NUMBER THEORY & DISCRETE MATHEMATICS

Editor

Mattias Raskob

scitus
academics

Number Theory & Discrete Mathematics

Edited by **Mattias Raskob**

ISBN: 9781681179643
Library of Congress Control Number: 2017949107

616, Corporate Way, Suite 2, 4766,
Valley Cottage, NY 10989

www.scitusacademics.com

Printed in United States of America.

Preface

Number Theory is one of the oldest and richest branches of mathematics. Number theory is concerned with the properties of numbers in general, particularly integers. It has applications to cryptography, cryptanalysis, and cryptology, particularly with regard to modular arithmetic, Diophantine equations, linear and quadratic congruence's, prime numbers and primarily testing. Other discrete aspects of number theory include geometry of numbers. In analytic number theory, techniques from continuous mathematics are also used. Ideas from combinatorics, algebra, and complex analysis all find their way and eventually become essential for understanding parts of number theory. Research in discrete mathematics increased in the latter half of the twentieth century partly due to the development of digital computers which operate in discrete steps and store data in discrete bits. Concepts and notations from discrete mathematics are useful in studying and describing objects and problems in branches of computer science, such as computer algorithms, programming languages, cryptography, automated theorem proving, and software development. Conversely, computer implementations are significant in applying ideas from discrete mathematics to real-world problems, such as in operations research.

This book 'Number Theory & Discrete Mathematics' is designed explicitly for mathematicians and computer scientists seeking a first approach to this important topic. It offers more comprehensive coverage of the fundamental theory, techniques, and applications of number theory. This approachable text studies discrete objects and the relationships that bind them.

Table of Contents

CHAPTER 1

Number in Mathematical Cryptography

Nathan Hamlin

Department of Mathematics and Statistics, Pullman, Washington, USA

ABSTRACT

With the challenge of quantum computing ahead, an analysis of number and representation adequate to the task is needed. Some clarifications on the combinatorial nature of representation are presented here; this is related to the foundations of digital representations of integers, and is thus also of interest in clarifying what numbers are and how they are used in pure and applied mathematics. The author hopes this work will help mathematicians and computer scientists better understand the nature of the Generalized Knapsack Code, a lattice-based code which the author believes to be particularly promising, and the use of number in computing in general.

Keywords: Number Theory, Quantum Computing, Public-Key Cryptography, Generalized Knapsack Code, Combinatorial Code

1. INTRODUCTION

The Generalized Knapsack Code (GKC), first introduced in a dissertation [1], was recently shown to be secure against basis reduction attacks [2]. With basis reduction eliminated as a threat, it was shown that the kinds of combinatorial disguising methods that are then possible would make the Shamir attack [3], an admittedly formidable problem for the knapsack of Merkle and Hellman [4], very manageable. Here I clarify some of the issues surrounding number generally that perhaps are necessary to understand and appreciate this code, as well as cryptography and computing in general.

My goal here is to arrive at a view of number and representation that is more or less adequate to conventional and quantum computing. The author is not attempting to challenge philosophical materialism, for which he has a great deal of respect, but he believes that something beyond the discussion of Dedekind [5] and Hilbert is necessary to grasp the role of number and representation in computing and cryptography, and that this can be accomplished without reverting to a form of Platonism. The traditional knapsack code [6] has been examined extensively. It is based on the possibility of identifying the binary representation of a number with a 0 - 1 knapsack problem. It seems that it has been assumed that the actual base of the digital representation of a number does not play a noticeable role in contemporary cryptography. The binary representation is particularly convenient for modern computation, and so it has been assumed that the binary representation is ideal, or at least neutral. While the actual information in a contemporary communication will be most likely transmitted in binary, as we shall see below, the selection of another more complicated base does have important cryptographic implications. In particular, the base that is chosen contributes to making certain properties of the number visible or not.

Cryptography is a distinct linguistic, mathematical, and representational process from computing, as can be seen by the fact that for most of its history it was done with paper and ink, and later, the telegraph. Thus there is at the very least no reason to assume that an account of number that is sufficient for computation or electronic communication will be adequate to cryptography as whole, even when done with the assistance of computers. And because of the linguistic nature of cryptography, there is every reason to believe that a complete account of the nature of representation is required to understand it fully. Since the practical cryptography we are dealing with today is largely a matter of disguising positive integers, something less than that full account, but, as we shall see, more than the account of binary and base 10, is needed to make sense of the field as it currently stands. That is what I am attempting to give here. To put things very simplistically, if you are trying to hide an integer from others in the computing process, it might be wise to store and transmit the number in a repre- sentation that does not reveal the essential properties of that number with respect to computation. The Hamlin-Webb representations make infinitely many such repre- sentations available for our use.

From my experience in presenting this material, I suspect this exposition will not be entirely satisfying to everyone, and especially to some who are more on the applied end of things with math and physics. I believe that, even so, some of the conclusions may be satisfactory and comprehensible to such readers. If the reader finds these arguments irritating, I suggest that he or she skip to the final paragraph of the paper.

2. A COMPARISON WITH OTHER LATTICE-BASED CODES

Unlike NTRU and related cryptosystems [7], the security of the GKC does not rest merely or even primarily on a closest or shortest vector problem. This makes it a fund- amentally different kind of code than many other lattice-based codes, and so a different kind of explanation regarding its security is needed.

Problems with the Classical Knapsack Code

A visualization of the problems with the original knapsack code will help prepare us to understand why the GKC need not have the same vulnerabilities of other lattice based codes.

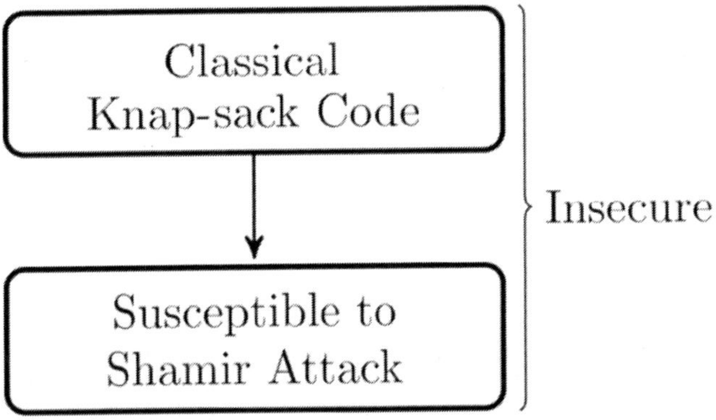

Attempts were made to defend against the Shamir attack, but the resulting change in density made another attack plausible:

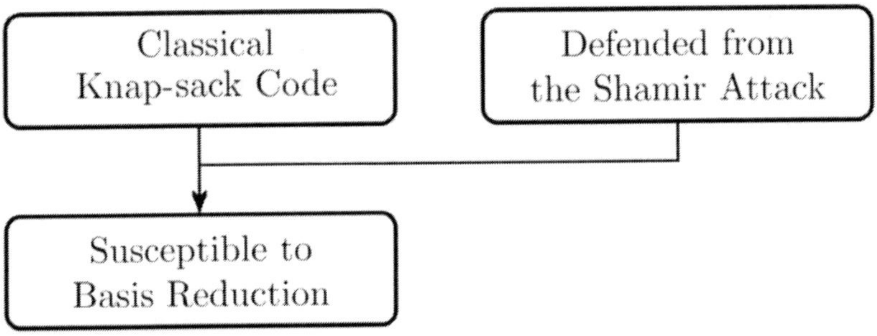

The situation with the Generalized Knapsack Code is much better [2]:

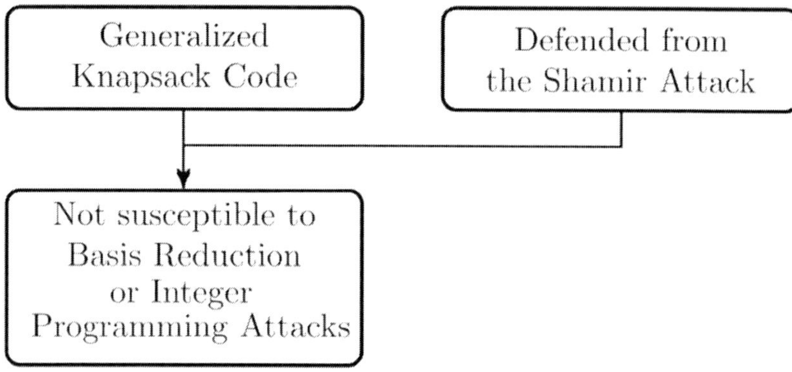

Basis reduction and the Shamir attack being the primary (mathematical) security concerns, this warrants the claim that the GKC is a secure public-key code.

3. NUMBERS, NUMERALS, IMAGES, AND REPRESENTATIONS

In order to make a compelling case for the security of the GKC within contemporary within the context of contemporary mathematical, computer science, and engineering understandings a few issues regarding representation will need to be clarified that are not always addressed in an article such as this. While some of what is said here might appear to be of use to a general audience, my goal is to clarify number for the people who use them the most, and pure mathematicians, applied mathematicians, computer scientists, and engineers all use number slightly differently; and people from each of these fields use them slightly differently. And people from all these fields are involved in mathematical cryptography. Nor, to my mind, would it be accurate to describe this a a philosophical approach to number. Philosophers have metaphysical and epistemo- logical commitments which I am I attempting to avoid even stepping on, and I believe their cultural responsibilities go beyond questions of mere accuracy. So, I am at- tempting to give a relatively pragmatic approach to an issue that many people care about for different reasons, in order to show how number works in mathematical cryptography, and hopefully provide a meeting place for the different concerns and people involved. Perhaps it might be helpful to illustrate this for one important case: the general pure mathematical approach to numbers versus the applied mathematical approach. Pure mathematicians must deal with infinite, and not merely infinitesimal, quantities or even sequences and we always have, practically speaking only finitely many symbols available to us. So a pure mathematician must leave open the possibility that numbers are not only non-identical with their symbols, but perhaps independent. The applied mathematician need only take into account quantities that impact his problems or kinds of problems, in the specific or general case, and so symbols with carefully defined meanings are of much greater importance, and these usually will ultimately refer to physical quantities, processes, and states, as opposed to entities that could be meaningfully described as abstract. By the time we get to physics, symbols can take on a life of their own, that perhaps a pure or applied mathematician might be un-comfortable with, but for the physicist extremely useful. I see no problem with any of these perspectives, but merely wish to point out that people from each area might be working together or making decisions on a problem that is primarily about number. In the case of cryptography we might be primarily interested in the natural numbers, and in my experience pure and applied mathematicians do not really agree about these. I don't wish to privilege one position over another, but to elucidate things to the point where there can some agreement here, even if that agreement cannot be elsewhere. This is important in making decisions about adopting specific public-key codes that work primarily on the natural numbers. I will include some things from the history of numbers and their representations, not as much as a survey, but rather to indicate some of the distinct sources of our perspectives on these issues. The philosophical view of number for elementary number theory which I favor is that of Husserl; but I very much believe it should not be regarded as giving the whole story. So, I am not attempting to fundamentally challenge the positions of Dedekind and Hilbert, which are very important in their own places, and which in my experience have a surprising amount of feeling attached to them in mathematical circles. Thus I will be treating this topic somewhat generally, but not for a general audience.

There is also, I believe, something very deep here, which has not generally been recognized within the mathematical and computing community, or at least been recognized as being of fundamental importance to cryptography. I will do my best to avoid being any more philosophical than necessary, but through talking with many people about this code and about numbers generally I have become convinced that an actual explanation is in fact needed here, even to talented and well-trained mathe- maticians. Husserl [8] argues that most human beings cannot perform basic operations with positive integers bigger than around 12 without a form of representation. This might shed some light on the dimensions of primary school multiplication tables. If you are new to thinking about this, you might think of an intelligent adult living in a community with no knowledge of mathematics, and words for numbers that do not go beyond the number 4 or 5. One could imagine that such an adult could work with sets of tools, weapons, etc. with clarity and accuracy as long as the quantities involved did not become too large. In the contemporary West, we typically teach our children how to count, and even sometimes to write numerals, before they really have need of them in normal life, so that does really work as effectively as a thought-exercise. Even if Husserl's estimate is substantially too low, which seems unlikely, or ultimately does not apply to certain people (savants, for example), a safe claim is that most of the numbers involved in the process of cryptography (and perhaps this holds true in a certain respect for computing generally) are too large for a human mind to think of and perform elementary operations upon without a system of representation (try to imagine mul- tiplying two integers on the order of 10^{10} without some way of representing either in symbols). It is well known of course that this can be very difficult or even impossible if you only have a poor system of numeration (e.g. Roman numerals).

The system of Roman numerals, which is often rightly ridiculed in contemporary life because of its extreme inconvenience in even basic calculations, actually had one important thing going for it, however: under Roman numerals, every natural number is given a unique name (here, a combination of letters) that does not depend essentially upon the number's mathematical properties, but only loosely on its relation to other natural numbers, and perhaps also to the idiosyncrasies of the Latin language. The base 10 system represents a number by its congruence classes modulo the powers of 10. The binary system, which does this modulo the powers of 2, is very useful for computation and communication with machines because of the easy ability to associate it with on/off switches and perhaps a high or low signal in a fiber-optic transmission. But strictly speaking, with respect to doing justice to each number, it is in many ways the worst representation possible, precisely because of this simplicity and machine-practicality. The very simplest non-constant recurrence is $u_n = 2u_{n-1}$ with $u_0 = 1$, which yields the powers of 2. Just like proving that a theorem, or finding good empirical evidence surrounding a theorem, that indicates it does or does not hold for the number 0 or 1 (or 2 for that matter) can be almost entirely irrelevant to its truth with respect to larger natural numbers or rational, algebraic, and transcendental numbers, so information about things in the very simplest case of recurrences may have no bearing at all on more complex or complicated sequences. And this simple fact seems to have been widely overlooked. The problem that the Leonardo de Pisa representation system has (and binary is just another species of this) is that it can bring you very close to making what is perhaps the fundamental intellectual error of the Middle Ages: to assume (in Aquinas's language) that one is convertible with being. It is surely no accident that the base 10 system was developed and popularized in a world already coming under the shadow of scholasticism. Saying that n is n copies of the number 1 $(n = n*1)$ does not necessarily give us any more information about n than we had before. Euclid's treatment of number theory in terms of "strips" in *the Elements*, which seems so foreign to those familiar with modern mathematics, probably contributed significantly to this confusion. Philosophically speaking, the natural numbers are individuals, as opposed to universals or images. The intellectual relationship that the greatest modern number theorists have had with them, to a thoughtful and knowledgeable mind, might be sufficient evidence of this fact, and while it is not within the realm of mathematical proof, it is within the general experience of mathematicians and students of mathe- matics. That *pi* more or less has an annual birthday celebration at this point is one of many things that could be adduced to this claim, which probably holds for real numbers in general, though the author will not insist on it here. Our usual way of writing the natural numbers in increasing order of cardinality is only natural in the sense that this is an extremely convenient order for our knowledge and purposes, and of course represents roughly the order in which we come to know them as children, and perhaps on a historical level to some degree as well. Cardinality is a property of number that is of at least moderate importance, of course. But the author believes it would perhaps be more entertaining than illuminating if someone were to attempt to argue that the most important property of the numbers 1 and 2, for example, was their cardinality. The number 5 is in some ways more interesting than 4, and of course such a direct comparison is only possible up to a point. There is surely no general result than a number's worth is inversely proportional to its cardinality, and this is a much more plausible proposition than direct proportion, and would be the main way one would have try to explain the order we have on the natural numbers in terms of cardinality. The number 1 is certainly not the least important natural number.

An ordinal interpretation has a place, but not so much in cryptography. In transmitting a message M (an integer) the processes of encoding, decoding, and cryptanalysis is surely made more and not less obscure if we interpret this as the Mth natural number. But to choose a simple example, say the number (and not numeral) 2: even if we could list (on paper, verbally, or in a computer database) all of its mathematical properties (and of course there are infinitely many of these) this list or these properties would not be identical with the number 2, nor would they perhaps even have exhausted what might be meant by saying "2". The base 10 system labels each number by a certain specific property that that number has in relation to the number 10. This property is not the same as the number, and it is arguably not even one of its most important mathematical properties. Our English system of appellations is in a sense, more natural than the base

10 number system, because it at least gives (somewhat generally) significant information about an individual human being (usually something about his or her history). The information the base 10 representation system gives about a number is roughly how it can be used for currency, or perhaps more generally, for measurement in a metric system. The base 2 system roughly gives the information as to how a number can be used for computing. The danger, which is less likely under Roman numeration, perhaps surprisingly, is for human beings through habituation to identify the number with this property. If we were to name roses and apples after the frequency of light they reflect in full sunlight, there would be much less danger in confusing that frequency with the particular (or even type) of rose or apple, than there is in confusing the integer represented in base 10 by 209,273,356 with its base 10 representation, since roses and apples are concrete visual objects and large integers are not. How a number can act or seem to act as a universal (say in the 15 apples left on a particular tree) is probably an important philosophical and practical question, but not directly relevant to mathematical cryptography at this time, as the interpretation of the meaning of the message number M (as apples, silver coins, time left in a time-share etc.) which perhaps does not currently play a direct role in the computational process of mathematical cryptography, but rather in the human inter- pretation of the decoded (or potentially decoded) message.

To be precise, so as to be able understand the general case, the base 10 system identifies the number with its properties in relationship to a sequence, and while we usually say this as the powers of 10, it is more accurate to say that it is in relation to the first order recurrence sequence $u_n = 10u_{n-1}$ with initial condition $u_0 = 1$, as the base 2 is more accurately related to the sequence $u_n = 10u_{n-1}$ with $u_0 = 1$. If you are having trouble seeing this, notice that we are generally trying to find how the number makes a certain finite sequence of place values in places and each place one to the left of another is 10 or 2 times that previous place, depending on the base. So, a mathematician of course will notice that this *amounts* to the powers of 10 or 2, which, perhaps sadly (in the author's opinion), are much more familiar to many mathematicians than even the simplest recurrence sequence. But the method of determining this is a recurrence relation, even though in these simple cases it yields an exponential sequence. And it is to Leonardo de Pisa that we owe both the Hindu-Arabic base 10 system (at least in its historical Western adoption) and also the Fibonacci sequence, which is enough to at least suggest that a recurrence interpretation of each may not be exactly "out of place" here.

I am going to simplify matters a bit for the sake of discussion, so let me give an example in full first so that it can be clear to the reader what I am leaving out. Let us interpret the basic information about the number that is represented in base 10 by 781 that is contained in its base 10 representation. When we write 781, in an implicitly base 10 situation, we are generally referring to the unique positive integer that is $\equiv 1 \pmod{10}$; after subtracting 1, is $\equiv 80 \pmod{100}$ (or $\equiv 8*10 \pmod{10*10}$); after subtracting another 80, is $\equiv 700 \pmod{1000}$ (or $\equiv 7*100 \pmod{10*100}$); and after subtracting another 700, is $\equiv$ 0 (mod every power of 10 (or $\equiv 0*10^n \pmod{10*10^n}$, for all n). In this last case we are in particular not interested in higher powers of 10 than 3. It might be convenient to call this a reverse-greedy algorithm. I am going to simplify things by identifying the digits of the base 10 representation (and similarly for base 2) with the equivalence classes of the integers mod the powers of 10. While this is not precise (781 (mod 10^2) is 81 and not 8, for example), the author believes it is a helpful way to keep what is actually happening in mind, and conveniently describes the information we are generally taking from the representation.

We perhaps need some new terminology. The representation classes with respect to a particular recurrence sequence are the registers in which the blocks are placed in a representation. For a first order recurrence these are just single places associated with the powers of the base. In the case of binary and base 10, these representation classes are roughly equivalence classes in each digit, and this brings a certain amount of freedom to the representation for the purposes they are used for (it is extremely convenient, for example, that adding a dime to a sum of money requires a relatively minimal change in the digits, the same perhaps could be said about shifting digits in certain conditions of the use of binary). That the representation classes for more complicated sequences *are not merely equivalence classes* is one thing that makes them superior for cryptography. The author is of course not advocating a return to classical mathematical notation. Nor does he think that our use of binary should be dis- continued. But he does think it is reasonable to claim that the immense success of binary for purposes of computation may have made us a little overconfident with regard to its mathematical status compared to other forms of representation. (I have yet to see someone wearing a Zeckendorf t-shirt yet). Recurrence Representations are not at all likely to do what binary can do in conventional computing. But the author believes these representations can do something that binary cannot do, namely, provide (actually, infinitely many) versions of the knapsack code that are secure from basis reduction and the Shamir attack. This is something very easy to overlook, and it appears that many people may have overlooked it.

Where does this leave us with cryptography and the knapsack code? To put it bluntly, what has actually been shown in the past by other researchers is that the knapsack code, using the worst possible mathematical representation of a number (binary), is not secure from basis reduction, and thus cannot be defended against the Shamir attack. If we have followed what has been said above, this may not have any implications whatsoever for more complex or cryptographically important representations. Namely, with respect to basis reduction, representations that do not fall prey to a shortest vector approach (and any binary vector, is going to be, relatively speaking, extremely short), or a closest vector attack (and again, a binary vector with the regularities of a first-order recurrence is probably in the very easy category to approximate). So with basis reduction and integer programming, most, if not all, of the work that has been placed prominently before the scholarly community, has been done on binary representations, and I think with the assumption that this is transparent, natural, general, or at least equivalent to other representations. This isn't the case, and so not only are there good mathematical arguments [2] [9], and data [2] to back-up the claims that the Gene- ralized Knapsack Code is secure from basis reduction, a simple inquiry into the nature of number and our usual representations would lead us to think that this is probably so, in light of what basis reduction and integer programming are recognized as able to accomplish. And if this is so, the arguments about providing security from the Shamir attack [1] [2] [9] with combinatorial disguising, after basis reduction is no longer a worry, are also sound and reliable. This puts us in a position of perhaps having available to us, a secure, public-key code. There seem to be good indication that this is the kind of code that is likely to also be safe under quantum computing, given at least one important inquiry into the 0 - 1 Knapsack under these conditions [10], and some analysis of the situation as a whole [6] which the author takes as suggesting, contrary to the situation with codes vulnerable to Schor's Algorithm (RSA, discrete log, elliptic curves), that the case may be that there are bounds on the possibility of finding algorithms for quantum computers that can effectively tackle some relevant lattice- based problems. To the extent which this can be currently tested, and to the extent which other concerns than security can be evaluated (for the code generally, as well as with respect to various kinds of choices of representations), I say that it is time for us to consider this code very seriously for adapting commerce and perhaps communication in light of the possibility of quantum computing. A code like this that can be im- plemented on conventional hardware, and yet that would be secure with a quantum computer somewhere "out there" would be much more convenient and cost effective than quantum cryptography, if the latter is to be understood as processes which require a quantum computer for one or both parties in communication.

3.1. Visualizing the Process

It may be helpful to see the previous and following content in terms of a diagram. We will examine the relationship between numerals, numbers, and computing with respect to both conventional and quantum computing. But first let us look at the general case:

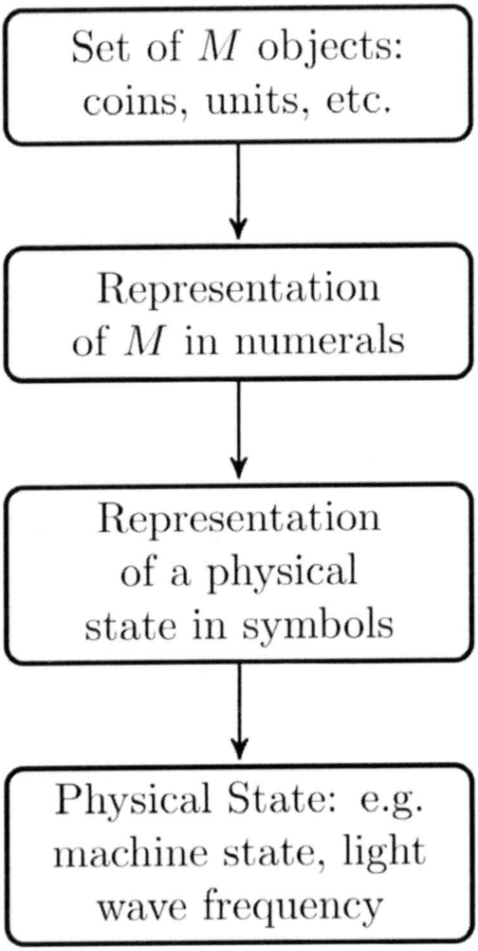

In the case of a conventional computer running an internal operation:

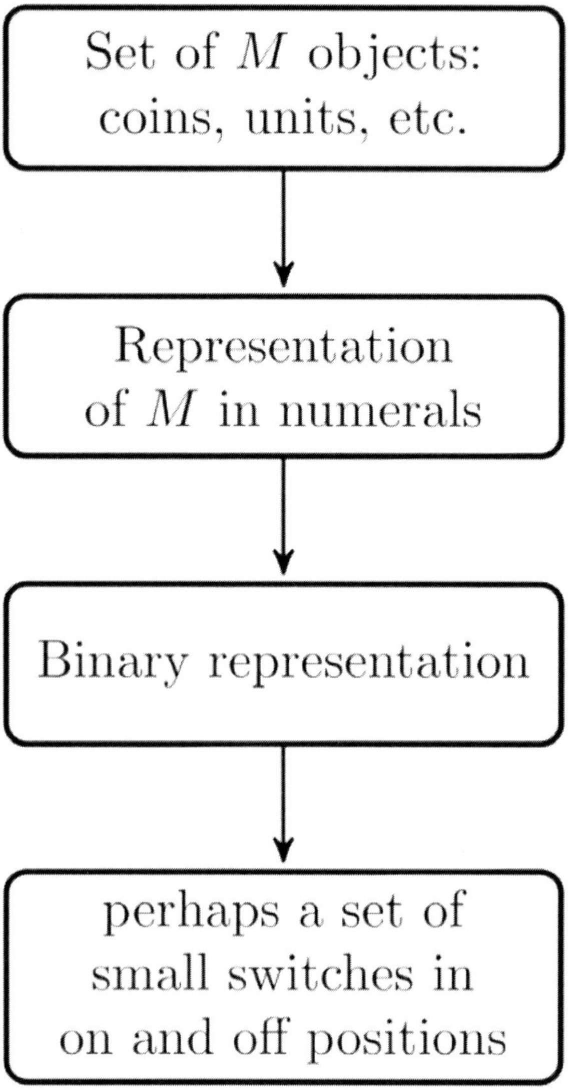

In a specific case of the latter describing the cardinality of a set of 5 coins:

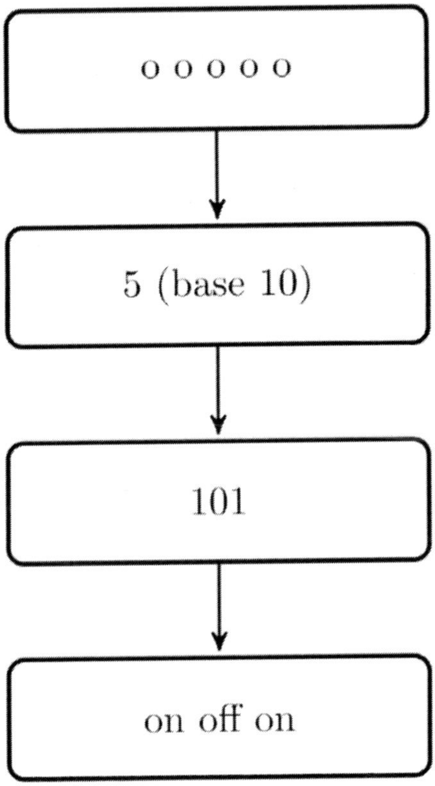

Of course this shows only the simplest possible cases, but it sufficiently illustrates that as we use number in computing, we have at least the numbered set (and the integer M could be handled as M units for the purpose of visualization as in Euclid), the common human representation (for us in the West at least, base 10), the representation for the machine (in conventional computing this is often binary), the actual state of the machine that corresponds with that representation. Binary is useful because of its correspondence with the last step, but the cryptographic process involves the first two steps as well, and is at least prior to the machine state that is transmitted over the wire or wirelessly.

3.2. Numerals

I think almost everyone can agree that for human purposes, numbers generally involve a visual way of expressing them, which I will describe as a numeral. I am trying to work very carefully through some general issues in this section so that it can be clear how we use numerals while avoiding a simplistic view of numbers as forms independent of matter, which are in turn instantiated somehow in the words or characters we use for them. I think many people know that this latter outlook could have been potentially fatal for the history of mathematics at more than one point, but at the same time, within cryptography some distinction must be made between the number and the representation, if only for the fact that the representation needs to hide the number at times, and this requires at the very least that the number and representation are non-identical. So again, some general material will be presented but in terms of what is likely to be useful to a mathematician, computer scientist, and perhaps engineer. I am using the term *numeral* to refer to the visual symbol that is used to represent the number for human purposes. In the discussion which follows, which attempts to go deeper on the representation side of things, I will use a relatively small number so that some measure of visualization will be possible, but I am primarily interested in integers that are very large for my general conclusions, and this will not pose a problem to the argument. The basic thing that needs to be explored and clarified is how *numerals* function in cryptography. The author believes there are certain considerations here that were not included in say, Dedekind's and Hilbert's important discussions of these matters, and which are essential to fully understanding the GKC, and numerical representation generally. In a typical encrypted communication today, the message is a number, say a positive integer like $M = 162$. Suppose this number represents something very concrete, as the number of silver coins the message sender has in a particular drawer in his or her office. The number itself is (logically at least) prior to any representation of it. If in the Western world we were still at a stage in the development of society where we could not easily represent numbers past the number 4, for example, there would still be a certain number of coins in that drawer (namely, 162; we probably need to ignore the role that numeration plays in the mint, but this has no fundamental importance here), but we might not be willing or able to record or communicate that fact, or to easily distinguish it from a situation where there were slightly more or fewer present. Even today, until someone has a great deal of familiarity with numbers, it can be difficult to meaningfully distinguish distinct large integers. Think, for example, how children will often skip from a million, to a billion, to infinite quantities so easily rather than include numbers in the sequence. $\left\{10^{10}, 10^{100}, 10^{1000} \cdots\right\}$.

If coins distract from the issue at hand, think of 162 smooth stones from a river bed. If we ignore concerns related to the computer (e.g having this discussion in 1600 C.E.), the base ten representation "162" does two things for us: 1) allows us keep a good record of the coins, for ourselves or others, and 2) enables us to perform mathematical operations that will then have some important implications for the coins. Operations on "162" will have implications for the number 162 apart from the representation. Roman numerals were excellent for task 1), but not so good for task 2). To reiterate, in a certain sense, and for the purposes of cryptography I do not think we have to be very metaphysical about this, numbers exist apart from their representations. The fact that base 10 and not base 2 was chosen for this example does not change this fact, but could illustrate it: 162 and 10,100,010 refer to the same number, and apart from human convention and the certain advantages one representation might have over another in a certain context (at a sporting goods store without a cash register vs. within a bank computer records system for safety deposit boxes) they are describing the same thing. Numbers which we use are in general present in the world without representations, but we need, as thinking and relating organisms, representations in order to keep track of, communicate, and operate on, number. The fact that as mathematicians we are often dealing with infinite, unending sequences of numbers in abstract settings can make this a bit cloudy, but the finite numbers mostly at use in daily communication and cryptography typically have a much more concrete significance and are integers or at least rationals, and so the important qualifications surrounding irrationals (Dedekind) and infinite sequences and series (Hilbert, see [8] for example) are not as essential to our understanding in this context.

So without claiming that all that I say here can be applied to all mathematical settings and entities, I will continue to argue that this is true in the most important settings relevant to cryptography. The introduction of the base representational 10 system in the Middle Ages by Leonardo de Pisa, vastly improved our relationship to number, but did not change how number was present in objects (our 162 coins in the drawer for example). This new relationship did in fact give Western Europeans a certain amount of power over number and numbered things, and so perhaps there was a change in the coins, at least in the expected number to be found in the drawer at the end of the day. I only intend to point out that 162 coins is a fact found in experience that can be real before someone comes along and counts them. When you first look in the drawer, and see the coins, there is a certain number them, even before you count them. In general practice at least, the act of counting is intended at to determine a fact that is already real but currently unknown and needed to be known. I suspect that most of what I am saying here will be relatively elementary to the reader, but I am trying to be thorough enough to handle some of the philosophical objections that have been raised in our history, and which I believe have a certain point to them.

Notice that a numeral, say “1”, or a collection of numerals e.g. “162” is actually an image. Its printed form, or the process of displaying it on paper or a computer screen might take the form of an algorithm involving numbers and numerals, but when we finally see it before us, we must recognize its meaning as a whole. The upper half of "1" does not mean 0.5 for example, and whether I draw a very small "1" in the bottom corner of a page, or make a large poster of it, its meaning, as a number anyway, is the same. (There might be a further meaning in the former as a *page* number, and in the latter as support for a favorite, as at maybe a football playoff, and for our purposes whether the number is *understood* to be cardinal or ordinal) In cryptography at least, we are usually dealing with cardinal numbers, and this is one of the main things that makes Husserl's treatment of greater use than Hilbert's. We may break up our larger digital representation (162) into parts, but when we finally arrive at the numeral "6" we do not break it down further but take it to represent the number 6, 6 things, etc. Similarly, when faced with a photo of an apple, we do not (educational purposes perhaps excepted) divide the photo into parts, but recognize the whole as in some way representing (say) a piece of fruit we once gave to a favorite teacher or happily found in our lunch bag on a daily basis when were much younger. If it is an image we recognize, in general, "cutting it up" either mentally or with a photo program, will not increase our understanding of the image if all we need is to recognize it's meaning as a whole. Incidentally, the fact that images are in some sense wholes is what makes them difficult to handle for internet search engines. Probably with our current algorithms, the best that can be done is to catalogue images with labels and then search for labels. But this is not to search for images themselves; it is only like what we do in our physical libraries where we use algorithms etc. to send the appropriate location for a cover with a particular number on it. If the glue is good, we will send of course the book that belongs to that cover, but we were actually searching for the number on the cover which probably has no ultimately meaningful connection to the book itself, except in a very general manner. One cannot after all get much of a real sense of Guizot's *History of France* (a rather long work) from the information in the card catalogue (physical or automated), even if this contains a substantial summary. Thus, at least until better algorithms for images can be found, people will probably still be seen browsing at bookstores and in libraries for the foreseeable future.

Please bear with the author as I take up one possible objection here. The point I am trying to make is essential in understanding the Generalized Knapsack Code, and the theoretical underpinnings of representation of number. Images as transmitted from computer to computer or printed on paper are, with some qualifications perhaps, large, many-valued arrays with a position in 2 or 3 space matched with a color. Without having to go into finite vs. infinite kinds of discussions which are in fact appropriate here, let us be very simple, since this paper is concerned mostly with representations of numbers. The numeral "6" on a printed page or computer monitor will indeed be describable in terms of pixels and dot matrices etc. And these are essential to producing a visual image such as "6".

But the image functions as a whole in at least our understanding. Even the neural network that produces the image in our imagination as we see it on the screen and compare it to other "6"'s we've seen is not the image itself, at least as far as it functions in our understanding.

Again, I don't think we have to be particularly metaphysical here. Think of the times you have seen people from different backgrounds and cultures write their numbers differently on checks or other documents. The important thing is recognizing the "6" however written, and if one has passed the early years of school, understanding its association with the number or a certain number of objects in a certain case.

A more rigorous treatment of this philosophically would probably be helpful for all this, but perhaps out of place in an article such as this. Here I only intend to make it clear that the numeral and the number are distinct; that the numeral functions as a whole, as an image, and is not reducible, at least in how we understand it in normal life, to an array in the sense that a machine stores it or produces it.

The author understands that the numeral "6" is probably simple enough to be handled exhaustively with algorithms etc., and will make a more a general point in this regard at a later time.

Let's apply all this to cryptography. The numeral "6" makes the number 6 present to our mind, if we have mastered our Hindu-Arabic numerals. Most of the time this is important so that we can know that there is six of something somewhere. If I am proud of my 162 silver coins in my drawer and have no fear of anyone stealing them, my message $M = 162$, or more likely, 10,100,010 (the numerals) can be sent in the clear and anyone can come to know about my collection. However, if I don't want others to know, I could send the number in a different set of numerals (perhaps resulting from a process of mathematical cryptography with algorithms) and these numerals would hide the true number of coins from anyone who intercepted the message (ideally speaking). So while for most other purposes of life, we use numerals to reveal numbers, or at least make them present, usable, or accessible, in cryptography we use them to hide, disguise, or secure. So we are using numerals for roughly speaking, the seemingly opposite purpose for which they were probably first developed. I see no problem with this of course, but I would like to give an account of how this happens so that we can understand how the representational system fits in. What is actually transmitted is a set of numerals (binary perhaps), and not a number. I understand that the coding of fiber-optic cables is designed so that these two can coincide in some way, but in at least an encrypted communication, this will not correspond to the message number. These images (numerals) either prevent someone from seeing (reading) the message, or if they have the proper knowledge to decipher (as in the case of the party I wish to communicate with), only delay, hopefully very briefly, the reading. In either case, the numerals obstruct the ordinary comprehension of unencrypted communication. So while it is indeed important to have difficult algorithms to reverse etc. the nature of representation plays a role in at least the first and last steps of all secure com- munication.

One could argue that it is the limitations of the strictly binary representation that made the original knapsack code insecure. It is the infinite possibilities of other representations using recurrence sequences, the basic building blocks of digital (in the most general sense of that word) representations, that makes the Generalized Knapsack Code secure. A system of representation involves more than the specific images (numerals) which represent numbers. In the base 10 system, we have numerals which represent the numbers 0 - 9 and then rules or conventions that extend the use of combinations of these numerals to represent (potentially at least) every positive integer. Thus representation is in essence a combinatorial problem, and since representation is essential to communicating (whether openly or securely) numbers to others, the combinatorial nature of representation is essential to any form of mathematical cryptography, which is a special way, type, or action upon communication. The Generalized Knapsack Code makes use what may be the most general theory of these representations in order to hide and reveal information. The vast combinatorial possibilities that result form the basis of its security.

Let us describe the Leonardo de Pisa's representation of positive integers in full (extending this to negatives or decimals is nearly trivial). There are ten images or numerals (0, 1, 2, 3, 4, 5, 6, 7, 8, 9), as well as an understanding of what each position (their are an infinite number of these potentially, but not in any actual representation of an integer) means, namely, the nth position to the left of the decimal point holds the symbol for the number of multiples of 10^{n-1}. Thus, the base 10 representation of an integer M is an ordered collection of numerals which are understood to have a certain interpretation in terms of the powers of 10. Describing binary would be even simpler. However, base ten and binary are among the very simplest of all representations [1], and thus we do not see the full complexity that can surround representation by merely looking at them. The powers of 2 and 10 have a very simple relationship, and each digit can seem to stand on its own, and provide separate information from the rest The same would be true to some extent even of the Zeckendorf representation. So let us take a representation of full complexity and examine it a little more fully.

We will of course use the Hindu-Arabic numerals, since these are not likely to go out use anytime soon, and have the great advantage of familiarity. But it is worth noting that the manufacture or adaptation of other numerals/symbols (e.g. letters from other non-Western languages) might have implications for public-key cryptography as well.

4. AN EXAMPLE REPRESENTATION

Let u_i satisfy the recurrence $u_n = u_{n-1} + 3u_{n-3} + 2u_{n-4}$ (the initial values are not vital, but the standard ones are generally more beautiful and convenient, i.e., $u_0 = u_1 = u_2 = 1$). The signature is $S = 1032$. The representation of any positive integer M is of the form

$$M = \sum_{i=0}^{n-1} d_i u_i, \tag{1}$$

where the string of digits $d_{n-1} d_{n-2} \cdots d_1 d_0$ must be composed of blocks of digits which are lexicographically smaller than S. In this case, the allowed blocks of digits are $0, 100, 101, 102, 1030, 1031$. Hence, for instance, $102|1030|0|101$ is a legitimate string, but $101|1102103$ is not. Notice that no allowed block begins with 11. The difference between the general situation and base ten or base 2 is thus two-fold: in terms of the actual position of Hindu-Arabic numerals, the number is read by finding the numeral in the nth position and multiplying it by u_{n-1}, and more importantly, not all possibilities of numerals and recurrence are used. This might seem like a truism, because the numerals we have are basically designed for base 10. But it means that these images are much more complicated. And if the recurrence is kept secret, and is one of very many possible recurrences, none of which reveal plainly useful mathematical properties of the integers, 10,210,300,101 may reveal very little if any information about the integer it represents, or even for how it could be used for mathematical com- putation (addition, multiplication etc.) This latter may be difficult or even virtually impossible, but at the very least will be far from transparent.

5. CONCLUSIONS

Representations are ordered combinations of images including numerals and sub- scripted letters which stand for members of recurrence sequences. These latter are implicit in base 10, base 2, and others, but not in general. Thus a representation of a positive integer is an image with ordered parts, and is understood in the context of a system of representation, namely a particular recurrence sequence with its properties, and the rules that develop from its intrinsic properties. If you are keeping numbers secret, it might be smart to represent them in such a way that very little meaningful information about what you are hiding is found in that representation. Recurrence sequences are an arbitrary choice except that they are the sequences that under certain conditions retain [1] the properties that are most useful to us in the binary and base ten representations. A paper is forthcoming which introduces a similar method of cryptography to the Generalized Knapsack Code, but using more arbitrary sequences, which, if secure, might have further advantages in keeping representations obscure.

This admittedly somewhat esoteric account which I have provided will perhaps not be entirely surprising for those who understand that quantum computing is a deeper scientific problem than conventional computing, and that a deeper understanding of number and representation and their role in mathematics and computing might be required for such an advanced project. At the bare minimum, the transition to qubits takes us away from the simplicity that the binary representation provided under conventional computing (in directly corresponding to bits), and so the relevance of Recurrence Representations to modern computing may transcend the cryptographic problem. As the binary and base 10 representations are effective for conventional computing and metric measurement, so other recurrences may have their own "natural" uses. But hopefully I have provided sufficient justification to make the claim that the problem and usefulness of number and representation goes far beyond what has traditionally been understood, and that the results that have been recently advanced in this area open up the field of lattice-based cryptography significantly more widely than has been generally recognized to be likely. For those who have followed the arguments presented here and elsewhere, the only thing that really remains to be done for the Generalized Knapsack Code is some heavy testing on the scale of computer systems that are only available in a corporate or governmental context, or perhaps at some leading intellectual institutions. There is every reason to believe this will lead to a working public-key code which can be implemented on conventional or quantum computers, and which will be secure in the presence of conventional or quantum computers.

If the numbers are individuals, then their properties are not exhausted by their relationships to one another, or perhaps even by a *priori* considerations generally. And thus, we come to know each number by experience; and the numbers can be and should be the proper objects of experimentation. Computer science allows us to come to know and understand numbers that are much too big for the human mind to appreciate directly, at least with our current modes of representation. Mathematical proof, which accomplishes a great deal in allowing us to understand and work with numbers, is not the only means to determining their properties and possibilities. And to require or be satisfied with an argument, when what is needed is actually *experiment* and *testing* is to put the cart before the horse. The arguments given above and in the citations indicate that the Generalized Knapsack Code, and perhaps the Needle-in-the-Haystack Code as well, are *very likely* to be secure public-key codes, but only through thorough experi- mentation with *actual numbers*, and *very large* ones at that, can we come to be reasonably satisfied that we can rely upon these codes. The test results regarding basis reduction are very favorable [2], but some large-scale implementation somewhere is needed, and one major purpose of this paper is to appeal for such an implementation.

ACKNOWLEDGEMENTS

Thanks to the Math Department and College of Arts and Sciences at Washington State University for making it possible for me to continue my work.

DEDICATION

For Ashley, who is often better about numbers, and about many other things as well. And for Thérèse, who already loves numbers, and many other good things too.

REFERENCES

1. Hamlin, N. (2016) Recurrence Representations: An Exploration of Number, Representation, and Public-Key Cryptography. Lambert Academic Publishing, Saarbrücken. (Originally published as a dissertation in 2012)
2. Hamlin, N., Krishnamoorthy, B. and Webb, W. (2015) A Knapsack-Like Code Using Recurrence Sequence Representations. The Fibonacci Quarterly, 1, 24-33.

3. Shamir, A. (1984) A Polynomial-Time Algorithm for Breaking the Basic Merkle-Hellman Cryptosystem. IEEE Transactions on Information Theory, 30, 699-704.
4. Merkle, R.C. and Hellman, M.E. (1978) Hiding Information and Signatures in Trap Door Knapsacks. IEEE Transactions on Information Theory, 24, 525-530.
5. Dedekind, R. and Beman, W., Trans. (1909) Essays on the Theory of Numbers. The Open Court Publishing Company, Chicago.
6. Bernstein, D., Buchmann, J. and Dahmen, E. (Eds.) (2009) Post-Quantum Cryptography. Springer, Heidelberg.
7. Hoffstein, J., Pipher, J. and Silverman, J.H. (2008) An Introduction to Mathematical Cryptography. Springer Science, New York.
8. Husserl, E. and Bernet, R. (Eds.), Willard, D. (Trans.) (2003) Philosophy of Arithmetic: Psychological and Logical investigations with Supplementary Texts from 1887-1901. Springer Science, Dordrecht.
9. Hamlin, N. and Webb, W.A. (2012) Representing Positive Integers as a Sum of Linear Recurrence Sequences. The Fibonacci Quarterly, 50, 99-105.
10. Arvind, V. and Schuler, R. (2003) The Quantum Query Complexity of 0-1 Knapsack and Associated Claw Problems. International Symposium on Algorithms and Computation, 2906, 168-177.

APPENDIX: THE GENERALIZED KNAPSACK CODE

This is some general material which appeared in [2] and earlier in [9], and, in a different form, in [1]. Please see these sources for more extensive explanations, data, and mathematical proofs.

1) The Traditional Knapsack Code

The plaintext message is assumed to be an integer $M,\ 0 \le M < 2^n$. We consider the representation of M in base 2:

$$M = \sum_{i=0}^{n-1} \epsilon_i 2^i,\ 0 \le \epsilon_i \le 1. \tag{2}$$

The creator of the code chooses a secret, superincreasing sequence $\{s_i\}$, i.e., where $s_i > s_{i-1} + \cdots + s_0$. The secret s_i are then disguised by one or more modular multi- plications of the form

$$w_i = ks_i \pmod{m}, \tag{3}$$

where k and m are kept secret. The w_i are made public. The sender of the message M computes

$$T = \sum_{i=0}^{n-1} \epsilon_i w_i. \tag{4}$$

The encoded message T is then sent over a possibly insecure channel. The hope was that only the creator of the code, who knows $k, m,$ and the s_i, can solve Equation (4) for the coefficients ϵ_i. In particular, the disguising step given in Equation (3) is easily reversed.

2) Generalized with More Generalized Representations

a) An Example Code

a) An Example Code

Let u_i satisfy the recurrence $u_{i+5} = u_{i+4} + u_{i+2} + 2u_{i+1} + 7u_i$ (the initial values are not vital, and we could take them to be the standard ones, i.e., $u_0 = 1, u_1 = 1, u_2 = 1, u_3 = 2$, and $u_4 = 4$). The signature is $S = 10127$. The representation of any positive integer M is of the form

$$M = \sum_{i=0}^{n-1} d_i u_i, \tag{5}$$

where the string of digits $d_{n-1} d_{n-2} \cdots d_1 d_0$ must be composed of blocks of digits which are lexicographically smaller than S. In this case, the allowed blocks of digits are

$0, 100, 1010, 1011, 10120, 10121, 10122, 10123, 10124, 10125$, and 10126.

Hence, for instance, $1011|1023|0|1010$ is a legitimate string, but $1011|1023|0|1010$ is not. Notice that no allowed block begins with 11. We now illustrate how to calculate this type of representation of any number M using a greedy approach. Although this calculation is straightforward, the fact that makes this code harder to model for the cryptanalyst is that so many strings of digits are not allowed in the representations, even though the strings appear similar to the allowed ones, and both classes of strings use digits of the same size.

b) Computing the Representation of M

An easy way to calculate the representation of any number M in the recurrence sequence $\{u_i\}$ is to calculate the augmented sequence $\{v_{j,i}\}$ for $1 \le j \le 10$, $0 \le i \le n$ given by

$$
\begin{aligned}
v_{1,i} &= u_i, \\
v_{2,i} &= u_i + u_{i-2}, \\
v_{3,i} &= u_i + u_{i-2} + u_{i-3}, \\
v_{4,i} &= u_i + u_{i-2} + 2u_{i-3}, \\
v_{5,i} &= u_i + u_{i-2} + 2u_{i-3} + u_{i-4}, \\
&\vdots \\
v_{10,i} &= u_i + u_{i-2} + 2u_{i-3} + 6u_{i-4}.
\end{aligned}
$$

The $v_{j,i}$ correspond to the allowed blocks of digits. In our example, the $v_{j,i}$ occur in groups of size 10. In other examples, the groups could be much larger. The correct expression for M is found simply by using the greedy algorithm on the $v_{j,i}$, and converting the sum into an expression in the u_i. The $v_{j,i}$ could be calculated and stored, or calculated from the u_i as needed.

We explore the memory requirements for storing all the $\{v_{j,i}\}$. The principal eigenvalue of the sequence $\{u_i\}$ is $\alpha \approx 1.9754$. We may assume that u_i is roughly α^i, or is close to 2^i. After disguising, the public weights w_i will be approximately 2^{n+40}, and the target T approximately 2^{n+50}. In other words, these quantities require 40 - 50 extra bits of memory to represent. If $n = 1000$, the memory required for the weights w_i is roughly $1,040,000$ bits (or 130 kilobytes). The memory required to store all the $v_{j,i}$ is hence 1.3 megabytes. Even if n is much larger, the memory needed for the w_i is negligible.

c) Encryption and Decryption

Let $\{u_i\}$ be a recurrence sequence that satisfies the following recurrence equation.

$$u_i = a_1 u_{i-1} + a_2 u_{i-2} + \cdots + a_h u_{i-h}, \tag{6}$$

where $a_1 > 0$ and all $a_i \geq 0$. The string $\S = a_1 a_2 \cdots a_h$ is its signature, and we let $A = a_1 + a_2 + \cdots + a_h$. Every natural number N has a unique representation in the form of Equation (5), where the digits are composed of blocks that are lexicographically smaller than S. Including the zero block, there are A such blocks. The auxiliary sequence $\{v_{j,i}\}$ is constructed as linear combinations of the u_i with coefficients same as the blocks other than the zero block. Hence there are $A-1$ of the $v_{j,i}$ in each group. The total number of $v_{j,i}$ numbers is hence $(A-1)n$ if there are n of the u_i.

The creator of the code chooses a secret sequence $\{s_i\}$ which has the property that $s_{i+1} > s_i(u_{i+1}/u_i)$ for all i. This property replaces the condition of $\{s_i\}$ being superincreasing as used in the traditional knapsack code.

The s_i are then disguised by any invertible mapping, some of which we describe below. The resulting quantities w_i are the public weights. If M is the original plaintext message, the user of the code expresses

$$M = \sum_{i=0}^{n-1} d_i u_i = \sum_{i=0}^{n-1} \epsilon_{j,i} v_{j,i}, \tag{7}$$

and computes

$$T = \sum_{i=0}^{n-1} d_i w_i = \sum_{i=0}^{n-1} \epsilon_{j,i} y_{j,i}, \tag{8}$$

which is the transmitted message.

It is relatively easy to see that this is appropriately reversible, and [2] explores the possibilities this version has with regard to protecting against the Shamir Attack, as well as explains why the code is secure from basis reduction.

CHAPTER 2

Competition Numbers of a Kind of Pseudo-Halin Graphs

Zhijun Cao, Yonggang Cui, Guoyan Ye, Yongqiang Zhao*

School of Mathematics and Information Science, Shijiazhuang University, Shijiazhuang, China

ABSTRACT

For any graph G,G together with sufficiently many isolated vertices is the competition graph of some acyclic digraph. The competition number k (G) of a graph G is defined to be the smallest number of such isolated vertices. In general, it is hard to compute the competition number k (G) for a graph G and characterizing a graph by its competition number has been one of important research problems in the study of competition graphs. A 2-connected planar graph G with minimum degree at least 3 is a pseudo-Halin graph if deleting the edges on the boundary of a single face 0 f yields a tree. It is a Halin graph if the vertices of 0 f all have degree 3 in G. In this paper, we compute the competition numbers of a kind of pseudo-Halin graphs.

Keywords: Competition Graph, Competition Number, Halin Graph, Generalized Halin Graph, Pseudo-Halin Graph

1. INTRODUCTION AND PRELIMINARY

Let G = (V, E) be a graph in which V is the vertex set and E the edge set. We always use V and E to denote the vertex number and the edge number of G, respectively. The notion of competition graph was introduced by Cohen [1] in connection with a problem in ecology. Let D = (V, A) be a digraph in which V is the vertex set and A the set of directed arcs. The competition graph $C(D)$ of D is the undirected graph G with the same vertex set as D and with an edge $uv \in E$ (G) if and only if there exists some vertex $x \in V(D)$ such that $(u, x),(v, x)\in A(D)$. We say that a graph G is a competition graph if there exists a digraph D such that $C(D) = G$. Roberts [2] observed that every graph together with sufficiently many isolated vertices is the competition graph of an acyclic digraph. The competition number $k(G)$ of a graph G is defined to be the smallest number k such that G together with k isolated vertices added is the competition graph of an acyclic digraph. It is difficult to compute the competition number of a graph in general as Opsut [3] has shown that the computation of the competition number of a graph is an NP-hard problem. But it has been one of important research problems in the study of competition graphs to characterize a graph by its competition number. Recently, many papers related to graphs' competition numbers appear. Kim et al. [4] studied the competition numbers of connected graphs with exactly one or two triangles. Sano [5] studied the competition numbers of regular polyhedra. Kim et al. [6] studied the competition numbers of Johnson graphs. Park and Sano [7] [8] studied the competition numbers of some kind of hamming graphs. Kim et al. [9] studied the competition numbers of the complement of a cycle. Furthermore, there are some papers (see [10] [11] [12] [13] [14]) focused on the competition numbers of the complete multipartite graphs, and some papers (see [15]-[21]) concentrated on the relationship between the competition number and the number of holes of a graph. A cycle of length at least 4 of a graph as an induced subgraph is called a hole of the graph. We use rI to denote the graph consisting only of r isolated vertices, and $rG\cup I$ the disjoint union of G and rI. The induced subgraph$G[S]$ of G is a subgraph of G whose vertex set is S and whose edge set is the set of those edges of G that have both ends in S.

All graphs considered in this paper are simple and connected. For a vertex v in a graph G, let the open neighborhood of v be denoted by $N_G(v)=\{u|u$ is adjacent to $v\}$. For any set S of vertices in G, we define the neighborhood of S in G to be the set of all vertices adjacent to vertices in S, this set is denoted by$N_G(S)$.An in-neighbor of a vertex v in digraph D is a vertex u such that $(u,v)\in A(D)$; an out-neighbor of a vertex v is a vertex w such that $(v,w)\in A(D)$. We denote the sets of in-neighbors and out-neighbors of v in D by $N_D^-(v)$and $N_D^+(v)$, respectively.

A subset S of the vertex set of a graph G is called a clique of G if $G[S]$ is a complete graph. For a clique S of a graph G and an edge e of G, we say e is covered by S if both of the endpoints of e are contained in S. An edge clique cover of a graph G is a family of cliques such that each edge of G is covered by some clique in the family. The edge clique cover number $\theta_e(G)$of a graph G is the minimum size of an edge clique cover of G. A vertex clique cover of a graph G is a family of cliques such that each vertex of G is contained in some clique in the family. The vertex clique cover number $\theta_v(G)$ of a graph G is the minimum size of a vertex clique cover of G

A generalized Halin graph $G = T \cup C$ is a plane graph consisting of a plane embedding of a tree T and a cycle C connecting the leaves (vertices of degree 1) of T such that C is the boundary of the exterior face. The tree T and the cycle C are called the characteristic tree and the adjoint cycle of G, respectively. For each $v \in V(C)$, if $N_T(v) \cap N_T(x) = \emptyset$ for any vertex $x \in N_C(v)$, then we called v a simple leaf of T, otherwise, a compound leaf of T. Denote all simple leaves of T by V C and all compound leaves of T by $V_2(C)$, respectively. It is obvious that $V(C) = V_1(C) \cup V_2(C)$, and $V_1(C) \cap V_2(C) = \emptyset$. Let $V_1' = N_T(V1(C))$, $V_1' = N_T(V_2(C))$. A generalized Halin graph G is a Halin graph when each interior vertex of G has degree at least 3.

A 2-connected planar graph G with minimum degree at least 3 is a pseudo-Halin graph if deleting the edges on the boundary of a single face $f_0$yields a tree. It is a Halin graph if the vertices of $f_0$all have degree 3 in G. The face $f_0$is the exterior face; the others are interior faces. Vertices of $f_0$are exterior vertices; the others are interior vertices. Vertices of $f_0$with degree 3 in G are regular vertices; the others are irregular vertices. Let $R(f_0)$ and $I(f_0)$denote the sets of regular and irregular vertices in f_0, respectively.

The main theme of this paper is to study the competition numbers of a kind of pseudo-Halin graphs with $|I(f_0)| = 1$, and gets the exact value or the upper bound of the competition number of this kind of pseudo-Halin graphs.

2. LEMMAS

We first introduce two useful Lemmas.

Lemma 1 (Harary et al. [22]). Let $D = (V, A)$ be a digraph. Then D is acyclic if and only if there exists an ordering of vertices, $\sigma = [v_1\ v_2, \cdots, v_n]$, such that one of the following two conditions holds:

1) For all $i, j \in \{1, 2, \cdots, n\}$, $(v_i, v_j) \in A$ implies that $i < j$;

2) For all $i, j \in \{1, 2, \cdots, n\}$, $(v_i, v_j) \in A$implies that $i > j$.

By this lemma, if D is an acyclic digraph, we can find a vertex labeling $\sigma: V \to \{1, 2, \cdots, |V|\}$ so that whenever $(x, y) \in A, \sigma(y) < \sigma(x)$. We call σ an acyclic labelling of D. Conversely, if D is a digraph with an acyclic labelling, then D is acyclic.

Lemma 2 (Kim and Roberts [4]). For a tree T and a vertex v of T, there is an acyclic digraph D so that $T \cup \{v_0\}$ is the competition graph of D for $v_0$not in T and so that v has only outgoing arcs in D:

Kim and Roberts [4] proved Lemma 2 by the following algorithm.

Let T_1= T, $V(D_1)$=$V(T)$, and $A(D_1)$= ∅. Choose a vertex V_1 of degree 1 from. If V_1 is adjacent to v_1 inT_1, let T_2=T-v_1, $V(D_2)$=$V(D_1)$∪= (V_0)for some vertex v_0 not in$T$, and$A(D_2)$={(v_1, v_0)(v_1, v_0)}. Having defined T_i andD_i, choose a vertex v_i of degree 1 fromT_i. If v^i is adjacent to v_i inT_i, then letT_{i+1}=$V(D_i)$, and$A(D_{i+1})$∪{(v_i, v_{i-1})(v^i, v_{i-1})}. Repeat this last step until D_n has been defined. Let$D$=($V(D_n)$, $A(D_n)$). In the procedure, we may avoid selecting v until we select all other vertices since there are at least two vertices of degree 1 in a tree with more than one vertex.

In fact, this algorithm provides an acyclic labelling σ=[v_0, v_1, v_2, ---, v_n] of D such thatv_n=v^{n-1}, $v_{n-1}v_n$∈$E(T)$, and v_{n-1} and v_n have only outgoing arcs in$D$, where |$V(T)$ |=n.

Opsut [3] gave the following two lower bounds for the competition number of a graph.

Theorem 1 (Opsut [3]). For any graph G,$k(G)\theta_e(G)$-|$V(G)$|+2.

Theorem 2 (Opsut [3]). For any graph G,$k(G)$≥ min{ $\theta_v(N_G(v))$|v∈$V(G)$.

Lemma 3. For any generalized Halin graph $G = T \cup C$,

$$k(G) \le \begin{cases} 2, & V_1(C) = \varnothing; \\ |V_1(C)| + 1, & V_1(C) \ne \varnothing \end{cases}.$$

Proof. Let G=T∪C be a generalized Halin graph, where T and C are the characteristic tree and the adjoint cycle of G, respectively. Suppose that along cycle C by clockwise order we can partition $V_2(C)$ (if$V_2(C)$≠∅) into k≥1 subsets $V_2^i(C)$ such that $V_2(C) = \cup_{i=1}^{k} V_2^i(C)$ and the vertices in each $V_2^i(C)$ are con- secutive on$C$, where1≤i≤k. Let u^i be the common neighbor of the vertices in, where $1 \le i \le k$. We assume that the vertices in $V_2^i(C)$ by clockwise order are $v_1^i, v_2^i, \cdots, v_{\alpha_i}^i$, where a_i≥2 and 1≤i≤k. Denote all vertices on Cbetween $v_{\alpha_i}^i$ and v_1^{i+1} by $V_1^i(C)$, where 1≤i≤k and $v_1^{k+1} = v_1^1$. We assume that the vertices in $V_1^i(C)$ (if $V_1^i(C) \ne \varnothing$) by clockwise order are $x_1^i, x_2^i, \cdots, x_{\beta_i}^i$, where β_i≥1 and $i \in \{1, 2, \cdots, k\}$. Note that $V_1(C) = \cup_{i=1}^{k} V_1^i(C)$, and if $V_1(C)$≠∅ then we always let $V_1^k(C) \ne \varnothing$. If$V_2(C)$≠∅, then let $V_1^1(C) = V(C)$ and arbitrarily select a vertex in $V(C)$as x_1^1.

By Lemma 1 and the algorithm in the proof of Lemma 2, we can construct an acyclic digraph D so that $T\cup\{v_0\}$ is the competition graph of D for $v_0$not in T, and get an acyclic labeling

$$\sigma : V \cup \{v_0\} \to \{1,2,\cdots,|V|+1\}$$

of$D$ so that

$\sigma(v_0)=1$;

$\sigma(v_j^i)=1+\sum_{k=1}^{i-1}(\alpha_k+\beta_k)+j$, where $1 \le i \le k$ and $1 \le j \le \alpha_i$;

$\sigma(x_j^i)=1+\sum_{k=1}^{i}\alpha_k+\sum_{k=1}^{i-1}\beta_k+j$, where $1 \le i \le k$ and $1 \le j \le \beta_i$;

Note that, if $V_2(C)=\emptyset$, then let $\alpha_1 = 0$, if $V_1^i = \emptyset$, then let $\beta_i = 0$, where $i\in\{1, 2,\cdots, k\}$, and we always have $\sum_{k=1}^{0}(\alpha_k+\beta_k)=0$ and $\sum_{k=1}^{0}\beta_k=0$. Label the other vertices of T arbitrarily.

Case 1. $V_1(C)=\emptyset$,

Let D_1be a digraph whose vertex set is

$V(G)\cup\{v_0, v_1\}$

and whose arc set is

$$A(D)\cup\bigcup_{i=1}^{k}\bigcup_{j=2}^{\alpha_i}\{(v_j^i, v_{j-2}^i)\}\cup\bigcup_{i=1}^{k-1}\{(v_1^{i+1}, v_{\alpha_i-1}^i)\}$$

$$\cup\{(v_1^1, v_1), (v_{\alpha_k}^k, v_1)\}-\bigcup_{i=1}^{k}\{(u^i, v_{\alpha_i-1}^i)\}$$

where $v_0^1 = v_0$, $v_0^i = v_{\alpha_{i-1}}^{i-1}$ for each $i\in\{2, 3, \cdots, k\}$, and v_1,$v_2$are new added vertices.

Case 2.$V_2(C)=\emptyset$.

Let D_2be a digraph whose vertex set is

$$V(G)\cup\{v_0\}\cup\{y_1^1, y_2^2, \cdots, y_{\beta_1}^1\}$$

and whose arc set is

$$A(D)\cup\bigcup_{j=1}^{\beta_1}\{(x_j^1,y_j^1),(x_{j+1}^1,y_j^1)\},$$

where $x_{\beta_1+1}^1 = x_1^1$, and all y_j^1 $(1\le j\le\beta_1)$ are new added vertices.

Case 3. $V_1(C)\neq\emptyset$ and $V_2(C)\neq\emptyset$

Let D_3 be a digraph whose vertex set is

$$V(G)\cup\{v_0\}\cup\bigcup_{1\le i\le k,\beta_i>0}\{y_1^i,y_2^i,\cdots,y_{\beta_i}^i\}$$

and whose arc set is

$$A(D)\cup\bigcup_{i=1}^{k}\bigcup_{j=2}^{\alpha_i}\{(v_j^i,v_{j-2}^i)\}\cup\bigcup_{1\le i\le k,\beta_i>0}\{(x_1^i,v_{\alpha_i-1}^i)\}\cup\bigcup_{1\le i\le k,\beta_i=0}\{(v_1^{i+1},v_{\alpha_i-1}^i)\}$$

$$\cup\bigcup_{1\le i\le k,\beta_i>0}\bigcup_{j=1}^{\beta_i}\{(x_j^i,y_j^i),(x_{j+1}^i,y_j^i)\}-\bigcup_{i=1}^{k}\{(u^i,v_{\alpha_i-1}^i)\},$$

where $v_0^1 = v_0$ and $v_0^i = x_{\beta_{i-1}}^{i-1}$ (if $\beta_{i-1}>0$) or $v_{\alpha_{i-1}}^{i-1}$ (if $\beta_{i-1}=0$) for any $i\in\{2, 3, ---, k\}$, $x_{\beta_k+1}^k = v_1^1$ and $x_{\beta_i+1}^i = v_1^{i+1}$ for any $i\in\{1, 2, ---, k-1\}$, such that $\beta_i>0$. All y_j^i $(1\le i\le k, 1\le j\le\beta_i)$ are new added vertices.

We note that D_1, D_2 and $D3$ are acyclic. This is because every arc added here goes either from a big label vertex to a small label vertex or from a vertex in $V(G)$ to a new added vertex not in $V(G)\cup\{v_0\}$. It is not difficult to check that

$$C(D_1)=G\cup\{v_0,v_1\},$$

$$C(D_2)=V(G)\cup\{v_0\}\cup\{y_1^1,y_2^2,\cdots,y_{\beta_1}^1\}\text{ and}$$

$$C(D_3)=G\cup\{v_0\}\cup\bigcup_{1\le i\le k,\beta_i>0}\{y_1^i,y_2^i,\cdots,y_{\beta_i}^i\}$$

And we know that $V_1(C)=\{x_1^1,x_2^2,\cdots,x_{\beta_1}^1\}$ if $V_2(C)=\emptyset$, and $V_1(C)=\bigcup_{1\le i\le k,\beta_i>0}\{x_1^i,x_2^i,\cdots,x_{\beta_i}^i\}$ if $V_1(C)\neq\emptyset$ and $V_2(C)\neq\emptyset$. So the result follows.

Lemma 4. For any not K_4 generalized Halin graph $G = T\cup C$,

$$|V_1(C)| = \theta_e(G) - |V(G)| + 1.$$

Proof. Let $G=T\cup C$ be a not K_4generalized Halin graph, where T and C are the characteristic tree and the adjoint cycle of G, respectively. Since $G[V(G)\setminus V(C)]$ is a tree, then $\theta_e(G[V(G)\setminus V(C)]) = |V(G)| - |V(C)| - 1$. Note that $G[V(C)\cup V_2']$ just includes all triangles in G and the edges in C,so $\theta_e(G[V(C)\cup V_2']) = |V(C)|$. It is easy to check that $\theta_e(G[V_1(C)\cup V_1']) = |V_1(C)|$. Furthermore, each pair of graphs $G[V(G)\setminus V(C)]$, $G[V(C)\cup V_2']$ and $G[V_1(C)\cup V_1']$ has not any common edges. So we have

$$\begin{aligned}\theta_e(G) &= \theta_e(G[V(G)\setminus V(C)]) + \theta_e(G[V(C)\cup V_2']) + \theta_e(G[V_1(C)\cup V_1']) \\ &= (|V(G)| - |V(C)| - 1) + |V(C)| + |V_1(C)| \\ &= |V(G)|| + |V_1(C)| - 1\end{aligned}$$

or

$$|V_1(C)| = \theta_e(G) - |V(G)| + 1$$

3. PSEUDO-HALIN GRAPH

Now we consider a pseudo-Halin graph G with the exterior face f_0 and $I(f_0)=\{x\}$. Let u and v be the neighbors of x on the boundary of f_0. Without lose of generality, we may always assume that u is on the left of x and v on the right of x. Let G' in a similar way as used in Lemma 3. By Lemma 3, $k(G')\leq 2$. Let $v = v_1^1$ and by the similar way used in the proof of Lemma 3, there is a digraph D' such that

$$C(D') = G' \cup \{v_0, v_1\},$$

and

$$\{(v,v_0),(v',v_0),(u,v_1),(v,v_1),(u,b)\} \subset A(D')$$

but

$$(u',b) \notin A(D')$$

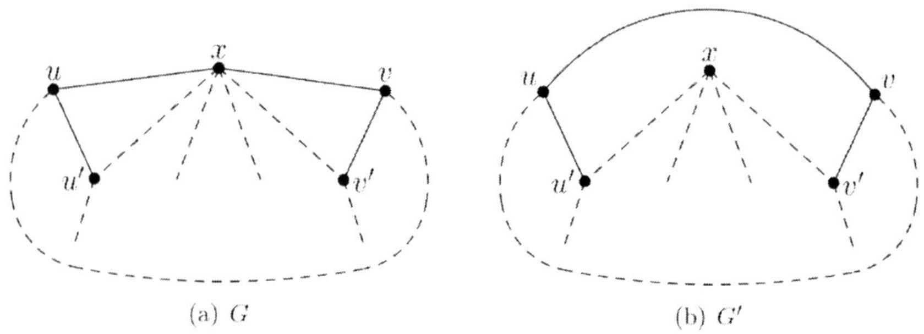

Figure 1. *G* and *G′*.

where v_0 and v_1 are two isolated vertices not in *G′*. Note that $N_{D'}^{-}(b)=\{u\}$. Let

$$V(D)=V(D')$$

and

$$A(D)=A(D')\cup\{(x,b),(x,v_1)\}\setminus\{(u,v_1)\}.$$

See Figure 2. Note that *D* is acyclic since every arc added here goes from a big label vertex to a small label vertex. It is easy to see that $C(D)\cup\{v_0, v_1\}$. So we have $k(G)\leq 2$. On the other hand, since for each vertex $v\in V(G)$, $d_G(v)\geq 3$, and note that the maximal clique in *G* is a triangle, so we have $\theta_v(N_G(v))\geq 2$. By Theorem 2, $k(G)\geq \min\{\theta_v(N_G(v)) \mid v\in V(G)\}\geq 2$. Therefore $k(G)=2$.

Lemma 5. Let *G* be a pseudo-Halin graph with $I(f_0)=\{x\}$, and *C′* the adjoint cycle of graph $G'=G-\{xu, xv\}+uv$. If $V_1(C')\neq\emptyset$, then we have $k(G)\leq |V_1(C')|+2$.

Proof. Suppose that *G* is a pseudo-Halin graph with $I(f_0)=\{x\}$ and $V_1(C')\neq\emptyset$, where *C′* is the adjoint cycle of graph $G'=G-\{xu, xv\}+uv$. Denote and labeling the vertices of *G′* in a similar way as used in Lemma 3. By Lemma 3, $k(G')\leq |V_1(C')|+1$, and there is a digraph *D′* such that $C(D')=G'\cup I_{|V_1(C')|+1}$, where $I_{|V_1(C')|+1}$ are $|V_1(C')|+1$ isolated vertices not in *G′*. By the similar way used in the proof of Lemma 3, there exists a vertex $y\in I_{|V_1(C')|+1}$ or $y\in V(C')$ such that $uv, uy\in A(D')$. Let

$$V(D)=V(D')\cup\{w\}$$

and

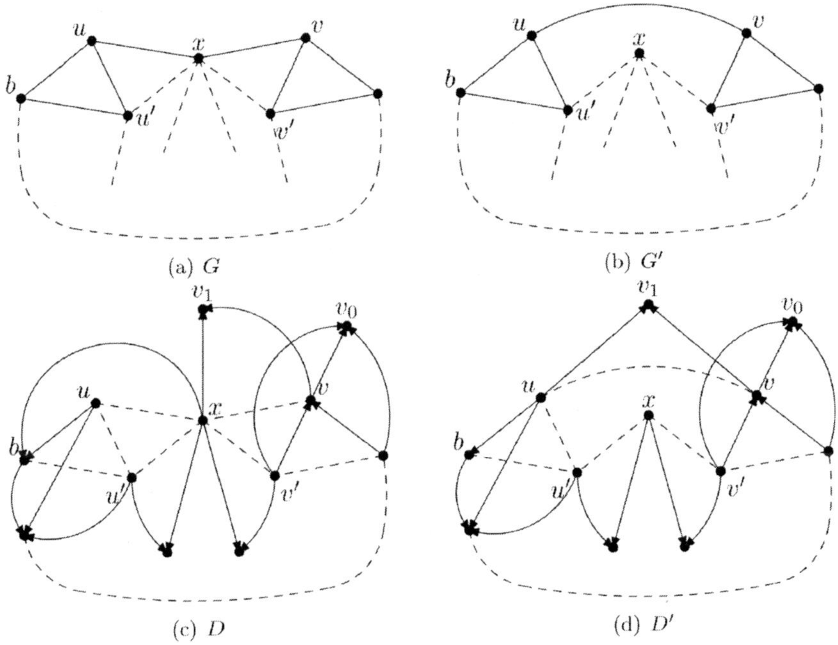

Figure 2. G and D ($V_1=\emptyset$).

$$A(D)=A(D')\cup\{(x,w),(u,w),(x,y)\}\setminus\{(u,y)\},$$

where w is a new added vertex to D'. Note that D is acyclic since every arc added here goes from a big label vertex to a small label vertex or to a new added vertex. It is easy to see that $C(D)=G\cup I_{|V_1(C')|+1}\cup\{w\}$. So we have $k(G)\leq|V_1(C')|+2$.

Lemma 6. Let G be a pseudo-Halin graph with $I(f_0)=\{x\}$, and C' the adjoint cycle of graph $G'=G-\{xu, xv\}+uv$.

1) If $xu', xv'\notin E(G)$, then $\theta_e(G')=\theta_e(G)-1$,

2) If $xu', xv'\in E(G)$, then

$$\theta_e(G')=\begin{cases}\theta_e(G)+3, & u,v\in V_1(C');\\ \theta_e(G)+1, & u,v\in V_2(C');\\ \theta_e(G)+2, & \text{otherwise},\end{cases}$$

3) If $xp' \in E(G)$, but $xq' \notin E(G)$, then $\theta_e(G') = \begin{cases} \theta_e(G)+1, & p \in V_1(C'); \\ \theta_e(G), & p \in V_2(C'), \end{cases}$ where

$\{p, q\}=\{u, v\}$.

Proof. 1) $xu', xv' \notin E(G)$.

Obviously, $\{x,u\}$ and $\{x,v\}$ are two maximal cliques of G. Since $\{u,v\}$ is a maximal clique of G', then $\theta_e(G') = \theta_e(G) - 1$.

2) $xu', xv' \in E(G)$.

It is easy to see that $\{x, u, u'\}$ and $\{x,v,v'\}$ are two maximal cliques of G. Note that $uv$is a maximal clique of G'. If $u, v \in V_1(C')$, then $\{x,u'\}$, $\{u,u'\}$, $\{x,v'\}$ and $\{v,v'\}$ are all maximal cliques of graph G'. $\theta_e(G') = \theta_e(G)+3$; If $u,v \in V_2(C')$, then $\{x. u'\}$ and $\{x. v'\}$ are two maximal cliques of graph G', but $\{u,u'\}$ and $\{v,v'\}$ not. Hence $\theta_e(G') = \theta_e(G)+1$; Otherwise, say $u \in V_1(C')$ and $u \in V_2(C')$ Then $\{x,u'\}$, $\{u,u'\}$ and $\{x,v'\}$ are all maximal cliques of graph G', but $\{v,v'\}$ not. Therefore $\theta_e(G')=\theta_e(G)+2$.

3) $xu' \in E(G)$ but $xv' \notin E(G)$ (or $xv' \in E(G)$ but $xu' \notin E(G)$).

Without lose of generality, we just need to consider the case $xu' \in E$ but $xv' \notin E$. By the proof of Case (1), $\theta_e(G\text{-}xu+uv)=\theta_e(G)$. If $v \in V_1(C')$, then $\{x,v'\}$ and $\{v,v'\}$ are all maximal cliques of graph G'. If $v \in V_2(C')$, then $\{x,v'\}$ is a maximal clique of graph G', but $\{v,v'\}$ not. So the result follows.

For a pseudo-Halin graph G, suppose that $f_0$be the exterior face of G and $I(f_0)=\{x\}$. Note that G' can not be K_4, so by Lemmas 4 we have

$|V_1(C')| = \theta_e(G') - |V(G)| + 1$, by Lemma 5 we have

$k(G) \le |V_1(C')| + 2 \le \theta_e(G') - |V(G)| + 3$. On the other hand, by Theorem 1 we have

$k(G) \ge \theta_e(G) - |V(G)| + 2$. So by lemmas 6, we have the following result.

Theorem 4. Let G be a pseudo-Halin graph with $I(f_0)=\{x\}$ and $V_1(C')\neq\emptyset$, where $C'$is the adjoint cycle of graph $G'=G-\{xu, xv\}+uv$.

1) If $xu', xv' \notin E$, then $k(G) = \theta_e(G) - |V(G)| + 2$,
2) If $xu', xv' \in E$, then

$$\theta_e(G)-|V(G)|+2\le k(G)\le\begin{cases}\theta_e(G)-|V(G)|+6, & u,v\in V_1(C');\\ \theta_e(G)-|V(G)|+4, & u,v\in V_2(C');\\ \theta_e(G)-|V(G)|+5, & \text{otherwise},\end{cases}$$

3) If $xp'\in E$, and $xq'\notin E$, then

$$\theta_e(G)-|V(G)|+2\le k(G)\le\begin{cases}\theta_e(G)-|V(G)|+4, & p\in V_1(C');\\ \theta_e(G)-|V(G)|+3, & p\in V_2(C'),\end{cases}$$

where $\{p, q\} = \{u,v\}$.

4. CONCLUDING REMARKS

In this paper, we study the competition numbers of a kind of pseudo-Halin graphs withjust one irregular vertex.

For a pseudo-Halin graph G with the exterior face $f_0$and $|I(f_0)|= 1$, we show that if all leaves of the characteristic tree of G are compound leaves, then $k(G) = 2$, otherwise, $\theta_e(G)-|V(G)|+2\le k(G)\le\theta_e(G)-|V(G)|+6$. Even we proved that $k(G)=\theta_e(G)-|V(G)|+2$ for some cases, but we can not provide the accurate value of the competition number of G for other cases. So it would be valuable to get the accurate value of the competition number of the pseudo-Halin graph with just one irregular vertex, and it may be interesting to study the competition numbers of general pseudo-Halin graphs.

For a digraph $D = (V, A)$, if we partition V into k types, then we may construct a undirected graph $C^k(D) = (V, E)$ of D as follows

1) $uv\in E$ if and only if there exists some vertex $x\in V$such that $(u, x),(v, x)\in A$ and $u,v$are of same type, or

2) $uv\in E$ if and only if there exists some vertex $x\in V$such that $(u, x),(v, x)\in A$ and $u,v$are of different types.

It is easy to see that $C^1(D) = C(D)$ for a given digraph D, and we note that multitype graphs can be used to study the multi-species in ecology and have been deeply studied, see [23] [24]. So these generalizations of competition graphs may be more realistic and more interesting.

ACKNOWLEDGEMENTS

We thank the editor and the referee for their valuable comments. The work was supported in part by the Natural Science Foundation of Hebei Province of China under Grant A2015106045, and in part by the Institute of Applied Mathematics of Shijiazhuang University.

REFERENCES

1. Cohen, J.E. (1968) Interval Graphs and Food Webs: A Finding and a Problem. Document 17696-PR, RAND Corporation, Santa Monica, CA.
2. Roberts, F.S. (1978) Food Webs, Competition Graphs, and the Boxicity of Ecological Phase Space. In: Alavi, Y. and Lick, D., Eds., Theory and Applications of Graphs, Lecture Notes in Mathematics, 642, 477-490. https://doi.org/10.1007/bfb0070404
3. Opsut, R.J. (1982) On the Computation of the Competition Number of a Graph. SIAM Journal on Algebraic Discrete Methods, 3, 420-428. https://doi.org/ 10.1137/0603043
4. Kim, S.-R. and Roberts, F.S. (1997) Competition Numbers of Graphs with a Small Number of Triangles. Discrete Applied Mathematics, 78, 153-162. https://doi.org/ 10.1016/s0166-218x(97)00026-7
5. Sano, Y. (2009) The Competition Numbers of Regular Polyhedra. Congressus Numerantium, 198, 211-219.
6. Kim, S.-R., Park, B. and Sano, Y. (2010) The Competition Numbers of Johnson Graphs. DiscussionesMathematicae Graph Theory, 30, 449-459. https://doi. org/10.7151/dmgt.1506
7. Park, B. and Sano, Y. (2011) The Competition Numbers of Hamming Graphs with Diameter at Most Three. Journal of the Korean Mathematical Society, 48, 691-702. https://doi.org/10.4134/jkms.2011.48.4.691
8. Park, B. and Sano, Y. (2011) The Competition Numbers of Ternary Hamming Graphs. Applied Mathematics Letters, 24, 1608-1613. https://doi.org/ 10.1016/ j.aml.2011.04.012
9. Kim, S.-R., Park, B. and Sano, Y. (2013) The Competition Number of the Complement of a Cycle. Discrete Applied Mathematics, 161, 1755-1760. https://doi.org/ 10.1016/j.dam.2011.10.034
10. Kim, S.-R., Park, B. and Sano, Y. (2012) The Competition Numbers of Complete Multipartite Graphs with Many Partite Sets. Discrete Applied Mathematics, 160, 1176-1182. https://doi.org/10.1016/j.dam.2011.12.017
11. Kim, S.-R. and Sano, Y. (2008) The Competition Numbers of Complete Tripartite Graphs. Discrete Applied Mathematics, 156, 3522-3524. https://doi.org/ 10.1016/j.dam.2008.04.009
12. Kuhl, J. (2013) Transversals and Competition Numbers of Complete Multipartite Graphs. Discrete Applied Mathematics, 161, 435-440. https://doi.org/ 10.1016/j.dam.2012.09.012
13. Li, B.-J.and Chang, G.J. (2012) Competition Numbers of Complete r -Partite Graphs. Discrete Applied Mathematics, 160, 2271-2276. https://doi.org/ 10.1016/ j.dam.2012.05.005
14. Park, B., Kim, S.-R. and Sano, Y. (2009) The Competition Numbers of Complete Multipartite Graphs and Mutually Orthogonal Latin Squares. Discrete Mathematics, 309, 6464-6469. https://doi.org/10.1016/j.disc.2009.06.016

15. Kamibeppu, A. (2012) A Sufficient Condition for Kim's Conjecture on the Competition Numbers of Graphs. Discrete Mathematics, 312, 1123-1127. https://doi.org/10.1016/j.disc.2011.11.035
16. Kamibeppu, A. (2010) An Upper Bound for the Competition Numbers of Graphs. Discrete Applied Mathematics, 158, 154-157. https://doi.org/ 10.1016/ j.dam.2009.09.007
17. Kim, S.-R., Lee, J.Y., Park, B. and Sano, Y. (2012) The Competition Number of a Graph and the Dimension of Its Hole Space. Applied Mathematics Letters, 25, 638-642. https://doi.org/10.1016/j.aml.2011.10.003
18. Kim, S.-R., Lee, J.Y. and Sano, Y. (2010) The Competition Number of a Graph Whose Holes Do Not Overlap Much. Discrete Applied Mathematics, 158, 1456-1460. https://doi.org/10.1016/j.dam.2010.04.004
19. Lee, J.Y., Kim, S.-R., Kim, S.-J.and Sano, Y. (2011) Graphs Having Many Holes but with Small Competition Numbers. Applied Mathematics Letters, 24, 1331-1335. https://doi.org/10.1016/j.aml.2011.03.003
20. Lee, J.Y., Kim, S.-R., Kim, S.-J.and Sano, Y. (2010) The Competition Number of a Graph with Exactly Two Holes. ArsCombinatoria, 95, 45-54.
21. Li, B.-J.and Chang, G.J. (2009) The Competition Number of a Graph with Exactly h Holes, All of Which Are Independent. Discrete Applied Mathematics, 157, 1337-1341. https://doi.org/10.1016/j.dam.2008.11.004
22. Harary, F., Norman, R.Z. and Cartwright, D. (1965) Structure Models: An Introduction to the Theory of Directed Graphs. Wiley, New York.
23. Barbour, A.D. and Reinert, G. (2011) The Shortest Distance in Random Multi-Type Intersection Graphs. Random Structures & Algorithms, 39, 179-209. https://doi.org/10.1002/rsa.20351
24. Shang, Y. (2016) Groupies in Multitype Random Graphs.SpringerPlus, 5, 989. https://doi.org/10.1186/s40064-016-2705-4

CHAPTER 3

Competition Numbers of Several Kinds of Triangulations of a Sphere

Yongqiang Zhao[1], Zhiming Fang[2,3], Yonggang Cui[1], Guoyan Ye[1], Zhijun Cao[1]

[1]*School of Science, Shijiazhuang University, Shijiazhuang, China*

[2]*Zhejiang Jinyun High School, Lishui, China*

[3]*School of Science, Hebei University of Technology, Tianjin, China*

ABSTRACT

It is hard to compute the competition number for a graph in general and characterizing a graph by its competition number has been one of important research problems in the study of competition graphs. Sano pointed out that it would be interesting to compute the competition numbers of some triangulations of a sphere as he got the exact value of the competition numbers of regular polyhedra. In this paper, we study the competition numbers of several kinds of triangulations of a sphere, and get the exact values of the competition numbers of a 24-hedron obtained from a hexahedron by adding a vertex in each face of the hexahedron and joining the vertex added in a face with the four vertices of the face, a class of dodecahedra constructed from a hexahedron by adding a diagonal in each face of the hexahedron, and a triangulation of a sphere with $3n(n \geq 2)$ vertices.

Keywords: Competition Graph, Competition Number, Edge Clique Cover, Vertex Clique Cover, Triangulation of a Sphere

1. INTRODUCTION AND PRELIMINARY

Let $G = (V, E)$ be a graph in which V is the vertex set and E the edge set. We always use $|V|$ and $|E|$ to to denote the vertex number and the edge number of G, respectively. The notion of competition graph was introduced by Cohen [1] in connection with a problem in ecology. Let $D = (V, A)$ be a digraph in which V is the vertex set and A the set of directed arcs. The competition graph $C(D)$ of D is the undirected graph G with the same vertex set as D and with an edge $uv \in E(G)$ if and only if there exists some vertex $x \in V(D)$ such that $(u, x),(v, x) \in A(D)$. We say that a graph G is a competition graph if there exists a digraph D such that $C(D) = G$.

Roberts [2] observed that every graph together with sufficiently many isolated vertices is the competition graph of an acyclic digraph. The competition number $k(G)$ of a graph G is defined to be the smallest number k such that G together with k isolated vertices added is the competition graph of an acyclic digraph. It is difficult to compute the competition number of a graph in general as Opsut [3] has shown that the computation of the competition number of a graph is an NP-hard problem. But it has been one of important research problems in the study of competition graphs to characterize a graph by its competition number. Recently, many papers related to graphs' competition numbers have appeared. Kim, et al., [4] studied the competition numbers of connected graphs with exactly one or two triangles. Sano [5] studied the competition numbers of regular polyhedra. Kim, et al., [6] studied the competition numbers of Johnson graphs. Park and Sano [7] [8] studied the competition numbers of some kind of Hamming graphs. Kim, et al., [9] studied the competition numbers of the complement of a cycle. Furthermore, there are some papers (see [10] [11] [12] [13] [14]) focused on the competition numbers of the complete multipartite graphs, and some papers (see [15] - [23]) concentrated on the relationship between the competition number and the number of holes of a graph. A cycle of length at least 4 of a graph as an induced subgraph is called a hole of the graph. We use I_r to denote the graph consisting only of r isolated vertices, and $G \cup I_r$ the disjoint union of G and I_r. The induced subgraph $G[S]$ of G is a subgraph of G whose vertex set is S and whose edge set is the set of those edges of G that have both ends in S.

All graphs considered in this paper are simple and connected. For a vertex v in a graph G, let the open neighborhood of v be defined by

$N_G(v)=\{u|u$ is adjacent to $v\}$. For any set U of vertices in G, we define the neighborhood of U in G to be the set of all vertices adjacent to vertices in U, this set is denoted by $N_G(U)$. Let $N_G[U]=N_G(U) \cup U$ and

$E_G[U] = \{e \in E(G) | e \text{ has an endpoint in } U\}$. We denote by the same symbol $N_G(U)$ the subgraph of G induced by $N_G(U)$. Note that $E_G[U]$ is con- tained in the edge set of the subgraph $N_G(U)$.

A subset U of the vertex set of a graph G is called a clique of G if $G[U]$ is a complete graph. For a clique U of a graph G and an edge e of G, we say e is covered by U if both of the endpoints of e are contained in U. An edge clique cover of a graph G is a family of cliques such that each edge of G is covered by some clique in the family. The edge clique cover number $\boldsymbol{\theta}_e(G)$ of a graph G is the minimum size of an edge clique cover of G. A vertex clique cover of a graph G is a family of cliques such that each vertex of G is contained in some clique in the family. The vertex clique cover number $\boldsymbol{\theta}_V(G)$ of a graph G is the minimum size of a vertex clique cover of G.

Let G be a graph and $F \subseteq E(G)$ be a subset of the edge set of G. An edge clique cover of F in G is a family of cliques of G such that each edge in F is covered by some clique in the family. The edge clique cover number $\boldsymbol{\theta}_e(F;G)$ of $F \subseteq E(G)$ in G is defined as the minimum size of an edge clique cover of F in G, i.e.,

$\theta_e(F;G) = \min\{|U| \,|\, U \text{ is an edge clique cover of } F \text{ in } G\}$ Note that the edge clique cover number $\boldsymbol{\theta}_e(E(G);G)$ of $E(G)$ in a graph G is equal to the edge clique cover number $\boldsymbol{\theta}_e(G)$ of the graph G.

Opsut [3] gave the following two lower bounds for the competition number of a graph.

Theorem 1 (Opsut [3]). For any graph G, G, $k(G) \geq \theta_e(G) - |V(G)| + 2$.

Theorem 2 (Opsut [3]). For any graph G, G, $k(G) \geq \min\{\theta_v(N_G(v)) | v \in V(G)\}$.

Recently, Sano [24] gave a generalization of the above two lower bounds as follows:

Theorem 3 (Sano [24]). Let $G = (V, E)$ be a graph. Let m be an integer such that $1 \leq m \leq |V|$. Then

$$k(G) \geq \min_{U \in \binom{V}{m}} \theta_e(E_G[U]; N_G[U]) - m + 1$$,

where $\binom{V}{m}$ denotes the set of all m-subsets of V.

The following results from [25] will be used in this paper.

Theorem 4 (Harary et al. [25]). Let D = (V, A) be a digraph. Then D is acyclic if and only if there exists an ordering of vertices, $\sigma = [v_1, v_2, \cdots, v_n]$, such that one of the following two conditions holds:

1) For all $i, j \in \{1, 2, \cdots, n\}$, $(v_i, v_j) \in A$ implies that $i < j$;

2) For all $i, j \in \{1, 2, \cdots, n\}$, $(v_i, v_j) \in A$ implies that $i > j$.

Sano [5] pointed out that it would be interesting to compute the competition numbers of some triangulations of a sphere as he obtained the exact value of the competition numbers of regular polyhedra. In this paper we try to study the competition numbers of several kinds of triangulations of a sphere. In Section 2, we study the competition number of a 24-hedron constructed from a hexahedron by adding a vertex in each face of the hexahedron and joining the vertex added in a face with the four vertices of the face. In Section 3, we study the competition numbers of a class of dodecahedra obtained from a hexahedron by adding a diagonal in each face of the hexahedron. In Section 4, we study the competition number of a triangulation of a sphere with $3n (n \geq 2)$ vertices.

In the following, "$S \rightarrow v$" means that we make an arc from each vertex in S to the verte v.

2. A 24-Hedron

In this section we study the competition number of a 24-hedron obtained from a hexahedron by adding a vertex in each face of the hexahedron and joining the vertex added in a face with the four vertices of the face. See Figure 1.

Theorem 5. *Let* T *be the 24-hedron shown in* Figure 1(b). Then k (T)=3.

Proof. Let $V(\mathrm{T})=\{v_1,v_2,...,v_{14})$ and suppose that the adjacencies between two vertices are given as Figure 2. Let

$$S_1 = \{v_2, v_7, v_9\}, S_2 = \{v_1, v_2, v_{13}\}, S_3 = \{v_1, v_4, v_{14}\}, S_4 = \{v_2, v_3, v_4\},$$
$$S_5 = \{v_3, v_6, v_7\}, S_6 = \{v_4, v_5, v_6\}, S_7 = \{v_5, v_{10}, v_{14}\}, S_8 = \{v_6, v_8, v_{10}\},$$
$$S_9 = \{v_7, v_8, v_{11}\}, S_{10} = \{v_9, v_{11}, v_{13}\}, S_{11} = \{v_{10}, v_{11}, v_{12}\}, S_{12} = \{v_{12}, v_{13}, v_{14}\}.$$

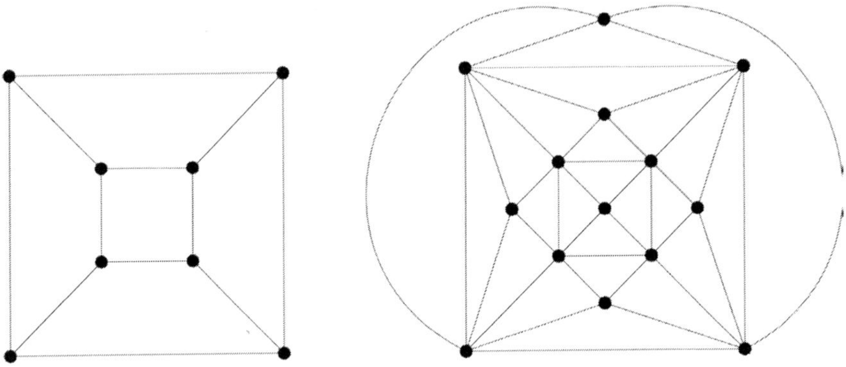

(a) A hexahedron (b) A 24-hedron

Figure 1. Planar embeddings of a hexahedron and a 24-hedron.

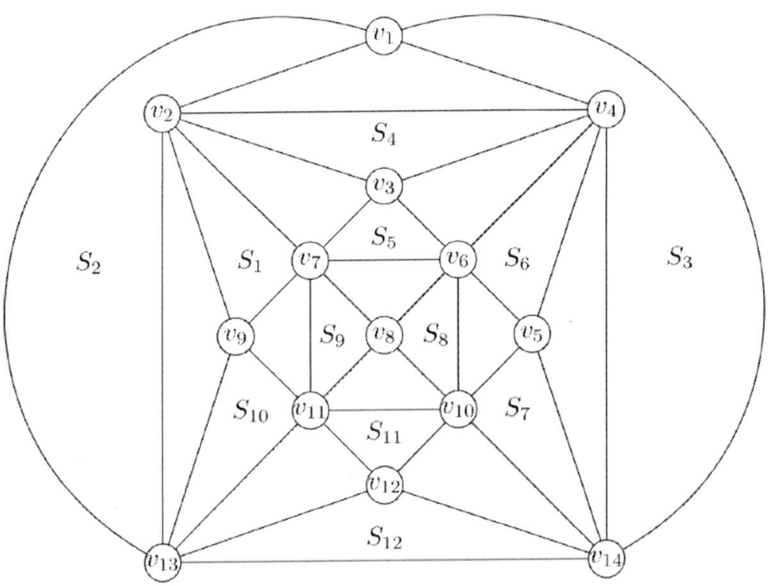

Figure 2. An edge clique cover of the 24-hedron T.

Then the family $\{S_1, S_2, \ldots, S_{12}\}$ is an edge clique cover of T.

Now, we define a digraph D as follows. Let

$$V(D) = V(\mathbb{T}) \cup \{a, b, c\},$$

$$S_1 \to a,\ \ S_2 \to b,\ \ S_3 \to c,\ \ S_4 \to v_1,\ \ S_5 \to v_2,\ \ S_6 \to v_3,$$
$$S_7 \to v_4,\ S_8 \to v_5,\ S_9 \to v_6,\ S_{10} \to v_7,\ S_{11} \to v_8,\ S_{12} \to v_9,$$

where a,b and c are new added vertices. Then by Theorem 4 the digraph D is acyclic and $C(D) = \mathrm{T} \cup \{a,b,c\}$. Hence we have

$$k(\mathbb{T}) \leq 3 \tag{1}$$

On the other hand, by Theorem 3 with m=2, we have

$$k(\mathbb{T}) \geq \min_{U \in \binom{V}{2}} \theta_e\left(E_{\mathbb{T}}[U]; N_{\mathbb{T}}[U]\right) - 1$$

.

There are 7 different cases for the set $E_{\mathrm{T}}[U]$ of edges in the subgraph $N_{\mathrm{T}}[U]$, where $U \in \binom{V}{2}$ (see Figure 3):

1) If $\mathbb{T}[U] = P_2$, then $\theta_e\left(E_{\mathbb{T}}[U]; N_{\mathbb{T}}[U]\right) = 5;$

2) If $\mathbb{T}[U] = I_2$, then $\theta_e\left(E_{\mathbb{T}}[U]; N_{\mathbb{T}}[U]\right) = 6;$

3) If $\mathbb{T}[U] = I_2$, then $\theta_e\left(E_{\mathbb{T}}[U]; N_{\mathbb{T}}[U]\right) = 6;$

4) If $\mathbb{T}[U] = I_2$, then $\theta_e\left(E_{\mathbb{T}}[U]; N_{\mathbb{T}}[U]\right) = 4;$

5) If $\mathbb{T}[U] = I_2$, then $\theta_e\left(E_{\mathbb{T}}[U]; N_{\mathbb{T}}[U]\right) = 4;$

6) If $\mathbb{T}[U] = P_2$, then $\theta_e\left(E_{\mathbb{T}}[U]; N_{\mathbb{T}}[U]\right) = 4;$

7) If $\mathbb{T}[U] = I_2$, then $\theta_e\left(E_{\mathbb{T}}[U]; N_{\mathbb{T}}[U]\right) = 5;$

Thus it holds that

$$\min_{U \in \binom{V}{2}} \theta_e\left(E_{\mathbb{T}}[U]; N_{\mathbb{T}}[U]\right) = 4$$

Hence we have

$$k(\mathbb{T}) \geq 3 \tag{2}$$

Combining inequalities (1) and (2) we have $k(\mathbb{T}) \geq 3$.

3. A CLASS OF DODECAHEDRA

In this section we study the competition numbers of a class of dodecahedra constructed from a hexahedron by adding a diagonal in each face of the hexahedron. It is not difficult to see that there are 6 nonisomorphic such dodecahedra. Denote the 6 different dodecahedra by H_1, H_2, H_3, H_4, H_5 and H_6, respectively. See Figure 4.

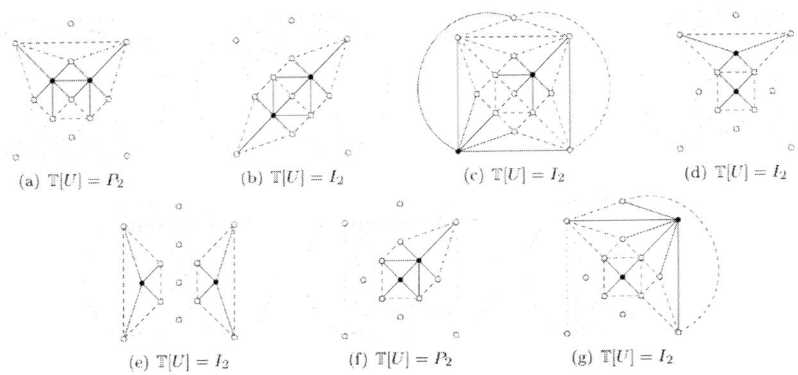

Figure 3. The set $E_T[U]$ of edges in the subgraph $N_T[U]$.

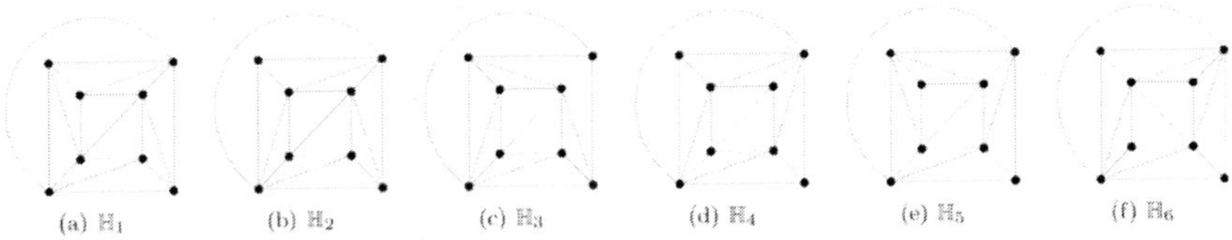

Figure 4. 6 nonisomorphic dodecahedra obtained from a hexahedron.

Theorem 6.
$$k(\mathbb{H}_i) = \begin{cases} 2, & i = 1,3; \\ 1, & i = 2,4,5,6 \end{cases}$$

Proof. Let $V(\mathbb{H}_i) = \{v_1, v_2, \cdots, v_8\}$ and suppose that the adjacencies between two vertices are given as Figure 5, where i=1,2,⋯,6. Now we define digraphs $D_1, D_2, \cdots, D_6$, respectively.

1) D_1.

Let $V(D_1) = V(\mathbb{H}_1) \cup \{a_1, b_1\}$, and $A(D_1)$ be defined as follows:

$$S_1=\{v_1,v_2,v_6\}\to a_1,\quad S_2=\{v_1,v_3,v_5\}\to b_1,\quad S_3=\{v_2,v_3,v_8\}\to v_1,$$
$$S_4=\{v_3,v_4,v_5\}\to v_2,\ S_5=\{v_4,v_7,v_8\}\to v_3,\ S_6=\{v_5,v_6,v_7\}\to v_4,$$
$$S_7=\{v_6,v_8\}\to v_5,$$

where a_1 and b_1 are new added vertices. Note that the family $\{S_1,S_2,\cdots,S_7\}$ is an edge clique cover of H_1.

2) D_2.

Let $V(D_2)=V(\mathbb{H}_2)\cup\{a_2\}$, and $A(D_2)$ be defined as follows:
$$S_1=\{v_1,v_2,v_3,v_5\}\to a_2,\ S_2=\{v_2,v_6,v_8\}\to v_1,\ S_3=\{v_3,v_4,v_6\}\to v_2,$$
$$S_4=\{v_4,v_5,v_7\}\to v_3,\quad S_5=\{v_5,v_7,v_8\}\to v_4,\ S_6=\{v_6,v_7\}\to v_5,$$

where a_2 is a new added vertex. Note that the family $\{S_1,S_2,\cdots,S_6\}$ is an edge clique cover of H_2.

3) D_3.

Let $V(D_3)=V(\mathbb{H}_3)\cup\{a_3,b_3\}$, and $A(D_3)$ be defined as follows:

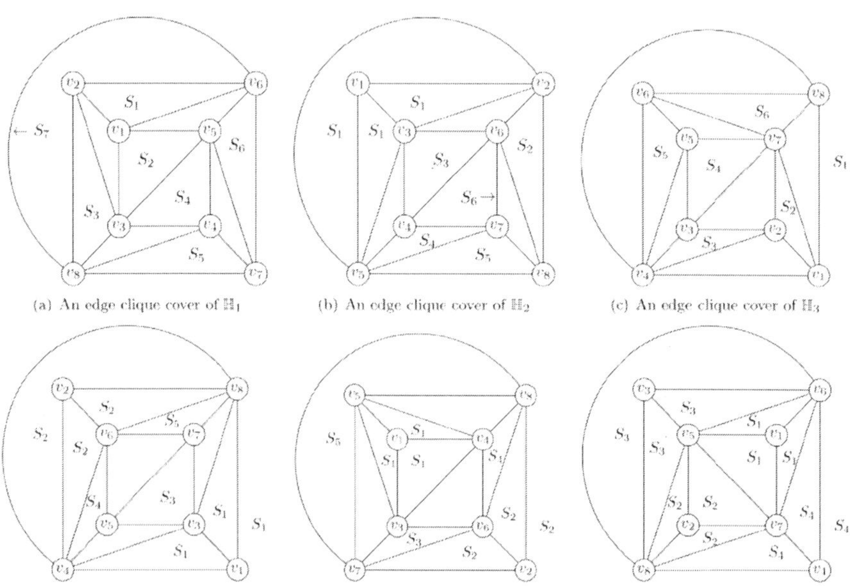

Figure 5. Edge clique covers for H_1, H_2 , H_3, H_4, $H_5$and H_6,.

$$S_1 = \{v_1, v_4, v_8\} \to a_3, \quad S_2 = \{v_1, v_2, v_7\} \to b_3, \quad S_3 = \{v_2, v_3, v_4\} \to v_1,$$
$$S_4 = \{v_3, v_5, v_7\} \to v_2, \quad S_5 = \{v_4, v_5, v_6\} \to v_3, \quad S_6 = \{v_6, v_7, v_8\} \to v_4,$$

where a_3 and b_3 are new added vertices. Note that the family $\{S1, S2, \cdots, S6\}$ is an edge clique cover of H_3.

4) D_4.

Let $V(D_4) = V(\mathbb{H}_4) \cup \{a_4\}$, and $A(D_4)$ be defined as follows:

$$S_1 = \{v_1, v_3, v_4, v_8\} \to a_4, \quad S_2 = \{v_2, v_4, v_6, v_8\} \to v_1, \quad S_3 = \{v_3, v_5, v_7\} \to v_2,$$
$$S_4 = \{v_4, v_5, v_6\} \to v_3, \quad S_5 = \{v_6, v_7, v_8\} \to v_4.$$

where a_4 is a new added vertex. Note that the family $\{S_1, S_2, \cdots, S_5\}$ is an edge clique cover of H_4.

5) D_5.

Let $V(D_5) = V(\mathbb{H}_5) \cup \{a_5\}$, and $A(D_5)$ be defined as follows:

$$S_1 = \{v_1, v_3, v_4, v_5\} \to a_5, \quad S_2 = \{v_2, v_6, v_7, v_8\} \to v_1, \quad S_3 = \{v_3, v_6, v_7\} \to v_2,$$
$$S_4 = \{v_4, v_6, v_8\} \to v_3, \quad S_5 = \{v_5, v_7, v_8\} \to v_4,$$

where a_5 is a new added vertex. Note that the family $\{S_1, S_2, \cdots, S_5\}$ is an edge clique cover of H_5.

6) D_6.

Let $V(D_6) = V(\mathbb{H}_6) \cup \{a_6\}$, and $A(D_6)$ be defined as follows:

$$S_1 = \{v_1, v_5, v_6, v_7\} \to a, \; S_2 = \{v_2, v_5, v_7, v_8\} \to v_1, \; S_3 = \{v_3, v_5, v_6, v_8\} \to v_2,$$
$$S_4 = \{v_4, v_6, v_7, v_8\} \to v_3,$$

where a_6 is a new added vertex. Note that the family $\{S_1, S_2, S_3, S_4\}$ is an edge clique cover of H_6.

By Theorem 4, each D_i is acyclic and

$$C(D_i) = \begin{cases} \mathbb{H}_i \cup \{a_i, b_i\}, & i = 1, 3; \\ \mathbb{H}_i \cup \{a_i\}, & i = 2, 4, 5, 6 \end{cases}.$$

Hence we have

$$k(\mathbb{H}_i)\begin{cases}\le 2, & i=1,3;\\ \le 1, & i=2,4,5,6\end{cases} \tag{3}$$

On the other hand, we note that

- $\delta(\mathbb{H}_1)=\delta(\mathbb{H}_3)=4$ and there is no clique with more than 3 vertices in H_1 and H_3, respectively;
- $N_{\mathbb{H}_2}(v_1)=\{v_2,v_3,v_5\}$ is covered by the clique $S_1=\{v_1,v_2,v_3,v_5\}$ in $\mathbb{H}_2$;
- $N_{\mathbb{H}_4}(v_1)=\{v_3,v_4,v_8\}$ is covered by the clique $S_1=\{v_1,v_3,v_4,v_8\}$ in $\mathbb{H}_4$;
- $N_{\mathbb{H}_5}(v_1)=\{v_3,v_4,v_5\}$ is covered by the clique $S_1=\{v_1,v_3,v_4,v_5\}$ in $\mathbb{H}_5$;
- $N_{\mathbb{H}_6}(v_1)=\{v_5,v_6,v_7\}$ is covered by the clique $S_1=\{v_1,v_5,v_6,v_7\}$ in $\mathbb{H}_6$.

Then we have

$$\theta_v\left(N_{\mathbb{H}_i}(v)\right)\ge 2, v\in V(\mathbb{H}_i), \quad \text{where} \quad i=1,3,$$

and

$$\theta_v\left(N_{\mathbb{H}_i}(v_1)\right)=1, \quad \text{where} \quad i=2,4,5,6.$$

By Theorem 2

$$k(\mathbb{H}_i)\ge \min\left\{\theta_v\left(N_{\mathbb{H}_i}(v)\right)\middle| v\in V(\mathbb{H}_i)\right\}\ge\begin{cases}2, & i=1,3;\\ 1, & i=2,4,5,6\end{cases} \tag{4}$$

Combining inequalities (3) and (4) we have

$$k(\mathbb{H}_i)=\begin{cases}2, & i=1,3;\\ 1, & i=2,4,5,6.\end{cases}$$

4. TRIANGULATION *G*(*N*) OF A SPHERE

In this section we study a graph $G(n)=(V,E)$ with $3n(n\ge 2)$ vertices, where

$$V=\bigcup_{i=1}^{n}\{x_i, y_i, z_i\}$$

and

$$E=\bigcup_{i=1}^{n}\{x_iy_i, y_iz_i, x_iz_i\}\cup\bigcup_{i=1}^{n-1}\{x_ix_{i+1}, x_iz_{i+1}, y_iy_{i+1}, y_iz_{i+1}, z_ix_{i+1}, z_iy_{i+1}\}$$

.

An example $G(3)$ is shown in Figure 6(a). In fact, we can draw $G(n)$ in the plane by the following way. First we draw the triangles $\Delta x_iy_iz_i\ (i=1,2,\cdots,n)$ such that $\Delta x_{i+1}y_{i+1}z_{i+1}$ includes $\Delta x_iy_iz_i$, and x_i, y_i, z_i are in clockwise order if i is even, or in counter-clockwise order if i is odd, respectively. Then we draw the other edges. It is easy to see that each face of $G(n)$ is a triangle. So for each $n\geq 2$, $G(n)$ is a triangulation of a sphere.

Theorem 7. For each $n\geq 2,\ k(G(n))=2$.

Proof. Let σ be a vertex ordering of $G(n)$ such that $\sigma(x_i)=v_{3i-2}$, $\sigma(y_i)=v_{3i-1}$ and $\sigma(z_i)=v_{3i}$, where $i=1,2,\cdots,n$. Let

$$S_1=\{v_1, v_2, v_3\},$$

$$S_{3i-1}=\{v_{3i-2}, v_{3i+1}, v_{3i+3}\},$$

$$S_{3i}=\{v_{3i-1}, v_{3i+2}, v_{3i+3}\}\ \text{and}$$

$$S_{3i+1}=\{v_{3i}, v_{3i+1}, v_{3i+2}\},\ \text{where}\ i=1,\cdots,n-1.$$

Then the family $\{S_1, S_2,\cdots, S_{3n-2}\}$ is an edge clique cover of $G(n)$. An edge clique cover of $G(3)$ is shown in Figure 6(b).

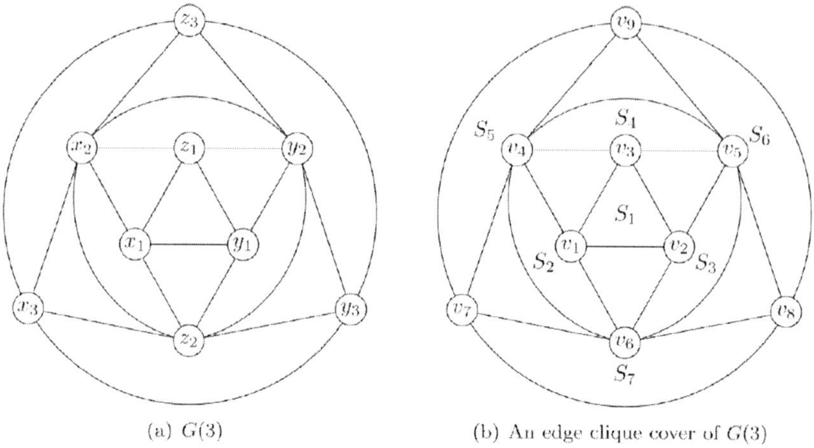

Figure 6. $G(3)$and an edge clique cover of $G(3)$.

Now, we define a digraph D by the following:

$$V(D)=V(G(n))\cup\{a,b\},$$

$$S_1=\{v_1,v_2,v_3\}\to v_4,$$

$$S_{3i-1}=\{v_{3i-2},v_{3i+1},v_{3i+3}\}\to v_{3i+4},$$

$$S_{3i}=\{v_{3i-1},v_{3i+2},v_{3i+3}\}\to v_{3i+5}, \text{ and}$$

$$S_{3i+1}=\{v_{3i},v_{3i+1},v_{3i+2}\}\to v_{3i+3}, \text{ where } i=1,\cdots,n-1$$

Note that $v_{3n+1}=a, v_{3n+2}=b$ are new vertices. Then by Theorem 4 the digraph D is acyclic and $C(D)=G(n)\cup\{a,b\}$. Hence we have

$$k(G(n))\le 2 \tag{5}$$

On the other hand, since $\delta(G(n))=4$ and there is no clique with more than 3 vertices in $G(n)$, then by Theorem 2

$$k(G(n))\ge \min\left\{\theta_v\left(N_{G(n)}(v)\right)\middle| v\in V(G(n))\right\}\ge 2 \tag{6}$$

Combining inequalities (5) and (6) we have $k(G(n))=2$.

5. CLOSING REMARKS

In this paper, we provide the exact values of the competition numbers of a 24-hedron, a class of dodecahedra and a triangulation of a sphere with

$3n(n \geq 2)$ vertices. It would be interesting to compute the competition numbers of some other triangulations of a sphere.

For a digraph$D=(V,A)$ if we partition V into k types, then we may construct a undirected graph $C^k(D)=(V,E)$ of D as follows:

1) $uv \in E$ if and only if there exists some vertex $x \in V$ such that

$(u, x),(v, x) \in A$ and u, v are of the same type, or

2) $uv \in E$ cif and only if there exists some vertex $x \in V$ such that

$(u, x),(v, x) \in A$ and u, v are of different types.

It is easy to see that $C^1(D)=C(D)$ for a given digraph D, and we note that multitype graphs can be used to study the multi-species in ecology and have been deeply studied (see [26] [27]). So these generalizations of competition graphs may be more realistic and more interesting.

ACKNOWLEDGEMENTS

We thank the editor and the referee for their valuable comments. The work was supported in part by the Natural Science Foundation of Hebei Province of China under Grant A2015106045, and in part by the Institute of Applied Mathematics of Shijiazhuang University.

REFERENCES

1. Cohen, J.E. (1968) Interval Graphs and Food Webs: A Finding and a Problem, Document 17696-PR, RAND Corporation, Santa Monica, CA.
2. Roberts, F.S. (1978) Food Webs, Competition Graphs, and the Boxicity of Ecological Phase Space. In: Alavi, Y. and Lick, D., Eds., Theory and Applications of Graphs (International Proceding Conference, Western Michigan University, Kalamazoo, Michigan, 1976), 477-490. https://doi.org/10.1007/bfb0070404
3. Opsut, R.J. (1982) On the Computation of the Competition Number of a Graph. SIAM Journal on Algebraic Discrete Methods, 3, 420-428. https://doi. org/ 10.1137/0603043
4. Kim, S. R. and Roberts, F.S. (1997) Competition Numbers of Graphs with a Small Number of Triangles. Discrete Applied Mathematics, 78, 153-162. https://doi.org /10.1016/S0166-218X(97)00026-7

5. Sano, Y. (2009) The Competition Numbers of Regular Polyhedra. Congressus Numerantium, 198, 211-219. https://doi.org/10.7151/dmgt.1506
6. Kim, S.R., Park, B. and Sano, Y. (2010) The Competition Numbers of Johnson Graphs. Discussiones Mathematicae Graph Theory, 30, 449-459.
7. Park, B. and Sano, Y. (2011) The Competition Numbers of Hamming Graphs with Diameter at Most Three. Journal of the Korean Mathematical Society, 48, 691-702. https://doi.org/10.4134/JKMS.2011.48.4.691
8. Park, B. and Sano, Y. (2011) The Competition Numbers of Ternary Hamming Graphs. Applied Mathematics Letters, 24, 1608-1613. https://doi.org/ 10.1016 /j.aml.2011.04.012
9. Kim, S.R., Park, B. and Sano, Y. (2013) The Competition Number of the Complement of a Cycle. Discrete Applied Mathematics, 161, 1755-1760. https://doi.org/ 10.1016/j.dam.2011.10.034
10. Kim, S.R., Park, B. and Sano, Y. (2012) The Competition Numbers of Complete Multipartite Graphs with Many Partite Sets. Discrete Applied Mathematics, 160, 1176-1182. https://doi.org/10.1016/j.dam.2011.12.017
11. Kim, S.R. and Sano, Y. (2008) The Competition Numbers of Complete Tripartite Graphs. Discrete Applied Mathematics, 156, 3522-3524. https://doi.org/ 10.1016/j.dam.2008.04.009
12. Kuhl, J. (2013) Transversals and Competition Numbers of Complete Multipartite Graphs. Discrete Applied Mathematics, 161, 435-440. https://doi.org/ 10.1016/j.dam.2012.09.012
13. Li, B.J. and Chang, G.J. (2012) Competition Numbers of Complete -Partite Graphs. Discrete Applied Mathematics, 160, 2271-2276. https://doi.org/ 10.1016 /j.dam.2012.05.005
14. Park, B., Kim, S.R. and Sano, Y. (2009) The Competition Numbers of Complete Multipartite Graphs and Mutually Orthogonal Latin Squares. Discrete Mathematics, 309, 6464-6469. https://doi.org/10.1016/j.disc.2009.06.016
15. Cho, H.H. and Kim, S.R. (2005) The Competition Number of a Graph Having Exactly One Hole. Discrete Mathematics, 303, 32-41. https://doi.org/ 10.1016/ j.disc.2004.12.016
16. Kamibeppu, A. (2012) A Sufficient Condition for Kim's Conjecture on the Competition Numbers of Graphs. Discrete Mathematics, 312, 1123-1127. https://doi.org/10.1016/j.disc.2011.11.035
17. Kamibeppu, A. (2010) An Upper Bound for the Competition Numbers of Graphs. Discrete Applied Mathematics, 158, 154-157. https://doi.org/ 10.1016/ j.dam.2009.09.007
18. Kim, S.R. (2005) Graphs with One Hole and Competition Number One. Journal of the Korean Mathematical Society, 42, 1251-1264. https://doi.org /10.4134/JKMS.2005.42.6.1251
19. Kim, S.R., Lee, J.Y., Park, B. and Sano, Y. (2012) The Competition Number of a Graph and the Dimension of Its Hole Space. Applied Mathematics Letters, 25, 638-642. https://doi.org/10.1016/j.aml.2011.10.003
20. Kim, S.R., Lee, J.Y. and Sano, Y. (2010) The Competition Number of a Graph Whose Holes Do Not Overlap Much. Discrete Applied Mathematics, 158, 1456-1460. https://doi.org/10.1016/j.dam.2010.04.004
21. Lee, J.Y., Kim, S.R., Kim, S.J. and Sano, Y. (2011) Graphs Having Many Holes But with Small Competition Numbers. Applied Mathematics Letters, 24, 1331-1335. https://doi.org/10.1016/j.aml.2011.03.003
22. Lee, J.Y., Kim, S.R., Kim, S.J. and Sano, Y. (2010) The Competition Number of a Graph with Exactly Two Holes. Ars Combinatoria, 95, 45-54.
23. Li, B.J. and Chang, G.J. (2009) The Competition Number of a Graph with Exactly h Holes, All of Which Are Independent. Discrete Applied Mathematics, 157, 1337-1341. https://doi.org/10.1016/j.dam.2008.11.004

24. Sano, Y. (2013) A Generalization of Opsut's Lower Bounds for the Competition Number of a Graph. Graphs and Combinatorics, 29, 1543-1547. https:// doi.org/10.1007/s00373-012-1188-5
25. Harary, F., Norman, R.Z. and Cartwright, D. (1965) Structure Models: An Introduction to the Theory of Directed Graphs, Wiley, New York.
26. Barbour, A.D. and Reinert, G. (2011) The Shortest Distance in Random Multi-Type Intersection Graphs. Random Structures & Algorithms, 39, 179-209. https://doi.org/10.1002/rsa.20351
27. Shang, Y. (2016) Groupies in Multitype Random Graphs. SpringerPlus, 5, 989. https://doi.org/10.1186/s40064-016-2705-4

CHAPTER 4

On the 2-Domination Number of Complete Grid Graphs

Ramy Shaheen, Suhail Mahfud, Khames Almanea

Department of Mathematics, Faculty of Science, Tishreen University, Lattakia, Syria

ABSTRACT

A set D of vertices of a graph $G = (V, E)$ is called k -dominating if every vertex $v \in V - D$ is adjacent to some k vertices of D. The k -domination number of a graph G, $\gamma_k(G)$, is the order of a smallest k -dominating set of G. In this paper we calculate the k -domination number (for $k = 2$) of the product of two paths $P_m \times P_n$ for $m = 1, 2, 3, 4, 5$ and arbitrary n. These results were shown an error in the paper [1].

Keywords: k-Dominating Set, k-Domination Number, 2-Dominating Set, 2-Domination Number, Cartesian Product Graphs, Paths

1. INTRODUCTION

Let $G = (V, E)$ be a graph. A subset of vertices $D \subseteq V$ is called a 2-dominating set of G if for every $v \in V$, either $v \in D$ or v is adjacent to at least two vertices of D. The 2-domination number $\gamma_2(G)$ is equal to min{|D|:D is a 2 – dominating set of G}.

The Cartesian product $G \times H$ of two graphs G and H is the graph with vertex set $V(G\times H) = V(G)\times V(H)$, where two vertices (v_1, v_2), $(u_1, u_2) \in G\times H$ are adjacent if and only if either $v_1u_1 \in E(G)$ and $v_2 = u_2$ or $v_2u_2 \in E(H)$ and $v_1 = u_1$.

Let G be a path of order n with vertex set $V(G) = \{1, 2,..., n\}$. Then for two paths of order m and n respectively, we have $P_m \times Pn = \{(i, j): 1 \le i \le m, 1 \le j \le n\}$. The jth column of $P_m + P_n$ is $k_j = \{(i, j): i = 1,...,m\}$. If D is a 2-dominating set for $P_m \times P_n$, then we put $W_j = D \cap K_j$. Let $s_j = |W_j|$. The sequence $(s_1, s_2,..., s_n)$ is called a 2-dominating sequence corresponding to D. For a graph G, we refer to minimum and maximum degrees by $\delta(G)$ and $\Delta(G)$, and for simplicity denoted those by δ and Δ, respectively. Also, we denote by V and E to order and size of graph G, respectively.

2. NOTATION AND TERMINOLOGY

Fink and Jacobson [2] [3] in 1985 to introduced the concept of multiple domination. A subset $D \subseteq V$ is k-dominating in G if every vertex of $V - D$ has at least k neighbours in D. The cardinality of a minimum k-dominating set is called the k-domination number $\gamma_k(G)$ of G. Clearly, $g_1(G) = g(G)$. Naturally, every k-dominating set of a graph G contains all vertices of degree less than k. Of course, every $(k+1)$ -dominating set is also a k-dominating set and so $\gamma_k(G) \le \gamma_{k+1}(G)$. Moreover, the vertex set V is the only $(\Delta+1)$ -dominating set but evidently it is not a minimum Δ-dominating set. Thus every graph G satisfies

$$\gamma_k(G) \le \gamma_{k+1}(G) \le \cdots \le \gamma_\Delta(G) < \gamma_{\Delta+1}(G) = |V|.$$

For a comprehensive treatment of domination in graphs, see the monographs by Haynes et al. [4]. Also, for more information see [5] [6]. Fink and Jacobson [2], introduced the following theorems:

Theorem 2.1 [2]. If $k \ge 2$, is an integer and G is a graph with $k \le \Delta(G)$, then $\gamma_k(G) \ge \gamma(G) + k - 2$.

Theorem 2.2 [2]. If T is a tree, then $\gamma_2(T) \ge \frac{|T|+1}{2}$.

In [6], Hansberg and Volkmann, proved the following theorem.

Theorem 2.4 [6]. Let Let $G = (V, E)$ be a graph of order n and minimum degree δ and let $k \in N$. If $\frac{\delta+1}{\ln(\delta+1)} \ge 2k$, then

$$\gamma_k(G) \le \frac{|V|}{\delta+1}\left(k\ln(\delta+1) + \sum_{i=0}^{k-1}\frac{\delta^i}{i!(\delta+1)^{k-1}}\right).$$

Cockayne, et al . [7], established an upper bound for the k-domination number of a graph G has minimum degree k, they gave the following result.

Theorem 2.3 [7]. Let G be a graph with minimum degree at least k , then $\gamma_k(G) \le \frac{k|V|}{(k+1)}$.

Favaron, et al . [9], gave new upper bounds of $\gamma_k(G)$.

Corollary 2.6 [9]. Let G be a graph of order n and minimum degree δ. If $k \le \delta$ is an integer, then $\gamma_k(G) \le \frac{\delta}{2\delta + 1 - k}|V|$.

In [4], Haynes et al . showed that the 2-domination number is bounded from below by the total domination number for every nontrivial tree.

Theorem 2.7 [4]. For every nontrivial tree, $\gamma_2(T) \ge \gamma_t(T)$.

Also, Volkmann [10] gave the important following result.

Theorem 2.8 [10]. Let G be a graph with minimum degree $\delta \ge k+1$, the $\gamma_{k+1}(G) \le \frac{|V| + \gamma_k(G)}{2}$.

Shaheen [11] considered the 2-domination number of Toroidal grid graphs and gave

an upper and lower bounds. Also, in [12], he introduced the following results.

Theorem 2.9 [12].

1) $\gamma_2(C_n) = [n/2]$.
2) $\gamma_2(C_3 \times C_n) = n : n \equiv 0 \pmod 3$, $\gamma_2(C_3 - C_n) = n + 1 : n \equiv 1, 2 \pmod 3$.
3) $\gamma_2(C_4 \times C_n) = n + [n/2] : n \equiv 0, 3, 5 \pmod 8$, $\gamma_2(C_4 - C_n) = n + [n/2] + 1$, $n \equiv 1,2,4,67 \pmod{14}$
4) $\gamma_2(C_5 \times C_n) = 2n$
5) $\gamma_2(C_6 \times C_n) = 2n : n \equiv 0 \pmod 3$
6) $\gamma_2(C_7 \times C_n) = [5n/2]$: $n \equiv 0,3,11 \pmod{14}$, $\gamma_2(C_7 \times C_n) = [5n/2] + 1$: $n \equiv 5,6,7,8,9,10 \pmod{14}$, $\gamma_2(C_7 \times C_n) = [5n/2] + 1$: $n \equiv 1,2,4,12,13 \pmod{14}$.

In this paper we calculate the k-domination number (for $k = 2$) of the product of two paths $P_m \times P_n$ for $m = 1, 2, 3, 4, 5$ and arbitrary n. These results were shown an error in the paper [1]. We believe that these results were wrong. In our paper we will provide improved and corrected her, especially for $m = 3, 4, 5$.

The following formulas appeared in [1],

$$\gamma_2(P_n) = \lceil (n+1)/2 \rceil \cdot \gamma_2(P_2 \times P_n) = n \cdot \gamma_2(P_3 \times P_n) = 2n - \lceil n/2 \rceil \cdot \gamma_2(P_4 \times P_n) = 2n.$$

$$\gamma_2(P_5 \times P_n) = 3n - \lceil n/2 \rceil \cdot \gamma_2(P_{2k+1} \times P_n) = (k+1)n - \lceil n/2 \rceil.$$

$$\gamma_2(P_m \times P_n) = \lceil m/2 \rceil n - \lceil n/2 \rceil : m \equiv 1 (\text{mod } 2),$$

$$\gamma_2(P_m \times P_n) = \lceil m/2 \rceil n : m \equiv 0 (\text{mod } 2).$$

In this paper, we correct the results in [1] and proves the following:

$$\gamma_2(P_n) = \lceil (n+1)/2 \rceil \cdot \gamma_2(P_2 \times P_n) = n \cdot \gamma_2(P_3 \times P_n) = n + \lceil n/3 \rceil.$$

$$\gamma_2(P_4 \times P_n) = 2n - \lfloor n/4 \rfloor : n \equiv 3, 7 (\text{mod } 8),$$

$$\gamma_2(P_4 \times P_n) = 2n - \lfloor n/4 \rfloor + 1 : n \equiv 0, 1, 2, 4, 5, 6 (\text{mod } 8).$$

$$\gamma_2(P_5 \times P_n) = 2n + \lceil n/7 \rceil : n \equiv 1, 2, 3, 5 (\text{mod } 7),$$

$$\gamma_2(P_5 \times P_n) = 2n + \lceil n/7 \rceil + 1 : n \equiv 0, 4, 6 (\text{mod } 7).$$

3. MAIN RESULTS

Our main results here are to establish the domination number of Cartesian product of two paths P_m and P_n for m = 1, 2, 3, 4, 5 and arbitrary n. We study 2-dominating sets in complete grid graphs using one technique: by given a minimum of upper 2-dominating set D of $P_m \times P_n$ and then we establish that D is a minimum 2-dominating set of $P_m \times P_n$ for several values of m and arbitrary n. Definitely we have $\gamma_2 (P_m \times P_n) = |D|$.

Let G be a path of order n with vertex set $V(G) = \{1,2,...,n\}$. For two paths of order m and n respectively is:

$(P_m \times P_n) = \{(i,j)$: $1 \leq i \leq m$, $1 \leq j \leq n,\}$. The jth column $P_m \times P_n$ is $K_j = \{(i,j)$: $i = 1,...,m\}$.

If D is a 2-dominating set for $P_m \times P_n$ then we put $W_j = D \cap K_j$. Let $s_j = |W_j|$. The sequence $(s_1, s_2,..., s_n)$ is called a 2-dominating sequence corresponding to D. Always we have $s_1, s_n \geq \lceil m/3 \rceil$. Suppose that $s_j = 0$ for some j (where $j \neq 1$ or n). The vertices of the jth column can only be 2-dominated by vertices of the (j – 1)st columns and (j + 1)st columns. Thus we have $S_{j-1} + S_{j+1} = m$, then . In general $S_{j-1} + 4S_j + S_{j+1} \geq 2m$.

Notice 3.1.

1) The study of 2-dominating sequence $(s_1, s_2,..., s_n)$ is the same as the study of the 2-dominating sequence $(s_n, s_{n-1},..., s_1)$.

2) If subsequence $(s_j, s_{j+1},..., s_{j+k})$ is not possible, then its reverse $(s_{j+k},..., s_{j+1}, s_j)$ is not possible.

3) We say that two subsequences $(s_j,..., s_{j+q})$,($s_{j+q+1},..., s_{j+r}$) are equivalent, if the sequence$(s_j,..., s_{j+q}, s_{j+q+1},..., s_{j+r})$ is possible.

We need the useful following lemma.

Lemma 3.1. There is a minimum 2-dominating set for $P_m \times P_n$ with 2-dominating sequence $(s_1, s_{2^-},..., s_n)$ such that, for all $j = 1,2,...,n$, is $\lfloor m/4 \rfloor \le s_j \le \lceil 3m/4 \rceil$.

Proof. Let D be a minimum 2-dominating set for $P_m \times P_n$ with 2-dominating sequence $(s_1, s_{2^-},..., s_n)$. Assume that for some j, s_j is large. Then we modify D by moving two vertices from column j, one to column $j-1$ and another one to column $j+1$, such that the resulting set is still 2-dominating set for $P_m \times P_n$. For $1 \le i \le m$ and $1 \le j \le n$, let $W = D \cap \{(i, j), (i+1, j), (i+2, j), (i+3, j)\}$. If $|W| = 4$, then we define $W = (D-W) \cup \{(i, j), (i+1, j-1), (i+2, j+1), (i+3, j)\}$, see Figure 1. We repeat this process if necessary eventually leads to a 2-dominating set with required properties. Also, we get D_1 is a 2-dominating set for $P_m \times P_n$ with $|D| = |D_1|$. Thus, we can assume that every four consecutive vertices of the jth column include at most three vertices of D. This implies that $s_j \le \lceil 3m/4 \rceil$, for all $1 \le j \le n$.

To prove the lower bound, we suppose that $|K_j \cap D|$ is be a maximum, i.e., $s_j = \lceil 3m/4 \rceil$. Then for each $(i, j) \notin D$, we have

$|\{(i-1,j+1), (i,j+1), (i+1,j+1)\} \cap D| \ge 1$. When $s_j \le \lceil 3m/4 \rceil$, there at must $m - \lceil 3m/4 \rceil = \lfloor m/4 \rfloor$ vertices does not in $K_j \cap D$. This implies that $s_{j+1} \ge \lfloor m/4 \rfloor$. So, the same as for $s_{j-1} \ge \lfloor m/4 \rfloor$.

By Lemma 3.1, always we have a minimum 2-dominating set D with 2-dominating sequence $(s_1, s_2,..., s_n)$, such that $\lfloor m/4 \rfloor \le s_j \le \lceil 3m/4 \rceil$, for all $j = 1,2,...,n$.

Lemma 3.2. $\gamma_2(P_n) = \left\lceil \frac{n+1}{2} \right\rceil$.

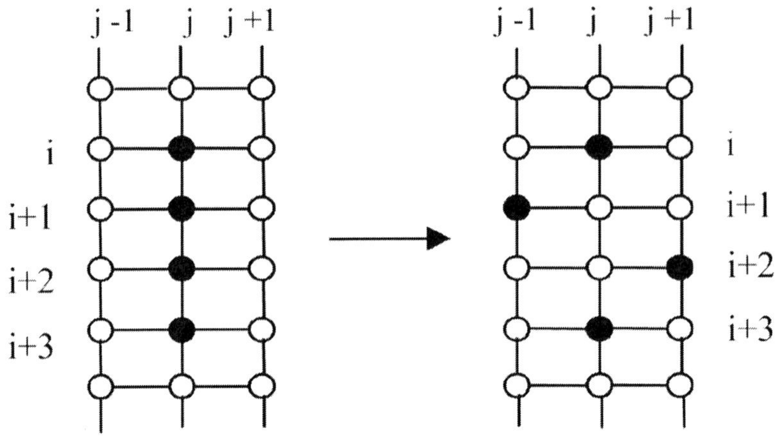

Figure 1. Modify D.

Proof. Let $D = \left\{(2k-1); 1 \le k \le \left\lceil \frac{n}{2} \right\rceil \right\}$.

We have D is a 2-dominating set of P_n for $n \equiv 1 \pmod 2$ with $|D| = \left\lceil \frac{n+1}{2} \right\rceil$, also $D \cup \{(n)\}$ is a 2-dominating set of P_n for $n \equiv 0 \pmod 2$ with $|D \cup \{(n)\}| = \left\lceil \frac{n+1}{2} \right\rceil$.

Let D_1 be a minimum 2-dominating set for P_n with $V(P_n) = \{x_1, x_2, \ldots, x_n\}$. Since $x_1 x_n \notin E(P_n)$, we need to $x_1, x_n \in D_1$, also if $x_j \notin D_1$ then x_{j-1}, x_{j+1} are belong to D_1, this implies that $x_{2j-1} \in D_1$ for $2 \le j \le \left\lfloor \frac{n}{2} \right\rfloor$. Thus implies that $|D_1| \ge 2 + \left\lfloor \frac{n}{2} \right\rfloor - 1 = \left\lceil \frac{n+1}{2} \right\rceil$. We result that $\gamma_2(P_n) = \left\lceil \frac{n+1}{2} \right\rceil$.

Theorem 3.1. $\gamma_2(P_2 \times P_n) = n$.

Proof. Let a set $D = \left\{(1, 2k-1) : 1 \le k \le \left\lceil \frac{n}{2} \right\rceil \right\} \cup \left\{(2, 2k) : 1 \le k \le \left\lfloor \frac{n}{2} \right\rfloor \right\}$.

It is clear that $|D| = n$. (1)

We can check that D is 2-dominating set for $P_2 \times P_n$, see Figure 2. Let D_1 be a minimum 2-dominating set for $P_2 \times P_n$ with dominating sequence $(s_1, \ldots, s_n)$. If $s_i \ge 1$ for all

$$j = 1, \cdots, n \text{ , then } |D_1| = \sum_{j=1}^{n} s_j \geq n \text{ .} \tag{2}$$

Let $s_j = 0$ for some j, then $s_{j-1} = s_{j+1} = 2$, also we have $s_1 \geq 1$ and $s_n \geq 1$. Now we define a new sequence $(s'_1, \cdots, s'_n)$, (not necessarily a 2-dominating sequence) as follows:

For $s_j = 2$, if $j = 1$ or n, we put $s'_j = s_j - 1$, $s'_2 = s_2 + 1/2$ and $s'_{n-1} = s_{n-1} + 1/2$.

If $j \neq 1$ or n, we put $s'_j = s_j - 1$, $s'_{j-1} = s_{j-1} + 1/2$ and $s'_{j+1} = s_{j+1} + 1/2$.

Otherwise $s'_j = s_j$.

We get a sequence $(s'_1, \cdots, s'_n)$ have property that each $s'_j \geq 1$ with

$$|D| = \sum_{j=1}^{n} s_j = \sum_{j=1}^{n} s'_j \geq n \text{ .} \tag{3}$$

By (1), (2) and (3) is γ_2 $(P_2 \times P_n) = n$. This completes the proof of the theorem.

Theorem 3.2. $\gamma_2(P_3 \times P_n) = n + \left\lceil \frac{n}{3} \right\rceil$.

Proof. Let

$$D = \left\{ (2, 3k-2) : 1 \leq k \leq \left\lceil \frac{n}{3} \right\rceil \right\} \cup \left\{ (2, 3k) : 1 \leq k \leq \left\lfloor \frac{n}{3} \right\rfloor \right\}$$
$$\cup \left\{ (1, 3k-1), (3, 3k-1) : 1 \leq k \leq \left\lceil \frac{n-1}{3} \right\rceil \right\}.$$

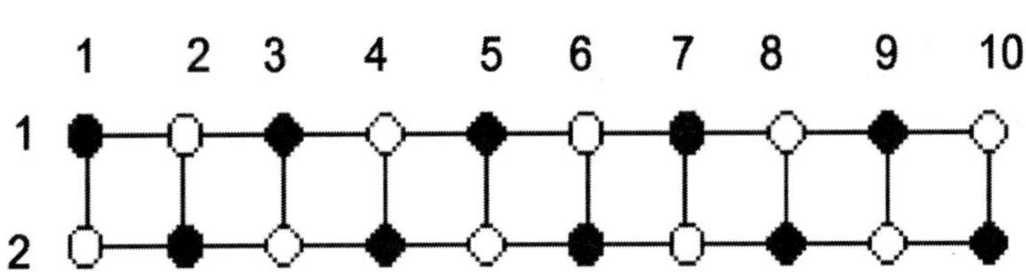

Figure 2. A 2-dominating set for $P_2 \times P_{10}$.

$$D' = \left\{(1,3k-2),(3,3k-2) : 1 \le k \le \left\lceil \frac{n}{3} \right\rceil\right\} \cup \left\{(2,3k-1) : 1 \le k \le \left\lceil \frac{n-1}{3} \right\rceil\right\}$$
$$\cup \left\{(2,3k) : 1 \le k \le \left\lfloor \frac{n}{3} \right\rfloor\right\}$$

.

We have $|D| = n + \left\lceil \frac{n}{3} \right\rceil$ and $|D'| = n + \left\lceil \frac{n}{3} \right\rceil$. (4)

By definition D and D' we note that

D is 2-dominating set for $P_3 \times P_n$ when $n = 0,2(\text{mod}3)$, (see Figure 3, for $P_3 \times P_{14}$).

D' is 2-dominating set for $P_3 \times P_n$ when $n = 1(\text{mod}3)$, (see Figure 4, for $P_3 \times P_{10}$).

Let D_1 be a minimum 2-dominating set for $P_3 \times P_n$ with 2-dominating sequence $(s_1,\ldots,s_n)$ we have $s_1, s_n \ge 1$ and

if $s_1, s_n \ge 1$ then $s_2, s_{n-1} \ge 2$,

if $s_1, s_n \ge 2$ then $s_2, s_{n-1} \ge 1$.

Also for $1 < j < n$, if $s_j = 0$ then $s_{j-1} = s_{j+1} = 3$,

$s_j = 1$ then $s_{j-1} + s_{j+1} \ge 3$,

$s_j = 2$ then $s_{j-1} + s_{j+1} \ge 2$,

If no one of $s_j = 0$ for all j, then $|D_1| = \sum_{j=1}^{n} s_j \ge n + \left\lceil \frac{n}{3} \right\rceil$.(5)

Let $s_j = 0$ ($j \ne 1$ or n) for some j, we define a sequence $(s'_1,\cdots,s'_n)$, (not necessarily a 2-dominating sequence) as follows:

If $s_j = 3$, then we put $s'_j = s_j - 1$, $s'_{j-1} = s_{j-1} + 1/2$ and $s'_{j+1} = s_{j+1} + 1/2$, otherwise

$s'_j = s_j$. We have $|D_1| = \sum_{j=1}^{n} s_j = \sum_{j=1}^{n} s'_j$. We note that the sequence $(s'_1,\cdots,s'_n)$ have the

property if $s'_j = 1$ then $s'_{j-1} + s'_{j+1} \ge 3$. Thus implies that

$$|D_1| = \sum_{j=1}^{n} s'_j \geq n + \left\lceil \frac{n}{3} \right\rceil . \tag{6}$$

From (4), (5) and (6) we get the required result. □

Theorem 3.3.

$$\gamma_2(P_4 \times P_n) = \begin{cases} 2n - \left\lfloor \dfrac{n}{4} \right\rfloor : n \equiv 3,7 \pmod 8, \\ 2n - \left\lfloor \dfrac{n}{4} \right\rfloor + 1 : n \equiv 0,1,2,4,5,6 \pmod 8. \end{cases}$$

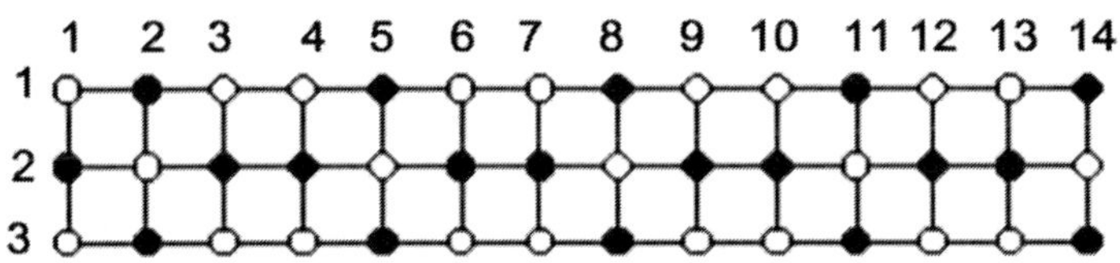

Figure 3. A 2-dominating set for $P_3 \times P_{14}$.

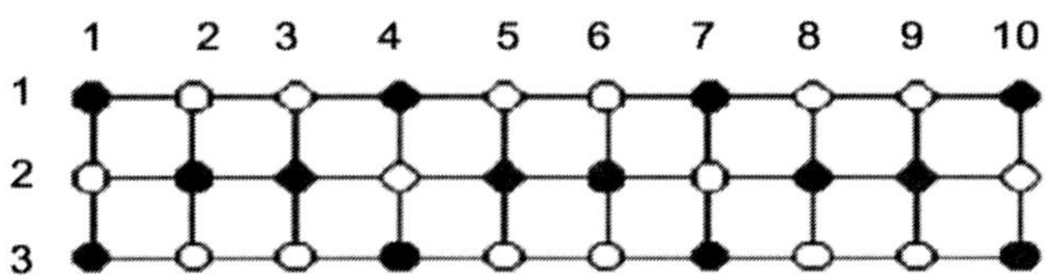

Figure 4. A 2-dominating set for $P_3 \times P_{10}$.

Proof. Let a set D defined as follows:

$$\begin{aligned} D = \Bigg\{ & \{(2,1),(3,1)\} \cup \left\{(1,4k-2),(4,4k-2); 1 \leq k \leq \left\lceil \frac{n-1}{4} \right\rceil \right\} \\ & \cup \left\{(2,8k-5); 1 \leq k \leq \left\lceil \frac{n-2}{8} \right\rceil \right\} \\ & \cup \left\{(1,8k-4),(3,8k-4),(4,8k-4); 1 \leq k \leq \left\lceil \frac{n-3}{8} \right\rceil \right\} \\ & \cup \left\{(2,8k-3); 1 \leq k \leq \left\lceil \frac{n-4}{8} \right\rceil \right\} \cup \left\{(3,8k-1); 1 \leq k \leq \left\lceil \frac{n-6}{8} \right\rceil \right\} \\ & \cup \left\{(1,8k),(2,8k),(4,8k); 1 \leq k \leq \left\lceil \frac{n-7}{8} \right\rceil \right\} \cup \left\{(3,8k+1); 1 \leq k \leq \left\lfloor \frac{n-1}{8} \right\rfloor \right\} \Bigg\} \end{aligned}$$

$$D' = \{(2,n)\}, \quad D'' = \{(3,n)\} .$$

We can check that the following sets are 2-dominating set for $P_4 \times P_n$ (see Figure 5, for $P_4 \times P_{11}$) as indicated:

D is 2-dominating set for $P_4 \times P_n$ when $n \equiv 0,4(\text{mod}8)$.

$D \cup D'$is 2-dominating set for $P_4 \times P_n$ when $n \equiv 1,2,7(\text{mod}8)$.

$D \cup D''$is 2-dominating set for $P_4 \times P_n$ when $n \equiv 3,5,6(\text{mod}8)$.

We have

$$|D| = \begin{Bmatrix} 2n - \left\lfloor \frac{n}{4} \right\rfloor - 1 : n \equiv 3,7 \pmod 8, \\ 2n - \left\lfloor \frac{n}{4} \right\rfloor : n \equiv 1,2,5,6 \pmod 8, \\ 2n - \left\lfloor \frac{n}{4} \right\rfloor + 1 : n \equiv 0,4 \pmod 8. \end{Bmatrix}$$

Let D_1 be a minimum 2-dominating set for $P_4 \times P_n$ with 2-dominating sequence $(s_1,\ldots,s_n)$ we shall show that

$$|D_1| = \begin{Bmatrix} 2n - \left\lfloor \frac{n}{4} \right\rfloor : n \equiv 3,7 \pmod 8, \\ 2n - \left\lfloor \frac{n}{4} \right\rfloor + 1 : n \equiv 0,1,2,4,5,6 \pmod 8. \end{Bmatrix}$$

By Lemma 3.1, we have $1 \leq s_j \leq 3$. Thus

If $s_j = 1$ then $s_{j-1} + s_{j+1} \geq 5$.

If $s_j = 2$ then $s_{j-1} + s_{j+1} \geq 2$.

If $s_j = 3$ then $s_{j-1} + s_{j+1} \geq 2$.

Also, we have s_1, $s_n \geq 2$. If s_1, $s_n = 2$ then s_1, $s_{n-1} \geq 2$, and if s_1, s_n = 3 then

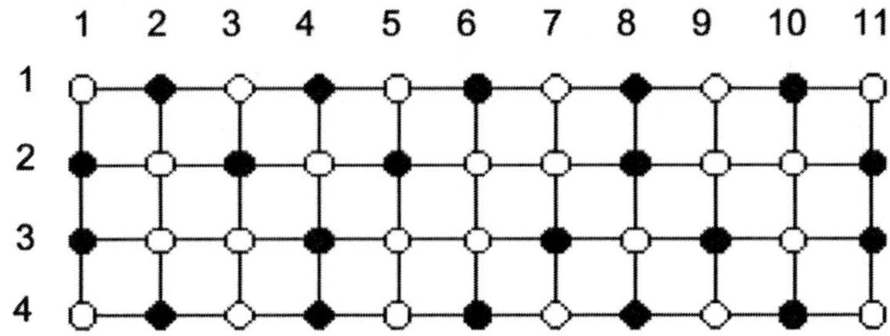

Figure 5. A 2-dominating set for $P_4 \times P_{11}$.

$s_2, s_{n-1} \geq 1$.

We define a new set D_1' with sequence $(s_1', \cdots, s_n')$, (not necessarily a 2-dominating sequence) as follows: if $s_j \geq 2$, let $M_j = s_j - \frac{7}{4}$. Now, for $j = 2$ to $j = n - 1$, if $s_j \geq 2$, then we put

$$s_j' = s_j - M_j, \ s_{j-1}' = s_{j-1} + \frac{M_j}{2} \text{ and } s_{j+1}' = s_{j+1} + \frac{M_j}{2}$$

Thus, for $3 \leq j \leq n - 2$, we have $s_j \geq \frac{7}{4}$. Since if $s_j \geq 2$ then $s_j' \geq \frac{7}{4}$ and if $s_j = 1$, then $s_{j-1} + s_{j+1} = 5$ this implies that $M_{j-1} + M_{j+1} = 5 - \frac{14}{4} = \frac{6}{4}$, which implies that $s_j' = s_j + \frac{M_{j-1}}{2} + \frac{M_{j+1}}{2} = 1 + \frac{3}{4} = \frac{7}{4}$.

We have three cases:

Case 1: $s_1, s_n \geq 2$, then $s_2, s_{n-1} \geq 2$, these implies that $s_1' \geq s_1 + \frac{1}{8}$ and $s_n' \geq s_n + \frac{1}{8}$ also

$$|D_1| = \sum_{j=1}^{n} s_j = \sum_{j=1}^{n} s_j' = s_1' + s_n' + \sum_{j=2}^{n-1} s_j' \geq 2 + \frac{1}{8} + 2 + \frac{1}{8} + \frac{7(n-2)}{4} = \frac{7n}{4} + \frac{3}{4}.$$

Case 2: S_1, s_n = 3 then S_1, s_{n-1} ≥ 2. Thus implies that S'_1, s'_n = 3 and $s'_2, s'_{n-1} \geq 1+\frac{1}{8}$ Then

$$|D_1| = \sum_{j=1}^{n} s_j = \sum_{j=1}^{n} s'_j = s'_1 + s'_2 + s'_{n-1} + s'_n + \sum_{j=2}^{n-2} s'_j \geq 3+1+\frac{1}{8}+3+1+\frac{1}{8}+\frac{7}{4} = \frac{7n}{4}+\frac{5}{4}$$

Case 3: s_1 = 2, s_n = 3 and s_2 ≥ 2, s_{n-1} ≥ 1 or s_1 = 3, s_n = 2 and s_2 ≥ 1, s_{n-1} ≥ 2. Two cases are similar by symmetry. We consider the first case:

s_1 = 2, s_2 ≥ 2and s_n = 3, s_{n-1} ≥ 1, this implies that

$s'_1 = 2+\frac{1}{8},\ s'_2 = \frac{7}{4},\ s'_n = 3,\ s'_{n-1} = 1+\frac{1}{8}$ and

$$|D_1| = \sum_{j=1}^{n} s_j = \sum_{j=1}^{n} s'_j = s'_1 + s'_2 + s'_{n-1} + s'_n + \sum_{j=3}^{n-2} s'_j \geq 2+\frac{1}{8}+\frac{7}{4}+3+1+\frac{1}{8}+\frac{7}{4}(n-4) = \frac{7n}{4}+1$$

But, we have the 2-domination number is positive integer number, also we have

$$2n-\left\lfloor\frac{n}{4}\right\rfloor = \frac{7n}{4}+\frac{3}{4} \quad \text{for} \quad n \equiv 3,7 \pmod 8$$
,

$$2n-\left\lfloor\frac{n}{4}\right\rfloor+1 = \begin{cases} \frac{7n}{4}+1 & \text{For} \quad n \equiv 0,4 \pmod 8, \\ \frac{7n}{4}+\frac{5}{4} & \text{For} \quad n \equiv 1,5 \pmod 8, \\ \frac{7n}{4}+\frac{6}{4} & \text{For} \quad n \equiv 2,6 \pmod 8, \end{cases}$$

Thus implies that

$$|D_1| \geq \left\{ \begin{array}{l} 2n-\left\lfloor\frac{n}{4}\right\rfloor; n \equiv 3,7 \pmod 8, \\ 2n-\left\lfloor\frac{n}{4}\right\rfloor+1; n \equiv 0,1,2,4,5,6 \pmod 8, \end{array} \right\}$$

Finally, we get

$$\gamma_2(P_4 \times P_n) = 2n - \left\lfloor \frac{n}{4} \right\rfloor : n \equiv 3,7 \pmod 8,$$

$$\gamma_2(P_4 \times P_n) = 2n - \left\lfloor \frac{n}{4} \right\rfloor + 1 : n \equiv 0,1,2,4,5,6 \pmod 8,$$

This complete the proof of the theorem.

Theorem 3.4.

$$\gamma_2(P_5 \times P_n) = \begin{cases} 2n + \left\lceil \frac{n}{7} \right\rceil : n \equiv 1,2,3,5 \pmod 7, \\ 2n + \left\lceil \frac{n}{7} \right\rceil + 1 : n \equiv 0,4,6 \pmod 7. \end{cases}$$

Proof. Let a set D defined as follows:

$$\begin{aligned} D = \{ & \{(2,1),(4,1)\} \cup \{(1,j),(2,j),(5,j) : j \equiv 2 \pmod 7\} \\ & \cup \{(3,j) : j \equiv 3 \pmod 7\} \cup \{(1,j),(4,j),(5,j) : j \equiv 4 \pmod 7\} \\ & \cup \{(2,j),(3,j) : j \equiv 5 \pmod 7\} \cup \{(2,j),(5,j) : j \equiv 6 \pmod 7\} \\ & \cup \{(1,j),(4,j) : j \equiv 0 \pmod 7\} \cup \{(3,j),(4,j) : j \equiv 1 \pmod 7\} \text{ and } j \neq 1\} \end{aligned}$$

We can check that the following sets are 2-dominating set for $P_5 \times P_n$ (see Figure 6, for $P_5 \times P_{23}$) as indicated:

$$\{D - \{K_n \cap D\}\} \cup \{(2,n),(3,n),(5,n)\} : n \equiv 1 \pmod 7.$$
$$D \cup \{(2,n)\} : n \equiv 0,4 \pmod 7.$$
$$D : n \equiv 2 \pmod 7.$$
$$\{D - \{K_n \cap D\}\} \cup \{(2,n),(4,n)\} : n \equiv 3,5 \pmod 7.$$
$$\{D - \{K_n \cap D\}\} \cup \{(1,n),(3,n),(5,n)\} : n \equiv 6 \pmod 7.$$

We have $D \leq 2n + \left\lceil \frac{n}{7} \right\rceil$ and

$$\gamma_2(P_5 \times P_n) \leq \begin{cases} 2n + \left\lceil \frac{n}{7} \right\rceil : n \equiv 1,2,3,5 \pmod 7, \\ 2n + \left\lceil \frac{n}{7} \right\rceil + 1 : n \equiv 0,4,6 \pmod 7. \end{cases}$$

This complete the proof of the theorem.

Lemma 3.3. The following cases are not possible:

1) (1, 2, 3, 1).

2) (1, 2, 1).

3) (1, 4, 1, 1).

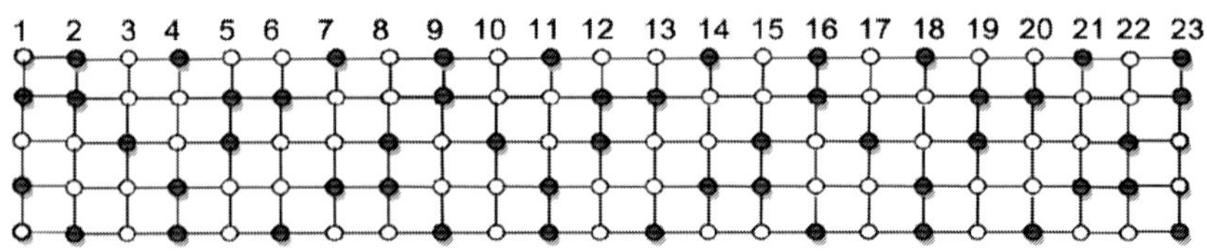

Figure 6. A 2-dominating set for $P_5 \times P_{23}$.

4) (1, 3, 1, 3, 1, 3).

5) (2, 1, 3).

6) (2, 2, 2, 2, 2, 2).

Proof. It follows directly from the drawing.

Lemma 3.4.

1) There is one case for subsequence $(s_j, s_{j+1}, s_{j+2}, s_{j+3}, s_{j+4}) = (2,2,2,2,2)$.

2) There is one case for subsequence $(s_j, s_{j+1}, s_{j+2}, s_{j+3}) = (1,3,1,3)$.

3) There is one case for subsequence $(s_j, s_{j+1}, s_{j+2}, s_{j+3}, s_{j+4}) = (1,3,1,3,1)$.

4) There is one case for subsequence $(s_j, s_{j+1}, s_{j+2}, s_{j+3}, s_{j+4}) = (1,2,3,2,1)$..

Proof. It follows directly from the drawing (see Figure 7).

Lemma 3.5.

1) $\sum_{j}^{j+3} s_j \geq 8$.

2) $\sum_{j}^{j+5} s_j \geq 12$.

3) $\sum_{j}^{j+6} s_j \geq 14$.

4) If $s_j = 3$ then $\sum_{j}^{j+6} s_j \geq 15$.

5) If $s_j = 4$ then $\sum_{j}^{j+6} s_j \geq 16$.

Proof. 1) By Lemma 3.3, imply that $\sum_{j}^{j+3} s_j \geq 8$.

2) By 1, we have $\sum_{j}^{j+3} s_j \geq 8$. If $\sum_{j}^{j+3} s_j \geq 8$, then we have the cases

$$(s_j, s_{j+1}, s_{j+2}, s_{j+3}) = (1,2,3,2),(1,3,1,3),(1,3,2,2),(1,4,1,2),(2,2,2,2).$$

From Lemma 3.3, we have $s_{j+4} + s_{j+5} \geq 4$, this implies that $\sum_{j}^{j+5} s_j \geq 12$.

If $\sum_{j}^{j+4} s_j \geq 9$ then $s_{j+4} + s_{j+5} \geq 3$. This implies that $\sum_{j}^{j+6} s_j \geq 12$.

3) We have $\sum_{j}^{j+2} s_j \geq 5$ and $\sum_{j+4}^{j+6} s_j \geq 5$. If $\sum_{j+4}^{j+6} s_j = 5$, then there is one case $(s_{j+4}, s_{j+5}, s_{j+6}) = (1,3,1)$ (where the cases (1,2,1), (1,2,2) are not possible). But the

case (1,3,1) is not compatible with any of the cases when $\sum_{j}^{j+3} s_j = 8$, this implies that

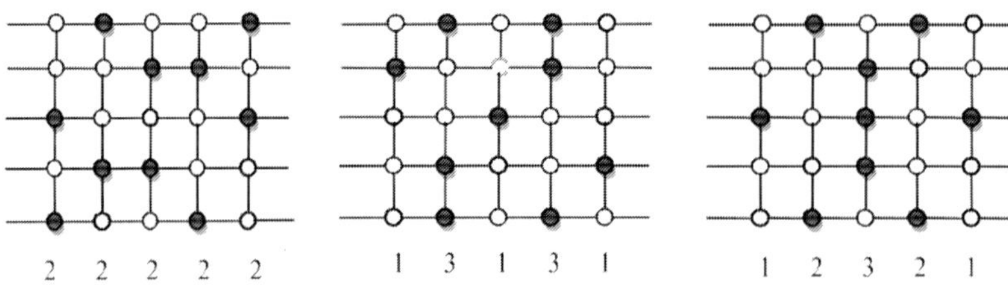

Figure 7. Cases 1, 2, 3 and 4 of Lemma 3.4.

$\sum_{j}^{j+3} s_j \geq 9$. Then $\sum_{j}^{j+6} s_j \geq 14$ (where the case (1,3,1,3,1,3) is not possible). If $\sum_{j+4}^{j+6} s_j \geq 6$ then $\sum_{j}^{j+6} s_j = \sum_{j}^{j+3} s_j + \sum_{j+4}^{j+6} s_j \geq 8 + 6 = 14$.

4) We have $s_j \geq 3$, then from 2 is $\sum_{j}^{j+6} s_j \geq 15$.

5) We have $s_j \geq 4$, then from 2 is $\sum_{j}^{j+6} s_j \geq 16$. This complete the proof of the Lemma. □

Lemma 3.6. If $\sum_{j}^{j+6} s_j = 14$, then $s_j = 1$ or $s_{j+6} = 1$.

Proof. We suppose the contrary s_j, $s_{j+6} \geq 2$. From Lemma 3.5, s_j, $s_{j+6} < 3$, else $\sum_{j}^{j+6} s_j \geq 15$. Now, we must study the case $s_j = s_{j+6} = 2$. We have $\sum_{j+2}^{j+5} s_j = 10$, by Lem-

ma 3.3, the case (2,2,2,2,2,2) is not possible, this implies that not all elements of the subsequence ($s_{j+1}, \ldots, s_{j+5}$) are equal to the value 2. If s_{j+1}, s_{j+2}, s_{j+3}, s_{j+4}, $s_{j+5} \geq 2$

where at least one of them is equal or greater than 3, then $\sum_{j}^{j+6} s_j \geq 15$, this is a contradiction with $\sum_{j}^{j+6} s_j = 14$. Now, we have $\sum_{j+2}^{j+5} s_j = 10$, where one of the subsequence ele-

ment ($s_{j+1}, \ldots, s_{j+5}$) is at most equal the value 1 (where $1 \leq s_j \leq 4$). We consider the cases $s_j = 1$ for $j+1 \leq j \leq j+5$:

1) $s_{j+1} = 1$ or $s_{j+5} = 1$ (where two cases are similar), we study the case $s_{j+1} = 1$

Then $s_{j+2} = 4$, these implies that $s_j + s_{j+1}\ 1 + s_{j+2} = 7$. By Lemma 3.5, we have $\sum_{j+3}^{j+6} s_j \geq 8$ then $\sum_{j}^{j+6} s_j \geq 15$, this is a contradiction.

2) $s_{j+2} = 1$ or $s_{j+4} = 1$ (where two cases are similar), we study the case $s_{j+2} = 1$ then $s_{j+1} \geq 3$, (because the case (2,2,1) is not possible). If $s_{j+1} = 3$ then $s_{j+1} \geq 3$

and we have $s_{j+6} = 2$ then $\sum_{j+4}^{j+5} s_j \geq 4$ (because two cases (1,2,2), (2,1,2) are not possible). Thus implies that $\sum_{j}^{j+6} s_j \geq 2+3+1+3+4+2 = 15$, this is a contradiction.

3) $s_{j+3} = 1$, then we have two subcases results from $s_{j+2} + s_{j+4} \geq 6$:

Subcase 1: $s_{j+2} = s_{j+4} = 3$,then $s_{j+1}, s_{j+5} \geq 2$ (because two cases $\left(s_j, s_{j+1}, s_{j+2}\right) = (2,1,3)$ and $\left(s_{j+4}, s_{j+5}, s_{j+6}\right) = (3,1,2)$ are not possible). Thus implies

That $\sum_{j}^{j+6} s_j \geq 15$, this is a contradiction.

Subcase 2: If $s_{j+2} = 2$, $s_{j+4} = 4$ or conversely (two cases are similar in studying), so

we will study case $s_{j+2} = 2$, $s_{j+4} = 4$ then $s_{j+5} \geq 1$, if $s_{j+5} \geq 2$, then $\sum_{j}^{j+6} s_j \geq 15$, because $s_{j+4} + s_{j+5} + s_{j+6} \geq 8$, we have $\sum_{j}^{j+3} s_j \geq 8$. Then $\sum_{j}^{j+6} s_j \geq 15$, this is a contradiction).

If $s_{j+5} = 1$, then $s_{j+4} + s_{j+5} + s_{j+6} = 7$. We have $\sum_{j}^{j+3} s_j \geq 8$. This implies that $\sum_{j}^{j+6} s_j \geq 15$ this is a contradiction.

Finally, we get if $\sum_{j}^{j+6} s_j = 14$, then $s_j = 1$ or $s_{j+6} = 1$. This completely the proof.

Result 3.1. If $\sum_{j}^{j+6} s_j = 14$, then from Lemma 3.6, we have the cases for subsequence

$(s_j, s_{j+1}, s_{j+2}, s_{j+3}, s_{j+4}, s_{j+5}, s_{j+6})$:

$$a_1:(1,2,3,2,1,4,1),\quad a_2:(1,2,3,2,2,2,2),\quad a_3:(1,2,3,2,2,3,1),$$
$$a_4:(1,2,3,3,1,3,1),\quad a_5:(1,3,1,3,1,4,1),\quad a_6:(1,3,1,3,2,2,2),$$
$$a_7:(1,3,1,3,2,3,1),\quad a_8:(1,3,1,3,3,2,1),\quad a_9:(1,3,1,4,1,3,1),$$
$$a_{10}:(1,3,2,2,2,2,2),\quad a_{11}:(1,3,2,2,2,3,1),\quad a_{12}:(1,3,2,2,3,2,1),$$
$$a_{13}:(1,3,2,3,1,3,1),\quad a_{14}:(1,4,1,2,3,2,1),\quad a_{15}:(1,4,1,3,1,3,1).$$

It is 15 cases (where $s_j = 1$ with $\sum_{j}^{j+6} s_j = 14$). We have three cases with $s_{j+1} =$ 2,

$s_{j+1} = 3$ and $s_{j+1} = 4$.

Case 1: $s_{j+1} = 2$ (including the cases $s_j = 1$ and $s_{j+1} = 2$ or $s_{j+6} = 1$ and $s_{j+5} = 2$). We have these cases are a_1, a_2, a_3, a_4, a_8, a_{12}, a_{14} and comes before these cases, $s_{j-1} = 4$ or comes after these cases $s_{j+7} = 4$, i.e., if $s_j = 1$, $s_{j+1} = 2$ then $s_{j-1} =$ 4 and if $s_{j+6} = 1$, $s_{j+5} = 2$ then $s_{j+7} = 4$.

Case 2: $s_{j+1} = 3$, $s_{j+1} = 4$ and these are the 8 remaining cases. We will study these cases after rejecting isomorphism cases when there is two cases or more, where $(s_j,\ldots,s_{j+6}) = (s_{j+6},\ldots,s_j)$, then we will study only one case. We have 8 cases as follows:

$$a_5:(1,3,1,3,1,4,1),\ a_6:(1,3,1,3,2,2,2),\ a_7:(1,3,1,3,2,3,1),\ a_9:(1,3,1,4,1,3,1),$$
$$a_{10}:(1,3,2,2,2,2,2),\ a_{11}:(1,3,2,2,2,3,1),\ a_{13}:(1,3,2,3,1,3,1),\ a_{15}:(1,4,1,3,1,3,1).$$

We note that two cases a_5, a_{15} are similar where one of them is contrary to the other one, so we study the case a_5. Also, two cases a_7, a_{13} are similar, so we study the case a_7. Then we study these cases: a_5, a_6, a_7, a_9, a_{10}, a_{11}. □

Notice 3.2. We note that all the possible cases in Result 3.1, do not begin or end with 3 or 4 and it do not begin or end with $s_j + s_{j+1} \geq 5$ or $s_{j+5} + s_{j+6} \geq$ 5 such that $s_j = 2$ or $s_{j+6} = 2$, and $s_{j+1} = 3$ or $s_{j+5} = 3$. Thus implies that if $s_j = 2$, $s_{j+1} = 3$,

Then $\sum_{j}^{j+6} s_j \geq 15$. Also, we note cases a_5, a_6, a_7 are beginning with (1, 3, 1, 3), but from

Lemma 3.4, we get $s_{j-1} = 4$. Now, remains our three cases for studying by the following lemma are:

$$a_9 : (1,3,1,4,1,3,1),\ a_{10} : (1,3,2,2,2,2,2),\ a_{11} : (1,3,2,2,2,3,1).$$

Result 3.2. If $s_{j+1} = 3$, $s_j = 1$ where $k_j \cap s = \{(1,j)\}$ or $k_j \cap s = \{(2,j)\}$ then $s_{j-1} = 4$, also for $k_j \cap s = \{(4,j)\}$ or $k_j \cap s = \{(5,j)\}$ because it are similar to two cases $k_j \cap s = \{(2,j)\}$ or $k_j \cap s = \{(1,j)\}$, respectively.

Lemma 3.7. If $\sum_{j}^{j+6} s_j = 14$, such that $s_{j+5} = 3$, $s_{j+6} = 1$, then $\sum_{j+7}^{j+13} s_j \geq 15$. Furthermore, if $\sum_{j+7}^{j+13} s_j \geq 15$ then $\sum_{j+14}^{j+20} s_j \geq 15$.

Proof. By Result 3.2, if $k_{j+6} \cap s = \{(1,j+6)\}$, $k_{j+6} \cap s = \{(2,j+6)\}$,

$k_{j+6} \cap s = \{(4,j+6)\}$or $k_{j+6} \cap s = \{(5,j+6)\}$ then $s_{j+7} = 4$. From Lemma 3.5, we get $\sum_{j+7}^{j+13} s_j \geq 16$. Assume $k_{j+6} \cap s = \{(3,j+6)\}$ then we have two cases for $k_{j+5} \cap s$:

Case 1. $k_{j+5} \cap s = \{(1,j+5),(3,j+5), (5,j+5)\}$. Then $s_{j+7} = 4$, by lemma 3.5,

$$\sum_{j+7}^{j+13} s_j \geq 16.$$

Case 2. $k_{j+5} \cap s = \{(1,j+5),(2,j+5), (5,j+5)\}$ or $k_{j+5} \cap s = \{(1,j+5),(4,j+5), (5,j+5)\}$and both cases are similar, so we will consider the first case. We have $3 \leq s_{j+7} \leq 4$ then by Lemma 3.5, $\sum_{j+7}^{j+13} s_j \geq 15$. If $s_{j+7} = 4$ then $\sum_{j+7}^{j+13} s_j \geq 16$. Assume $s_{j+7} = 3$, if $\sum_{j+7}^{j+13} s_j \geq 16$ the proof is finish. Assume $\sum_{j+7}^{j+13} s_j \geq 15$ then we have cases $s_{j+8} = 1,2,3$ or 4.

Subcase 2.1. If $s_{j+8} = 4$ then $s_{j+9} \geq 1$. This implies that

$$\sum_{j+7}^{j+13} s_j \geq 3+4+1+\sum_{j+10}^{j+13} s_j = 8+8 = 16$$

{By Lemma 3.5, $\sum_{j}^{j+3} s_j \geq 8$}.

Subcase 2.2. If s_{j+8} = 3 then $\sum_{j+9}^{j+13} s_j \geq 9$. If $\sum_{j+9}^{j+13} s_j \geq 9$ then $\sum_{j+7}^{j+13} s_j \geq 16$.

Assume that $\sum_{j+9}^{j+13} s_j \geq 9$ then we have only one case $(s_{j+9},\dots, s_{j+13})$ = (1,3,1,3,1) or $(s_{j+9},\dots, s_{j+13})$ = (1,2,3,2,1). For any case we have s_{j+8} = 4. So, we get $\sum_{j+9}^{j+13} s_j \geq 9$. Which implies that $\sum_{j+7}^{j+13} s_j \geq 16$.

Subcase 2.3. If s_{j+8} = 1 then s_{j+9} = 4 {because the case $(s_{j+5},s_{j+6},s_{j+7},s_{j+8},s_{j+9})$ = (3,1,3,1,3)is not possible, by Lemma 3.3}. Then

$$\sum_{j+7}^{j+13} s_j \geq 3+1+4+\sum_{j+10}^{j+13} s_j \geq 8+8=16.$$

Subcase 2.4. If s_{j+8} = 2 then s_{j+7} = 3, s_{j+8} = 2, we have the following cases:

2.4.1. $s_{j+9} \geq 3$ then $\sum_{j+7}^{j+13} s_j \geq 3+2+3+\sum_{j+10}^{j+13} s_j \geq 8+8=16$.

2.4.2. $s_{j+9} \neq 1$ {because there is only one case for $(s_{j+7},s_{j+8},s_{j+9})$ = (3,2,1) such that

$$\{K_{j+7} \cup K_{j+8} \cup K_{j+9}\} \cap S = \{(2,j+7),(3,j+7),(4,j+7),(1,j+8),(5,j+8),(3,j+9)\}$$

But according to distribution vertices $k_{j+5} \cap s$ and $k_{j+6} \cap s$ we have

$$k_{j+5} \cap 7 \neq \{(2,j+7),(3,j+7),(4,j+7)\}.$$

2.4.3. s_{j+9} = 2 then $s_{j+7} + s_{j+8} + s_{j+9}$ = 7. This implies that $(s_{j+7},s_{j+8},s_{j+9})$ = (3,2,2). We will study the cases that leads to $\sum_{j+7}^{j+13} s_j = 15$, i.e., $\sum_{j+10}^{j+13} s_j = 8$, {because the cases which leads to $\sum_{j+7}^{j+13} s_j \geq 16$ the proof will be done}. Now, we have the fixed case $(s_{j+7},s_{j+8},s_{j+9})$ = (3,2,2) We will consider the vertices $k_{j+10} \cap s$ which imply the following:

2.4.3.1. If s_{j+10} = 4 then $(3,2,2,4,s_{j+11},s_{j+12},s_{j+13})$, this implies that $\sum_{j+11}^{j+13} s_j = 4$

and $(s_{j+11}, s_{j+12}, s_{j+13}) = (1,2,1)$ is not possible.

2.4.3.2. If $s_{j+10} = 3$ then $(3,2,2,3,s_{j+11},s_{j+12},s_{j+13})$ and $\sum_{j+11}^{j+13} s_j = 5$ which imply that $(s_{j+11},s_{j+12},s_{j+13}) = (2,1,2)(2,2,1),(1,2,2)$ or $(1, 3, 1)$, and the only possible case is $(1, 3, 1)$. Thus implies that $(s_{j+11},\ldots,s_{j+13}) = (3,2,2,3,1,3,1)$. By Lemma 3.4 and Lemma 3.5 is $s_{j+14} = 4$, these implies that $\sum_{j+14}^{j+20} s_j \geq 16$.

2.4.3.3. If $s_{j+10} = 2$ then $(3,2,2,2,s_{j+11},s_{j+12},s_{j+13})$, i.e., $\sum_{j+11}^{j+13} s_j = 6$. We have $s_{j+11} \neq 1$ {because the case (2,2,1) is not possible}. Then we have the following cases for $s_{j+11},s_{j+12},s_{j+13}$:

1). If $s_{j+11} = 4$ then $s_{j+12} = 1$ and $s_{j+13} = 1$, but the case (4,1,1) is not possible.

2). If $s_{j+11} = 3$ and $s_{j+12} = 1$ then $s_{j+13} = 2$, also the case (3,1,2) is not possible.

3). If $s_{j+11} = 3$, $s_{j+12} = 2$ and $s_{j+13} = 1$ then $(s_j,\ldots, s_{j+6}) = (3,2,2,2,3,2,1)$ which gets $s_{j+7} = 4$ and $\sum_{j+7}^{j+13} s_j \geq 16$.

4). If $s_{j+11} = 2$ and $s_{j+12} = 2$ then $s_{j+13} = 2$, but the case (3,2,2,2,2,2,2) is not possible. If $s_{j+11} = 2$, $s_{j+12} = 3$ and $s_{j+13} = 1$ then we gets $(s_j,\ldots, s_{j+6}) = (3,2,2,2,2,3,1)$ During the proof of Lemma, we notice that if $s_j = 3$ And $s_{j+1} = 1$, then $\sum_{j+2}^{j+8} s_j \geq 15$. This complete the proof.

Result 3.3. Based on the Lemma 3.6, and the other Lemmas and results precede it.

We see that when we have case of $\sum_{j}^{j+6} s_j = 14$, then the only case that comes after it, is $\sum_{j+7}^{j+13} s_j = 15$ such that $(s_{j+7},\ldots, s_{j+13}) = (3,2,2,2,2,3,1)$ which continues in the same

way or it is followed by 7 columns contain 16 vertices from S {by Lemma 3.6,

$\sum_{j+14}^{j+20} s_j \geq 15$, because $s_{j+12} = 3, s_{j+13} = 1\}$. When this case is repeated then $\sum_{j=n-6}^{n} s_j \geq 15$ and then when the case $\sum_{j}^{j+6} s_j = 14$ it is necessary, the case $\sum_{j+6+q}^{j+6+q-1+7r} s_j \geq 16$ exists as well {where $j + 6 + q - 1 + 7r \leq n$} these implies that $\sum_{j=1}^{n} s_j \geq \left\lceil \frac{15n}{7} \right\rceil$ then

$$\gamma_2(P_5 \times P_n) = \sum_{j=1}^{n} s_j \geq 2n + \left\lceil \frac{n}{7} \right\rceil$$

Lemma 3.8. Let S be 2-dominating set for $P_5 \times P_n$ then:

1) $s_1 \geq 2$ and $s_1 + s_2 \geq 4$ ($s_{n-1} + s_n \geq 4$, $s_n \geq 2$).

2) If $s_1 + s_2 = 4$ then $s_1 + s_2 + s_3 = 8$ ($s_{n-1} + s_n = 4$ then $s_{n-2} + s_{n-1} + s_n = 8$).

3) $s_1 + s_2 + s_3 \geq 6$ ($s_{n-2} + s_{n-1} + s_n \geq 6$).

4) $\sum_{j=1}^{4} s_j \geq 9 \left(\sum_{j=n-3}^{n} s_j \geq 9 \right)$.

5) $\sum_{j=1}^{5} s_j \geq 10 \left(\sum_{j=n-4}^{n} s_j \geq 10 \right)$ and if $\sum_{j=1}^{5} s_j = 10$ then $\sum_{j=1}^{6} s_j \geq 14$, also if $\sum_{j=n-4}^{n} s_j = 10$ then $\sum_{j=n-5}^{n} s_j \geq 14$

6) $\sum_{j=1}^{6} s_j \geq 13 \left(\sum_{j=n-5}^{n} s_j \geq 13 \right)$.

7) $\sum_{j=1}^{7} s_j \geq 15 \left(\sum_{j=n-6}^{n} s_j \geq 15 \right)$.

8) If $s_1 + s_2 = 5$ then either $\sum_{j=1}^{5} s_j \geq 11$ or $\sum_{j=1}^{6} s_j \geq 14$, also if $s_{n-1} + s_n = 5$ then either $\sum_{j=n-4}^{n} s_j \geq 11$ or $\sum_{j=n-5}^{n} s_j \geq 14$.

Proof. The study of dominating sequence $(s_1,s_2,...,s_n)$ is the same as the study of the dominating sequence $(s_n,s_{n-1},...,s_1)$, so we study one case $(s_1,s_2,...,s_n)$. Also, the study of $\sum_{j=1}^{r} s_j$ is the same as the study of $\sum_{j=n-r+1}^{n} s_j$.

1) We have $s_1 \geq 2$, if $s_1 = 2$ then $s_2 \geq 3$ thus, $s_1 + s_2 \geq 5$ if $s_1 \geq 3$ then $s_2 \geq 1(1 \leq s_j \leq 4)$these implies that $s_1 + s_2 \geq 4$.

2) If $s_1 + s_2 = 4$, then we have only one the case $k_1 \cap s = \{(1,1),(3,1),(5,1)\}$ these implies that $k_1 \cap s = \{(3,2)\}$ and $s_3 = 4$ then $s_1 + s_2 + s_3 = 8$.

3) If $s_1 + s_2 \geq 5$, then $\sum_{j=1}^{3} s_j \geq 6$ {because $1 \leq s_j \leq 4$} and if $s_1 + s_2 = 4$ then by 2, is $\sum_{j=1}^{3} s_j = 8$.

4) If $s_1 + s_2 = 4$ then $\sum_{j=1}^{4} s_j = 8$ these implies that $\sum_{j=1}^{4} s_j \geq 9$ and if $s_1 + s_2 \geq 6$ then $\sum_{j=1}^{4} s_j \geq 9$ {because $s_3 + s_4 \geq 3$}. Assume that $s_1 + s_2 = 5$, then we have three cases:

4.1) $s_1 = 2$, $s_2 = 3$ then $s_3 + s_4 \geq 4$, because the case $(s_2,s_3,s_4) = (3,1,2)$ is not possible. Also the case $(s_2,s_3,s_4) = (3,1,2)$ is not possible, else when $k_2 \cap s = \{(2,2),(3,2),(4,2)\}$and this is not possible.

4.2) $s_1 = 3$, $s_2 = 2$ then $s_3 + s_4 \geq 4$ because the cases $(s_2,s_3,s_4) = (2,2,1)$, $(s_2,s_3,s_4) = (2,1,2)$ are not possible.

4.3) $s_1 = 4$, $s_2 = 1$ then $s_3 + s_4 \geq 4$, because the cases $(s_1,s_2,s_3,s_4) = (4,1,2,1)$, $(s_1,s_2,s_3,s_4) = (4,1,2,2)$ are not possible. Thus implies that we have $\sum_{j=1}^{4} s_j \geq 9$.

5) By Lemma 3.4, we have two cases for $\sum_{j=1}^{4} s_j \geq 9$ and these two cases are (1,2,3,2,1), (1,3,1,3,1), furthermore these cannot be shown here because $s_1 \geq 2$. Thus implies that we $\sum_{j=1}^{5} s_j \geq 10$.

6). If $s_1 + s_2 \geq 5$ then $\sum_{j=1}^{6} s_j = s_1 + s_2 + \sum_{j=3}^{6} s_j \geq 5 + 8 = 13$.

(where by Lemma 3.5, we have $\sum_{j}^{j+3} s_j \geq 8)$). Let $s_1 + s_2 = 4$ then $\sum_{j=1}^{3} s_j = 8$ these implies that $\sum_{j=1}^{6} s_j \geq 8 + \sum_{j=4}^{6} s_j$. Thus implies that $\sum_{j=1}^{6} s_j \geq 8 + 5 = 13$ {because $\sum_{j}^{j+2} s_j \geq 5$ }.

7) If $s_1 \geq 3$ then from Lemma 3.5, $\sum_{j=1}^{7} s_j \geq 15$. Let $s_1 = 2$ {because $s_1 > 1$} then $s_2 \geq 3$. This implies that $\sum_{j=1}^{7} s_j \geq 15$ {by Notice 3.2}.

8) If $s_1 + s_2 = 5$ then either $\sum_{j=1}^{5} s_j \geq 11$ or $\sum_{j=1}^{6} s_j \geq 14$. We have $s_1 + s_2 = 5$, then we have three cases:

8.1) $s_1 = 4$, $s_2 = 1$, then $s_3 + s_4 + s_4 \geq 7$ because the cases

(s_1, s_2, s_3, s_4, s_5) = (4,1,2,2,2),(4,1,3,2,1),(4,1,2,3,1) or (4,1,3,1,2) are not possible.

Thus implies that $\sum_{j=1}^{5} s_j \geq 11$.

8.2) $s_1 = 2$, $s_2 = 3$, then $\sum_{j=1}^{5} s_j \geq 10$ and if $\sum_{j=1}^{5} s_j \geq 10$ then (s_1, s_2, s_3, s_4, s_5) = (2,3,1,3,1). By Lemma 3.4, $s_6 = 4$. Thus implies that $\sum_{j=1}^{6} s_j \geq 14$.

8.3) $s_1 = 3$, $s_2 = 2$, then (s_1, s_2, s_3, s_4, s_5) it has minimal numerals in the following cases (s_1, s_2, s_3, s_4, s_5) = (3,2,2,2,2),(3,2,1,4,1) or (3,2,3,1,3) and for the case (s_3, s_4, s_5) = (1,3,1)is not compatible with the case (s_1, s_2) = (3,2) Thus implies that $\sum_{j=1}^{5} s_j \geq 11$. This completes the proof. □

Theorem 3.5.

$$\gamma_2(P_5 \times P_n) = \begin{cases} 2n + \left\lceil \frac{n}{7} \right\rceil : n \equiv 1,2,3,5 \pmod 7, \\ 2n + \left\lceil \frac{n}{7} \right\rceil + 1 : n \equiv 0,4,6 \pmod 7. \end{cases}$$

Proof. By Result 3.3, we have $\gamma_2(p_5 \times p_n) = \sum_{j=1}^{n} s_j \geq \left\lceil \frac{15n}{7} \right\rceil$. By Theorem 3.4, we get $\gamma_2(p_5 \times p_n) = 2n + \left\lceil \frac{n}{7} \right\rceil : n \equiv 1,2,3,5 \pmod 7.$

Now, for $n \equiv 0,4,6 \pmod 7$, by Theorem 3.4, we have $\gamma_2(p_5 \times p_n) \leq 2n + \left\lceil \frac{n}{7} \right\rceil + 1$. From Result 3.3, we have $\gamma_2(p_5 \times p_n) \leq 2n + \left\lceil \frac{n}{7} \right\rceil$. We will study the cases:

1) $n \equiv 0 \pmod 7$. We have $\gamma_2(p_5 \times p_n) = \sum_{j=1}^{n} s_j$. So, we consider the following:

a) $s_1 + s_2 = 4$ then $s_1 + s_2 + s_3 = 8$ and by Lemma 3.8,

$$\gamma_2(p_5 \times p_n) = \sum_{j=1}^{n} s_j = s_1 + s_2 + s_3 + \sum_{j=4}^{n-4} s_j + \sum_{j=n-3}^{n} s_j \geq 8 + 2(n-2) + \frac{n-7}{7} + 9,$$

$$\gamma_2(p_5 \times p_n) \geq 17 + 2n - 14 + \frac{n-7}{7} = 2n + \frac{n+14}{7} = 2n + \left\lceil \frac{n}{7} \right\rceil + 2 \geq 2n + \left\lceil \frac{n}{7} \right\rceil + 1.$$

b) $s_1 + s_2 \geq 5$ if $s_1 + s_2 \geq 6$ then

$$\gamma_2(p_5 \times p_n) = \sum_{j=1}^{n} s_j = s_1 + s_2 + \sum_{j=3}^{n-5} s_j + \sum_{j=n-4}^{n} s_j \geq 6 + 2(n-7) + \frac{n-7}{7} + 10$$

$$= 2n + \frac{n-7+14}{7} = 2n + \left\lceil \frac{n}{7} \right\rceil + 1.$$

Let $s_1 + s_2 = 5$ then by Lemma 3.8, $\sum_{j=1}^{5} s_j \geq 11$ or $\sum_{j=1}^{6} s_j \geq 14$. If $\sum_{j=1}^{5} s_j \geq 11$ then

$$\gamma_2(p_5\times p_n)=\sum_{j=1}^{n}s_j=\sum_{j=1}^{5}s_j+\sum_{j=6}^{n-2}s_j+s_{n-1}+s_n\geq 11+2(n-7)+\frac{n-7}{7}+5$$
$$=2n+\frac{n}{7}+1=2n+\left\lceil\frac{n}{7}\right\rceil+1.$$

{where the case $s_{n-1}+s_n=4$ is the same as $s_1+s_2=4$}. If $\sum_{j=1}^{5}s_j\geq 11$ then by Lemma 3.8, we have $\sum_{j=1}^{6}s_j\geq 14$

$$\gamma_2(p_5\times p_n)=\sum_{j=1}^{n}s_j=s_1+s_2+\sum_{j=3}^{n-5}s_j+\sum_{j=n-4}^{n}s_j\geq 6+2(n-7)+\frac{n-7}{7}+10$$
$$=2n+\frac{n-7+14}{7}=2n+\left\lceil\frac{n}{7}\right\rceil+1.$$

And with Theorem 3.4, we get $\gamma_2(p_5\times p_n)=2n+\left\lceil\frac{n}{7}\right\rceil+1: n\equiv 0(\mathrm{mod}\,7)$.

2) When $n\equiv 4(\mathrm{mod}7)$ we have two cases:

a) $s_1+s_2=4$. Thus implies that $s_1+s_2+s_3=8$ then

$$\gamma_2(p_5\times p_n)=\sum_{j=1}^{n}s_j=\sum_{j=1}^{3}s_j+\sum_{j=4}^{n-1}s_j+s_n\geq 8+\frac{15(n-4)}{7}+2$$
$$=2n+\frac{n+10}{7}=2n+\left\lceil\frac{n}{7}\right\rceil+1.$$

b) $s_1+s_2\geq 5$ {where $s_{n-1}+s_n\geq 5$} then

$$\gamma_2(p_5\times p_n)=\sum_{j=1}^{n}s_j=s_1+s_2+\sum_{j=3}^{n-2}s_j+s_{n-1}+s_n\geq 5+2(n-4)+\frac{n-4}{7}+5$$
$$=2n+\frac{n+10}{7}=2n+\left\lceil\frac{n}{7}\right\rceil+1.$$

Then by Theorem 3.4, we get $\gamma_2(p_5\times p_n)=2n+\left\lceil\frac{n}{7}\right\rceil+1: n\equiv 4(\mathrm{mod}\,7)$.

3) $n\equiv 6(\mathrm{mod}7)$. We have two cases:

a) If $s_1 + s_2 = 4$ then $s_1 + s_2 + s_3 = 8$. Thus implies that

$$\gamma_2(p_5 \times p_n) = \sum_{j=1}^{n} s_j = s_1 + s_2 + s_3 + \sum_{j=4}^{n-3} s_j + s_{n-2} + s_{n-1} + s_n \geq 8 + 2(n-6) + \frac{n-6}{7} + 6$$
$$= 2n + \left\lceil \frac{n}{7} \right\rceil + 1.$$

b) If $s_1 + s_2 \geq 5$ then $s_{n-1} + s_n \geq 5$. Thus implies that

$$\gamma_2(p_5 \times p_n) = \sum_{j=1}^{n} s_j = \sum_{j=1}^{4} s_j + \sum_{j=5}^{n-2} s_j + s_{n-1} + s_n \geq 9 + 2(n-6) + \frac{n-6}{7} + 5$$
$$= 2n + \frac{n+8}{7} = 2n + \left\lceil \frac{n}{7} \right\rceil + 1.$$

By Theorem 3.4, we get $\gamma_2(p_5 \times p_n) = 2n + \left\lceil \frac{n}{7} \right\rceil + 1 : n \equiv 6 \pmod 7$ Finally, we get

$$\gamma_2(p_5 \times p_n) = \begin{cases} 2n + \left\lceil \frac{n}{7} \right\rceil : n \equiv 1,2,3,5 \pmod 7, \\ 2n + \left\lceil \frac{n}{7} \right\rceil + 1 : n \equiv 0,4,6 \pmod 7. \end{cases}$$

This completes the proof.

REFERENCES

1. Mohan, J.J. and Kelkar, I. (2012) Restrained 2-Domination Number of Complete Grid Graphs. International Journal of Applied Mathematics and Computation, 4, 352-358.
2. Fink, J.F. and Jacobson, M.S. (1985) n-Domination in graphs, in: Graph Theory with Application to Algorithms and Computer Science. John Wiley and Sons, New York, 282-300.
3. Fink, J.F. and Jacobson, M.S. (1985) On n-Domination, n-Dependence and Forbidden Subgraphs. In: Graph Theory with Application to Algorithms and Computer Science, John Wiley and Sons, New York, 301-311.
4. Haynes, T.W., Hedetniemi, S.T., Henning, M.A. and Slater, P.J. (2003) H-Forming Sets in Graphs. Discrete Mathematics, 262, 159-169. https://doi.org/10.1016/S0012-365X(02)00496-X
5. Haynes, T.W., Hedetniemi, S.T. and Slater, P.J. (1998) Fundamentals of Domination in Graphs. Marcel Dekker, Inc., New York.
6. Hansberg, A. and Volkmann, L. (2009) Upper Bounds on the k-Domination Number and the k-Roman Domination Number. Discrete Applied Mathematics, 157, 1634-1639. https://doi.org/10.1016/j.dam.2008.10.011
7. Cockayne, E.J., Gamble, B. and Shepherd, B. (1985) An Upper Bound for the k-Domination Number of a Graph. Journal of Graph Theory, 9, 533-534. https://doi.org/10.1002/jgt.3190090414

8. Blidia, M., Chellali, M. and Volkmann, L. (2006) Some Bounds on the p-Domination Number in Trees. Discrete Mathematics, 306, 2031-2037. https://doi.org/10.1016/j.disc.2006.04.010
9. Favaron, O., Hansberg, A. and Volkmann, L. (2008) On k-Domination and Minimum Degree in Graphs. Journal of Graph Theory, 57, 33-40. https://doi.org/10.1002/jgt.20279
10. Volkmann, L. (2010) A Bound on the k-Domination Number of a Graph. Czechoslovak Mathematical Journal, 60, 77-83. https://doi.org/10.1007/s10587-010-0019-1
11. Shaheen, R. (2009) Bounds for the 2-Domination Number of Toroidal Grid Graphs. International Journal of Computer Mathematics, 86, 584-588. https://doi.org/10.1080/00207160701690284
12. Shaheen, R. (2013) On the 2-Domination Number of Cartesian Product of Two Cycles. Advances and Applications in Discrete Mathematics, 12, 83-108.

CHAPTER 5

Dislocated Soft Metric Space with Soft Fixed Point Theorems

Balaji Raghunath Wadkar[1], Vishnu Narayan Mishra[2,3*], Ramakant Bhardwaj[4], Basant Singh[1]

[1]*Department of Mathematics, AISECT University, Bhopal, India*

[2]*Applied Mathematics & Humanities Department, S.V. National Institute of Technology, Surat, India*

[3]*Department of Mathematics, Indira Gandhi National Tribal University, Amarkantak, India*

[4]*Department of Mathematics, TIT Group of Institutes, Bhopal, India*

ABSTRACT

In the present paper, we define Dislocated Soft Metric Space and discuss about the existence and uniqueness of soft fixed point of a cyclic mapping in soft dislocated metric space. We also prove the unique soft fixed point theorems of a cyclic mapping in the context of dislocated soft metric space. Examples are given for support of the results.

Keywords: Soft Set, Soft Fixed Point, Soft Point of Soft Mapping, Soft Metric Space, Dislocated Soft Metric Space

1. INTRODUCTION

The soft set theory is one of the branches of mathematics, which aims to describe phenomena and concepts of an ambiguous, undefined vague and imprecise meaning, which was initiated by Molodtsov [1]. This theory is applicable where there are no clearly defined mathematical models. Recently many papers concerning soft sets have been published (see [2] - [8]). In many aspects of Mathematics, fixed point theory has wonderful applications. Shabir and Naz [9] presented soft topological spaces and they investigated some properties of soft topological spaces. Later many researchers were studied about soft topological spaces. In these studies, concepts of soft fixed point are expressed by different approaches. Das and Samanta [10] [11] introduced a different notion of soft metric space by using different concept of soft point and investigated some basic properties of these. In 2000, Hitzler and Seda [12] introduced the notion of dislocated metric space, in which self distance of a point need not be equal to zero. Aage and Salunke [13] established some important fixed point theorem in single and pair of mappings in dislocated metric space. Later Karapnar and Salimi [14] discussed the existence and uniqueness of fixed point of a cyclic mapping in the context of metric like space. The study of common fixed point of mapping in dislocated metric space satisfying certain contractive condition has been at the Centre of vigorous research activity (see [15] [16] [17] [18]). Dislocated metric space plays very important role in topology, logical programming and in electronic engineering. Recently Wadkar et al. [19], Mishra et al. [20] [21] [22] [23], Deepmala and pathak [24], Wadkar et al. [25], [26] discussed and proved fixed point theorems by employing different concepts.

In the present paper, we discuss about the investigations concerning the existence and uniqueness of soft fixed point of a cyclic mapping in soft dislocated metric space. We also prove the unique soft fixed point theorems of a cyclic mapping in the context of dislocated metric space. To check the validity of the result we give the examples. Before starting to prove main result, some basic definitions are required.

Definition 1.1: Let X and E are respectively an initial inverse set and a parameter set. A soft set over X is pair denoted by (Y, E) if and only if Y is a mapping from E into the set of all subsets of the set X, i.e. $Y: E \rightarrow P(x)$, where $P(x)$ is the power set of X.

Example of Soft Set: Let $(\breve{C}, \bar{D})$ be soft set, which describes the Nature of

workers at Industry. Suppose that $U = \{u_1, u_2, u_3, u_4\}$, i.e. universe of four workers. Let a set of decision parameters be $\breve{D}$ = {s_1,s_2,s_2,s_4,s_5}. Now s_i (i = 1,2,3,4,5) stand for the parameters: high working speed, slow working speed, average working speed, work delay working speed and no working speed properly respectively.

Now consider $\tilde{C}(s_1)=\{u_1,u_3\}$, $\tilde{C}(s_2)=\{u_1,u_3,u_4\}$, $\tilde{C}(s_3)=U$, and $\tilde{C}(s_4)=\{u_2,u_4\}$. By consisting of the following collection of approximations, the soft set ($\breve{C}$, $\breve{D}$) can be viewed.

$$\begin{aligned}(\tilde{C},\tilde{D}) = \{&(\text{high working speed}, \{u_1,u_3\}), (\text{slow working speed}, \{u_1,u_3,u_4\}),\\ &(\text{average working speed}, U), (\text{work delay working speed}, \{u_2,u_4\}),\\ &(\text{no working speed, empty})\}.\end{aligned}$$

Definition 1.2: The intersection of two soft sets (Y, A) and (Z, B) over X is a soft set over X denoted by (I, C) and is given by $(Y,A)\tilde{\cap}(Z,B)=(I,C)$, where $C = A \cup B$ and $\forall \varepsilon \in C$, $I(\varepsilon)=Y(\varepsilon)\cap Z(\varepsilon)$.

Definition 1.3: The union of two soft sets (Y, A) and (Z, B) over X is the soft set (I, C), where $C = A \cup B$ and for all k in C,

$$I(k)=\begin{cases} Y(k), & \text{if } k \text{ is an element of } A-B,\\ Z(k), & \text{if } k \text{ is an element of } B-A,\\ Y(k)\cup Z(k), & \text{if } k \text{ is an element of } A\cap B.\end{cases}$$

This relationship is denoted by $(Y,A)\tilde{\cup}(Z,B)=(I,C)$

Definition 1.4: A soft set (Y, A) over X is said to be a null soft set if $Y(K)$ = empty, for all k in A and is denoted by Φ.

Definition 1.5: For all $k \in A$, if $Y(k) = X$ then (Y, A) is called an absolute soft set over X.

Definition 1.6: The difference of two soft sets (F, E) and (G, E) over X is a soft set (H, E) over X, denoted by (F, E)\(G, E) and is defined as $H(x)=F(x)\setminus G(x),\ \forall x \in E$.

Definition 1.7: The complement of soft set (Y, A) is denoted by $(Y, A)^C$ and is defined as $(Y,A)^c=(Y^c,A)$, where $Y^c : A \to P(X)$ is a mapping given by $Y^c(\beta)=X-Y(\beta)$, for all β.

Definition 1.8: Let B(R) be the collection of all non-empty bounded subsets of R and E taken as a set of parameters. Then the mapping $Y: R \rightarrow B(R)$ is called a soft real set. It is denoted by (Y, E).

Definition 1.9: For two soft real numbers $\tilde{\tilde{m}}$ and $\tilde{n}$ the following conditions

holds:

(i) $\tilde{\tilde{m}} \leq \tilde{n}$ if $\tilde{\tilde{m}}(s) \leq \tilde{n}(s)$, for all $s \in E$;

(ii) $\tilde{\tilde{m}} \geq \tilde{n}$ if $\tilde{\tilde{m}}(s) \geq \tilde{n}(s)$, for all $s \in E$;

(iii) $\tilde{\tilde{m}} < \tilde{n}$ if $\tilde{\tilde{m}}(s) < \tilde{n}(s)$, for all $s \in E$;

(iv) $\tilde{\tilde{m}} > \tilde{n}$ if $\tilde{\tilde{m}}(s) > \tilde{n}(s)$, for all $s \in E$.

Definition 1.10: A soft set (P, E) over X is said to have a soft point if there is exactly one $s \in E$ such that $P(s) = \{x\}$, for some $x \in X$, also $P(s') = \varphi \ \forall s' \in E / \{s\}$,. It will be denoted by $\tilde{x}_s$.

Definition 1.11: Two soft points $\tilde{x}_i$, $\tilde{y}_j$ are said to be equal if $i = j$ and $P(i) =$

$P(j)$ i.e. $x = y$. Hence $\tilde{x}_i \neq \tilde{y}_i \Leftrightarrow x \neq y$ or $i \neq j$.

Definition 1.12: A mapping $\tilde{\rho}: SP(\tilde{X}) \times SP(\tilde{X}) \rightarrow R(E)^*$ is soft metric on soft set $\tilde{X}$ with following properties.

SM1. for all $\tilde{x}_{s_1}, \tilde{y}_{s_2} \in \tilde{X}, \quad \tilde{\rho}(\tilde{x}_{s_1}, \tilde{y}_{s_2}) \tilde{\geq} \overline{0}$;

SM2. $\tilde{\rho}(\tilde{x}_{s_1}, \tilde{y}_{s_2}) = \overline{0}$, if and only if $\tilde{x}_{s_1} = \tilde{y}_{s_2}$;

SM3. for all $\tilde{x}_{s_1}, \tilde{y}_{s_2} \in \tilde{X}, \quad \tilde{\rho}(\tilde{x}_{s_1}, \tilde{y}_{s_2}) = \tilde{\rho}(\tilde{y}_{s_2}, \tilde{x}_{s_1})$;

SM4. for all $\tilde{x}_{s_1}, \tilde{y}_{s_2}, \tilde{z}_{s_3} \in \tilde{X}, \quad \tilde{\rho}(\tilde{x}_{s_1}, \tilde{z}_{s_3}) \leq \tilde{\rho}(\tilde{x}_{s_1}, \tilde{y}_{s_2}) + \tilde{\rho}(\tilde{y}_{s_2}, \tilde{z}_{s_3})$

The soft set $\tilde{X}$ with a soft metric $\tilde{\rho}$ defined on $\tilde{X}$ is called a soft metric space and denoted by $(\tilde{X}, \tilde{\rho}, \tilde{E})$.

Definition 1.13: Let us consider a soft metric $(\tilde{X}, \tilde{\rho}, \tilde{E})$ and $\tilde{\alpha}$ be a non-negative soft real number. The soft open ball with center at $\tilde{x}_s$ and radius $\tilde{\alpha}$ is given by

$$B\left(\tilde{x}_s, \tilde{\alpha}\right) = \left\{\tilde{y}_{s'} \in \tilde{X} : \tilde{\rho}\left(\tilde{x}_s, \tilde{y}_{s'}\right) \leq \tilde{\alpha}\right\} \subset SP\left(\tilde{X}\right)$$

and the soft closed ball with center at $\tilde{x}_s$ and radius $\tilde{\alpha}$ is given by

$$B\left(\tilde{x}_s, \tilde{\alpha}\right) = \left\{\tilde{x}_{s'} \in \tilde{X} : \tilde{\rho}\left(\tilde{x}_{s'}, \tilde{y}_{s'}\right) \leq \tilde{\alpha}\right\} \subset SP\left(\tilde{X}\right)$$

Definition 1.14: A sequence $\left\{\tilde{x}^n_{\lambda_n}\right\}$ of soft points in soft metric space $(\tilde{X}, \tilde{\rho}, \tilde{E})$ is said to be convergent in $(\tilde{X}, \tilde{\rho}, \tilde{E})$ if there is a soft point $\tilde{y}_\mu \tilde{\in} \tilde{X}$ such that $\tilde{\rho}\left(\tilde{x}^n_{\lambda_n}, \tilde{y}_\mu\right) \to \tilde{0}$ as $n \to \infty$, that is for every $\tilde{\varepsilon} > \bar{0}$, there is a natural number N = N($\tilde{\varepsilon}$) such that $\bar{0} \leq \tilde{\rho}\left(\tilde{x}^n_{\lambda_n}, \tilde{y}_\mu\right) \tilde{<} \tilde{\varepsilon}$, whenever $n > N$.

Definition 1.15: Let $(\tilde{X}, \tilde{\rho}, \tilde{E})$ be a soft metric space, then the sequence $\left\{\tilde{x}^n_{\lambda_n}\right\}$ of soft points in $(\tilde{X}, \tilde{\rho}, \tilde{E})$ is said to be a Cauchy sequence in $\tilde{X}$, if corresponding to every $\tilde{\varepsilon} \tilde{\geq} \overline{0}$ there exist $m \in N$ such that $\tilde{\rho}\left(\tilde{x}^i_{\lambda_i}, \tilde{y}^j_{\lambda_j}\right) \tilde{\leq} \tilde{\varepsilon}, \forall i, j \geq m$ i.e. $\tilde{\rho}\left(\tilde{x}^i_{\lambda_i}, \tilde{y}^j_{\lambda_j}\right) \to \overline{0}$ as $i, j, \to \infty$.

Definition 1.16: The soft metric space $(\tilde{X}, \tilde{\rho}, \tilde{E})$ is called complete, if every Cauchy sequence in $\tilde{X}$ converges to some point of $\tilde{X}$.

Definition 1.17: Let $(\tilde{X}, \tilde{\rho}, \tilde{E})$ be a soft metric space. A function (f, φ): $(\tilde{X}, \tilde{\rho}, E) \to (\tilde{X}, \tilde{\rho}, \tilde{E})$ is called a soft contraction mapping if there is a soft real number $a \in R$, $0 \leq \alpha < \bar{1}$ such that for every point $\tilde{x}_\lambda, \tilde{y}_\mu \in SP(X)$ we have

$$\tilde{\rho}\left((f,\varphi)(\tilde{x}_\lambda),(f,\varphi)(\tilde{y}_\mu)\right) \leq \alpha\tilde{\rho}(\tilde{x}_\lambda, \tilde{y}_\mu).$$

Definition 1.18: A mapping $\tilde{\rho} : SP(\tilde{X}) \times SP(\tilde{X}) \to R(E)^*$ is said to be dislocated soft metric on the soft set $\tilde{X}$ if $\tilde{\rho}$ $\tilde{\rho}$ satisfies the following conditions:

(d1) $\tilde{\rho}(\tilde{x}_{s_1}, \tilde{y}_{s_2}) = \bar{0}$ then $\tilde{x}_{s_1} = \tilde{y}_{s_2}$,

(d2) $\tilde{\rho}(\tilde{x}_{s_1}, \tilde{y}_{s_2}) = \tilde{\rho}(\tilde{y}_{s_2}, \tilde{x}_{s_1})$, for all $\tilde{x}_{s_1}, \tilde{y}_{s_2} \in \tilde{X}$,

(d3) $\tilde{\rho}(\tilde{x}_{s_1}, \tilde{z}_{s_3}) \leq \tilde{\rho}(\tilde{x}_{s_1}, \tilde{y}_{s_2}) + \tilde{\rho}(\tilde{y}_{s_2}, \tilde{z}_{s_3})$, for all $\tilde{x}_{s_1}, \tilde{y}_{s_2}, \tilde{z}_{s_3} \in \tilde{X}$

The soft set $\tilde{X}$ with soft dislocated metric $\tilde{\rho}$ defined on $\tilde{X}$ is called a dislocated soft metric space and denoted by $(\tilde{X}, \tilde{\rho}, E)$.

2. MAIN RESULTS

Theorem 2.1: Let A and B be two non-empty closed subsets of a complete dislocated soft metric space $(\tilde{X}, \tilde{\rho}, E)$.

Suppose $(f,\varphi): A \cup B \to A \cup B$ is cyclic and satisfy the following:

(c) There exist a constant $k \in (0, 1)$ such that $\tilde{\rho}\left((f,\varphi)x_{\lambda},(f,\varphi)y_{\mu}\right) \leq k\tilde{\rho}\left(x_{\lambda},y_{\mu}\right)$ for all $x_{\lambda} \in A,\ y_{\mu} \in B$

Then (f, φ) has a unique soft fixed point that belongs to $A \cap B$.

Definition 2.2: Let $(\tilde{X}, \tilde{\rho}, E)$ be dislocated soft metric space and U be a subset of $\tilde{X}$. We say that U is $\tilde{\rho}$-open subset of X if $\forall x \in \tilde{X}$ there exists $r > 0$ such that $B_p(x,r) \subseteq U$. Also, $V \subseteq \tilde{X}$ is a $\tilde{\rho}$-closed subset of $\tilde{X}$ if $(\tilde{X}/V)$ is $\tilde{\rho}$-open subset of $\tilde{X}$.

Lemma 2.3: Let $(\tilde{X}, \tilde{\rho}, E)$ be dislocated soft metric space and V be a $\tilde{\rho}$-closed subset of X. Let $\left\{x^{n}_{\lambda_n}\right\}$ be a sequence in V. If $x^{n}_{\lambda_n} \to x_{\lambda}$ as $n \to \infty$ then $x_\lambda \in V$.

Proof: Let $x_\lambda \notin V$, by definition 2.2, $X \setminus V$ is a $\tilde{\rho}$-open set then there exist $r > 0$ such that $B_{\tilde{\rho}}\left(x_{\lambda},r\right) \subseteq X \setminus V$ On the other hand, we have $\lim_{n\to\infty}\left|\tilde{\rho}\left(x^{n}_{\lambda_n},x_{\lambda}\right)-\tilde{\rho}\left(x_{\lambda},x_{\lambda}\right)\right|=0$, Since $x^{n}_{\lambda_n} \to x_{\lambda}$ as $n \to \infty$ So there exist $n_0 \in N$ such that

$$\left|\tilde{\rho}\left(x^{n}_{\lambda_n},x_{\lambda}\right)-\tilde{\rho}\left(x_{\lambda},x_{\lambda}\right)\right|<r$$

for all $n > n_0$. Hence we conclude that $\left\{x^{n}_{\lambda_n}\right\} \subseteq B_{\rho}\left(x_{\lambda},r\right) \subseteq X \setminus V$, for all $n > n_0$. This is contradiction since $\left\{x^{n}_{\lambda_n}\right\} \subseteq V$ for all n in N.

Lemma 2.4: Let $(\tilde{X}, \tilde{\rho}, E)$ be dislocated soft metric space and $\{x_{\lambda_n}^n\}$ be a sequence in $(\tilde{X}, \tilde{\rho}, E)$ such that $\{x_{\lambda_n}^n\} \to x_\lambda$ as $n \to \infty$ and $\tilde{\rho}(x_\lambda, x_\lambda) = 0$ Then

$$\lim_{n\to\infty} \tilde{\rho}(x_{\lambda_n}^n, y_\mu) = \tilde{\rho}(x_\lambda, y_\mu), \text{ for all } y_\mu \in (\tilde{X}, \tilde{\rho}, E).$$

Lemma 2.5: If $(\tilde{X}, \tilde{\rho}, E)$ be soft dislocated metric space, then the following conditions holds

A. If $\tilde{\rho}(x_{\lambda_n}^n, x_{\lambda_n}^n) = 0$ then $\tilde{\rho}(x_\lambda x_\lambda) = \tilde{\rho}(y_\mu, y_\mu) = 0$;

B. If $\{x_{\lambda_n}^n\}$ be a sequence such that $\lim_{n\to\infty} \tilde{\rho}(x_{\lambda_n}^n, x_{\lambda_{n+1}}^{n+1}) = 0$ then we have $\lim_{n\to\infty} \tilde{\rho}(x_{\lambda_n}^n, x_{\lambda_n}^n) = \lim_{n\to\infty} \tilde{\rho}(x_{\lambda_{n+1}}^{n+1}, x_{\lambda_{n+1}}^{n+1}) = 0$;

C. If $x_\lambda \neq y_\mu$ then $\tilde{\rho}(x_\lambda, y_\mu) > 0$;

$\tilde{\rho}(x_\lambda x_\lambda) = \frac{2}{n}\sum_{i=1}^{n} \tilde{\rho}(x_\lambda, x_{\lambda_i}^i)$, holds for all $x_\lambda, x_{\lambda_i}^i \in \tilde{X}$ where $1 \le i \le n$.

D. At first we define the class of Φ and Ψ by the following ways:

$$\Psi = \{\psi : [0,\infty) \to [0,\infty) \text{ such that } \psi \text{ is non decreasing and continuous}\}$$

and

$$\Phi = \{\phi : [0,\infty) \to [0,\infty) \text{ such that } \phi \text{ is lower semicontinuous}\}.$$

Definition 2.6: Let $(\tilde{X}, \tilde{\rho}, E)$ be a soft dislocated metric space, $m \in N$, let A_1, A_2,.., A_m be $\tilde{\rho}$-closed non empty subsets of $\tilde{X}$ and let $\tilde{Y} = \bigcup_{i=1}^{i=m} A_i$. We say that (f, φ) is a cyclic generalized ϕ - ψ contractive mapping if

i. $\tilde{Y} = \bigcup_{i=1}^{i=m} A_i$ is a cyclic representation of $\tilde{Y}$ with respect to (f, φ),

ii. $\psi(t) - \psi(s) + \phi(s) > 0$ for all $t > 0$ and $s = t$ or $s = 0$ and

$$\psi\left(\tilde{\rho}\left((f,\varphi)x_\lambda,(f,\varphi)y_\mu\right)\right) \prec \psi\left(M_{\tilde{\rho}}\left(x_\lambda,y_\mu\right)\right) - \phi\left(M_{\tilde{\rho}}\left(x_\lambda,y_\mu\right)\right) \quad (1)$$

for any $x_\lambda \in A_i, y_\mu \in A_{i+1}, \quad i = 1,2,3,\cdots,m$, where $A_{i+1} = A_1, \phi \in \Phi, \psi \in \Psi$ and

$$M_{\tilde{\rho}}\left(x_\lambda,y_\mu\right) = \max\left\{\tilde{\rho}\left(x_\lambda,y_\mu\right), \tilde{\rho}\left(x_\lambda,(f,\varphi)x_\lambda\right), \tilde{\rho}\left(y_\mu,(f,\varphi)y_\mu\right), \frac{\tilde{\rho}\left(y_\mu,(f,\varphi)x_\lambda\right) + \tilde{\rho}\left(y_\mu,(f,\varphi)y_\mu\right) + \tilde{\rho}\left(x_\lambda,(f,\varphi)x_\lambda\right)}{6}\right\}$$

Let $\tilde{X}$ be a non-empty set and (f, φ): $(\tilde{X},\tilde{\rho},\tilde{E}) \rightarrow (\tilde{X},\tilde{\rho},\tilde{E})$ be given map. The set of all soft fixed points of (f,φ) will be denoted by Fix((f,φ)) i.e.

$$Fix\left((f,\varphi)\right) = \left\{x_\lambda \in \tilde{X} : x_\lambda = (f,\varphi)x_\lambda\right\}.$$

Theorem 2.7: Let $(\tilde{X},\tilde{\rho},\tilde{E})$ be a complete dislocated soft metric space, $m \in N$, let $A_1, A_2, \cdots, A_m$ be non-empty $\tilde{\rho}$-closed subsets of $(\tilde{X},\tilde{\rho},\tilde{E})$ and

Let $\tilde{Y} = \bigcup_{i=1}^{i=m} A_i$. Suppose that $(f,\varphi): \tilde{Y} \rightarrow \tilde{Y}$ is a cyclic generalized $\phi - \psi$ con-

tractive mapping. Then (f,φ) has fixed point in $\bigcap_{i=1}^{n} A_i$. Moreover if $\tilde{\rho}\left(x_\lambda,y_\mu\right) \geq \tilde{\rho}\left(x_\lambda,x_\lambda\right)$ for all $x, y \in Fix\left(f,\varphi\right)$, then (f,φ) has a unique fixed point in $\bigcap_{i=1}^{n} A_i$

Proof: Let $x_{\lambda_0}^0$ be an arbitrary point of $\tilde{Y}$, so there exists some i_0 such

that $x_{\lambda_0}^0 \in A_{i_0}$. We know that $(f,\varphi)(A_{i_0}) \subseteq A_{i_0+1}$ we conclude that $(f,\varphi)(x_{\lambda_0}^0) \in A_{i_0+1}$. Thus there exist $x_{\lambda_1}^1$ in A_{i_0+1} such that $(f,\varphi)x_{\lambda_0}^0 = x_{\lambda_1}^1$ Recursive $(f,\varphi)x_{\lambda_n}^n = x_{\lambda_{n+1}}^{n+1}$, where $x_{\lambda_n}^n \in A_{i_n}$. Hence for $n \geq 0$ there exist $i_n \in \{1,2,3,\cdots,m\}$ such that $x_{\lambda_n}^n \in A_{i_n}$. In case $x_{\lambda_{n_0}}^{n_0} = x_{\lambda_{n_0+1}}^{n_0+1}$ for some $n_0 = 0,1,2,\cdots$, then it is clear that $x_{\lambda_{n_0}}^{n_0}$ is a soft fixed point of (f,φ). Now assume that $x_{\lambda_n}^n \neq x_{\lambda_{n+1}}^{n+1}$ for all n. Hence by lemma 2.5(c) we have $\tilde{\rho}(x_{\lambda_{n-1}}^{n-1}, x_{\lambda_n}^n) > 0$ for all n. We shall show that the sequence $\{d_n\}$ is non-in- creasing, where $d_n = \tilde{\rho}(x_{\lambda_n}^n, x_{\lambda_{n+1}}^{n+1})$. Assume that there exists some $n_0 \in N$ such that

$$\tilde{\rho}\left(x_{\lambda_{n_0-1}}^{n_0-1}, x_{n_0}^{n_0}\right) \leq \tilde{\rho}\left(x_{\lambda_{n_0}}^{n_0}, x_{n_0+1}^{n_0+1}\right)$$

Hence we get

$$\psi\left(\tilde{\rho}\left(x_{n_0-1}^{n_0-1}, x_{n_0}^{n_0}\right)\right) \leq \psi\left(\tilde{\rho}\left(x_{n_0}^{n_0}, x_{n_0+1}^{n_0+1}\right)\right) \tag{2}$$

Set $\tilde{x}_{\lambda_1}^1 = (f,\varphi)(\tilde{x}_\lambda^0) = \left(f(\tilde{x}_\lambda^0)\right)_{\varphi(\lambda)}, \tilde{x}_{\lambda_2}^2 = (f,\varphi)(\tilde{x}_{\lambda_1}^1) = \left(f^2(\tilde{x}_\lambda^0)\right)_{\varphi^2(\lambda)}, \cdots,$

$$\tilde{x}_{\lambda_{n+1}}^{n+1} = (f,\varphi)\left(\tilde{x}_{\lambda_n}^n\right) = \left(f^{n+1}\left(\tilde{x}_\lambda^0\right)\right)_{\varphi^{n+1}(\lambda)}, \cdots.$$

Using conditions (1) together with (2), we get

$$\psi\left(\tilde{\rho}\left(x^{n}_{\lambda_n},x^{n+1}_{\lambda_{n+1}}\right)\right)=\psi\left(\tilde{\rho}\left((f,\varphi)x^{n-1}_{\lambda_{n-1}},(f,\varphi)x^{n}_{\lambda_n}\right)\right)\leq\tilde{\rho}\left((f,\varphi)x^{n-1}_{\lambda_{n-1}},(f,\varphi)x^{n}_{\lambda_n}\right)$$

$$\leq\psi\left(\max\left\{\tilde{\rho}\left(x^{n-1}_{\lambda_{n-1}},x^{n}_{\lambda_n}\right),\tilde{\rho}\left(x^{n-1}_{\lambda_{n-1}},(f,\varphi)x^{n-1}_{\lambda_{n-1}}\right),\tilde{\rho}\left(x^{n}_{\lambda_n},(f,\varphi)x^{n}_{\lambda_n}\right),\right.\right.$$
$$\left.\left.\frac{\tilde{\rho}\left(x^{n}_{\lambda_n},(f,\varphi)x^{n-1}_{\lambda_{n-1}}\right)+\tilde{\rho}\left(x^{n}_{\lambda_n},(f,\varphi)x^{n}_{\lambda_n}\right)+\tilde{\rho}\left(x^{n-1}_{\lambda_{n-1}},(f,\varphi)x^{n-1}_{\lambda_{n-1}}\right)}{6}\right\}\right)$$

$$-\phi\left(\max\left\{\tilde{\rho}\left(x^{n-1}_{\lambda_{n-1}},x^{n}_{\lambda_n}\right),\tilde{\rho}\left(x^{n-1}_{\lambda_{n-1}},(f,\varphi)x^{n-1}_{\lambda_{n-1}}\right),\tilde{\rho}\left(x^{n}_{\lambda_n},(f,\varphi)x^{n}_{\lambda_n}\right),\right.\right.$$
$$\left.\left.\frac{\tilde{\rho}\left(x^{n}_{\lambda_n},(f,\varphi)x^{n-1}_{\lambda_{n-1}}\right)+\tilde{\rho}\left(x^{n}_{\lambda_n},(f,\varphi)x^{n}_{\lambda_n}\right)+\tilde{\rho}\left(x^{n-1}_{\lambda_{n-1}},(f,\varphi)x^{n-1}_{\lambda_{n-1}}\right)}{6}\right\}\right)$$

$$\leq\psi\left(\max\left\{\tilde{\rho}\left(x^{n-1}_{\lambda_{n-1}},x^{n}_{\lambda_n}\right),\tilde{\rho}\left(x^{n-1}_{\lambda_{n-1}},x^{n}_{\lambda_n}\right),\tilde{\rho}\left(x^{n}_{\lambda_n},x^{n+1}_{\lambda_{n+1}}\right),\frac{\tilde{\rho}\left(x^{n}_{\lambda_n},x^{n}_{\lambda_n}\right)+\tilde{\rho}\left(x^{n}_{\lambda_n},x^{n+1}_{\lambda_{n+1}}\right)+\tilde{\rho}\left(x^{n-1}_{\lambda_{n-1}},x^{n}_{\lambda_n}\right)}{6}\right\}\right)$$

$$-\phi\left(\max\left\{\tilde{\rho}\left(x^{n-1}_{\lambda_{n-1}},x^{n}_{\lambda_n}\right),\tilde{\rho}\left(x^{n-1}_{\lambda_{n-1}},x^{n}_{\lambda_n}\right),\tilde{\rho}\left(x^{n}_{\lambda_n},x^{n+1}_{\lambda_{n+1}}\right),\frac{\tilde{\rho}\left(x^{n}_{\lambda_n},x^{n}_{\lambda_n}\right)+\tilde{\rho}\left(x^{n}_{\lambda_n},x^{n+1}_{\lambda_{n+1}}\right)+\tilde{\rho}\left(x^{n-1}_{\lambda_{n-1}},x^{n}_{\lambda_n}\right)}{6}\right\}\right)$$

$$\leq\psi\left(\max\left\{\tilde{\rho}\left(x^{n-1}_{\lambda_{n-1}},x^{n}_{\lambda_n}\right),\tilde{\rho}\left(x^{n}_{\lambda_n},x^{n+1}_{\lambda_{n+1}}\right),\frac{\tilde{\rho}\left(x^{n}_{\lambda_n},x^{n}_{\lambda_n}\right)+\tilde{\rho}\left(x^{n}_{\lambda_n},x^{n+1}_{\lambda_{n+1}}\right)+\tilde{\rho}\left(x^{n-1}_{\lambda_{n-1}},x^{n}_{\lambda_n}\right)}{6}\right\}\right)$$

$$-\phi\left(\max\left\{\tilde{\rho}\left(x^{n-1}_{\lambda_{n-1}},x^{n}_{\lambda_n}\right),\tilde{\rho}\left(x^{n}_{\lambda_n},x^{n+1}_{\lambda_{n+1}}\right),\frac{\tilde{\rho}\left(x^{n}_{\lambda_n},x^{n}_{\lambda_n}\right)+\tilde{\rho}\left(x^{n}_{\lambda_n},x^{n+1}_{\lambda_{n+1}}\right)+\tilde{\rho}\left(x^{n-1}_{\lambda_{n-1}},x^{n}_{\lambda_n}\right)}{6}\right\}\right).$$

(3)

On the other hand, from lemma 2.5(D) we have $\tilde{\rho}\left(x^{n}_{\lambda_n},x^{n}_{\lambda_n}\right)\leq\tilde{\rho}\left(x^{n-1}_{\lambda_{n-1}},x^{n}_{\lambda_n}\right)+\tilde{\rho}\left(x^{n}_{\lambda_n},x^{n+1}_{\lambda_{n+1}}\right)$

and $$\frac{\tilde{\rho}\left(x^{n}_{\lambda_n},x^{n}_{\lambda_n}\right)+\tilde{\rho}\left(x^{n}_{\lambda_n},x^{n+1}_{\lambda_{n+1}}\right)+\tilde{\rho}\left(x^{n-1}_{\lambda_{n-1}},x^{n}_{\lambda_n}\right)}{6}\leq\frac{\tilde{\rho}\left(x^{n-1}_{\lambda_{n-1}},x^{n}_{\lambda_n}\right)+\tilde{\rho}\left(x^{n}_{\lambda_n},x^{n+1}_{\lambda_{n+1}}\right)}{3}$$

That

is $$\max\left\{\tilde{\rho}\left(x^{n-1}_{\lambda_{n-1}},x^{n}_{\lambda_n}\right),\tilde{\rho}\left(x^{n}_{\lambda_n},x^{n+1}_{\lambda_{n+1}}\right),\frac{\tilde{\rho}\left(x^{n-1}_{\lambda_{n-1}},x^{n}_{\lambda_n}\right)+\tilde{\rho}\left(x^{n}_{\lambda_n},x^{n+1}_{\lambda_{n+1}}\right)}{3}\right\}$$

$$\leq\max\left\{\tilde{\rho}\left(x^{n-1}_{\lambda_{n-1}},x^{n}_{\lambda_n}\right),\tilde{\rho}\left(x^{n}_{\lambda_n},x^{n+1}_{\lambda_{n+1}}\right)\right\}$$

Therefore from (3) we get

$$\psi\left(\tilde{\rho}\left(x_{\lambda_n}^{n}, x_{\lambda_{n+1}}^{n+1}\right)\right) \leq \psi\left(\max\left\{\tilde{\rho}\left(x_{\lambda_{n-1}}^{n-1}, x_{\lambda_n}^{n}\right), \tilde{\rho}\left(x_{\lambda_n}^{n}, x_{\lambda_{n+1}}^{n+1}\right)\right\}\right) - \phi\left(\max\left\{\tilde{\rho}\left(x_{\lambda_{n-1}}^{n-1}, x_{\lambda_n}^{n}\right), \tilde{\rho}\left(x_{\lambda_n}^{n}, x_{\lambda_{n+1}}^{n+1}\right)\right\}\right).$$

Now,

if $\max\left\{\tilde{\rho}\left(x_{\lambda_{n-1}}^{n-1}, x_{\lambda_n}^{n}\right), \tilde{\rho}\left(x_{\lambda_n}^{n}, x_{\lambda_{n+1}}^{n+1}\right)\right\} = \tilde{\rho}\left(x_{\lambda_n}^{n}, x_{\lambda_{n+1}}^{n+1}\right)$ then $\tilde{\rho}\left(x_{\lambda_n}^{n}, x_{\lambda_{n+1}}^{n+1}\right) \leq \alpha\tilde{\rho}\left(x_{\lambda_n}^{n}, x_{\lambda_{n+1}}^{n+1}\right) - \beta\tilde{\rho}\left(x_{\lambda_n}^{n}, x_{\lambda_{n+1}}^{n+1}\right)$.

This is contradiction. Hence we have

$$\psi\tilde{\rho}\left(x_{\lambda_n}^{n}, x_{\lambda_{n+1}}^{n+1}\right) \leq \psi\left\{\tilde{\rho}\left(x_{\lambda_{n-1}}^{n-1}, x_{\lambda_n}^{n}\right)\right\} - \phi\left\{\tilde{\rho}\left(x_{\lambda_{n-1}}^{n-1}, x_{\lambda_n}^{n}\right)\right\}, \tag{4}$$

for all $n \in N$. By taking $x_\lambda = x_{\lambda_{n_0-1}}^{n_0-1}$ and $y_\mu = x_{\lambda_{n_0}}^{n_0}$ in (4) and keeping (2) in

mind, we deduce that

$$\psi\tilde{\rho}\left(x_{\lambda_{n_0-1}}^{n_0-1}, x_{\lambda_{n_0}}^{n_0}\right) \leq \psi\left\{\tilde{\rho}\left(x_{\lambda_{n_0-1}}^{n_0-1}, x_{\lambda_{n_0}}^{n_0}\right)\right\} - \phi\left\{\tilde{\rho}\left(x_{\lambda_{n_0-1}}^{n_0-1}, x_{\lambda_{n_0}}^{n_0}\right)\right\}.$$

This is a contradiction. Hence we conclude that $d_n < d_{n-1}$ i.e.

$\tilde{\rho}\left(x_{\lambda_n}^{n}, x_{\lambda_{n+1}}^{n+1}\right) \leq \tilde{\rho}\left(x_{\lambda_{n-1}}^{n-1}, x_{\lambda_n}^{n}\right)$ hold for all $n \in N$. Thus there exist $r \leq$ 0 such that $\lim\limits_{n\to\infty} d_n = \lim\limits_{n\to\infty} \tilde{\rho}\left(x_{\lambda_n}^{n}, x_{\lambda_{n+1}}^{n+1}\right) = r$. We shall show that r = 0 by the method of

reductio ad absurdum. For this purpose, we assume that $r > 0$. By (4) together with the property of ϕ and ψ we have

$$\psi(r) = \limsup_{n\to\infty} \psi(d_n) \leq \limsup_{n\to\infty}\left[\psi(d_{n-1}) - \phi(d_{n-1})\right] \leq \psi(r) - \phi(r).$$

This yields that $\varphi(r) \leq 0$. This is contradiction. Hence we obtain that

$$\lim_{n\to\infty} d_n \le \lim_{n\to\infty}\left(x_{\lambda_n}^{n}, x_{\lambda_{n+1}}^{n+1}\right) = 0 \quad (5)$$

We shall show that $\left\{x_{\lambda_n}^{n}\right\}$ is a $\tilde{\rho}$-Cauchy sequence. To reach this goal, first we prove the followings claim:

(k) For every $\epsilon > 0$, there exists $n \in N$ such that if $r, q \ge n$ with $r - q = 1(m)$ then $\tilde{\rho}\left(x_{\lambda_{r_n}}^{r_n}, x_{\lambda_{q_n}}^{q_n}\right) < \epsilon$.

Suppose, on the contrary that there exist $\epsilon > 0$ such that for any $n \in N$, we can find $r_n > q_n \ge n$ with $r_n - q_n \equiv 1(m)$ satisfying

$$\tilde{\rho}\left(x_{\lambda_{q_n}}^{q_n}, x_{\lambda_{r_n}}^{r_n}\right) \ge \epsilon \quad (6)$$

Now we consider $n > 2m$. Then, corresponding to $q_n \ge n$, we can choose r_n in such a way that it is the smallest integer with $r_n > q_n$ satisfying

$r_n - q_n \equiv 1(m)$ and $\tilde{\rho}\left(x_{\lambda_{q_n}}^{q_n}, x_{\lambda_{r_n}}^{r_n}\right) \ge \epsilon$ Therefore $\tilde{\rho}\left(x_{\lambda_{q_n}}^{q_n}, x_{\lambda_{r_n-m}}^{r_n-m}\right) \le \epsilon$. By using triangular inequality, we obtain

$$\begin{aligned}
\epsilon &\le \tilde{\rho}\left(x_{\lambda_{q_n}}^{q_n}, x_{\lambda_{r_n}}^{r_n}\right) \\
&\le \tilde{\rho}\left(x_{\lambda_{q_n}}^{q_n}, x_{\lambda_{q_n-m}}^{q_n-m}\right) + \sum_{i=1}^{m} \tilde{\rho}\left(x_{\lambda_{r_n-i}}^{r_n-i}, x_{\lambda_{r_n-i+1}}^{r_n-i+1}\right) \\
&\le \epsilon + \sum_{i=1}^{m} \tilde{\rho}\left(x_{\lambda_{r_n-i}}^{r_n-i}, x_{\lambda_{r_n-i+1}}^{r_n-i+1}\right)
\end{aligned}$$

Passing to the limit as $n \to \infty$ in the last inequality and taking (5) into account, we obtain that

$$\lim_{n\to\infty} \tilde{\rho}\left(x_{\lambda_{q_n}}^{q_n}, x_{\lambda_{r_n}}^{r_n}\right) = \epsilon \quad . (7)$$

Again by (d_3) we derive that

$$\begin{aligned}
\epsilon &\le \tilde{\rho}\left(x_{\lambda_{q_n}}^{q_n}, x_{\lambda_{r_n}}^{r_n}\right) \\
&\le \tilde{\rho}\left(x_{\lambda_{q_n}}^{q_n}, x_{\lambda_{q_{n+1}}}^{q_{n+1}}\right) + \tilde{\rho}\left(x_{\lambda_{q_{n+1}}}^{q_{n+1}}, x_{\lambda_{r_n+1}}^{r_n+1}\right) + \tilde{\rho}\left(x_{\lambda_{r_n+1}}^{r_n+1}, x_{\lambda_{r_n}}^{r_n}\right) \\
&\le \tilde{\rho}\left(x_{\lambda_{q_n}}^{q_n}, x_{\lambda_{q_{n+1}}}^{q_{n+1}}\right) + \tilde{\rho}\left(x_{\lambda_{q_{n+1}}}^{q_{n+1}}, x_{\lambda_{q_n}}^{q_n}\right) + \tilde{\rho}\left(x_{\lambda_{q_n}}^{q_n}, x_{\lambda_{r_n}}^{r_n}\right) \\
&\quad + \tilde{\rho}\left(x_{\lambda_{r_n}}^{r_n}, x_{\lambda_{r_n}+1}^{r_n+1}\right) + \tilde{\rho}\left(x_{\lambda_{r_n}+1}^{r_n+1}, x_{\lambda_{r_n}}^{r_n}\right) \\
&\le 2\tilde{\rho}\left(x_{\lambda_{q_n}}^{q_n}, x_{\lambda_{q_{n+1}}}^{q_{n+1}}\right) + \tilde{\rho}\left(x_{\lambda_{q_n}}^{q_n}, x_{\lambda_{r_n}}^{r_n}\right) + 2\tilde{\rho}\left(x_{\lambda_{r_n}+1}^{r_n+1}, x_{\lambda_{r_n}}^{r_n}\right)
\end{aligned}$$

Taking (5) and (7) in account we get

$$\lim_{n\to\infty} \tilde{\rho}\left(x_{\lambda_{q_{n+1}}}^{q_{n+1}}, x_{\lambda_{r_n+1}}^{r_n+1}\right) = \epsilon \quad . (8)$$

By (d_3) we have the following inequality

$$\tilde{\rho}\left(x_{\lambda_{q_n}}^{q_n}, x_{\lambda_{r_n+1}}^{r_n+1}\right) \le \tilde{\rho}\left(x_{\lambda_{q_n}}^{q_n}, x_{\lambda_{r_n}}^{r_n}\right) + \tilde{\rho}\left(x_{\lambda_{r_n}}^{r_n}, x_{\lambda_{r_n+1}}^{r_n+1}\right) \quad (9)$$

and $\tilde{\rho}\left(x_{\lambda_{q_n}}^{q_n}, x_{\lambda_{r_n}}^{r_n}\right) \leq \tilde{\rho}\left(x_{\lambda_{q_n}}^{q_n}, x_{\lambda_{r_n+1}}^{r_n+1}\right) + \tilde{\rho}\left(x_{\lambda_{r_n}}^{r_n}, x_{\lambda_{r_n+1}}^{r_n+1}\right)$. (10)

Letting lim $n \to \infty$ in (9) and (10), we have

$$\lim_{n\to\infty} \tilde{\rho}\left(x_{\lambda_{q_n}}^{q_n}, x_{\lambda_{r_n+1}}^{r_n+1}\right) = \epsilon \quad . (11)$$

Again by (d_3) we have

$$\tilde{\rho}\left(x_{\lambda_{r_n}}^{r_n}, x_{\lambda_{q_n+1}}^{q_n+1}\right) \leq \tilde{\rho}\left(x_{\lambda_{r_n}}^{r_n}, x_{\lambda_{r_n+1}}^{r_n+1}\right) + \tilde{\rho}\left(x_{\lambda_{r_n+1}}^{r_n+1}, x_{\lambda_{q_n+1}}^{q_n+1}\right) \quad (12)$$

and

$$\tilde{\rho}\left(x_{\lambda_{r_n+1}}^{r_n+1}, x_{\lambda_{q_n+1}}^{q_n+1}\right) \leq \tilde{\rho}\left(x_{\lambda_{r_n+1}}^{r_n+1}, x_{\lambda_{r_n}}^{r_n}\right) + \tilde{\rho}\left(x_{\lambda_{r_n}}^{r_n}, x_{\lambda_{q_n+1}}^{q_n+1}\right) \quad . (13)$$

Letting $n \to \infty$ in (12) and (13), we derive that

$$\lim_{n\to\infty} \tilde{\rho}\left(x_{\lambda_{r_n}}^{r_n}, x_{\lambda_{q_n+1}}^{q_n+1}\right) = \epsilon \quad . \quad (14)$$

Since $x_{\lambda_{q_n}}^{q_n}$ and $x_{\lambda_{r_n}}^{r_n}$ lie in different adjacently labeled sets and for certain $1 \leq i \leq m$. By using (5), (7), (8), (11) and (14) together with the fact that (f, φ) is a generalized cyclic $\phi - \psi$ contractive mappings, we find that

$$\psi\left(\tilde{\rho}\left(x_{\lambda_{q_n+1}}^{q_n+1}, x_{\lambda_{r_n+1}}^{r_n+1}\right)\right) = \psi\left(\tilde{\rho}\left((f,\varphi)x_{\lambda_{q_n}}^{q_n}, (f,\varphi)x_{\lambda_{r_n}}^{r_n}\right)\right)$$

$$\leq \psi\left(\max\left\{\tilde{\rho}\left(x_{\lambda_{q_n}}^{q_n}, x_{\lambda_{r_n}}^{r_n}\right), \tilde{\rho}\left(x_{\lambda_{q_n}}^{q_n}, (f,\varphi)x_{\lambda_{q_n}}^{q_n}\right), \tilde{\rho}\left(x_{\lambda_{r_n}}^{r_n}, (f,\varphi)x_{\lambda_{r_n}}^{r_n}\right), \frac{\tilde{\rho}\left(x_{\lambda_{r_n}}^{r_n}, (f,\varphi)x_{\lambda_{q_n}}^{q_n}\right) + \tilde{\rho}\left(x_{\lambda_{r_n}}^{r_n}, (f,\varphi)x_{\lambda_{r_n}}^{r_n}\right) + \tilde{\rho}\left(x_{\lambda_{q_n}}^{q_n}, (f,\varphi)x_{\lambda_{q_n}}^{q_n}\right)}{6}\right\}\right)$$

$$-\phi\left(\max\left\{\tilde{\rho}\left(x_{\lambda_{q_n}}^{q_n}, x_{\lambda_{r_n}}^{r_n}\right), \tilde{\rho}\left(x_{\lambda_{q_n}}^{q_n}, (f,\varphi)x_{\lambda_{q_n}}^{q_n}\right), \tilde{\rho}\left(x_{\lambda_{r_n}}^{r_n}, (f,\varphi)x_{\lambda_{r_n}}^{r_n}\right), \frac{\tilde{\rho}\left(x_{\lambda_{r_n}}^{r_n}, (f,\varphi)x_{\lambda_{q_n}}^{q_n}\right) + \tilde{\rho}\left(x_{\lambda_{r_n}}^{r_n}, (f,\varphi)x_{\lambda_{r_n}}^{r_n}\right) + \tilde{\rho}\left(x_{\lambda_{q_n}}^{q_n}, (f,\varphi)x_{\lambda_{q_n}}^{q_n}\right)}{6}\right\}\right)$$

$$\leq \psi\left(\max\left\{\tilde{\rho}\left(x_{\lambda_{q_n}}^{q_n}, x_{\lambda_{r_n}}^{r_n}\right), \tilde{\rho}\left(x_{\lambda_{q_n}}^{q_n}, x_{\lambda_{q_{n+1}}}^{q_{n+1}}\right), \tilde{\rho}\left(x_{\lambda_{r_n}}^{r_n}, x_{\lambda_{r_n+1}}^{r_{n+1}}\right), \frac{\tilde{\rho}\left(x_{\lambda_{r_n}}^{r_n}, x_{\lambda_{q_{n+1}}}^{q_{n+1}}\right) + \tilde{\rho}\left(x_{\lambda_{r_n}}^{r_n}, x_{\lambda_{r_n+1}}^{r_{n+1}}\right) + \tilde{\rho}\left(x_{\lambda_{q_n}}^{q_n}, x_{\lambda_{q_{n+1}}}^{q_{n+1}}\right)}{6}\right\}\right)$$

$$-\phi\left(\max\left\{\tilde{\rho}\left(x_{\lambda_{q_n}}^{q_n}, x_{\lambda_{r_n}}^{r_n}\right), \tilde{\rho}\left(x_{\lambda_{q_n}}^{q_n}, x_{\lambda_{q_{n+1}}}^{q_{n+1}}\right), \tilde{\rho}\left(x_{\lambda_{r_n}}^{r_n}, x_{\lambda_{r_n+1}}^{r_{n+1}}\right), \frac{\tilde{\rho}\left(x_{\lambda_{r_n}}^{r_n}, x_{\lambda_{q_{n+1}}}^{q_{n+1}}\right) + \tilde{\rho}\left(x_{\lambda_{r_n}}^{r_n}, x_{\lambda_{r_n+1}}^{r_{n+1}}\right) + \tilde{\rho}\left(x_{\lambda_{q_n}}^{q_n}, x_{\lambda_{q_{n+1}}}^{q_{n+1}}\right)}{6}\right\}\right).$$

Regarding the property of ϕ and ψ in the last inequality, we obtain that $\psi(\epsilon) \leq \psi(\epsilon) - \phi(\epsilon)$, which is a contradiction. Hence the condition (k) is a satisfied. Fix $\epsilon > 0$. By the claim we find $n_0 \in N$ such that if $r,q \geq n_0$ with $r - q \equiv 1(m)$,

$$\tilde{\rho}\left(x_{\lambda_r}^{r}, x_{\lambda_q}^{q}\right) \leq \frac{\epsilon}{2} \tag{15}$$

Since $\lim_{n\to\infty} \tilde{d}\left(x_{\lambda_n}^{n}, x_{\lambda_{n+1}}^{n+1}\right) = 0$, we also find $n_1 \in N$ such that

$$\tilde{\rho}\left(x_{\lambda_n}^{n}, x_{\lambda_{n+1}}^{n+1}\right) \leq \frac{\epsilon}{2m} \tag{16}$$

for any $n \geq n_1$. Suppose that $r, s \geq \max\{n_0, n_1\}$ and $s > r$. There exist $k \in \{1, 2, 3, \cdots, m\}$ such that $s - r \equiv k(m)$ Therefore $s - r + \phi \equiv 1(m)$, for $\phi = m - k + 1$.

So we have for $j \in \{1,2,3,\ldots,m\}$ and $s + j - r \equiv 1(m)$,

$$\tilde{\rho}\left(x_{\lambda_r}^{r}, x_{\lambda_s}^{s}\right) \leq \tilde{\rho}\left(x_{\lambda_r}^{r}, x_{\lambda_{s+j}}^{s+j}\right) + \tilde{\rho}\left(x_{\lambda_{s+j}}^{s+j}, x_{\lambda_{s+j-1}}^{s+j-1}\right) + \cdots + \tilde{\rho}\left(x_{\lambda_{s+1}}^{s+1}, x_{\lambda_s}^{s}\right)$$

.

By (15) & (16) and from the last inequality, we get

$$\tilde{\rho}\left(x_{\lambda_r}^{r}, x_{\lambda_s}^{s}\right) \leq \frac{\epsilon}{2} + j \times \frac{\epsilon}{2m} \leq \frac{\epsilon}{2} + m \times \frac{\epsilon}{2m} = \epsilon$$

.

This proves that $\{x_{\lambda_n}^{n}\}$ is a $\tilde{\rho}$-Cauchy sequence. Since ϵ is arbitrary,

$\{x_{\lambda_n}^{n}\}$ is a Cauchy sequence. Since Y is $\tilde{\rho}$-closed in $(\tilde{X},\tilde{\rho},E)$, then $(\tilde{Y},\tilde{\rho},E)$ is also complete, there exists $x_\lambda \in \tilde{Y} = \bigcup_{i=1}^{m} A_i$ such that $\lim_{n\to\infty} x_{\lambda_n}^{n} = x_\lambda$ in $(\tilde{Y},\tilde{\rho},E)$; equivalently

$$\tilde{\rho}(x_\lambda, x_\lambda) = \lim_{n\to\infty} \tilde{\rho}\left(x_\lambda, x_{\lambda_n}^{n}\right) = \lim_{m,n\to\infty} \tilde{\rho}\left(x_{\lambda_n}^{n}, x_{\lambda_m}^{m}\right) = 0 \tag{17}$$

.

In what follows, we prove that x_λ is a soft fixed point of (f, φ). In fact, since $\lim_{n\to\infty} x^n_{\lambda_n} = x_\lambda$ and $\tilde{Y} = \bigcup_{i=1}^{m} A_i$ is a cyclic representation of Y with respect to (*f*,*φ*). The sequence $\{x^n_{\lambda_n}\}$ has infinite terms in each A_i, for $i \in \{1,2,3,\cdots,m\}$. Suppose that $x_\lambda \in A_i$, $(f,\varphi)x_\lambda \in A_{i+1}$ and we take a subsequence $\{x^{n_k}_{\lambda_{n_k}}\}$ of $\{x^n_{\lambda_n}\}$ with $x^{n_k}_{\lambda_{n_k}} \in A_{i-1}$ (the existence of this subsequence is guaranteed by above- mentioned comment). By using the contractive condition we can obtain

$$\psi\left(\tilde{\rho}\left((f,\varphi)x_\lambda,(f,\varphi)x^{n_k}_{\lambda_{nk}}\right)\right)$$

$$\leq \psi\left(\max\left\{\tilde{\rho}\left(x_\lambda,x^{n_k}_{\lambda_{nk}}\right),\tilde{\rho}\left(x_\lambda,(f,\varphi)x_\lambda\right),\tilde{\rho}\left(x^{n_k}_{\lambda_{nk}},(f,\varphi)x^{n_k}_{\lambda_{nk}}\right),\right.\right.$$
$$\left.\left.\frac{\tilde{\rho}\left(x^{n_k}_{\lambda_{nk}},(f,\varphi)x_\lambda\right)+\tilde{\rho}\left(x^{n_k}_{\lambda_{nk}},(f,\varphi)x^{n_k}_{\lambda_{nk}}\right)+\tilde{\rho}\left(x_\lambda,(f,\varphi)x_\lambda\right)}{6}\right\}\right)$$

$$-\phi\left(\max\left\{\tilde{\rho}\left(x_\lambda,x^{n_k}_{\lambda_{nk}}\right),\tilde{\rho}\left(x_\lambda,(f,\varphi)x_\lambda\right),\tilde{\rho}\left(x^{n_k}_{\lambda_{nk}},(f,\varphi)x^{n_k}_{\lambda_{nk}}\right),\right.\right.$$
$$\left.\left.\frac{\tilde{\rho}\left(x^{n_k}_{\lambda_{nk}},(f,\varphi)x_\lambda\right)+\tilde{\rho}\left(x^{n_k}_{\lambda_{nk}},(f,\varphi)x^{n_k}_{\lambda_{nk}}\right)+\tilde{\rho}\left(x_\lambda,(f,\varphi)x_\lambda\right)}{6}\right\}\right).$$

Passing to the limit as $n \to \infty$ and using $x^{n_k}_{\lambda_{nk}} \to x_\lambda$, lower semi-continuity of ϕ we have

$$\psi\left(\tilde{\rho}\left(x_\lambda,(f,\varphi)x_\lambda\right)\right) \leq \psi\left(\tilde{\rho}\left(x_\lambda,(f,\varphi)x_\lambda\right)\right) - \phi\left(\tilde{\rho}\left(x_\lambda,(f,\varphi)x_\lambda\right)\right)$$

.

So, $\tilde{\rho}\left(x_\lambda,(f,\varphi)x_\lambda\right)=0$. Therefore x_λ is a soft fixed point of (*f*, *φ*). Finally to prove the uniqueness of soft fixed point, suppose that $y_\mu, z_w \in \left(\tilde{X},\tilde{\rho},E\right)$ are two distinct soft fixed points of (*f*, *φ*). The cyclic character of (*f*, *φ*) and the fact that $y_\mu, z_w \in \left(\tilde{X},\tilde{\rho},E\right)$ are soft fixed points of (*f*, *φ*) implies that

$x_\lambda, y_\mu \in \bigcap_{i=1}^{m} A_i$. Suppose that $x_\lambda \neq y_\mu$ and for all $p_\lambda, q_\theta \in Fix((f,\varphi))$

$\tilde{\rho}(p_\lambda, q_\theta) \geq \tilde{\rho}(p_\lambda, p_\lambda)$. Using the contractive condition, we obtain

$$\begin{aligned}
&\psi\left(\tilde{\rho}((f,\varphi)x_\lambda,(f,\varphi)y_\mu)\right) \\
&\leq \psi\left(\max\left\{\tilde{\rho}(x_\lambda, y_\mu), \tilde{\rho}(x_\lambda,(f,\varphi)x_\lambda), \tilde{\rho}(y_\mu,(f,\varphi)y_\mu),\right.\right. \\
&\left.\left.\frac{\tilde{\rho}(y_\mu,(f,\varphi)x_\lambda)+\tilde{\rho}(y_\mu,(f,\varphi)y_\mu)+\tilde{\rho}(x_\lambda,(f,\varphi)x_\lambda)}{6}\right\}\right) \\
&-\phi\left(\max\left\{\tilde{\rho}(x_\lambda, y_\mu), \tilde{\rho}(x_\lambda,(f,\varphi)x_\lambda), \tilde{\rho}(y_\mu,(f,\varphi)y_\mu),\right.\right. \\
&\left.\left.\frac{\tilde{\rho}(y_\mu,(f,\varphi)x_\lambda)+\tilde{\rho}(y_\mu,(f,\varphi)y_\mu)+\tilde{\rho}(x_\lambda,(f,\varphi)x_\lambda)}{6}\right\}\right)
\end{aligned}$$

Then we have $\psi\left(\tilde{\rho}(x_\lambda, y_\mu)\right) \leq \psi\left(\tilde{\rho}(x_\lambda, y_\mu)\right) - \phi\left(\tilde{\rho}(x_\lambda, y_\mu)\right)$.

This is a contradiction. Thus we derive that $\psi\left(\tilde{\rho}(x_\lambda, y_\mu)\right) \leq \psi\left(\tilde{\rho}(x_\lambda, y_\mu)\right) - \phi\left(\tilde{\rho}(x_\lambda, y_\mu)\right)$. Hence proved.

In the theorem 2.7, if we take $(\tilde{X}, \tilde{\rho}, E) = A_i$, for all $0 \leq i \leq m$ then we deduce the following theorem.

Theorem 2.8: Let $(\tilde{X}, \tilde{\rho}, E)$ be a complete soft dislocated metric space and (f, φ) be self map on $(\tilde{X}, \tilde{\rho}, E)$. Assume that there exist $\varphi \in \Phi$, $\psi \in \Psi$ such

that $\psi\left(\tilde{\rho}((f,\varphi)x_\lambda,(f,\varphi)y_\mu)\right) \leq \psi\left(M_{\tilde{\rho}}(x_\lambda, y_\mu)\right) - \phi\left(M_{\tilde{\rho}}(x_\lambda, y_\mu)\right)$, for all $x_\lambda, y_\mu \in (\tilde{X}, \tilde{\rho}, E)$, where

$$M_{\tilde{\rho}}\left(x_{\lambda},y_{\mu}\right)=\max\left\{\tilde{\rho}\left(x_{\lambda},y_{\mu}\right),\tilde{\rho}\left(x_{\lambda},(f,\varphi)x_{\lambda}\right),\tilde{\rho}\left(y_{\mu},(f,\varphi)y_{\mu}\right),\right.$$
$$\left.\frac{\tilde{\rho}\left(y_{\mu},(f,\varphi)x_{\lambda}\right)+\tilde{\rho}\left(y_{\mu},(f,\varphi)y_{\mu}\right)+\tilde{\rho}\left(x_{\lambda},(f,\varphi)x_{\lambda}\right)}{6}\right\}.$$

Then (f, φ) has a soft fixed point. Moreover if $\tilde{\rho}\left(x_{\lambda},y_{\mu}\right)\geq\tilde{\rho}\left(x_{\lambda},x_{\lambda}\right)$ for all $x_{\lambda},y_{\mu}\in Fix\left(\left(f,\varphi\right)\right)$ then (f, φ) has a unique soft fixed point.

If in theorem (2.7) we take $\psi(t) = t$ and $\phi(t) = (1 - r)t$ where $r \in [0,1)$ then we deduce the following corollary.

Corollary 2.9: Let $(\tilde{X},\tilde{\rho},E)$ be a complete soft dislocated metric space, $m \in N$,

let $A1, A2,\ldots, Am$ be non empty $\tilde{\rho}$-closed subsets of $(\tilde{X},\tilde{\rho},E)$ and let

$Y=\bigcup_{i=1}^{i=m}A_i$. Suppose that $(f,\varphi):Y\rightarrow Y$ is an operator such that

i. $\bigcup_{i=1}^{i=m}A_i$ is a cyclic representation of $(f,\varphi):Y\rightarrow Y$ with respect to (f, φ)

ii. there exist $r \in [0,1)$ such that

$$\tilde{\rho}\left((f,\varphi)x_{\lambda},(f,\varphi)y_{\mu}\right)$$
$$=r\max\left\{\tilde{\rho}\left(x_{\lambda},y_{\mu}\right),\tilde{\rho}\left(x_{\lambda},(f,\varphi)x_{\lambda}\right),\tilde{\rho}\left(y_{\mu},(f,\varphi)y_{\mu}\right),\right.$$
$$\left.\frac{\tilde{\rho}\left(y_{\mu},(f,\varphi)x_{\lambda}\right)+\tilde{\rho}\left(y_{\mu},(f,\varphi)y_{\mu}\right)+\tilde{\rho}\left(x_{\lambda},(f,\varphi)x_{\lambda}\right)}{6}\right\},$$

for any $x \in A_i$ and $y \in A_{i+1}$, $i = 1,2,3,\ldots, m$. Where $A_{m+1}=A_1$, then (f, φ) has a soft fixed point $z_w\in\bigcap_{i=1}^{m}A_i$. Moreover if $\tilde{\rho}\left(x_{\lambda},y_{\mu}\right)\geq\tilde{\rho}\left(x_{\lambda},x_{\lambda}\right)$ for all $x_{\lambda}\in Fix\left(f,\varphi\right)$ then (f, φ) has a unique soft fixed point.

Example 2.10: Let X = R with soft dislocated metric

$\tilde{\rho}(x_\lambda, y_\mu) = \max\left\{\left|\frac{x_\lambda}{2}\right|, \left|\frac{y_\mu}{2}\right|\right\}$, for all $x_\lambda, y_\mu \in \overline{X}$. Suppose $A_1 = [-2,0]$, $A_2 = [0,2]$ & $Y = \bigcup_{i=1}^{i=2} A_i$. Define $(f,\varphi): Y \to Y$ by

$$(f,\varphi) = \begin{cases} \dfrac{(x_\lambda)^2}{8}, & \text{if } x_\lambda \in [-2,0] \\ \dfrac{-x_\lambda}{5}, & \text{if } x_\lambda \in [0,2] \end{cases}$$

It is clear that $\bigcup_{i=1}^{i=2} A_i$ is a cyclic representation of Y with respect to (f, φ).

Let $x_\lambda \in A_1 = [-2,0]$ and $y_\mu \in A_2 = [0,2]$ then

$$\tilde{\rho}((f,\varphi)x_\lambda, (f,\varphi)y_\mu)$$
$$= \tilde{\rho}\left(\frac{x_\lambda^2}{8}, \frac{-y_\mu}{5}\right) = \max\left\{\left|\frac{x_\lambda^2/8}{2}\right|, \left|\frac{-y_\mu/5}{2}\right|\right\} = \max\left\{\left|\frac{x_\lambda^2}{16}\right|, \left|\frac{-y_\mu}{10}\right|\right\}$$
$$\leq \max\left\{\frac{x_\lambda}{4}, \frac{y_\mu}{4}\right\} \leq \frac{1}{2}\max\left\{\frac{x_\lambda}{2}, \frac{y_\mu}{2}\right\} \leq \frac{1}{2}\tilde{\rho}(x_\lambda, y_\mu).$$

and so

$$\tilde{\rho}((f,\varphi)x_\lambda, (f,\varphi)y_\mu)$$
$$= r\max\left\{\tilde{\rho}(x_\lambda, y_\mu), \tilde{\rho}(x_\lambda, (f,\varphi)x_\lambda), \tilde{\rho}(y_\mu, (f,\varphi)y_\mu),\right.$$
$$\left.\frac{\tilde{\rho}(y_\mu, (f,\varphi)x_\lambda) + \tilde{\rho}(y_\mu, (f,\varphi)y_\mu) + \tilde{\rho}(x_\lambda, (f,\varphi)x_\lambda)}{6}\right\}.$$

Hence the condition of corollary (2.9) (theorem 2.7) holds and (f, φ) has a fixed point in $A_1 \cap A_2$. Here $x_\lambda = 0$ is a fixed point of (f, φ).

Example 2.11: Let X = R with soft dislocated metric

$\tilde{\rho}(x_\lambda, y_\mu) = \max\{|x_\lambda|, |y_\mu|\}$, for all $x_\lambda, y_\mu \in X$.

Suppose $A_1 = [-1,0]$ and $A_2 = [0,1]$ & $Y = \bigcup_{i=1}^{i=2} A_i$ we define $(f,\varphi): Y \to Y$ by

$$(f,\varphi) = \begin{cases} \dfrac{(x_\lambda)^2}{2}, & \text{if } x_\lambda \in [-1,0], \\ \dfrac{-x_\lambda}{4}, & \text{if } x_\lambda \in [0,1]. \end{cases}$$

It is clear that $\bigcup_{i=1}^{i=2} A_i$ is a cyclic representation of Y with respect to (f, φ).

Let $x_\lambda \in A_1 = [-1,0]$ and $y_\mu \in A_2 = [0,1]$, then

$$\tilde{\rho}((f,\varphi)x_\lambda, (f,\varphi)y_\mu)$$

$$= \tilde{\rho}\left(\frac{x_\lambda^2}{2}, \frac{-y_\mu}{2}\right) = \max\left\{\left|\frac{x_\lambda^2}{2}\right|, \left|\frac{-y_\mu}{4}\right|\right\} \leq \max\left\{\frac{x_\lambda}{2}, \frac{-y_\mu}{2}\right\}$$

$$\leq \frac{1}{2}\max\{x_\lambda, -y_\mu\} \leq \frac{1}{2}\max\{|x_\lambda|, |y_\mu|\} \leq \frac{1}{2}\tilde{\rho}(x_\lambda, y_\mu),$$

$$\tilde{\rho}((f,\varphi)x_\lambda, (f,\varphi)y_\mu)$$

$$= r\max\left\{\tilde{\rho}(x_\lambda, y_\mu), \tilde{\rho}(x_\lambda, (f,\varphi)x_\lambda), \tilde{\rho}(y_\mu, (f,\varphi)y_\mu), \frac{\tilde{\rho}(y_\mu, (f,\varphi)x_\lambda) + \tilde{\rho}(y_\mu, (f,\varphi)y_\mu) + \tilde{\rho}(x_\lambda, (f,\varphi)x_\lambda)}{6}\right\}.$$

Hence the condition of corollary (2.9) (theorem 2.7) holds and (f, φ) has a soft fixed point in $A_1 \cap A_2$. Here $x_\lambda = 0$ is a soft fixed point of (f, φ).

In the above corollary we take $A_i = \left(\tilde{X}, \tilde{\rho}, E\right)$ for all $0 \le i \le m$, then we deduce the following corollary.

Corollary 2.12: Let $\left(\tilde{X}, \tilde{\rho}, E\right)$ be a complete soft dislocated metric space and let (f, φ) be a self map on X. Assume that there exist $r \in [0, 1)$ such that

$$\tilde{\rho}\left((f,\varphi)x_\lambda,(f,\varphi)y_\mu\right)$$
$$\le r\max\left\{\tilde{\rho}\left(x_\lambda,y_\mu\right),\tilde{\rho}\left(x_\lambda,(f,\varphi)x_\lambda\right),\tilde{\rho}\left(y_\mu,(f,\varphi)y_\mu\right),\right.$$
$$\left.\frac{\tilde{\rho}\left(y_\mu,(f,\varphi)x_\lambda\right)+\tilde{\rho}\left(y_\mu,(f,\varphi)y_\mu\right)+\tilde{\rho}\left(x_\lambda,(f,\varphi)x_\lambda\right)}{6}\right\}$$

holds for all $x_\lambda, y_\mu \in \left(X, \tilde{\rho}, E\right)$. Then (f, φ) has a soft fixed point. Moreover if $\tilde{\rho}\left(x_\lambda, y_\mu\right) \ge \tilde{\rho}\left(x_\lambda, x_\lambda\right)$ for all $x_\lambda, y_\mu \in Fix\left((f,\varphi)\right)$, then (f, φ) has a unique soft fixed point.

Example 2.13: Let X = R with soft dislocated metric space and $\tilde{\rho}\left(x_\lambda, y_\mu\right) = \max\left\{x_\lambda, y_\mu\right\}$. For any x_λ let $(f,\varphi):\left(X,\tilde{\rho},E\right)\to\left(X,\tilde{\rho},E\right)$ be defined by

$$(f,\varphi)x_\lambda = \begin{cases} \dfrac{x_\lambda}{6}, & \text{if } 0 \le x_\lambda < \dfrac{1}{2} \\ \dfrac{x_\lambda^2}{5}, & \text{if } \dfrac{1}{2} \le x_\lambda \le 1 \\ \dfrac{x_\lambda}{7}, & \text{if } x > 1. \end{cases}$$

Proof: To show that the existence and uniqueness of soft point of (f, φ), we need to consider the following cases

Let $0 \le x_\lambda, y_\mu < \frac{1}{2}$ then

$$\tilde{\rho}\left((f,\varphi)x_\lambda,(f,\varphi)y_\mu\right)=\frac{1}{6}\max\{x_\lambda,y_\mu\}\le\frac{1}{2}\max\{x_\lambda,y_\mu\}=\frac{1}{2}\tilde{\rho}(x_\lambda,y_\mu)$$

.

Let $\frac{1}{2} \le x_\lambda, y_\mu \le 1$ then

$$\tilde{\rho}\left((f,\varphi)x_\lambda,(f,\varphi)y_\mu\right)=\frac{1}{5}\max\{x_\lambda^2,y_\mu^2\}\le\frac{1}{5}\max\{x_\lambda,y_\mu\}$$
$$\le\frac{1}{2}\max\{x_\lambda,y_\mu\}=\frac{1}{2}\tilde{\rho}(x_\lambda,y_\mu).$$

Let $x_\lambda, y_\mu > 1$ then

$$\tilde{\rho}\left((f,\varphi)x_\lambda,(f,\varphi)y_\mu\right)=\frac{1}{7}\max\{x_\lambda,y_\mu\}\le\frac{1}{2}\max\{x_\lambda,y_\mu\}=\frac{1}{2}\tilde{\rho}(x_\lambda,y_\mu)$$

.

Let $0 \le x_\lambda < \frac{1}{2}$ and $\frac{1}{2} \le y_\mu \le 1$ then

$$\tilde{\rho}\left((f,\varphi)x_\lambda,(f,\varphi)y_\mu\right)=\max\left\{\frac{x_\lambda}{6},\frac{y_\mu^2}{5}\right\}\le\frac{1}{2}\max\{x_\lambda,y_\mu\}=\frac{1}{2}\tilde{\rho}(x_\lambda,y_\mu)$$

.

Let $0 \le x_\lambda < \frac{1}{2}$ and $y_\mu > 1$ then

$$\tilde{\rho}\left((f,\varphi)x_\lambda,(f,\varphi)y_\mu\right)=\max\left\{\frac{x_\lambda}{6},\frac{y_\mu}{7}\right\}\le\frac{1}{2}\max\{x_\lambda,y_\mu\}=\frac{1}{2}\tilde{\rho}(x_\lambda,y_\mu)$$

.

Let $\frac{1}{2} \le x_\lambda \le 1$ and $0 \le y_\mu \le \frac{1}{2}$ then

$$\tilde{\rho}((f,\varphi)x_\lambda,(f,\varphi)y_\mu) = \max\left\{\frac{x_\lambda}{5},\frac{y_\mu}{6}\right\} \le \frac{1}{2}\max\{x_\lambda,y_\mu\} = \frac{1}{2}\tilde{\rho}(x_\lambda,y_\mu)$$

.

and so

$$\tilde{\rho}((f,\varphi)x_\lambda,(f,\varphi)y_\mu)$$

$$\le r\max\left\{\tilde{\rho}(x_\lambda,y_\mu),\tilde{\rho}(x_\lambda,(f,\varphi)x_\lambda),\tilde{\rho}(y_\mu,(f,\varphi)y_\mu),\right.$$

$$\left.\frac{\tilde{\rho}(y_\mu,(f,\varphi)x_\lambda)+\tilde{\rho}(y_\mu,(f,\varphi)y_\mu)+\tilde{\rho}(x_\lambda,(f,\varphi)x_\lambda)}{6}\right\}$$

Hence we conclude that all the condition of corollary (2.12) (theorem 2.7) holds and (f, φ) has a soft fixed point in $[0, \infty)$. By corollary 2.9 we deduce the following result.

Corollary 2.14: Let $(\tilde{X}, \tilde{\rho}, E)$ be a complete soft dislocated metric, $m \in N$,

let $A_1, A_2, \cdots, A_m$ be non empty $\tilde{\rho}$-closed subsets of X and $Y = \bigcup_{i=1}^{i=m} A_i$.

Suppose that $(f,\varphi): Y \to Y$ is an operator such that

i. $Y = \bigcup_{i=1}^{i=m} A_i$ is a cyclic representation of $(\tilde{X}, \tilde{\rho}, E)$ with respect to (f, φ).

ii. there exist $r \in [0,1)$ such that

$$\int_0^{\tilde{\rho}((f,\varphi)x_\lambda,(f,\varphi)y_\mu)} \rho(t)\,dt \le r\int_0^m \rho(t)\,dt,$$

where
$$m = \max\left\{\tilde{\rho}(x_\lambda,y_\mu),\tilde{\rho}(x_\lambda,(f,\varphi)x_\lambda),\tilde{\rho}(y_\mu,(f,\varphi)y_\mu),\right.$$
$$\left.\frac{\tilde{\rho}(y_\mu,(f,\varphi)x_\lambda)+\tilde{\rho}(y_\mu,(f,\varphi)y_\mu)+\tilde{\rho}(x_\lambda,(f,\varphi)x_\lambda)}{6}\right\},$$
for any $x \in A_i$

$y \in A_{i+1},\ i = 1,2,3,\cdots,m$, where A_{m+1} = A_1 and $\rho:[0,\infty)\to[0,\infty)$ be Lebesgue-integrable mapping satisfying $\int_0^{\epsilon}\rho(t)\,\mathrm{d}t > 0,\ \text{for } \epsilon > 0$ for $\epsilon > 0$. Then (f, φ) has a fixed point $z_\omega \in \bigcap_{i=1}^{m} A_i$. Moreover if $\tilde{\rho}(x_\lambda, y_\mu) \geq \tilde{\rho}(x_\lambda, x_\lambda)$ for all $x_\lambda, y_\mu \in Fix((f,\varphi))$, then (f, φ) has a unique soft fixed point.

Definition 2.15: Let $(f,\varphi):(X,\tilde{\rho},E)\to(X,\tilde{\rho},E)$ with $\psi:(X,\tilde{\rho},E)\to[0,\infty)$ and $\gamma \in [0,1]$. A mapping (f, φ) is said to be a $\gamma-\psi$ sub admissible soft mapping if $\psi(x_\lambda)\leq\gamma \Rightarrow \psi((f,\varphi)x_\lambda)\leq\gamma$ for $x_\lambda \in (X,\tilde{\rho},E)$.

Example 2.16: Let $(f,\varphi):[-\pi,\pi]\to[-\pi,\pi]$ and $\psi:[-\pi,\pi]\to R_+$ be defined by $(f,\varphi)x_\lambda = \frac{\pi}{4}(\tan x_\lambda)$ and $\psi(x_\lambda) = \left|x_\lambda - \frac{1}{4}\pi\right| + \frac{1}{2}$ then (f, φ) is a $\gamma-\psi$ sub admissible mapping, where $\gamma = \frac{1}{2}$, indeed if $\psi(x_\lambda) = \left|x_\lambda - \frac{1}{4}\pi\right| + \frac{1}{2} \leq \frac{1}{2}$ then $x_\lambda = \frac{1}{4}\pi$ then $x_\lambda = \frac{1}{4}\pi$.

Hence $(f,\varphi)(x_\lambda) = \frac{1}{4}\pi$ and $\psi((f,\varphi)x_\lambda) = \frac{1}{2}$.

Let Λ be the class of all the functions $\phi:[0,\infty)^3\to[0,\infty)$ that are a continuous with the property: $\phi(x, y, z) = 0$ if and only if $x = y = z = 0$.

Definition 2.17: Let $(X,\tilde{\rho},E)$ be a soft dislocated metric space, $m \in N$, let $A_1, A_2, \cdots, A_m$ be $\tilde{\rho}$-closed non empty subsets of $(X,\tilde{\rho},E)$ and let $Y = \bigcup_{i=1}^{i=m} A_i$. Assume that $(f,\varphi):Y\to Y$ is γ-ψ-sub admissible mapping, where $\gamma = \frac{1}{8}$. Then (f, φ) is called ψ-cyclic generalized weakly C-contraction if

i. $Y = \bigcup_{i=1}^{i=m} A_i$ is a cyclic representation of Y with respect to

$$\begin{aligned}
&\tilde{\rho}\left((f,\varphi)x_\lambda,(f,\varphi)y_\mu\right) \\
&\le \psi(x_\lambda)\tilde{\rho}\left(y_\mu,(f,\varphi)x_\lambda\right)+\psi\left((f,\varphi)x_\lambda\right)\tilde{\rho}\left(x_\lambda,(f,\varphi)y_\mu\right) \\
&\quad +\psi\left((f,\varphi)^2 x_\lambda\right)\tilde{\rho}\left(y_\mu,(f,\varphi)y_\mu\right)+\psi\left((f,\varphi)^3 x_\lambda\right)\tilde{\rho}\left(x_\lambda,(f,\varphi)x_\lambda\right) \\
&\quad +\psi\left((f,\varphi)^4 x_\lambda\right)\frac{\tilde{\rho}\left(x_\lambda,(f,\varphi)x_\lambda\right)\tilde{\rho}\left(y_\mu,(f,\varphi)y_\mu\right)}{1+\tilde{\rho}\left(x_\lambda,y_\mu\right)} \\
&\quad +\psi\left((f,\varphi)^5 x_\lambda\right)\frac{\tilde{\rho}\left(x_\lambda,(f,\varphi)x_\lambda\right)\tilde{\rho}\left(x_\lambda,(f,\varphi)y_\mu\right)}{1+\tilde{\rho}\left(y_\mu,(f,\varphi)x_\lambda\right)} \\
&\quad -\phi\Big\{\tilde{\rho}\left(x_\lambda,(f,\varphi)x_\lambda\right),\tilde{\rho}\left(y_\mu,(f,\varphi)y_\mu\right), \\
&\quad \frac{1}{2}\left[\tilde{\rho}\left(x_\lambda,(f,\varphi)y_\mu\right)+\tilde{\rho}\left(y_\mu,(f,\varphi)x_\lambda\right)\right]\Big\}
\end{aligned} \tag{18}$$

for any $x \in A_i$, $y \in A_{i+1}$, $i = 1,2,3,\ldots m$ and $A_{m+1} = A_1$ and $\phi \in \Lambda$.

Theorem 1.18: Let $(\tilde{X},\tilde{\rho},E)$ be a complete soft dislocated metric space, $m \in N$, let $A_1, A_2, \cdots, A_m$, be $\tilde{\rho}$-closed non-empty subsets of $(\tilde{X},\tilde{\rho},E)$ and let $Y = \bigcup_{i=1}^{i=m} A_i$. Assume that (f, φ): $Y \to Y$ is a ψ-cyclic generalized weakly C-contraction. If there exists $x_{\lambda_0}^0 \in Y$ such that $\psi(x_\lambda^0) \le \frac{1}{8}$, then (f, φ) has a soft fixed point $z_\omega \in \bigcap_{i=1}^n A_i$ Moreover if $\psi\left(z_{w_0}^0\right) \le \frac{1}{8}$ then $z_{w_0}^0$ is unique.

Proof: Let $x_{\lambda_0}^0 \in Y$ be such that $\psi(x_{\lambda_0}^0) \le \frac{1}{8}$. Since (f, φ) is a sub ψ-ad- missible mapping with respect to $\frac{1}{8}$,

then $\psi((f,\varphi)x_{\lambda_0}^0) \le \frac{1}{8}$. $\psi((f,\varphi)^n x_{\lambda_0}^0) \le \frac{1}{8}$ for all $n \in N \cup 0$. Also, there exist some i_0 such that $x_{\lambda_0}^0 \in A_{i_0}$ Now $(f,\varphi)(A_{i_0}) \subseteq A_{i_0+1}$ implies that $(f,\varphi)(x_{\lambda_0}^0) \in A_{i_0+1}$. Thus there exist $x_{\lambda_1}^1$ in A_{i_0+1} such that $(f,\varphi)x_{\lambda_0}^0 = x_{\lambda_1}^1$. Similarly $(f,\varphi)x_{\lambda_n}^n = x_{\lambda_{n+1}}^{n+1}$, where $x_{\lambda_n}^n \in A_{i_n}$. Hence for $n \ge 0$ there exist $i_n \in \{1,2,3,\cdots,m\}$ such that $x_{\lambda_n}^n \in A_{i_n}$ and $x_{\lambda_{n+1}}^{n+1} \in A_{i_n+1}$. In case $x_{\lambda_{n_0}}^{n_0} = x_{\lambda_{n_0+1}}^{n_0+1}$, for some n_0 = 0,1,2,3,.., then it is clear that $x_{\lambda_{n_0}}^{n_0}$ is a fixed point of (f, φ). Now assume that $x_{\lambda_n}^n \ne x_{\lambda_{n+1}}^{n+1}$, for all n.

Since (f, φ): $Y \to Y$ is a cyclic generalized weak C-contraction, we have for all $n \in N^*$ we have

$$\tilde{\rho}\left(x_{\lambda_n}^{n}, x_{\lambda_{n+1}}^{n+1}\right)=\tilde{\rho}\left(\left(f,\varphi\right)x_{\lambda_{n-1}}^{n-1},\left(f,\varphi\right)x_{\lambda_n}^{n}\right)$$
$$\leq\psi\left(x_{\lambda_{n-1}}^{n-1}\right)\tilde{\rho}\left(x_{\lambda_n}^{n},\left(f,\varphi\right)x_{\lambda_{n-1}}^{n-1}\right)+\psi\left(\left(f,\varphi\right)x_{\lambda_{n-1}}^{n-1}\right)\tilde{\rho}\left(x_{\lambda_{n-1}}^{n-1},\left(f,\varphi\right)x_{\lambda_n}^{n}\right)$$
$$+\psi\left(\left(f,\varphi\right)^{2}x_{\lambda_{n-1}}^{n-1}\right)\tilde{\rho}\left(x_{\lambda_n}^{n},\left(f,\varphi\right)x_{\lambda_n}^{n}\right)+\psi\left(\left(f,\varphi\right)^{3}x_{\lambda_{n-1}}^{n-1}\right)\tilde{\rho}\left(x_{\lambda_{n-1}}^{n-1},\left(f,\varphi\right)x_{\lambda_{n-1}}^{n-1}\right)$$
$$+\psi\left(\left(f,\varphi\right)^{4}x_{\lambda_{n-1}}^{n-1}\right)\frac{\tilde{\rho}\left(x_{\lambda_{n-1}}^{n-1},\left(f,\varphi\right)x_{\lambda_{n-1}}^{n-1}\right)\tilde{\rho}\left(x_{\lambda_n}^{n},\left(f,\varphi\right)x_{\lambda_n}^{n}\right)}{1+\tilde{\rho}\left(x_{\lambda_{n-1}}^{n-1},x_{\lambda_n}^{n}\right)}$$
$$+\psi\left(\left(f,\varphi\right)^{5}x_{\lambda_{n-1}}^{n-1}\right)\frac{\tilde{\rho}\left(x_{\lambda_{n-1}}^{n-1},\left(f,\varphi\right)x_{\lambda_{n-1}}^{n-1}\right)\tilde{\rho}\left(x_{\lambda_{n-1}}^{n-1},\left(f,\varphi\right)x_{\lambda_n}^{n}\right)}{1+\tilde{\rho}\left(x_{\lambda_n}^{n},\left(f,\varphi\right)x_{\lambda_{n-1}}^{n-1}\right)}$$
$$-\phi\left\{\tilde{\rho}\left(x_{\lambda_n}^{n},\left(f,\varphi\right)x_{\lambda_n}^{n}\right),\tilde{\rho}\left(x_{\lambda_{n-1}}^{n-1},\left(f,\varphi\right)x_{\lambda_{n-1}}^{n-1}\right),\right.$$
$$\left.\frac{1}{2}\left[\tilde{\rho}\left(x_{\lambda_{n-1}}^{n-1},\left(f,\varphi\right)x_{\lambda_n}^{n}\right)+\tilde{\rho}\left(x_{\lambda_n}^{n},\left(f,\varphi\right)x_{\lambda_{n-1}}^{n-1}\right)\right]\right\}$$
$$\leq\frac{1}{8}\left\{\tilde{\rho}\left(x_{\lambda_n}^{n},x_{\lambda_n}^{n}\right)+\tilde{\rho}\left(x_{\lambda_{n-1}}^{n-1},x_{\lambda_{n+1}}^{n+1}\right)+\tilde{\rho}\left(x_{\lambda_n}^{n},x_{\lambda_{n+1}}^{n+1}\right)+\tilde{\rho}\left(x_{\lambda_{n-1}}^{n-1},x_{\lambda_n}^{n}\right)\right.$$
$$\left.+\frac{\tilde{\rho}\left(x_{\lambda_{n-1}}^{n-1},x_{\lambda_n}^{n}\right)\tilde{\rho}\left(x_{\lambda_n}^{n},x_{\lambda_{n+1}}^{n+1}\right)}{1+\tilde{\rho}\left(x_{\lambda_{n-1}}^{n-1},x_{\lambda_n}^{n}\right)}+\frac{\tilde{\rho}\left(x_{\lambda_{n-1}}^{n-1},x_{\lambda_n}^{n}\right)\tilde{\rho}\left(x_{\lambda_{n-1}}^{n-1},x_{\lambda_{n+1}}^{n+1}\right)}{1+\tilde{\rho}\left(x_{\lambda_n}^{n},x_{\lambda_n}^{n}\right)}\right\}$$
$$-\phi\left\{\tilde{\rho}\left(x_{\lambda_n}^{n},x_{\lambda_{n+1}}^{n+1}\right),\tilde{\rho}\left(x_{\lambda_{n-1}}^{n-1},x_{\lambda_n}^{n}\right),\frac{1}{2}\left[\tilde{\rho}\left(x_{\lambda_{n-1}}^{n-1},x_{\lambda_{n+1}}^{n+1}\right)+\tilde{\rho}\left(x_{\lambda_n}^{n},x_{\lambda_n}^{n}\right)\right]\right\},$$

And so

$$\tilde{\rho}\left(x_{\lambda_n}^{n},x_{\lambda_{n+1}}^{n+1}\right)$$
$$\leq\frac{1}{8}\left\{\tilde{\rho}\left(x_{\lambda_n}^{n},x_{\lambda_n}^{n}\right)+\tilde{\rho}\left(x_{\lambda_{n-1}}^{n-1},x_{\lambda_{n+1}}^{n+1}\right)+\tilde{\rho}\left(x_{\lambda_n}^{n},x_{\lambda_{n+1}}^{n+1}\right)+\tilde{\rho}\left(x_{\lambda_{n-1}}^{n-1},x_{\lambda_n}^{n}\right)\right.$$
$$\left.+\tilde{\rho}\left(x_{\lambda_n}^{n},x_{\lambda_{n+1}}^{n+1}\right)+\frac{\tilde{\rho}\left(x_{\lambda_{n-1}}^{n-1},x_{\lambda_n}^{n}\right)\tilde{\rho}\left(x_{\lambda_{n-1}}^{n-1},x_{\lambda_{n+1}}^{n+1}\right)}{1+\tilde{\rho}\left(x_{\lambda_n}^{n},x_{\lambda_n}^{n}\right)}\right\}. \tag{19}$$

On the other hand from (d3) we have

$$\tilde{\rho}\left(x_{\lambda_{n-1}}^{n-1}, x_{\lambda_{n+1}}^{n+1}\right) \leq \tilde{\rho}\left(x_{\lambda_{n-1}}^{n-1}, x_{\lambda_n}^{n}\right) + \tilde{\rho}\left(x_{\lambda_n}^{n}, x_{\lambda_{n+1}}^{n+1}\right).$$

And by lemma (2.5D) we have

$$\tilde{\rho}\left(x_{\lambda_n}^{n}, x_{\lambda_n}^{n}\right) \leq \tilde{\rho}\left(x_{\lambda_{n-1}}^{n-1}, x_{\lambda_n}^{n}\right) + \tilde{\rho}\left(x_{\lambda_n}^{n}, x_{\lambda_{n+1}}^{n+1}\right).$$

Also

$$\begin{aligned}
\frac{\tilde{\rho}\left(x_{\lambda_{n-1}}^{n-1}, x_{\lambda_n}^{n}\right)\tilde{\rho}\left(x_{\lambda_{n-1}}^{n-1}, x_{\lambda_{n+1}}^{n+1}\right)}{1+\tilde{\rho}\left(x_{\lambda_n}^{n}, x_{\lambda_n}^{n}\right)} &\leq \frac{\tilde{\rho}\left(x_{\lambda_{n-1}}^{n-1}, x_{\lambda_n}^{n}\right)\tilde{\rho}\left(x_{\lambda_{n-1}}^{n-1}, x_{\lambda_{n+1}}^{n+1}\right)}{\tilde{\rho}\left(x_{\lambda_n}^{n}, x_{\lambda_n}^{n}\right)} \\
&\leq \frac{\tilde{\rho}\left(x_{\lambda_{n-1}}^{n-1}, x_{\lambda_n}^{n}\right)\left[\tilde{\rho}\left(x_{\lambda_{n-1}}^{n-1}, x_{\lambda_n}^{n}\right) + \tilde{\rho}\left(x_{\lambda_n}^{n}, x_{\lambda_{n+1}}^{n+1}\right)\right]}{\tilde{\rho}\left(x_{\lambda_{n-1}}^{n-1}, x_{\lambda_n}^{n}\right) + \tilde{\rho}\left(x_{\lambda_n}^{n}, x_{\lambda_{n+1}}^{n+1}\right)} \\
&\leq \tilde{\rho}\left(x_{\lambda_{n-1}}^{n-1}, x_{\lambda_{n+1}}^{n+1}\right)
\end{aligned}$$

From (19) we have

$$\begin{aligned}
&\tilde{\rho}\left((f,\varphi)x_{\lambda_{n-1}}^{n-1}, (f,\varphi)x_{\lambda_n}^{n}\right) \\
&\leq \frac{1}{8}\left\{\tilde{\rho}\left(x_{\lambda_{n-1}}^{n-1}, x_{\lambda_n}^{n}\right) + \tilde{\rho}\left(x_{\lambda_n}^{n}, x_{\lambda_{n+1}}^{n+1}\right) + \tilde{\rho}\left(x_{\lambda_{n-1}}^{n-1}, x_{\lambda_n}^{n}\right) + \tilde{\rho}\left(x_{\lambda_n}^{n}, x_{\lambda_{n+1}}^{n+1}\right)\right. \\
&\quad \left. + \tilde{\rho}\left(x_{\lambda_n}^{n}, x_{\lambda_{n+1}}^{n+1}\right) + \tilde{\rho}\left(x_{\lambda_{n-1}}^{n-1}, x_{\lambda_n}^{n}\right) + \tilde{\rho}\left(x_{\lambda_n}^{n}, x_{\lambda_{n+1}}^{n+1}\right) + \tilde{\rho}\left(x_{\lambda_{n-1}}^{n-1}, x_{\lambda_n}^{n}\right)\right\} \\
&\leq \frac{1}{8}\left\{4\tilde{\rho}\left(x_{\lambda_{n-1}}^{n-1}, x_{\lambda_n}^{n}\right) + 4\tilde{\rho}\left(x_{\lambda_n}^{n}, x_{\lambda_{n+1}}^{n+1}\right)\right\} \\
&\leq \frac{1}{2}\left\{\tilde{\rho}\left(x_{\lambda_{n-1}}^{n-1}, x_{\lambda_n}^{n}\right) + \tilde{\rho}\left(x_{\lambda_n}^{n}, x_{\lambda_{n+1}}^{n+1}\right)\right\}.
\end{aligned}$$

Therefore we have

$$\tilde{\rho}\left(x_{\lambda_n}^{n}, x_{\lambda_{n+1}}^{n+1}\right) \leq \tilde{\rho}\left(x_{\lambda_{n-1}}^{n-1}, x_{\lambda_n}^{n}\right). \tag{20}$$

For any $n \geq 1$, set $t_n = \tilde{\rho}\left(x_{\lambda_n}^{n}, x_{\lambda_{n-1}}^{n-1}\right)$. On the occasion of the facts above $\{t_n\}$

is a non-increasing sequence of non negative real numbers. Consequently, there exist $L \geq 0$ such that

$$\lim_{n\to\infty} \tilde{\rho}\left(x_{\lambda_n}^{n}, x_{\lambda_{n-1}}^{n-1}\right) = L \tag{21}$$

We shall prove that $L = 0$. Since $\tilde{\rho}\left(x_{\lambda_n}^{n}, x_{\lambda_n}^{n}\right) = 2\phi\tilde{\rho}\left(x_{\lambda_n}^{n}, x_{\lambda_{n+1}}^{n+1}\right)$ then we get $\lim_{n\to\infty} \tilde{d}\left(x_{\lambda_n}^{n}, x_{\lambda_n}^{n}\right) = 2L$. Similarly, $\lim_{n\to\infty} \tilde{\rho}\left(x_{\lambda_{n-1}}^{n-1}, x_{\lambda_{n+1}}^{n+1}\right) = 2L$ Then

$$\lim_{n\to\infty}\left[\tilde{\rho}\left(x_{\lambda_n}^{n}, x_{\lambda_n}^{n}\right) + \tilde{\rho}\left(x_{\lambda_{n-1}}^{n-1}, x_{\lambda_{n+1}}^{n+1}\right)\right] \leq 4L.$$

On the other hand, by taking limit as $n \to \infty$ in (19), we have,

$$L \leq \frac{1}{8}\left[4L + \left\{\tilde{\rho}\left(x_{\lambda_n}^{n}, x_{\lambda_n}^{n}\right) + \tilde{\rho}\left(x_{\lambda_{n-1}}^{n-1}, x_{\lambda_{n+1}}^{n+1}\right)\right\}\right],$$

This implies $$4L \leq \lim_{n\to\infty}\left\{\tilde{\rho}\left(x_{\lambda_n}^{n}, x_{\lambda_n}^{n}\right) + \tilde{\rho}\left(x_{\lambda_{n-1}}^{n-1}, x_{\lambda_{n+1}}^{n+1}\right)\right\}.$$

Hence $$\lim_{n\to\infty}\left\{\tilde{\rho}\left(x_{\lambda_n}^{n}, x_{\lambda_n}^{n}\right) + \tilde{\rho}\left(x_{\lambda_{n-1}}^{n-1}, x_{\lambda_{n+1}}^{n+1}\right)\right\} = 4L$$

Now from (18) we have

$$\begin{aligned} t_{n+1} \leq\ & \psi\left(x_{\lambda_{n-1}}^{n-1}\right)\tilde{\rho}\left(x_{\lambda_n}^{n}, x_{\lambda_n}^{n}\right) + \psi\left(x_{\lambda_n}^{n}\right)\tilde{\rho}\left(x_{\lambda_{n-1}}^{n-1}, x_{\lambda_{n+1}}^{n+1}\right) + \psi\left(x_{\lambda_{n+1}}^{n+1}\right)t_{n+1} \\ & + \psi\left(x_{\lambda_{n+2}}^{n+2}\right)t_n + \psi\left(x_{\lambda_{n+3}}^{n+3}\right)t_{n+1} + \psi\left(x_{\lambda_{n+4}}^{n+4}\right)t_{n-1}\frac{\tilde{\rho}\left(x_{\lambda_{n-1}}^{n-1}, x_{\lambda_{n+1}}^{n+1}\right)}{1+\tilde{\rho}\left(x_{\lambda_n}^{n}, x_{\lambda_n}^{n}\right)} \\ & - \phi\left\{t_n, t_{n+1}, \frac{1}{2}\left[\tilde{\rho}\left(x_{\lambda_{n-1}}^{n-1}, x_{\lambda_{n+1}}^{n+1}\right) + \tilde{\rho}\left(x_{\lambda_n}^{n}, x_{\lambda_n}^{n}\right)\right]\right\}. \end{aligned}$$

$$\begin{aligned} t_{n+1} \leq \frac{1}{8}\Bigg\{ & \tilde{\rho}\left(x_{\lambda_n}^{n}, x_{\lambda_n}^{n}\right) + \tilde{\rho}\left(x_{\lambda_{n-1}}^{n-1}, x_{\lambda_{n+1}}^{n+1}\right) + t_{n+1} + t_n + t_{n+1} + t_{n-1}\frac{\tilde{\rho}\left(x_{\lambda_{n-1}}^{n-1}, x_{\lambda_{n+1}}^{n+1}\right)}{1+\tilde{\rho}\left(x_{\lambda_n}^{n}, x_{\lambda_n}^{n}\right)}\Bigg\} \\ & - \phi\left\{t_n, t_{n+1}, \frac{1}{2}\left[\tilde{\rho}\left(x_{\lambda_{n-1}}^{n-1}, x_{\lambda_{n+1}}^{n+1}\right) + \tilde{\rho}\left(x_{\lambda_n}^{n}, x_{\lambda_n}^{n}\right)\right]\right\} \end{aligned}$$

By taking limit as $n \to \infty$ in the above inequality, we deduce that, $L \leq L - \phi(L, L, 2L)$.

So $\phi(L,L,2L)=0$ since $\phi\left(x_\lambda, y_\mu, z_\omega\right)=0 \Leftrightarrow x_\lambda = y_\mu = z_\omega = 0$ we get $L = 0$.

Due to $\lim_{n\to\infty} \tilde{\rho}\left(x^n_{\lambda_n}, x^n_{\lambda_n}\right) \leq 2L$ and $\lim_{n\to\infty} \tilde{\rho}\left(x^{n-1}_{\lambda_{n-1}}, x^{n+1}_{\lambda_{n+1}}\right) \leq 2L$ we have

$$\lim_{n\to\infty} \tilde{\rho}\left(x^n_{\lambda_n}, x^n_{\lambda_n}\right) = \lim_{\substack{n\to\infty \\ n\to\infty}} \tilde{\rho}\left(x^{n-1}_{\lambda_{n-1}}, x^{n+1}_{\lambda_{n+1}}\right) = \lim_{n\to\infty} \tilde{\rho}\left(x^n_{\lambda_n}, x^{n+1}_{\lambda_{n+1}}\right) = 0 . \tag{22}$$

We shall show that $\left\{x^n_{\lambda_n}\right\}$ is a $\tilde{\rho}$-Cauchy sequence. At first, we prove the following fact:

(K) for every $\epsilon > 0$, there exist $n \in N$ such that if $r, q \geq n$ with $r - q \equiv 1(m)$ then $\tilde{\rho}\left(x^r_{\lambda_r}, x^q_{\lambda_q}\right) < \epsilon$.

Suppose to the contrary that there exist $\epsilon > 0$, such that for any n in N, we can find $r_n > q_n \geq n$ with $r_n - q_n \equiv 1(m)$ satisfying

$$\tilde{\rho}\left(x^{r_n}_{\lambda_{r_n}}, x^{q_n}_{\lambda_{q_n}}\right) \geq \epsilon \tag{23}$$

Following the related lines of the proof of theorem (1.8) we have

$$\lim_{n\to\infty} \tilde{\rho}\left(x^{q_n}_{\lambda_{q_n}}, x^{r_n}_{\lambda_{r_n}}\right) = \epsilon,$$

$$\lim_{n\to\infty} \tilde{\rho}\left(x^{q_n+1}_{\lambda_{q_n+1}}, x^{r_n+1}_{\lambda_{r_n+1}}\right) = \epsilon,$$

$$\lim_{n\to\infty} \tilde{\rho}\left(x^{q_n}_{\lambda_{q_n}}, x^{r_n+1}_{\lambda_{r_n+1}}\right) = \epsilon \text{ and } \lim_{n\to\infty} \tilde{\rho}\left(x^{r_n}_{\lambda_{r_n}}, x^{q_n+1}_{\lambda_{q_n+1}}\right) = \epsilon. \tag{24}$$

Since $x^{q_n}_{\lambda_{q_n}}$ & $x^{r_n}_{\lambda_{r_n}}$ lie in different adjacently labeled sets A_i and A_{i+1}, for a certain $1 \le i \le m$. Using the fact that (f, φ) is ψ-cyclic generalized weakly C- contraction, we have

$$\begin{aligned}
&\tilde{\rho}\left(x^{q_n+1}_{\lambda_{q_n+1}}, x^{r_n+1}_{\lambda_{r_n+1}}\right) = \tilde{\rho}\left((f,\varphi)x^{q_n}_{\lambda_{q_n}}, (f,\varphi)x^{r_n}_{\lambda_{r_n}}\right) \\
&\le \psi\left(x^{q_n}_{\lambda_{q_n}}\right)\tilde{\rho}\left(x^{r_n}_{\lambda_{r_n}}, (f,\varphi)x^{q_n}_{\lambda_{q_n}}\right) + \psi\left((f,\varphi)x^{q_n}_{\lambda_{q_n}}\right)\tilde{\rho}\left(x^{q_n}_{\lambda_{q_n}}, (f,\varphi)x^{r_n}_{\lambda_{r_n}}\right) \\
&\quad + \psi\left((f,\varphi)^2 x^{q_n}_{\lambda_{q_n}}\right)\tilde{\rho}\left(x^{r_n}_{\lambda_{r_n}}, (f,\varphi)x^{r_n}_{\lambda_{r_n}}\right) + \psi\left((f,\varphi)^3 x^{q_n}_{\lambda_{q_n}}\right)\tilde{\rho}\left(x^{q_n}_{\lambda_{q_n}}, (f,\varphi)x^{q_n}_{\lambda_{q_n}}\right) \\
&\quad + \psi\left((f,\varphi)^4 x^{q_n}_{\lambda_{q_n}}\right)\frac{\tilde{\rho}\left(x^{q_n}_{\lambda_{q_n}}, (f,\varphi)x^{q_n}_{\lambda_{q_n}}\right)\tilde{\rho}\left(x^{r_n}_{\lambda_{r_n}}, (f,\varphi)x^{r_n}_{\lambda_{r_n}}\right)}{1+\tilde{\rho}\left(x^{q_n}_{\lambda_{q_n}}, x^{r_n}_{\lambda_{r_n}}\right)} \\
&\quad + \psi\left((f,\varphi)^5 x^{q_n}_{\lambda_{q_n}}\right)\frac{\tilde{\rho}\left(x^{q_n}_{\lambda_{q_n}}, (f,\varphi)x^{q_n}_{\lambda_{q_n}}\right)\tilde{\rho}\left(x^{q_n}_{\lambda_{q_n}}, (f,\varphi)x^{r_n}_{\lambda_{r_n}}\right)}{1+\tilde{\rho}\left(x^{r_n}_{\lambda_{r_n}}, (f,\varphi)x^{q_n}_{\lambda_{q_n}}\right)} \\
&\quad - \phi\Bigg\{\tilde{\rho}\left(x^{q_n}_{\lambda_{q_n}}, (f,\varphi)x^{q_n}_{\lambda_{q_n}}\right), \tilde{\rho}\left(x^{r_n}_{\lambda_{r_n}}, (f,\varphi)x^{r_n}_{\lambda_{r_n}}\right), \\
&\qquad \frac{1}{2}\left[\tilde{\rho}\left(x^{q_n}_{\lambda_{q_n}}, (f,\varphi)x^{r_n}_{\lambda_{r_n}}\right) + \tilde{\rho}\left(x^{r_n}_{\lambda_{r_n}}, (f,\varphi)x^{q_n}_{\lambda_{q_n}}\right)\right]\Bigg\} \\
&\le \frac{1}{8}\Bigg\{\tilde{\rho}\left(x^{r_n}_{\lambda_{r_n}}, x^{q_n+1}_{\lambda_{q_n+1}}\right) + \tilde{\rho}\left(x^{q_n}_{\lambda_{q_n}}, x^{r_n+1}_{\lambda_{r_n+1}}\right) + \tilde{\rho}\left(x^{r_n}_{\lambda_{r_n}}, x^{r_n+1}_{\lambda_{r_n+1}}\right) + \tilde{\rho}\left(x^{q_n}_{\lambda_{q_n}}, x^{q_n+1}_{\lambda_{q_n+1}}\right) \\
&\quad + \frac{\tilde{\rho}\left(x^{q_n}_{\lambda_{q_n}}, x^{q_n+1}_{\lambda_{q_n+1}}\right)\tilde{\rho}\left(x^{r_n}_{\lambda_{r_n}}, x^{r_n+1}_{\lambda_{r_n+1}}\right)}{1+\tilde{\rho}\left(x^{q_n}_{\lambda_{q_n}}, x^{r_n}_{\lambda_{r_n}}\right)} + \frac{\tilde{\rho}\left(x^{q_n}_{\lambda_{q_n}}, x^{q_n+1}_{\lambda_{q_n+1}}\right)\tilde{\rho}\left(x^{q_n}_{\lambda_{q_n}}, x^{r_n+1}_{\lambda_{r_n+1}}\right)}{1+\tilde{\rho}\left(x^{r_n}_{\lambda_{r_n}}, x^{q_n+1}_{\lambda_{q_n+1}}\right)}\Bigg\} \\
&\quad - \phi\left\{\tilde{\rho}\left(x^{q_n}_{\lambda_{q_n}}, x^{q_n+1}_{\lambda_{q_n+1}}\right), \tilde{\rho}\left(x^{r_n}_{\lambda_{r_n}}, x^{r_n+1}_{\lambda_{r_n+1}}\right), \frac{1}{2}\left[\tilde{\rho}\left(x^{q_n}_{\lambda_{q_n}}, x^{r_n+1}_{\lambda_{r_n+1}}\right) + \tilde{\rho}\left(x^{r_n}_{\lambda_{r_n}}, x^{q_n+1}_{\lambda_{q_n+1}}\right)\right]\right\}.
\end{aligned}$$

Now, by taking limit as $n \to \infty$ in the above inequality, we derive that

$$\epsilon \le \frac{1}{8}[\epsilon + \epsilon + 0 + 0 + 0 + 0] - \phi[0, 0, \epsilon] \le \frac{1}{4}\epsilon$$

.

This is a contradiction. Hence condition (k) holds. We are ready to show that the sequence $\left\{x_{\lambda_n}^{n}\right\}$ is a Cauchy. Fix $\epsilon > 0$ By the claim, we find $n_0 \in N$ such that if $r, q \geq n_0$ with $r - q \equiv 1(m)$ then

$$\tilde{\rho}\left(x_{\lambda_{r_n}}^{r_n}, x_{\lambda_{q_n}}^{q_n}\right) \leq \frac{\epsilon}{4} \leq \frac{\epsilon}{2} \tag{25}$$

Since $\lim_{n\to\infty} \tilde{\rho}\left(x_{\lambda_n}^{n}, x_{\lambda_{n+1}}^{n+1}\right) = 0$, we find $n_1 \in N$ such that

$$\tilde{\rho}\left(x_{\lambda_n}^{n}, x_{\lambda_{n+1}}^{n+1}\right) \leq \frac{\epsilon}{2m} \tag{26}$$

for any $n \geq n_1$. Suppose that $r, s \geq \max\left\{n_0, n_1\right\}$ with $s > r$. Then there exist $k \in \{1, 2, \cdots, m\}$ such that $s - r \equiv k(m)$. Therefore, $s - r + \phi \equiv 1(m)$, for $\phi = m - k + 1$.

So we have, for $j \in \{1, 2, 3, \cdots, m\}, \quad s + j - r \equiv 1(m)$,

$$\tilde{\rho}\left(x_{\lambda_r}^{r}, x_{\lambda_s}^{s}\right) \leq \tilde{\rho}\left(x_{\lambda_r}^{r}, x_{\lambda_{s+j}}^{s+j}\right) + \tilde{\rho}\left(x_{\lambda_r}^{r}, x_{\lambda_{s+j}}^{s+j}\right)$$
$$+ \tilde{\rho}\left(x_{\lambda_{s+j}}^{s+j}, x_{\lambda_{s+j-1}}^{s+j-1}\right) + \cdots + \tilde{\rho}\left(x_{\lambda_{s+1}}^{s+1}, x_{\lambda_s}^{s}\right).$$

By (25) and (26) and from the last inequality, we get

$$\tilde{\rho}\left(x_{\lambda_r}^{r}, x_{\lambda_s}^{s}\right) \leq \frac{\epsilon}{2} + j \times \frac{\epsilon}{2m} \leq \frac{\epsilon}{2} + m \times \frac{\epsilon}{2m} = \epsilon.$$

This proves that $\left\{x_{\lambda_n}^{n}\right\}$ is a $\tilde{\rho}$-Cauchy sequence.

Since Y is $\tilde{\rho}$-closed in (X, $\tilde{\rho}$, *E*) then (Y, $\tilde{\rho}$, *E*) is also complete, there exists $z_\omega \in Y = \bigcup_{i=1}^{m} A_i$ such that $\lim_{n\to\infty} x_{\lambda_n}^{n} = z_\omega$ in (Y, $\tilde{\rho}$, *E*), equivalently

$$\tilde{\rho}\left(z_\omega, z_\omega\right) = \lim_{n\to\infty} \tilde{\rho}\left(z_\omega, x_{\lambda_n}^{n}\right) = \lim_{m,n\to\infty} \tilde{\rho}\left(x_{\lambda_n}^{n}, x_{\lambda_m}^{m}\right) = 0 \tag{27}$$

In what follows, we prove that x_λ is a soft fixed point of (*f*, φ). In fact, since $\lim_{n\to\infty} x_{\lambda_n}^{n} = z_w$ and as $Y = \bigcup_{i=1}^{m} A_i$ is cyclic representation of Y with respect to (*f*, φ), the sequence $\left\{x_{\lambda_n}^{n}\right\}$ has infinite terms in each A_i for $i \in \{1,2,3,\ldots,m$. Suppose that $x_\lambda \in A_i,\ (f,\varphi)x_\lambda \in A_{i+1}$ and we take a subsequence $\left\{x_{\lambda_{nk}}^{n_k}\right\}$ of $\left\{x_{\lambda_n}^{n}\right\}$ with $x_{\lambda_{n_k}}^{n_k} \in A_{i-1}$ (the existence of this subsequence is guaranteed by the above mentioned comment). By using the contractive condition, we can obtain

$$\begin{aligned}
&\tilde{\rho}\left(x_{\lambda_{n_k+1}}^{n_k+1}, (f,\varphi)x_\lambda\right)x_{\lambda_{n_k}}^{n_k} = \tilde{\rho}\left(\left(f,\varphi\right)x_{\lambda_{n_k}}^{n_k}, (f,\varphi)x_\lambda\right) \\
&\le \psi\left(x_{\lambda_{n_k}}^{n_k}\right)\tilde{\rho}\left(x_\lambda, (f,\varphi)x_{\lambda_{n_k}}^{n_k}\right) + \psi\left(\left(f,\varphi\right)x_{\lambda_{n_k}}^{n_k}\right)\tilde{\rho}\left(x_{\lambda_{n_k}}^{n_k}, (f,\varphi)x_\lambda\right) \\
&\quad + \psi\left(\left(f,\varphi\right)^2 x_{\lambda_{n_k}}^{n_k}\right)\tilde{\rho}\left(x_\lambda, (f,\varphi)x_\lambda\right) + \psi\left(\left(f,\varphi\right)^3 x_{\lambda_{n_k}}^{n_k}\right)\tilde{\rho}\left(x_{\lambda_{n_k}}^{n_k}, (f,\varphi)x_{\lambda_{n_k}}^{n_k}\right) \\
&\quad + \psi\left(\left(f,\varphi\right)^4 x_{\lambda_{n_k}}^{n_k}\right)\frac{\tilde{\rho}\left(x_{\lambda_{n_k}}^{n_k}, (f,\varphi)x_{\lambda_{n_k}}^{n_k}\right)\tilde{\rho}\left(x_\lambda, (f,\varphi)x_\lambda\right)}{1+\tilde{\rho}\left(x_{\lambda_{n_k}}^{n_k}, x_\lambda\right)}
\end{aligned}$$

$$+\psi\left((f,\varphi)^5 x_{\lambda_{n_k}}^{n_k}\right)\frac{\tilde{\rho}\left(x_{\lambda_{n_k}}^{n_k},(f,\varphi)x_{\lambda_{n_k}}^{n_k}\right)\tilde{\rho}\left(x_{\lambda_{n_k}}^{n_k},(f,\varphi)x_\lambda\right)}{1+\tilde{\rho}\left(x_\lambda,(f,\varphi)x_{\lambda_{n_k}}^{n_k}\right)}$$

$$-\phi\left\{\tilde{\rho}\left(x_{\lambda_{n_k}}^{n_k},(f,\varphi)x_{\lambda_{n_k}}^{n_k}\right),\tilde{\rho}\left(x_\lambda,(f,\varphi)x_\lambda\right),\right.$$

$$\left.\frac{1}{2}\left[\tilde{\rho}\left(x_{\lambda_{n_k}}^{n_k},(f,\varphi)x_\lambda\right)+\tilde{\rho}\left(x_\lambda,(f,\varphi)x_{\lambda_{n_k}}^{n_k}\right)\right]\right\},$$

$$\leq\frac{1}{8}\left\{\tilde{\rho}\left(x_\lambda,x_{\lambda_{n_k+1}}^{n_k+1}\right)+\tilde{\rho}\left(x_{\lambda_{n_k}}^{n_k},(f,\varphi)x_\lambda\right)+\tilde{\rho}\left(x_\lambda,(f,\varphi)x_\lambda\right)+\tilde{\rho}\left(x_{\lambda_{n_k}}^{n_k},x_{\lambda_{n_k+1}}^{n_k+1}\right)\right.$$

$$\left.+\frac{\tilde{\rho}\left(x_{\lambda_{n_k}}^{n_k},x_{\lambda_{n_k+1}}^{n_k+1}\right)\tilde{\rho}\left(x_\lambda,(f,\varphi)x_\lambda\right)}{1+\tilde{\rho}\left(x_{\lambda_{n_k}}^{n_k},x_\lambda\right)}+\frac{\tilde{\rho}\left(x_{\lambda_{n_k}}^{n_k},x_{\lambda_{n_k+1}}^{n_k+1}\right)\tilde{\rho}\left(x_{\lambda_{n_k}}^{n_k},(f,\varphi)x_\lambda\right)}{1+\tilde{\rho}\left(x_\lambda,x_{\lambda_{n_k+1}}^{n_k+1}\right)}\right\}$$

$$-\phi\left\{\tilde{\rho}\left(x_{\lambda_{n_k}}^{n_k},x_{\lambda_{n_k+1}}^{n_k+1}\right),\tilde{\rho}\left(x_\lambda,(f,\varphi)x_\lambda\right),\frac{1}{2}\left[\tilde{\rho}\left(x_{\lambda_{n_k}}^{n_k},(f,\varphi)x_\lambda\right)+\tilde{\rho}\left(x_\lambda,x_{\lambda_{n_k+1}}^{n_k+1}\right)\right]\right\}.$$

Passing to the limit as lim $n \to \infty$ and using $x_{\lambda_{n_k}}^{n_k} \to x_\lambda$ lower semi-conti- nuity of ϕ, we have

$$\tilde{\rho}\left(x_\lambda,(f,\varphi)x_\lambda\right)$$

$$\leq\frac{1}{8}\left\{0+\tilde{\rho}\left(x_\lambda,(f,\varphi)x_\lambda\right)+\tilde{\rho}\left(x_\lambda,(f,\varphi)x_\lambda\right)+0+0+0\right\}$$

$$-\phi\left\{0,\tilde{\rho}\left(x_\lambda,(f,\varphi)x_\lambda\right),\frac{1}{2}\left[\tilde{\rho}\left(x_\lambda,(f,\varphi)x_\lambda\right)\right]\right\}$$

$$\leq\frac{1}{4}\tilde{\rho}\left(x_\lambda,(f,\varphi)x_\lambda\right).$$

So $\tilde{\rho}\left(x_\lambda,(f,\varphi)x_\lambda\right)=0$ and therefore x_λ is fixed point of (f, φ).

Finally to prove the uniqueness of soft fixed point theorem, suppose that $y_\mu, z_\omega \in (X,\tilde{\rho},E)$ are soft fixed points of (f, φ). The cyclic character of (f, φ) and the fact that $y_\mu, z_\omega \in (X,\tilde{\rho},E)$ are soft fixed points of $(f,$

φ) implies that $y_\mu, z_\omega \in \bigcap_{i=1}^{m} A_i$. Also suppose that $\psi(y_\mu) \le \frac{1}{8}$. By using contractive condition we derive that

$$
\begin{aligned}
\tilde{\rho}(y_\mu, z_\omega) &= \tilde{\rho}((f,\varphi) y_\mu, (f,\varphi) z_\omega) \\
&\le \psi(y_\mu)\tilde{\rho}(z_\omega, (f,\varphi) y_\mu) + \psi((f,\varphi) y_\mu)\tilde{\rho}(y_\mu, (f,\varphi) z_\omega) \\
&\quad + \psi((f,\varphi)^2 y_\mu)\tilde{\rho}(z_\omega, (f,\varphi) z_\omega) + \psi((f,\varphi)^3 y_\mu)\tilde{\rho}(y_\mu, (f,\varphi) y_\mu) \\
&\quad + \psi((f,\varphi)^4 y_\mu)\frac{\tilde{\rho}(y_\mu, (f,\varphi) y_\mu)\tilde{\rho}(z_w, (f,\varphi) z_w)}{1+\tilde{\rho}(y_\mu, z_w)} \\
&\quad + \psi((f,\varphi)^5 y_\mu)\frac{\tilde{\rho}(y_\mu, (f,\varphi) y_\mu)\tilde{\rho}(y_\mu, (f,\varphi) z_w)}{1+\tilde{\rho}(z_w, (f,\varphi) y_\mu)} \\
&\quad - \phi\Big(\tilde{\rho}(y_\mu, (f,\varphi) y_\mu), \tilde{\rho}(z_w, (f,\varphi) z_w), \\
&\qquad \frac{1}{2}\big[\tilde{\rho}(y_\mu, (f,\varphi) z_w) + \tilde{\rho}(z_w, (f,\varphi) y_\mu)\big]\Big)
\end{aligned}
$$

$$
\begin{aligned}
&\le \frac{1}{8}\Bigg\{\tilde{\rho}(z_\omega, y_\mu) + \tilde{\rho}(y_\mu, z_\omega) + \tilde{\rho}(z_\omega, z_\omega) + \tilde{\rho}(y_\mu, y_\mu) \\
&\quad + \frac{\tilde{\rho}(y_\mu, y_\mu)\tilde{\rho}(z_w, z_w)}{1+\tilde{\rho}(y_\mu, z_w)} + \frac{\tilde{\rho}(y_\mu, y_\mu)\tilde{\rho}(y_\mu, z_w)}{1+\tilde{\rho}(z_w, y_\mu)}\Bigg\} \\
&\quad - \phi\rho\Big(\tilde{\rho}(y_\mu, y_\mu), \tilde{\rho}(z_w, z_w), \frac{1}{2}\big[\tilde{\rho}(y_\mu, z_w) + \tilde{\rho}(z_w, y_\mu)\big]\Big) \\
&\le \frac{1}{8}\{2\tilde{\rho}(z_\omega, y_\mu)\} - \phi\Big(0, \tilde{\rho}(z_w, z_w), \frac{1}{2}\big[\tilde{\rho}(y_\mu, z_w) + \tilde{\rho}(z_w, y_\mu)\big]\Big) \\
&\le \frac{1}{4}\tilde{\rho}(z_\omega, y_\mu).
\end{aligned}
$$

This gives us $\tilde{\rho}(z_\omega, y_\mu) = 0$ that is $z_\omega = y_\mu$. This finishes the proof.

Corollary 2.19: Let (X, $\tilde{\rho}$, E) be a complete soft dislocated metric space, $m \in N$, let $A_1, A_2, \ldots, A_m$ be non empty $\tilde{\rho}$-closed subsets of (X, d, E) and

Let $Y = \bigcup_{i=1}^{i=m} A_i$. Suppose that $(f,\varphi): Y \to Y$ is an operator such that

i) $Y = \bigcup_{i=1}^{i=m} A_i$ is cyclic representation of (X, $\tilde{\rho}$, E) with respect to (f, φ);

ii) there exist $\beta \in \left[0, \frac{1}{8}\right)$ such that

$$\begin{aligned}
&\tilde{\rho}\left((f,\varphi)x_\lambda,(f,\varphi)y_\mu\right) \\
&\leq \beta\left\{\tilde{\rho}\left(y_\mu,(f,\varphi)x_\lambda\right)+\tilde{\rho}\left(x_\lambda,(f,\varphi)y_\mu\right)+\tilde{\rho}\left(y_\mu,(f,\varphi)y_\mu\right)\right. \\
&\quad +\tilde{\rho}\left(x_\lambda,(f,\varphi)x_\lambda\right)+\frac{\tilde{\rho}\left(x_\lambda,(f,\varphi)x_\lambda\right)\tilde{\rho}\left(y_\mu,(f,\varphi)y_\mu\right)}{1+\tilde{\rho}\left(x_\lambda,y_\mu\right)} \\
&\quad \left.+\frac{\tilde{\rho}\left(x_\lambda,(f,\varphi)x_\lambda\right)\tilde{\rho}\left(x_\lambda,(f,\varphi)y_\mu\right)}{1+\tilde{\rho}\left(y_\mu,(f,\varphi)x_\lambda\right)}\right\},
\end{aligned} \tag{28}$$

for any $x_\lambda \in A_i,\ y_\mu \in A_{i+1},\ i = 1,2,3,\cdots,m$ where $A_{m+1} = A_1$. Then (f, φ) has fixed point $z_w \in \bigcap_{i=1}^{i=m} A_i$.

Example 2.20: Let (X, $\tilde{\rho}$, E) = R with soft dislocates metric space

$\tilde{\rho}\left(x_\lambda, y_\mu\right) = \max\left\{\left|x_\lambda\right|, \left|y_\mu\right|\right\}$, for any $x_\lambda, y_\mu \in (X, \tilde{\rho}, E)$

suppose $A_1 = [-2, 0]$

and $A_2 = [0,2]$ & $Y = \bigcup_{i=1}^{i=2} A_i$ we define $(f, \varphi): Y \to Y$ by

$$(f,\varphi)x_\lambda = \begin{cases} \dfrac{-x_\lambda}{32}, & \text{if } x_\lambda \in [-2,0], \\ \dfrac{-x_\lambda}{16}, & \text{if } x_\lambda \in [0,2], \end{cases}$$

It clear that $\bigcup_{i=1}^{i=2} A_i$ is cyclic representation of Y with respect to (f, φ).

Proof: Let $x_{\lambda_1}^1 \in A_1 = [-2,0]$ and $x_{\lambda_2}^2 \in A_2 = [0,2]$

$$\begin{aligned}
&\tilde{\rho}((f,\varphi)x_\lambda,(f,\varphi)y_\mu) \\
&= \max\left\{\left|\frac{-x_\lambda}{32}\right|,\left|\frac{-y_\mu}{16}\right|\right\} \le \max\left\{\left|\frac{-x_\lambda}{32}\right|,\frac{y_\mu}{16}\right\} \le \max\left\{\frac{-x_\lambda}{16},\frac{y_\mu}{16}\right\} \\
&\le \frac{1}{16}\max\{|x_\lambda|,|y_\mu|\} = \frac{1}{16}\tilde{\rho}(x_\lambda,y_\mu).
\end{aligned}$$

$$\begin{aligned}
&\tilde{\rho}((f,\varphi)x_\lambda,(f,\varphi)y_\mu) \\
&\le \frac{1}{16}\Bigg\{\tilde{\rho}(y_\mu,(f,\varphi)x_\lambda)+\tilde{\rho}(x_\lambda,(f,\varphi)y_\mu)+\tilde{\rho}(y_\mu,(f,\varphi)y_\mu) \\
&\quad+\tilde{\rho}(x_\lambda,(f,\varphi)x_\lambda)+\frac{\tilde{\rho}(x_\lambda,(f,\varphi)x_\lambda)\tilde{\rho}(y_\mu,(f,\varphi)y_\mu)}{1+\tilde{\rho}(x_\lambda,y_\mu)} \\
&\quad+\frac{\tilde{\rho}(x_\lambda,(f,\varphi)x_\lambda)\tilde{\rho}(x_\lambda,(f,\varphi)y_\mu)}{1+\tilde{\rho}(y_\mu,(f,\varphi)x_\lambda)}\Bigg\}.
\end{aligned}$$

Hence the condition of corollary (2.19) (theorem 2.18) holds and (f, φ) has a soft fixed point in $A_1 \cap A_2$. Here $x_\lambda = 0$ is a soft fixed point of (f, φ).

In the above theorem 2.18, if we take $A_i = (X,\tilde{\rho},E)$ for all $0 \le i \le m$ then we deduce the following theorem.

Theorem 2.21: Let $(X, \tilde{\rho}, E)$ be a complete soft dislocated metric space and $(f,\varphi):(X,\tilde{\rho},E)\to(X,\tilde{\rho},E)$ be a sub ψ-admissible mapping such that

$$\begin{aligned}
&\tilde{\rho}\left((f,\varphi)x_{\lambda},(f,\varphi)y_{\mu}\right)\\
&\leq\psi(x_{\lambda})\tilde{\rho}\left(y_{\mu},(f,\varphi)x_{\lambda}\right)+\psi\left((f,\varphi)x_{\lambda}\right)\tilde{\rho}\left(x_{\lambda},(f,\varphi)y_{\mu}\right)\\
&\quad+\psi\left((f,\varphi)^{2}x_{\lambda}\right)\tilde{\rho}\left(y_{\mu},(f,\varphi)y_{\mu}\right)+\psi\left((f,\varphi)^{3}x_{\lambda}\right)\tilde{\rho}\left(x_{\lambda},(f,\varphi)x_{\lambda}\right)\\
&\quad+\psi\left((f,\varphi)^{4}x_{\lambda}\right)\frac{\tilde{\rho}\left(x_{\lambda},(f,\varphi)x_{\lambda}\right)\tilde{\rho}\left(y_{\mu},(f,\varphi)y_{\mu}\right)}{1+\tilde{\rho}\left(x_{\lambda},y_{\mu}\right)}\\
&\quad+\psi\left((f,\varphi)^{5}x_{\lambda}\right)\frac{\tilde{\rho}\left(x_{\lambda},(f,\varphi)x_{\lambda}\right)\tilde{\rho}\left(x_{\lambda},(f,\varphi)y_{\mu}\right)}{1+\tilde{\rho}\left(y_{\mu},(f,\varphi)x_{\lambda}\right)}\\
&\quad-\phi\left\{\tilde{\rho}\left(x_{\lambda},(f,\varphi)x_{\lambda}\right),\tilde{\rho}\left(y_{\mu},(f,\varphi)y_{\mu}\right),\right.\\
&\quad\left.\frac{1}{2}\left[\tilde{\rho}\left(x_{\lambda},(f,\varphi)y_{\mu}\right)+\tilde{\rho}\left(y_{\mu},(f,\varphi)x_{\lambda}\right)\right]\right\};
\end{aligned}$$

for any $x_{\lambda}, y_{\mu}\in(X,\tilde{\rho},E)$, where $\phi\in\wedge$, $\psi\in\Psi$. Then (f, φ) has unique soft fixed point in $(X, \tilde{\rho}, E)$.

Corollary 2.22: Let $(X, \tilde{\rho}, E)$ be a complete soft dislocated metric space and $(f,\varphi):(X,\tilde{\rho},E)\to(X,\tilde{\rho},E)$ be a sub ψ-admissible mapping such that

$$\tilde{\rho}\left(\left(f,\varphi\right)x_\lambda,\left(f,\varphi\right)y_\mu\right)$$

$$\leq \beta\left\{\tilde{\rho}\left(y_\mu,\left(f,\varphi\right)x_\lambda\right)+\tilde{\rho}\left(x_\lambda,\left(f,\varphi\right)y_\mu\right)+\tilde{\rho}\left(y_\mu,\left(f,\varphi\right)y_\mu\right)\right.$$

$$+\tilde{\rho}\left(x_\lambda,\left(f,\varphi\right)x_\lambda\right)+\frac{\tilde{\rho}\left(x_\lambda,\left(f,\varphi\right)x_\lambda\right)\tilde{\rho}\left(y_\mu,\left(f,\varphi\right)y_\mu\right)}{1+\tilde{\rho}\left(x_\lambda,y_\mu\right)}$$

$$\left.+\frac{\tilde{\rho}\left(x_\lambda,\left(f,\varphi\right)x_\lambda\right)\tilde{\rho}\left(x_\lambda,\left(f,\varphi\right)y_\mu\right)}{1+\tilde{\rho}\left(y_\mu,\left(f,\varphi\right)x_\lambda\right)}\right\};$$

for any $x_\lambda, y_\mu \in \left(X,\tilde{\rho},E\right)$, where $\beta \in \left[0,\frac{1}{8}\right)$. Then (f, φ) has unique soft

fixed point in $(X, \tilde{\rho}, E)$.

Example 2.23: Let $(X, \tilde{\rho}, E) = R$ with soft dislocated metric space

$\tilde{\rho}\left(x_\lambda,y_\mu\right)=\max\left\{x_\lambda,y_\mu\right\}$, for any $x_\lambda, y_\mu \in \left(X,\tilde{\rho},E\right)$.
Let $\left(f,\varphi\right):\left(X,\tilde{\rho},E\right)\rightarrow\left(X,\tilde{\rho},E\right)$ be defined by

$$\left(f,\varphi\right)x_\lambda=\begin{cases}\dfrac{\left(x_\lambda\right)^2+x_\lambda}{18}, & \text{if } 0\leq x_\lambda\leq 1,\\[2ex] \dfrac{x_\lambda}{12}, & \text{if } x_\lambda\geq 1,\end{cases}$$

Proof: To show the existence and uniqueness soft point of (f, φ), we investigate the following cases

Let $0\leq x_\lambda, y_\mu < 1$ then

$$\tilde{\rho}\left((f,\varphi)x_\lambda,(f,\varphi)y_\mu\right)$$
$$=\max\left\{\frac{x_\lambda^2+x_\lambda}{18},\frac{y_\mu^2+y_\mu}{18}\right\}\le\frac{1}{10}\max\left\{x_\lambda,y_\mu\right\}=\frac{1}{10}\tilde{\rho}\left(x_\lambda,y_\mu\right)$$

Let $x_\lambda, y_\mu \ge 1$ then

$$\tilde{\rho}\left((f,\varphi)x_\lambda,(f,\varphi)y_\mu\right)=\frac{1}{12}\max\left\{x_\lambda,y_\mu\right\}\le\frac{1}{10}\max\left\{x_\lambda,y_\mu\right\}=\frac{1}{10}\tilde{\rho}\left(x_\lambda,y_\mu\right)$$

Let $0 \le x_\lambda < 1$ and $y_\mu = 1$ then

$$\tilde{\rho}\left((f,\varphi)x_\lambda,(f,\varphi)y_\mu\right)=\max\left\{\frac{x_\lambda^2+x_\lambda}{18},\frac{y_\mu}{12}\right\}\le\frac{1}{12}\max\left\{x_\lambda,y_\mu\right\}=\frac{1}{10}\tilde{\rho}\left(x_\lambda,y_\mu\right).$$

Hence

$$\tilde{\rho}\left((f,\varphi)x_\lambda,(f,\varphi)y_\mu\right)$$
$$\le\frac{1}{10}\left\{\tilde{\rho}\left(y_\mu,(f,\varphi)x_\lambda\right)+\tilde{\rho}\left(x_\lambda,(f,\varphi)y_\mu\right)+\tilde{\rho}\left(y_\mu,(f,\varphi)y_\mu\right)\right.$$
$$+\tilde{\rho}\left(x_\lambda,(f,\varphi)x_\lambda\right)+\frac{\tilde{\rho}\left(x_\lambda,(f,\varphi)x_\lambda\right)\tilde{\rho}\left(y_\mu,(f,\varphi)y_\mu\right)}{1+\tilde{\rho}\left(x_\lambda,y_\mu\right)}$$
$$\left.+\frac{\tilde{\rho}\left(x_\lambda,(f,\varphi)x_\lambda\right)\tilde{\rho}\left(x_\lambda,(f,\varphi)y_\mu\right)}{1+\tilde{\rho}\left(y_\mu,(f,\varphi)x_\lambda\right)}\right\}.$$

Hence all the condition of corollary (2.22) (theorem 2.21) are satisfied. Thus (f, φ) has a unique soft fixed point in (X, d, E) indeed 0 is the unique soft fixed point of (f, φ).

Corollary 2.24: Let in (X, d, E) be a complete soft dislocated metric space, $m \in N$, let $A_1, A_2, \cdots, A_m$ be non empty $\tilde{\rho}$-closed subsets of $(X, \tilde{\rho}, E)$ and let

$Y=\bigcup_{i=1}^{i=m}A_i$. Suppose that $(f, \varphi): Y \to Y$ is an operator such that

i) $Y=\bigcup_{i=1}^{i=m}A_i$ is cyclic representation of $(X, \tilde{\rho}, E)$ with respect to (f, φ)

ii) $\int_0^{\tilde{\rho}((f,\varphi)x_\lambda,(f,\varphi)y_\mu)} \rho(t)\,dt \le \beta \int_0^m \rho(t)\,dt$.

where

$$m = \tilde{\rho}(y_\mu,(f,\varphi)x_\lambda) + \tilde{\rho}(x_\lambda,(f,\varphi)y_\mu) + \tilde{\rho}(y_\mu,(f,\varphi)y_\mu) + \tilde{\rho}(x_\lambda,(f,\varphi)x_\lambda) + \frac{\tilde{\rho}(x_\lambda,(f,\varphi)x_\lambda)\tilde{\rho}(y_\mu,(f,\varphi)y_\mu)}{1+\tilde{\rho}(x_\lambda,y_\mu)} + \frac{\tilde{\rho}(x_\lambda,(f,\varphi)x_\lambda)\tilde{\rho}(x_\lambda,(f,\varphi)y_\mu)}{1+\tilde{\rho}(y_\mu,(f,\varphi)x_\lambda)}$$

for any $x_\lambda \in A_i, y_\mu \in A_{i+1}$, $i = 1,2,\cdots,m$. Where A_{m+1} = A_1 and $\rho:[0,\infty)\to[0,\infty)$ is

Lebsegue-integrable mapping satisfying $\int_0^\epsilon \rho(t)\,dt$, for $\epsilon > 0$ and the constant $\beta \in \left[0,\frac{1}{8}\right)$. Then (f, φ) has unique soft fixed point $z_\omega \in \bigcap_{i=1}^{i=m} A_i$

In corollary 2.24, if we take Ai = $(X, \tilde{\rho}, E)$, for i = 1,2,..., m We obtain the

following result.

Theorem 2.25: Let $(X, \tilde{\rho}, E)$ be a complete soft dislocated metric space

and $(f,\varphi):(X,\tilde{\rho},E)\to(X,\tilde{\rho},E)$ be a mapping such that for any $\int_0^{\tilde{\rho}((f,\varphi)x_\lambda,(f,\varphi)y_\mu)} \rho(t)\,dt \le \beta \int_0^m \rho(t)\,dt$ then

where

$$m = \tilde{\rho}(y_\mu,(f,\varphi)x_\lambda) + \tilde{\rho}(x_\lambda,(f,\varphi)y_\mu) + \tilde{\rho}(y_\mu,(f,\varphi)y_\mu) + \tilde{\rho}(x_\lambda,(f,\varphi)x_\lambda) + \frac{\tilde{\rho}(x_\lambda,(f,\varphi)x_\lambda)\tilde{\rho}(y_\mu,(f,\varphi)y_\mu)}{1+\tilde{\rho}(x_\lambda,y_\mu)} + \frac{\tilde{\rho}(x_\lambda,(f,\varphi)x_\lambda)\tilde{\rho}(x_\lambda,(f,\varphi)y_\mu)}{1+\tilde{\rho}(y_\mu,(f,\varphi)x_\lambda)},$$

Let $\rho:[0,\infty)\to[0,\infty)$ is Lebsegue-integrable mapping satisfying $\int_0^\epsilon \rho(t)\,dt$ for $\epsilon > 0$ and the constant $\beta \in \left[0,\frac{1}{8}\right)$ Then (f, φ) has unique soft fixed point.

3. CONCLUSION

In this paper, the investigations concerning the existence and uniqueness of soft fixed point of a cyclic mapping in soft dislocated metric space are established. Examples are given in the support of established results. These results can be extended to any directions, and can also be extended to fixed point theory of non-expansive multivalued mappings. These proved results lead to different directions and aspect of soft metric fixed point theory.

ACKNOWLEDGEMENTS

The authors would like their sincere thanks to the editor and the anonymous referees for their valuable comments and useful suggestions in improving the article.

REFERENCES

1. Molodtsov, D. (1999) Soft Set-Theory-First Results. Computers & Mathematics with Applications, 37, 19-31. https://doi.org/10.1016/S0898-1221(99)00056-5
2. Ali, M.I., Feng, F., Liu, X., Min, W.K. and Shabir, M. (2011) On Some New Operations in Soft Set Theory. Computers & Mathematics with Applications, 62, 351-358.
3. Chen, D. (2005) The Parameterization Reduction of Soft Sets and Its Applications. Computers & Mathematics with Applications, 49, 757-763. https://doi.org/10.1016/j.camwa.2004.10.036
4. Gunduz, C., Sonmez, A. and Çakalli, H. (2013) On Soft Mappings. arXiv:1305.4545v1 [math.GM]
5. Hussain, S. and Ahmad, B. (2011) Some Properties of Soft Topological Spaces. Computers and Math with Applications, 62, 4058-4067. https://doi.org/10.1016/j.camwa.2011.09.051
6. Maji, P.K., Roy, A.R. and Biswas, R. (2002) An Application of Soft Sets in a Decision Making Problem. Computers & Mathematics with Applications, 44, 1077-1083. https://doi.org/10.1016/S0898-1221(02)00216-X
7. Maji, P.K., Biswas, R. and Roy, A.R. (2003) Soft Set Theory. Computers and Mathematics with Applications, 45, 555-562. https://doi.org/10.1016/S0898-1221(03)00016-6
8. Majumdar, P. and Samanta, S.K. (2010) On Soft Mappings. Computers and Mathematics with Applications, 60, 2666-2672. https://doi.org/ 10.1016 /j.camwa.2010.09.004
9. Shabir, M. and Naz, M. (2011) On Soft Topological Spaces. Computers and Mathematics with Applications, 61, 1786-1799. https://doi.org/ 10.1016/ .camwa.2011.02.006
10. Das, S. and Samanta, S.K. (2012) Soft Real Sets, Soft Real Numbers and Their Properties. Journal of Fuzzy Mathematics, 20, 551-576.
11. Das, S. and Samanta, S.K. (2013) On Soft Metric Spaces. The Journal of Fuzzy Mathematics, 21, 207-213.

12. Hitzler, P. and Seda, A.K. (2000) Dislocated Topologies. Journal of Electrical Engineering, 51, 3-7.
13. Aage, C.T. and Salunke, J.N. (2008) Some Results of Fixed Point Theorem in Dislocated Quasi-Metric Spaces. Bulletin of the Marathwada Mathematical Society, 9, 1-5.
14. Karapinar, E. and Salimi, P. (2013) Dislocated Metric Space to Metric Space with Some Fixed Point Theorem. Fixed Point Theory and Application, 2013, 222. https://doi.org/10.1186/1687-1812-2013-222
15. Bayramov, S. and Gunduz, C. (2013) Soft Locally Compact and Soft Para Compact Spaces. Journal of Mathematics and System Science, 3, 122-130.
16. Rhoades, B.E. (1977) A Comparison of Various Definitions of Contractive Mappings. Transactions of the American Mathematical Society, 266, 257-290. https://doi.org/10.1090/S0002-9947-1977-0433430-4
17. Huang, L.-G. and Zhang, X. (2007) Cone Metric Spaces and Fixed Point Theorems of Contractive Mappings. Journal of Mathematical Analysis and Applications, 332, 1468-1476. https://doi.org/10.1016/j.jmaa.2005.03.087
18. Yazar, M.I., Gunduz, C. and Bayramov, S. (2013) Fixed Point for Soft Contractive Mappings. Cornell University Library, Thurston Ave Ithaca, NY.
19. Wadkar, B.R., Bhardwaj, R.K. and Singh, B. (2012) A Common Fixed Point Theorem in Dislocated Metric Space. International Journal of Engineering Research and Development, 10, 14-17.
20. Mishra, L.N., Mishra, V.N., Khatri, K. and Deepmala (2014) On the Trigonometric Approximation of Signals (Functions) Belonging to Generalized Weighted Lipschitz Class by Matrix Operator of Conjugate Series of Its Fourier Series. Applied Mathematics and Computation, 237, 252-263. https://doi.org/10.1016/j.amc.2014.03.085
21. Mishra, V.N., Khatri, K. and Mishra, L.N. (2013) Using Linear Operator to Approximate Signal of Class. Filomat, 27, 355-365. https://doi.org/10.2298/FIL1302353M
22. Mishra, L.N., Tiwari, S.K., Mishra, V.N. and Khan, I.A. (2015) Unique Fixed Point Theorems for Generalized Contractive Mappings in Partial Metric Spaces. Journal of Function Spaces, 2015, Article ID: 960827, 8 p. https://doi.org/10.1155/2015/960827
23. Mishra, L.N., Tiwari, S.K. and Mishra, V.N. (2015) Fixed Point Theorems for Generalized Weakly S-Contractive Mappings in Partial Metric Spaces. Journal of Applied Analysis and Computation, 5, 600-612.
24. Deepmala and Pathak, H.K. (2013) A Study on Some Problems on Existence of Solutions for Nonlinear Functional-Integral Equations. Acta Mathematica Scientia, 33, 1305-1313. https://doi.org/10.1016/S0252-9602(13)60083-1
25. Wadkar, B.R., Bhardwaj, R., Mishra, V.N. and Singh, B. (2017) Coupled Fixed Point Theorems in Soft Metric and B-Soft Metric Space. Ser. A: Appl. Math. Inform. And Mech., 9, 59-73. http://www.np.ac.rs/downloads/publications/vol9_br_1/rad6.pdf
26. Wadkar, B.R., Bhardwaj, R., Singh, B. and Mishra, V.N. (2017) Coupled Fixed Point Theorems with Monotone Property in Soft Metric and Soft B-Metric Space. International Journal of Mathematical Analysis, 11, 363-375. https://doi.org/10.12988/ijma.2017.7222

CHAPTER 6

An Alternative Proof of the Largest Number of Maximal Independent Sets in Connected Graphs Having at Most Two Cycles

Min-Jen Jou[1], Jenq-Jong Lin[2]

[1]*Department of Information Technology, Ling Tung University, Taiwan*

[2]*Department of Finance, Ling Tung University, Taiwan*

ABSTRACT

G. C. Ying, Y. Y. Meng, B. E. Sagan, and V. R. Vatter [1] found the maximum number of maximal independent sets in connected graphs which contain at most two cycles. In this paper, we give an alternative proof to determine the largest number of maximal independent sets among all connected graphs of order $n \geq 12$, which contain at most two cycles. We also characterize the extremal graph achieving this maximum value.

Keywords: Maximal Independent Set, Connected Graph Having at Most Two Cycle

1. INTRODUCTION

Let $G=(V,E)$ be a simple undirected graph. An independent set is a subset S of V such that no two vertices in S are adjacent. A maximal independent set is an independent set that is not a proper subset of any other independent set. The set of all maximal independent sets of a graph G is denoted by $MI(G)$ and its cardinality by $mi(G)$.

The problem of determining the largest value of $mi(G)$ in a general graph of order n and those graphs achieving the largest number was proposed by Erdös and Moser, and solved by Moon and Moser [2]. It was then extensively studied for various classes of graphs in the literature, including trees, forests, (connected) graphs with at most one cycle, bipartite graphs, connected graphs, k-connected graphs, (connected) triangle-free graphs; for a survey see [3]. Recently, Jin and Li [4] determined the second largest number of maximal independent sets among all graphs of order n.

There are results on independent sets in graphs from a different point of view. The Fibonacci number of a graph is the number of independent vertex subsets. The concept of the Fibonacci number of a graph was introduced in [5] and discussed in several papers [6] [7]. In addition, Jou and Chang [8] showed a linear-time algorithm for counting the number of maximal independent sets in a tree.

Jou and Chang [9] determined the largest number of maximal independent sets among all graphs and connected graphs of order n, which contain at most one cycle. Later B. E. Sagan and V. R. Vatter [10] found the largest number of maximal independent sets among all graphs of order n, which contain at most r cycles. In 2012, Jou [11] settled the second largest number of maximal independent sets in graphs with at most one cycle. G. C. Ying, Y. Y. Meng, B. E. Sagan, and V. R. Vatter [1] found the maximum number of maximal independent sets in connected graphs which contain at most two cycles. In this paper, we give an alternative proof to determine the largest number of maximal independent sets among all connected graphs of order $n\geq 12$, which contain at most two cycles. We also characterize the extremal graph achieving this maximum value.

For a graph $G=(V,E)$ the cardinality of $V(G)$ is called the order, and it is denoted by $|G|$. The neighborhood $N(x)$ of a vertex $x\in V(G)$ is the set of vertices adjacent to x in G and the closed neighborhood $N[x]$ is $\{x\}\cup N(x)$. The degree of x is the cardinality of $N(x)$, and it is denoted by $\deg(x)$. A vertex x is said to be a leaf if $\deg(x) = 1$. For a set $A \subseteq V(G)$, the deletion of A from G is the graph $G - A$ obtained from G by removing all vertices in A and their incident edges. Two graphs G_1 and G_2 are disjoint if $V(G_1)\cap V(G_2)=\emptyset$. The union of two disjoint graphs G_1 and G_2 is the graph $G_1\cup G_2$ with vertex set $V(G_1 \cup G_2)=V(G_1) \cup (G_2)$ and edge set $E(G_1 \cup G_2) = E(G_1) \cup E(G_2)$. If a graph G is isomorphic to another graph H, we denote $G=H$. Denote K_n a complete graph of order n and C_n a cycle of order n. The join of two disjoint graphs G_1 and G_2 is the graph G_1+G_2 with vertex set $V(G_1 \cup G_2) = V(G_1) \cup V(G_2)$ and edge set

$$E(G_1 \cup G_2) = E(G_1) \cup E(G_2) \cup \{uv : u \in V(G_1) \text{ and } v \in V(G_2)\}$$

. The star-product of two disjoint graphs G_1 and G_2 is the graph $G_1 \times G_2$ with vertex set $V(G_1 \cup G_2) = V(G_1) \cup V(G_2)$ and edge set $E(G_1 \cup G_2) = E(G_1) \cup E(G_2) \cup \{v_1 v_2\}$, where v_i is a vertex with maximum degree in G_i for $i = 1, 2$.

2. PRELIMINARY

The following results are needed.

Lemma 1. ([12] [13]) *For any vertex x in a graph G, the following hold.*

1) $mi(G) \le mi(G-x) + mi(G - N_G[x])$.

2) If x is a leaf adjacent to y, then $mi(G) = mi(G - N_G[x]) + mi(G - N_G[y])$.

Lemma 2. ([13]) If *G is the union of two disjoint graphs G_1 and G_2, then* $mi(G) = mi(G_1)mi(G_2)$.

Lemma 3. Let *a,b,n* be integers such that $0 < a < \frac{n}{2} < b$ and let $f(x) = 2^x + 2^{n-x}$.

Then $f(x) \le \max\{f(a), f(b)\}$.

Proof. The derivative of $f(x)$ is

$$f'(x) = 2^x(\ln 2) - (2^{n-x})(\ln 2) = (\ln 2)\left[2^x - 2^{n-x}\right].$$

So $f'(x)<0$ for any $x\in\left(a,\frac{n}{2}\right)$ and $f'(x)>0$ for any $x\in\left(\frac{n}{2},b\right)$.

Then *f*(*x*) is decreasing on $\left(a,\frac{n}{2}\right)$ and *f*(*x*) is increasing on $\left(\frac{n}{2},b\right)$. Hence $f(x)\le\max\{f(a),f(b)\}$.□

Theorem 1. ([9]) If T is a tree with $n\ge 1$ vertices, then $mi\ (G)\le t\ (n)$, where

$$t(n)=\begin{cases}2^{\frac{n-1}{2}}, & \text{if } n \text{ is odd;}\\ 2^{\frac{n-2}{2}}+1, & \text{if } n \text{ is even.}\end{cases}$$

Furthermore, mi(*T*)=*t*(*n*)if *and only if T*∈*T*(*n*),*where*

$$T(n)=\begin{cases}B\left(1,\frac{n-1}{2}\right), & \text{if } n \text{ is odd;}\\ B\left(2,\frac{n-2}{2}\right)\text{) or } B\left(4,\frac{n-4}{2}\right), & \text{if } n \text{ is even,}\end{cases}$$

where B(i,j)is the set of batons, which are the graphs obtained from a basic path P of i≥1 vertices by attaching j≥0 paths of length two to the endpoints of P in all possible ways (see Figure 1).

Theorem 2. ([9]) *If F is a forest with n≥1 vertices, then mi*(*G*) ≤ *f* (*n*), *where*

$$f(n)=\begin{cases}2^{\frac{n-1}{2}}, & \text{if } n \text{ is odd ;}\\ 2^{\frac{n}{2}}, & \text{if } n \text{ is even.}\end{cases}$$

Furthermore, *mi*(*F*) ≤ *f* (*n*)if and only if *F* ∈ *F* (*n*), where

$$F(n)=\begin{cases}B\left(1,\frac{n-1-2s}{2}\right)\cup sP_2, & \text{if } n \text{ is odd ;}\\ \frac{n}{2}P_2, & \text{if } n \text{ is even,}\end{cases}$$

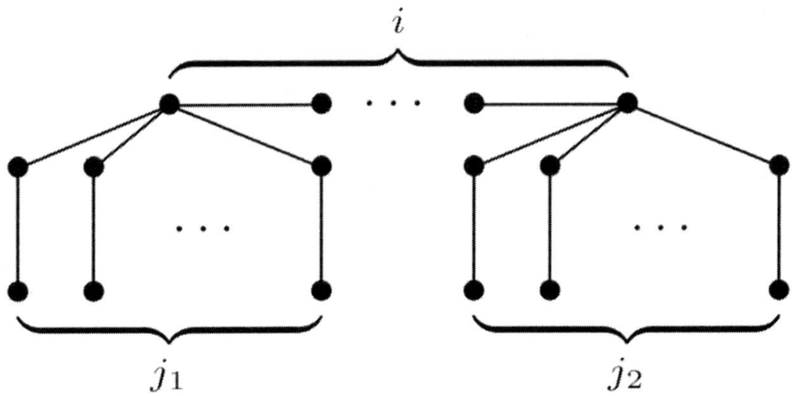

Figure 1. The baton $B(i,j)$ with$j=j_1=j_2$..

where $0 \le s \le \frac{n-1}{2}$.

Theorem 3. ([9]) *If G is a graph of order n≥2 vertices with at most one cycle, then mi (G) ≤ g (n,1), where*

$$g(n,1) = \begin{cases} 2^{\frac{n}{2}}, & \text{if } n \text{ is even;} \\ 3 \cdot 2^{\frac{n-3}{2}}, & \text{if } n \text{ is odd.} \end{cases}$$

Furthermore, mi (G) = g (n,1)if and only if G=G(n,1) where

$$G(n,1) = \begin{cases} \frac{n}{2} K_2, & \text{if } n \text{ is even;} \\ K_3 \cup \frac{n-3}{2} K_2, & \text{if } n \text{ is odd.} \end{cases}$$

Theorem 4. ([11]) *If G is a graph of order n≥4 with at most one cycle such that G ≠ G(n,1), then mi (G) ≤ g' (n,1), where*

$$g'(n,1) = \begin{cases} 3 \cdot 2^{\frac{n-4}{2}}, & \text{if } n \ge 4 \text{ is even;} \\ 5 \cdot 2^{\frac{n-5}{2}}, & \text{if } n \ge 5 \text{ is odd.} \end{cases}$$

Furthermore, $mi(G) = g'(n,1)$ if and only if $G \in G'(n,1)$, where

$$G'(n,1)=\begin{cases} P_4 \cup \frac{n-4}{2}P_2,\ C_3 \cup T(2s-3) \cup \frac{n-2s}{2}P_2 \text{ or } C(2s,1) \cup \frac{n-2s}{2}P_2, & \text{if } n \geq 4 \text{ is even } (s \geq 2); \\ C_5 \cup \frac{n-5}{2}P_2 \text{ or } C(5,1) \cup \frac{n-5}{2}P_2, & \text{if } n \geq 5 \text{ is odd.} \end{cases}$$

Theorem 5. ([9]) *If H is a connected graph of order n≥3 with at most one cycle, then $mi(H) \leq h(n,1)$, where*

$$h(n,1)=\begin{cases} 2^{\frac{n-1}{2}}+1, & \text{if } n \geq 3 \text{ is odd;} \\ 3 \cdot 2^{\frac{n-4}{2}}, & \text{if } n \geq 4 \text{ is even.} \end{cases}$$

Furthermore, $mi(H) = h(n,1)$ if and only if $H \in H(n,1)$ (see Figure 2), where

$$H(n,1)=\begin{cases} K_3 * \frac{n-3}{2}K_2 \text{ or } C_5, & \text{if } n \geq 3 \text{ is odd;} \\ K_1 * \left(K_3 \cup \frac{n-4}{2}K_2\right) \text{ or } P_4, & \text{if } n \geq 4 \text{ is even.} \end{cases}$$

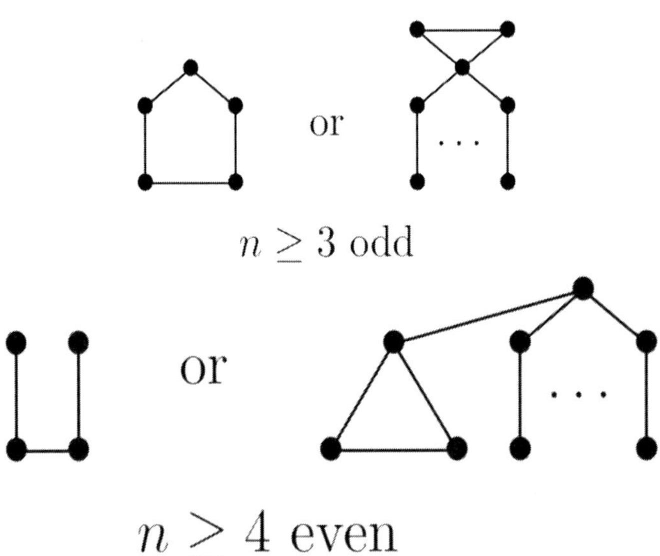

Figure 2. The extremal graphs $H(n,1)$ for $n \geq 3$.

3. THE ALTERNATIVE PROOF

In this section, we give an alternative proof to determine the largest number of maximal independent sets among all connected graphs of order $n \geq 12$, which contain at most two cycles. We also characterize the extremal graph achieving this maximum value.

Theorem 6. *If H is a connected graph of order n ≥ 12 with at most two cycles, then mi(H)≤h(n,2), where*

$$h(n,2)=\begin{cases} 3\cdot 2^{\frac{n-4}{2}}+2, & \text{if } n\geq 12 \text{ is even;} \\ 9\cdot 2^{\frac{n-7}{2}}, & \text{if } n\geq 13 \text{ is odd.} \end{cases}$$

Furthermore, mi(H)≤h(n,2) if and only if H ∈ H (n, 2), where

$$H(n,2)=\begin{cases} K_3\times\left(K_3\cup\frac{n-6}{2}K_2\right), & \text{if } n\geq 12 \text{ is even;} \\ K_1\times\left(2K_3\cup\frac{n-7}{2}K_2\right), & \text{if } n\geq 13 \text{ is odd.} \end{cases}$$

A unicyclic graph is a connected graph having one cycle. The order of a unicyclic graph is at least three. The following lemmas will be needed in the proof of main theorem.

Lemma 4. *Suppose that G = T ∪ H is the union of a tree T and a unicyclic graph H, where |G|= n ≥ 6. Then mi (G) ≤ 2 h(n − 2,1). The equality holds if and only if*

$$G=K_2\cup H(n-2,1) \text{ or } G=K_3\cup B\left(1,\frac{n-4}{2}\right).$$

Proof. Let$|T|=k$. Note that H has one cycle, then $|H|=n-k\geq 3$. We consider two cases.

Case 1. $n\geq 6$is even.

By Lemma 2, Theorem 1 and Theorem 5, we have

$$mi(G) = mi(T)\cdot mi(H) \le t(k)\cdot h(n-k,1)$$
$$= \begin{cases} 2^{\frac{k-1}{2}}\cdot\left(2^{\frac{n-k-1}{2}}+1\right), & \text{if } 1\le k\le n-3 \text{ is odd;} \\ \left(2^{\frac{k-2}{2}}+1\right)\cdot\left(3\cdot 2^{\frac{n-k-4}{2}}\right), & \text{if } 2\le k\le n-4 \text{ is even;} \end{cases}$$
$$= \begin{cases} 2^{\frac{n-2}{2}}+2^{\frac{k-1}{2}}, & \text{if } 1\le k\le n-3 \text{ is odd;} \\ 3\left(2^{\frac{n-6}{2}}+2^{\frac{n-k-4}{2}}\right), & \text{if } 2\le k\le n-4 \text{ is even;} \end{cases}$$
$$\le \begin{cases} 2^{\frac{n-2}{2}}+2^{\frac{(n-3)-1}{2}}, & \text{if } 1\le k\le n-3 \text{ is odd;} \\ 3\left(2^{\frac{n-6}{2}}+2^{\frac{n-(2)-4}{2}}\right), & \text{if } 2\le k\le n-4 \text{ is even;} \end{cases}$$
$$= \begin{cases} 3\cdot 2^{\frac{n-4}{2}}, & \text{if } 1\le k\le n-3 \text{ is odd;} \\ 3\cdot 2^{\frac{n-4}{2}}, & \text{if } 2\le k\le n-4 \text{ is even;} \end{cases} = \begin{cases} t(n-3)\cdot 3, & \text{if } 1\le k\le n-3 \text{ is odd;} \\ 2\cdot h(n-2,1), & \text{if } 2\le k\le n-4 \text{ is even;} \end{cases}$$
$$= 2\cdot h(n-2,1)$$

If the equality holds, then $mi(G) = t(n-3)\,h(3,1)$ or $mi(G) = t(2)\,h(n-2,1)$. Hence the equality holds if and only if $G=K_2 \cup H(n\text{-}2,1)$ or $G = K_3 \cup B\left(1,\frac{n-4}{2}\right)$.

Case 2. n≥7 is odd.

By Lemma 3 and since n≥7,

$$2^{\frac{k-2}{2}}+2^{\frac{n-k-1}{2}} \le \max\left\{2^{\frac{(2)-2}{2}}+2^{\frac{n-(2)-1}{2}},\ 2^{\frac{(n-3)-2}{2}}+2^{\frac{n-(n-3)-1}{2}}\right\} = 2^{\frac{n-3}{2}}+1$$

for $2 \le k \le n\text{-}3$.

By Theorem 1 and Theorem 5, we have

$$mi(G) = mi(T)\cdot mi(H) \le t(k)\cdot h(n-k,1)$$

$$= \begin{cases} 2^{\frac{k-1}{2}}\cdot\left(3\cdot 2^{\frac{n-k-4}{2}}\right), & \text{if } 1 \le k \le n-4 \text{ is odd;} \\ \left(2^{\frac{k-2}{2}}+1\right)\cdot\left(2^{\frac{n-k-1}{2}}+1\right), & \text{if } 2 \le k \le n-3 \text{ is even;} \end{cases}$$

$$= \begin{cases} 3\cdot 2^{\frac{n-5}{2}}, & \text{if } 1 \le k \le n-4 \text{ is odd;} \\ 2^{\frac{n-3}{2}}+2^{\frac{k-2}{2}}+2^{\frac{n-k-1}{2}}+1, & \text{if } 2 \le k \le n-3 \text{ is even;} \end{cases}$$

$$\le \begin{cases} 3\cdot 2^{\frac{n-5}{2}}, & \text{if } 1 \le k \le n-4 \text{ is odd;} \\ 2^{\frac{n-3}{2}}+1+\left(2^{\frac{n-3}{2}}+1\right), & \text{if } 2 \le k \le n-3 \text{ is even;} \end{cases}$$

$$\le 2^{\frac{n-1}{2}}+2 = 2\cdot h(n-2,1).$$

If the equality holds, then $mi\ (G) = t\ (2)\ h(n-2,1)$. Hence the equality holds if and only if $G{=}K_2{\cup}H(n\text{-}2{,}1)$.

By case 1 and case 2, we have that $mi\ (G) \le t\ (2)\ h(n-2,1)$. The equality holds if and

only if $G{=}K_2{\cup}H(n\text{-}2{,}1)$ or $G = K_3 \cup B\left(1,\dfrac{n-4}{2}\right)$. □

Lemma 5. *Suppose that F is a forest of order $n{\ge}4$ having at most two components. Then $mi\ (F) \le 2\ t\ (n-2)$ and the equality holds if and only if $F{=}K_2 \cup T(n\text{-}2)$ or*

$$F = B\left(1,\frac{n-1}{2}\right).$$

Proof. Let F be a forest of order $n{\ge}4$ having at most two components such that $m_i(F)$ as large as possible. Then

$mi(F) \geq mi(K_2 \cup T(n-2)) = 2 \cdot t(n-2)$. If F has one component, then *F* is a tree and, by Theorem 1, $2 \cdot t(n-2) \leq mi(F) \leq t(n)$. Then n is odd and $m_i(F)$=$t(n)$. By Theorem 1, $F = B\left(1, \frac{n-1}{2}\right)$. Now we assume that *F*

have two components. Let T_1 and T_2 be the components of *F*, where $|T_1|$=*s*. We consider two cases.

Case 1. *n*≥*4* is even.

By Lemma 3,

$$2^{\frac{s-2}{2}} + 2^{\frac{n-s-2}{2}} \leq \max\left\{2^{\frac{(2)-2}{2}} + 2^{\frac{n-(2)-2}{2}}, 2^{\frac{(n-2)-2}{2}} + 2^{\frac{n-(n-2)-2}{2}}\right\} = 2^{\frac{n-4}{2}} + 1$$

for

2 ≤ *s* ≤ *n*-2. By Lemma 2 and Theorem 1, then

$$2 \cdot t(n-2) = 2\left(2^{\frac{n-4}{2}} + 1\right) \leq mi(F) = mi(T_1) \cdot mi(T_2) \leq t(s) \cdot t(n-s)$$

$$= \begin{cases} 2^{\frac{s-1}{2}} \cdot \left(2^{\frac{n-s-1}{2}}\right), & \text{if } 1 \leq s \leq n-1 \text{ is odd;} \\ \left(2^{\frac{s-2}{2}} + 1\right) \cdot \left(2^{\frac{n-s-2}{2}} + 1\right), & \text{if } 2 \leq s \leq n-2 \text{ is even;} \end{cases}$$

$$= \begin{cases} 2^{\frac{n-2}{2}}, & \text{if } 1 \leq s \leq n-1 \text{ is odd;} \\ 2^{\frac{n-4}{2}} + 1 + \left(2^{\frac{s-2}{2}} + 2^{\frac{n-s-2}{2}}\right), & \text{if } 2 \leq s \leq n-2 \text{ is even;} \end{cases}$$

$$\leq \begin{cases} 2^{\frac{n-2}{2}}, & \text{if } 1 \leq s \leq n-1 \text{ is odd;} \\ 2^{\frac{n-4}{2}} + 1 + \left(2^{\frac{n-4}{2}} + 1\right), & \text{if } 2 \leq s \leq n-2 \text{ is even;} \end{cases}$$

$$= \begin{cases} 2^{\frac{n-2}{2}}, & \text{if } 1 \le s \le n-1 \text{ is odd;} \\ 2^{\frac{n-2}{2}} + 2, & \text{if } 2 \le s \le n-2 \text{ is even;} \end{cases}$$

$$\le 2\left(2^{\frac{n-4}{2}} + 1\right) = 2 \cdot t(n-2).$$

The equalities hold andmi $(F) = t$ (2) t $(n - 2)$. By Theorem 1,$F=K_2 \cup T(n\text{-}2)$. The equality holds if and only if $F=K_2 \cup T(n\text{-}2)$.

Case 2. $n \geq 5$ is odd.

Then F has exactly one even component, we assume that $s \geq 2$ is even. By Theorem 1, then

$$2 \cdot t(n-2) = 2^{\frac{n-1}{2}} \le mi(F) = mi(T_1) \cdot mi(T_2) \le t(s) \cdot t(n-s) = \left(2^{\frac{s-2}{2}} + 1\right) \cdot \left(2^{\frac{n-s-1}{2}}\right)$$

$$= 2^{\frac{n-3}{2}} + 2^{\frac{n-s-1}{2}} \le 2^{\frac{n-3}{2}} + 2^{\frac{n-(2)-1}{2}} = 2^{\frac{n-1}{2}} = 2 \cdot t(n-2)$$

. The equalities hold and mi $(F) = t$ (2) t $(n - 2)$. The equality holds if and only if $F=K_2 \cup T(n\text{-}2)$. □

The following is the proof of Theorem 6.

Proof. Let H be a connected graph of order $n \geq 12$ with at most two cycles such that $mi(H)$ as large as possible. Then mi $(H) \geq mi$ $(H(n, 2)) = h(n, 2)$. Since mi $(H) \geq h(n, 2) > h(n,1)$, by Theorem 5, H have at least two cycles. That means that H have exactly two cycles and $H \neq C_n$. Let v be a vertex lying on some cycle such that $deg(v)$ is as large as possible. Since $H \neq C_n$, we can see that $deg(v)=s \geq 3$. The graph $H\text{-}v$ is a graph of order $n\text{-}1$ with at most one cycle. We consider two cases.

Case 1. $H - v = G(n-1,1)$.

Then $H - v = K_3 \cup \frac{n-4}{2} P_2$ or $H - v = \frac{n-1}{2} P_2$. Since H is connected, this means

that v connects to every component of $G(n-1,1)$. Then mi $(H - N[v])$ has at most one edge, then mi $(H - N[v]) \leq 2$. So we have

$$h(n,2) \le mi(H) \le mi(H-v) + mi(H-N[v]) \le g(n-1,1) + 2$$

$$= \begin{cases} 3\cdot 2^{\frac{n-4}{2}} + 2, & \text{if } n \ge 12 \text{ is even;} \\ 2^{\frac{n-1}{2}} + 2, & \text{if } n \ge 13 \text{ is odd;} \end{cases} \le h(n,2).$$

Then $mi\ (G - N\ [v]) = 2$ and $n \ge 12$ is even. So $H - v = K_3 \cup \frac{n-4}{2} P_2$. Note that H has two cycles, hence $H = K_3 \times \left(K_3 \cup \frac{n-6}{2} K_2 \right) = H(n,2)$ for even $n \ge 12$.

Case 2. $H - v \ne G(n-1,1)$.

Let $deg\ (v) = d$. Then the subgraph $H - N\ [v]$ is a graph of order $n - d - 1$ having at most one cycle. By Theorem 3, $mi\ (H - N\ [v]) \le g\ (n - d - 1)$. By Theorem 4, $mi\ (H - v) \le g'(n-1,1)$. By Lemma 1 and Theorem 4, then

$$g(n-d-1) \ge mi(H - N[v]) \ge mi(H) - mi(H-v) \ge h(n,2) - g'(n-1,1)$$

$$= \begin{cases} \left(3\cdot 2^{\frac{n-4}{2}} + 2\right) - \left(5\cdot 2^{\frac{n-6}{2}}\right), & \text{if } n \ge 12 \text{ is even;} \\ \left(9\cdot 2^{\frac{n-7}{2}}\right) - \left(3\cdot 2^{\frac{n-5}{2}}\right), & \text{if } n \ge 13 \text{ is odd;} \end{cases}$$

$$= \begin{cases} 2^{\frac{n-6}{2}} + 2, & \text{if } n \ge 12 \text{ is even;} \\ 3\cdot 2^{\frac{n-7}{2}}, & \text{if } n \ge 13 \text{ is odd;} \end{cases} = \begin{cases} g(n-6,1) + 2, & \text{if } n \ge 12 \text{ is even;} \\ g(n-4,1), & \text{if } n \ge 13 \text{ is odd;} \end{cases}$$

Then

$$deg(v) = d \le \begin{cases} 4, & \text{if } n \ge 12 \text{ is even;} \\ 3, & \text{if } n \ge 13 \text{ is odd.} \end{cases}$$

Claim. $deg\ (v) = 3$.

Suppose that $deg\ (v) = 4$, then n is even. By Theorem 3, $mi\ (H - N\ [v]) \le g\ (n - 5,1)$

and

$$mi(H-v) \ge mi(H) - mi(H-N[v]) \ge h(n,2) - g(n-5,1) = \left(3\cdot 2^{\frac{n-4}{2}} + 2\right) - 3\cdot 2^{\frac{n-8}{2}}$$

$$= 9\cdot 2^{\frac{n-8}{2}} + 2 > 2^{\frac{n-2}{2}} = f(n-1)$$

.

By Theorem 2, $H - v$ is not a forest and $H - v$ has one cycle. Let H'' be the component of $H - v$ having one cycle and $F = H - v - V(H'')$, where $|H''| = s \geq 3$. Note that H has two cycles and v is lying on some cycle. Thus v has two edges incident to some component of H-v. Since *deg* $(v) = 4$, the number of the components of H-v is at most three. Thus F is either a tree or the union of two trees. By Lemma 5, $mi\,(F) \leq 2\,t\,(n - s - 3)$. By Lemma 3,

$$2^{\frac{s+1}{2}} + 2^{\frac{n-s-3}{2}} \leq \max\left\{2^{\frac{(3)+1}{2}} + 2^{\frac{n-(3)-3}{2}}, 2^{\frac{(n-5)+1}{2}} + 2^{\frac{n-(n-5)-3}{2}}\right\} \leq 2^{\frac{n-4}{2}} + 2$$

for $3 \leq s \leq n-5$. By

Theorem 5, $mi\,(H'') \leq h(s,1)$. Note that $mi(H-v) \geq 9\cdot 2^{\frac{n-8}{2}} + 2$. By Lemma 2 and

Lemma 5,

$$9\cdot 2^{\frac{n-8}{2}} + 2 \leq (H-v) = mi(F)\cdot mi(H'') \leq \begin{cases} 2\cdot h(n-3,1), & \text{if } s = n-3; \\ (2\cdot t(n-s-3))\cdot h(s,1), & \text{if } s \neq n-3; \end{cases}$$

$$= \begin{cases} 2\cdot(2^{\frac{n-4}{2}}+1), & \text{if } s = n-3; \\ \left(2\cdot\left(2^{\frac{n-s-5}{2}}+1\right)\right)\cdot\left(2^{\frac{s-1}{2}}+1\right), & \text{if } 3 \leq s \leq n-5 \text{ is odd;} \\ \left(2\cdot 2^{\frac{n-s-4}{2}}\right)\cdot\left(3\cdot 2^{\frac{s-4}{2}}\right), & \text{if } 2 \leq s \leq n-4 \text{ is even;} \end{cases}$$

$$= \begin{cases} 2^{\frac{n-2}{2}}+2, & \text{if } s = n-3; \\ 2^{\frac{n-4}{2}}+2+\left(2^{\frac{s+1}{2}}+2^{\frac{n-s-3}{2}}\right), & \text{if } 3 \leq s \leq n-5 \text{ is odd;} \\ 3\cdot 2^{\frac{n-6}{2}}, & \text{if } 2 \leq s \leq n-4 \text{ is even;} \end{cases}$$

$$\leq \begin{cases} 2^{\frac{n-2}{2}}+2, & \text{if } s = n-3\,; \\ 2^{\frac{n-4}{2}}+2+\left(2^{\frac{n-4}{2}}+2\right), & \text{if } 3 \leq s \leq n-5 \text{ is odd;} \quad \leq 2^{\frac{n-2}{2}}+4 < 9\cdot 2^{\frac{n-8}{2}} \\ 3\cdot 2^{\frac{n-6}{2}}, & \text{if } 2 \leq s \leq n-4 \text{ is even;} \end{cases}$$

where $n \geq 12$. This is a contradiction, so *deg* $(v) = 3$.

By Claim, *deg* $(v) = 3$. Note that H has two cycles and v is lying on some cycle. Thus v has two edges incident to some component of $H - v$. Since *deg* $(v) = 3$, the number of the components of $H - v$ is at most two. Thus $H - v = T \cup H'$, where T is a tree and H' is a unicyclic graph. By Lemma 4 and Theorem 3, then $mi\,(H - v) = mi\,(T)\, mi\,(H') \le 2\, h(n-3,1) = 2\, h(n-3,1)$ and $mi\,(H - N[v]) \le g\,(n-4,1)$. So

$$\begin{aligned} h(n,2) &\le mi(H) \le mi(H-v) + mi(H-N[v]) \\ &\le 2\cdot h(n-3,1) + g(n-4,1) \\ &= \begin{cases} 2\left(2^{\frac{n-4}{2}}+1\right)+2^{\frac{n-4}{2}}, & \text{if } n \ge 12 \text{ is even;} \\ 2\left(3\cdot 2^{\frac{n-7}{2}}\right)+3\cdot 2^{\frac{n-7}{2}}, & \text{if } n \ge 13 \text{ is odd;} \end{cases} \\ &= \begin{cases} 3\cdot 2^{\frac{n-4}{2}}+2, & \text{if } n \ge 12 \text{ is even;} \\ 9\cdot 2^{\frac{n-7}{2}}, & \text{if } n \ge 13 \text{ is odd.} \end{cases} \\ &= h(n,2). \end{aligned}$$

The equalities hold. Then $H - N\,[v] = G(n-4,1)$ and, by Lemma 4, $H - v = K_2 \cup H(n-3,1)$ or $K_3 \cup B\left(1, \frac{n-5}{2}\right)$. If $H - v = K_3 \cup B\left(1, \frac{n-5}{2}\right)$, then n is odd and $H - N[v] \ne K_3 \cup \frac{n-7}{2} K_2$. That is $H - N[v] \ne G(n-4,1)$, this is a contra-

diction. Thus $H - v = K_2 \cup H(n-3,1)$. If n is even, where $n \ge 12$, then

$$H - v = K_2 \cup H(n-3,1) = K_2 \cup \left(K_3 \times \cup \frac{n-6}{2} K_2\right)$$ and

$H - N[v] = G(n-4,1) = \frac{n-4}{2}K_2$. Then, there exists a vertex $u \in V(H-v)$ lying on some cycle such that $deg(u) \geq \frac{n}{2} > 3$. This contradicts to the claim, so n is odd.

Thus $H - v = K_2 \cup H(n-3,1) = K_2 \cup \left(K_1 * \left(K_3 \cup \frac{n-7}{2}K_2 \right) \right)$ and

$H - N[v] = G(n-4,1) = K_3 \cup \frac{n-7}{2}K_2$.

Hence $H = K_1 \times \left(2K_3 \cup \frac{n-7}{2}K_2 \right) = H(n,2)$ for odd $n \geq 13$.

ACKNOWLEDGEMENTS

The authors are grateful to the referees for the helpful comments.

REFERENCES

1. Ying, G.C., Meng, Y.Y., Sagan, B.E. and Vatter, V.R. (2006) Maximal Independent Sets in Graphs with at Most r Cycles. Journal of Graph Theory, 53, 270-282. http://dx.doi.org/10.1002/jgt.20185
2. Moon, J.W. and Moser, L. (1965) On Cliques in Graphs. Israel Journal of Mathematics, 3, 23-28. http://dx.doi.org/10.1007/BF02760024
3. Jou, M.J. and Chang, G.J. (1995) Survey on Conunting Maximal Independent Sets. In: Tangmance, S. and Schulz, E., Eds., Proceedings of the Second Asian Mathematical Conference, World Scientific, Singapore City, 265-275.
4. Jin, Z. and Li, X. (2008) Graphs with the Second Largest Number of Maximal Independent Sets. Discrete Mathematics, 308, 5864-5870. http://dx.doi.org/10.1016/j.disc.2007.10.032
5. Prodinger, H. and Tichy, R.F. (1982) Fibonacci Numbers of Graphs. The Fibonacci Quarterly, 20, 16-21.
6. Knopfmacher, A., Tichy, R.F., Wagner, S. and Ziegler, V. (2007) Graphs, Partitions and Fibonacci Numbers. Discrete Applied Mathematics, 155, 1175-1187. http://dx.doi.org/10.1016/j.dam.2006.10.010
7. Wagner, S.G. (2006) The Fibonacci Number of Generalized Petersen Graphs. The Fibonacci Quarterly, 44, 362-367.

8. Jou, M.J. and Chang, G.J. (2002) Algorithmic Aspects of Counting Independent Sets. Ars Combinatoria, 65, 265-277.
9. Jou, M.J. and Chang, G.J. (1997) Maximal Independent Sets in Graphs with at Most One Cycle. Discrete Applied Mathematics, 79, 67-73. http://dx.doi.org/10.1016/S0166-218X(97)00033-4
10. Sagan, B.E. and Vatter, V.R. (2006) Maximal and Maximum Independent Sets in Graphs with at Most r Cycles. Journal of Graph Theory, 53, 283-314. http://dx.doi.org/10.1002/jgt.20186
11. Jou, M.J. (2012) The Second Largest Number of Maximal Independent Sets in Connected Graphs with at Most One Cycle. Journal of Combinatorial Optimization, 24, 192-201. http://dx.doi.org/10.1007/s10878-011-9376-4
12. Hujter, M. and Tuza, Z. (1993) The Number of Maximal Independent Sets in Triangle-Free Graphs. SIAM: SIAM Journal on Discrete Mathematics, 6, 284-288. http://dx.doi.org/10.1137/0406022
13. Jou, M.J. (1991) The Number of Maximal Independent Sets in Graphs. Master Thesis, Department of Mathematics, National Central University, Taoyuan.

CHAPTER 7

On the Number of Cycles in a Graph

Nazanin Movarraei, Samina A. Boxwala

Department of Mathematics, University of Pune, Pune, India

ABSTRACT

In this paper, we obtain explicit formulae for the number of 7-cycles and the total number of cycles of lengths 6 and 7 which contain a specific vertex v_i in a simple graph G, in terms of the adjacency matrix and with the help of combinatorics.

Keywords: Adjacency Matrix, Cycle, Graph Theory, Path, Subgraph, Walk

1. INTRODUCTION

In a simple graph G, a walk is a sequence of vertices and edges of the form $v_0, e_1, v_1, \ldots, e_k, v_k$ such that the edge e_i has ends v_{i-1} and v_i. A walk is called closed if $v_0 = v_k$. In graph theory, a path in a graph is a finite or infinite sequence of edges which connect a sequence of vertices which, are all distinct from one another. A closed path (with the common end points) is called a cycle.

It is known that if a graph G has adjacency matrix $A=[a_{ij}]$, then for $k=0,1,\cdots$, the ij-entry of A^k is the number of v_i=v_j walks of length k in G. It is also known that $tr(A^n)$ is the sum of the diagonal entries of A^n and d_i is the degree of the vertex v_i.

In 1971, Frank Harary and Bennet Manvel [1], gave formulae for the number of cycles of lengths 3 and 4 in simple graphs as given by the following theorems:

Theorem 1. [1] *If G is a simple graph with adjacency matrix A, then the number of 3-cycles in G is* $\frac{1}{6}tr\left(A^3\right)$. *(It is known that* $tr\left(A^3\right)=\sum_{i=1}^{n}a_{ii}^{(3)}=\sum_{j\neq i}a_{ij}^{(2)}a_{ij}$ *).*

Theorem 2. *[1] If G is a simple graph with adjacency matrix A, then the number of 4-cycles in G is*

$$\frac{1}{8}\left[tr\left(A^4\right)-2q-2\sum_{j\neq i}a_{ij}^{(2)}\right]$$

, where q is the number of edges in G. (It is obvious that the above formula is also equal to

$$\frac{1}{8}\left[trA^4-trA^2-2\sum_{j\neq i}a_{ij}^{(2)}\right]$$
)

Theorem 3. [1] *If G is a simple graph with n vertices and the adjacency matrix* $A=\left[a_{ij}\right]$, *then the number of 5-cycles in G is*

$$\frac{1}{10}\left[tr\left(A^5\right)+5tr\left(A^3\right)-5\sum_{i=1}^{n}d_i a_{ii}^{(3)}\right]$$

In 2003, V. C. Chang and H. L. Fu [2], found a formula for the number of 6-cycles in a simple graph which is stated below:

Theorem 4. [2] *If G is a simple graph with adjacency matrix A, then the number of 6-cycles in G is*

$$\frac{1}{12}\Bigg[tr\left(A^6\right)-6tr\left(A^4\right)++5tr\left(A^3\right)-4tr\left(A^2\right)-3\sum_{i,j=1}^{p}a_{ij}^{(3)}+12\sum_{i,j=1}^{p}a_{ij}^{(2)}-3\sum_{i=1}^{p}\left[a_{ii}^{(3)}\right]^2+9\sum_{j\neq i}a_{ij}^{(2)}\left(a_{ij}^{(2)}-1\right)a_{ij}$$
$$-6\sum_{j\neq i}a_{ij}^{(2)}\left(a_{ij}^{(2)}-1\right)\left(a_{ij}^{(2)}-2\right)-2\sum_{i=1}^{p}a_{ii}^{(2)}\left(a_{ii}^{(2)}-1\right)\left(a_{ii}^{(2)}-2\right)\Bigg]$$

, where p isthe number of vertices in G.

Their proofs are based on the following fact:

The number of n-cycles (n=3,4,5,6)in a graph G is equal to $\frac{1}{2n}\left(tr\left(A^n\right)-x\right)$ where x is the number of closed walks of length n, which are not n-cycles.

In 1997, N. Alon, R. Yuster and U. Zwick [3], gave number of 7-cyclic graphs. They also gave some for- mulae for the number of cycles of lengths 5, which contains a specific vertex v_i in a graph G.

In [3] - [9], we have also some bounds to estimate the total time complexity for finding or counting paths and cycles in a graph.

In our recent works [10] [11], we obtained some formulae to find the exact number of paths of lengths 3, 4 and 5, in a simple graph G, given below:

Theorem 5. [11] *Let G be a simple graph with n vertices and the adjacency matrix* $A = [a_{ij}]$. *The number of paths of length 3 in G is* $\sum_{j \neq i} a_{ij}^{(2)} (d_j - a_{ij} - 1)$.

Theorem 6. [11] *Let G be a simple graph with n vertices and the adjacency matrix* $A = [a_{ij}]$. *The number of*

paths of length 4 in G is

$$\sum_{j \neq i} \left[a_{ij}^{(4)} - 2a_{ij}^{(2)} (d_j - a_{ij}) \right] - \sum_{i=1}^{n} \left[(2d_i - 1) a_{ii}^{(3)} + 6 \binom{d_i}{3} \right].$$

Theorem 7. [11] *Let G be a simple graph with n vertices and the adjacency matrix* $A = [a_{ij}]$. *The number of*

paths of length 3 in G, each of which starts from a specific vertex v_i *is*

$$\sum_{j \neq i} a_{ij}^{(2)} (d_j - a_{ij} - 1).$$

Theorem 8. [11] *Let G be a simple graph with n vertices and the adjacency matrix* $A = [a_{ij}]$. *The number of paths of length 4 in G, each of which starts from a specific vertex* v_i *is*

$$\sum_{j \neq i} \left[a_{ij}^{(4)} - (d_i + d_j - 3a_{ij}) a_{ij}^{(2)} - \left(a_{ii}^{(3)} + a_{jj}^{(3)} + 2\binom{d_j - 1}{2} \right) a_{ij} \right]$$

.

Theorem 9. [10] *Let G be a simple graph with n vertices and the adjacency matrix* $A = \left[a_{ij} \right]$. *The number of*

paths of length 5 in G is $\sum_{j \neq i} a_{ij}^{(5)} - 2\sum_{j \neq i} a_{ij}^{(4)} + 2\sum_{i=1}^{n} a_{ii}^{(3)} (d_i - 2) + 4\sum_{j \neq i} a_{ij}^{(2)} - 2\sum_{j \neq i} a_{ij}^{(2)} (d_j - a_{ij} - 1)$

$$- 4\sum_{j \neq i} a_{ij}^{(2)} \binom{d_i - a_{ij} - 1}{2} + 6\sum_{j \neq i} a_{ij} \binom{a_{ij}^{(2)}}{2} - 2\sum_{j \neq i} a_{ii}^{(3)} a_{ij}^{(2)} - 2\sum_{i=1}^{n} a_{ii}^{(3)} \binom{d_i - 2}{2}$$

$$- 2\sum_{i=1}^{n} \left(a_{ii}^{(4)} - a_{ii}^{(2)} - 2\binom{d_i}{2} - \sum_{j=1, j \neq i}^{n} a_{ij}^{(2)} \right) (d_i - 2) - \sum_{j \neq i} a_{ij} - 3\, tr\, A^4 + 6\, tr\, A^3 + 3\, tr\, A^2$$

.

Theorem 10. [10] *If G is a simple graph with n vertices and the adjacency matrix* $A = \left[a_{ij} \right]$, *then the number of 4-cycles each of which contains a specific vertex* v_i *of G is*

$$\frac{1}{2}\left[a_{ii}^{(4)} - a_{ii}^{(2)} - 2\binom{d_i}{2} - \sum_{j=1, j \neq i}^{n} a_{ij}^{(2)} \right].$$

In [3] we can also see a formula for the number of 5-cycles each of which contains a specific vertex v_i but, their formula has some problem in coefficients. In [12] we gave the correct formula as considered below:

Theorem 11. [12] *If G is a simple graph with n vertices and the adjacency matrix* $A = \left[a_{ij} \right]$, *then the number of 5-cycles each of which contains a specific vertex* v_i *of G is*

$$\frac{1}{2}\left[a_{ii}^{(5)} - 5a_{ii}^{(3)} - 2(d_i - 2)a_{ii}^{(3)} - 2\sum_{j=1, j \neq i}^{n} a_{ij}^{(2)} a_{ij} (d_j - 2) - 2\sum_{j=1, j \neq i}^{n} a_{ij} \left(\frac{1}{2} a_{jj}^{(3)} - a_{ij} a_{ij}^{(2)} \right) \right].$$

In this paper, we give a formula to count the exact number of cycles of length 7 and the number of cycles of lengths 6 and 7 containing a specific vertex v_i in a simple graph G, in terms of the adjacency matrix of *G* and with the help of combinatorics.

2. NUMBER OF 7-CYCLES

In 1997, N. Alon, R. Yuster and U. Zwick [3], gave number of 7-cyclic graphs. The n-cyclic graph is a graph that contains a closed walk of length n and these walks are not necessarily cycles. In this section we obtain a formula for the number of cycles of length 7 in a simple graph G with the helps of [3].

Method: To count N in the cases considered below, we first count $tr(A^7)$ for the graph of first con- figuration. This will give us the number of all closed walks of length 7 in the corresponding graph. But, some of these walks do not pass through all the edges and vertices of that configuration and to find N in each case, we have to include in any walk, all the edges and the vertices of the corresponding subgraphs at least once. So, we delete the number of closed walks of length 7 which do not pass through all the edges and vertices. To find these kind of walks we also have to count $tr(A^7)$ for all the subgraphs of the corresponding graph that can contain a closed walk of length 7.

Theorem 12. *If G is a simple graph with n vertices and the adjacency matrix* $A=\left[a_{ij}\right]$*, then the number of 7-cycles in G is* $\frac{1}{14}\left(tr\left(A^7\right)-x\right)$*, where x is equal to* $\sum_{i=1}^{11} F_i$ *in the cases that are considered below.*

Proof: The number of 7-cycles of a graph G is equal to $\frac{1}{14}\left(tr\left(A^7\right)-x\right)$, where x is the number of closed

walks of length 7 that are not 7-cycles. To find x, we have 11 cases as considered below; the cases are based on the configurations-(subgraphs) that generate all closed walks of length 7 that are not 7-cycles.

In each case, N denotes the number of closed walks of length 7 that are not 7-cycles in the corresponding subgraph, M denotes the number of subgraphs of G of the same configuration and F_n, (n = 1, 2,…) denote the total number of closed walks of length 7 that are not cycles in all possible subgraphs of G of the same configurations. However, in the cases with more than one figure (Cases 5, 6, 8, 9, 11), N, M and F_n are based on the first graph in case n of the respective figures and P_1,P_2,… denote the number of subgraphs of G which don't have the same configuration as the first graph but are counted in M. It is clear that F_n is equal to $N \times M - P - P -$…). To find N in each case, we have to include in any walk, all the edges and the vertices of the corresponding subgraphs at least once.

Case 1: For the configuration of Figure 1, N=126, $M=\frac{1}{6}tr\left(A^3\right)$ and $F_1=21\,tr\left(A^3\right)$. (See Theorem 1)

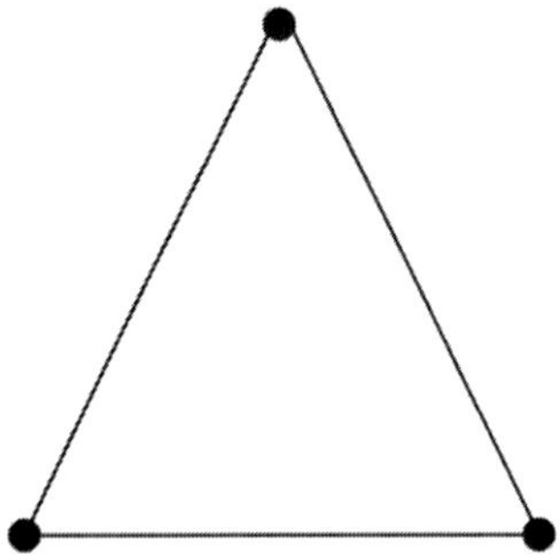

Figure 1. Closed walks of length 7 type 1.

Case 2: For the configuration of Figure 2, N=84,

$$M = \frac{1}{2}\sum_{i=1}^{n} a_{ii}^{(3)}\left(d_i - 2\right) \text{ and } F_2 = 42\sum_{i=1}^{n} a_{ii}^{(3)}\left(d_i - 2\right).$$

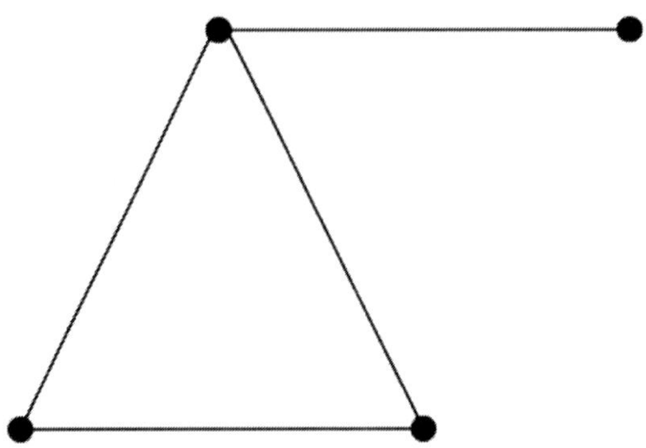

Figure 2. Closed walks of length 7 type 2.

Case 3: For the configuration of Figure 3, N=28,

$$M = \frac{1}{2}\sum_{i=1}^{n} a_{ii}^{(3)}\binom{d_i - 2}{2} \text{ and } F_3 = 14\sum_{i=1}^{n} a_{ii}^{(3)}\binom{d_i - 2}{2}.$$

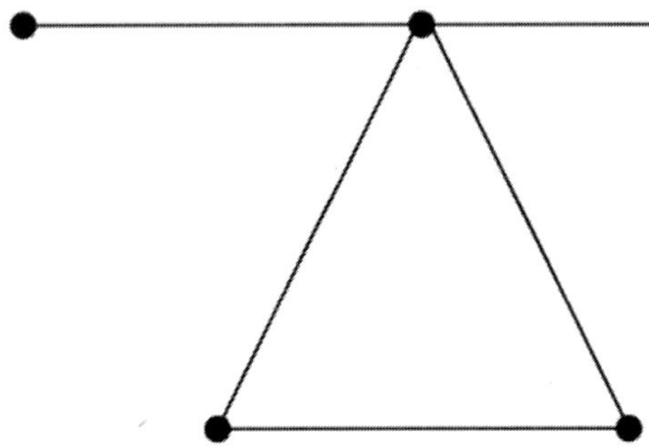

Figure 3. Closed walks of length 7 type 3.

Case 4: For the configuration of Figure 4 $N=112$, $M=\frac{1}{2}\sum_{j\neq i}\binom{a_{ij}^{(2)}}{2}a_{ij}$, and $F_4=56\sum_{j\neq i}\binom{a_{ij}^{(2)}}{2}a_{ij}$.

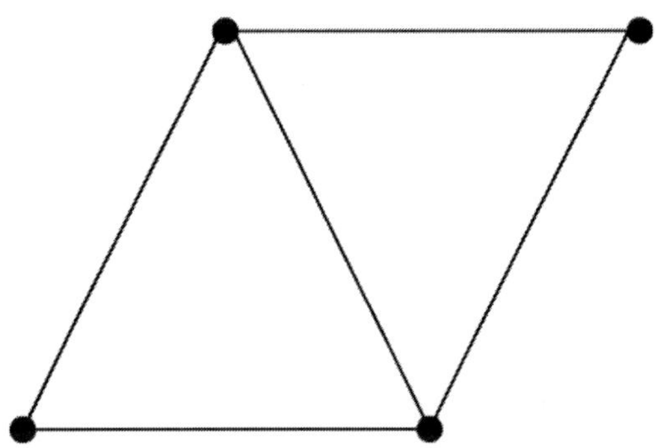

Figure 4. Closed walks of length 7 type 4.

Case 5: For the configuration of Figure 5(a) $N=14$, $M=\frac{1}{2}\sum_{j\neq i}a_{ii}^{(3)}a_{ij}^{(2)}$,. Let P_1 denote the number of subgraphs of *G* that have the same configuration as the graph of Figure 5(b) and are counted in *M*. Thus $P_1=6\times\left(\frac{1}{126}F_1\right)$, where $\frac{1}{126}F_1$ is the number of subgraphs of *G* that have the same configuration as the graph of Figure 5(b) and 6 is the number of times that this subgraph is counted in *M*. Let P_2denote the number of subgraphs of *G* that have the same configuration as the graph of Figure 5(c) and are counted in *M*. nThus

$P_2 = 2\times\left(\frac{1}{84}F_2\right)$, where $\frac{1}{84}F_2$ is the number of subgraphs of G that have the same configuration as the graph of Figure 5(c) and 2 is the number of times that this subgraph is counted in M. Let P_3 denote the number of subgraphs of G that have the same configuration as the graph of Figure 5(d) and are counted in M.

Thus $P_3 = 4\times\left(\frac{1}{112}F_4\right)$, where $\frac{1}{112}F_4$ is the number of subgraphs of G that have the same configuration as the graph of Figure 5(d) and 4 is the number of times that this subgraph is counted in M. Consequently,

$$F_5 = 7\sum_{j\neq i} a_{ii}^{(3)} a_{ij}^{(2)} - \frac{2}{3}F_1 - \frac{1}{3}F_2 - \frac{1}{2}F_4.$$

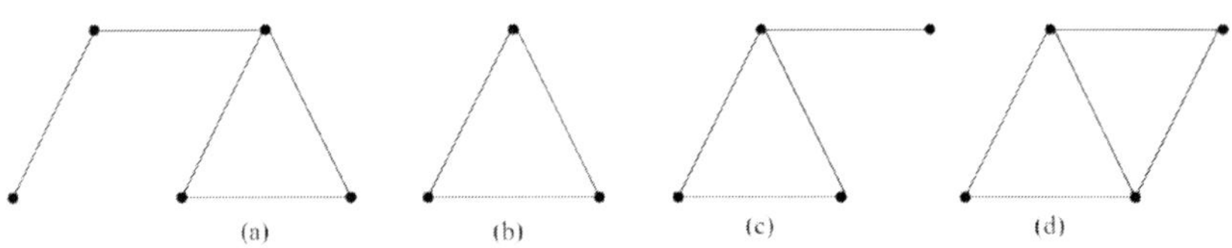

Figure 5. Closed walks of length 7 type 5.

Case 6: For the configuration of Figure 6(a)

$N = 14,\quad M = \frac{1}{2}\sum_{j\neq i} a_{ij}^{(2)} a_{ij}\left(d_j - 2\right)\left(d_i - 2\right)$. Let P_1 denote the number of subgraphs of G that have the same configuration as the graph of Figure 6(b) and are counted in M.

Thus $P_1 = 2\times\left(\frac{1}{112}F_4\right)$, where $\frac{1}{112}F_4$ is the number of subgraphs of G that have the same configuration as the graph of Figure 6(b) and 2 is the number of times that this subgraph is counted in M. Consequently,

$$F_6 = 7\sum_{j\neq i} a_{ij}^{(2)} a_{ij}\left(d_j - 2\right)\left(d_i - 2\right) - \frac{1}{4}F_4.$$

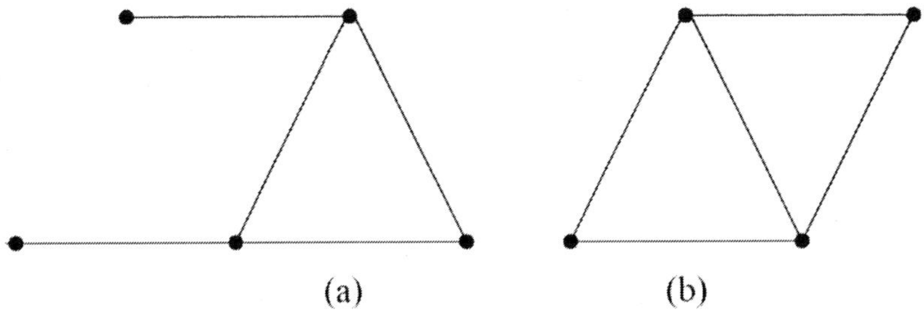

Figure 6. Closed walks of length 7 type 6.

Case 7: For the configuration of Figure 7 $N=70$, $M=\frac{1}{10}\left[tr\left(A^5\right)+5tr\left(A^3\right)-5\sum_{i=1}^{n}d_i a_{ii}^{(3)}\right]$ (see Theorem 3) and

$$F_7=7\,tr\left(A^5\right)+35\,tr\left(A^3\right)-35\sum_{i=1}^{n}d_i a_{ii}^{(3)}.$$

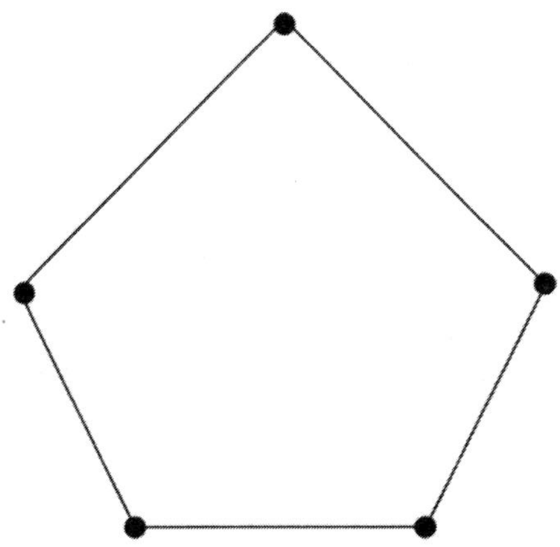

Figure 7. Closed walks of length 7 type 7.

Case 8: For the configuration of Figure 8(a)

$N=42$, $M=\frac{1}{2}\sum_{j\neq i}a_{ij}^{(2)}\left(d_j-a_{ij}-1\right)\left(a_{ij}a_{ij}^{(2)}\right)$ (see Theorem 5).

Let P_1 denote the number of subgraphs of G that have the same configuration as the graph of Figure 8(b) and are counted in M. Thus

$P_1 = 4\times\left(\frac{1}{112}F_4\right)$, where $\frac{1}{112}F_4$ is the number of subgraphs of G that have the same configuration as the graph of Figure 8(b) and 4 is the number of times that this subgraph is counted in M. Consequently, $F_8 = 21\sum_{j\neq i} a_{ij}^{(2)}\left(d_j - a_{ij} - 1\right)\left(a_{ij}a_{ij}^{(2)}\right) - \frac{3}{2}F_4$.

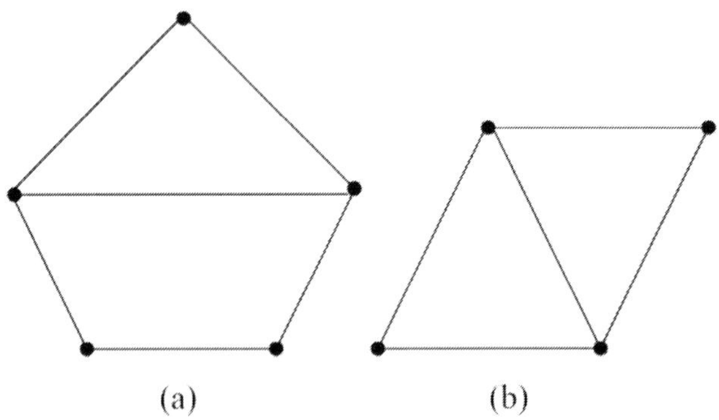

Figure 8. Closed walks of length 7 type 8.

Case 9: For the configuration of Figure 9(a), N=14,

$$M = \frac{1}{2}\Bigg[\sum_{i=1}^{n}\left(a_{ii}^{(5)} - 5a_{ii}^{(3)} - 2\left(d_i - 2\right)a_{ii}^{(3)}\right)\left(d_i - 2\right)$$

$$-2\sum_{j\neq i} a_{ij}^{(2)}a_{ij}\left(d_j - 2\right)\left(d_i - 2\right) - 2\sum_{j\neq i} a_{ij}\left(d_i - 2\right)\left(\frac{1}{2}a_{jj}^{(3)} - a_{ij}a_{ij}^{(2)}\right)\Bigg]$$

(see Theorem 11). Let P_1 denote the number

of subgraphs of G that have the same configuration as the graph of Figure 9(b) and are counted in M. Thus $P_1 = 2\times\left(\frac{1}{42}F_8\right)$, where $\frac{1}{42}F_8$ is the number subgraphs of G that have the same configuration as the graph of Figure 9(b) and 2 is the number of times that this subgraph is counted in M. Consequently,

$$F_9 = 7\sum_{i=1}^{n}\left(a_{ii}^{(5)} - 5a_{ii}^{(3)} - 2(d_i - 2)a_{ii}^{(3)}\right)(d_i - 2) - 14\sum_{j\neq i} a_{ij}^{(2)} a_{ij}(d_j - 2)(d_i - 2)$$
$$-14\sum_{j\neq i} a_{ij}(d_i - 2)\left(\frac{1}{2}a_{ij}^{(3)} - a_{ij}a_{ij}^{(2)}\right) - \frac{2}{3}F_8$$
.

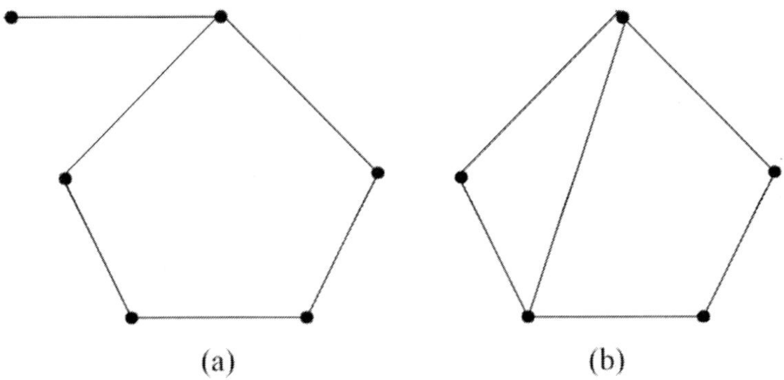

Figure 9. Closed walks of length 7 type 9.

Case 10: For the configuration of Figure 10, $N = 84$, $M = \frac{1}{2}\sum_{j\neq i}\binom{a_{ij}^{(2)}}{3}a_{ij}$ and

$$F_{10} = 42\sum_{j\neq i}\binom{a_{ij}^{(2)}}{3}a_{ij}$$
.

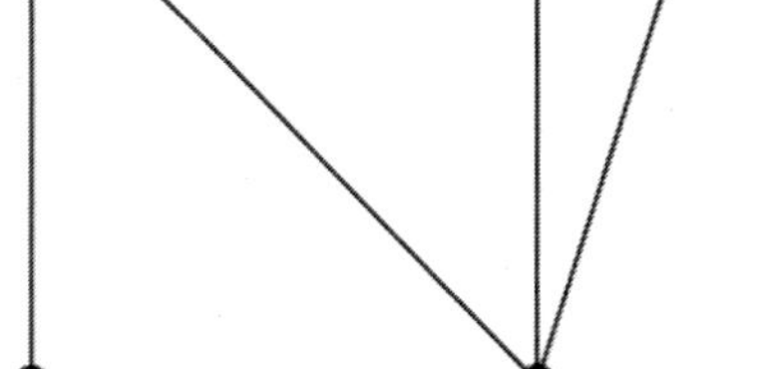

Figure 10. Closed walks of length 7 type 10.

Case 11: For the configuration of Figure 11(a) $N = 28$, $M = \frac{1}{2}\sum_{j\neq i}\binom{a_{ij}^{(2)}}{2}a_{ii}^{(3)}$. Let P_1 denote the number of subgraphs of G that have the same configuration as the graph of Figure 11(b) and are counted in M. Thus $P_1 = 2\times\left(\frac{1}{42}F_8\right)$, where $\frac{1}{42}F_8$ is the number of subgraphs of G that have the same configuration as the graph of Figure 11(b) and 2 is the number of times that this subgraph is counted in M. Let P_2 denote the number of subgraphs of G that have the same configuration as the graph of Figure 11(c) and are counted in M. Thus $P_2 = 6\times\left(\frac{1}{112}F_4\right)$, where $\frac{1}{112}F_4$ is the number of subgraphs of G that have the same configuration as the graph of Figure 11(c) and 6 is the number of times that this subgraph is counted in M. Let P_3 denote the number of subgraphs of G that have the same configuration as the graph of Figure 11(d) and are counted in M. Thus $P_3 = 6\times\left(\frac{1}{84}F_{10}\right)$, where $\frac{1}{84}F_{10}$ is the number of subgraphs of G that have the same configuration as the graph of Figure 11(d) and 6 is the number of times that this subgraph is counted in nM. Consequently,

$$F_{11} = 14\sum_{j\neq i}\binom{a_{ij}^{(2)}}{2}a_{ii}^{(3)} - \frac{4}{3}F_8 - \frac{3}{2}F_4 - 2F_{10}$$

.

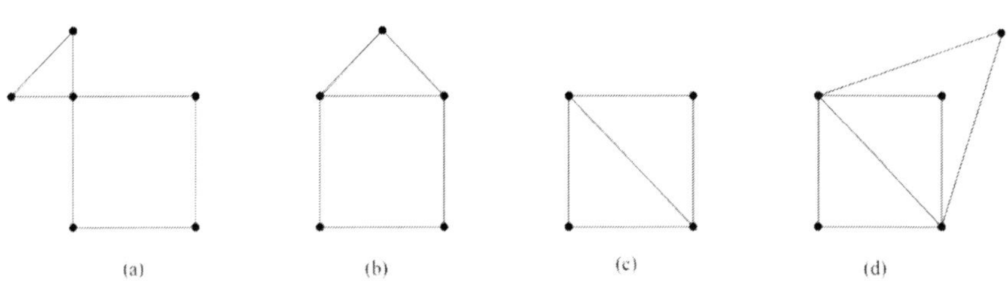

Figure 11. Closed walks of length 7 type 11.

Now, we add the values of F_n arising from the above cases and determine x. By putting the value of x in

$\frac{1}{14}\left(tr\left(A^7\right)-x\right)$ we get the desired formula. □

Example 1. *In* K_7, $F_1=4410$, $F_2=35280$, $F_3=17640$, $F_4=23520$, $F_5=17640$, $F_6=17640$ $F_7=17640$, $F_8=52920$, $F_9=35280$, $F_{10}=17640$, $F_{11}=35280$, and $trA^7=279930$. So $x=274890$ and by Theorem 12, the number of cycles of length 7 in K_7 is $\frac{1}{14}(279930-274890)=360$.

3. Number of Cycles Passing the Vertex v_i

In this section we give formulae to count the number of cycles of lengths 6 and 7, each of which contain a specific vertex v_i of the graph G.

Theorem 13. *If G is a simple graph with n vertices and the adjacency matrix* $A=\left[a_{ij}\right]$*, then the number of 6-cycles each of which contains a specific vertex* v_i *of G is* $\frac{1}{2}\left(a_{ii}^{(6)}-x\right)$*, where x is equal to* $\sum_{i=1}^{17}F_i$ *in the cases that are considered below.*

Proof: The number of 6-cycles each of which contain a specific vertex v_i of the graph G is equal t $\frac{1}{2}\left(a_{ii}^{(6)}-x\right)$, where x is the number of closed walks of length 6 form the vertex v_i to v_i that are not 6-cycles.

To find x, we have 17 cases as considered below; the cases are based on the configurations-(subgraphs) that generate v_i- v_i walks of length 6 that are not cycles. In each case, N denotes the number of walks of length 6 from v_i to v_i that are not cycles in the corresponding subgraph, M denotes the number of subgraphs of G of the same configuration and F_n, (n=1,2,···) denote the total number of v_i- v_i walks of length 6 that are not cycles in all possible subgraphs of G of the same configuration. However, in the cases with more than one figure (Cases 11, 12, 13, 14, 15, 16, 17), N, M and F_n are based on the first graph of the respective figures and P_1,P_2,··· denote the number of subgraphs of G which don't have the same configuration as the first graph but are counted in M. It is clear that F_n is equal to N×(M- P_1,P_2,-···). To find N in each case, we have to include in any walk, all the edges and the vertices of the corresponding subgraphs at least once.

Case 1: For the configuration of Figure 12, N=1, $M = a_{ii}^{(2)}$ and $F_1 = a_{ii}^{(2)}$.

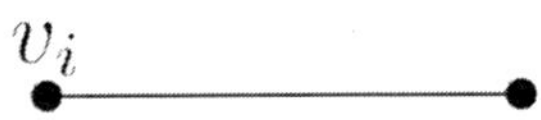

Figure 12. v_i-v_i walks of length 6 type 1.

Case 2: For the configuration of Figure 13, $N = 3$, $M = \sum_{j \neq i} a_{ij}^{(2)}$ and $F_2 = 3\sum_{j \neq i} a_{ij}^{(2)}$.

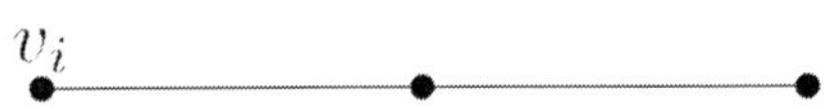

Figure 13. v_i-v_i walks of length 6 type 2.

Case 3: For the configuration of Figure 14, N=6, $M = \binom{d_i}{2}$ and $F_3 = 6\binom{d_i}{2}$.

Figure 14. v_i-v_i walks of length 6 type 3.

Case 4: For the configuration of Figure 15 $N = 1$, $M = \sum_{j \neq i} a_{ij}^{(2)} \left(d_j - a_{ij} - 1\right)$, and $F_4 = \sum_{j \neq i} a_{ij}^{(2)} \left(d_j - a_{ij} - 1\right)$. (See Theorem 7)

Figure 15. v_i-v_i walks of length 6 type 4.

Case 5: For the configuration of Figure 16 $N=2$, $M=\sum_{j\neq i} a_{ij}^{(2)}\left(d_i-a_{ij}-1\right)$, and $F_5=2\sum_{j\neq i} a_{ij}^{(2)}\left(d_i-a_{ij}-1\right)$.

Figure 16. v_i-v_i walks of length 6 type 5.

Case 6: For the configuration of Figure 17, $N=2$, $M=\sum_{j\neq i} a_{ij}\binom{d_j-1}{2}$ and $F_6=2\sum_{j\neq i} a_{ij}\binom{d_j-1}{2}$.

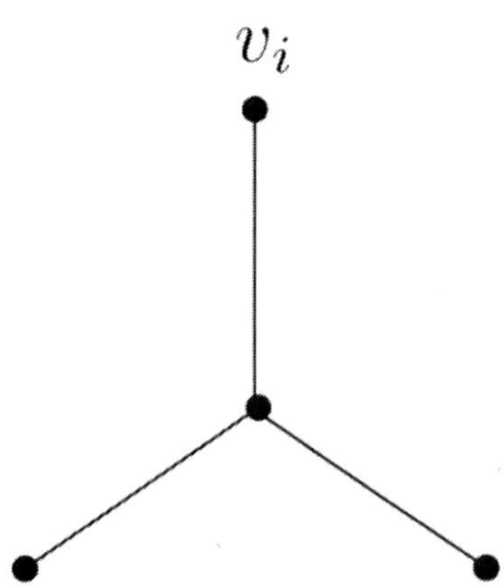

Figure 17. v_i-v_i walks of length 6 type 6.

Case 7: For the configuration of Figure 18, $N=6$, $M=\binom{d_i}{3}$ and $F_7=6\binom{d_i}{3}$.

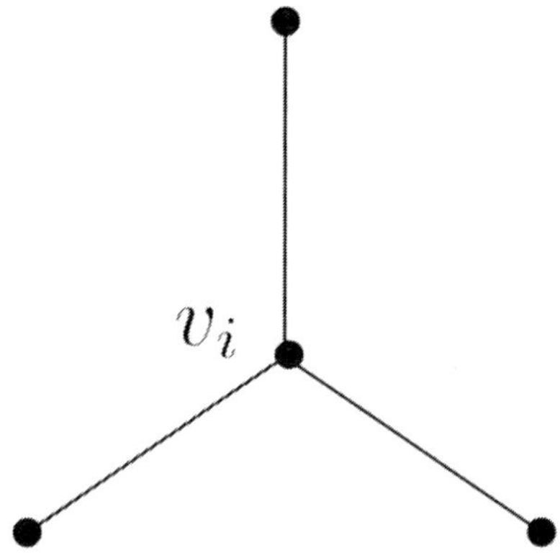

Figure 18. v_i-v_i walks of length 6 type 7.

Case 8: For the configuration of Figure 19, $N = 8$, $M = \frac{1}{2}\sum_{j \neq i} a_{ij}^{(2)} a_{ij}$ and $F_8 = 4\sum_{j \neq i} a_{ij}^{(2)} a_{ij}$.

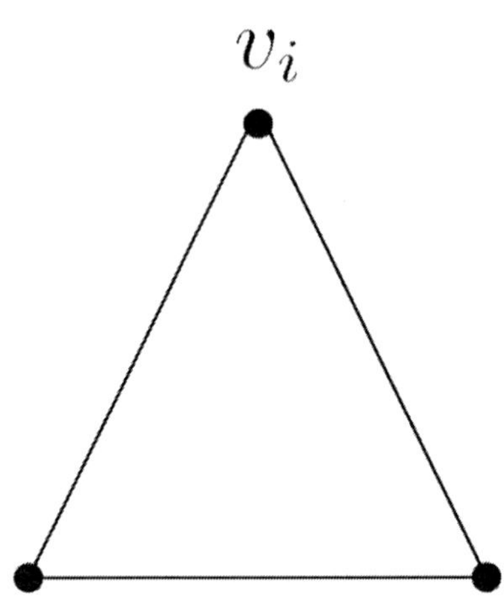

Figure 19. v_i-v_i walks of length 6 type 8.

Case 9: For the configuration of Figure 20, $N = 12$, $M = \sum_{j \neq i} \binom{a_{ij}^{(2)}}{2}$ and $F_9 = 12\sum_{j \neq i} \binom{a_{ij}^{(2)}}{2}$.

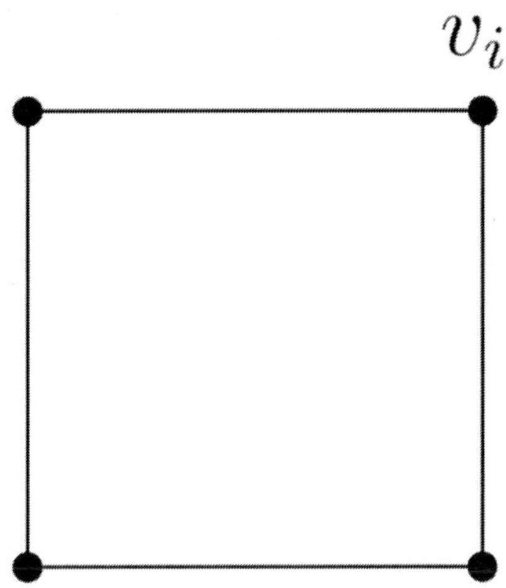

Figure 20. v_i-v_i walks of length 6 type 9.

Case 10: For the configuration of Figure 21, $N = 12$, $M = \sum_{j \neq i} \binom{a_{ij}^{(2)}}{2} a_{ij}$ and $F_{10} = 12 \sum_{j \neq i} \binom{a_{ij}^{(2)}}{2} a_{ij}$.

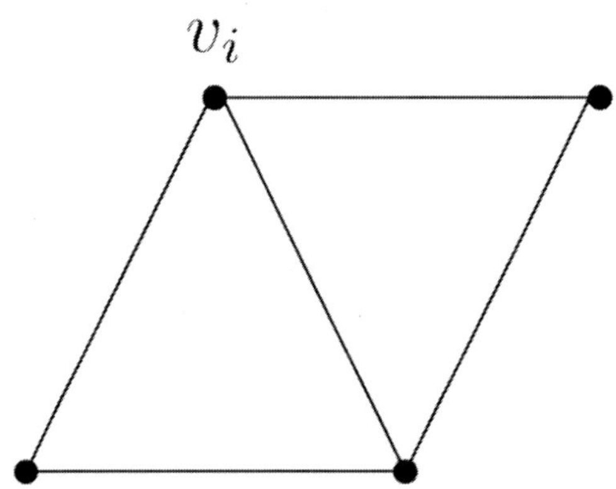

Figure 21. v_i-v_i walks of length 6 type 10.

Case 11: For the configuration of Figure 22(a), $N = 6$, $M = \frac{1}{2} \sum_{k \neq j, j, k \neq i} a_{jk}^{(2)} a_{ij} a_{ik} a_{jk}$. Let P_1 denote the number of all subgraphs of G that have the same configuration as the graph of Figure 22(b) and are counted in M. Thus $P_1 = 1 \times \left(\frac{1}{8} F_8 \right)$, where $\frac{1}{8} F_8$ is the number of subgraphs of G that have the same configuration as the graph of Figure 22(b) and this subgraph is counted only once in M. Consequently,

$$F_{11} = 3 \sum_{k \neq j, j, k \neq i} a_{jk}^{(2)} a_{ij} a_{ik} a_{jk} - \frac{3}{4} F_8$$

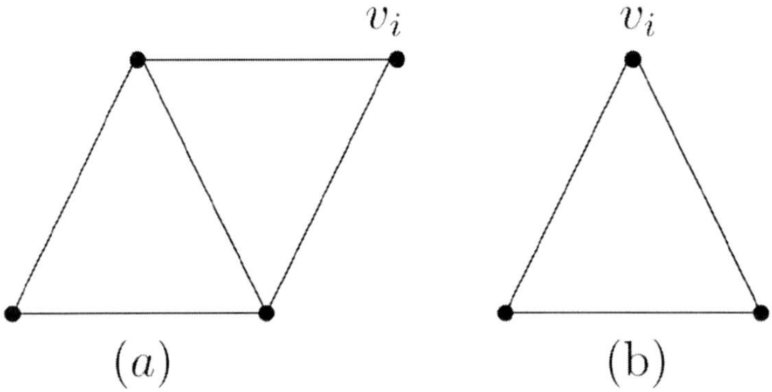

Figure 22. v_i-v_i walks of length 6 type 11.

Case 12: For the configuration of Figure 23(a),N=2 , $M = \sum_{k \neq j, j, k \neq i}^{n} \binom{a_{jk}^{(2)}}{2} a_{ij}$. Let P_1 denote the number of all subgraphs of G that have the same configuration as the graph of Figure 23(b) and are counted in M. Thus $P_1 = 2 \times \left(\frac{1}{12} F_9 \right)$, where $\frac{1}{12} F_9$ is the number of subgraphs of G that have the same configuration as the graph of Figure 23(b) and 2 is the number of times that this subgraph is counted in M. Consequently,

$$F_{12} = 2 \sum_{k \neq j, j, k \neq i} \binom{a_{jk}^{(2)}}{2} a_{ij} - \frac{1}{3} F_9$$

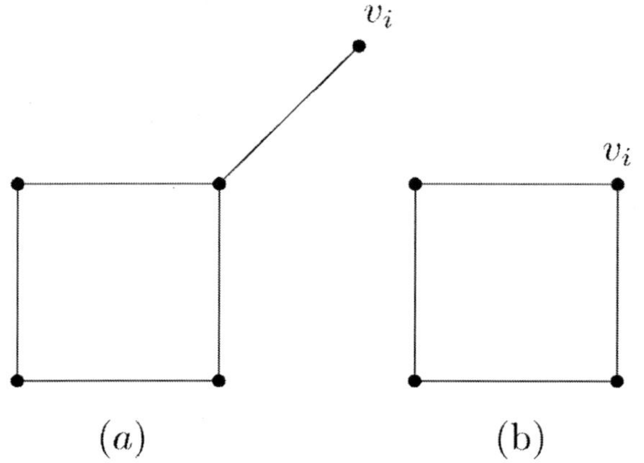

Figure 23. v_i-v_i walks of length 6 type 12.

Case 13: For the configuration of Figure 24(a) $N = 4$, $M = \sum_{j \neq i} \binom{a_{ij}^{(2)}}{2}(d_i - 2)$,. Let P_1 denote the number of all subgraphs of G that have the same configuration as the graph of Figure 24(b) and are counted in M. Thus $P_1 = 1 \times \left(\frac{1}{12} F_{10}\right)$, where $\frac{1}{12} F_{10}$ is the number of subgraphs of G that have the same configuration as the graph of Figure 24(b) and this subgraph is counted only once in M. Consequently,

$$F_{13} = 4\sum_{j \neq i} \binom{a_{ij}^{(2)}}{2}(d_i - 2) - \frac{1}{3} F_{10}$$

.

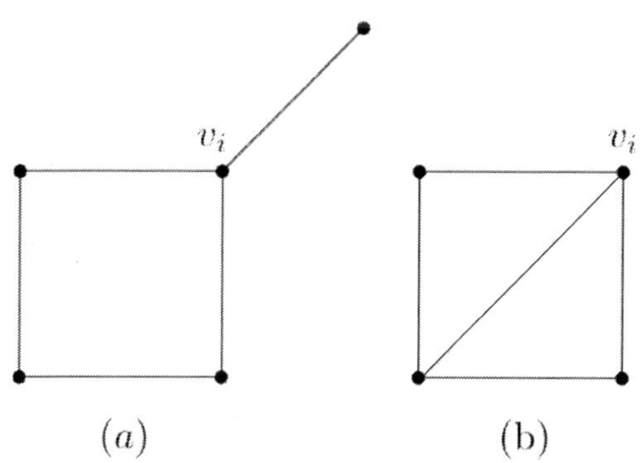

Figure 24. v_i-v_i walks of length 6 type 13.

Case 14: For the configuration of Figure 25(a), $N=2$, $M=\sum_{j\neq i}\binom{a_{ij}^{(2)}}{2}(d_j-2)$. Let P_1 denote the number of all subgraphs of G that have the same configuration as the graph of Figure 25(b) and are counted in M. Thus $P_1=1\times\left(\frac{1}{12}F_{10}\right)$, where $\frac{1}{12}F_{10}$ is the number of subgraphs of G that have the same configuration as the graph of Figure 25(b) and this subgraph is counted only once in M. Consequently, $F_{14}=2\sum_{j\neq i}\binom{a_{ij}^{(2)}}{2}(d_j-2)-\frac{1}{6}F_{10}$.

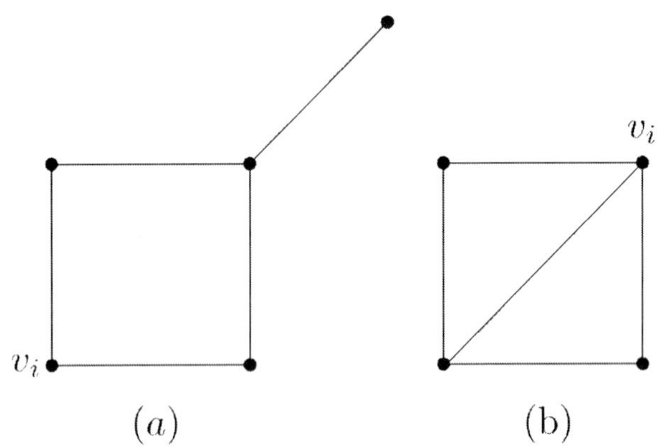

Figure 25. v_i-v_i walks of length 6 type 14.

Case 15: For the configuration of Figure 26(a), N=2, $M=\sum_{j\neq i}a_{ij}^{(2)}(d_j-a_{ij}-1)a_{ij}(d_j-2)$. Let P_1 denote the number of all subgraphs of G that have the same configuration as the graph of Figure 26(b) and are counted in M. Thus $P_1=2\times\left(\frac{1}{6}F_{11}\right)$, where $\frac{1}{6}F_{11}$ is the number of subgraphs of G that have the same configuration as the graph of Figure 26(b) and 2 is the number of times that this subgraph is counted in M. Consequently, $F_{15}=2\sum_{j\neq i}a_{ij}^{(2)}(d_j-a_{ij}-1)a_{ij}(d_j-2)-\frac{2}{3}F_{11}$.

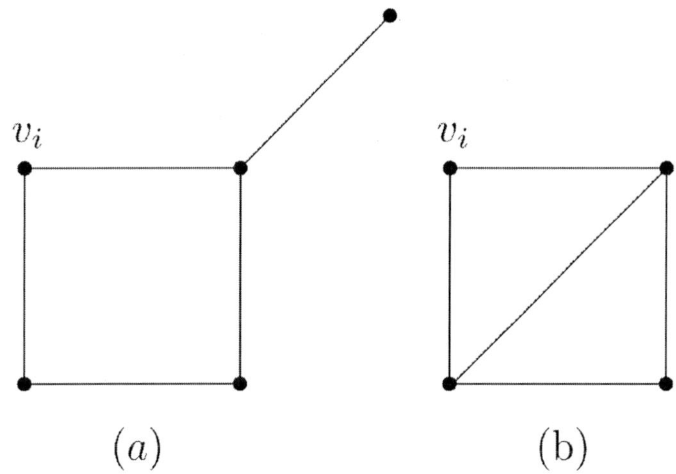

Figure 26. v_i-v_i walks of length 6 type 15.

Case 16: For the configuration of Figure 27(a), N=4, $M=\frac{1}{2}\sum_{j\neq i}a_{jj}^{(3)}a_{ij}^{(2)}a_{ij}$.

Let P_1 denote the number of all subgraphs of G that have the same configuration as the graph of Figure 27(b) and are counted in M. Thus $P_1=2\times\left(\frac{1}{8}F_8\right)$, where $\frac{1}{8}F_8$ is the number of subgraphs of G that have the same configuration as the graph of Figure 27(b) and 2 is the number of times that this subgraph is counted in M. Let P_2 denote the number of all subgraphs of G that have the same configuration as the graph of Figure 27(c) and are counted in M. Thus $P_2=2\times\left(\frac{1}{6}F_{11}\right)$, where $\frac{1}{6}F_{11}$ is the number of subgraphs of G that have the same configuration as the graph of Figure 27(c) and 2 is the number of times that this subgraph is counted in M. Let P_3denote the number of all subgraphs of G that have the same configuration as the graph of Figure 27(d) and are counted in M. Thus $P_3=2\times\left(\frac{1}{12}F_{10}\right)$, where $\frac{1}{12}F_{10}$ is the number of subgraphs of G that have the same configuration as the graph of Figure 27(d) and 2 is the number of times that this subgraph is counted in M. Consequently,

$$F_{16}=2\sum_{j\neq i}a_{jj}^{(3)}a_{ij}^{(2)}a_{ij}-F_8-\frac{4}{3}F_{11}-\frac{2}{3}F_{10}.$$

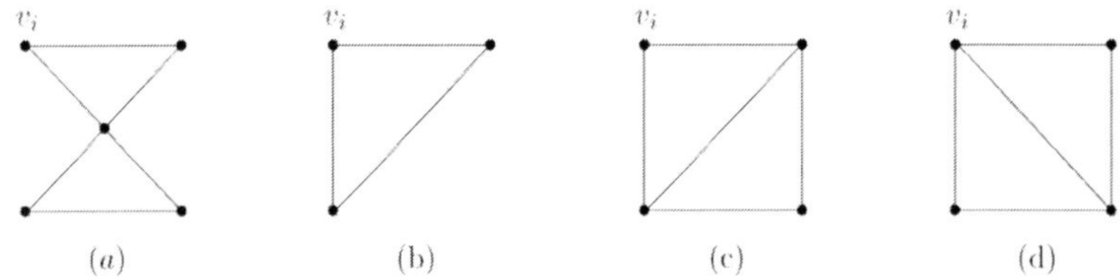

Figure 27. v_i-v_i walks of length 6 type 16.

Case 17: For the configuration of Figure 28(a), N=8, $M = \binom{\frac{1}{2}a_{ii}^{(3)}}{2}$.
Let P_1 denote the number of all subgraphs of G that have the same configuration as the graph of Figure 28(b) and are counted in M. Thus $P_1 = 1 \times \left(\frac{1}{12}F_{10}\right)$, where $\frac{1}{12}F_{10}$ is the number of subgraphs of G that have the same configuration as the graph of Figure 28(b) and this subgraph is counted only once in M. Consequently, $F_{17} = 8\binom{\frac{1}{2}a_{ii}^{(3)}}{2} - \frac{2}{3}F_{10}$.

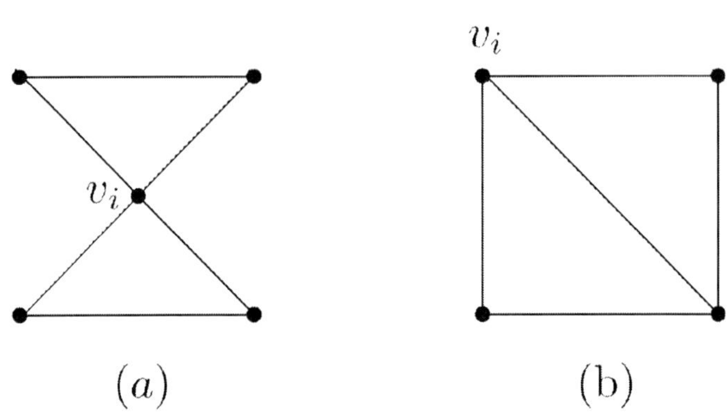

Figure 28. v_i-v_i walks of length 6 type 17.

Now we add the values of F_n arising from the above cases and determine x. Substituting the value of x in $\frac{1}{2}\left(a_{ii}^{(6)} - x\right)$ and simplifying, we get the number of 6-cycles each of which contains a specific vertex v_i of G. □

Example 2. *In the graph of* Figure 29 *we have,* $F_1 = 5,\quad F_2 = 60,\quad F_3 = 60,\quad F_4 = 60,\quad F_5 = 120,\quad F_6 = 60,$ $F_7 = 60,\quad F_8 = 80,\quad F_9 = 360,\quad F_{10} = 360,\quad F_{11} = 180,\quad F_{12} = 120,\quad F_{13} = 240,\quad F_{14} = 120,\quad F_{15} = 240,$ $F_{16} = 240,\quad F_{17} = 120$. *So, we have x=2485. Consequently, by Theorem 13, the number of 6-cycles each of which contains the vertex* v_1 *in the graph of* Figure 29*is* 60.

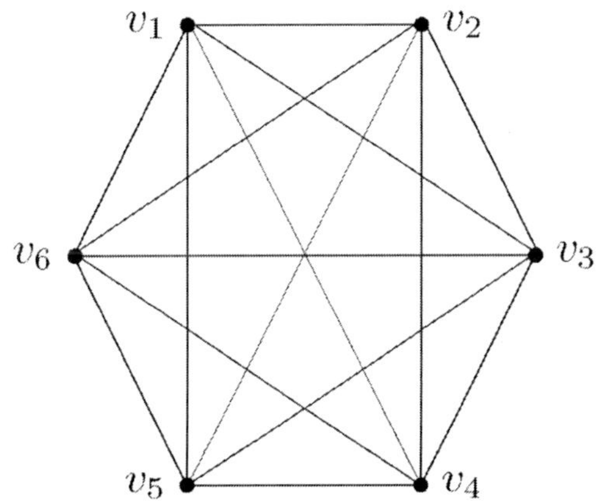

Figure 29. Complete graph with 7 vertices.

Theorem 14. *If G is a simple graph with n vertices and the adjacency matrix* $A = \left[a_{ij}\right]$, *then the number of 7-cycles each of which contains a specific vertex* v_i *of G is* $\frac{1}{2}\left(a_{ii}^{(7)} - x\right)$, *where x is equal to* $\sum_{i=1}^{30} F_i$ *in the cases that are considered below.*

Proof: The number of 7-cycles each of which contains a specific vertex v_i of the graph G is equal to $\frac{1}{2}\left(a_{ii}^{(7)} - x\right)$, where x is the number of closed walks of length 7 form the vertex v_i to v_i that are not 7-cycles.

To find x, we have 30 cases as considered below; the cases are based on the configurations-(subgraphs) that generate v_i- v_i walks of length 7 that are not cycles. In each case, N denotes the number of walks of length 7 from v_i to v_i that are not cycles in the corresponding subgraph, M denotes the number of subgraphs of G of the same configuration and F_n, (n=1,2,···) denote the total number of v_i- v_i walks of length 7 that are not cycles in all possible subgraphs of G of the same configuration. However, in the cases with more than one figure (Cases 9, 10, ⋯, 18, 20, ⋯, 30), N, M and F_n are based on the first graph of the respective figures and P, P , ⋯ denote the number of subgraphs of G which do not have the same configuration as the first graph but are counted in M. It is clear that F_n is equal to $N\times(M-P_1-P_2-\cdots)$. To find N in each case, we have to include in any walk, all the edges and the vertices of the corresponding subgraphs at least once.

Case 1: For the configuration of Figure 30, N=42, $M=\frac{1}{2}\sum_{j\neq i} a_{ij}^{(2)}a_{ij}$ and $F_1=21\sum_{j\neq i} a_{ij}^{(2)}a_{ij}$.

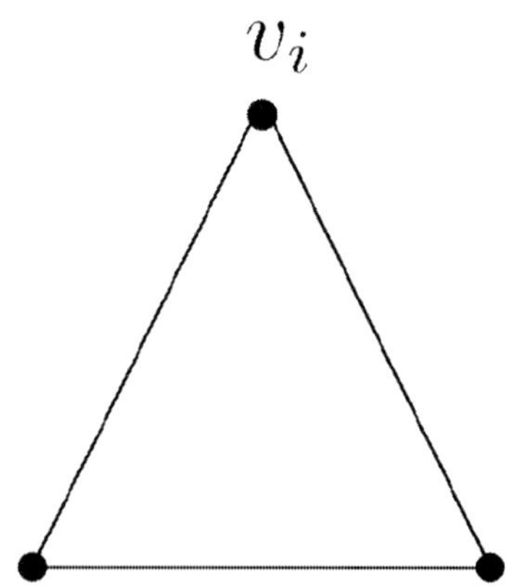

Figure 30. v_i-v_i walks of length 7 type 1.

Case 2: For the configuration of Figure 31, N=34, $N=34$, $M=\frac{1}{2}a_{ii}^{(3)}(d_i-2)$ and $F_2=17a_{ii}^{(3)}(d_i-2)$.

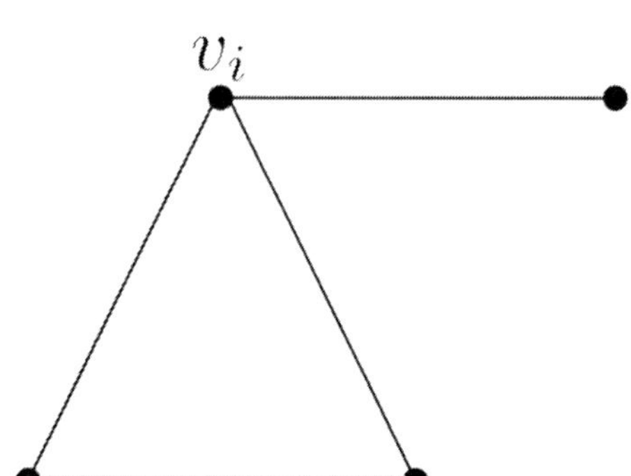

Figure 31. v_i-v_i walks of length 7 type 2.

Case 3: For the configuration of Figure 32, N=18, $M=\sum_{j\neq i} a_{ij}^{(2)} a_{ij}\left(d_j-2\right)$ and $F_3=18\sum_{j\neq i} a_{ij}^{(2)} a_{ij}\left(d_j-2\right)$.

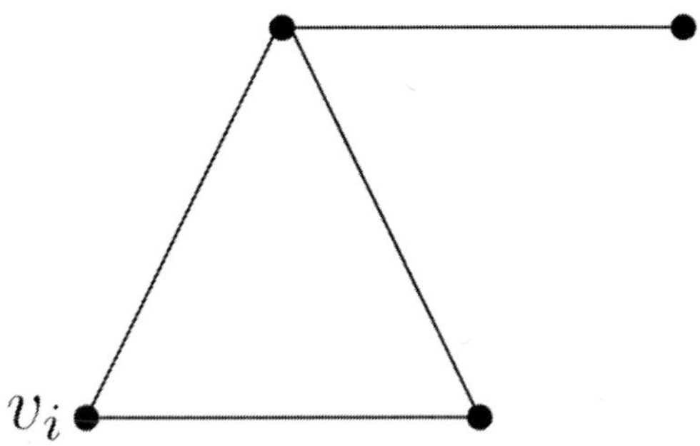

Figure 32. v_i-v_i walks of length 7 type 3.

Case 4: For the configuration of Figure 33, N=4, $M=\sum_{j\neq i}\left(\frac{1}{2}a_{jj}^{(3)}a_{ij}-a_{ij}^{(2)}a_{ij}\right)\left(d_j-3\right)$ and

$F_4=4\sum_{j\neq i}\left(\frac{1}{2}a_{jj}^{(3)}a_{ij}-a_{ij}^{(2)}a_{ij}\right)\left(d_j-3\right)$.

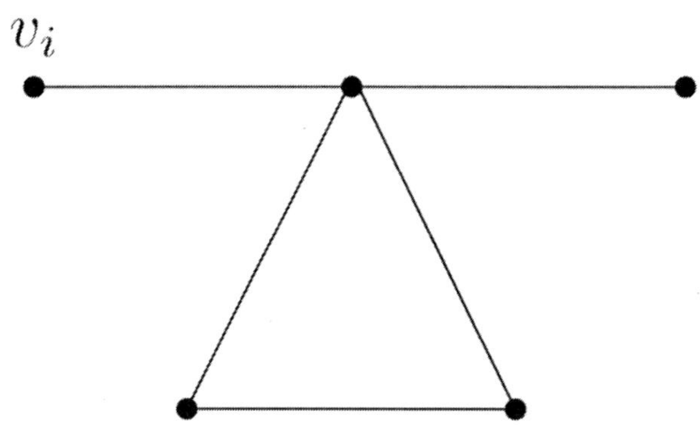

Figure 33. v_i-v_i walks of length 7 type 4.

Case 5: For the configuration of Figure 34, N=12, $M=\frac{1}{2}a_{ii}^{(3)}\binom{d_i-2}{2}$ and $F_5=6a_{ii}^{(3)}\binom{d_i-2}{2}$.

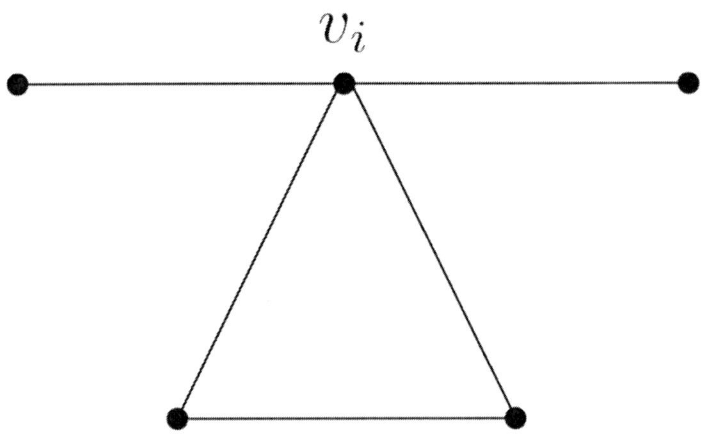

Figure 34. v_i-v_i walks of length 7 type 5.

Case 6: For the configuration of Figure 35, N=4, $M=\sum_{j\neq i} a_{ij}^{(2)} a_{ij} \binom{d_j-2}{2}$ and $F_6 = 4\sum_{j\neq i} a_{ij}^{(2)} a_{ij} \binom{d_j-2}{2}$.

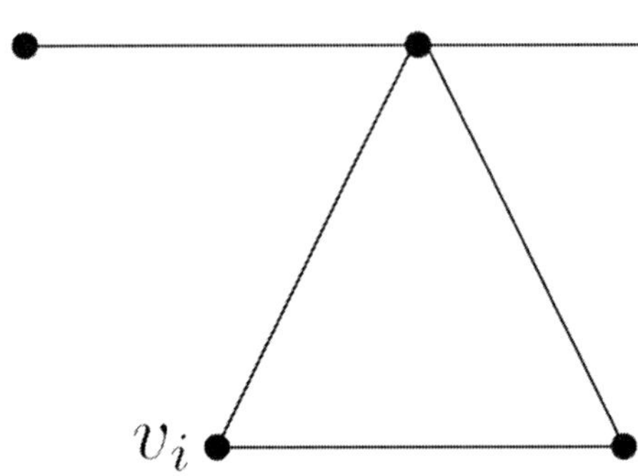

Figure 35. v_i-v_i walks of length 7 type 6.

Case 7: For the configuration of Figure 36, N=32, $M=\sum_{j\neq i} \binom{a_{ij}^{(2)}}{2} a_{ij}$ and $F_7 = 32\sum_{j\neq i} \binom{a_{ij}^{(2)}}{2} a_{ij}$.

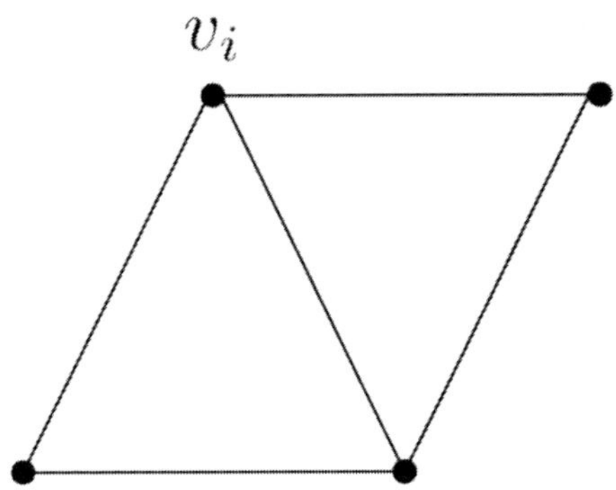

Figure 36. v_i-v_i walks of length 7 type 7.

Case 8: For the configuration of Figure 37, N=24, $M=\sum_{j\neq i}\binom{a_{ij}^{(2)}}{3}a_{ij}$, $F_8=24\sum_{j\neq i}\binom{a_{ij}^{(2)}}{3}a_{ij}$.

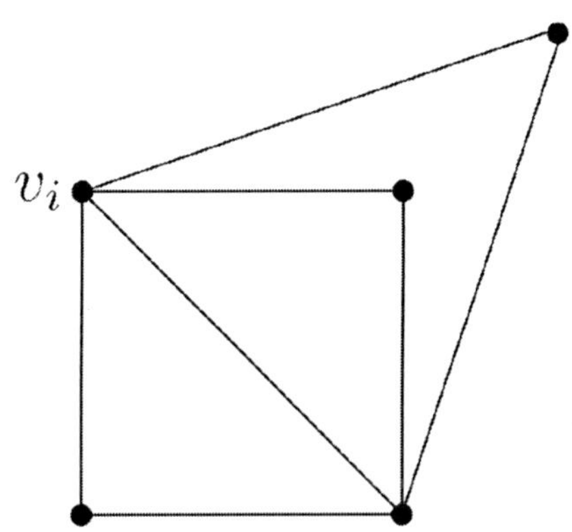

Figure 37. v_i-v_i walks of length 7 type 8.

Case 9: For the configuration of Figure 38(a), N=12, $M=\frac{1}{2}\sum_{k\neq j,j,k\neq i}\binom{a_{jk}^{(2)}}{2}a_{jk}a_{ij}a_{ik}$. Let P_1 denote the number of all subgraphs of G that have the same configuration as the graph of Figure 38(b) and are counted in M. Thus $P_1=1\times\left(\frac{1}{24}F_{10}\right)$, where $\frac{1}{24}F_{10}$ is the number of subgraphs of G that have the same configuration as the graph of Figure

38(b) and this subgraph is counted only once in *M*. Consequently,

$$F_9 = 6\sum_{k\neq j,j,k\neq i}\binom{a_{jk}^{(2)}}{2}a_{jk}a_{ij}a_{ik} - \frac{1}{2}F_{10}$$

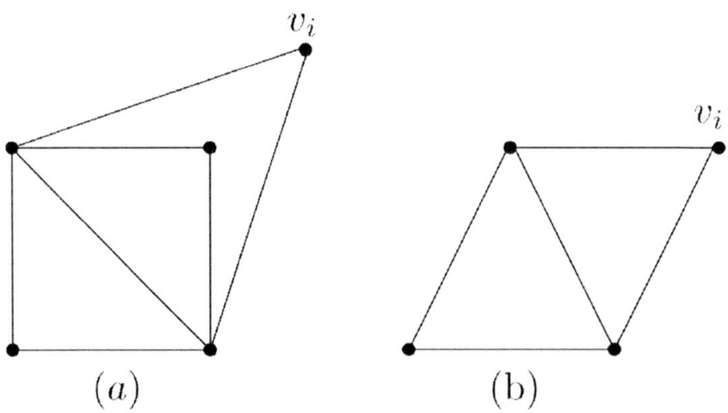

Figure 38. v_i-v_i walks of length 7 type 9.

Case 10: For the configuration of Figure 39(a), *N*=24, $M = \frac{1}{2}\sum_{k\neq j,j,k\neq i} a_{jk}^{(2)}a_{ij}a_{ik}a_{jk}$. Let P_1 denote the number of all subgraphs of G that have the same configuration as the graph of Figure 39(b) and are counted in *M*. Thus $P_1 = 1\times\left(\frac{1}{42}F_1\right)$, where $\frac{1}{42}F_1$ is the number of subgraphs of *G* that have the same configuration as the graph of Figure 39(b) and this subgraph is counted only once in *M*. Consequently,

$$F_{10} = 12\sum_{k\neq j,j,k\neq i} a_{jk}^{(2)}a_{ij}a_{ik}a_{jk} - \frac{4}{7}F_1$$

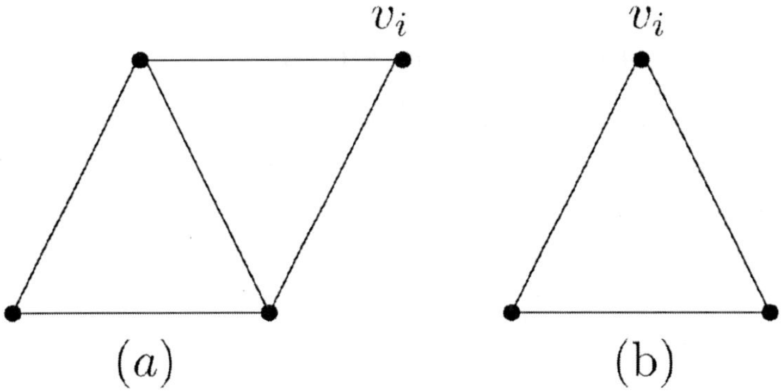

Figure 39. v_i-v_i walks of length 7 type 10.

Case 11: For the configuration of Figure 40(a), N=14, $M=\frac{1}{2}\sum_{j\neq i} a_{jj}^{(3)}a_{ij}$.
Let P_1 denote the number of all subgraphs of G that have the same configuration as the graph of Figure 40(b) and are counted in M. Thus $P_1 = 2\times\left(\frac{1}{42}F_1\right)$, where $\frac{1}{42}F_1$ is the number of subgraphs of G that have the same configuration as the graph of Figure 40(b) and 2 is the number of times that this subgraph is counted in M. Consequently,

$$F_{11} = 7\sum_{j\neq i} a_{jj}^{(3)}a_{ij} - \frac{2}{3}F_1.$$

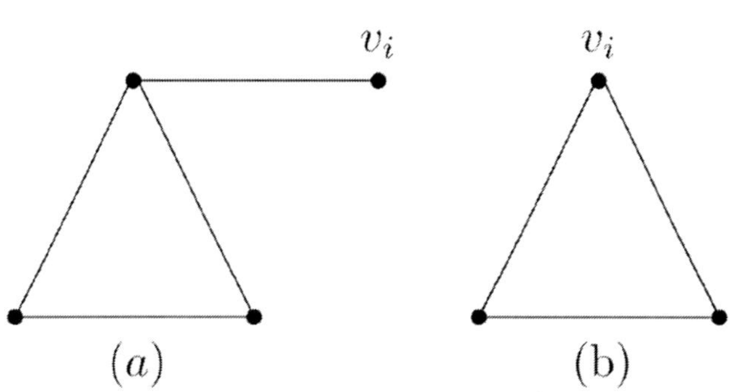

Figure 40. v_i-v_i walks of length 7 type 11.

Case 12: For the configuration of Figure 41(a), N=2, $M=\frac{1}{2}\sum_{j\neq i} a_{ij}^{(2)} a_{jj}^{(3)}$. Let P_1 denote the number of all subgraphs of G that have the same configuration as the graph of Figure 41(b) and are counted in M. Thus $P_1 = 2\times\left(\frac{1}{42}F_1\right)$, where $\frac{1}{42}F_1$ is the number of subgraphs of G that have the same configuration as the graph of Figure 41(b) and 2 is the number of times that this subgraph is counted in M. Let P_2 denote the number of all subgraphs of G that have the same configuration as the graph of Figure 41(c) and are counted in M. Thus $P_2 = 2\times\left(\frac{1}{14}F_{11}\right)$, where $\frac{1}{14}F_{11}$ is the number of subgraphs of G that have the same configuration as the graph of Figure 41(c) and 2 is the number of times that this subgraph is counted in M. Let P_2denote the number of all subgraphs of G that have the same configuration as the graph of Figure 41(d) and are counted in M. Thus $P_2 = 2\times\left(\frac{1}{14}F_{11}\right)$, where $\frac{1}{14}F_{11}$ is the number of subgraphs of G that have the same configuration as the graph of Figure 41(d) and 2 is the number of times that this subgraph is counted in M. Consequently, $F_{12}=\sum_{j\neq i} a_{ij}^{(2)} a_{jj}^{(3)} - \frac{2}{21}F_1 - \frac{2}{7}F_{11} - \frac{1}{8}F_7$.

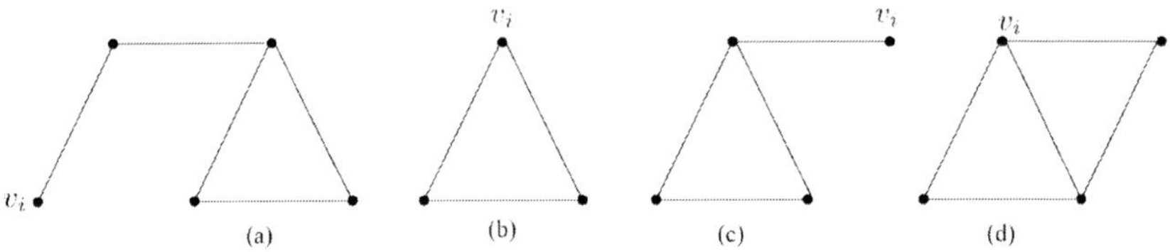

Figure 41. v_i-v_i walks of length 7 type 12.

Case 13: For the configuration of Figure 42(a), N=4, $M=\frac{1}{2}\sum_{j\neq i} a_{jj}^{(3)} a_{ij}\left(d_i-1\right)$. Let P_1 denote the number of all subgraphs of G that have the same configuration as the graph of Figure 42(b) and are counted in M. Thus $P_1 = 2\times\left(\frac{1}{42}F_1\right)$, where $\frac{1}{42}F_1$ is the number of subgraphs of G that have the same configuration as the graph of Figure 42(b) and 2 is the number of times that this subgraph is counted in M. Let P_2 denote the number of all subgraphs of G that have the same configuration as the graph of Figure 42(c) and are counted in M. Thus $P_2 = 2\times\left(\frac{1}{34}F_2\right)$, where $\frac{1}{34}F_2$ is the number of subgraphs of G that have the sameconfiguration as the graph of Figure 42(c) and 2 is the number of times that this subgraph is counted in M. Let P_3 denote the number of all subgraphs of G that have the same configuration as the graph of Figure 42(d) and are counted in M. Thus $P_3 = 2\times\left(\frac{1}{24}F_{10}\right)$, where $\frac{1}{24}F_{10}$ is the number of subgraphs of G that have the same configuration as the graph of Figure 42(d) and 2 is the number of times that this subgraph is counted in M. Consequently, $F_{13} = 2\sum_{j\neq i} a_{jj}^{(3)} a_{ij}\left(d_i - 1\right) - \frac{4}{21}F_1 - \frac{4}{17}F_2 - \frac{1}{3}F_{10}$.

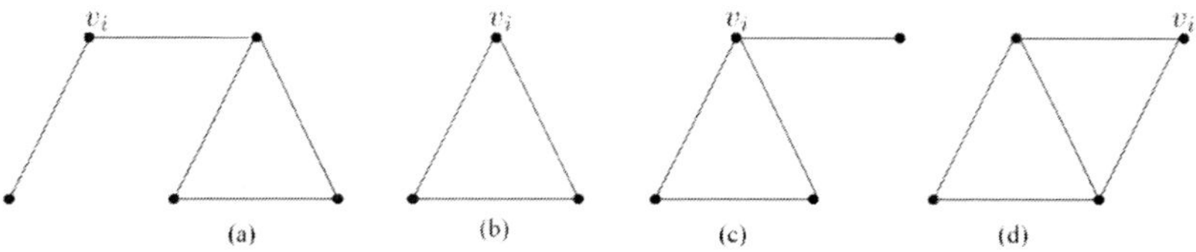

Figure 42. v_i-v_i walks of length 7 type 13.

Case 14: For the configuration of Figure 43(a), N=4, $M = \frac{1}{2}\sum_{j\neq i} a_{ii}^{(3)} a_{ij}^{(2)}$. Let P_1 denote the number of all subgraphs of G that have the same configuration as the graph of Figure 43(b) and are counted in M. Thus $P_1 = 2\times\left(\frac{1}{42}F_1\right)$, where $\frac{1}{42}F_1$ is the number of subgraphs of G that have the same configuration as the graph of Figure 43(b) and 2 is the number of times that this subgraph is counted in M. Let P_2 denote the number of all subgraphs of G that have the same configuration as the graph of Figure 43(c) and are counted in M. Thus $P_2 = 1\times\left(\frac{1}{18}F_3\right)$, where $\frac{1}{18}F_3$ is the number of subgraphs of G that have the same configuration as the graph of Figure 43(c) and this subgraph is counted only once in M. Let P_3 denote the number of all subgraphs of G that have the same configuration as the graph of Figure 43(d) and are counted in M. Thus $P_3 = 2\times\left(\frac{1}{32}F_7\right)$, where $\frac{1}{32}F_7$ is the number of subgraphs of G that have the same configuration as the graph of Figure 43(d) and 2 is the number of times that this subgraph is counted in M. Consequently, $F_{14} = 2\sum_{j\neq i} a_{ii}^{(3)}a_{ij}^{(2)} - \frac{4}{21}F_1 - \frac{2}{9}F_3 - \frac{1}{4}F_7$.

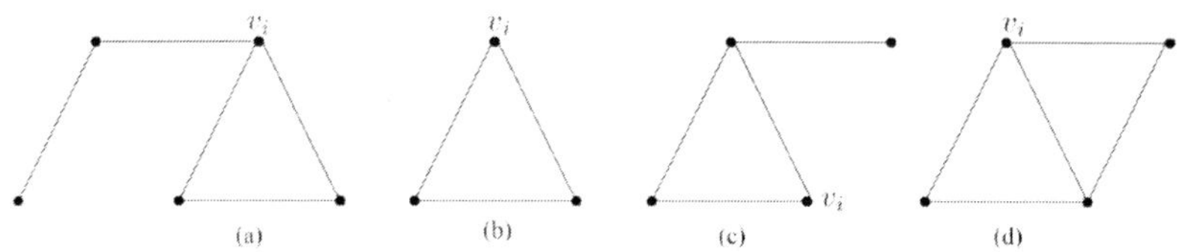

Figure 43. v_i-v_i walks of length 7 type 14.

Case 15: For the configuration of Figure 44(a), N=2, $M = \sum_{j\neq k,\, j,k\neq i} a_{ij}^{(2)}a_{jk}^{(2)}a_{ij}$. Let P_1 denote the number of all subgraphs of G that have the same configuration as the graph of Figure 44(b) and are counted in M. Thus $P_1 = 2\times\left(\frac{1}{42}F_1\right)$, where $\frac{1}{42}F_1$ is the number of subgraphs of G that have the same configuration as the graph of Figure 44(b) and 2 is the number of times that this subgraph is counted in M. Let P_2 denote the number of all subgraphs of G that have the same configuration as the graph of Figure 44(c) and are counted in M. Thus $P_2 = 2\times\left(\frac{1}{34}F_2\right)$, where $\frac{1}{34}F_2$ is the number of subgraphs of G that have the same configuration as the graph of Figure 44(c) and 2 is the number of times that this subgraph is counted in M. Let P_3 denote the number of all subgraphs of G that have the same configuration as the graph of Figure 44(d) and are counted in M. Thus $P_3 = 2\times\left(\frac{1}{24}F_{10}\right)$, where $\frac{1}{24}F_{10}$ is the number of subgraphs of G that have the same configuration as the graph of Figure 44(d) and 2 is the number of times that this subgraph is counted in M. Let P_4 denote the number of all subgraphs of G that have the same configuration as the graph of Figure 44(e) and are counted in M. Thus $P_4 = 1\times\left(\frac{1}{18}F_3\right)$, where $\frac{1}{18}F_3$ is the number of subgraphs of G that have the same configuration as the graph of Figure 44(e) and 1 is the number of times that this subgraph is counted in M. Consequently,

$$F_{15} = 2\sum_{j\neq k,\, j,k\neq i} a_{ij}^{(2)} a_{jk}^{(2)} a_{ij} - \frac{2}{21}F_1 - \frac{2}{17}F_2 - \frac{1}{6}F_{10} - \frac{1}{9}F_3.$$

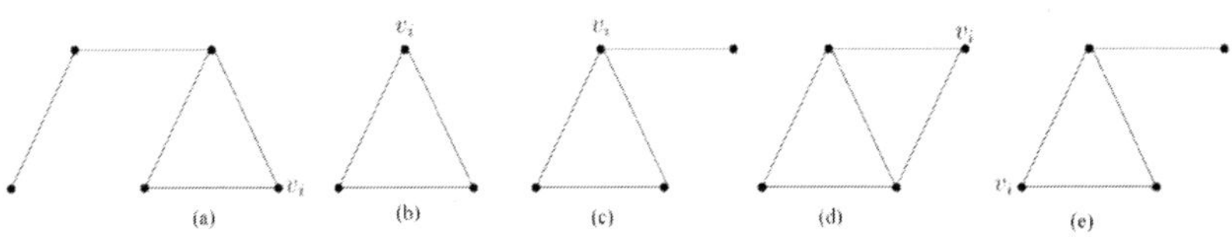

Figure 44. v_i-v_i walks of length 7 type 15.

Case 16: For the configuration of Figure 45(a), N=2, $M = \sum_{j\neq k,\, j,k\neq i} a_{jk}^{(2)} a_{jk} a_{ij}(d_k - 2)$. Let P_1 denote the number of all subgraphs of G that have the same configuration as the graph of Figure 45(b) and are counted in M. Thus $P_1 = 2\times\left(\frac{1}{24}F_{10}\right)$, where $\frac{1}{24}F_{10}$ is the number of subgraphs of G that have the same configuration as the graph of Figure 45(b) and 2 is the number of times that this subgraph is counted in M. Let P_2 denote the number of all subgraphs of G that have the same configuration as the graph of Figure 45(c) and are counted in M. Thus $P_2 = 1\times\left(\frac{1}{18}F_3\right)$, where $\frac{1}{18}F_3$ is the number of subgraphs of G that have the same configuration as the graph of Figure 45(c) and 1 is the number of times that this subgraph is counted in M. Consequently,

$$F_{16} = 2\sum_{j\neq k,\, j,k\neq i} a_{jk}^{(2)} a_{jk} a_{ij}(d_k - 2) - \frac{1}{6}F_{10} - \frac{1}{9}F_3.$$

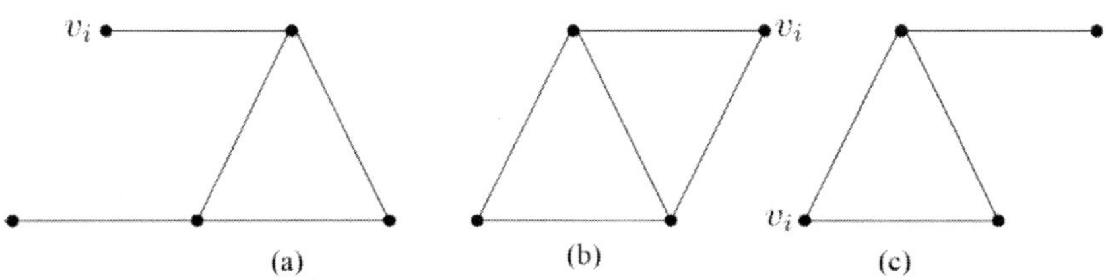

Figure 45. v_i-v_i walks of length 7 type 16.

Case 17: For the configuration of Figure 46(a), N=4, $M=\sum_{j\neq i} a_{ij}^{(2)} a_{ij}(d_j-2)(d_i-2)$. Let P_1 denote the number of all subgraphs of G that have the same configuration as the graph of Figure 46(b) and are counted in M. Thus $P_1=2\times\left(\frac{1}{32}F_7\right)$, where $\frac{1}{32}F_7$ is the number of subgraphs of G that have the same configuration as the graph of Figure 46(b) and 2 is the number of times that this subgraph is counted in M. Consequently,

$$F_{17}=4\sum_{j\neq i} a_{ij}^{(2)} a_{ij}(d_j-2)(d_i-2)-\frac{1}{4}F_7.$$

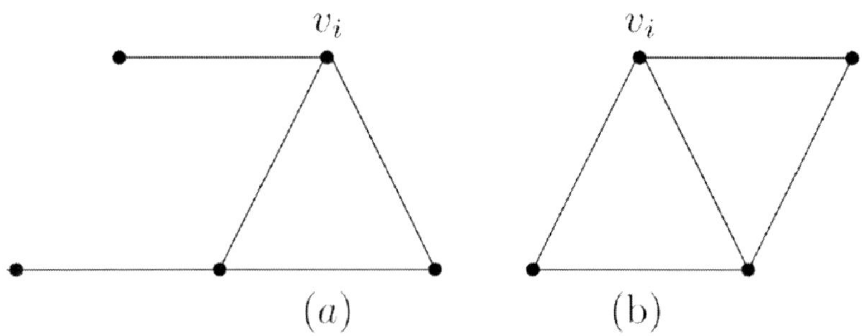

Figure 46. v_i-v_i walks of length 7 type 17.

Case 18: For the configuration of Figure 47(a), N=2, $M=\frac{1}{2}\sum_{k\neq j, j, k\neq i}(d_j-2)(d_k-2)a_{ij}a_{ik}a_{jk}$. Let P_1 denotes the number of all subgraphs of G that have the same configuration as the graph of Figure 47(b) and are counted in M. Thus $P_1=1\times\left(\frac{1}{24}F_{10}\right)$, where $\frac{1}{24}F_{10}$ is the number of subgraphs of G that have the same configuration as the graph of Figure 47(b) and 1 is the number of times that this subgraph is counted in M. Consequently,

$$F_{18}=4\sum_{k\neq j, j, k\neq i}(d_j-2)(d_k-2)a_{ij}a_{ik}a_{jk}-\frac{1}{12}F_{10}.$$

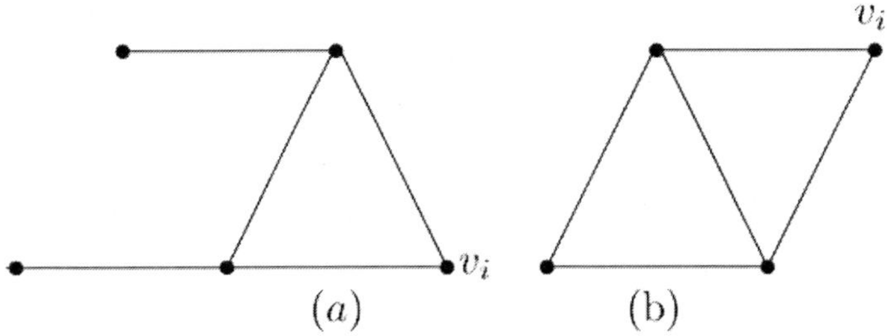

Figure 47. v_i-v_i walks of length 7 type 18.

Case 19: For the configuration of Figure 48, N=14,

$$M = \frac{1}{2}\left[a_{ii}^{(5)} - 5a_{ii}^{(3)} - 2(d_i - 2)a_{ii}^{(3)} - 2\sum_{j=1, j\neq i}^{n} a_{ij}^{(2)}a_{ij}(d_j - 2) - 2\sum_{j=1, j\neq i}^{n} a_{ij}\left(\frac{1}{2}a_{jj}^{(3)} - a_{ij}a_{ij}^{(2)}\right)\right]$$

$$F_{19} = 7\left[a_{ii}^{(5)} - 5a_{ii}^{(3)} - 2(d_i - 2)a_{ii}^{(3)} - 2\sum_{j=1, j\neq i}^{n} a_{ij}^{(2)}a_{ij}(d_j - 2) - 2\sum_{j=1, j\neq i}^{n} a_{ij}\left(\frac{1}{2}a_{jj}^{(3)} - a_{ij}a_{ij}^{(2)}\right)\right]$$

(see Theorem 11) and.

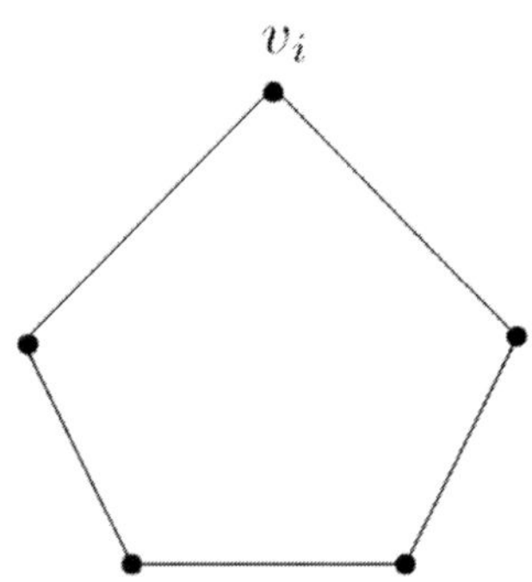

Figure 48. v_i-v_i walks of length 7 type 19.

Case 20: For the configuration of Figure 49(a), N=6, $M = \frac{1}{2}\sum_{k\neq j, j, k\neq i} a_{jk}^{(2)}(d_k - a_{jk} - 1)(a_{jk}a_{ij}a_{ik})$ (see Theorem 5). Let P_1 denote the number of all subgraphs of G that have the same configuration as the graph of Figure 49(b) and are counted in M. Thus $P_1 = 2\times\left(\frac{1}{32}F_7\right)$, where $\frac{1}{32}F_7$ is the number of subgraphs of G that have the same configuration as the graph of Figure 49(b) and 2 is the number of times that this subgraph is counted in *M*. Consequently, $F_{20} = 3\sum_{k\neq j,j,k\neq i} a_{jk}^{(2)}\left(d_k - a_{jk} - 1\right)\left(a_{jk}a_{ij}a_{ik}\right) - \frac{3}{8}F_7$.

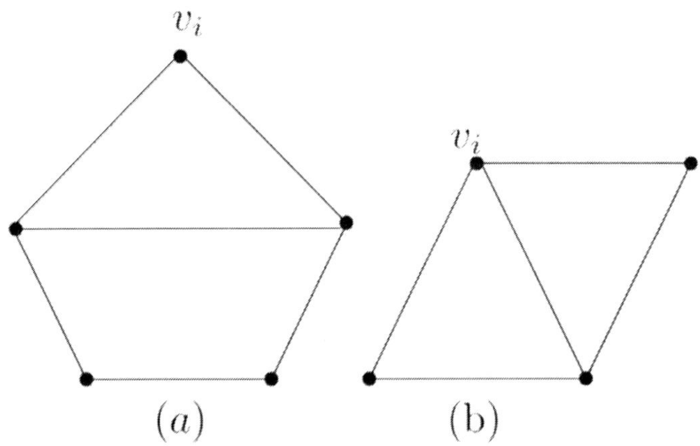

Figure 49. v_i-v_i walks of length 7 type 20.

Case 21: For the configuration of Figure 50(a), *N*=12, $M = \sum_{j\neq i} a_{ij}^{(2)}\left(d_j - a_{ij} - 1\right)\left(a_{ij}^{(2)}a_{ij}\right)$ (see Theorem 7).

Let P_1 denote the number of all subgraphs of *G* that have the same configuration as the graph of Figure 50(b) and are counted in *M*. Thus $P_1 = 2\times\left(\frac{1}{32}F_7\right)$, where $\frac{1}{32}F_7$ is the number of subgraphs of *G* that have the same configuration as the graph of Figure 50(b) and 2 is the number of times that this subgraph is counted in *M*. Let P_2 denote the number of all subgraphs of *G* that have the same configuration as the graph of Figure 50(c) and are counted in *M*. Thus $P_2 = 2\times\left(\frac{1}{24}F_{10}\right)$, where $\frac{1}{24}F_{10}$ is the number of subgraphs of *G* that have the same configuration as the graph of Figure 50(c) and 2 is the number of times that this subgraph is counted in *M*. Consequently, $F_{21} = 12\sum_{j\neq i} a_{ij}^{(2)}\left(d_j - a_{ij} - 1\right)\left(a_{ij}^{(2)}a_{ij}\right) - \frac{3}{4}F_7 - F_{10}$.

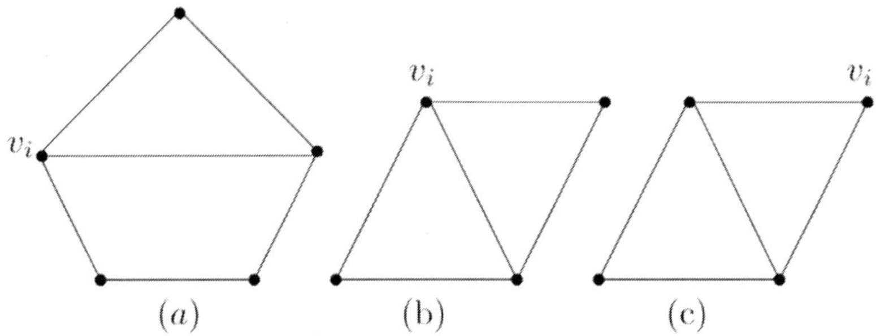

Figure 50. v_i-v_i walks of length 7 type 21.

Case 22: For the configuration of Figure 51(a), N=6, $M=\frac{1}{2}\sum_{j\neq i} a_{ij}^{(2)}\left(d_j-a_{ij}-1\right)\left(a_{jj}^{(3)}a_{ij}\right)$ (see Theorem 7). Let P_1 denote the number of all subgraphs of G that have the same configuration as the graph of Figure 51(b) and are counted in M. Thus $P_1=2\times\left(\frac{1}{24}F_{10}\right)$, where $\frac{1}{24}F_{10}$ is the number of subgraphs of G that have the same configuration as the graph of Figure 51(b) and 2 is the number of times that this subgraph is counted in M. Let P_2 denote the number of all subgraphs of G that have the same configuration as the graph of Figure 51(c) and are counted in M. Thus $P_2=6\times\left(\frac{1}{12}F_9\right)$, where $\frac{1}{12}F_9$ is the number of subgraphs of G that have the same configuration as the graph of Figure 51(c) and 6 is the number of times that this subgraph is counted in M. Let P_3 denotes the number of all subgraphs of G that have the same configuration as the graph of Figure 51(d) and are counted in M. Thus $P_3=1\times\left(\frac{1}{12}F_{21}\right)$, where $\frac{1}{12}F_{21}$ is the number of subgraphs of G that have the same configuration as the graph of Figure 51(d) and 1 is the number of times that this subgraph is counted in M. Let P_4 denote the number of all subgraphs of G that have the same configuration as the graph of Figure 51(e) and are counted in M. Thus $P_4=2\times\left(\frac{1}{32}F_7\right)$, where $\frac{1}{32}F_7$ is the number of subgraphs of G that have the same configuration as the graph of Figure 51(e) and 2 is the number of times that this subgraph is counted in M. Let P_5 denote the number of all subgraphs of G that have the same configuration as the graph of Figure 51(f) and are counted in M. Thus $P_5 = 1 \times \left(\frac{1}{4} F_{29} \right)$, where $\frac{1}{4} F_{29}$ is the number of subgraphs of G that have the same configuration as the graph of Figure 51(f) and 1 is the number of times that this subgraph is counted in M. Consequently,

$$F_{22} = 3\sum_{j \neq i} a_{ij}^{(2)} \left(d_j - a_{ij} - 1 \right)\left(a_{jj}^{(3)} a_{ij} \right) - \frac{1}{2} F_{10} - 3F_9 - \frac{1}{2} F_{21} - \frac{3}{8} F_7 - \frac{3}{2} F_{29}.$$

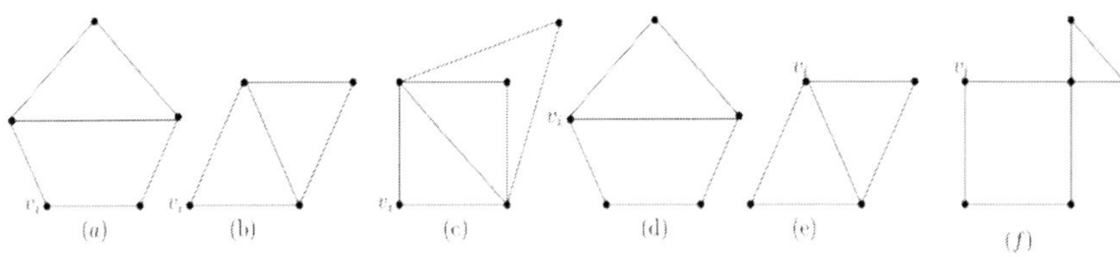

Figure 51. v_i-v_i walks of length 7 type 22.

Case 23: For the configuration of Figure 52(a), N=2,

$$M = \frac{1}{2}\left[\sum_{j \neq i} \left[a_{jj}^{(5)} - 5a_{jj}^{(3)} - 2\left(d_j - 2 \right) a_{jj}^{(3)} - 2\sum_{k \neq j} a_{jk}^{(2)} a_{jk} \left(d_k - 2 \right) - 2\sum_{k \neq j} a_{jk} \left(\frac{1}{2} a_{kk}^{(3)} - a_{jk} a_{jk}^{(2)} \right) \right] a_{ij} \right]$$

(See Theorem 11).

Let P_1 denote the number of all subgraphs of G that have the same configuration as the graph of Figure 52(b)and are counted in M. Thus $P_1 = 2 \times \left(\frac{1}{14} F_{19} \right)$, where $\frac{1}{14} F_{19}$ is the number of subgraphs of G that have the same configuration as the graph of Figure 52(b) and 2 is the number of times that this subgraph is counted in M. Let P_2 denote the number of all subgraphs of G that have the same configuration as the graph of Figure 52(c) and are counted in M. Thus $P_2 = 1 \times \left(\frac{1}{12} F_{21} \right)$, where $\frac{1}{12} F_{21}$ is the number of subgraphs of G that have the same configuration as the graph of Figure 52(c) and 1 is the number of times that this subgraph is counted in M. Consequently,

$$F_{23} = \left[\sum_{j \neq i} \left[a_{jj}^{(5)} - 5a_{jj}^{(3)} - 2\left(d_j - 2 \right) a_{jj}^{(3)} - 2\sum_{k \neq j} a_{jk}^{(2)} a_{jk} \left(d_k - 2 \right) - 2\sum_{k \neq j} a_{jk} \left(\frac{1}{2} a_{kk}^{(3)} - a_{jk} a_{jk}^{(2)} \right) \right] a_{ij} \right] - \frac{2}{7} F_{19} - \frac{1}{6} F_{21}.$$

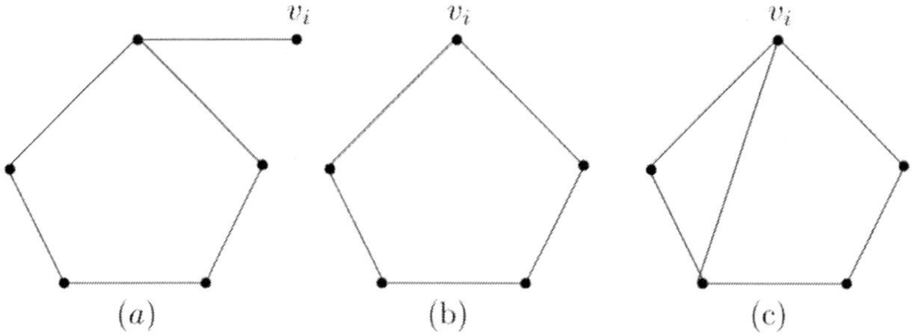

Figure 52. v_i-v_i walks of length 7 type 23.

Case 24: For the configuration of Figure 53(a), N=4,

$$M=\frac{1}{2}\left[\left(a_{ii}^{(5)}-5a_{ii}^{(3)}-2(d_i-2)a_{ii}^{(3)}\right)(d_i-2)-2\sum_{j\neq i}a_{ij}^{(2)}a_{ij}\left(d_j-2\right)(d_i-2)-2\sum_{j\neq i}a_{ij}\left(d_i-2\right)\left(\frac{1}{2}a_{ij}^{(3)}-a_{ij}a_{ij}^{(2)}\right)\right]$$

(See Theorem 11). Let P_1 denote the number of all subgraphs of G that have the same configuration as thegraph of Figure 53(b) and are counted in M. Thus $P_1=1\times\left(\frac{1}{12}F_{21}\right)$, where $\frac{1}{12}F_{21}$ is the number of subgraphsof G that have the same configuration as the graph of Figure 53(b) and 1 is the number of times that this figure is counted in M. Consequently,

$$F_{24}=2\left[\left(a_{ii}^{(5)}-5a_{ii}^{(3)}-2(d_i-2)a_{ii}^{(3)}\right)(d_i-2)-2\sum_{j\neq i}a_{ij}^{(2)}a_{ij}\left(d_j-2\right)(d_i-2)\right.$$
$$\left.-2\sum_{j\neq i}a_{ij}\left(d_i-2\right)\left(\frac{1}{2}a_{ij}^{(3)}-a_{ij}a_{ij}^{(2)}\right)\right]-\frac{1}{3}F_{21}.$$

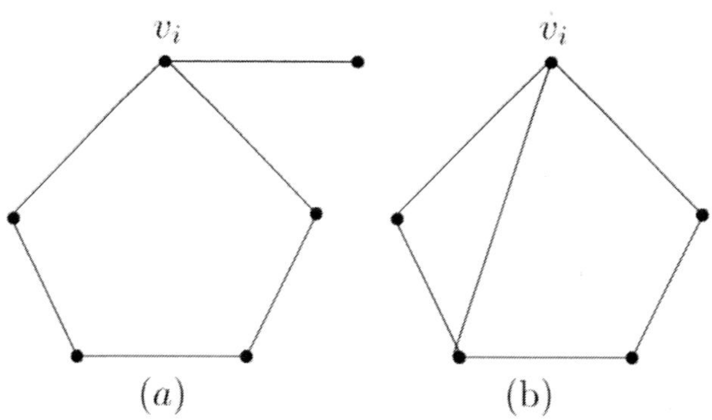

Figure 53. v_i-v_i walks of length 7 type 24.

Case 25: For the configuration of Figure 54(a), N=2,

$$M=\sum_{j\neq i}\left[a_{ij}^{(4)}-\left(d_i+d_j-3a_{ij}\right)a_{ij}^{(2)}-\left(a_{ii}^{(3)}+a_{jj}^{(3)}+2\binom{d_j-1}{2}\right)a_{ij}\right]\left(d_j-2\right)a_{ij}$$

(See Theorem 8). Let P_1 denote the number of all subgraphs of G that have the same configuration as the graph of Figure 54(b) and are counted in M. Thus $P_1=2\times\left(\frac{1}{6}F_{20}\right)$, where $\frac{1}{6}F_{20}$ is the number of subgraphs of G that have the same configuration as the graph of Figure 54(b) and 2 is the number of times that this subgraph is counted in M. Let P_2denote the number all subgraphs of G that have the same configuration as the graph of Figure 54(c) and are counted in M. Thus $P_2=1\times\left(\frac{1}{6}F_{22}\right)$, where $\frac{1}{6}F_{22}$ is the number of subgraphs of G that have the same configuration as the graph of Figure 54(c) and 1 is the number of times that this subgraph is counted in M. Consequently,

$$F_{25}=2\sum_{j\neq i}\left[a_{ij}^{(4)}-\left(d_i+d_j-3a_{ij}\right)a_{ij}^{(2)}-\left(a_{ii}^{(3)}+a_{jj}^{(3)}+2\binom{d_j-1}{2}\right)a_{ij}\right]\left(d_j-2\right)a_{ij}-\frac{2}{3}F_{20}-\frac{1}{3}F_{22}$$

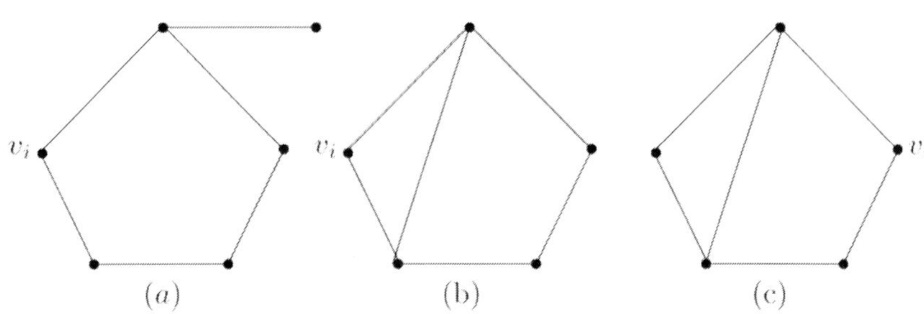

Figure 54. v_i-v_i walks of length 7 type 25.

Case 26: For the configuration of Figure 55(a), N=2,

$$M=\sum_{j\neq i}\left[\left(a_{ij}^{(3)}-\left(d_i+d_j-1\right)a_{ij}\right)a_{ij}^{(2)}-\sum_{k\neq i,j}\left(a_{ik}^{(2)}a_{ik}-a_{ik}a_{ij}a_{jk}\right)a_{jk}-\sum_{k\neq i,j}\left(a_{jk}^{(2)}a_{jk}-a_{ik}a_{ij}a_{jk}\right)a_{ik}\right]\left(d_j-2\right).$$

Let P_1 denote the number of all subgraphs of G that have the same configuration as the graph of Figure 55(b) and are counted in M. Thus $P_1=1\times\left(\frac{1}{12}F_{21}\right)$, where $\frac{1}{12}F_{21}$ is the number of subgraphs of G that have the same configuration as the graph of Figure 55(b) and 1 is the number of times that this subgraph is counted in *M*. Let P_2 denote the number of all subgraphs of *G* that have the same configuration as the graph of Figure 55(c) and are counted in *M*. Thus $P_2 = 1\times\left(\frac{1}{6}F_{22}\right)$, where $\frac{1}{6}F_{22}$ is the number of subgraphs of *G* that have the same configuration as the graph of Figure 55(c) and 1 is the number of times that this subgraph is counted in *M*. Consequently,

$$F_{26} = 2\sum_{j\neq i}\Bigg[\left(a_{ij}^{(3)} - (d_i + d_j - 1)a_{ij}\right)a_{ij}^{(2)} - \sum_{k\neq i,j}\left(a_{ik}^{(2)}a_{ik} - a_{ik}a_{ij}a_{jk}\right)a_{jk} - \sum_{k\neq i,j}\left(a_{jk}^{(2)}a_{jk} - a_{ik}a_{ij}a_{jk}\right)a_{ik}\Bigg](d_j - 2) - \frac{1}{6}F_{21} - \frac{1}{3}F_{22}.$$

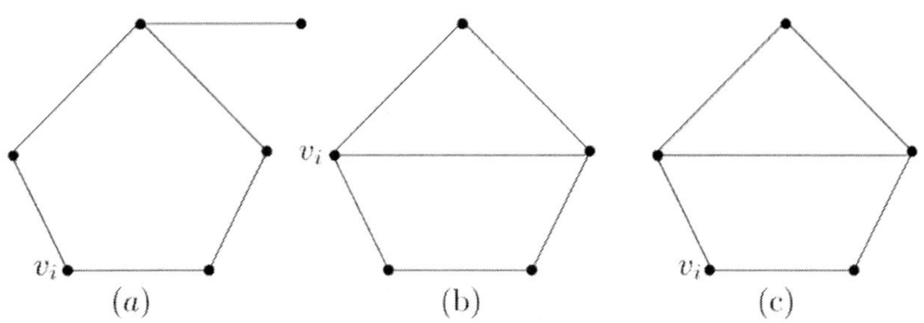

Figure 55. v_i-v_i walks of length 7 type 26.

Case 27: For the configuration of Figure 56(a), *N*=4, $M = \sum_{j\neq i}\left[a_{ij}^{(2)}a_{ij} \times \sum_{k\neq i,j}\binom{a_{ij}^{(2)}}{2}\right]$. Let P_1 denote the Nnumber of all subgraphs of *G* that have the same configuration as the graph of Figure 56(b) and are counted in *M*. Thus $P_1 = 1\times\left(\frac{1}{12}F_{21}\right)$, where $\frac{1}{12}F_{21}$ is the number of subgraphs of *G* that have the same configuration as the graph of Figure 56(b) and 1 is the number of times that this subgraph is counted in *M*. Let P_2 denote the number of all subgraphs of *G* that have the same configuration as the graph of Figure 56(c) and are counted in *M*. Thus $P_2 = 2\times\left(\frac{1}{24}F_{10}\right)$, where $\frac{1}{24}F_{10}$ is the number of subgraphs of *G* that have the same configuration as the graph of Figure 56(c) and 2 is the number of times that this subgraph is counted in M. Let P_3 denote the number of all subgraphs of G that have the same configuration as the graph of Figure 56(d) and are counted in M. Thus $P_3 = 2\times\left(\frac{1}{12}F_9\right)$, where $\frac{1}{12}F_9$ is the number of subgraphs of G that have the same configuration as the graph of Figure 56(d) and 2 is the number of times that this subgraph is counted in M. Let P_4 denote the number of all subgraphs of G that have the same configuration as the graph of Figure 56(e) and are counted in M. Thus $P_4 = 2\times\left(\frac{1}{6}F_{20}\right)$, where $\frac{1}{6}F_{20}$ is the number of subgraphs of G that have the same configuration as the graph of Figure 56(e) and 2 is the number of times that this subgraph is counted in M. Let P_5 denote the number of all subgraphs of G that have the same configuration as the graph of Figure 56(f) and are counted in M. Thus $P_5 = 2\times\left(\frac{1}{32}F_7\right)$, where $\frac{1}{32}F_7$ is the number of subgraphs of G that have the same configuration as the graph of Figure 56(f) and 2 is the number of times that this subgraph is counted in M.

Consequently,

$$F_{27} = 4\sum_{j\neq i}\left[a_{ij}^{(2)}a_{ij}\times\sum_{k\neq i,j}\binom{a_{ij}^{(2)}}{2}\right]-\frac{1}{3}F_{21}-\frac{1}{3}F_{10}-\frac{2}{3}F_9-\frac{4}{3}F_{20}-\frac{1}{4}F_7$$

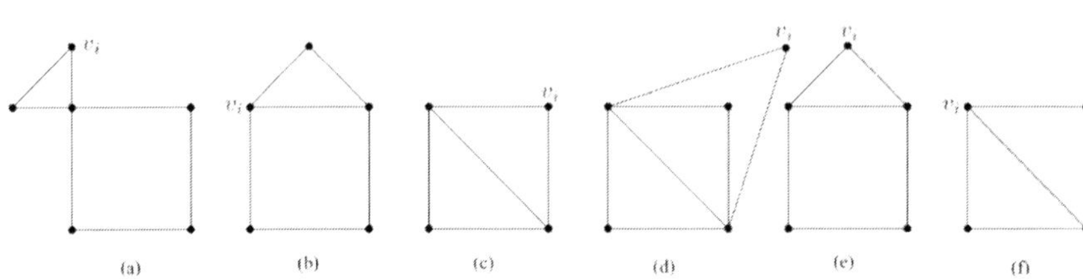

Figure 56. v_i-v_i walks of length 7 type 27.

Case 28: For the configuration of Figure 57(a) $N=8$, $M=\left(\frac{1}{2}a_{ii}^{(3)}\right)\sum_{j\neq i}\binom{a_{ij}^{(2)}}{2}$. Let P_1 denote the number of all subgraphs of G that have the same configuration as the graph of Figure 57(b) and are counted in M. Thus $P_1 = 1\times\left(\frac{1}{12}F_{21}\right)$, where $\frac{1}{12}F_{21}$ is the number of subgraphs of G that have the same configuration as the graph of Figure 57(b) and 1 is the number of times that this subgraph is counted in M. Let P_2 denote the number of all subgraphs of G that have the same configuration as the graph of Figure 57(c) and are counted in *M*. Thus $P_2 = 1\times\left(\frac{1}{24}F_{10}\right)$, where $\frac{1}{24}F_{10}$ is the number of subgraphs of *G* that have the same configuration as the graph of Figure 57(c) and 1 is the number of times that this subgraph is counted in *M*. Let P_3 denote the number of all subgraphs of *G* that have the same configuration as the graph of Figure 57(d) and are counted in *M*. Thus $P_3 = 3\times\left(\frac{1}{24}F_8\right)$, where $\frac{1}{24}F_8$ is the number of subgraphs of *G* that have the same configuration as the graph of Figure 57(d) and 3 is the number of times that this subgraph is counted in *M*. Let P_4 denote the number of all subgraphs of *G* that have the same configuration as the graph of Figure, 57(e) and are counted in *M*. Thus $P_4 = 2\times\left(\frac{1}{32}F_7\right)$, where $\frac{1}{32}F_7$ is the number of subgraphs of *G* that have the same configuration as the graph of Figure 57(e) and 2 is the number of times that this subgraph is

counted in *M*. Consequently,

$$F_{28} = \left(4a_{ii}^{(3)}\right)\sum_{j\neq i}\binom{a_{ij}^{(2)}}{2} - \frac{2}{3}F_{21} - \frac{1}{3}F_{10} - F_8 - \frac{1}{2}F_7$$

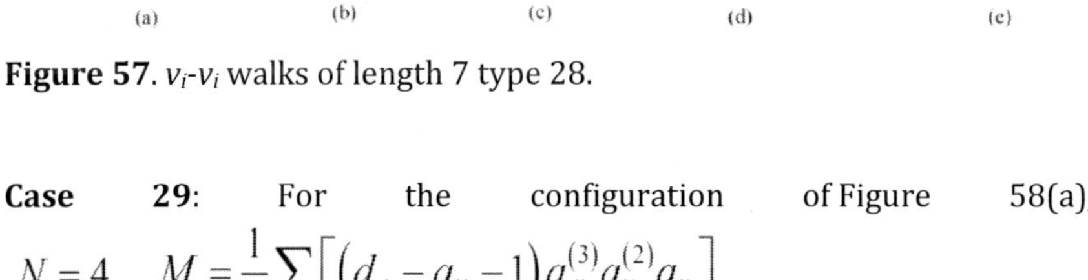

Figure 57. v_i-v_i walks of length 7 type 28.

Case 29: For the configuration of Figure 58(a), $N = 4$, $M = \frac{1}{2}\sum_{j\neq i}\left[\left(d_j - a_{ij} - 1\right)a_{jj}^{(3)}a_{ij}^{(2)}a_{ij}\right]$. Let P_1 denote the number of all subgraphs of *G* that have the same configuration as the graph of Figure 58(b) and are counted in *M*. Thus $P_1 = 1\times\left(\frac{1}{12}F_{21}\right)$, where $\frac{1}{12}F_{21}$ is the number of subgraphs of *G* that have the same configuration as the graph of Figure 58(b) and 1 is the number of times that this subgraph is counted in *M*. Let P_2 denote the number of all subgraphs of *G* that have the same configuration as the graph of Figure 58(c) and are counted in M. Thus $P_2 = 4\times\left(\frac{1}{24}F_{10}\right)$, where $\frac{1}{24}F_{10}$ is the number of subgraphs of G that have the same configuration as the graph of Figure 58(c) and 4 is the number of times that this subgraph is counted in M. Let P_3 denote the number of all subgraphs of G that have the same configuration as the graph of Figure 58(d) and are counted in M. Thus $P_3 = 4\times\left(\frac{1}{12}F_9\right)$, where $\frac{1}{12}F_9$ is the number of subgraphs of G that have the same configuration as the graph of Figure 58(d) and 4 is the number of times that this subgraph is counted in M. Let P_4 denote the number of all subgraphs of G that have the same configuration as the graph of Figure 58(e) and are counted in M. Thus $P_4 = 1\times\left(\frac{1}{6}F_{22}\right)$, where $\frac{1}{6}F_{22}$ is the number of subgraphs of G that have the same configuration as the graph of Figure 58(e) and 1 is the number of times that this subgraph is counted in M. Let P_5 denote the number of all subgraphs of G that have the same configuration as the graph of Figure 58(f) and are counted in M. Thus $P_5 = 2\times\left(\frac{1}{32}F_7\right)$, where $\frac{1}{32}F_7$ is the number of subgraphs of G that have the same configuration as the graph of Figure 58(f) and 2 is the number of times that this subgraph is counted in M. Consequently,

$$F_{29} = 2\sum_{j\neq i}\left[\left(d_j - a_{ij} - 1\right)a_{jj}^{(3)}a_{ij}^{(2)}a_{ij}\right] - \frac{1}{3}F_{21} - \frac{2}{3}F_{10} - \frac{4}{3}F_9 - \frac{2}{3}F_{22} - \frac{1}{4}F_7 .$$

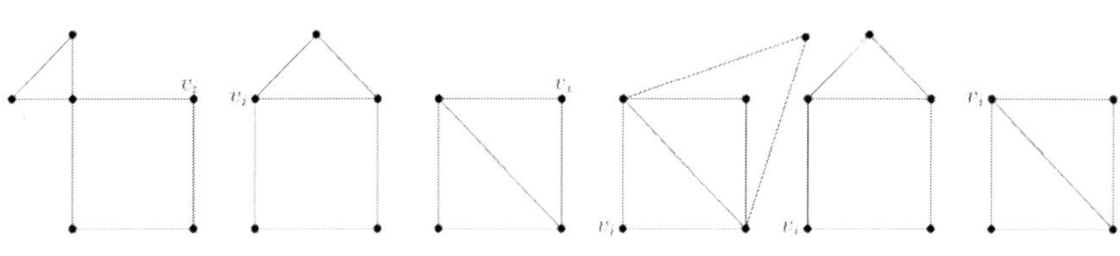

Figure 58. v_i-v_i walks of length 7 type 29.

Case 30: For the configuration of Figure 59(a), $N = 4$, $M = \frac{1}{2}\sum_{j\neq i}\binom{a_{ij}^{(2)}}{2}a_{jj}^{(3)}$. Let P_1 denote the number of all subgraphs of G that have the same configuration as the graph of Figure 59(b) and are counted in M. Thus $P_1 = 1\times\left(\frac{1}{6}F_{22}\right)$, where $\frac{1}{6}F_{22}$ is the number of subgraphs of G that have the same configuration as the graph of Figure 59(b) and 1 is the number of times that this subgraph is counted in M. Let P_2 denote the number of all subgraphs of G that have the same configuration as the graph of Figure 59(c) and are counted in M. Thus $P_2 = 1\times\left(\frac{1}{24}F_{10}\right)$, where $\frac{1}{24}F_{10}$ is the number of subgraphs of G that have the same configuration as the graph of Figure 59(c) and 1 is the number of times that this subgraph is counted in M. Let P_3 denote the number of all subgraphs of G that have the same configuration as the graph of Figure 59(d) and are counted in M. Thus $P_3 = 3\times\left(\frac{1}{24}F_8\right)$, where $\frac{1}{24}F_8$ is the number of subgraphs of G that have the same configuration as the graph of Figure 59(d) and 3 is the number of times that this subgraph is counted in M. Let P_4 denote the number of all subgraphs of G that have the same configuration as the graph of Figure 59(e) and are counted in M. Thus $P_4 = 2\times\left(\frac{1}{32}F_7\right)$, where $\frac{1}{32}F_7$ is the number of subgraphs of G that have the same configuration as the graph of Figure 59(e) and 2 is the number of times that this subgraph is counted in M.

Consequently, $F_{30} = 2\sum_{j\neq i}\binom{a_{ij}^{(2)}}{2}a_{jj}^{(3)} - \frac{2}{3}F_{22} - \frac{1}{6}F_{10} - \frac{1}{2}F_8 - \frac{1}{4}F_7$.

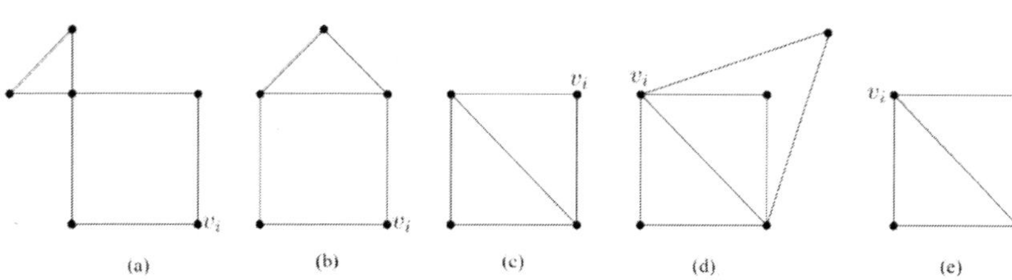

Figure 59. v_i-v_i walks of length 7 type 30.

Now, we add the values of F_n arising from the above cases and determine x.

Substituting the value of x in $\frac{1}{2}\left(a_{ii}^{(7)} - x\right)$ and simplifying, we get the number of 7-cycles each of which contains a specific vertex v_i of G. □

Example 3 *In the graph of* Figure 29 *we have* $F_1 = 420,\ F_2 = 1020,\ F_3 = 1080,\ F_4 = 240,\ F_5 = 360,$

$F_6 = 240$, $F_7 = 960$, $F_8 = 480$, $F_9 = 360$, $F_{10} = 720$, $F_{11} = 420$, $F_{12} = 120$, $F_{13} = 240$, $F_{14} = 240$, $F_{15} = 240$, $F_{16} = 240$, $F_{17} = 480$, $F_{18} = 120$, $F_{19} = 840$, $F_{20} = 360$, $F_{21} = 1440$, $F_{22} = 720$, $F_{23} = 120$, $F_{24} = 240$, $F_{25} = 240$, $F_{26} = 240$, $F_{27} = 240$, $F_{28} = 240$, $F_{29} = 240$, $F_{30} = 120$. *So, we have.x*=13020. *Consequently, by Theorem 14, the number of 7-cycles each of which contains the vertex* v_i *in the graph of* Figure 29*is* 0.

REFERENCES

1. Harary, F. and Manvel, B. (1971) On the Number of Cycles in a Graph. Mat. Casopis Sloven. Akad. Vied, 21, 55-63.
2. Chang, Y.C. and Fu, H.L. (2003) The Number of 6-Cycles in a Graph. Bulletin of the ICA, 39, 27-30.
3. Alon, N., Yuster, R. and Zwick, U. (1997) Finding and Counting Given Length Cycles. Algorithmica, 17, 209-223. http://dx.doi.org/10.1007/BF02523189
4. Bjorklund, A., Husfeldt, T., Kaski, P. and Koivisto, M. (2009) Counting Paths and Packing in Halves. Lecture Notes in Computer Science, 5757, 578-586. http://dx.doi.org/10.1007/978-3-642-04128-0_52
5. Bjorklund, A., Husfeldt, T., Kaski, P. and Koivisto, M. (2008) The Fast Intersection Transform with Applications to Counting Paths, CoRR, abs/0809.2489.
6. Chen, J., Lu, S., Sze, S.H. and Zhang, F. (2007) Improved Algorithms for Path, Matching and Packing Problems. 18th Annual ACM-SIAM Symposium on Discrete Algorithms (SODA 2007), Philadelphia, 298-307.
7. Koutis, I. (2008) Faster Algebraic Algorithm for Path and Packing Problems. CALP, LNCS 5125, 575-586. Springer, Berlin.
8. Kroese, D.P. and Roberts, B. (2007) Estimating the Number of s-t Paths in a Graph. Journal of Graph Algorithms and Applications, 11, 195-214. http://dx.doi .org/10.7155/jgaa.00142
9. Williams, R. (2009) Finding a Path of Length k in $O^*(2^K)$Time. Information Processing Letters, 109, 315-318. http://dx.doi.org/10.1016/j.ipl.2008.11.004
10. Boxwala, S.A. and Movarraei, N. (2015) On the Number of Paths of Length 5 in a Graph. International Journal of Applied Mathematical Research, 4, 30-51. http://dx.doi.org/10.14419/ijamr.v4i1.3874
11. Movarraei, N. and Shikare, M.M. (2014) On the Number of Paths of Lengths 3 and 4 in a Graph. International Journal of Applied Mathematical Research, 3, 178-189.
12. Boxwala, S.A. and Movarraei, N. (2015) On the Number of Paths of Length 6 in a Graph. International Journal of Applied Mathematical Research, 4, 267-280. http://dx.doi.org/10.14419/ijamr.v4i2.4314

CHAPTER 8

Dominating Sets and Domination Polynomials of Square of Paths

A. Vijayan[1], K. Lal Gipson[2]

[1]*Department of Mathematics, Nesamony Memorial Christian College, Marthandam, India*

[2]*Department of Mathematics, Mar Ephraem College of Engineering and Technology, Kanayakumari District, India*

ABSTRACT

Let $G = (V, E)$ be a simple graph. A set $S \subseteq V$ is a dominating set of G, if every vertex in V-S is adjacent to at least one vertex in S. Let P_n^2 be the square of the Path P_n and let $D(P_n^2, i)$ denote the family of all dominating sets of P_n^2 with cardinality i. Let $d(P_n^2, i) = |D(P_n^2, i)|$. In this paper, we obtain a recursive formula for $d(P_n^2, i)$. Using this recursive formula, we construct the polynomial, $D(P_n^2, x) = \sum_{i=\lceil \frac{n}{5} \rceil}^{n} d(P_n^2, i) x^i$, which we call domination polynomial of P_n^2 and obtain some properties of this polynomial.

Keywords: Domination Set; Domination Number; Domination Polynomials

1. INTRODUCTION

Let $G = (V, E)$ be a simple graph of order $|V| = n$. For any vertex $v \in V$, the open neighbourhood of v is the set $N(v) = \{u \in V / uv \in E\}$ and the closed neighbourhood of v is the set $N[v] = N(v) \cup \{v\}$. For a set $S \subseteq V$, the open neighbourhood of S is $N(S) = U_{n \in S} N(v)$ and the closed neighbourhood of S is $N[S] = N(S) \cup S$. A set $S \subseteq V$ is a dominating set of G, if $N[S] = V$, or equivalently, every vertex in V-S is adjacent to at least one vertex in S. The domination number of a graph G is defined as the minimum size of a dominating set of vertices in G and it is denoted as $\gamma(G)$. A simple path is a path in which all its internal vertices have degree two, and the end vertices have degree one and is denoted by P_n.

Definition 1.1.: Let $D(G,i)$ be the family of dominating sets of a graph G with Cardinality i and let

$$d(G,i) = |D(G,i)|.$$

Then the domination polynomial $D(G,x)$ of G is defined as

$$D(G,x) = \sum_{i=\gamma(G)}^{|V(G)|} d(G,i)x^i.$$

where $\gamma(G)$ is the domination number of G.

Result 1.2. [1]: If a graph G consists of m components $G_1, \cdots, G_m$, then

$$D(G,x) = D(G_1,x) \cdots D(G_m,x).$$

Result 1.3. [1]: Let G_1 and G_2 be graphs of order n_1 and n_2 respectively. Then

$$D(G_1 \vee G_2, x) = \left((1+x)^{n_1} - 1\right)\left((1+x)^{n_2} - 1\right) + D(G_1,x) + D(G_2,x).$$

Result 1.4. [1]: Let G be a bipartite graph with bipartition (X,Y). Then G contains a matching that saturates every vertex in X if and only if for all $S \subseteq V |N(S)| \geq |S|$.

Result 1.5. [1]: Let G be a graph of order n. Then for every $0 \le i \le \frac{n}{2}$, we have $d(G,i) \le d(G,i+1)$.

Result 1.6. [1]: For any graph G of order n, $\gamma(G \circ K_1) = n$.

Result 1.7. [1]: For any graph G of order n and $n \le m \le 2n$, we have

$$d(G \circ K_1, m) = \binom{n}{m-n} 2^{2n-m}.$$

Hence $D(G \circ K_1, x) = x^n (x+2)^n$.

Definition 1.8: The Square of a graph with the same set of vertices as G and an edge between two vertices if and only if there is a path of length at most 2 between them. The second power of a graph is also called its Square.

Let P_n^2 be the square of the path P_n (2nd power)with n vertices. Let $D(P_n^2, i)$ be the family of dominating sets of the graph P_n^2 with cardinality i and let

$$d(P_n^2, i) = \left| D(P_n^2, i) \right|.$$

We call the polynomial

$$D(P_n^2, x) = \sum_{i=\left\lceil \frac{n}{5} \right\rceil}^{n} d(P_n^2, i) x^i$$

the domination polynomial of the graph P_n^2 [2].

In the next section, we construct the families of the dominating sets of the square of paths by recursive method.

As usual we use $\lfloor x \rfloor$ for the largest integer less than or equal to x and $\lceil x \rceil$ for the smallest integer greater than or equal to x. Also, we denote the set $\{1, 2, \cdots, n\}$ by $[n]$, throughout this paper.

2. MAIN RESULT

Let $D(P_n^2, i)$ be the family of dominating sets of P_n^2 with cardinality *i*. We investigate the dominating sets of P_n^2. We need the following lemma to prove our main results in this section.

Lemma 2.1. [3]: $\gamma(P_n^2) = \left\lceil \frac{n}{5} \right\rceil$ By Lemma 2.1 and the definition of domination number, one has the following Lemma:

Lemma 2.2. $D(P_n^2, i) = \Phi$ if and only if $i \succ n$ or $i \prec \left\lceil \frac{n}{5} \right\rceil$. A simple path is a path in which all its internal vertices have degree two, and the end vertices have degree one. The following Lemma follows from observation [2].

Lemma 2.3. [2]: If a graph *G* contains a simple path of length 5*k* – 1, then every dominating set of *G* must contain at least *k* vertices of the path.

In order to find a dominating set of P_n^2, with cardinality i, we only need to consider

$D(P_{n-1}^2, i-1)$, $D(P_{n-2}^2, i-1)$,

$D(P_{n-3}^2, i-1)$, $D(P_{n-4}^2, i-1)$

and $D(P_{n-5}^2, i-1)$. The families of these dominating sets can be empty or otherwise. Thus, we have eight combinations, whether these five families are empty or not. Two of these combinations are not **possible** (Lemma 2.5(i), & (ii)). Also, the combination that

$$D(P_{n-1}^2, i-1) = D(P_{n-2}^2, i-1) = D(P_{n-3}^2, i-1) = \Phi$$

$$D(P_{n-4}^2, i-1) = D(P_{n-5}^2, i-1) = \Phi$$

does not need to be considered because it implies that $D(P_n^2, i) = \Phi$ (See Lemma 2.5 (iii)). Thus we only need to consider five combinations or cases. We consider those cases in Theorem 2.7.

Lemma 2.4.: If $Y \in D(P_{n-6}^2, i-1)$, and there exists $x \in [n]$ such that $Y \cup \{x\} \in D(P_n^2, i)$, then

$Y \in D(P_{n-5}^2, i-1)$.

Proof: Suppose that $Y \notin D(P_{n-5}^2, i-1)$. Since

$Y \in D(P_{n-6}^2, i-1)$,

Y contains at least one vertex labeled $n-6$, $n-7$ or $n-8$.

If $n-6 \in Y$, then $Y \in D(P_{n-5}^2, i-1)$, a contradiction. Hence $n-7$ or $n-8 \in Y$, but then in this case, $Y \cup \{x\} \notin D(P_n^2, i)$, for any $x \in [n]$, also a contradiction.

Lemma 2.5.:

1) If

$$D(P_{n-1}^2, i-1) = D(P_{n-3}^2, i-1) = \Phi$$

then

$$D(P_{n-2}^2, i-1) = \Phi$$

2) If

$$D(P_{n-1}^2, i-1) \neq \Phi,\ D(P_{n-3}^2, i-1) \neq \Phi$$

then

$$D(P_{n-2}^2, i-1) \neq \Phi,$$

3) If

$D(P_{n-1}^2, i-1) = \Phi$, $D(P_{n-2}^2, i-1) = \Phi$,

$D(P_{n-3}^2, i-1) = \Phi$, $D(P_{n-4}^2, i-1) = \Phi$,

$D(P_{n-5}^2, i-1) = \Phi$, then $D(P_n^2, i) = \Phi$

Proof:

1) Since $D(P_{n-1}^2, i-1) = D(P_{n-3}^2, i-1) = \Phi$, by Lemma 2.2, $i-1 \succ n-1$ (or) $i-1 \prec \left\lceil \frac{n-3}{5} \right\rceil$. In either case, we have $D(P_{n-2}^2, i-1) = \Phi$.

2) Suppose that $D(P_{n-2}^2, i-1) = \Phi$, so by Lemma 2.2we have $i-1 \succ n-2$ or $i-1 \prec \left\lceil \frac{n-2}{5} \right\rceil$ If $i-1 \succ n-2$then $i-1 \succ n-3$. Therefore $D(P_{n-3}^2, i-1) = \Phi$, a contradiction.

3) Suppose that $D(P_n^2, i) \neq \Phi$, Let $Y \in D(P_n^2, i)$. Thus, at least one vertex labeled *n* or *n* – 1 or *n* – 2 is in *Y*. If *n* ∈ *Y*, then by Lemma (2.3), at least one vertex labeled *n* – 1, *n* – 2, *n* – 3, *n* – 4 or *n* – 5 is in *Y*. If *n* – 1 ∈ Y or *n* – 2 ∈ *Y*, then $Y - \{n\} \in D(P_{n-1}^2, i-1)$, a contradiction.

If *n* – 3 ∈ *Y*, then $Y - \{n\} \in D(P_{n-2}^2, i-1)$, a contradiction. Now, suppose that *n* – 1 ∈ *Y*. Then, by Lemma 2.3, at least one vertex labeled *n* – 2, *n* – 3, *n* – 4, *n* – 5 or *n* – 6 is in *Y*. If *n* – 2 ∈ *Y* or *n* – 3 ∈ *Y*, then $Y - \{n-1\} \in D(P_{n-2}^2, i-1)$a contradiction. If *n* – 4 ∈ *Y*then $Y - \{n-1\} \in D(P_{n-3}^2, i-1)$, a contradiction.

If *n* – 5 ∈ *Y*, then $Y - \{n-1\} \in D(P_{n-2}^2, i-1)$, a contradiction If *n* – 6 ∈ *Y*, then $Y - \{n-1\} \in D(P_{n-5}^2, i-1)$, a contradiction. Therefore, $D(P_n^2, i) = \Phi$.

Lemma 2.6.: If $D(P_n^2, i) \neq \Phi$, then

1) $D\left(P_{n-1}^2, i-1\right) = D\left(P_{n-2}^2, i-1\right) = D\left(P_{n-3}^2, i-1\right)$

$= D\left(P_{n-4}^2, i-1\right) = \Phi$ and $D\left(P_{n-5}^2, i-1\right) \neq \Phi$ if and only if $n = 5k$ and $i = k$ for some $k \in N$;

2) $D\left(P_{n-2}^2, i-1\right) = D\left(P_{n-3}^2, i-1\right) = D\left(P_{n-4}^2, i-1\right)$

$= D\left(P_{n-5}^2, i-1\right) = \Phi$ then $D\left(P_{n-1}^2, i-1\right) \neq \Phi$ if and only if $i = n$;

3) $D\left(P_{n-1}^2, i-1\right) = \Phi$, $D\left(P_{n-2}^2, i-1\right) \neq \Phi$,

$D\left(P_{n-3}^2, i-1\right) \neq \Phi$, $D\left(P_{n-4}^2, i-1\right) \neq \Phi$,

$D\left(P_{n-5}^2, i-1\right) \neq \Phi$, if and only if n = $5k + 2$ and

$i = \left\lceil \frac{5k+2}{5} \right\rceil$ for some $k \in N$;

4) $D\left(P_{n-1}^2, i-1\right) \neq \Phi$, $D\left(P_{n-2}^2, i-1\right) \neq \Phi$,

$D\left(P_{n-3}^2, i-1\right) \neq \Phi$, $D\left(P_{n-4}^2, i-1\right) \neq \Phi$, and

$D\left(P_{n-5}^2, i-1\right) = \Phi$ if and only if $i = n - 3$.

5) $D\left(P_{n-1}^2, i-1\right) \neq \Phi$, $D\left(P_{n-2}^2, i-1\right) \neq \Phi$,

$D\left(P_{n-3}^2, i-1\right) \neq \Phi$, $D\left(P_{n-4}^2, i-1\right) \neq \Phi$, and

$D\left(P_{n-5}^2, i-1\right) \neq \Phi$ if and only if

$\left\lceil \frac{n-1}{5} \right\rceil + 1 \leq i \leq n-4$.

Proof:

1) ($\Rightarrow$)

Since

$$D\left(P_{n-1}^2, i-1\right) = D\left(P_{n-2}^2, i-1\right)$$
$$= D\left(P_{n-3}^2, i-1\right) = D\left(P_{n-4}^2, i-1\right) = \Phi,$$

by Lemma 2.2,

$$i-1 \succ n-1 \text{ or } i-1 \prec \left\lceil \frac{n-2}{5} \right\rceil$$

If $i-1 \succ n-1$, then $i \succ n$ and by Lemma 2.2, $D\left(P_n^2, i\right) = \Phi$, a contradiction.

So $i \prec \left\lceil \frac{n-2}{5} \right\rceil + 1$, and since $D\left(P_n^2, i\right) \neq \Phi$, we have

$\left\lceil \frac{n}{5} \right\rceil \leq i \leq \left\lceil \frac{n-2}{5} \right\rceil + 1$, which implies that $n = 5k$ and $i = k$ for some $k \in N$.

($\Leftarrow$) If $n = 5k$ and $i = k$ for some $k \in N$, then by Lemma 2.2,

$$D\left(P_{n-1}^2, i-1\right) = D\left(P_{n-2}^2, i-1\right)$$
$$= D\left(P_{n-3}^2, i-1\right) = D\left(P_{n-4}^2, i-1\right) = \Phi,$$

and $D\left(P_{n-5}^2, i-1\right) \neq \Phi$.

2) ($\Rightarrow$) Since

$$D\left(P_{n-2}^2, i-1\right) = D\left(P_{n-3}^2, i-1\right)$$
$$= D\left(P_{n-4}^2, i-1\right) = D\left(P_{n-5}^2, i-1\right) = \Phi,$$

by Lemma 2.2,

$$i-1 \succ n-2 \quad \text{or} \quad i-1 < \left\lceil \frac{n-3}{5} \right\rceil.$$

If

$i-1 \prec \left\lceil \frac{n-3}{5} \right\rceil$, then $i-1 \prec \left\lceil \frac{n-1}{5} \right\rceil$

and hence $D\left(P_{n-2}^{2},i-1\right)=\Phi$, a contradiction.

So $i-1 \succ n-2$. Also, $D\left(P_{n-1}^{2},i-1\right)\neq\Phi$.

Therefore $i-1 \succ n-1$. Therefore $i \geq n$. But $i \leq n$. Hence $i = n$.

($\Leftarrow$) If $i = n$, then by lemma 2.2,

$$D\left(P_{n-2}^{2},i-1\right)=D\left(P_{n-3}^{2},i-1\right)$$
$$=D\left(P_{n-4}^{2},i-1\right)=D\left(P_{n-5}^{2},i-1\right)=\Phi$$

then $D\left(P_{n-1}^{2},i-1\right)\neq\Phi$.

3) ($\Rightarrow$) Since $D\left(P_{n-1}^{2},i-1\right)=\Phi$, by Lemma 2.2,

$i-1 \succ n-1$ or $i-1 \prec \left\lceil \frac{n-1}{5} \right\rceil$.

If $i-1 \succ n-1$, then $i-1 \succ n-2$, by Lemma 2.2,

$$D\left(P_{n-2}^{2},i-1\right)=D\left(P_{n-3}^{2},i-1\right)$$
$$=D\left(P_{n-4}^{2},i-1\right)=D\left(P_{n-5}^{2},i-1\right)=\Phi,$$

a contradiction.

Therefore $i \prec \left\lceil \frac{n-1}{5} \right\rceil + 1$.

But $i-1 \geq \left\lceil \frac{n-2}{5} \right\rceil$, because $D\left(P_{n-2}^{2},i-1\right)\neq\Phi$.

Therefore $i \geq \left\lceil \frac{n-2}{5} \right\rceil + 1$.

Hence, $\left\lceil \frac{n-2}{5} \right\rceil + 1 \leq i \prec \left\lceil \frac{n-1}{5} \right\rceil + 1$.

This holds only if $n = 5k + 2$ and

$i = k+1 = \left\lceil \frac{5k+2}{5} \right\rceil$,

for some $k \in N$.

($\Leftarrow$) If $n = 5k + 2$ and $i = \left\lceil \frac{5k+2}{5} \right\rceil$ for some $k \in N$, then by Lemma 2.2,

$$D\left(P_{n-1}^2, i-1\right) = \Phi, \quad D\left(P_{n-2}^2, i-1\right) \neq \Phi,$$

$$D\left(P_{n-3}^2, i-1\right) \neq \Phi, \quad D\left(P_{n-4}^2, i-1\right) \neq \Phi$$

and

$D\left(P_{n-5}^2, i-1\right) \neq \Phi$.

4) ($\Rightarrow$) Since $D\left(P_{n-5}^2, i-1\right) = \Phi$, by Lemma 2.2,

$$i-1 \succ n-5 \quad \text{or} \quad i-1 \prec \left\lceil \frac{n-5}{5} \right\rceil.$$

Since $D\left(P_{n-2}^2, i-1\right) \neq \Phi$, by Lemma 2.2,

$\left\lceil \frac{n-2}{5} \right\rceil + 1 \leq i \leq n-1$.

Therefore $i-1 \prec \left\lceil \frac{n-5}{5} \right\rceil$ is not possible. Therefore $i-1 \succ n-5$. Therefore $i \succ n-4$.

Therefore $i \geq n-3$.

Since $D\left(P_{n-3}^2, i-1\right) = \Phi$, $i-1 \leq n-4$.

Therefore $i \leq n-3$. Hence $i = n-3$.

($\Leftarrow$) if $i = n-3$ then

By Lemma 2.2,

$$D\left(P_{n-1}^2, i-1\right) \neq \Phi, \quad D\left(P_{n-2}^2, i-1\right) \neq \Phi,$$

$$D\left(P_{n-3}^2, i-1\right) = \Phi, \quad D\left(P_{n-4}^2, i-1\right) \neq \Phi$$

and

$D\left(P_{n-5}^2, i-1\right) = \Phi$.

5) (⟹) Since $D\left(P_{n-1}^2, i-1\right) \neq \Phi$, $D\left(P_{n-2}^2, i-1\right) \neq \Phi$

$D\left(P_{n-3}^2, i-1\right) \neq \Phi$, and $D\left(P_{n-5}^2, i-1\right) = \Phi$, then by applying Lemma (2.2),

$\left\lceil \frac{n-1}{5} \right\rceil \leq i-1 \leq n-1$, $\left\lceil \frac{n-2}{5} \right\rceil \leq i-1 \leq n-2$,

$\left\lceil \frac{n-3}{5} \right\rceil \leq i-1 \leq n-3$, $\left\lceil \frac{n-3}{5} \right\rceil \leq i-1 \leq n-4$

and

$\left\lceil \frac{n-3}{5} \right\rceil \leq i-1 \leq n-5$.

So

$\left\lceil \frac{n-1}{5} \right\rceil \leq i-1 \leq n-5$

and hence $\left\lceil \frac{n-1}{5} \right\rceil + 1 \leq i \leq n-4$.

(⟸) If $\left\lceil \frac{n-1}{5} \right\rceil + 1 \leq i \leq n-4$, then the result follows from Lemma 2.2.

Theorem 2.7.: For every $n \geq 6$ and $i \geq \left\lceil \frac{n}{5} \right\rceil$,

1) If

$$D\left(P_{n-1}^2, i-1\right) = D\left(P_{n-2}^2, i-1\right)$$
$$= D\left(P_{n-3}^2, i-1\right) = D\left(P_{n-4}^2, i-1\right) = \Phi$$

and

$$D\left(P_{n-5}^2, i-1\right) \neq \Phi$$

then

$$D\left(P_n^2, i\right) = \{\{3, 8, \cdots, n-7, n-2\}\}.$$

2) If

$$D\left(P_{n-2}^2, i-1\right) = D\left(P_{n-3}^2, i-1\right)$$

$$= D\left(P_{n-4}^2, i-1\right) = D\left(P_{n-5}^2, i-1\right) = \Phi$$

and

$$D\left(P_{n-1}^2, i-1\right) \neq \Phi$$

then

$$D\left(P_n^2, i\right) = \{[n]\}.$$

3) If

$$D\left(P_{n-1}^2, i-1\right) = \Phi, \quad D\left(P_{n-2}^2, i-1\right) \neq \Phi,$$

$$D\left(P_{n-3}^2, i-1\right) \neq \Phi, \quad D\left(P_{n-4}^2, i-1\right) \neq \Phi$$

and

$$D\left(P_{n-5}^2, i-1\right) \neq \Phi$$

then

$$\begin{aligned} &D\left(P_n^2, i\right) \\ &= \{\{3, 8, \cdots, n-4, n-2\}, X_1 \cup \{n\}, X_2 \cup \{n-1\}, \\ &X_3 \cup \{n-2\} / X_1 \in D\left(P_{n-3}^2, i-1\right), \\ &X_2 \in D\left(P_{n-4}^2, i-1\right), X_3 \in D\left(P_{n-5}^2, i-1\right)\} \end{aligned} \quad (2.1)$$

4) If $D\left(P_{n-3}^2, i-1\right) = \Phi$, $D\left(P_{n-2}^2, i-1\right) \neq \Phi$, $D\left(P_{n-1}^2, i-1\right) \neq \Phi$,

then

$$D\left(P_n^2, i\right) = \{[n] - \{x\}, \ x \in [n]\}.$$

5) If

$D\left(P_{n-1}^2, i-1\right) \neq \Phi$, $D\left(P_{n-2}^2, i-1\right) \neq \Phi$,

$D\left(P_{n-3}^2, i-1\right) \neq \Phi$, $D\left(P_{n-4}^2, i-1\right) \neq \Phi$, $D\left(P_{n-5}^2, i-1\right) \neq \Phi$

then

$$D\left(P_n^2, i\right) = \left\{X_1 \cup \{n\}, X_2 \cup \{n-1\}, X_3 \cup \{n-2\} / X_1 \in D\left(P_{n-1}^2, i-1\right), X_2 \in D\left(P_{n-2}^2, i-1\right), X_3 \in D\left(P_{n-3}^2, i-1\right), \left\{\{n-2\} \cup (X_4 - X_3)\right\}, \left\{\{n\} \cup (X_4 \cap X_3)\right\}, \left\{\{n-1\} \cup (X_5 - X_4)\right\}. / X_4 \in D\left(P_{n-4}^2, i-1\right), X_5 \in D\left(P_{n-5}^2, i-1\right)\right\}$$

(2.2)

Proof:

1) Since

$D\left(P_{n-1}^2, i-1\right) = D\left(P_{n-2}^2, i-1\right)$

$= D\left(P_{n-3}^2, i-1\right) = D\left(P_{n-4}^2, i-1\right) = \Phi$

and $D\left(P_{n-5}^2, i-1\right) \neq \Phi$ by Lemma 2.6 i) $n = 5k$, and $i = k$ for some $k \in N$.

Clearly the set $\{3, 8, \cdots, n-7, n-2\}$ has $\frac{n}{5}$ elements.

By the definition of P_n^2, 3 has joining with 1, 2, 4, 5, and 8 has joining with 6, 7, 9, 10. Therefore 3 and 8 covers all the vertices up to10for n = 10. Proceeding like this, we obtain that $\{3, 8, \cdots, n-7, n-2\}$ covers all Vertices up to n. The other sets with cardinality $\frac{n}{5}$ are $\{2, 7, 12, \cdots, n-3\}$, $\{4, 9, \cdots, n-1\}$ etc. In the first set, $n-3$ does not cover the vertex n. The second set does not cover the vertex 1and so on.

Therefore $\{3, 8, \cdots, n-7, n-2\}$ is the only dominating set of cardinality $\frac{n}{5} = k = i$.

2) We have

$D\left(P_{n-2}^2, i-1\right) = D\left(P_{n-3}^2, i-1\right)$

$= D\left(P_{n-4}^2, i-1\right) = D\left(P_{n-5}^2, i-1\right) = \Phi$

and

$$D\left(P_{n-1}^2, i-1\right) \neq \Phi.$$

By Lemma 2.6 (ii), we have $i = n$. So,

$$D\left(P_n^2, i\right) = \{[n]\}.$$

3) We have

$$D\left(P_{n-1}^2, i-1\right) = \Phi, \quad D\left(P_{n-2}^2, i-1\right) \neq \Phi,$$

$$D\left(P_{n-3}^2, i-1\right) \neq \Phi, \quad D\left(P_{n-4}^2, i-1\right) \neq \Phi$$

and

$$D\left(P_{n-5}^2, i-1\right) \neq \Phi.$$

By Lemma 2.6 (iii), $n = 5k + 2$ and

$$D\left(P_{n-5}^2, i-1\right) \neq \Phi.$$

for some $k \in N$.

Since

$$X = \{3, 8, \cdots, 5k-2\} \in D\left(P_{5k}^2, k\right),$$

$$X \cup \{5k+2\} \in D\left(P_{5k+2}^2, k+1\right).$$

Also, if $X \in D\left(P_{5k+1}^2, k\right)$ then

$$X \cup \{5k+1\} \in D\left(P_{5k+2}^2, k+1\right)$$

Therefore, we have

$$\left\{\{3, 8 \cdots 5k-2, 5k+1\}, X_1 \cup \{n\}, X_2 \cup \{n-1\}, X_3 \cup \{n-2\} \middle/ X_1 \in D\left(P_{n-3}^2, i-1\right),\right.$$
$$\left. X_2 \in D\left(P_{n-4}^2, i-1\right), X_3 \in D\left(P_{n-5}^2, i-1\right)\right\} \subseteq D\left(P_{5k+2}^2, k+1\right) \tag{2.3}$$

Now let $Y \in D\left(P_{5k+2}^2, k+1\right)$. Then $5k + 2$, or $5k + 1$ is in Y. If $5k + 2 \in Y$, then by Lemma 3, at least one vertex labeled $5k + 1$, $5k$ or $5k - 1$ is in Y. If $5k + 1$ or $5k$ is in Y, then

$$Y-\{5k+2\}\in D\left(P_{5k+1}^{2},k\right),$$

a contradiction; because $D\left(P_{5k+1}^{2},k\right)=\Phi$.

Hence $5k-1\in Y$, $5k\notin Y$ and $5k+1\notin Y$.

Therefore, $Y=X\cup\{5k+2\}$ for some $X\in D\left(P_{5k}^{2},k\right)$;

that is $Y=\{3,8,\cdots,5k-2,5k+1\}$. Now suppose that $5k+1\in Y$ and $5k+2\notin Y$. By Lemma 2.3 at least one vertex labeled $5k$, $5k-1$ or $5k-2$ is in Y. If $5k\in Y$ then

$$Y-\{5k+1\}\in D\left(P_{5k}^{2},k\right)=\{3,8,\cdots,5k-2\},$$

a contradiction because $5k\notin X$ for all $X\in D\left(P_{5k}^{2},k\right)$.

Therefore $5k-1$, or $5k-2$ is in Y, but $5k\notin Y$.

Thus $Y=X\cup\{5k+1\}$ for some $X\in D\left(P_{5k+1}^{2},k\right)$. So

$$D\left(P_{5k+2}^{2},k+1\right)\subseteq\left\{\{3,8,\cdots,5k-2,5k+1\},X_1\cup\{n\},X_2\cup\{n-1\},X_3\cup\{n-2\}/X_1\in D\left(P_{n-3}^{2},i-1\right),\right.$$
$$\left.X_2\in D\left(P_{n-4}^{2},i-1\right),X_3\in D\left(P_{n-5}^{2},i-1\right)\right\}\subseteq D\left(P_{5k+2}^{2},k+1\right) \tag{2.4}$$

From (2.3) and (2.4),

$$D\left(P_n^2,i\right)=\left\{\{3,8,\cdots,n-4,n-2\},\right.$$
$$\left.X_1\cup\{n\},X_2\cup\{n-1\},X_3\cup\{n-2\}/X_1\in D\left(P_{n-3}^{2},i-1\right),X_2\in D\left(P_{n-4}^{2},i-1\right),X_3\in D\left(P_{n-5}^{2},i-1\right)\right\}.$$

4) If

$D\left(P_{n-3}^{2},i-1\right)=\Phi$,

$D\left(P_{n-2}^{2},i-1\right)\neq\Phi$,

$D\left(P_{n-1}^{2},i-1\right)\neq\Phi$,

by Lemma 2.6 (iv) $i=n-1$.

Therefore $D\left(P_n^2,i\right)=\{[n]-\{x\}\ \ x\in[n]\}$.

5) $D\left(P_{n-1}^{2},i-1\right)\neq\Phi$, $D\left(P_{n-2}^{2},i-1\right)\neq\Phi$,

$$D\left(P_{n-3}^2, i-1\right) \neq \Phi, \quad D\left(P_{n-4}^2, i-1\right) \neq \Phi,$$

$$D\left(P_{n-5}^2, i-1\right) \neq \Phi.$$

Let $X_1 \in D\left(P_{n-2}^2, i-1\right)$, so at least one vertex labeled $n-1$, $n-2$ or $n-3$ is in X_1. If $n-1$, $n-2$ or $n-3 \in X_1$, then

$$X_1 \cup \{n\} \in D\left(P_n^2, i\right).$$

Let $X_2 \in D\left(P_{n-2}^2, i-1\right)$, then $n-2$ or $n-3$ or $n-4$ is in X_2.

If $n-2$, $n-3$ or $n-4 \in X_2$, then

$$X_2 \cup \{n-1\} \in D\left(P_n^2, i\right).$$

Now let $X_3 \in D\left(P_{n-3}^2, i-1\right)$ then n $-$ 3, $n-4$ or $n-5$ is in X_3. If $n-3$ or $n-4$ or $n-5 \in X_3$, then

$$X_3 \cup \{n-2\} \in D\left(P_n^2, i\right).$$

Now let $X_4 \in D\left(P_{n-4}^2, i-1\right)$, then $n-4$, or $n-5$ or $n-6$ is in X_4.

If $n-4 \in X_4$, then $X_4 \cup \{n\} \in D\left(P_n^2, i\right)$. If $n-5 \in X_4$, then

$$X_4 \cup \{n-1\} \in D\left(P_n^2, i\right).$$

If $n-6 \in X_4$, then $X_4 \cup \{n-2\} \in D\left(P_n^2, i\right)$. Thus we have

$$\left\{X_1 \cup \{n\}, X_2 \cup \{n-1\}, X_3 \cup \{n-2\} / X_1 \in D\left(P_{n-1}^2, i-1\right), X_2 \in D\left(P_{n-2}^2, i-1\right), X_3 \in D\left(P_{n-3}^2, i-1\right)\right\}$$
$$\cup \left\{\left\{\{n-2\} \cup (X_4 - X_3)\right\}, \left\{\{n\} \cup (X_4 \cap X_3)\right\}, \left\{\{n-1\} \cup (X_5 - X_4)\right\} / X_4 \in D\left(P_{n-4}^2, i-1\right), X_5 \in D\left(P_{n-5}^2, i-1\right)\right\} \subseteq D\left(P_n^2, i\right) \quad (2.5)$$

Now suppose that $n-1 \in Y$, $n \in Y$ then by Lemma 2.3, at least one vertex labeled $n-2$, $n-3$ or in Y. If $n-2 \in Y$ then $Y = (X_4 - X_3) \cup \{n-2\}$ for some

$$X_4 \in D\left(P_{n-4}^2, i-1\right), \quad X_3 \in D\left(P_{n-3}^2, i-1\right).$$

Now suppose that, $n-5 \in Y$ & $n-4 \in Y$, then by Lemma 2.3 one vertex labeled $n-6$, $n-7$, in Y.

If n $-$ 4 $\in$ Y, then $Y = (X_5 - X_4) \cup \{n-1\}$ for

$X_4 \in D(P_{n-4}^2, i-1)$, $X_5 \in D(P_{n-5}^2, i-1)$,

So

$$D(P_n^2, i) \subseteq \{X_1 \cup \{n\}, X_2 \cup \{n-1\}, X_3 \cup \{n-2\} / X_1 \in D(P_{n-1}^2, i-1), X_2 \in D(P_{n-2}^2, i-1), X_3 \in D(P_{n-3}^2, i-1)\}$$
$$\cup \{\{\{n-2\} \cup (X_4 - X_3)\}, \{\{n\} \cup (X_4 \cap X_5)\}, \{\{n-1\} \cup (X_5 - X_4)\} / X_4 \in D(P_{n-4}^2, i-1), X_5 \in D(P_{n-5}^2, i-1)\}$$

(2.6)

From (2.5) & (2.6)

$$D(P_n^2, i) = \{X_1 \cup \{n\}, X_2 \cup \{n-1\}, X_3 \cup \{n-2\} / X_1 \in D(P_{n-1}^2, i-1), X_2 \in D(P_{n-2}^2, i-1), X_3 \in D(P_{n-3}^2, i-1)\}$$
$$\cup \{\{\{n-2\} \cup (X_4 - X_3)\}, \{\{n\} \cup (X_4 \cap X_5)\}, \{\{n-1\} \cup (X_5 - X_4)\} / X_4 \in D(P_{n-4}^2, i-1), X_5 \in D(P_{n-5}^2, i-1)\}$$

3. DOMINATION POLYNOMIAL OF P_n^2

Let $D(P_n^2, x) = \sum_{i=\lceil \frac{n}{5} \rceil}^{n} d(P_n^2, i) x^i$ be the domination polynomial of a path P_n^2. In this section, we derive the expression for $D(P_n^2, x)$.

Theorem 3.1.

a) If $D(P_n^2, i)$ is the family of dominating sets with cardinality i of P_n^2, then

$$d(P_n^2, i) = d(P_{n-1}^2, i-1) + d(P_{n-2}^2, i-1) + d(P_{n-3}^2, i-1) + d(P_{n-4}^2, i-1) + d(P_{n-5}^2, i-1)$$

where $d(P_n^2, i) = |D(P_n^2, i)|$.

b) For every $n \geq 6$,

$$D(P_n^2, x) = x\left[D(P_{n-1}^2, x) + D(P_{n-2}^2, x) + D(P_{n-3}^2, x) + D(P_{n-4}^2, x) + D(P_{n-5}^2, x)\right]$$

with the initial values

$$D(P_1^2, x) = x, \quad D(P_2^2, x) = x^2 + 2x,$$

$$D(P_3^2, x) = x^3 + 3x^2 + 3x,$$

$$D(P_4^2, x) = x^4 + 4x^3 + 6x^2 + 2x,$$

$$D(P_5^2, x) = x^5 + 5x^4 + 10x^3 + 8x^2 + x.$$

Proof:

a) Using (i), (ii), (iii), (iv) and (v) of Theorem 2.7, we prove (a) part.

1) Suppose (i) of Theorem 2.7 holds.

From (v), we have

$$D(P_n^2,i)=\{\Phi \cup \{n-1\} \cup (X_5) \ \ X_5 \in D(P_{n-5}^2,i-1)\}.$$

Since

$$D(P_{n-1}^2,i-1)=D(P_{n-2}^2,i-1)=\Phi,$$

$$D(P_{n-3}^2,i-1)=D(P_{n-4}^2,i-1)=\Phi$$

$$|D(P_{n-1}^2,i-1)|=|D(P_{n-2}^2,i-1)|=0$$

$$|D(P_{n-3}^2,i-1)|=|D(P_{n-4}^2,i-1)|=0$$

Therefore $|D(P_n^2,i)|=|D(P_{n-5}^2,i-1)|$.

Therefore $d(P_n^2,i)=d(P_{n-5}^2,i-1)$.

Therefore, i[illegible]s case $d(P_n^2,i)=d(P_{n-5}^2,i-1)$ holds.

2) Suppose (ii) of Theorem 2.7 holds.

From (v), we have

$$D(P_n^2,i)=\{\Phi \cup \{n\} \cup (X_1) \ \ X_1 \in D(P_{n-1}^2,i-1)\}.$$

Since $D(P_{n-1}^2,i-1)=D(P_{n-3}^2,i-1)=\Phi$

$$D(P_{n-4}^2,i-1)=D(P_{n-5}^2,i-1)=\Phi$$

$$|D(P_{n-2}^2,i-1)|=|D(P_{n-3}^2,i-1)|=0$$

$$|D(P_{n-4}^2,i-1)|=|D(P_{n-5}^2,i-1)|=0$$

Therefore $|D(P_n^2,i)|=|D(P_{n-1}^2,i-1)|$

Therefore $d\left(P_n^2,i\right)=d\left(P_{n-1}^2,i-1\right)$.

Therefore, In this case $d\left(P_n^2,i\right)=d\left(P_{n-1}^2,i-1\right)$ holds.

3) Suppose (iii) of Theorem 2.7 holds.

From (v), we have

$$D\left(P_n^2,i\right)=\{\Phi\cup\{n-1\}\cup(X_2),\{n-2\}\cup(X_3),\cup(X_4-X_3)\cup\{n-2\},(X_4\cap X_3)\cup\{n\},$$
$$(X_5-X_4)\cup\{n-1\}/X_2\in D\left(P_{n-2}^2,i-1\right),X_3\in D\left(P_{n-3}^2,i-1\right),X_4\in D\left(P_{n-4}^2,i-1\right)\text{ and }X_5\in D\left(P_{n-5}^2,i-1\right)\}$$

Since

$$D\left(P_{n-1}^2,i-1\right)=\Phi,\ \left|D\left(P_{n-1}^2,i-1\right)\right|=0.$$

Therefore

$$\left|D\left(P_n^2,i\right)\right|=\left|D\left(P_{n-2}^2,i-1\right)\right|\cup\left|D\left(P_{n-3}^2,i-1\right)\right|$$
$$\cup\left|D\left(P_{n-4}^2,i-1\right)\right|\cup\left|D\left(P_{n-5}^2,i-1\right)\right|.$$

Therefore

$$d\left(P_n^2,i\right)=d\left(P_{n-2}^2,i-1\right)$$
$$+d\left(P_{n-3}^2,i-1\right)+d\left(P_{n-4}^2,i-1\right).$$
$$+d\left(P_{n-5}^2,i-1\right).$$

4) Suppose (v) of Theorem 2.7 holds.

From (v), we have

$$D\left(P_n^2,i\right)=\{\{n\}\cup(X_1),\{n-1\}\cup(X_2),\{n-2\}\cup(X_3),\cup(X_4-X_3)\cup\{n-2\},(X_4\cap X_3)\cup\{n\},$$
$$(X_5-X_4)\cup\{n-1\}/X_1\in D\left(P_{n-1}^2,i-1\right),X_2\in D\left(P_{n-2}^2,i-1\right),X_3\in D\left(P_{n-3}^2,i-1\right),$$
$$X_4\in D\left(P_{n-4}^2,i-1\right)\text{ and }X_5\in D\left(P_{n-5}^2,i-1\right)\}.$$

Therefore

$$\left|D\left(P_n^2,i\right)\right|=\left|D\left(P_{n-1}^2,i-1\right)\right|\cup\left|D\left(P_{n-2}^2,i-1\right)\right|\cup\left|D\left(P_{n-3}^2,i-1\right)\right|\cup\left|D\left(P_{n-4}^2,i-1\right)\right|\cup\left|D\left(P_{n-5}^2,i-1\right)\right|.$$

Therefore

$$\Sigma d\left(P_n^2,i\right)=d\left(P_{n-1}^2,i-1\right)+d\left(P_{n-2}^2,i-1\right)+d\left(P_{n-3}^2,i-1\right)+d\left(P_{n-4}^2,i-1\right)+d\left(P_{n-5}^2,i-1\right).$$

Therefore, we have the theorem.

b)

$$d(P_n^2,i)x^i = d(P_{n-1}^2,i-1)x^i + d(P_{n-2}^2,i-1)x^i + d(P_{n-3}^2,i-1)x^i + d(P_{n-4}^2,i-1)x^i + d(P_{n-5}^2,i-1)x^i$$

$$\sum d(P_n^2,i)x^i = \sum d(P_{n-1}^2,i-1)x^i + \sum d(P_{n-2}^2,i-1)x^i + \sum d(P_{n-3}^2,i-1)x^i + \sum d(P_{n-4}^2,i-1)x^i + \sum d(P_{n-5}^2,i-1)x^i$$

$$d(P_n^2,i)x^i = x\sum d(P_{n-1}^2,i-1)x^{i-1} + x\sum d(P_{n-2}^2,i-1)x^{i-1} + x\sum d(P_{n-3}^2,i-1)x^{i-1}$$
$$+x\sum d(P_{n-4}^2,i-1)x^{i-1} + x\sum d(P_{n-5}^2,i-1)x^{i-1}$$

$$\sum d(P_n^2,i)x^i = x\Big[\sum d(P_{n-1}^2,i-1)x^{i-1} + \sum d(P_{n-2}^2,i-1)x^{i-1} + \sum d(P_{n-3}^2,i-1)x^{i-1}$$
$$+\sum d(P_{n-4}^2,i-1)x^{i-1} + \sum d(P_{n-5}^2,i-1)x^{i-1}\Big]$$

$$D(P_n^2,x) = x\Big[D(P_{n-1}^2,x) + D(P_{n-2}^2,x) + D(P_{n-3}^2,x) + D(P_{n-4}^2,x) + D(P_{n-5}^2,x)\Big]$$

with the initial values

$$D(P_1^2,x) = x,\ D(P_2^2,x) = x^2 + 2x,$$

$$D(P_3^2,x) = x^3 + 3x^2 + 3x$$

$$D(P_4^2,x) = x^4 + 4x^3 + 6x^2 + 2x,$$

$$D(P_5^2,x) = x^5 + 5x^4 + 10x^3 + 8x^2 + x.$$

We obtain $d(P_n^2,i)$ for $1 \le n \le 10$ as shown in **Table 1**.

Table 1. $d(P_n^2,i)$, the number of dominating set of P_n^2 with cardinality i.

i / n	1	2	3	4	5	6	7	8	9	10
1	1									
2	2	1								
3	3	3	1							
4	2	6	4	1						
5	1	8	10	5	1					
6	0	9	18	15	6	1				
7	0	8	27	33	21	7	1			
8	0	6	34	60	54	28	8	1		
9	0	3	37	93	114	82	36	9	1	
10	0	1	34	126	206	196	118	45	10	1

In the following Theorem, we obtain some properties of $d(P_n^2, i)$

Theorem 3.2. The following properties hold for the coefficients of $D(P_n^2, x)$;

1) $d(P_{5n}^2, n) = 1$ for every $n \in N$.

2) $d(P_n^2, n) = 1$ for every n ∈ N

3) $d(P_n^2, n-1) = n$ for every $n \geq 2$

4) $d(P_n^2, n-2) = \frac{n(n-1)}{2}$ for every $n \geq 3$

5) $d(P_n^2, n-3) = nC_3 - 2$ for every $n \geq 4$

6) $d(P_n^2, n-4) = nC_4 - (2n-6)$ for every $n \geq 5$

7) $d(P_{5n-1}^2, n) = n+1$ for every n ∈ N.

Proof:

1) Since $D(P_{5n}^2, n) = \{3, 8, \cdots, 5k-2\}$, we have $d(P_{5n}^2, n) = 1$.

2) Since $D(P_n^2, n) = \{[n]\}$, we have the result

3) Since $D(P_n^2, n-1) = \{[n] - \{x\} \; x \in [n]\}$, we have $d(P_n^2, n-1) = n$.

4) By induction on n. The result is true for n = 3.

$\text{L.H.S.} = d(P_3^2, 1) = 3$ (from table)

$\text{R.H.S} = \left(\frac{3 \times 2}{2}\right) = 3$

Therefore, the result is true for n = 4.

Now suppose that the result is true for all numbers less than "n" and we prove it for n.

By Theorem 3.1,

$$d\left(P_n^2, n-2\right) = d\left(P_{n-1}^2, n-3\right) + d\left(P_{n-2}^2, n-3\right)$$
$$+d\left(P_{n-3}^2, n-3\right) + d\left(P_{n-4}^2, n-3\right) + d\left(P_{n-5}^2, n-3\right)$$
$$= \frac{(n-1)(n-2)}{2} + (n-2) + 1$$
$$= \frac{n^2 - 3n + 2 + 2n - 4 + 2}{2} = \frac{n^2 - n}{2} = \frac{n(n-1)}{2}.$$

5) By induction on n. The result is true for n = 4.

$$\text{L.H.S} = d\left(P_4^2, 1\right) = 2 \text{ (from table)}$$

$$\text{R.H.S} = d\left(P_4^2, 1\right) = \frac{4\times 3\times 2}{6} - 2$$

Therefore the result is true for n = 1. Now suppose the result is for all natural numbers less than n.

By Theorem 3.1,

$$d\left(P_n^2, n-3\right) = d\left(P_{n-1}^2, n-4\right) + d\left(P_{n-2}^2, n-4\right) + d\left(P_{n-3}^2, n-4\right) + d\left(P_{n-4}^2, n-4\right) + d\left(P_{n-5}^2, n-4\right)$$
$$= \frac{(n-1)(n-2)(n-3)}{6} - 2 + \frac{(n-1)(n-2)}{2} + (n-3) + 1$$
$$= \frac{1}{6}\left[(n-1)(n-2)(n-3) + 3(n-2)(n-3) + 6(n-3) - 12 + 6\right]$$
$$= \frac{1}{6}\left[(n-3)\left[(n-2)(n+2) + 6\right] - 6\right] = \frac{1}{6}\left[(n-3)\left[n^2 - 4 + 6\right] - 6\right]$$
$$= \frac{1}{6}\left[n^3 + 2n - 3n^2 - 6 - 6\right] = \frac{1}{6}\left[n^3 - 3n^2 + 2n - 12\right]$$

6) By induction on n. Let n = 5

$$\text{L.H.S} = d\left(P_5^2, 1\right) = 1 \text{ (from table)}$$

$$\text{R.H.S} = \frac{n(n-1)(n-2)(n-3)}{1\times 2\times 3\times 4} - (2n-6) = \frac{5\times 4\times 3\times 2}{1\times 2\times 3\times 4} - (2\times 5 - 6) = 5 - 4 = 1$$

Therefore the result is true for n =1.

Now suppose that the result is true for all natural numbers less than or equal to n.

$$d\left(P_n^2, n-4\right) = d\left(P_{n-1}^2, n-5\right) + d\left(P_{n-2}^2, n-5\right) + d\left(P_{n-3}^2, n-5\right) + d\left(P_{n-4}^2, n-5\right) + d\left(P_{n-5}^2, n-5\right)$$
$$= \frac{n(n-1)(n-2)(n-3)(n-4)}{24}(2(n-1)-6) + \frac{(n-2)(n-3)(n-4)}{6}2 + \frac{(n-3)(n-4)}{2} + (n-4) + 1$$
$$= \frac{1}{24}\left[(n-1)(n-2)(n-3)(n-4) - 12(n-2)(n-3)(n-4) + 24(n-4) + 4(n-2)(n-3)(n-4) + 48n + 192 - 48 + 24\right]$$
$$= \frac{1}{24}\left[(n-2)(n-3)(n-4)[n-1+4] + 12(n-4)[n-3+2] - 48n + 168\right]$$
$$= \frac{1}{24}\left[(n-2)(n-3)(n-4)(n+3) + 12(n-4)(n-1) - 48n + 168\right]$$
$$= \frac{1}{24}\left[(n-4)\left[(n-2)(n-3)(n+3) + 12(n-1) - 48n + 168\right]\right]$$
$$= \frac{1}{24}\left[(n-4)\left[(n-2)\left(n^2-9\right) + 12n - 12 - 48n\right]\right] = \frac{1}{24}\left[(n-4)\left(n^3 - 9n - 2n^2 + 18\right) - 36n + 156\right]$$
$$= \frac{1}{24}\left[n^4 - 9n^2 - 2n^3 + 18n - 4n^3 + 36n + 8n^2 + 72 - 36n + 156\right] = \frac{1}{24}\left[n^4 - 6n^3 + 11n^2 - 54n + 144\right]$$

7) From the table it is true.

Theorem 3.3.

1) $\sum_{i=n}^{5n} d\left(P_i^2, n\right) = 5\sum_{i=2}^{5n-5} d\left(P_i^2, n-1\right)$

2) For every $j \geq \left\lceil \frac{n}{5} \right\rceil$,

$$d\left(P_{n+1}^2, j+1\right) - d\left(P_n^2, j+1\right) = d\left(P_n^2, j\right) - d\left(P_{n-5}^2, j\right)$$

3) If $S_n = \sum_{i=\left\lceil \frac{n}{5} \right\rceil}^{n} d\left(P_n^2, j\right)$, then for every $n \geq 6$,

$S_n = S_{n-1} + S_{n-2} + S_{n-3} + S_{n-4} + S_{n-5}$ with initial values $S_1 = 1$, $S_2 = 3$, $S_3 = 7$, $S_4 = 13$ and $S_5 = 25$.

Proof:

1) Proof by induction on n First suppose that $n = 2$ then,

$$\sum_{i}^{10} d\left(P_i^2, 2\right) = 45 = 5\sum_{i=2}^{5} d\left(P_i^2, 1\right)$$
$$\sum_{i=k}^{5k} d\left(P_i^2, k\right) = \sum_{i=k}^{5k} d\left(P_{i-1}^2, k-1\right) + \sum_{i=k}^{5k} d\left(P_{i-2}^2, k-1\right) + \sum_{i=k}^{5k} d\left(P_{i-3}^2, k-1\right) + \sum_{i=k}^{5k} d\left(P_{i-4}^2, k-1\right) + \sum_{i=k}^{5k} d\left(P_{i-5}^2, k-1\right)$$
$$= 5\sum_{i=k-1}^{5(k-1)} d\left(P_{i-1}^2, k-2\right) + 5\sum_{i=k-1}^{5(k-1)} d\left(P_{i-2}^2, k-2\right) + 5\sum_{i=k-1}^{5(k-1)} d\left(P_{i-3}^2, k-2\right) + 5\sum_{i=k-1}^{5(k-1)} d\left(P_{i-4}^2, k-2\right) + 5\sum_{i=k-1}^{5(k-1)} d\left(P_{i-5}^2, k-2\right)$$
$$= 5\sum_{i=k-1}^{5k-5} d\left(P_i^2, k-1\right)$$

We have the result.

2) By Theorem 3.1, We have

$$d(P_{n+1}^2, j+1) - d(P_n^2, j+1) = d(P_n^2, j) + d(P_{n-1}^2, j) + d(P_{n-2}^2, j) + d(P_{n-3}^2, j)$$
$$+ d(P_{n-4}^2, j) + d(P_{n-1}^2, j) - d(P_{n-1}^2, j) - d(P_{n-2}^2, j) - d(P_{n-3}^2, j) - d(P_{n-4}^2, j) - d(P_{n-5}^2, j)$$
$$d(P_{n+1}^2, j+1) - d(P_n^2, j+1) = d(P_n^2, j) - d(P_{n-5}^2, j)$$

Therefore we have the result 3) By theorem (3.1), we have

$$S_n = \sum_{i=\left\lceil \frac{n}{5} \right\rceil}^{n} d(P_n^2, j)$$

$$S_n = \sum_{i=\left\lceil \frac{n}{5} \right\rceil}^{n} \left[d(P_{n-1}^2, j-1) + d(P_{n-2}^2, j-1) + d(P_{n-3}^2, j-1) + d(P_{n-4}^2, j-1) + d(P_{n-5}^2, j-1) \right]$$

$$= \sum_{i=\left\lceil \frac{n}{5} \right\rceil-1}^{n-1} d(P_{n-1}^2, j-1) + \sum_{i=\left\lceil \frac{n}{5} \right\rceil-1}^{n-1} d(P_{n-2}^2, j-1) + \sum_{i=\left\lceil \frac{n}{5} \right\rceil-1}^{n-1} d(P_{n-3}^2, j-1) + \sum_{i=\left\lceil \frac{n}{5} \right\rceil-1}^{n-1} d(P_{n-4}^2, j-1) + \sum_{i=\left\lceil \frac{n}{5} \right\rceil-1}^{n-1} d(P_{n-5}^2, j-1)$$

$$S_n = S_{n-1} + S_{n-2} + S_{n-3} + S_{n-4} + S_{n-5}.$$

4. CONCLUSIONS

In [2], the domination polynomial of path was studied and obtained the very important property,

$$d(P_n^2, i) = d(P_{n-1}^2, i-1) + d(P_{n-2}^2, i-1) + d(P_{n-3}^2, i-1).$$

It is interesting that we have derived an analogues relation for the square of path of the form

$$d(P_n^2, i) = d(P_{n-1}^2, i-1) + d(P_{n-2}^2, i-1) + d(P_{n-3}^2, i-1)$$
$$+ d(P_{n-4}^2, i-1) + d(P_{n-5}^2, i-1)$$

One can characterise the roots of the polynomial $D(P_n^2, x)$ and identify whether they are real or complex.

Another interesting character r to be investigated is whether $D(P_n^2, x)$ is log concave are not.

REFERENCES

1. S. Alikhani and Y.-H. Peng, "Introduction to Domination Polynomial of a Graph," .arXiv:0905.2251v1[math.co], 2009.
2. S. Alikhani and Y.-H. Peng, "Domination Sets and Domination Polynomials of Paths," International Journal of Mathematics and Mathematical Sciences, Vol. 2009, 2009, Article ID: 542040.
3. G. Chartand and P. Zhang, "Introduction to Graph Theory," McGraw-Hill, Boston, 2005.

CHAPTER 9

On powers that are Sums of Consecutive Like Powers

Vandita Patel and Samir Siksek*

Mathematics Institute, University of Warwick, Coventry CV4 7AL, UK

ABSTRACT

Let $k \geq 2$ be even, and let r be a non-zero integer. We show that for almost all $d \geq 2$ (in the sense of natural density), the equation

$x^k + (x + r)^k + \cdots + (x + (d - 1)r)^k = y^n$, x, y, n $\in$ Z,n $\geq$ 2,

has no solutions.

Keywords: Exponential equation Bernoulli polynomial Newton polygon

1 BACKGROUND

The problem of cubes that are sums of consecutive cubes goes back to Euler ([10] art. 249) who noted the remarkable relation $3^3 + 4^3 + 5^3 = 6^3$. Similar problems were considered by several mathematicians during the nineteenth and early twentieth century as surveyed in Dickson's *History of the Theory of Numbers* ([7] p. 582–588). These questions are still of interest today. For example, both Cassels [5] and Uchiyama [17] determined the squares that can be written as sums of three consecutive cubes. Stroeker [16] determined all squares that are expressible as the sum of $2 \leq d \leq 50$ consecutive cubes, using a method based on linear forms in elliptic logarithms. More recently, Bennett, Patel and Siksek [2] determined all perfect powers that are expressible as sums of $2 \leq d \leq 50$ consecutive cubes, using linear forms in logarithms, sieving and Frey curves. There has been some interest in powers that are sums of k-th powers for other exponents k. For example, the solutions to the equation

$$x^k + (x + 1)k + (x + 2)k = y^n, x, y, n \in Z, n \geq 2,$$

have been determined by Zhongfeng Zhang [18] for $k = 2, 3, 4$ and by Bennett, Patel and Siksek [1] for $k = 5, 6$, and similar problems are considered by Soydan [15].

In view of the above, it is natural to consider the equation

$$x^k + (x + 1)k^+ \cdots + (x + d - 1)k = y^n, x, y, n \in Z, \mathrm{n} \geq 2 \quad (1)$$

with $k, d \geq 2$. This was studied by Zhang and Bai [19] for $k = 2$. They show that if q is a prime $\equiv \pm 5(\mathrm{mod}12) \equiv \pm 5\ (\mathrm{mod}12)$ and $\upsilon_q(d) = 1$ then Eq. (1) has no solutions for $k = 2$; here $\upsilon_q(d)$ denotes the q-adic valuation of d. It follows from a standard result in analytic number theory (as we shall see later) that the set of d for which there is a solution with $k = 2$ has natural density 0. We prove the following generalization to all even exponents k.

Theorem 1 Let $k \geq 2$ be even and let r be a non-zero integer. Write $A_{k,r}$ for the set of integers $d \geq 2$ such that the equation

$$x^k + (x + r)^k + \cdots + (x + (d - 1)r)^k = y^n, x, y, n \in Z, \mathrm{n} \geq 2, \quad (2)$$

has a solution (x, y, n). Then $A_{k,r}$ has natural density 0; by this we mean

$$\lim_{X \to \infty} \frac{\#\{d \in \mathcal{A}_{k,r}\ :\ d \leq X\}}{X} = 0.$$

If k is odd, then $A_{k,r}$ contains all of the odd d: we can take $(x, y, n) = (r(1 - d)/2, 0, n)$. Thus the conclusion of the theorem does not hold for odd k.

2 SOME PROPERTIES OF BERNOULLI NUMBERS AND POLYNOMIALS

In this section we summarise some classical properties of Bernoulli numbers and polynomials. These are found in many references, including [8]. The Bernoulli numbers BkBk are defined via the expansion

$$\frac{x}{e^x - 1} = \sum_{k=0}^{\infty} B_k \frac{x^k}{k!}$$

The first few Bernoulli numbers are

$B_0 = 1$, $B_1 = -1/2$, $B_2 = 1/6$, $B_3 = 0$, $B_4 = -1/30$, $B_5 = 0$, $B_6 = 1/42$.

It is easy to show that $B_{2k+1} = 0$ for all $k \geq 1$. The B_k are rational numbers, and the von Staudt–Clausen theorem asserts that for $k \geq 2$ even

$$B_k + \sum_{(p-1)|k} \frac{1}{p} \in \mathbb{Z}$$

where the sum ranges over primes p such that $(p - 1)|k$.

The k-th Bernoulli polynomial can be defined by

$$B_k(x) = \sum_{m=0}^{k} \binom{k}{m} B_m x^{k-m} \tag{3}$$

Thus it is a monic polynomial with rational coefficients, and all primes appearing in the denominators are bounded by k+1k+1. It satisfies the symmetry

$$B_k(1 - x) = (-1)^k B_k(x), \tag{4}$$

the identity

$$B_k(x + 1) - B_k(x) = kx^{k-1}, \tag{5}$$

and the recurrence

$$B'_k(x) = kB_{k-1}(x). \tag{6}$$

Whilst all the above results have been known since at least the nineteenth century, we also make use of the following far more recent and difficult theorem due to Brillhart [3] and Dilcher [9].

Theorem 2 (Brillhart and Dilcher) The Bernoulli polynomials are squarefree.

2.1 Relation to sums of consecutive like powers

Lemma 2.1 Let r be a non-zero integer and $k, d \geq 1$. Then

$$x^k + (x+r)^k + \cdots + (x + r(d-1))^k = \frac{r^k}{k+1}\left(B_{k+1}\left(\frac{x}{r} + d\right) - B_{k+1}\left(\frac{x}{r}\right)\right)$$

This formula can be found in ([8] Section 24.4), but is easily deduced from the identity (5).

Lemma 2.2 Let $q \geq k + 3$ be a prime. Let a, r, d be integers with $d \geq 2$, and $r \neq 0$. Suppose $q|d$ and $q \nmid r$. Then

$a^k + (a + r)k + \cdots + (a + r(d - 1))k \equiv r^k \cdot d \cdot B^k(a/r) (\bmod q^2)$.

Proof By Taylor's Theorem

$$B_{k+1}(x + d) = B_{k+1}(x) + d \cdot B'_{k+1}(x) + \frac{d^2}{2} B^{(2)}_{k+1}(x) + \cdots + \frac{d^{k+2}}{(k+2)!} \cdot B^{(k+2)}_{k+1}(x)$$

It follows from the assumption $q \geq k + 3$ that the coefficients of $B_{k+1}(x)$ are q-adic integers. Thus the coefficients of the polynomials $B^{(i)}_{k+1}(x)/i!$ are also q-adic integers. As $q|d$ and $q \nmid r$ we have

$$B_{k+1}\left(\frac{a}{r} + d\right) - B_{k+1}\left(\frac{a}{r}\right) \equiv d \cdot B'_{k+1}(a/r) \pmod{q^2}$$

The lemma follows from (6) and Lemma 2.1. □

Lemma 2.3 Let k, r be integers with $k \geq 2$ and $r \neq 0$. Let $q \geq k + 3$ be a prime not dividing r such that the congruence $B_k(x) \equiv 0 \pmod q$ has no solutions. Let d be a positive integer such that $\upsilon q(d) = 1$. Then Eq. (2) has no solutions (i.e. $d \notin A_{k,r}$).

Proof Suppose $(x, y, n) = (a,b,n)$ is a solution to (2). By Lemma 2.2,

$r^k \cdot d \cdot B^k(a/r) \equiv b^n (\bmod q^2)$.

However, the hypotheses of the lemma ensure that the left-hand side has q-adic valution 1. Thus $\upsilon_q(b^n) = 1$ giving a contradiction. □

Remarks For $k \geq 3$ odd, the k-th Bernoulli polynomial has known rational roots 0, 1 / 2, 1. Thus the criterion in the lemma fails to hold for all primes q. For even $k \geq 2$ we shall show that there is a positive density of primes q such that $B^k(x)$ has no roots modulo q.

- The second Bernoulli polynomial is $B_2(x) = x^2 - x + 1/6$. By quadratic reciprocity, this has a root modulo $q \nmid 6$ if and only if $q \equiv \pm 1 (\bmod 12)$. We thus recover the result of Bai and Zhang mentioned in the introduction: if $q \equiv \pm 5 (\bmod 12)$ and $\upsilon_q(d) = 1$ then (1) has no solutions with $k = 2$.

3 A GALOIS PROPERTY OF EVEN BERNOULLI POLYNOMIALS

Proposition 3.1 Let $k \geq 2$ be even, and let G_k be the Galois group of the Bernoulli polynomial $B_k(x)$. Then there is an element $\mu \in G_k$ that acts freely on the roots of $B_k(x)$.

There is a long-standing conjecture that the even Bernoulli polynomials are irreducible; see for example [3, 4, 14]. One can easily deduce Proposition 3.1 from this conjecture. We give an unconditional proof of Proposition 3.1 in Sect. 5. As noted previously, if k is odd, then $B_k(x)$ has rational roots 0, 1 / 2, 1, so the conclusion of the proposition certainly fails for odd k.

3.1 A density result

Let A be a set of positive integers. For X positive, define

$A(X) = \#\{d \in A : d \leq X\}$.

The *natural density* of AA is defined as the limit (if it exists)

$$\delta(\mathcal{A}) = \lim_{X \to \infty} \frac{\mathcal{A}(X)}{X}$$

For a given prime q, define

$A^{(q)} = \{d \in A : v_q(d) = 1\}$.

We shall need the following result of Niven ([13] Corollary 1).

Theorem 3 (Niven) Let $\{q_i\}$ be a set of primes such that $\delta(A(q_i)) = 0$ and $\sum qi^{-1} = \infty$. Then $\delta(A) = 0$.

3.2 Proposition 3.1 implies Theorem 1

We now suppose Proposition 3.1 and use it to deduce Theorem 1. Let $k \geq 2$ be an even integer. Write G_k for the Galois group of the Bernoulli polynomial $B_k(x)$. Let $\mu \in G_k$ be the element acting freely on the roots of $B_k(x)$ whose existence is asserted by Proposition 3.1. By the Chebotarev density theorem ([6] Chapter VIII) there is a set of primes $\{q_i\}^{\infty}{}_{i=1}$ having positive Dirichlet density such that for each $q = q_i$, the Frobenius element Frob$q \in G_k$ is conjugate to μ. We omit from $\{q_i\}$ (without affecting the density) the following:

- primes $q \leq k + 2$;

- primes q dividing r;
- primes q dividing the numerator of the discriminant of $B_k(x)$ (which is non-zero by Theorem 2).

As μ acts freely on the roots of $B_k(x)$, it follows that the polynomial $B_k(x)$ has no roots modulo any of the q_i. Now let $A = A_{k,r}$ be as in the statement of Theorem 1. By Lemma 2.3, if $\upsilon_{q_i}(d) = 1$ then $d \notin A$. It follows that $A(q_i) = \emptyset$. By Theorem 3, we have $\delta(A) = 0$ as required.

4 THE 2-ADIC NEWTON POLYGONS OF EVEN BERNOULLI POLYNOMIALS

Lemma 4.1 Let $k \geq 2$ be even and write $k = 2^s t$ where t is odd and $s \geq 1$. The 2-adic Newton polygon of $B_k(x)$ consists two segments:

(i) a horizantal segment joining the points $(0, -1)$ and $(k - 2^s, -1)$;
(ii) a segment joining the points $(k - 2^s, -1)$ and $(k, 0)$ of slope $1/2^s$.

Proof Consider the definition of $B_k(x)$ in (3). We know that $B_0 = 1, B_1 = -1/2$ and $B_m = 0$ for all odd $m \geq 3$. From the von Staudt–Clausen theorem, we know that $\upsilon_2(B_m) = -1$ for even $m \geq 2$. It follows that the Newton polygon is bounded below by the Horizontal line $y = -1$.

We shall need to make use of the following result of Kummer (see [11]): if p is a prime, and u, v are positive integers then

$$\binom{u}{v} \equiv \binom{u_0}{v_0}\binom{u_1}{v_1} \pmod{p},$$

where u_0, u_1 are respectively the remainder and quotient on dividing u by p, and likewise v_0, v_1 are respectively the remainder and quotient on dividing v by p. Here we adopt the convention $\binom{r}{s} = 0$ if $r < s$. Applying this with $p = 2$ we see that

$$\binom{k}{2^s} = \binom{2^s t}{2^s} \equiv \binom{t}{1} \equiv t \equiv 1 \pmod{2}.$$

Thus the coefficient of x^{k-2} in $B_k(x)$ has 2-adic valuation -1. Since the constant coefficient of $B_k(x)$ also has valuation -1, we obtain the segment (i) as part of the Newton polygon. We also see that for $0 < v < 2^s$,

$$\binom{k}{\nu} \equiv 0 \pmod{2}$$

and so the valuation of the coefficient of $x^{k-\nu}$ is ≥ 0. Finally the coefficient of x^k is $B_0 = 1$ and so has valuation 0. This gives segment (ii) and completes the proof. □

Remarks Inkeri [12] showed that $B_k(x)$ has no rational roots for k even. His proof required very precise (and difficult) estimates for the real roots of $B_k(x)$. Lemma 4.1 allows us to give a much simpler proof of the following stronger result.

Theorem 4

Let k be even. Then $B_k(x)$ has no roots in Q_2.

Proof Indeed, suppose $\alpha \in Q_2$ is a root of $B_k(x)$. From the slopes of the Newton polygon segments we see that $\upsilon_2(\alpha) = 0$ or $-1/2^s$. As υ_2 takes only integer values on Q_2, we see that $\upsilon_2(\alpha) = 0$ and so $\alpha \in Z_2$. Let $f(x) = 2B_k(x) \in Z_2[x]$. Thus $f(\alpha) = 0$ and so $f(\bar{\alpha}) = \bar{0} \in F_2$. However, $\bar{\alpha} \in F_2 = \{\bar{0},\bar{1}\}$. Now $f(\bar{0}) = \overline{(2Bk)} = \bar{1}$, and from (4) we know that $f(\bar{1}) = f(\bar{0}) = \bar{1}$. This gives a contradiciton. □

Although Theorem 4 is not needed by us, its proof helps motivate part of the proof of Proposition 3.1.

5 COMPLETING THE PROOF OF THEOREM 1

5.1 A little group theory

Lemma 5.1 Let H be a finite group acting transitively on a finite set $\{\beta_1,\ldots,\beta_n\}$. Let $H_i \subseteq H$ be the stabilizer of β_i, and suppose $H_1 = H_2$. Let $\pi : H \to C$ be a surjective homomorphism from H onto a cyclic group C. Then there is some $\mu \in H$ acting freely on $\{\beta_1,\ldots,\beta_n\}$ such that $\pi(\mu)$ is a generator of C.

Proof Let $m = \#C$ and write $C = \langle\sigma\rangle$. Consider the subset

$$C' = \{\sigma^r : \gcd(r,m) = 1\};$$

this is the set of elements that are cyclic generators of C, and has cardinality $\varphi(m)$, where φ is the Euler totient function. As π is surjective we see that

$$\#\pi^{-1}(C') = \frac{\varphi(m)}{m} \cdot \#H. \tag{7}$$

As H acts transitively on the β_i, the stabilizers H_i are conjugate and so have the same image $\pi(H_i)$ in C. If this image is a proper subgroup of C, then take μ to be any preimage of σ. Thus $\pi(\mu) = \sigma$ is a generator of C, and moreover, μ does not belong to any of the stabilizers H_i and so acts freely on $\{\beta_1,\ldots,\beta_n\}$, completing the proof in this case. Thus we suppose that $\pi(H_i) = C$ for all i. It follows that

$$\#\pi^{-1}(C') \cap H_i = \frac{\varphi(m)}{m} \cdot \#H_i = \frac{\varphi(m)}{m} \cdot \frac{\#H}{n}, \tag{8}$$

where the second equality follows from the Orbit-Stabilizer Theorem. The lemma states that there is some element μ belonging to $\pi^{-1}(C')$ but not to $\cup H_i$. Suppose otherwise. Then

$$\pi^{-1}(C') \subseteq \bigcup_{i=1}^{n} H_i,$$

and therefore

$$\pi^{-1}(C') = \bigcup_{i=1}^{n} \pi^{-1}(C') \cap H_i. \tag{9}$$

Now (7), (8) and (9) together imply that the $\pi^{-1}(C') \cap H_i$ are pairwise disjoint. This contradicts the hypothesis that $H_1 = H_2$ completing the proof. □

Proof of Propostion 3.1 We now complete the proof of Theorem 1 by proving Proposition 3.1. Fix an even $k \geq 2$, and let L be the splitting field of $B_k(x)$. Let $G_k = \mathrm{Gal}(L/Q) = \mathrm{Gal}(B_k(x))$ be the Galois group of $B_k(x)$. Let $\mathfrak{P}$ be a prime of L above 2. The 2-adic valuation v_2 on Q_2 has a unique extension to $L_{\mathfrak{P}}$ which we continue to denote by v_2. We let $H = \mathrm{Gal}(L_{\mathfrak{P}}/Q_2) \subseteq G_k$ be the decomposition subgroup corresponding to $\mathfrak{P}$.

From Lemma 4.1 we see that $B_k(x)$ factors as $B_k(x) = g(x)h(x)$ over Q_2 where the factors g, h correspond respectively to the segments (i), (ii) in the lemma. Thus g, h have degree $k - 2^s$ and 2^s respectively. We denote the roots of g by $\{\alpha_1,\ldots,\alpha_{k-2s}\} \subset L_{\mathfrak{P}}\{\alpha_1,\ldots,\alpha_{k-2s}\} \subset L_{\mathfrak{P}}$ and the roots of h by $\{\beta_1,\ldots,\beta_{2S}\} \subset L_{\mathfrak{P}}$. From the slopes of the segments we see that $v_2(\alpha_i) = 0$ and $v_2(\beta_j) = -1/2^s$. It clearly follows that h is irreducible and therefore that H acts transitively on the β_j. Moreover, from the symmetry (4) we see that $1 - \beta_1$ is a root of $B_k(x)$, and by appropriate relabelling we can suppose that $\beta_2 = 1 - \beta_1$. In the notation of Lemma 5.1, we have $H_1 = H_2$. Now let $C = \mathrm{Gal}(F_{\mathfrak{P}}/F_2)$, where $F_{\mathfrak{P}}$ is the residue field of $\mathfrak{P}$. This group is cyclic generated by the Frobenius map: $\bar{\gamma} \mapsto \bar{\gamma}^2$. We let $\pi : H \to C$ be the induced surjection. By Lemma 5.1 there is some $\mu \in H$ that acts freely on the β_i and such that $\pi(\mu)$ generates C. To complete the proof of Proposition 3.1 it is enough to show that μ also acts freely on the α_i. Suppose otherwise, and let α be one of the α_i that is fixed by μ. As $\upsilon_2(\alpha) = 0$, we can write $\bar{\alpha} \in F_{\mathfrak{P}}$ for the reduction of α modulo $\mathfrak{P}$. Now α is fixed by μ, and so $\bar{\alpha} \in F_{\mathfrak{P}}$ is fixed by $\langle \pi(\mu) \rangle = C$. Thus $\bar{\alpha} \in \mathrm{F}_2$ and so $\tilde{\alpha} = \bar{0}$ or $\bar{1}$. Now let $f(x) = 2B_k(x) \in \mathrm{Z}_2[x]$. Thus $f(\bar{\alpha}) = \bar{0}$. But $f(\bar{0}) = \overline{(2Bk)} = \bar{1}$, and from (4) we know that $f(\bar{1})$=f($\bar{0}$) = $\bar{1}$. This contradiction completes the proof.

ACKNOWLEDGEMENTS

The first-named author is supported by an EPSRC studentship. The second-named author is supported by the EPSRC *LMF: L-Functions and Modular Forms* Programme Grant EP/K034383/1.

REFERENCES

1. Bennett, M.A., Patel, V., Siksek, S.: Perfect powers that are sums of consecutive cubes. Mathematika **63**, 230–249 (2016)
2. Bennett, M.A., Patel V., Siksek S.: Superelliptic equations arising from sums of consecutive powers, Acta Arithmetica, to appear
3. Brillhart, J.: On the Euler and Bernoulli polynomials. J. Reine Angew. Math. **234**, 45–64 (1969)
4. Carlitz, L.: Note on irreducibility of the Bernoulli and Euler polynomial. Duke Math. J. **19**, 475–481 (1952)
5. Cassels, J.W.S.: A Diophantine equation. Glasgow Math. J. **27**, 11–88 (1985)
6. Cassels, J.W.S., Fröhlich, A. (eds.): Algebraic Number Theory. Academic Press, New York (1967)
7. Dickson, L.E.: History of the theory of numbers, vol. II. Chelsea, New York (1971)
8. Dilcher, K.: Chapter 24: Bernoulli and Euler polynomials, in digital library of mathematical functions. http://dlmf.nist.gov/24
9. Dilcher, K.: On multiple zeros of Bernoulli polynomials. Acta Arithmetica **134**(2), 149–155 (2008)
10. Euler L.: Vollständige Anleitung zur Algebra, vol. 2, St. Petersburg (1770)
11. Granville, A.: Arithmetic properties of binomial coefficients. I. Binomial coefficients modulo prime powers. CMS Conf. Proc. Amer. Math. Soc. **20**, 253–276 (1997)
12. Inkeri, K.: The real roots of Bernoulli polynomials. Ann. Univ. Turku. Ser. A **I**(37), 1–20 (1959)

13. Niven, I.: The asymptotic density of sequences. Bull. Amer. Math. Soc. **57**, 420–434 (1951)
14. Kimura, N.: On the degree of an irreducible factor of the Bernoulli polynomials. Acta Arith. **50**, 243–249 (1988)
15. Soydan G., On the Diophantine equation
16. $(x+1)^k + (x+2)^k + \cdots + (lx)^k = y^n$ preprint (2016)
17. Stroeker, R.J.: On the sum of consecutive cubes being a square. Compositio Mathematica **97**, 295–307 (1995)
18. Uchiyama, S.: On a Diophantine equation. Proc. Japan Acad. Ser. A Math. Sci. **55**(9), 367–369 (1979)
19. Zhang, Z.: On the Diophantine equation $(x-1)^k + x^k + (x+1)^k = y^n$. Publ. Math. Debrecen **85**, 93–100 (2014)
20. Zhang, Z., Bai, M.: On the Diophantine equation $(x+1)^2 + (x+2)^2 + \ldots + (x+d)^2 = y^n$. Funct. Approx. Comment. Math. **49**, 73–77 (2013)

CHAPTER 10

Some Edge Product Cordial Graphs in the Context of Duplication of Some Graph Elements

Udayan M. Prajapati[1], Prakruti D. Shah[2]

[1]St. Xavier's College, Ahmedabad, India

[2]Shankersinh VaghelaBapu Institute of Technology, Gandhinagar, India

1. INTRODUCTION

We begin with a simple, finite, undirected graph $G = (V(G), E(G))$ where $V(G)$ and $E(G)$ denote the vertex set and the edge set respectively. For all other terminology, we follow Gross [1]. We will provide a brief summary of definitions and other information which are necessary for the present investigations.

Definition 1. A graph labeling is an assignment of integers to the vertices or edges or both subject to certain condition. If the domain of the mapping is the set of vertices , edges or both then the labeling is called a vertex labeling , an edge labeling or a total labeling.

Definition 2. For a graph G, an edge labeling function is defined as $f : E(G) \rightarrow \{0,1\}$ and the induced vertex labeling function $f^* : V(G) \rightarrow \{0,1\}$ is given by $f^* = f(e_1)f(e_2)\text{---}f(e_k)$ if $e_1, e_2, \text{---}, e_k$ are the edges incident with the vertex v.

We denote the number of vertices of G having label i under f^* by $v_f(i)$ and the number of edges of G having label i under f by $e_f(i)$ for $i = 0,1$.

The function f is called an edge product cordial labeling of G if $|v_f(0)-v_f(1)|\le 1$ and $|e_f(0)-e_f(1)|\le 1$. A graph G is called edge product cordial if it admits edge product cordial labeling.

The concept of edge product cordial labeling was introduce by Vaidya and Barasara [2] in which they proved that C_n for n odd, trees of order greater than 2, unicyclic graphs of odd order , crowns , armed crowns , helms, closed helms, webs , flowers graph are edge product cordial . They also proved that wheel and gear for even are not edge product cordial . They also [3] proved that T_n, DT_n for odd, Q_n for odd, DQ_n for odd are edge product cordial labeling . They also proved that DT_n for even, Q_n for even, DQ_n for even, DF_n are not edge product cordial labeling.

Definition 3. The graph $W_n= C_n+ K_1$ is called wheel graph, the vertex corresponding to 1 K is called apex vertex and vertices corresponding to C_n are called rim vertices.

Definition 4. The helm H_n is the graph obtained from a wheel W_n by attaching a pendent edge at each vertex of the n-cycle.

Definition 5. A gear graph is obtained from the wheel graph W_n by adding a vertex between every pair of adjacent vertices of the n-cycle.

Definition 6. The crown $C_n \odot K_1$ is obtained by joining a single pendent edge to each vertex of C_n.

Definition 7. The neighborhood of a vertex v of a graph is the set of all vertices adjacent to v. It is denoted by $N(v)$.

Definition 8. Duplication of a vertex of the graph G is the graph G' obtained from G by adding a new vertex v' to G such that $N(v') = N(v)$.

Definition 9. Duplication of a vertex v_k by a new edge $e = v'_k v''_k$ in a graph G produces a new graph G' such that $N(v'_k) = \{v_k, v''_k\}$ and $N(v''_k) = \{v_k, v'_k\}$.

The concept of duplication of vertex by edge was introduce by Vaidya and Barasara [4].

Definition 10. Duplication of an edge $e = uv$ by a new vertex w in a graph G produces a new graph G' such that $N(w) = \{u, v\}$.

The concept of duplication of edge by vertex was introduce by Vaidya and Dani [5].

2. MAIN RESULTS

Theorem 1. The graph obtained by duplication of an arbitrary vertex of the cycle in a crown graph is an edge product cordial graph.

Proof. Let C_n be a cycle with consecutive vertices v_1, v_2, ---, v_n and edges $e_i=v_iv_{i+1}$ for each $i\in\{1, 2,3, ---, n\}$. Let u_i be a new vertex adjacent to v_i with $e_i' = u_iv_i$ for each $i\in\{1, 2,3, ---, n\}$. Resulting graph is a crown graph G_1.

Let G be the graph obtained by duplication of the vertex n v by a new vertex v' of $G1$ such that $e_1'' = v'v_1$, $e_2'' = v'v_{n-1}$ and $e_3'' = v'u_n$.

Thus $|V(G)| = 2n+1$ and $|E(G)| = 2n+3$.

Now for n = 3 and n = 4 Figure 1 shows that the graphs are edge product cordial as follows:

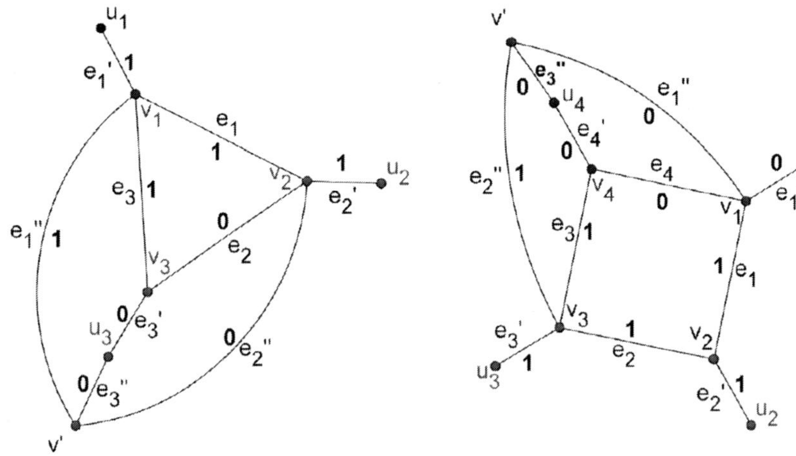

Figure 1. n = 3 and n = 4 .

Case 1: When n is odd. For $n \geq 5$, define f: $E(G)\rightarrow\{0,1\}$ as follows:

$$f(e)=\begin{cases} 1, \text{ if } e=e_i, \text{ for } 1\le i\le \left\lceil \frac{n}{2} \right\rceil; \\ 0, \text{ if } e=e_i, \text{ for } \left\lceil \frac{n}{2} \right\rceil+1\le i\le n; \\ 1, \text{ if } e=e_i', \text{ for } 1\le i\le \left\lceil \frac{n}{2} \right\rceil; \\ 0, \text{ if } e=e_i', \text{ for } \left\lceil \frac{n}{2} \right\rceil+1\le i\le n; \\ 0, \text{ if } e=e_i'', \text{ for } i=1,2,3. \end{cases}$$

In the view of above labeling pattern we have,

$V_f(1)=v_f(0)-1=n$ and $e_f(1)=e_f(0)-1=n+1$.

Case 2: When n is even. For $n\ge 6$, define $f: E(G)\rightarrow\{0,1\}$ as follows:

$$f(e)=\begin{cases} 1, \text{ if } e=e_i, \text{ for } 1\le i\le \frac{n}{2}+1; \\ 0, \text{ if } e=e_i, \text{ for } \frac{n}{2}+2\le i\le n; \\ 1, \text{ if } e=e_i', \text{ for } 1\le i\le \frac{n}{2}+1; \\ 0, \text{ if } e=e_i', \text{ for } \frac{n}{2}+2\le i\le n; \\ 0, \text{ if } e=e_i'', \text{ for } i=1,2,3. \end{cases}$$

In the view of the above labeling pattern we have,

$V_f(0)=v_f(1)-1=n$ and $e_f(0)=e_f(1)-1=n+1$.

Thus, from both the cases we have $|v_f(0)-v_f(1)|\le 1$ and $|e_f(0)-e_f(1)|\le 1$.

Hence, graph G admits edge product cordial labeling. Thus, G is an edge product cordial graph.

Illustration 1. The graph obtained by duplication of an arbitrary vertex of the cycle $C_n$in a crown graph is an edge product cordial graph as shown in Figure 2 as follows:

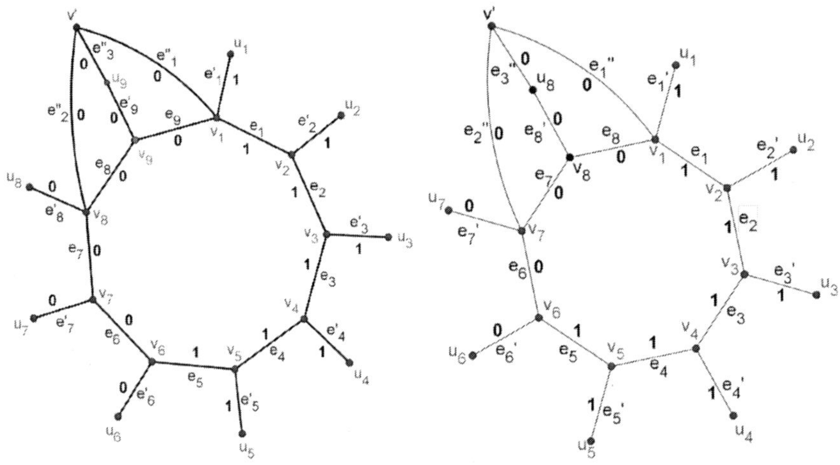

Figure 2. $n = 9$ and $n = 8$.

Theorem 2. The graph obtained by duplication of an arbitrary vertex of the cycle by a new edge in a crown graph is edge product cordial graph.

Proof. Let $C_n$be a cycle with consecutive vertices v_1, v_2,---, $v_n$and edges e_i=$v_iv_{i+1}$for each $i\in\{1, 2,---, n\}$. Let $u_i$be a new vertex adjacent to $v_i$with $e_i= u_i v_i$for each $i\in\{1, 2,---, n\}$. Resulting graph is a crown graph G_1.

Let G be the graph obtained by duplication of the vertex $v_n$by an edge $e_1'' = v'v''$ of G_1such that $e_2'' = v'v_n$ and $e_3'' = v''v_n$. Thus $|V(G)| = 2n+2$ and $|E(G)| = 2n+3$.

Define f: $E(G)\rightarrow\{0,1\}$ as follows:

$$f(e) = \begin{cases} 1, \text{ if } e = e_i, \text{ for } i = 1,2; \\ 0, \text{ if } e = e_i, \text{ for } 3 \le i \le n; \\ 1, \text{ if } e = e_i', \text{ for } i \in \{1,2,\cdots,n\} \\ 0, \text{ if } e = e_i'', \text{ for } i = 1,2,3. \end{cases};$$

In the view of the above labeling pattern we have, $v_f(0)=v_f(1)=n+1$ and $e_f(0)=e_f(1)-1=n+1$. Thus, we have $|v_f(0) - v_f(1)| \le 1$ and $|e_f(0) - e_f(1)| \le 1$.

Hence, graph G admits edge product cordial labeling. Thus, G is an edge product cordial graph.

Illustration 2. The graph obtained by duplication of an arbitrary vertex of the cycle $C_n$by a new edge in a crown graph is edge product cordial graph as shown in Figure 3.

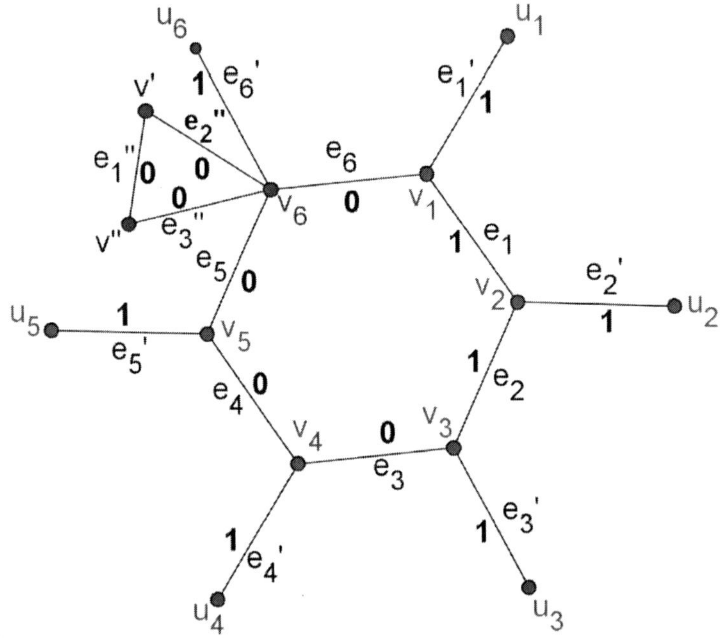

Figure 3.n = 6.

Theorem 3. The graph obtained by duplication of an arbitrary edge of the cycle $C_n$by a new vertex in a crown graph is edge product cordial

Proof. Let n C be a cycle with the consecutive vertices v_1, v_2, ---, $v_n$and edges $e_i=v_iv_{i+1}$for each $i\in\{1, 2,---, n\}$. Let $u_i$be a new vertex adjacent to $v_i$with $e_i' = u_iv_i$for each $i\in\{1, 2,---, n\}$. Resulting graph is a crown graph G_1.

Let G be the graph obtained by duplication of an edge $e_1=v_1v_2$by a vertex v'in G_1such that $e_1'' = v'v_1$ and $e_2'' = v'v_2$. Thus $|V(G)| = 2n+1$ and $|E(G)| = 2n+2$.

Define f: $E(G)\rightarrow\{0,1\}$ as follows:

$$f(e) = \begin{cases} 1, \text{ if } e = e_i, \text{ for } i = 1; \\ 0, \text{ if } e = e_i, \text{ for } 2 \le i \le n; \\ 1, \text{ if } e = e_i', \text{ for } i \in \{1, 2, \cdots, n\} \\ 0, \text{ if } e = e_i'', \text{ for } i = 1, 2. \end{cases}$$

In the view of above labeling pattern we have, $v_f(0)-1=v_f(1)=n$ and$e_f(0)=e_f(1)=n+1$. Thus, we have$\left|v_f(0)-v_f(1)\right|\le 1$and $\left|e_f(0)-e_f(1)\right|\le 1$.

Hence, graph G admits edge product cordial labeling. Thus, G is an edge product cordial graph.

Illustration 3. The graph obtained by duplication of an arbitrary edge of the cycle $C_n$by a new vertex in a crown graph is edge product cordial graph as shown in Figure 4.

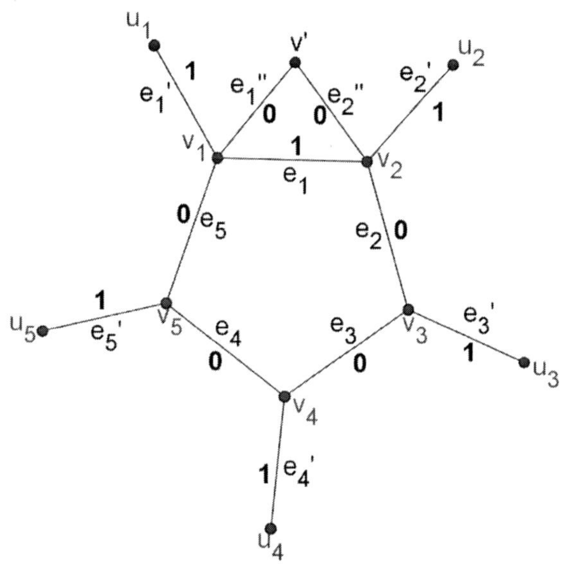

Figure 4.$n = 5$.

Theorem 4. The graph obtained by duplication of each pendent vertex by a new vertex in a crown graph is edge product cordial graph.

Proof. Let $C_n$be a cycle with consecutive vertices $v_1, v_2, ---, v_n$and edges $e_i=v_iv_{i+1}$for each $i\in\{1, 2, ---, n\}$. Let $u_i$be a new vertex adjacent to $v_i$with $e_i' = u_iv_i$for each $i\in\{1, 2, ---, n\}$. Resulting graph is a crown graph G_1.

Let G be the graph obtained by duplication of each pendent vertex $u_i$by a new vertex v_i' of G_1 such that $e_i'' = v_i'v_i$for $i\in\{1, 2, ---, n\}$. Thus $\left|V(G)\right|=3n$and $\left|E(G)\right|=3n$.

Case 1: When n is odd, define f: $E(G)\to\{0,1\}$ as follows:

$$f(e)=\begin{cases}0, \text{ if } e=e_i, \text{ for } i\in\{1,2,\cdots,n\};\\ 1, \text{ if } e=e_i', \text{ for } i\in\{1,2,\cdots,n\};\\ 1, \text{ if } e=e_i'', 1\le i\le\left\lceil\frac{n}{2}\right\rceil;\\ 0, \text{ if } e=e_i'', \left\lceil\frac{n}{2}\right\rceil+1\le i\le n.\end{cases}$$

In the view of above labeling pattern we have,

$v_f(0)=v_f(1)-1=\left\lfloor\frac{3n}{2}\right\rfloor$ and $e_f(0)=e_f(1)-1=\left\lfloor\frac{3n}{2}\right\rfloor$.

Case 2: When n is even, define $f:E(G)\to\{0,1\}$ as follows:

$$f(e)=\begin{cases}0, \text{ if } e=e_i, \text{ for } i\in\{1,2,\cdots,n\};\\ 1, \text{ if } e=e_i', \text{ for } i\in\{1,2,\cdots,n\};\\ 1, \text{ if } e=e_i'', 1\le i\le\frac{n}{2};\\ 0, \text{ if } e=e_i'', \frac{n}{2}+1\le i\le n.\end{cases}$$

In the view of above labeling pattern we have, $v_f(0)=v_f(1)=\frac{3n}{2}$ and $e_f(0)=e_f(1)=\frac{3n}{2}$.

Thus, from both the cases we have $|v_f(0)-v_f(1)|\le 1$ and $|e_f(0)-e_f(1)|\le 1$. Hence, graph G admits edge product cordial labeling. Thus, G is an edge product cordial graph.

Illustration 4. The graph obtained by duplication of each pendent vertex by a new vertex in a crown graph is edge product cordial graph as shown in Figure 5 as follows :

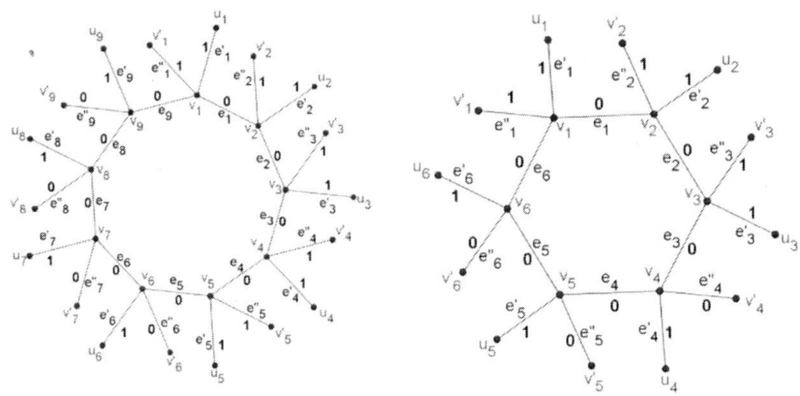

Figure 5. $n = 9$ and $n = 6$.

Theorem 5. The graph obtained by duplication of each of the vertices of degree three by an edge in a gear graph is an edge product cordial graph.

Proof. Let W_n be the wheel graph with apex vertex v and consecutive rim vertices $v_1, v_2, ---, v_n$. To obtained the gear graph G_n subdivide each of the rim edges $v_1v_2, v_2v_3, ---, v_{n-1}v_n, v_nv_1$ of the wheel graph by the vertices $u_1, u_2, ---, u_n$ respectively such that $e_i=vv_i$, $e_{li}=v_iu_i$ and $e_{ri}=u_iv_{i+1}$ for $i\in\{1, 2, ---, n\}$. Thus $\left|V\left(G\right)\right| = 4n+1$ and $\left|E\left(G\right)\right| = 6n$.

Define $f:E(G)\to\{0,1\}$ as follows:

$$f(e)=\begin{cases} 0, \text{ if } e=e_i, \text{ for } i=\{1,2,3,\cdots,n\}; \\ 0, \text{ if } e=e_{l_i}, \text{ for } i=\{1,2,3,\cdots,n\}; \\ 0, \text{ if } e=e_{r_i}, \text{ for } i\in\{1,2,3,\cdots,n\}; \\ 1, \text{ if } e=f_i, \text{ for } i=\{1,2,3,\cdots,n\}; \\ 1, \text{ if } e=f_i', \text{ for } i=\{1,2,3,\cdots,n\}; \\ 1, \text{ if } e=f_i'', \text{ for } i=\{1,2,3,\cdots,n\}. \end{cases}$$

In the view of the above labeling pattern we have, $v_f(0)-1=v_f(1)=2n$ and $e_f(0)=e_f(1)=3n$. Thus, we have $\left|v_f(0)-v_f(1)\right|\le 1$ and $\left|e_f(0)-e_f(1)\right|\le 1$. Hence, graph G admits edge product cordial labeling. Thus, G is edge product cordial graph.

Illustration 5. The graph obtained by duplication of each vertex of degree three by an edge in a gear graph is an edge product cordial graph as shown in Figure 6.

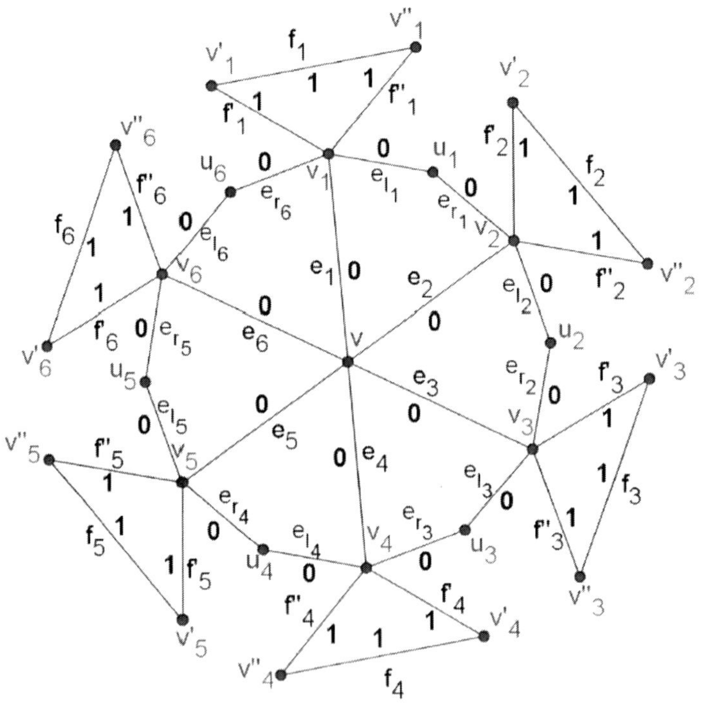

Figure 6. $n = 6$.

Theorem 6. The graph obtained by duplication of each of the pendent vertices by anew vertex in a helm graph is edge product cordial graph.

Proof. Let v be the apex vertex and $v_1, v_2, ---, v_n$ be the consecutive rim vertices of the wheel W_n with edges $e_i = v_i v_{i+1}$ and $e'_i = vv_i$ for $i \in \{1, 2, ---, n\}$. Let u_i be anew vertex adjacent to v_i with edges $e''_i = u_i v_i$ for $i \in \{1,2, ---,n\}$. Resulting graph is helm graph H_n.

Let G be the graph obtained from H_n by duplication of each pendent vertex u_i by a new vertex v'_i such that $f_i = v'_i v_i$ for $i \in \{1, 2, ---, n\}$. Thus $|V(G)| = 3n + 1$ and $|E(G)| = 4n$

Case 1: When n is odd, define $f : E(G) \to \{0,1\}$ as follows:

$$f(e)=\begin{cases}1, \text{ if } e=e_i,\ 1\le i\le\left\lfloor\frac{n}{2}\right\rfloor; \\ 0, \text{ if } e=e_i,\ \left\lfloor\frac{n}{2}\right\rfloor+1\le i\le n; \\ 0, \text{ if } e=e_i',\ \text{for } i\in\{1,2,\cdots,n\}; \\ 1, \text{ if } e=e_i'',\ 1\le i\le n-1; \\ 0, \text{ if } e=e_i'',\ \text{for } i=n; \\ 1, \text{ if } e=f_i,\ 1\le i\le\left\lfloor\frac{n}{2}\right\rfloor+2; \\ 0, \text{ if } e=f_i,\ \left\lfloor\frac{n}{2}\right\rfloor+3\le i\le n.\end{cases}$$

In the view of above labeling pattern we have, $v_f(0)=v_f(1)=\left\lceil\frac{3n}{2}\right\rceil$ and $e_f(0)=e_f(1)=2n$.

Case 2: When n is even, define $f:E(G)\rightarrow\{0,1\}$ as follows:

$$f(e)=\begin{cases}1, \text{ if } e=e_i,\ 1\le i\le\frac{n}{2}-1; \\ 0, \text{ if } e=e_i,\ \frac{n}{2}\le i\le n; \\ 0, \text{ if } e=e_i',\ \text{for } i\in\{1,2,\cdots,n\}; \\ 1, \text{ if } e=e_i'',\ 1\le i\le\frac{n}{2}+1; \\ 0, \text{ if } e=e_i'',\ \frac{n}{2}+2\le i\le n; \\ 1, \text{ if } e=f_i,\ \text{for } i\in\{1,2,\cdots,n\}.\end{cases}$$

In the view of above labeling pattern we have, $v_f(0)=v_f(1)-1=\frac{3n}{2}$ and $e_f(0)=e_f(1)=2n$.

Thus, from both the cases we have $|v_f(0)-v_f(1)| \leq 1$ and $|e_f(0)-e_f(1)| \leq 1$. Hence,graph G admits edge product cordial labeling. Thus, G is an edge product cordialgraph.

Illustration 6. The graph obtained by duplication of each pendent vertex by a newvertex in a helm graph is edge product cordial graph as shown in Figure 7.

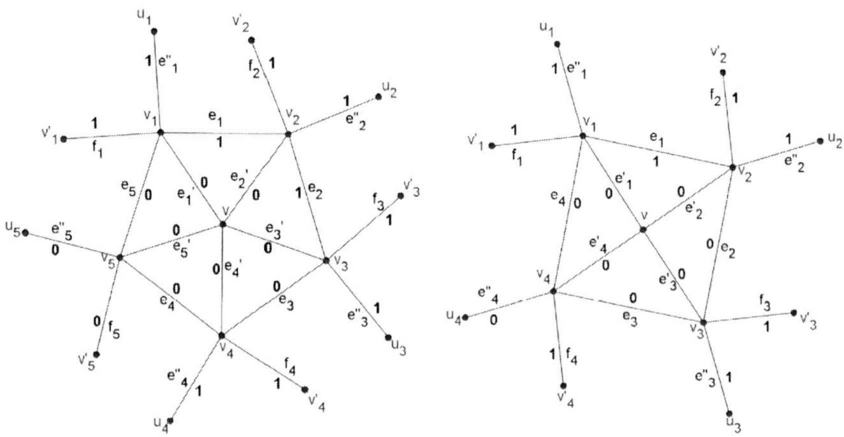

Figure 7.$n = 5$ and $n = 4$.

3. CONCLUSION

We have derived six results for edge product cordial related to crown graph, gear graphand helm graph in the context of duplication of various graph elements. Similar problemcan be discussed for other graph family for edge product cordial labeling.

ACKNOWLEDGEMENTS

The authors are highly thankful to the anonymous referee for valuable comments andconstructive suggestions. The First author is thankful to the University Grant Commission,India for supporting him with Minor Research Project under No. F. 47-903/ 14(WRO) dated 11th March, 2015.

REFERENCES

1. Gross, J.L. and Yellen, J. (Eds.) (2004) Handbook of Graph Theory. CRC Press, Boca Raton.
2. Vaidya, S.K. and Barasara, C.M. (2012) Edge Product Cordial Labeling of Graphs. Journalof Mathematics and Computer Science , 2, 1436-1450.http://scik.org /index.php/jmcs/article/view/420/189
3. Vaidya, S.K. and Barasara, C.M. (2013) Some New Families of Edge Product CordialGraphs. Advanced Modeling Optimization, 15, 103-111.http://camo. ici.ro/ journal/vol15/v15a9.pdf
4. Vaidya, S.K. and Barasara, C.M. (2011) Product Cordial Graphs in the Context of SomeGraph Operations. International Journal of Mathematics and Computer Science , 1, 1-6.
5. Vaidya, S.K. and Dani, N.A. (2011) Cordial and 3-Equitable Graphs Induced by Duplicationof Edge. Mathematics Today , 27, 71-82.

CHAPTER 11

Mean Square Numerical Methods for Initial Value Random Differential Equations

Magdy A. El-Tawil[1*], Mohammed A. Sohaly[2]

[1] *Engineering Mathematics Department, Faculty of Engineering, Cairo University, Giza, Egypt*

[2] *Mathematic Department, Faculty of Science, Mansoura University, Mansoura, Egypt*

ABSTRACT

In this paper, the random Euler and random Runge-Kutta of the second order methods are used in solving random differential initial value problems of first order. The conditions of the mean square convergence of the numerical solutions are studied. The statistical properties of the numerical solutions are computed through numerical case studies.

Keywords: Random Differential Equations, Mean Square Sense, second random variable, Initial Value Problems, Random Euler Method, Random Runge Kutta-2 Method

1. INTRODUCTION

Random differential equations (RDE) are defined as differential equations involving random inputs. In recent years, increasing interest in the numerical solution of (RDE) has led to the progressive development of several numerical methods. This paper is interested in studying the following random differential initial value problem (RIVP) of the form:

$$\frac{dX}{dt} = f(t, X),\ X(t_0) = X_0 \tag{1.1}$$

Randomness may exist in the initial value or in the differential operator or both. In [1,2], the authors discussed the general order conditions and a global convergence proof is given for stochastic Runge-Kutta methods applied to stochastic ordinary differential equations (SODEs) of Stratonovich type. In [3,4], the authors discussed the random Euler method and the conditions for the mean square convergence of this problem. In [5], the authors considered a very simple adaptive algorithm based on controlling only the drift component of a time step. Platen, E. [6] discussed discrete time strong and weak approximation methods that are suitable for different applications. Other numerical methods are discussed in [7-12].

In this paper the random Euler and random RungeKutta of the second order methods are used to obtain an approximate solution for equation (1.1). This paper is organized as follows. In Section 2, some important preliminaries are discussed. In Section 3, the existence and uniqueness of the solution of random differential initial value problem is discussed and the convergence of random Euler and random Runge-Kutta of the second order methods is discussed. In Section 4, the statistical properties for the exact and numerical solutions are studied. Section 5 presents the solution of some numerical examples of first order random differential equations using random Euler and random Runge-Kutta of the second order methods showing the convergence of the numerical solutions to the exact ones (if possible). The general conclusions are presented in the end section.

2. PRELIMINARIES

2.1. Mean Square Calculus [13]

Definition1. Let us consider the properties of a class of real r.v.'s $X_1, X_2, \cdots, X_n$ whose second moments, $E\{X_1^2\}, E\{X_2^2\}, \cdots$ are finite. In this case, they are called"second order random variables", (2.r.v's).

Definition 2. The linear vector space of second order random variables with inner product, norm and distance, is called an L_2-space.

A s.p. $\{X(t), t \in T\}$ is called a "second order stochastic process" (2.s.p) if for $t_1, t_2, \cdots t_n$, the r.v's $\{X(t_1), X(t_2), \cdots, X(t_n)\}$ are elements of L_2-space.

A second order s.p. $\{X(t), t \in T\}$ is characterized by

$\|X(t)\|^2 = E\{X^2(t)\} < \infty, \quad t \in T$.

2.1.1. The Convergence in Mean Square

A sequence of r.v's $\{X_n\}$ converges in mean square (m.s) to a random variable X

if $\lim_{n\to\infty} \|X_n - X\| = 0$ i.e. $X_n \xrightarrow{m.s} X$ or $\lim_{n\to\infty} X_n = X$

where lim is the limit in mean square sense.

2.1.2. Mean-Square Differentiability

The random process $\{X(t)\}$ is mean-square differentiable at t if $\lim_{h\to 0} \frac{X_{t+h} - X_t}{h}$ exists, and is denoted by $\lim_{h\to 0} \frac{X_{t+h} - X_t}{h} = \dot{X}_t$

3. RANDOM INITIAL VALUE PROBLEM (RIVP)

3.1. Existence and Uniqueness

Let us have the random initial value problem

$$\frac{dX}{dt} = b(t, X), \; t \in T = [t_0, t], \; X(t_0) = X_0 \tag{3.1}$$

where $X(t)$ is second order random process. This equation is equivalent to integral equation

$$X(t) = X_0 + \int_{t0}^{t} b(X(s), s)\, ds \tag{3.2}$$

Theorem (3.1.1)

If we have the random initial value problem (3.1) and suppose the right-hand side function $b(t, X)$ is continuous and satisfies a mean square (m.s) Lipschitz condition in its second argument:

$$\left\| b(t, X) - b(t, Y) \right\| \leq c \left\| X - Y \right\| \tag{3.3}$$

where C is a constant or

$$\left\| b(t, X) - b(t, Y) \right\| \leq c(t) \left\| X - Y \right\| \leq c \left\| X - Y \right\| \tag{3.4}$$

where c(t) is a continuous function {because in every finite interval $c(t) \leq$ constant}.

then the solution of equation (3.1) exists and is unique.

The proof

The existence can be proved by using successive approximations. Let

$$X_t^0 = X_0 \tag{3.5}$$

and for $n \geq 1$

$$X_t^n = X_0 + \int_{t0}^{t} b\left(X_s^{n-1}, s \right) \mathrm{d}s. \tag{3.6}$$

For $n = 1$ we obtain:

$$\left\| X_t^{(1)} - X_t^{(0)} \right\| = \left\| \int_{t0}^{t} b(s, X_0) \mathrm{d}s \right\| \leq k \cdot \left| t - t_0 \right|$$

where

$$\left\| b(t, X) \right\| \leq k \tag{3.7}$$

For $n > 1$ we obtain:

$$\begin{aligned} \left\| X_t^{(n)} - X_t^{(n-1)} \right\| &= \left\| \int_{t0}^{t} b\left(s, X_s^{n-1}\right) - b\left(s, X_s^{n-2}\right) \mathrm{d}s \right\| \\ &\leq \int_{t0}^{t} c \cdot \left\| X_s^{(n-1)} - X_s^{(n-2)} \right\| \mathrm{d}s \end{aligned} \tag{3.8}$$

Successively, we can obtain the following:

$$\left\|X_t^{(n)} - X_t^{(n-1)}\right\| \le \int_{t_0}^{t} c\left\|X_s^{n-1} - X_s^{n-2}\right\| \mathrm{d}s$$

$$\le \int_{t_0}^{t} c\int_{t_0}^{t} c\left\|X_s^{n-2} - X_s^{n-3}\right\| \mathrm{d}s\mathrm{d}s \le c^3 \int_{t_0}^{t}\int_{t_0}^{t}\int_{t_0}^{t}\left\|X_s^{n-3} - X_s^{n-4}\right\| \mathrm{d}s\mathrm{d}s\mathrm{d}s$$

$$\le c^{n-1}\int_{t_0}^{t}\int_{t_0}^{t}\cdots\int_{t_0}^{t}\left\|X_s^{1} - X_s^{0}\right\| \mathrm{d}s\cdots\mathrm{d}s \le c^{n-1}\int_{t_0}^{t}\int_{t_0}^{t}\cdots\int_{t_0}^{t} k\left|s-t_0\right|\mathrm{d}\cdots\mathrm{d}s.$$

Hence

$$\left\|X^n - X^{n-1}\right\| \le kc^{n-1}\int_{t_0}^{t}\left[\int_{t_0}^{t}\cdots\int_{t_0}^{t}\left[\int_{t_0}^{t}\left|s-t_0\right|\mathrm{d}s]\mathrm{d}s\right]\cdots\mathrm{d}s\right]\mathrm{d}s = kc^{n-1}\frac{\left|t-t_0\right|^n}{n!}$$

(3.9)

Since:

$$\sum_{n=1}^{\infty} k.c^{(n-1)}\frac{\left|t-t_0\right|}{n!} = \frac{k}{c}\mathrm{e}^{c\left\|t-t_0\right\|}$$ is convergent for finite t, (3.10)

hence we can have the following

$$\left\|X_t^1 - X_t^0\right\| + \left\|X_t^2 - X_t^1\right\| + \cdots + \left\|X_t^{n-1} - X_t^{n-2}\right\| + \cdots \le \left(\frac{k\left|t-t_0\right|}{1!}\right) + \left(\frac{kc\left|t-t_0\right|^2}{2!}\right) + \cdots + \left(\frac{kc^{n-2}\left|t-t_0\right|^{n-1}}{(n-1)!}\right) + \cdots = \sum_{n=1}^{\infty} k.c^{(n-1)}\frac{\left|t-t_0\right|^n}{n!}$$

(3.11)

Accordingly,

$$\left\|X_t^1 - X_t^0 + X_t^2 - X_t^1 + \cdots + X_t^{n-1} - X_t^{n-2} + X_t^n - X_t^{n-1}\right\| = \lim_{n\to\infty}\left\|X_t^n - X_t^0\right\| = \left\|X_t\right\|$$

Hence:

$$\left\|X_t^1 - X_t^0 + X_t^2 - X_t^1 + \cdots X_t^{n-1} - X_t^{n-2} + X_t^n - X_t^{n-1}\right\| \le \left\|X_t^1 - X_t^0\right\| + \left\|X_t^2 - X_t^1\right\| + \cdots + \left\|X_t^{n-1} - X_t^{n-2}\right\| + \left\|X^n - X^{n-1}\right\| = \frac{k}{c}\mathrm{e}^{c\left\|t-t_0\right\|}$$

This yield $\lim_{n\to\infty}\left\|X_t^n - X_t^0\right\| = \left\|X_t\right\| \le \frac{k}{c}\mathrm{e}^{c\left\|t-t_0\right\|}$

Then $\lim_{n\to\infty} X_n$ exists. i.e.

$$X_t = \lim_{n\to\infty} X_t^n \tag{3.12}$$

Since X_t^n is the general solution of equation (3.6) and X_t is the general solution of equation (3.2).

To prove the uniqueness of the solution, let X_t is a solution of the initial-value problem (3.1), or, which is the same, of the integral equation (3.2), and Y_t is the solution of

$$\frac{\mathrm{d}y}{\mathrm{d}t} = b\left(Y(t),t\right), t \in T = \left[t_0, t\right],\ Y(t_0) = Y_0 \tag{3.13}$$

to prove the uniqueness of the solution we want to prove that

$X_t = Y_t$ (3.14)

By subtraction (3.2) and the corresponding integral equation for Y_t

$$X_t - Y_t = X_0 - Y_0 + \int_{t_0}^{t} \left(b\left(X_s, s\right) - b\left(Y_s, s\right)\right)\mathrm{d}s$$

Since $X_0 = Y_0$ then: (3.15)

$$\left\| X_t - Y_t \right\| \le \int_{t_0}^{t} c \cdot \left\| X_s - Y_s \right\| \mathrm{d}s \tag{3.16}$$

i.e; $U' \le \int_{t0}^{t} c \cdot U_s \, \mathrm{d}s$ (3.17)

where $U_t = \left\| X_t - Y_t \right\|$. (3.18)

From equation (3.17) we have:

$$\left\| U_t \right\| \le c \left| t - t_0 \right| \left\| U_t \right\| \tag{3.19}$$

Note that: at $t = t_0$ we obtain $\| U_0 \| \le 0$ then: $\| U_0 \| = 0$

From (3.19) c must satisfy the following condition:

$$c \ge \frac{1}{\left| t - t_0 \right|} \tag{3.20}$$

which is in contradiction with being an independent free constant, hence the only solution of the integral equation (3.17) is

$$U^t = 0 \tag{3.21}$$

Hence $X_t = Y_t$ i.e., the solution of equation (3.1) exists and is unique.

3.2. The Convergence of Euler Scheme for Random Differential Equations in (m.s.) Sense

Let us have the random differential equation

$$\dot{X}(t) = f(X(t),t),\ t \in T = [t_0,t_1],\ X(t_0) = X_0 \quad (3.22)$$

where X_0 is a random variable and the unknown $X(t)$ as well as the right-hand side $f(X,t)$ are stochastic processes defined on the same probability space.

Definitions [6,7]

- Let g: $T \rightarrow L_2$ is an m.s. bounded function and let h > 0 then The "m.s. modulus of continuity of g" is the function

$$W(g,h) = \sup_{|t-t^*|\le h} \left\| g(t) - g(t^*) \right\|,\ t,t^* \in T$$

- The function g is said to be m.s uniformly continuous in T if:

$$\lim_{h\to 0} W(g,h) = 0$$

Note that:

(The limit depends on h because g is defined at every t so we can write $W(g,h) = W(h)$

In the problem (3.22), we find that the convergence of this problem depends on the right hand side (i.e. $f(X(t),t)$ then we want to apply the previous definition on $f(X(t),t)$ hence:

Let $f(X(t),t)$ be defined on $S \times T$ where S is bounded set in L_2

Then we say that f is "randomly bounded uniformly continuous" in S, if

$$\lim_{h\to 0} W(f(x,.),h) = 0$$

(note that $W(f(X(.),h)) = W(h)$)

3.2.1. Random Mean Value Theorem for Stochastic Processes

The aim of this section is to establish a relationship between the increment $X(t) = X(t_0)$ of a 2-s.p. and its m.s. derivative $\dot{X}(\xi)$ for some ξ lying in the interval $[t_0,t]$ for $t > t_0$. This result will be used in the next section to prove the convergence of the random Euler method.

Lemma (3.3.2) [6,7]

Let $Y(t)$ is a 2-s.p., m.s. continuous on interval $T = [t_0,t]$. Then, there exists $\xi \in [t_0,t]$ such that

$$\int_{t_0}^{t} Y(s)ds = Y(\xi)(t-t_0), \; t_0 < t_1 < t \tag{3.25}$$

The proof

Since $Y(t)$ is m.s. continuous, the integral process

$\int_{t_0}^{t} Y(s)ds$ is well defined and the correlation function

$\Gamma_y(r,s)$ is well defined, is a deterministic continuous function on $T \times T$.

For each fixed r, the function $\Gamma_y(r,s)$ is continuous and by the classic mean value theorem for integrals, it follows that:

$$\int_{t_0}^{t} \Gamma_y(r,s)ds = \Gamma_y(r,\xi)(t-t_0)$$

$$\forall \xi \in [t_0,t] \tag{3.26}$$

Note that by definition of $\Gamma_y(r,s)$ expression (3.26) can be written in the form

$$\int_{t_0}^{t} E[y(r)y(s)]ds = E[y(r)y(\xi)](t-t_0)$$

Since $\Gamma_y(r,s) = E[y(r)y(s)]$

We must prove that for the value ξ satisfying (3.26) one get:

$$\left\| \int_{t_0}^{t} y(s)ds - y(\xi)(t-t_0) \right\|^2 = E\left[\left(\int_{t_0}^{t} y(s)ds - y(\xi)(t-t_0) \right)^2 \right] = 0 \tag{3.27}$$

The proof of (3.27)

As

$$E\left[\left(\int_{t_0}^{t} y(s)\,\mathrm{d}s - y(\xi)(t-t_0)\right)^2\right]$$
$$= E\left[\left(\int_{t_0}^{t} y(s)\,\mathrm{d}s\right)^2\right] - 2E\left[\left(\int_{t_0}^{t} y(s)\,\mathrm{d}s\right) y(\xi)\right](t-t_0)$$
$$+ E\left[y(\xi)^2 (t-t_0)^2\right] \tag{3.28}$$

and since:

$$E\left[\left(\int_{t_0}^{t} y(s)\,\mathrm{d}s\right)^2\right] = \int_{t_0}^{t}\int_{t_0}^{t} E\left[y(s)\,y(r)\right]\mathrm{d}r\mathrm{d}s$$

then by substituting in (3.28)

$$E\left[\left(\int_{t_0}^{t} y(s)\,\mathrm{d}s - y(\xi)(t-t_0)\right)^2\right]$$
$$= \int_{t_0}^{t}\int_{t_0}^{t} E\left[y(s)\,y(r)\right]\mathrm{d}r\mathrm{d}s - \int_{t_0}^{t} E\left[y(s)\,y(\xi)\,\mathrm{d}s\right)(t-t_0)$$
$$- \int_{t_0}^{t} E\left[y(s)\,y(\xi)\,\mathrm{d}s\right)(t-t_0) + E[y(\xi)\,y(\xi)(t-t_0)^2 \tag{3.29}$$

And since:

$$\int_{t_0}^{t} E\left[y(r)\,y(s)\right]\mathrm{d}\,s = E\left[y(r)\,y(\xi)\right](t-t_0)$$

then by substituting in (3.28) we have:

$$E\left[\left(\int_{t_0}^{t} y(s)\,\mathrm{d}s - y(\xi)(t-t_0)\right)^2\right]$$
$$= \int_{t_0}^{t} E\left[y(s)\,y(\xi)\,\mathrm{d}s\right)(t-t_0) - \int_{t_0}^{t} E\left[y(s)\,y(\xi)\,\mathrm{d}s\right)(t-t_0)$$
$$- E[y(\xi)\,y(\xi)(t-t_0)^2 + E[y(\xi)\,y(\xi)(t-t_0)^2 = 0$$

i.e. $\left\|\int_{t_0}^{t} y(s)\,\mathrm{d}s - y(\xi)(t-t_0)\right\|^2 = 0$ we obtain

$$\int_{t_0}^{t} y(s)\,\mathrm{d}s = y(\xi)(t-t_0)$$

Theorem (3.3.1) [6,7]

Let $X(s)$ be a m.s. differentiable 2-s.p. in $[t_0,t_1]$ and m.s. continuous in $T = [t_0,t]$. Then, there exists $\xi \in [t_0,t_1]$ such that $X(t) - X(t_0) = \dot{X}(\xi)(t - t_0)$,

$t_0 < t_1 < t$

The proof

The result is a direct consequence of Lemma (3.3.2) applied to the 2-s.p. $Y(t) = \dot{X}(t)$

$$\int_{t_0}^{t} \dot{X}(s)\,\mathrm{d}s = \dot{X}(\xi)(t-t_0)$$

And the integral formula

$$\int_{t_0}^{t} \dot{X}(s)\,\mathrm{d}s = X(t) - X(t_0) \tag{3.30}$$

The proof of (3.30)

Let $X(t)$ be a m.s. differentiable on T and let the ordinary function $f(t,s)$ be continuous on $T \times T$ whose partial derivative $\frac{\partial f(t,s)}{\partial s}$ exist

If $$Y(t) = \int_{a}^{t} f(t,s)\dot{X}(s)\,\mathrm{d}s \tag{3.31}$$

Then $$Y(t) = f(t,s)X(s)\Big|_{a}^{t} - \int_{a}^{t} \frac{\partial f(t,s)}{\partial s} X(s)\,\mathrm{d}s \tag{3.32}$$

Let $f(t,s) = 1$ in Equations (3.31) and (3.32) we have the useful result that:

If $\dot{X}(t)$ is m.s. Riemann integrable on T then:

$$\int_{a}^{t} \dot{X}(s)\,ds = X(t) - X(a), \ [a,t] \subset T$$

Then we have:

$$X(t) - X(t_0) = \dot{X}(\xi)(t-t_0)$$

3.2.2. The Convergence of Random Euler Scheme

In this section we are interested in the mean square convergence, in the fixed station sense, of the random Euler method defined by

$$X_{n+1} = X_n + hf\left(X_n, t_n\right),\ X\left(t_0\right) = X_0,\ n \geq 0 \quad (3.33)$$

where X_n and $f(X_n, t_n)$ are 2-r.v.'s $h = t_n - t_{n-1}$, $t_n = t_0 + nh$ and f: $S \times T \rightarrow L_2$ satisfies the following conditions:

C1: $f(X,t)$ is randomly bounded uniformly continuous C2: $f(X,t)$ satisfies the m.s. Lipschitz condition

$$\left\| f(x,t) - f(y,t) \right\| \leq k(t) \left\| x - y \right\|$$

where

$$\int_{t_0}^{t_1} k(t) \leq \infty \quad (3.34)$$

Note that under hypothesis C1 and C2, we are interested in the m.s. convergence to zero of the error

$$e_n = X_n - X(t) \quad (3.35)$$

where $X(t)$ is the theoretical solution 2-s.p. of the problem (3.22), $t = t_n = t_0 + nh$.

Taking into account (3.22), and Theorem (3.3.1), one getsSince from (3.22) we have at $t = t_\xi$ then

$$\dot{X}\left(t_\xi\right) = f\left(x\left(t_\xi\right), t_\xi\right)$$

Note $\xi \in \left[t_0, t_1\right]$ and we can use ξ instead of t_ξ and from Theorem (3.3.1) at $t = t_\xi$ then we have:

$X\left(t_\xi\right) - X\left(t_0\right) = \dot{X}(t)\left(t_\xi - t_0\right)$ $X(t_\xi) - X(t_0) = X(t)(t_\xi - t_0)$ then

Note that we deal with the interval $\left(t_n, t_{n+1}\right) \ni t_\xi \in \left(t_n, t_{n+1}\right)$ and hence t_0 was the starting in the problem (3.22) and here t_n is the starting and since Euler method deal with solution depend on previous solution and if we have $X(t_n)$ instead of $X(t_0)$ then we can use $X(t_{n+1})$ instead of $X(t_\xi)$.

Then the final form of the problem (3.22) is

$$X(t_{n+1}) = X(t_n) + hf\left(X\left(t_\xi, t_\xi\right)\right), \text{ for some } t_\xi \in \left(t_n, t_{n+1}\right) \tag{3.36}$$

Now we have the solution of problem (3.22) is $X(t_n)$

At $t = t_n$ then $X(t_n) = X(t)$ and the solution of Euler method (3.33) is X_n

Then we can define the error

$$e_n = X_n - X(t_n)$$

$$e_n = X_n - X(t)$$

By (3.33) and (3.36) it follows that

$$X_{n+1} - X(t_{n+1})$$
$$= X_n + hf(X_n, t_n) - X(t_n) - hf\left(X\left(t_\xi, t_\xi\right)\right)$$

This implies

$$e_{n+1} = X_n - X(t_n) + h\left\{f(X_n, t_n) - f\left(X\left(t_\xi\right), t_\xi\right)\right\}$$

Hence

$$\|e_{n+1}\| = \left\|X_n - X(t_n) + h\left\{f(X_n, t_n) - f\left(X\left(t_\xi\right), t_\xi\right)\right\}\right\|$$
$$\leq \left\|X_n - X(t_n)\right\| + h\left\|f\left(X\left(t_\xi\right), t_\xi\right) - f(X_n, t_n)\right\| \tag{3.37}$$

Since:

$$\left\|f\left(X\left(t_\xi\right),t_\xi\right)-f\left(X_n,t_n\right)\right\|$$
$$=\left\|f\left(X\left(t_\xi\right),t_\xi\right)-f\left(X\left(t_\xi\right),t_n\right)+f\left(X\left(t_\xi\right),t_n\right)\right.$$
$$\left.+f\left(X\left(t_n\right),t_n\right)-f\left(X\left(t_n\right),t_n\right)-f\left(X_n,t_n\right)\right\|$$
$$\leq\left\|f(X(t_\xi),t_\xi)-f(X(t_\xi),t_n)\right\|$$
$$+\left\|f(X(t_\xi),t_n)-f(X(t_n),t_n)\right\|$$
$$+\left\|f(X(t_n),t_n)-f(X_n,t_n)\right\| \qquad (3.38)$$

Since the theoretical solution X is m.s. bounded in $[t_0,t_1]$, $\sup_{t_0\leq t\leq t_1}\left\|X\left(t\right)\right\|\leq M<\infty$ and Under hypothesis C1, C2 We obtain

- $\left\|f\left(X\left(t_\xi\right),t_\xi\right)-f\left(X\left(t_\xi\right),t_n\right)\right\|=w\left(h\right)$
- $\left\|f\left(X\left(t_\xi\right),t_n\right)-f\left(X\left(t_n\right),t_n\right)\right\|\leq k\left(t_n\right)Mh$ (*)

Since $k(t_n)$ is Lipschitz constant (from C2) and from Theorem (3.3.1) we have $X(t)$ – $X(t_0)$ = $\dot{X}(\xi)(t$ – $t_0)$ and note that the two points are $X(t_\xi)$ and $X(t_n)$ in (*) then we have:

$$\left\|X\left(t_\xi\right)-X\left(t_n\right)\right\|=\left\|\dot{X}\left(\xi\right)\right\|\left|t_\xi-t_n\right|\leq Mh$$

Since $\left|t_\xi-t_n\right|=h$ and $M=\sup_{t_0\leq t\leq t_1}\left\|\dot{X}\left(t\right)\right\|$

- $\left\|f\left(X\left(t_n\right),t_n\right)-f\left(X_n,t_n\right)\right\|$
$\leq k\left(t_n\right)\left\|X\left(t_n\right)-X_n\right\|=k\left(t_n\right)\left\|e_n\right\|$

Then by substituting in (3.38) we have

$$\left\| f\left(X\left(t_{\xi}\right), t_{\xi}\right) - f\left(X_n, t_n\right) \right\| \le w(h) + k\left(t_n\right) Mh + k\left(t_n\right) \left\| e_n \right\| \tag{3.39}$$

Then by substituting in (3.37) we have

$$\begin{aligned}
\left\|e_{n+1}\right\| &\le \left\|e_n\right\| + h\left[w(h) + k\left(t_n\right)Mh + k\left(t_n\right)\left\|e_n\right\|\right] = \left(1 + k\left(t_n\right)h\right)\left\|e_n\right\| + h\left[w(h) + k\left(t_n\right)Mh\right] \\
&\le \left(1 + K\left(t_n\right)h\right)[\left(1 + k\left(t_n\right)h\right)\left\|e_{n-1}\right\| + h\left[w(h) + k\left(t_n\right)Mh\right] + h\left[w(h) + k\left(t_n\right)Mh\right] \\
&= \left(1 + K\left(t_n\right)h\right)^2 \left\|e_{n-1}\right\| + h\left[w(h) + k\left(t_n\right)Mh\right]\left[1 + \left(1 + k\left(t_n\right)h\right)\right] \\
&\le \left(1 + K\left(t_n\right)h\right)^3 \left\|e_{n-2}\right\| + h\left[w(h) + k\left(t_n\right)Mh\right]\left[1 + \left(1 + k\left(t_n\right)h\right) + \left(1 + k\left(t_n\right)h\right)^2\right] \\
&\le \left(1 + K\left(t_n\right)h\right)^{n+1} \left\|e_0\right\| + h\left[w(h) + k\left(t_n\right)Mh\right]\left[1 + \left(1 + k\left(t_n\right)h\right) + \left(1 + k\left(t_n\right)h\right)^2 + \cdots + \left(1 + k\left(t_n\right)h\right)^n\right]
\end{aligned}$$

Since:

$$\left[1 + \left(1 + k\left(t_n\right)h\right) + \left(1 + k\left(t_n\right)h\right)^2 + \cdots + \left(1 + k\left(t_n\right)h\right)^n\right]$$

is geometrical sequence.

Then:

$$\left[1 + \left(1 + k\left(t_n\right)h\right) + \left(1 + k\left(t_n\right)h\right)^2 + \cdots + \left(1 + k\left(t_n\right)h\right)^n\right]$$

$$= \frac{\left(1 + k\left(t_n\right)h\right)^n - 1}{k\left(t_n\right)h}$$

Then we get

$$\left\|e_{n+1}\right\| \le \left(1 + K\left(t_n\right)h\right)^{n+1} \left\|e_0\right\|$$

$$+ \left[w(h) + k\left(t_n\right)Mh\right] \frac{[\left(1 + k\left(t_n\right)h\right)^n - 1]}{k\left(t_n\right)}$$

Taking into account that $e_0 = 0$ where

$$e_0 = X_0 - X\left(t_0\right) = 0.$$

$$\left\|e_{n+1}\right\| \le \left[w(h) + k\left(t_n\right)Mh\right] \frac{\left[\left(1 + k\left(t_n\right)h\right)^n - 1\right]}{k\left(t_n\right)}$$

$$\lim_{h\to 0}\|e_{n+1}\| \le \lim_{h\to 0}\left[w(h)+k(t_n)Mh\right]\frac{\left[\left(1+k(t_n)h\right)^n-1\right]}{k(t_n)}$$

$$=\lim_{h\to 0}\frac{\left[w(h)+k(t_n)Mh\right]}{k(t_n)}\left[\left(1+K(t_n)h\right)^n\right]$$

$$-\lim_{h\to 0}\frac{\left[w(h)+k(t_n)Mh\right]}{k(t_n)} \tag{3.40}$$

Note that:

The term: $\lim_{h\to 0}\frac{\left[w(h)+k(t_n)Mh\right]}{k(t_n)}=0$ as

$$h \to 0 \ (w(h) \to 0 \ni \text{as } h \to 0) \tag{3.41}$$

And the second term:

$$\lim_{h\to 0}\frac{\left[w(h)+k(t_n)Mh\right]}{k(t_n)}\left[\left(1+K(t_n)h\right)^n\right]$$

we have:

$$\lim_{h\to 0}\frac{\left[w(h)+k(t_n)Mh\right]}{k(t_n)}[\left(1+K(t_n)h\right)^n]$$

$$=\lim_{h\to 0}\frac{\left[w(h)+k(t_n)Mh\right]}{k(t_n)}\lim_{h\to 0}\left[\left(1+K(t_n)h\right)^n\right] \tag{3.42}$$

The first limit in (3.42) equal zero and:

The computation of $\lim_{h\to 0}\left[\left(1+K(t_n)h\right)^n\right]$ as follows:

Let $y=\lim_{h\to 0}\left[\left(1+K(t_n)h\right)^n\right]$ then by tacking the logarithm of the two sides we have:

$$\ln y = \ln \lim_{h\to 0}\left[\left(1+K\left(t_n\right)h\right)^n\right]$$
$$= \lim_{h\to 0} \ln\left[\left(1+K\left(t_n\right)h\right)^n\right]$$
$$= \lim_{h\to 0} n \ln\left[\left(1+K\left(t_n\right)h\right)\right]$$
$$= \lim_{h\to 0} \frac{t_n - t_0}{h} \ln\left[\left(1+K\left(t_n\right)h\right)\right]$$
$$= \lim_{h\to 0} \frac{t_n - t_0}{h} \ln\left[\left(1+K\left(t_n\right)h\right)\right]$$
$$= \lim_{h\to 0} \frac{\left(t_n - t_0\right)\left[\ln\left(1+k\left(t_n\right)h\right)\right]}{h}$$

By using the (L'Hospital's Rule):

$$\lim_{h\to 0} \frac{\left(t_n - t_0\right)\left[\ln\left(1+k\left(t_n\right)h\right)\right]}{h}$$
$$= \lim_{h\to 0} \frac{\left(t_n - t_0\right)\dfrac{1}{\left(1+k\left(t_n\right)h\right)}k\left(t_n\right)}{1}$$
$$= \lim_{h\to 0} \frac{\left(t_n - t_0\right)k\left(t_n\right)}{1+k\left(t_n\right)h} = \left(t - t_0\right)k\left(t\right) \tag{3.43}$$

Then $\ln y = \left(t - t_0\right)k\left(t_n\right)$ which implies that

$y = e^{\left(t-t_0\right)k\left(t_n\right)}$ hence

$$\lim_{h\to 0}\left[\left(1+K\left(t_n\right)h\right)^n\right] = e^{\left(t-t_0\right)k\left(t_n\right)}$$

By substituting in (3.42):

$$\lim_{h\to 0}\frac{\left[w(h)+k(t_n)Mh\right]}{k(t_n)}\left[\left(1+K(t_n)h\right)^n\right]$$

$$=0\times e^{(t-t_0)k(t_n)}=0 \tag{3.44}$$

By substituting from (3.44) and (3.42) in (3.40) hence $\lim_{h\to 0}\left\|e_{n+1}\right\|\to 0 \;\; i.e., \left\{e_n\right\}$ i.e., $\{e_n\}$ converge in m.s to zero as $h\to 0$ hence $X_n \xrightarrow{m.s} X(t_n)=X(t)$.

3.3. The Convergence of Runge-Kutta of Second Order Scheme for Random Differential Equations in Mean Square Sense

In this section we are interested in the mean square convergence, in the fixed station sense, of the random Runge-Kutta of second order method defined by

$$X_{n+1}=X_n+\frac{h}{2}[f\left(X_n,t_n\right)+f\left(X_n+f\left(X_n,t_n\right),t_{n+1}\right),$$

$$X(t_0)=X_0,\ n\geq 0 \tag{3.45}$$

where X_n and $f(X_n,t_n)$ are 2-r.v.'s, $h=t_n-t_{n-1}$, $t_n=t_0+nh$ and f: $S\times T\to L_2$, $S\subset L_2$ satisfies the following conditions:

C1: $f(X,t)$ is randomly bounded uniformly continuousC2: $f(X,t)$ satisfies the m.s. Lipschitz condition

$$\left\|f(x,t)-f(y,t)\right\|\leq k(t)\left\|x-y\right\|$$

where

$$\int_{t_0}^{t_1}k(t)\leq\infty \tag{3.46}$$

Note that under hypothesis C1 and C2, we are interested in the m.s. convergence to zero of the error

$$e_n=X_n-X(t) \tag{3.47}$$

where $X(t)$ is the theoretical solution 2-s.p. of the problem (3.22), $t = t_n = t_0 + n_h$.

Taking into account (3.22), and Theorem (3.3.1), one gets,

Since from (3.22) we have at $t = t_\xi$ then

$$\dot{X}(t_\xi) = f(x(t_\xi), t_\xi)$$

Note $\xi \in [t_0, t_1]$ and we can use ξ instead of t_ξ

And from Theorem (3.3.1) at $t = t_\xi$ then we obtain

$$X(t_\xi) - X(t_0) = \dot{X}(t)(t_\xi - t_0) \Rightarrow$$
$$X(t_\xi) - X(t_0) = f(X(t_\xi, t_\xi)(t_\xi - t_0))$$

Note that we deal with the interval $(t_n, t_{n+1}) \ni t_\xi \in (t_n, t_{n+1})$ and hence t_0 was the starting in the problem (3.22) and here t_n is the starting and since Euler method deal with solution depend on previous solution and if we have $X(t_n)$ instead of $\Rightarrow X(t_0)$ we can use $X(t_n + 1)$ instead of $X(t_\xi)$ then the final form of the problem (3.22) is

$$X(t_{n+1}) = X(t_n) + hf(X(t_\xi, t_\xi)), \text{ for some } t_\xi \in (t_n, t_{n+1}) \tag{3.48}$$

Now we have the solution of problem (3.22) is $X(t_n)$

At $t = t_n$ then $X(t_n) = X(t)$ and the solution of Runge-Kutta of 2 order method (3.45) is X_n

Then we can define the error

$e_n = X_n - X(t)$

By (3.45) and (3.48) it follows that

$$X_{n+1} - X(t_{n+1})$$
$$= X_n + \frac{h}{2}\left[f(X_n, t_n) + f(X_n + f(X_n, t_n), t_{n+1})\right] - X(t_n)$$
$$-\frac{h}{2} f(X(t_\xi), t_\xi) - \frac{h}{2} f(X(t_\xi), t_\xi)$$

Then we obtain:

$$e_{n+1} = X_n - X(t_n) + \frac{h}{2}\left\{f(X_n, t_n) - f\left(X(t_\xi), t_\xi\right)\right\}$$
$$+\frac{h}{2}\left\{f\left(X_n + f(X_n, t_n), t_{n+1}\right) - f\left(X(t_\xi), t_\xi\right)\right\}$$

By taking the norm for the two sides:

$$\left\|e_{n+1}\right\| = \left\| X_n - X(t_n) + \frac{h}{2}\left\{f(X_n, t_n) - f\left(X(t_\xi), t_\xi\right)\right\}\right.$$
$$\left. +\frac{h}{2}\left\{f\left(X_n + f(X_n, t_n), t_{n+1}\right) - f\left(X(t_\xi), t_\xi\right)\right\}\right\|$$
$$\leq \left\|X_n - X(t_n)\right\| + \frac{h}{2}\left\|f\left(X(t_\xi), t_\xi\right) - f(X_n, t_n)\right\|$$
$$+\frac{h}{2}\left\|f\left(X_n + f(X_n, t_n), t_{n+1}\right) - f\left(X(t_\xi), t_\xi\right)\right\| \quad (3.49)$$

Since:

$$\left\|f\left(X(t_\xi), t_\xi\right) - f(X_n, t_n)\right\|$$
$$= \left\|f\left(X(t_\xi), t_\xi\right) - f\left(X(t_\xi), t_n\right) + f\left(X(t_\xi), t_n\right)\right.$$
$$\left. + f\left(X(t_n), t_n\right) - f\left(X(t_n), t_n\right) - f(X_n, t_n)\right\|$$
$$\leq \left\|f\left(X(t_\xi), t_\xi\right) - f\left(X(t_\xi), t_n\right)\right\| + \left\|f\left(X(t_\xi), t_n\right) - f\left(X(t_n), t_n\right)\right\|$$
$$+ \left\|f\left(X(t_n), t_n\right) - f(X_n, t_n)\right\| \quad (3.50)$$

Since the theoretical solution *X* is m.s. bounded in

$[t_0, t_1]$, $\sup_{t_0 \leq t \leq t_1} \left\|X(t)\right\| \leq M < \infty$ and Under hypothesis C1, C2 We have

- $\left\|f\left(X(t_\xi), t_\xi\right) - f\left(X(t_\xi), t_n\right)\right\| = w(h)$

- $\left\| f\left(X\left(t_{\xi}\right),t_{n}\right)-f\left(X\left(t_{n}\right),t_{n}\right)\right\| \leq k\left(t_{n}\right)Mh$ (*)

where $k(t_n)$ Is Lipschitz constant (from C2) and:

From Theorem (3.3.1) we have

$X(t)-X(t_0)=\dot{X}(\xi)(t-t_0)$ and note that the two points Are $X(t_\xi)$ and $X(t_n)$ in (*) then

$$\left\|X\left(t_{\xi}\right)-X\left(t_{n}\right)\right\|=\left\|\dot{X}(\xi)\right\|\left|t_{\xi}-t_{n}\right| \leq Mh$$

where $\left|t_{\xi}-t_{n}\right|=h$ and $M=\sup_{t_0\leq t\leq t_1}\left\|\dot{X}(t)\right\|$

- $$\begin{aligned}&\left\|f\left(X\left(t_{n}\right),t_{n}\right)-f\left(X_{n},t_{n}\right)\right\|\\&\leq k\left(t_{n}\right)\left\|X\left(t_{n}\right)-X_{n}\right\|=k\left(t_{n}\right)\left\|e_{n}\right\|\end{aligned}$$

Then by substituting in (3.50) we have

$$\left\|f\left(X\left(t_{\xi}\right),t_{\xi}\right)-f\left(X_{n},t_{n}\right)\right\| \leq w(h)+k\left(t_{n}\right)Mh+k\left(t_{n}\right)\left\|e_{n}\right\| \tag{3.51}$$

And another term:

$$\begin{aligned}&\left\|f\left(X_{n}+f\left(X_{n},t_{n}\right),t_{n+1}\right)-f\left(X\left(t_{\xi}\right),t_{\xi}\right)\right\|\\
&=\left\|f\left(X\left(t_{\xi}\right),t_{\xi}\right)-f\left(X_{n}+f\left(X_{n},t_{n}\right),t_{n+1}\right)\right\|\\
&=\left\|f\left(X\left(t_{\xi}\right),t_{\xi}\right)-f\left(X\left(t_{\xi}\right),t_{n+1}\right)+f\left(X\left(t_{\xi}\right),t_{n+1}\right)\right.\\
&\quad\left.+f\left(X\left(t_{n}\right),t_{n+1}\right)-f\left(X\left(t_{n}\right),t_{n+1}\right)-f\left(X_{n}+f\left(X_{n},t_{n}\right),t_{n+1}\right)\right\|\\
&\leq\left\|f\left(X\left(t_{\xi}\right),t_{\xi}\right)-f\left(X\left(t_{\xi}\right),t_{n+1}\right)\right\|+\left\|f\left(X\left(t_{\xi}\right),t_{n+1}\right)-f\left(X\left(t_{n}\right),t_{n+1}\right)\right\|\\
&\quad+\left\|f\left(X\left(t_{n}\right),t_{n+1}\right)-f\left(X_{n}+f\left(X_{n},t_{n}\right),t_{n+1}\right)\right\|\\
&\leq w(h)+k\left(t_{n+1}\right)Mh+k\left(t_{n+1}\right)\left[\left\|e_{n}\right\|-M\right]\end{aligned}$$

Since:

- $\left\|f\left(X\left(t_{\xi}\right),t_{\xi}\right)-f\left(X\left(t_{\xi}\right),t_{n+1}\right)\right\|\leq w(h)$
- $\left\|f\left(X\left(t_{\xi}\right),t_{n+1}\right)-f\left(X\left(t_{n}\right),t_{n+1}\right)\right\|\leq k\left(t_{n+1}\right)Mh$

where $k(t_n)$. Is Lipschitz constant (from C2) and:

From Theorem (3.3.1) we have

$X(t) - X(t_0) = \dot{X}(\xi)(t - t_0)$ and note that the two points are $X(t_\xi)$ and $X(t_n)$ in (*) then we have:

$$\left\|X\left(t_{\xi}\right)-X\left(t_{n}\right)\right\|=\left\|\dot{X}\left(\xi\right)\right\|\left|t_{\xi}-t_{n}\right|\leq Mh$$

where $\left|t_{\xi}-t_{n}\right|=h$ and $M=\sup_{t_0\leq t\leq t_1}\left\|\dot{X}\left(t\right)\right\|$

And the last term:

$$\begin{aligned}
&\left\|f\left(X\left(t_{n}\right),t_{n+1}\right)-f\left(X_{n}+f\left(X_{n},t_{n}\right),t_{n+1}\right)\right\|\\
&\leq k\left(t_{n+1}\right)\left\|X\left(t_{n}\right)-X_{n}-f\left(X_{n},t_{n}\right)\right\|\\
&\leq k\left(t_{n+1}\right)\left\|X\left(t_{n}\right)-X_{n}\right\|-\left\|f\left(X_{n},t_{n}\right)\right\|\\
&=k\left(t_{n+1}\right)\left[\left\|e_{n}\right\|-M\right]
\end{aligned}$$

Then by substituting in (3.49) we have

$$\|e_{n+1}\| \le \|e_n\| + \frac{h}{2}\left[w(h) + k(t_n)Mh + k(t_n)\|e_n\|\right] + \frac{h}{2}\left[w(h) + k(t_{n+1})Mh + k(t_{n+1})\left[\|e_n\| - M\right]\right]$$

$$= \|e_n\|\left(1 + \frac{h}{2}k(t_n) + k(t_{n+1})\right) + \frac{h}{2}\left[2w(h) + hk(t_n)M + hk(t_{n+1})M - K(t_{n+1})M\right]$$

$$\le \left\{\|e_{n-1}\|\left(1 + \frac{h}{2}k(t_n) + k(t_{n+1})\right) + \frac{h}{2}\left[2w(h) + hk(t_n)M + hk(t_{n+1})M - K(t_n)M\right]\right\}$$

$$\left\{\left(1 + \frac{h}{2}k(t_n) + k(t_{n+1})\right)\right\} + \frac{h}{2}\left[2w(h) + hk(t_n)M + hk(t_{n+1})M - K(t_{n+1})M\right]$$

$$= \|e_{n-1}\|\left(1 + \frac{h}{2}k(t_n) + k(t_{n+1})\right)^2 + \frac{h}{2}\left[2w(h) + hk(t_n)M + hk(t_{n+1})M - K(t_{n+1})M\right]\left(2 + \frac{h}{2}k(t_n) + k(t_{n+1})\right)$$

$$\le \left\{\|e_{n-2}\|\left(1 + \frac{h}{2}k(t_n) + k(t_{n+1})\right) + \frac{h}{2}\left[2w(h) + hk(t_n)M + hk(t_{n+1})M - K(t_n)M\right]\right\}$$

$$\left\{\left(1 + \frac{h}{2}k(t_n) + k(t_{n+1})\right)^2\right\} + \frac{h}{2}\left[2w(h) + hk(t_n)M + hk(t_{n+1})M - K(t_{n+1})M\right]\left(2 + \frac{h}{2}k(t_n) + k(t_{n+1})\right)$$

$$= \|e_{n-2}\|\left(1 + \frac{h}{2}k(t_n) + k(t_{n+1})\right)^3 + \frac{h}{2}\left[2w(h) + hk(t_n)M + hk(t_{n+1})M - K(t_{n+1})M\right]$$

$$\left[1 + \left(1 + \frac{h}{2}k(t_n) + k(t_{n+1})\right) + \left(1 + \frac{h}{2}k(t_n) + k(t_{n+1})\right)^2\right]$$

Then we have:

$$\|e_{n+1}\| \le \|e_0\|\left(1 + \frac{h}{2}k(t_n) + k(t_{n+1})\right)^{n+1} + \frac{h}{2}\left[2w(h) + hk(t_n)M + hk(t_{n+1})M - K(t_{n+1})M\right]$$

$$\left[1 + \left(1 + \frac{h}{2}k(t_n) + k(t_{n+1})\right) + \left(1 + \frac{h}{2}k(t_n) + k(t_{n+1})\right)^2 + \cdots + \left(1 + \frac{h}{2}k(t_n) + k(t_{n+1})\right)^n\right]$$

Since:

$$\left[1 + \left(1 + \frac{h}{2}k(t_n) + k(t_{n+1})\right) + \left(1 + \frac{h}{2}k(t_n) + k(t_{n+1})\right)^2 + \cdots + \left(1 + \frac{h}{2}k(t_n) + k(t_{n+1})\right)^n\right]$$

is geometrical sequence then we have:

$$[1 + \left(1 + \frac{h}{2}k(t_n) + k(t_{n+1})\right) + \left(1 + \frac{h}{2}k(t_n) + k(t_{n+1})\right)^2 + \cdots + \left(1 + \frac{h}{2}k(t_n) + k(t_{n+1})\right)^n] = \frac{\left[\left(1 + \frac{h}{2}k(t_n) + k(t_{n+1})\right)^n\right] - 1}{\frac{h}{2}k(t_n) + k(t_{n+1})}$$

Then we get:

$$\|e_{n+1}\| \le \|e_0\|\left(1 + \frac{h}{2}k(t_n) + k(t_{n+1})\right)^{n+1} + \frac{h}{2}\left[2w(h) + hk(t_n)M + hk(t_{n+1})M - K(t_{n+1})M\right]\frac{\left[\left(1 + \frac{h}{2}k(t_n) + k(t_{n+1})\right)^n\right] - 1}{\frac{h}{2}k(t_n) + k(t_{n+1})}$$

Taking into account that $e_0 = 0$ where

$$e_0 = X_0 - X(t_0) = 0$$

$$\|e_{n+1}\| \le \frac{h}{2}\left[2w(h) + hk(t_n)M + hk(t_{n+1})M - k(t_{n+1})M\right]\frac{\left[\left(1+\frac{h}{2}k(t_n)+k(t_{n+1})\right)^n\right]-1}{\frac{h}{2}k(t_n)+k(t_{n+1})}$$

$$\lim_{h\to 0}\|e_{n+1}\| \le \lim_{h\to 0}\frac{h}{2}\left[2w(h) + k(t_n)Mh + hk(t_{n+1})M - K(t_{n+1})M\right]\frac{\left[\left(1+\frac{h}{2}k(t_n)+k(t_{n+1})\right)^n\right]-1}{\frac{h}{2}k(t_n)+k(t_{n+1})}$$

$$= \lim_{h\to 0}\frac{\frac{h}{2}\left[2w(h)+k(t_n)Mh+k(t_{n+1})Mh-k(t_{n+1})M\right]}{\frac{h}{2}k(t_n)+k(t_{n+1})}\left[\left(1+\frac{h}{2}K(t_n)+k(t_{n+1})\right)^n\right]$$

$$-\lim_{h\to 0}\frac{\frac{h}{2}\left[2w(h)+k(t_n)Mh+k(t_{n+1})Mh-k(t_{n+1})M\right]}{\frac{h}{2}k(t_n)+K(t_{n+1})}$$

(3.52)

Note that:

The term:

$$\lim_{h\to 0}\frac{\frac{h}{2}\left[2w(h)+k(t_n)Mh+k(t_{n+1})Mh-k(t_{n+1})M\right]}{\frac{h}{2}k(t_n)+K(t_{n+1})} = \frac{0}{k(t_{n+1})} = 0$$

and the second term:

$$\lim_{h\to 0}\frac{\frac{h}{2}\left[2w(h)+k(t_n)Mh+k(t_{n+1})Mh-k(t_{n+1})M\right]}{\frac{h}{2}k(t_n)+k(t_{n+1})}\left[\left(1+\frac{h}{2}K(t_n)+k(t_{n+1})\right)^n\right]$$

we have:

$$\lim_{h\to 0}\frac{\frac{h}{2}\left[2w(h)+k(t_n)Mh+k(t_{n+1})Mh-k(t_{n+1})M\right]}{\frac{h}{2}k(t_n)+k(t_{n+1})}\left[\left(1+\frac{h}{2}K(t_n)+k(t_{n+1})\right)^n\right]$$

$$=\lim_{h\to 0}\frac{\frac{h}{2}\left[2w(h)+k(t_n)Mh+k(t_{n+1})Mh-k(t_{n+1})M\right]}{\frac{h}{2}k(t_n)+k(t_{n+1})}\lim_{h\to 0}\left[\left(1+\frac{h}{2}K(t_n)+k(t_{n+1})\right)^n\right] \quad (3.53)$$

The first limit in (3.53) equals zero and:

The computation of

is as follows:

Let $\lim_{h\to 0}\left[\left(1+\frac{h}{2}K(t_n)+k(t_{n+1})\right)^n\right]$ then by tacking the logarithm of the two sides we have:

$$\ln y=\ln\lim_{h\to 0}\left[\left(1+\frac{h}{2}K(t_n)+k(t_{n+1})\right)^n\right]=\lim_{h\to 0}\ln\left[\left(1+\frac{h}{2}K(t_n)+k(t_{n+1})\right)^n\right]$$

$$=\lim_{h\to 0}n\left\{\ln\left[\left(1+\frac{h}{2}K(t_n)+k(t_{n+1})\right)\right]\right\}=\lim_{h\to 0}\frac{t_n-t_0}{h}\left[\ln[\left(1+\frac{h}{2}K(t_n)+k(t_{n+1})\right)\right]$$

$$=\lim_{h\to 0}\frac{(t_n-t_0)\left[\ln(1+\frac{h}{2}k(t_n)+k(t_{n+1})\right]}{h}$$

By using the (L'Hospital's Rule):

$$\lim_{h\to 0}\frac{(t_n-t_0)\left[\ln\left(1+\frac{h}{2}k(t_n)+k(t_{n+1})\right)\right]}{h}$$

$$=\lim_{h\to 0}\frac{(t_n-t_0)\frac{1}{\left(1+\frac{h}{2}k(t_n)+k(t_{n+1})\right)}\left[\frac{1}{2}k(t_n)\right]}{1} \quad (3.54)$$

$$=\lim_{h\to 0}\frac{\frac{1}{2}(t_n-t_0)k(t_n)}{1+\frac{h}{2}k(t_n)+k(t_{n+1})}=\frac{\frac{1}{2}(t-t_0)k(t)}{1+k(t)}$$

$\ln y = \dfrac{(t-t_0)k(t)}{2[1+k(t)]}$ Then $y = e^{\frac{(t-t_0)k(t)}{2[1+k(t)]}}$ hence:

$$\lim_{h\to 0}\left[\left(1+\frac{h}{2}K(t_n)h+k(t_{n+1})\right)^n\right] = e^{\frac{(t-t_0)k(t)}{2[1+k(t)]}}$$

By substituting in (3.53):

$$\lim_{h\to 0}\frac{\frac{h}{2}\left[2w(h)+k(t_n)Mh+k(t_{n+1})Mh-k(t_{n+1})M\right]}{\frac{h}{2}k(t_n)+k(t_{n+1})}$$

$$\left[\left(1+\frac{h}{2}K(t_n)+k(t_{n+1})\right)^n\right] = 0\times e^{\frac{(t-t_0)k(t)}{2[1+k(t)]}} = 0 \tag{3.55}$$

By substituting from (3.55) and (3.53) in (3.51) then we obtain $\lim_{h\to 0}\|e_{n+1}\|\to 0$ i.e. $\{e_n\}$ converges in m.s to zero as $h\to 0$ hence $X_n \xrightarrow{m.s} X(t_n) = X(t)$

4. SOME RESULTS

Theorem 4.1

Let $\{X_n, n = 0, 1, \cdots\}$, $\{Y_n, n = 0, 1, \cdots\}$ be sequences of 2-r.v's over the same probability space and let a and b be deterministic real numbers.

Suppose: $X = \lim_{n\to\infty} X_n$ and $Y = \lim_{n\to\infty} Y_n$

Then:

1) $(aX+bY) = \lim_{n\to\infty}(aX_n+bY_n)$

2) $E(X) = \lim_{n\to\infty} E\{X_n\}$

3) $E(XY) = \lim_{n\to\infty} E\{X_n Y_n\}$

4) $\lim_{n\to\infty} E(X_n^2) = E(X^2)$

5) $\lim_{n\to\infty} Var(X_n) == Var(X)$

Definition 4.1 [13]. "The convergence in probability"

A sequence of r.v's $\{X_n\}$ converges in probability to a random variable X as $n \to \infty$ if

$$\lim_{n\to\infty} p\{|X_n - X| > \varepsilon\} = 0 \ \forall \varepsilon > 0$$

Definition 4.2 [13]. "The convergence in distribution"

A sequence of r.v's $\{X_n\}$ converge in distribution to a random variable X as $n \to \infty$ if

$$\lim_{n\to\infty} F_{x_n}(x) = F_x(x)$$

Lemma (4.1) [13]

The convergence in m.s implies convergence in probability

Lemma (4.2) [13]

The convergence in probability implies convergence in distribution

Theorem 4.2

If $X_n \xrightarrow{m.s} X$ then PDF of $\{X_n\} \xrightarrow{m.s}$ PDF PDF of $\{X\}$ i.e.; $\lim_{n\to\infty} f_{X_n}(x) = f_X(x)$

Proof Since we have shown that If $X_n \xrightarrow{m.s} X$ then

$$X_n \xrightarrow{d} X$$

i.e., if $X_n \xrightarrow{m.s} X$ then $\lim_{n\to\infty} F_{X_n}(x) = F_X(x)$

Then we have: $\lim_{n\to\infty} \frac{d}{dx} F_{X_n}(x) = \frac{d}{dx} F_X(x)$ then

$$\lim_{n\to\infty} f_{x_n}(x) = f_X(x)$$

5. NUMERICAL EXAMPLES

Example (5.1)

The differential equation with random term in it and random initial condition

$$y' = Kx,\ y(x_0) = D,\ x \in [x_0, x_n],$$

K, D are independent Poisson random variables with joint PDF

$$f_{K,D}(K,D) = \frac{\mathrm{e}^{-4}\,2^{k+D}}{k!D!},\ K,D = 0,1,2,\cdots$$

1) The exact solution,

$$y = D + \frac{K\left(x^2 - x_0^2\right)}{2}$$

2) The numerical solution

Using the Random Euler Method:

$$y_n = y_{n-1} + hf\left(y_{n-1}, x_{n-1}\right),\ y(x_0) = y_0$$

at $n = 1$

$$y_1 = y_0 + hf\left(y_0, x_0\right) = D + hKx_0$$

at $n = 2$

$$y_2 = y_1 + hf\left(y_1, x_1\right) = D + hKx_0 + hKx_1$$
$$= D + hKx_0 + hK\left(x_0 + h\right)$$

at $n = 3$

$$y_3 = y_2 + hf(y_2, x_2) = D + hKx_0 + hKx_1 + hKx_2$$
$$= D + hKx_0 + hK(x_0 + h) + hK(x_0 + 2h)$$

at $n = 4$

$$y_4 = y_3 + hf(y_3, x_3) = D + hKx_0 + hKx_1 + hKx_2 + hKx_3$$
$$= D + hKx_0 + hK(x_0 + h) + hK(x_0 + 2h) + hK(x_0 + 3h)$$

and so on...

Then the general numerical solution is

$$y_n = D + hKx_0 + hK(x_0 + h) + hK(x_0 + 2h)$$
$$+ hK(x_0 + 3h) + \cdots + hK(x_0 + (n-1)h)$$

i.e., $$y_n = D + hK\sum_{i=0}^{n-1}(x_0 + ih).$$

This can be written in another form:

$$y_n = D + nhKx_0 + [n(n-1)/2]Kh^2.$$

We can prove that:

1) $\lim_{n\to\infty} y_n = y$

Proof Since $\lim_{n\to\infty} y_n = y$ (if and only if) $\lim_{n\to\infty} E|y_n - y|^2 = 0$

Then:

$$y_n - y = nhKx_0 + \frac{n(n-1)}{2}\left[h^2K\right] - \frac{K\left(x^2 - x_0^2\right)}{2}$$

$$\left|y_n - y\right|^2$$

$$= n^2h^2K^2x_0^2 + 2nhK^2x_0\left[\frac{n(n-1)}{2}h^2 - \frac{\left(x^2 - x_0^2\right)}{2}\right]$$

$$+ K^2\left[\frac{n(n-1)}{2}h^2 - \frac{\left(x^2 - x_0^2\right)}{2}\right]^2$$

where $h = \dfrac{x_n - x_0}{n}$

$$\left|y_n - y\right|^2 = K^2x_0^2\left(x_n - x_0\right)^2 + K^2x_0\left(x_n - x_0\right)\left[\frac{(n-1)}{n}\left(x_n - x_0\right)^2 - \left(x^2 - x_0^2\right)\right]$$

$$+\frac{K^2}{4}\left[\frac{(n-1)}{n}\left(x_n - x_0\right)^2 - \left(x^2 - x_0^2\right)\right]^2$$

$$= K^2x_0^2x_n^2 - 2K^2x_nx_0^3 + K^2x_0^4 + K^2x_nx_0\left(\frac{n-1}{n}\right)\left(x_n - x_0\right)^2 - K^2x_0^2\left(\frac{n-1}{n}\right)\left(x_n - x_0\right)^2$$

$$-K^2x_nx_0\left(x^2 - x_0^2\right) + K^2x_0^2\left(x^2 - x_0^2\right) + \frac{K^2}{4}\left(\frac{n-1}{n}\right)^2\left(x_n - x_0\right)^4$$

$$-\frac{K^2}{2}\left(\frac{n-1}{n}\right)\left(x_n - x_0\right)^2\left(x^2 - x_0^2\right) + \frac{K^2}{4}\left(x^2 - x_0^2\right)^2$$

$$\lim_{n\to\infty} E|y_n - y|^2 = 18x_0^2x^2 - 36x_0^3x + 18x_0^4 + 18xx_0(x-x_0)^2 - 18x_0^2(x-x_0)^2$$
$$-18xx_0(x^2 - x_0^2) + 18x_0^2(x^2 - x_0^2) + \frac{18}{4}x^4 - 18x^3x_0 + 27x^2x_0^2 - 18xx_0^3$$
$$+\frac{18}{4}x_0^4 - 9x^4 + 18x^3x_0 - 18xx_0^3 + 9x_0^4 + \frac{18}{4}x^4 - 9x^2x_0^2 + \frac{18}{4}x_0^4$$
$$= 18x_0^2x^2 - 36x_0^3x + 18x_0^4 + 72xx_0^3 - 36x^2x_0^2 - 36x_0^4 + \frac{18}{4}x^4 - 18x^3x_0$$
$$+27x^2x_0^2 - 18xx_0^3 + \frac{18}{4}x_0^4 - 9x^4 + 18x^3x_0 - 18xx_0^3 + 9x_0^4$$
$$+\frac{18}{4}x^4 - 9x^2x_0^2 + \frac{18}{4}x_0^4 = 0$$

i.e; $\lim_{n\to\infty} y_n = y$

We can verify theorem (4.1) as follows

2) $\lim_{n\to\infty} E\{y_n\} = E\{y\}$

Proof

$$y_n = D + hK\sum_{i=0}^{n-1}(x_0 + ih) = D + hK\sum_{j=1}^{n}(x_0 + (j-1)h) = D + hK\left[x_0 + (x_0 + h) + (x_0 + 2h) + (x_0 + 3h) + \cdots (x_0 + (n-1)h)\right]$$
$$= D + hK\left[nx_0 + h\left[1 + 2 + 3 + \cdots + (n-1)\right]\right] = D + hK\left[nx_0 + h\left(\frac{n(n-1)}{2}\right)\right]$$
$$= D + K\left[nhx_0 + h^2\left(\frac{n(n-1)}{2}\right)\right] \quad \text{where } h = \frac{x_n - x_0}{n}$$
$$= D + K\left[nx_0 + \frac{(x_n - x_0)}{n} + \left(\frac{(x_n - x_0)}{n}\right)^2\left(\frac{n(n-1)}{2}\right)\right] = D + K\left[x_0(x_n - x_0) + \frac{(n-1)}{2n}(x_n - x_0)^2\right]$$
$$E(y_n) = E(D) + E\left\{K\left[x_0(x_n - x_0) + \frac{(n-1)}{2n}(x_n - x_0)^2\right]\right\} = 2 + 2\left\{E(x_0x_n) - E(x_0^2) + \frac{(n-1)}{2n}E(x_n - x_0)^2\right\}$$

Then:

$$\lim_{n\to\infty} E\{y_n\} = 2 + 2\left\{E(x_0 x) - E(x_0^2) + \frac{1}{2}E(x - x_0)^2\right\}$$

$$= 2 + 2\left\{E(x_0 x) - E(x_0^2) + \frac{1}{2}E(x^2) - E(x_0 x) + \frac{1}{2}E(x_0^2)\right\}$$

$$= 2 + 2\left\{\frac{1}{2}E(x^2) - \frac{1}{2}E(x_0^2)\right\} = 2 + \left\{E(x^2) - E(x_0^2)\right\} = E\{y\}$$

i.e.; $\lim_{n\to\infty} E\{y_n\} = E\{y\}$.

3) $\lim_{n\to\infty} E\{y_n^2\} = E\{y^2\}$

Proof Since $y_n = D + nhKx_0 + [n(n-1)/2]Kh^2$

Then we have:

$$E\{y_n^2\} = E\left[D + Kx_0(x_n - x_0) + Kn(n-1)/2\right]\left[(x_n - x_0)^2/n^2\right]^2$$

$$= E\left[D + K(x_n - x_0)x_0\right]^2 + 2E\left\{\left[D + K(x_n - x_0)x_0\right]\left[K\left[(n-1)/n\right]\left[(x_n - x_0)^2/2\right]\right]\right\}$$

$$+ E\left[K\left[n(n-1)/n\right]\left[(x_n - x_0)^2/2\right]\right]^2$$

$$= E\left[D^2\right] + E\left[K(x_n - x_0)x_0\right]^2 + 2E\left[DK(x_n x_0 - x_0^2)\right] + 2E\left\{\left[D + K(x_n - x_0)x_0\right]\left[K\left[(n-1)/n\right]\left[(x_n - x_0)^2/2\right]\right]\right\}$$

$$+ E[K^2\left[n(n-1)/n\right]^2 E\left[(x_n - x_0)^2/2\right]^2$$

$$= E\left[D^2\right] + E\left[K^2\right]E\left[x_n x_0\right]^2 - 2E\left[K^2\right]E\left[x_n x_0^3\right] + E\left[K^2\right]E\left[x_0\right]^4 + 2E[D]E[K]E\left[x_n x_0\right] - 2E[D]E[K]E\left[x_0^2\right]$$

$$+ 2E[D]E[K]\left[(n-1)/n\right]E\left[(x_n - x_0)^2/2\right] + 2\left[(n-1)/n\right]E\left[K^2\right]E\left[x_0(x - x_0)^3/2\right]$$

$$+ [n(n-1)/n]^2 E\left[K^2\right]E\left[(x_n - x_0)^2/2\right]^2$$

$$= E[D]^2 + E\left[K^2\right]E\left[x_n x_0\right]^2 - 2E\left[K^2\right]E\left[x_n x_0^3\right] + E\left[K^2\right]E\left[x_0\right]^4 + 2E[D]E[K]E\left[x_n x_0\right] - 2E[D]E[K]E\left[x_0^2\right]$$

$$+ E[D]E[K]\left[(n-1)/n\right]E\left[(x_n - x_0)^2\right] + \left[(n-1)/n\right]E\left[K^2\right]E\left[x_0(x - x_0)^3\right] + \frac{1}{4}\left[n(n-1)/n\right]^2 E\left[K^2\right]E\left[(x_n - x_0)^4\right]$$

Then by taking the limit:

$$\lim_{n\to\infty} E\{y_n^2\} = 6 + 6E[xx_0]^2 - 12E[xx_0^3] + 6E[x_0]^4 + 8E[xx_0] - 8E[x_0^2]$$
$$+4E(x-x_0)^2 + 6E[x_0(x-x_0)^3] + \frac{6}{4}E(x-x_0)^4$$
$$= 6 + 6E[xx_0]^2 - 12E[xx_0^3] + 6E[x_0]^4 + 8E[xx_0] - 8E[x_0^2] + 4E[x^2]$$
$$-8E[xx_0] + 4E[x_0^2] + 6E[x^3x_0] - 18E[x^2x_0^2] + 18E[xx_0^3] - 6E[x_0^4]$$
$$+\frac{6}{4}E[x^4] - 6E[x^3x_0] - 6E[xx_0^3] + 9E[x^2x_0^2] + \frac{6}{4}E[x_0^4] = E(y^2)$$

i.e. $\lim_{n\to\infty} E\{y_n^2\} = E\{y^2\}$.

4) $\lim_{n\to\infty} \text{var}\{y_n\} = \text{var}\{y\}$

Proof

$$\lim_{n\to\infty} Var\{y_n\} = \lim_{n\to\infty}\left[E(y_n^2) - [E(y_n)]^2\right]$$
$$= \lim_{n\to\infty} E(y_n^2) - \lim_{n\to\infty}[E(y_n)]^2$$
$$= E(y^2) - [E(y)]^2 = Var(y)$$

i.e., $\lim_{n\to\infty} Var(y_n) = Var(y)$.

5) $\lim_{n\to\infty} PDF(y_n) = PDF(y)$

Proof Since $y = D + \frac{K(x^2 - x_0^2)}{2}$

Let us define Z = D. Then the inverse transformation is:

$D = Y - \frac{K(x^2 - x_0^2)}{2}$, $D = Z$ then we have $D = Z$ and

$$K=\frac{2(y-Z)}{x^2-x_0^2}$$

$$J=\begin{vmatrix}\frac{\partial k}{\partial y} & \frac{\partial k}{\partial z}\\ \frac{\partial D}{\partial y} & \frac{\partial D}{\partial z}\end{vmatrix}=\begin{vmatrix}\frac{2}{x^2-x_0^2} & \frac{-2}{x^2-x_0^2}\\ 0 & 1\end{vmatrix}=\frac{2}{x^2-x_0^2}$$

Then:

$$f_{y,z}(y,z)=f_{K,D}\left(\frac{2(y-Z)}{x^2-x_0^2},Z\right)|J|=\frac{e^{-4}2^{Z+\frac{2(Y-Z)}{X^2-X_0^2}}}{Z!\left[\frac{2(y-Z)}{x^2-x_0^2}\right]!}$$

Since D ≥ 0 then Z ≥ 0 hence

$$y\geq Z+\frac{K\left(x^2-x_0^2\right)}{2}$$

$$f_y(y)=\sum_{Z=0}^{y}\frac{e^{-4}2^{Z+\frac{2(Y-Z)}{x^2-x_0^2}}}{Z!\left[\frac{2(y-Z)}{x^2-x_0^2}\right]!}=PDF(y)$$

For a numerical solution:

since $y_n=D+nhKx_0+\left[n(n-1)/2\right]Kh^2$

Let $z_n = D$ $z_n=D$ then $K=\frac{(y_n-z_n)}{nhx_0+\left[n(n-1)/2\right]h^2}$

$$J = \begin{vmatrix} \dfrac{\partial K}{\partial y_n} & \dfrac{\partial K}{\partial Z_n} \\ \dfrac{\partial D}{\partial y_n} & \dfrac{\partial D}{\partial Z_n} \end{vmatrix}$$

$$= \begin{vmatrix} \dfrac{1}{nhx_0 + \left[n(n-1)/2\right]h^2} & \dfrac{-1}{nhx_0 + \left[n(n-1)/2\right]h^2} \\ 0 & 1 \end{vmatrix}$$

$$= \frac{1}{nhx_0 + \left[n(n-1)/2\right]h^2}$$

Then:

$$f_{y_n, Z_n}(y_n, z_n) = f_{K,D}\left(\frac{(y_n - z_n)}{nhx_0 + \left[n(n-1)/2\right]h^2}, z_n\right)|J|$$

$$= \frac{e^{-4} 2^{Z_n + \frac{(Y_n - Z_n)}{nhx_0 + [n(n-1)/2]h^2}}}{z_n! \left[\dfrac{(y_n - z_n)}{nhx_0 + \left[n(n-1)/2\right]h^2}\right]!}$$

$$f_{yn}(y_n) = \sum_{Z_n=0}^{y_n} \frac{e^{-4} 2^{Z_n + \frac{(Y_n - z_n)}{nhx_0 + [n(n-1)/2]h^2}}}{z_n! \left[\dfrac{(y_n - z_n)}{nhx_0 + \left[n(n-1)/2\right]h^2}\right]!}$$

where $h = \dfrac{x_n - x_0}{n}$

$$f_{yn}(y_n) = \sum_{z_n}^{y_n} \frac{e^{-4} 2^{z_n + \frac{(Y_n - z_n)}{(x_n - x_0)x_0 + [(n-1)/2n](x_n - x_0)^2}}}{z_n![\frac{(y_n - z_n)}{(x_n - x_0)x_0 + [(n-1)/2n](x_n - x_0)^2}]!}$$

$$= PDF(y_n)$$

Then by taking the limit we have

$$\lim_{n\to\infty} f_{y_n}(y_n) = \sum_{Z=0}^{y} \frac{e^{-4} 2^{Z + \frac{2(Y-z)}{x^2 - x_0^2}}}{z!\left[\frac{2(y-z)}{x^2 - x_0^2}\right]!} = f_y(y)$$

i.e.; $\lim_{n\to\infty} PDF(y_n) = PDF(y)$

B. Using the Random Runge-Kutta method:

$$y_{n+1} = y_n + \frac{h}{2}\left[f(y_n, x_n) + f(y_n + f(y_n, x_n), x_{n+1})\right]$$

at $n = 0$

$$y_1 = y_0 + \frac{h}{2}\left[f(y_0, x_0) + f(y_0 + f(y_0, x_0), x_1)\right]$$

$$= D + \frac{h}{2}[Kx_0 + Kx_1] = D + \frac{hK}{2}[x_0 + x_1].$$

At $n = 1$

$$y_2 = y_1 + \frac{h}{2}\left[f(y_1, x_1) + f(y_1 + f(y_1, x_1), x_2)\right]$$

$$= D + \frac{h}{2}[Kx_0 + Kx_1] + \frac{h}{2}[Kx_1 + Kx_2]$$

$$= D + \frac{hK}{2}[x_0 + 2x_1 + x_2].$$

At n = 2

$$y_3 = y_2 + \frac{h}{2}\left[f(y_2, x_2) + f(y_2 + f(y_2, x_2), x_3)\right]$$

$$= D + \frac{hK}{2}[x_0 + 2x_1 + x_2] + \frac{h}{2}[Kx_2 + Kx_3]$$

$$= D + \frac{hK}{2}[x_0 + 2x_1 + 2x_2 + x_3].$$

Then the general solution is:

$$y_n = D + \frac{hK}{2}[x_0 + 2x_1 + 2x_2 + 2x_3 + \ldots\ldots + 2x_{n-1} + x_n],$$

$$y_n = D + hK\left[\frac{1}{2}x_0 + (x_0 + h) + (x_0 + 2h) + \cdots + (x_0 + (n-1)h) + \frac{1}{2}(x_0 + nh)\right],$$

$$y_n = D + hK\left[\frac{1}{2}x_0 + (x_0 + h) + (x_0 + 2h) + \cdots + (x_0 + (n-1)h) + \frac{1}{2}(x_0 + nh)\right]$$

$$y_n = D + hK\sum_{i=0}^{n-1}(x_0 + ih) + \frac{1}{2}nh^2K.$$

This can be written in another form:

$$y_n = D + nhKx_0 + \frac{n^2 h^2 K}{2}.$$

We can prove that:

1) $\lim_{n\to\infty} y_n = y$

Proof

Since $\lim_{n\to\infty} y_n = y$ (if and only if) $\lim_{n\to\infty} E\left|y_n - y\right|^2 = 0$

$$y_n - y = nhKx_0 + \frac{n^2}{2}\left[h^2 K\right] - \frac{K\left(x^2 - x_0^2\right)}{2}$$

$$\left|y_n - y\right|^2$$

$$= n^2 h^2 K^2 x_0^2 + 2nhK^2 x_0 \left[\frac{n^2}{2}h^2 - \frac{\left(x^2 - x_0^2\right)}{2}\right]$$

$$+ K^2 \left[\frac{n^2}{2}h^2 - \frac{\left(x^2 - x_0^2\right)}{2}\right]^2$$

where $h = \dfrac{x_n - x_0}{n}$

$$
\begin{aligned}
\left|y_n - y\right|^2 &= K^2 x_0^2 \left(x_n - x_0\right)^2 + K^2 x_0 \left(x_n - x_0\right)\left[\left(x_n - x_0\right)^2 - \left(x^2 - x_0^2\right)\right] \\
&\quad + \frac{K^2}{4}\left[\left(x_n - x_0\right)^2 - \left(x^2 - x_0^2\right)\right]^2 \\
&= K^2 x_0^2 x_n^2 - 2K^2 x_n x_0^3 + K^2 x_0^4 + K^2 x_n x_0 \left(x_n - x_0\right)^2 - K^2 x_0^2 \left(x_n - x_0\right)^2 \\
&\quad - K^2 x_n x_0 \left(x^2 - x_0^2\right) + K^2 x_0^2 \left(x^2 - x_0^2\right) + \frac{K^2}{4}\left(x_n - x_0\right)^4 \\
&\quad - \frac{K^2}{2}\left(x_n - x_0\right)^2 \left(x^2 - x_0^2\right) + \frac{K^2}{4}\left(x^2 - x_0^2\right)^2
\end{aligned}
$$

$$
\begin{aligned}
&\lim_{n\to\infty} E\left|y_n - y\right|^2 \\
&= 18x_0^2x^2 - 36x_0^3x + 18x_0^4 + 18xx_0\left(x - x_0\right)^2 - 18x_0^2\left(x - x_0\right)^2 - 18xx_0\left(x^2 - x_0^2\right) + 18x_0^2\left(x^2 - x_0^2\right) \\
&\quad + \frac{18}{4}x^4 - 18x^3x_0 + 27x^2x_0^2 - 18xx_0^3 + \frac{18}{4}x_0^4 - 9x^4 + 18x^3x_0 - 18xx_0^3 + 9x_0^4 + \frac{18}{4}x^4 - 9x^2x_0^2 + \frac{18}{4}x_0^4 \\
&= 18x_0^2x^2 - 36x_0^3x + 18x_0^4 + 72xx_0^3 - 36x^2x_0^2 - 36x_0^4 + \frac{18}{4}x^4 - 18x^3x_0 + 27x^2x_0^2 - 18xx_0^3 \\
&\quad + \frac{18}{4}x_0^4 - 9x^4 + 18x^3x_0 - 18xx_0^3 + 9x_0^4 + \frac{18}{4}x^4 - 9x^2x_0^2 + \frac{18}{4}x_0^4
\end{aligned}
$$

i.e.; $\lim_{n\to\infty} y_n = y$.

Verification of Theorem (4.1):

2) $\lim_{n\to\infty} E\left\{y_n\right\} = E\left\{y\right\}$

Proof

$$y_n = D + hK\sum_{i=0}^{n-1}(x_0 + ih) + \frac{nh^2K}{2} = D + hK\sum_{j=1}^{n}(x_0 + (j-1)h) + \frac{nh^2K}{2}$$

$$= D + hK\left[x_0 + (x_0 + h) + (x_0 + 2h) + (x_0 + 3h) + \cdots (x_0 + (n-1)h)\right] + \frac{nh^2K}{2}$$

$$= D + hK\left[nx_0 + h\left[1 + 2 + 3 + \cdots + (n-1)\right]\right] + \frac{nh^2K}{2}$$

$$= D + hK\left[nx_0 + h\left(\frac{n(n-1)}{2}\right)\right] + \frac{nh^2K}{2}$$

$$= D + K\left[nhx_0 + h^2\left(\frac{n(n-1)}{2}\right)\right] + \frac{nh^2K}{2} \quad \text{where } h = \frac{x_n - x_0}{n}$$

$$= D + K\left[nx_0\frac{(x_n - x_0)}{n} + \left(\frac{(x_n - x_0)}{n}\right)^2\left(\frac{n(n-1)}{2}\right)\right] + \frac{(x_n - x_0)^2 K}{2n}$$

$$= D + K\left[x_0(x_n - x_0) + \frac{(n-1)}{2n}(x_n - x_0)^2\right] + \frac{(x_n - x_0)^2 K}{2n}$$

$$E(y_n) = E(D) + E\left\{K\left[x_0(x_n - x_0) + \frac{(n-1)}{2n}(x_n - x_0)^2\right] + \frac{(x_n - x_0)^2 K}{2n}\right\}$$

$$= 2 + 2\left\{E(x_0x_n) - E(x_0^2) + \frac{(n-1)}{2n}E(x_n - x_0)^2\right\} + \frac{2E(x_n - x_0)^2}{2n}$$

$$\lim_{n\to\infty} E\{y_n\} = 2 + 2\left\{E(x_0x) - E(x_0^2) + E(x - x_0)^2/2\right\}$$

$$= 2 + 2\left\{E(x_0x) - E(x_0^2) + \frac{1}{2}E(x^2) - E(x_0x) + \frac{1}{2}E(x_0^2)\right\}$$

$$= 2 + E(x^2) - E(x_0^2) = E\{y\}$$

i.e.; $\lim_{n\to\infty} E\{y_n\} = E\{y\}$.

3) $\lim_{n\to\infty} E\{y_n^2\} = E\{y^2\}$

Proof Since $y_n = D + nhKx_0 + \frac{n^2h^2K}{2}$ then:

$$E\left\{y_n^2\right\} = E\left[D + Kx_0\left(x_n - x_0\right) + \frac{\left(x_n - x_0\right)^2 K}{2}\right]^2$$

$$= E\left[D + K\left(x_n - x_0\right)x_0\right]^2 + 2E\left\{\left[D + K\left(x_n - x_0\right)x_0\right]\left[\frac{K\left(x_n - x_0\right)^2}{2}\right]\right\} + E\left[\frac{K\left(x_n - x_0\right)^2}{2}\right]^2$$

$$= E\left[D^2\right] + E\left[K\left(x_n - x_0\right)x_0\right]^2 + 2E\left[DK\left(x_n x_0 - x_0^2\right)\right]$$

$$+ 2E\left\{\left[D + K\left(x_n - x_0\right)x_0\right]\left[\frac{K\left(x_n - x_0\right)^2}{2}\right]\right\} + E\left[\frac{K\left(x_n - x_0\right)^2}{2}\right]^2$$

$$= E\left[D^2\right] + E\left[K^2\right]E\left[x_n x_0\right]^2 - 2E\left[K^2\right]E\left[x_n x_0^3\right] + E\left[K^2\right]E\left[x_0\right]^4$$

$$+ 2E[D]E[K]E\left[x_n x_0\right] - 2E[D]E[K]E\left[x_0^2\right] + 2E[D]E[K]E\left[\left(x_n - x_0\right)^2/2\right]$$

$$+ 2E\left[K^2\right]E\left[x_0\left(x - x_0\right)^3/2\right] + E\left[K^2\right]E\left[\left(x_n - x_0\right)^2/2\right]^2$$

$$= E[D]^2 + E\left[K^2\right]E\left[x_n x_0\right]^2 - 2E\left[K^2\right]E\left[x_n x_0^3\right] + E\left[K^2\right]E\left[x_0\right]^4$$

$$+ 2E[D]E[K]E\left[x_n x_0\right] - 2E[D]E[K]E\left[x_0^2\right] + E[D]E[K]E\left[\left(x_n - x_0\right)^2\right]$$

$$+ E\left[K^2\right]E\left[x_0\left(x_n - x_0\right)^3\right] + \frac{1}{4}E\left[K^2\right]E\left[\left(x_n - x_0\right)^4\right]$$

$$\lim_{n\to\infty} E\left\{y_n^2\right\} = 6 + 6E\left[xx_0\right]^2 - 12E\left[xx_0^3\right] + 6E\left[x_0\right]^4 + 8E\left[xx_0\right] - 8E\left[x_0^2\right]$$

$$+ 4E\left[\left(x - x_0\right)^2\right] + 6E\left[x_0\left(x - x_0\right)^3\right] + \frac{6}{4}E\left[\left(x - x_0\right)^4\right]$$

$$= 6 + 6E\left[xx_0\right]^2 - 12E\left[xx_0^3\right] + 6E\left[x_0\right]^4 + 8E\left[xx_0\right] - 8E\left[x_0^2\right] + 4E\left[x^2\right]$$

$$- 8E\left[xx_0\right] + 4E\left[x_0^2\right] + 6E\left[x^3 x_0\right] - 18E\left[x^2 x_0^2\right] + 18E\left[xx_0^3\right] - 6E\left[x_0^4\right]$$

$$+ \frac{6}{4}E\left[x^4\right] - 6E\left[x^3 x_0\right] - 6E\left[xx_0^3\right] + 9E\left[x^2 x_0^2\right] + \frac{6}{4}E\left[x_0^4\right] = E\left(y^2\right)$$

i.e.; $\lim_{n\to\infty} E\left\{y_n^2\right\} = E\left\{y^2\right\}$.

4) $\lim_{n\to\infty} Var\left(y_n\right) = Var(y)$

Proof

$$\lim_{n\to\infty} Var\left\{y_n\right\} = \lim_{n\to\infty}\left[E\left(y_n^2\right) - \left[E\left(y_n\right)\right]^2\right]$$

$$= \lim_{n\to\infty} E\left(y_n^2\right) - \lim_{n\to\infty}\left[E\left(y_n\right)\right]^2$$

$$= E\left(y^2\right) - \left[E(y)\right]^2 = Var(y)$$

i.e. $\lim_{n\to\infty} Var(y_n) = Var(y)$

5) $\lim_{n\to\infty} PDF(y_n) = PDF(y)$

Since $y = D + \dfrac{K(x^2 - x_0^2)}{2}$

Let us define $Z = D$. Then the inverse transformation is:

$$D = y - \frac{K(x^2 - x_0^2)}{2}$$

$D = z$ then we have $D = z$

$$K = \frac{2(y-z)}{x^2 - x_0^2}$$

$$J = \begin{vmatrix} \frac{\partial K}{\partial y} & \frac{\partial K}{\partial z} \\ \frac{\partial D}{\partial y} & \frac{\partial D}{\partial z} \end{vmatrix} = \begin{vmatrix} \frac{2}{x^2 - x_0^2} & \frac{-2}{x^2 - x_0^2} \\ 0 & 1 \end{vmatrix} = \frac{2}{x^2 - x_0^2}$$

and $K = \dfrac{2(y-z)}{x^2 - x_0^2}$

$$f_{y,z}(y,z) = f_{K,D}\left(\frac{2(y-z)}{x^2 - x_0^2}, z\right)|J| = \frac{e^{-4}2^{Z + \frac{2(Y-Z)}{X^2 - X_0^2}}}{z!\left[\frac{2(y-z)}{x^2 - x_0^2}\right]!}$$

Since $D \geq 0$ then $z \geq 0$ this implies

$$y \geq z + \frac{K(x^2 - x_0^2)}{2}$$

$$f_y(y) = \sum_{z=0}^{y} \frac{e^{-4}2^{Z + \frac{2(Y-Z)}{X^2 - X_0^2}}}{z!\left[\frac{2(y-z)}{x^2 - x_0^2}\right]!} = PDF(y)$$

Numerically

since $$y_n = D + nhKx_0 + \frac{n^2h^2K}{2}$$

$$K = \frac{(y_n - z_n)}{nhx_0 + \frac{n^2h^2}{2}}$$

Let $z_n = D$ then

$$J = \begin{vmatrix} \frac{\partial k}{\partial y_n} & \frac{\partial k}{\partial z_n} \\ \frac{\partial D}{\partial y_n} & \frac{\partial D}{\partial z_n} \end{vmatrix} = \begin{vmatrix} \frac{1}{nhx_0 + \frac{n^2h^2}{2}} & \frac{-1}{nhx_0 + \frac{n^2h^2}{2}} \\ 0 & 1 \end{vmatrix}$$

$$= \frac{1}{nhx_0 + \frac{n^2h^2}{2}}$$

$$f_{y_n, z_n}(y_n, z_n) = f_{k.D}\left(\frac{(y_n - z_n)}{nhx_0 + \frac{n^2h^2}{2}}, z_n\right)|J|$$

$$= \frac{e^{-4}2^{Z_n + \frac{(Y_n - Z_n)}{nhx_0 + \frac{n^2h^2}{2}}}}{z_n!\left[\frac{(y_n - z_n)}{nhx_0 + \frac{n^2h^2}{2}}\right]!}$$

$$f_{yn}(y_n) = \sum_{z_n=0}^{y_n} \frac{e^{-4}2^{Z_n + \frac{(Y_n - Z_n)}{nhx_0 + \frac{n^2h^2}{2}}}}{z_n!\left[\frac{(y_n - z_n)}{nhx_0 + \frac{n^2h^2}{2}}\right]!}$$ where $$h = \frac{x_n - x_0}{n}$$

$$f_{yn}(y_n) = \sum_{z_n}^{y_n} \frac{e^{-4} 2^{Z_n + \frac{(Y_n - Z_n)}{(x_n - x_0)x_0 + (x_n - x_0)^2/2}}}{z_n!\left[\frac{(y_n - z_n)}{(x_n - x_0)x_0 + (x_n - x_0)^2/2}\right]!}$$

$$= PDF(y_n)$$

$$\lim_{n\to\infty} f_{y_n}(y_n) = \sum_{z=0}^{y} \frac{e^{-4} 2^{Z + \frac{2(Y-Z)}{X^2 - X_0^2}}}{z!\left[\frac{2(y-z)}{x^2 - x_0^2}\right]!} = f_y(y)$$

i.e.; $\lim_{n\to\infty} PDF(y_n) = PDF(y)$.

Example (5.2)

Solve the problem

$$\frac{dy}{dt} = y^2 - Ky,\ y(0) = K,\ t \in [0, t_n],\ K \sim \exp(1)$$

The exact solution

$y = K$

The numerical solution by the Euler method:

$$y_n = y_{n-1} + hf(y_{n-1}, t_{n-1}), y(0) = K.$$

AT $n = 1$

$$y_1 = y_0 + hf(y_0, t_0) = K + h(y_0^2 - Ky_0) = K.$$

At $n = 2$

$$y_2 = y_1 + hf(y_1, t_1) = K + h(y_1^2 - ky_1) = K.$$

At $n = 3$

$$y_3 = y_2 + hf\left(y_2, t_2\right) = K + h\left(y_2^2 - ky_2\right) = K$$

And so on....

Then the general numerical solution: $y_n = K$.

It is clear that:

1) $$\lim_{n\to\infty} y_n = y$$

Since $$\lim_{n\to\infty} E\left|y_n - y\right|^2 = \lim_{n\to\infty} E\left|K - K\right|^2 = 0$$

Verification of Theorem (4.1)

It is clear that:

$$\lim_{n\to\infty} E\left\{y_n\right\} = E\left\{y\right\}$$

2) $$E\left(y_n\right) = E\left(K\right) = K = E\left(y\right)$$

3) $$\lim_{n\to\infty} E\left\{y_n^2\right\} = E\left\{y^2\right\}$$

Since $$y_n = K \Rightarrow E\left\{y_n^2\right\} = E\left[K\right]^2 = K^2 = E\left\{y^2\right\}$$

$$\lim_{n\to\infty} Var\left(y_n\right) = Var\left(y\right)$$

$$\begin{aligned}\lim_{n\to\infty} Var\left\{y_n\right\} &= \lim_{n\to\infty}\left[E\left(y_n^2\right) - \left[E\left(y_n\right)\right]^2\right] \\ &= \lim_{n\to\infty} E\left(y_n^2\right) - \lim_{n\to\infty}\left[E\left(y_n\right)\right]^2 \\ &= E\left(y^2\right) - \left[E\left(y\right)\right]^2 = Var\left(y\right)\end{aligned}$$

4)

5) $$\lim_{n\to\infty} PDF\left(y_n\right) = PDF\left(y\right)$$

$y = K$ Then $|J| = 1$ which implies

$PDF(y) = |J|\,PDF(k) = e^{-y}$

$y_n - k$ Then $|J| = 1$ which implies

$$PDF(y_n) = |J|\,PDF(K) = e^{-y_n k}$$

Then: $\lim_{n \to \infty} PDF(y_n) = PDF(y)$

6. CONCLUSIONS

The initially valued first order random differential equations can be solved numerically using the random Euler and random Runge-Kutta methods in mean square sense. The existence and uniqueness of the solution have been proved. The convergence of the presented numerical techniques has been proven in mean square sense. The results of the paper have been illustrated through some examples.

7. REFERENCES

1. K. Burrage and P. M. Burrage, "High Strong Order Explicit Runge-Kutta Methods for Stochastic Ordinary Differential Equations," Applied Numerical Mathematics, Vol. 22, No. 1-3, 1996, pp. 81-101. doi:10.1016/S0168-9274(96)00027-X
2. K. Burrage and P. M. Burrage, "General Order Conditions for Stochastic Runge-Kutta Methods for Both Commuting and Non-commuting Stochastic Ordinary Equations," Applied Numerical Mathematics, Vol. 28, No. 2-4, 1998, pp. 161-177. doi:10.1016/S0168-9274(98)00042-7
3. J. C. Cortes, L. Jodar and L. Villafuerte, "Numerical Soluion of Random Differential Equations, a Mean Square Approach," Mathematical and Computer Modelling, Vol. 45, No. 7, 2007, pp. 757-765. doi:10.1016/j.mcm.2006.07.017
4. J. C. Cortes, L. Jodar and L.Villafuerte, "A Random Euler Method for Solving Differential Equations with Uncertainties," Progress in Industrial Mathematics at ECMI, Madrid, 2006.
5. H. Lamba, J. C. Mattingly and A. Stuart, "An adaptive Euler-Maruyama Scheme for SDEs, Convergence and Stability," IMA Journal of Numerical Analysis, Vol. 27, No. 3, 2007, pp. 479-506. doi:10.1093/imanum/drl032
6. E. Platen, "An Introduction to Numerical Methods for Stochastic Differential Equations," Acta Numerica, Vol. 8, 1999, pp. 197-246. doi:10.1017/S0962492900002920
7. D. J. Higham, "An Algorithmic Introduction to Numerical Simulation of SDE," SIAM Review, Vol. 43, No. 3, 2001, pp. 525-546. doi:10.1137/S0036144500378302
8. D. Talay and L. Tubaro, "Expansion of the Global Error for Numerical Schemes Solving Stochastic Differential Equation," Stochastic Analysis and Applications, Vol. 38, No. 4, 1990, pp. 483-509. doi:10.1080/07362999008809220

9. P. M. Burrage, "Numerical Methods for SDE," Ph.D. Thesis, The University of Queensland, 1999.
10. P. E. Kloeden, E. Platen and H. Schurz, "Numerical Solution of SDE Through Computer Experiments," Second Edition, Springer, Berlin, 1997.
11. M. A. El-Tawil, "The Approximate Solutions of Some Stochastic Differential Equations Using Transformation," Journal of Applied Mathematics and Computing, Vol. 164, No. 1, 2005, pp. 167-178. doi:10.1016/j.amc.2004.04.062
12. P. E. Kloeden and E. Platen, "Numeical Solution of Stochastic Differential Equations," Springer, Berlin, 1999.
13. T. T. Soong, "Random Differential Equations in Science and Engineering," Academic Press, New York, 1973.

CHAPTER 12

A Combinatorial Analysis of Tree-Like Sentences

Gilbert Labelle[1], Louise Laforest[2]

[1]*Département de Mathématiques, Université du Québec à Montréal, Montreal, Canada*

[2]*Département d'Informatique, Université du Québec à Montréal, Montreal, Canada*

ABSTRACT

A sentence over a finite alphabet A, is a finite sequence of non-empty words over A. More generally, we define a graphical sentence over A by attaching a non-empty word over A to each arrow and each loop of a connected directed graph (digraph, for short). Each word is written according to the direction of its corresponding arrow or loop. Graphical sentences can be used to encode sets of sentences in a compact way: the readable sentences of a graphical sentence being the sentences corresponding to directed paths in the digraph. We apply combinatorial equations on enriched trees and rooted trees, in the context of combinatorial species and Pólya theories, to analyze parameters in classes of tree-like sentences. These are graphical sentences constructed on tree-like digraphs.

Keywords: Pólya Theory, Combinatorial Species, Digraphs, Tree-Like Sentences

1. INTRODUCTION

Figure 1 (left) shows a completely unlabelled[1] connected digraph. We define a graphical sentence over a finite alphabet A by attaching a non-empty word over A to each arrow and each loop of a completely unlabelled connected digraph. Each word must be written according to the direction of its corresponding arrow or loop, from source to target. Figure 1 (middle) shows a graphical sentence over alphabet {A, C, G, T} and Figure 1 (right) shows another over alphabet {♪, 𝅗𝅥, ♫, 𝅝, ♩}.

Graphical sentences can be used to encode sets of ordinary sentences in a compact way: The readable sentences of a graphical sentence being the sentences corresponding to directed paths in its digraph. For example,

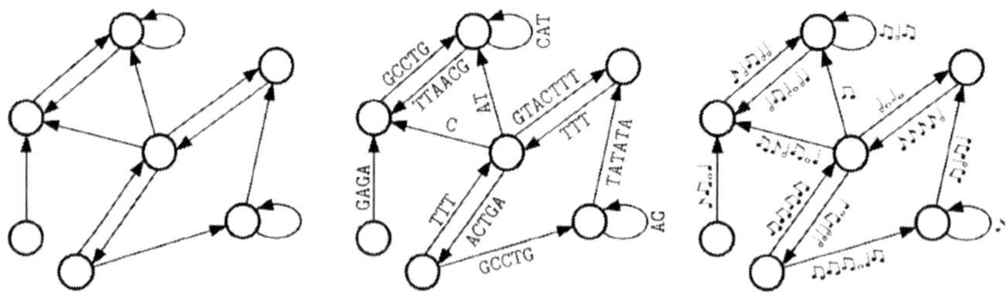

Figure 1. Unlabelled digraph and graphical sentences over alphabets {A, C, G, T} and {♪, 𝅗𝅥, ♫, 𝅝, ♩}.

TTT C GCCTG CAT CAT GCAATT, is a readable sentence arising from the graphical sentence of Figure 1 (middle).

In the present paper we focus our attention on the structure of graphical sentences as combinatorial objects using methods from the theory of combinatorial species [1] [2] and classical Pólya theory [3]. We leave aside the generation of the readable sentences of a graphical sentence since this is easily done via the computation of powers of incidence matrices[2]. Of course, special sentences among the readable sentences can be selected by adding extra structure to graphical sentences (such as source points, sink points, STOP points, counters, extensions of the alphabet by adding special characters such as ⊔, !, ?, etc). We also leave aside this aspect in our analysis of graphical sentences.

Various descriptive parameters can be attached to each graphical sentence over a given alphabet A. For example, the graphical sentence of Figure 1 (middle) is made of 7 vertices, 11 arrows, 2 loops, 52 letters, letter A appears 13 times, letter C appears 9 times, letter G appears 10 times and letter T appears 20 times.

As usual in enumerative combinatorics, families of parameters associated to structures are conveniently encoded by weight-monomials.

Definition 1.1. The weight of a graphical sentence s over an alphabet A is the (commutative) formal monomial[3]

$$\mathbf{w}(s) = x^{\#\text{vertices in } s} y^{\#\text{arrows in } s} z^{\#\text{loops in } s} t^{\#\text{letters in } s} \prod_{a \in A} a^{\nu_a(s)} \tag{1}$$

where $\nu_a(s)$ is the number of occurrences of letter a in s.

In (1), each letter $a \in A$ is reinterpreted as a formal variable. For example, the weight of the graphical sentence s of Figure 1 (middle) is given by

$$\mathbf{w}(s) = x^7 y^{11} z^2 t^{52} A^{13} C^9 G^{10} T^{20}. \tag{2}$$

Definition 1.2. Let C be any class of totally unlabelled connected digraphs and $\mathbb{L} \subseteq \mathbb{N}^+ = \{1,2,3,\cdots\}$. Let $\mathfrak{S}$ be the (countable) set of all graphical sentences over alphabet A arising from digraphs in C, the word on each arrow or loop having a length $\in$L. The inventory of $\mathfrak{S}$ is the formal sum of the weights of all graphical sentences in $\mathfrak{S}$:

$$\mathbf{w}\mathfrak{S} = \mathbf{w}\mathfrak{S}\left(x, y, z, t, (a)_{a \in A}\right) = \sum_{s \in \mathfrak{S}} \mathbf{w}(s). \tag{3}$$

As usual in enumeration problems, the (explicit or recursive) computation of an inventory of a class of structures provides a great deal of information about the structures to which it is associated. This information is extracted from the inventory through *expansion, collection of terms, specialization/confluence of variables, algebraic/differential manipulations and coefficient extraction.*

For example, in the present situation, expanding and collecting terms in (3) gives, of course,

$$\mathbf{w}\mathfrak{S}\left(x, y, z, t, (a)_{a \in A}\right) = \sum_{m,n,p,q,(\nu_a)_{a \in A}} c_{m,n,p,q,(\nu_a)_{a \in A}} x^m y^n z^p t^q \prod a^{\nu_a}, \tag{4}$$

where the coefficient $c_{m,n,p,q,(\nu_a)_{a \in A}}$ is the total number of graphical sentences $s \in \mathfrak{S}$ having m vertices, n ar- rows, p loops, a total number of q letters, ν_a of which are letter a, for each $a \in A$.

Assigning the value 1 to each letter a and collecting terms gives

$$\mathbf{w}\mathfrak{S}\left(x,y,z,t,(1)_{a\in A}\right)=\sum_{m,n,p,q} b_{m,n,p,q}x^m y^n z^p t^q \quad ,(5)$$

where $b_{m,n,p,q}$ is the number of $s\in\mathfrak{S}$ having m vertices, n arrows, p loops and q letters. Letting $x=y=z=1$, gives

$$\mathbf{w}\mathfrak{S}\left(1,1,1,t,(a)_{a\in A}\right)=\sum_{q,(v_a)_{a\in A}} d_{q,(v_a)_{a\in A}} t^q \prod a^{v_a} \quad ,(6)$$

where $d_{q,(v_a)_{a\in A}}$ is the number of $s\in\mathfrak{S}$ made of q letters occurring with frequencies $(v_a)_{a\in A}$.

Letting $x=y=z=1$ and assigning the value 1 to each letter a gives

$$\mathbf{w}\mathfrak{S}\left(1,1,1,t,(1)_{a\in A}\right)=\sum_q e_q t^q \quad , \quad (7)$$

where e_q is the number of $s\in\mathfrak{S}$ made of q letters.

Let $f_{p,q}$ be the number of graphical sentences $s\in\mathfrak{S}$ made of p words (*i.e.*, p is the total number of arrows and loops in s) and q letters. Then

$$\mathbf{w}\mathfrak{S}\left(1,z,z,t,(1)_{a\in A}\right)=\sum_{p,q} f_{p,q} z^p t^q \quad (8)$$

Moreover, if we let t=1 in (8) and if $\mathfrak{S}$ is a finite set, then

$$\mathbf{w}\mathfrak{S}\left(1,z,z,1,(1)_{a\in A}\right)=\sum_p h_p z^p \quad , \quad (9)$$

is a polynomial, where h_p is the number of $s\in\mathfrak{S}$ made of p words. Differentiation gives

$$\left[\frac{\frac{\mathrm{d}}{\mathrm{d}z}\mathbf{w}\mathfrak{S}\left(1,z,z,1,(1)_{a\in A}\right)}{\mathbf{w}\mathfrak{S}\left(1,z,z,1,(1)_{a\in A}\right)}\right]_{z=1}=\frac{\sum_p p h_p}{\sum_p h_p}=\text{Expected number of words in a random } s\in\mathfrak{S} \quad .(10)$$

Of course, a variety of other similar manipulations of the inventory $\mathfrak{S}\left(x,y,z,t,(a)_{a\in A}\right)$ are possible. In Section 2 we apply methods from the theory of species and Pólya theory, to express inventories of general classes of graphical sentences in terms of cycle index series. Section 3 deals with specific classes of graphical sentences: linear sentences (corresponding to path-like digraphs) and general tree-like sentences (corresponding to classes of tree-like digraphs). We conclude (Section 4) by giving suggestions for possible extensions and generalizations of our results. Various explicit examples are given and to make the text easier to read, the proofs of the main results are collected in Section 5. In a previous paper, [4], we studied the distribution of runs in arborescent words.

We assume that the reader is familiar with Pólya theory [3] and with the basic concepts of the theory of combinatorial species [1] [2].

2. INVENTORY OF GRAPHICAL SENTENCES VIA CYCLE INDEX SERIES

In order to give a rigorous meaning to the notion of a totally unlabelled digraph and to be able to take into account the possible symmetries within graphical sentences, we must recall first some definitions concerning labelled digraphs. *A digraph* on (or labelled by) a finite set V of *vertices*, a finite set V_1 of arrows, and a finite set V_0 of loops is an ordered pair g=(g_0,g_1) of njections,

$$g_0 : V_0 \to V,\ \ g_1 : V_1 \to V\times V\setminus \mathrm{diag}(V) \qquad ,(11)$$

where $\mathrm{diag}(V)=\{(v,v): v\in V\}$ and $V\times V\setminus \mathrm{diag}(V)=\{(v,w): v\in V, w\in V, v\neq w\}$. Given any loop $i_0 \in V_0$ and any vertex $v \in V$, the equality $g_0(i_0)$ v = means that i_0 is a loop at v in the digraph g. Given any arrow $j_1 \in V_1$ and any ordered pair of distinct vertices $(v,w) \in V \times V \setminus \mathrm{diag}(V)$, the equality $g_0(j_1)= (v,w)$ means that arrow j_1 is going from v to w in the digraph g. In other words

$$g_0(i_0)=v \Leftrightarrow \underset{i_0}{\overset{}{\underset{\circlearrowleft}{v}}}\ \text{in}\ g,\quad g_1(j_1)=(v,w)\Leftrightarrow v\xrightarrow{j_1} w\ \text{in}\ g \qquad (12)$$

Figure 2(a) shows a digraph on the sets $V=\{1,2,3,\cdots,8\}$ of vertices, $V_1=\{1_1,2_1,3_1,\cdots,12_1\}$ of arrows and $V_0=\{1_0,2_0\}$ of loops, with loop 1_0 at vertex 4 and loop 2_0 at vertex 1. Let $g=(g_0,g_1)$ be a digraph on (V,V_1,V_0) and $g'=(g'_0,g'_1)$ be a digraph on (V',V'_1,V'_0). An isomorphism $\theta : g \tilde{\rightarrow} g'$ from digraph g to digraph g', is an ordered triple, $\theta=(\alpha,\beta,\gamma)$, of bijections $\alpha : V \tilde{\rightarrow} V'$, $\beta : V_1 \tilde{\rightarrow} V'_1$, $\gamma : V_0 \tilde{\rightarrow} V'_0$, such that

$$\underset{i_0}{\overset{\circlearrowleft}{v}} \text{ in } g \Leftrightarrow \underset{\gamma(i_0)}{\overset{\circlearrowleft}{\alpha(v)}} \text{ in } g' \quad , \tag{13}$$

$$v \xrightarrow{j_1} w \text{ in } g \Leftrightarrow \alpha(v) \xrightarrow{\beta(j_1)} \alpha(w) \text{ in } g' \; . \tag{14}$$

If $g=g'$ then θ is called an automorphism (or symmetry) of the digraph g. For example, the triple of permutations (α,β,γ), defined (in cyclic notation) by

$$\alpha=(3,8)(7,6), \quad \beta=(4_1,12_1)(9_1,1_1)(2_1,5_1), \quad \gamma=\mathrm{id} \;, \tag{15}$$

is an automorphism of the digraph of Figure 2(a). Note that labelled digraphs are elastic and not considered as embedded in the plane. Only the incidence relations between vertices, arrows and loops are taken into account. A totally unlabelled digraph (see Figure 2(b)), is simply an isomorphism class of labelled digraphs. The class of a labelled digraph g is denoted $[g]$; so that $[g]$ is a totally unlabelled digraph with representative g.

We now give a rigorous definition of the notion of a graphical sentence.

Definition 2.1. Let A be a finite alphabet and $A^+ = A + A^2 + A^3 + \cdots$, be the set of non-empty words over A. A graphical sentence over A is an equivalence class, s, of ordered triples, (g,σ_0,σ_1), where g is a connected labelled digraph on (V,V_1,V_0) and $\sigma_0 : V_0 \rightarrow A^+$, $\sigma_1 : V_1 \rightarrow A^+$ are arbitrary functions assigning a non-empty word to each loop and each arrow of g. Two such triples (g,σ_0,σ_1) and (g',σ'_0,σ'_1) being equivalent if there exists an isomorphism $\theta=(\alpha,\beta,\gamma) : g \tilde{\rightarrow} g'$ such that $\sigma'_0 \circ \gamma = \sigma_0$ and $\sigma'_1 \circ \beta = \sigma_1$. We write $s=[(g,\sigma_0,\sigma_1)]$ to mean that s is a graphical sentence with representative (g,σ_0,σ_1).

Now take any species $\mathcal{G}$ of connected labelled digraphs[4]. Our goal is to compute the inventory (3) of the class $\mathfrak{S}$ of all graphical sentences $s = [(g, \sigma_0, \sigma_1)]$ where $g \in \mathcal{G}$. To emphasize the fact that digraphs are made of three sorts of elements, vertices, arrows and loops, the given species $\mathcal{G}$ of digraphs can be written in the form $\mathcal{G} = \mathcal{G}(X, Y, Z)$, where X is the sort of vertices, Y is the sort of arrows, and Z is the sort of loops[5]. Any digraph $g \in \mathcal{G}$ is called a $\mathcal{G}$-structure for short.

Following standard notations from the theory of species, the set of all $\mathcal{G}$-structures on a set V of vertices, a set V_1 of arrows and a set V_0 of loops is denoted by $\mathcal{G}=[V,V_1,V_0]$ (note the square brackets). Given bijections $\alpha : V \tilde{\rightarrow} V'$, $\beta : V_1 \tilde{\rightarrow} V_1'$, $\gamma : V_0 \tilde{\rightarrow} V_0'$ we denote by

$$\mathcal{G}[\alpha, \beta, \gamma] : \mathcal{G}[V, V_1, V_0] \tilde{\rightarrow} \mathcal{G}[V', V_1', V_0'], \quad (16)$$

the bijection $(g \mapsto g')$ that transforms (or transports) each digraph $g \in \mathcal{G}[V, V_1, V_0]$ into a corresponding isomorphic digraph $g' \in \mathcal{G}[V', V_1', V_0']$, as described above. Note that if, $V = V'$, $V_1 = V_1'$ and $V_0 = V_0'$, then $\mathcal{G}[\alpha, \beta, \gamma]$ is a permutation of the set $\mathcal{G}[V, V_1, V_0]$ (i.e., a bijection of $\mathcal{G}[V, V_1, V_0]$ into itself).

Many power series can be associated to any species F. An important one is the Pólya-Joyal cycle index series $\mathbf{Z}_F$. In the context of a species $\mathcal{G} = \mathcal{G}(X, Y, Z)$ of digraphs, this is a power series in a triple infinity of variables, $x_1, x_2, x_3, \cdots; y_1, y_2, y_3, \cdots; z_1, z_2, z_3, \cdots$, defined by

$$\mathbf{Z}_{\mathcal{G}} = \mathbf{Z}_{\mathcal{G}}(x_1, x_2, x_3, \cdots; y_1, y_2, y_3, \cdots; z_1, z_2, z_3, \cdots) \quad (17)$$

$$= \sum_{m,n,p} \frac{1}{m!n!p!} \sum_{\alpha \in S_m, \beta \in S_n, \gamma \in S_p} \operatorname{fix} \mathcal{G}[\alpha, \beta, \gamma] x_1^{\alpha_1} x_2^{\alpha_2} x_3^{\alpha_3} \cdots y_1^{\beta_1} y_2^{\beta_2} y_3^{\beta_3} \cdots z_1^{\gamma_1} z_2^{\gamma_2} z_3^{\gamma_3} \cdots, \quad (18)$$

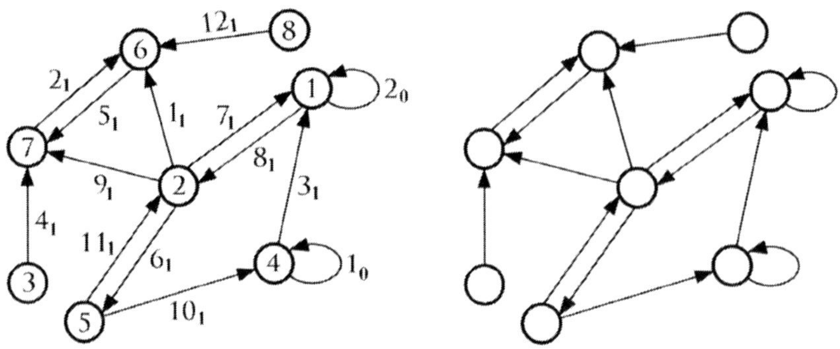

(a) (b)

Figure 2. (a) A labelled digraph g; (b) $[g]$ = totally unlabelled g.

where, for each permutations $\alpha \in S_m$, $\beta \in S_n$ and $\gamma \in S_p$, $\mathrm{fix}\,\mathcal{G}[\alpha,\beta,\gamma]$ is the number[6] of all digraphs $g \in \mathcal{G}[V,V_1,V_0]$, on the sets $V=\{1,2,\cdots,m\}$ of vertices, $V_1=\{1_1,2_1,\cdots,n_1\}$ of arrows and $V=\{1_0,2_0,\cdots,p_0\}$ of loops, for which (α,β,γ) is an automorphism (i.e., $\mathrm{fix}\,\mathcal{G}[\alpha,\beta,\gamma]$ is the cardinality (or total weight) of the set of fixed points of the permutation $\mathcal{G}[\alpha,\beta,\gamma]$). For a permutation $\sigma \in S_k$, the notation σ_i is used to denote the number of cycles[7] of length i in the cyclic decomposition of σ.

The sequence of integers $(\sigma_1,\sigma_2,\sigma_3,\cdots)$ is called the cyclic type of σ and it is well known that the number of $\sigma \in S_k$ having cyclic type $\mathbf{k}=(k_1,k_2,k_3,\cdots)$, with $k_1+2k_2+3k_3+\cdots=k$, is

$$\frac{k!}{\mathbf{k}\!\downarrow}, \quad \text{where } \mathbf{k}\!\downarrow = 1^{k_1}k_1!\,2^{k_2}k_2!\,3^{k_3}k_3!\cdots \tag{19}$$

Note that each sequence $\mathbf{k}$ has a finite number of nonzero terms and will be considered, in the present text, as a finite sequence with $c=c(\mathbf{k})=\max\{j : k_j \neq 0\}$ components. By this convention, $\mathbf{k}$ can be viewed as a partition of the integer $k=k_1+2k_2+\cdots+ck_c$, having k_i parts of length i, for $i = 1, 2, \cdots, c$ and we use the classical notation[8]

$$\mathbf{k} \vdash k, \tag{20}$$

to express this fact. Note also that $\mathrm{fix}\mathcal{G}[\alpha, \beta, \gamma]$, in (18), only depends on the cyclic types of the three permutations α, β, γ. Hence, every three permutations α, β, γ having given cyclic types $\mathbf{m} = (m_1, m_2, m_3, \cdots)$, $\mathbf{n} = (n_1, n_2, n_3, \cdots)$, $\mathbf{p} = (p_1, p_2, p_3, \cdots)$ contribute to the same monomial

$$\mathbf{x}^{\mathbf{m}}\mathbf{y}^{\mathbf{n}}\mathbf{t}^{\mathbf{p}} = x_1^{m_1} x_2^{m_2} x_3^{m_3} \cdots y_1^{n_1} y_2^{n_2} y_3^{n_3} \cdots z_1^{p_1} z_2^{p_2} z_3^{p_3} \cdots, \tag{21}$$

in (18). In order to eliminate this redundancy, we regroup monomials which correspond to each of these types, and taking (19) into account we obtain the following more compact variant expression for the cycle index series of $\mathcal{G}$:

$$\mathbf{Z}_{\mathcal{G}} = \mathbf{Z}_{\mathcal{G}}(\mathbf{x}, \mathbf{y}, \mathbf{z}) = \sum_{\mathbf{m}, \mathbf{n}, \mathbf{p}} \mathrm{fix}\mathcal{G}[\mathbf{m}, \mathbf{n}, \mathbf{p}] \frac{\mathbf{x}^{\mathbf{m}}}{\mathbf{m}\downarrow} \frac{\mathbf{y}^{\mathbf{n}}}{\mathbf{n}\downarrow} \frac{\mathbf{z}^{\mathbf{p}}}{\mathbf{p}\downarrow}, \tag{22}$$

in which each monomial appears only once and $\mathrm{fix}\mathcal{G}[\mathbf{m}, \mathbf{n}, \mathbf{p}]$ is the number (or total weight) of all $\mathcal{G}$-struc- tures for which any given (α, β, γ) of types $(\mathbf{m}, \mathbf{n}, \mathbf{p})$ is an automorphism.

The following proposition is a consequence of general principles from the theory of species and Pólya theory. It shows that the computation of $\mathbf{Z}_{\mathcal{G}}$ is an essential step in the determination of the inventory series (3) of classes of graphical sentences arising from digraphs $g \in \mathcal{G}$.

Proposition 2.1. *Let* $\mathcal{G} = \mathcal{G}(X, Y, Z)$ *be any species of connected digraphs, let* $\mathbb{L} \subseteq \mathbb{N}^+$, *and let* $\mathfrak{S} = \mathfrak{S}_{A, \mathcal{G}, \mathbb{L}}$ *be the set of all graphical sentences over alphabet A arising from* $\mathcal{G}$, *the word on each arrow or loop having a length* $\in \mathbb{L}$. *Then, the inventory of* $\mathfrak{S}$ *is given by*

$$\mathrm{w}\mathfrak{S}\left(x, y, z, t, (a)_{a \in A}\right) = \mathbf{Z}_{\mathcal{G}}\left(x, x^2, x^3, \cdots; \lambda_1 y, \lambda_2 y^2, \lambda_3 y^3, \cdots; \lambda_1 z, \lambda_2 z^2, \lambda_3 z^3, \cdots\right), \tag{23}$$

where $\lambda_k = \Lambda\left(t^k \Sigma a^k\right),\ k = 1,2,3,\cdots,\ \Lambda(u) = \sum_{i\in\mathbb{L}} u^i$,

and $\Sigma a^k = \sum_{a\in A} a^k$ *is the formal k-th power sum of the letters of the alphabet.*

Proof. See Section 5. □

Making use of the compact expression (22) for the cycle index series and collecting terms, inventory (23) can be rewritten in the following more explicit form.

Corollary 2.2. *The inventory,* $\mathbf{w}\mathfrak{S} = \mathbf{w}\mathfrak{S}\left(x,y,z,t,(a)_{a\in A}\right)$, *of the set* $\mathfrak{S} = \mathfrak{S}_{A,\mathcal{G},\mathbb{L}}$ *of all graphical sentences arising from a species* $\mathcal{G} = \mathcal{G}(X,Y,Z)$ *of digraphs is given by*

$$\mathbf{w}\mathfrak{S} = \sum_{m\geq 0, n\geq 0, p\geq 0} \omega_{m,n,p}\left(\lambda_1,\lambda_2,\lambda_3,\cdots\right)x^m y^n z^p \qquad ,(24)$$

Where $\lambda_k = \Lambda\left(t^k \Sigma a^k\right),\ k = 1,2,3,\cdots,\ \Lambda(u) = \sum_{i\in\mathbb{L}} u^i$, and

$$\omega_{m,n,p}\left(\lambda_1,\lambda_2,\lambda_3,\cdots\right) = \sum_{\mathbf{m}\vdash m, \mathbf{n}\vdash n, \mathbf{p}\vdash p} \frac{\mathrm{fix}\,\mathcal{G}[\mathbf{m},\mathbf{n},\mathbf{p}]}{\mathbf{m}\!\downarrow \mathbf{n}\!\downarrow \mathbf{p}\!\downarrow} \lambda_1^{m_1+p_1}\lambda_2^{n_2+p_2}\lambda_3^{n_3+p_3}\cdots \qquad (25)$$

Note. If $\mathbb{L} = \mathbb{N}^+$ in Proposition 2.1, then $\Lambda(u) = u/(1-u)$ and no restrictions are put on the lengths of the words in inventory (23). If $\mathbb{L} = \{1,3,5,\cdots\}$, then $\Lambda(u)=u/(1=u^2)$ and the lengths of the words are all odd. If $\mathbb{L} = \{1,2,\cdots,N\}$, then $\Lambda(u) = \sum_{k=1}^{N} u^k = u\left(1-u^N\right)/(1-u)$ and the lengths of the words are bounded by N, etc.

3. ANALYSIS OF CLASSES OF TREE-LIKE SENTENCES

As shown in the preceding section, the computation of the inventory of a class $\mathfrak{S}$ of graphical sentences can be reduced to the computation of the cycle index series $\mathbf{Z}_{\mathcal{G}}$ provided that $\mathfrak{S}$ arises from a 3-sort species $\mathcal{G} = \mathcal{G}(X,Y,Z)$ of connected digraphs. However, the explicit or recursive computation of the cycle index series of most species of graphical structures is a very difficult (or intractable) task. For example, even in the ordinary one-sort case[9], the complete cycle index series of the species of all ordinary plane digraphs and all transitive digraphs are still unknown[10].

For this reason, we focus our study on the following basic classes of graphical sentences:

1) Linear sentences (arising from the species of path-shaped digraphs).

2) General tree-like sentences (arising from various species of tree-like digraphs).

Note that linear sentences are special kinds of tree-like sentences. Due to their close relationship with ordinary sentences, we have chosen to present first a separate subsection devoted to their study. Our methods will use the fact that species of tree-like digraphs can be built from simpler species by making use of basic combinatorial operations and that cycle index series behave well with respect to these operations. For example, if F, G and H are species, then

$$\mathbf{Z}_{F+G} = \mathbf{Z}_F + \mathbf{Z}_G,\ \mathbf{Z}_{F\cdot G} = \mathbf{Z}_F \cdot \mathbf{Z}_G,\ \mathbf{Z}_{F(G)} = \mathbf{Z}_F \circ \mathbf{Z}_G,\ \mathbf{Z}_{F(G,H)} = Z_F \circ (Z_G, Z_H),\ \text{etc,} \tag{26}$$

where ° denotes the classical plethystic substitution of cycle index series (see [1]).

3.1. Linear Sentences

We say that a digraph $g=(g_0,g_1)$ is path-shaped if its underlying simple graph is a simple path. A graphical sentence is linear if it comes from a path-shaped digraph. Figure 3 shows a path-shaped digraph, together with its underlying simple path and a linear sentence over alphabet A={a,b,c}. Note that a path-shaped digraph can have non-trivial automorphisms. For example, the 180° rotation, (α,β,γ), where the cyclic decompositions of the permutations α,β,γ are given by

$$\alpha = (3,8)(5,2)(6,4)(1,9),\ \beta = (3_0,4_0)(5_0,1_0) \tag{27}$$

$$\gamma = (7_1,3_1)(9_1,4_1)(1_1,12_1)(5_1,6_1)(10_1,2_1)(8_1,11_1) \tag{28}$$

is an automorphism of the path-shaped digraph of Figure 3.

Special kinds of linear sentences over an alphabet include ordinary sentences (Figure 4 top), corresponding to directed paths without loops, and ordinary sentences with (possible) loops (Figure 4 bottom), corresponding to directed paths with (possible) loops.

For example, the sentence

MY TAYLOR IS RICH RICH AND MY COUSIN IS POOR POOR POOR (29)

is one of the readable sentences in Figure 4 bottom.

Proposition 3.1 *Let* $\mathcal{O}$ *be the set of all ordinary sentences,* $\mathcal{O}^{\circlearrowright}$*, the set of all ordinary sentences with loops,* $\mathfrak{L}$*, the set of all linear sentences without loops, and* $\mathfrak{L}^{\circlearrowright}$*, the set of all linear sentences with loops over an alphabet A and a set* $\mathbb{L} \subseteq \mathbb{N}^+$ *of allowed word-lengths. Then, the following inventories hold*

$$\mathbf{w}\mathcal{O} = \frac{x}{1-\lambda_1 xy}, \quad \mathbf{w}\mathcal{O}^{\circlearrowright} = \frac{(1+\lambda_1 z)x}{1-(1+\lambda_1 z)\lambda_1 xy}, \tag{30}$$

$$\mathbf{w}\mathfrak{L} = \frac{1}{2}\left(\frac{x}{1-(2+\lambda_1 y)\lambda_1 xy} + \frac{x+\lambda_2 x^2 y^2}{1-(2+\lambda_2 y^2)\lambda_2 x^2 y^2}\right), \tag{31}$$

$$\mathbf{w}\mathfrak{L}^{\circlearrowright} = \frac{1}{2}\left(\frac{(1+\lambda_1 z)x}{1-(1+\lambda_1 z)(2+\lambda_1 y)\lambda_1 xy} + \frac{(1+\lambda_1 z)x+(1+\lambda_2 z^2)\lambda_2 x^2 y^2}{1-(1+\lambda_2 z^2)(2+\lambda_2 y^2)\lambda_2 x^2 y^2}\right), \tag{32}$$

Where $\lambda_1 = \Lambda(t\Sigma a), \lambda_2 = \Lambda(t^2 \Sigma a^2), \quad \Lambda(u) = \sum_{i\in\mathbb{L}} u^i$.

Proof. See Section 5. □

In view of (30)-(32), the alternate general inventory formula (24) in the case of any class of linear sentences, does not involve $\lambda_3, \lambda_4, \cdots$. We have the following explicit expansions.

Corollary 3.2. *For the classes* $\mathcal{O}, \mathcal{O}^{\circlearrowright}, \mathfrak{L}, \mathfrak{L}^{\circlearrowright}$ *of linear sentences, we have*

$$\mathbf{w}\mathcal{O}^{\circlearrowright} = \sum_{m\geq 1, 0\leq p\leq m} \binom{m}{p} \lambda_1^{m+p-1} x^m y^{m-1} z^p \tag{33}$$

$$\mathbf{w}\mathcal{O} = \mathbf{w}\mathcal{O}^{\circlearrowleft}\Big|_{z=0} \tag{34}$$

$$\mathbf{w}\mathfrak{L}^{\circlearrowleft} = \sum_{(m,n,p)\in W}\left(a_{m,n,p}\lambda_1^{n+p} + b_{m,n,p}\lambda_2^{\frac{n+p}{2}} + c_{m,n,p}\lambda_1\lambda_2^{\frac{n+p-1}{2}}\right)x^m y^n z^p \tag{35}$$

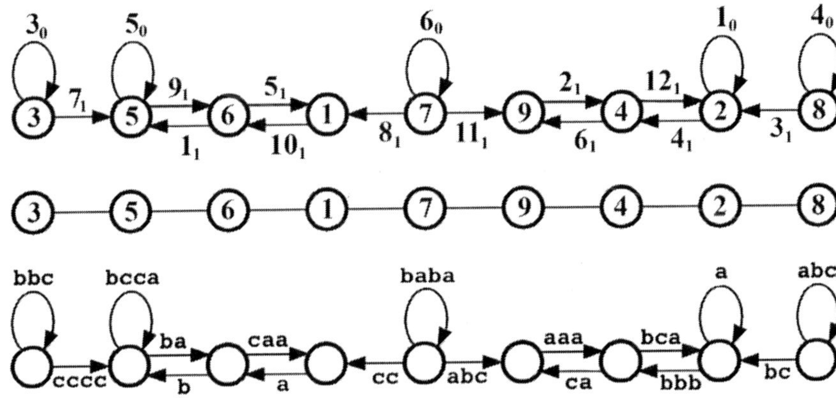

Figure 3. Path-shaped digraph, its underlying simple path and a linear sentence.

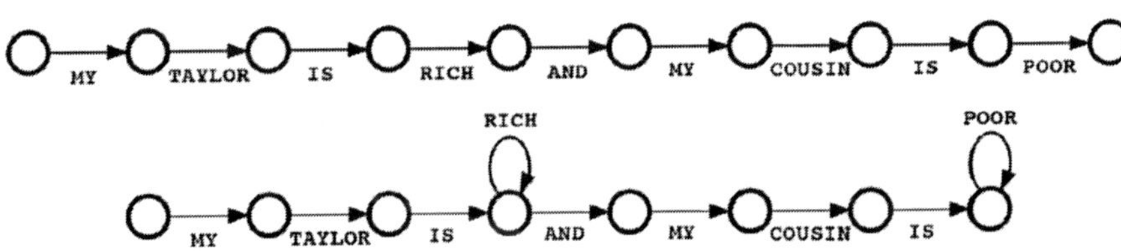

Figure 4. An ordinary sentence and an ordinary sentence with loops.

$$\mathbf{w}\mathfrak{L} = \mathbf{w}\mathfrak{L}^{\circlearrowleft}\Big|_{z=0}, \tag{36}$$

where

$$W = \left\{(m,n,p)\,\middle|\, m \geq 1, m-1 \leq n \leq 2(m-1), 0 \leq p \leq m\right\}, \tag{37}$$

$$a_{m,n,p} = \binom{m-1}{n-m+1}\binom{m}{p}2^{2m-n-3}, \tag{38}$$

$$b_{m,n,p} = \binom{\frac{m-1}{2}}{\frac{n-m+1}{2}} \binom{\frac{m-1}{2}}{\frac{p}{2}} 2^{m-\frac{n}{2}-2} + \binom{\frac{m}{2}-1}{\frac{n-m}{2}} \binom{\frac{m}{2}}{\frac{p}{2}} 2^{m-\frac{n}{2}-2} \quad ,(39)$$

$$c_{m,n,p} = \binom{\frac{m-1}{2}}{\frac{n-m+1}{2}} \binom{\frac{m-1}{2}}{\frac{p-1}{2}} 2^{m-\frac{n}{2}-2} \quad , \qquad (40)$$

using the convention $\binom{\alpha}{\beta} = 0$ *if α and β are not both integers.*

Proof. (Sketch) Formulas (34) and (36) are immediate since no loops are involved. Careful computations, starting from (30) and (32) using geometric series, the binomial theorem, and manipulation or indices lead to expansions (33) and (35). □

Sample of explicit examples of computations.

Example 3.1. General shape of the inventory of the class $\mathbf{W}\mathfrak{L}^{\circlearrowleft}$.

The first few terms of $\mathbf{W}\mathfrak{L}^{\circlearrowleft}$ read as follows

$$\begin{aligned}
\mathbf{w}\mathfrak{L}^{\circlearrowleft} &= x + \lambda_1 xz + \lambda_1 x^2 y + 2\lambda_1^2 x^2 yz + \lambda_1^3 x^2 yz^2 + \left(\frac{1}{2}\lambda_1^2 + \frac{1}{2}\lambda_2\right) x^2 y^2 + \lambda_1^3 x^2 y^2 z \\
&+ \left(\frac{1}{2}\lambda_1^4 + \frac{1}{2}\lambda_2^2\right) x^2 y^2 z^2 + \left(2\lambda_1^2 + \lambda_2\right) x^3 y^2 + \left(6\lambda_1^3 + \lambda_1\lambda_2\right) x^3 y^2 z \\
&+ \left(6\lambda_1^4 + \lambda_2^2\right) x^3 y^2 z^2 + \left(2\lambda_1^5 + \lambda_1\lambda_2^2\right) x^3 y^2 z^3 + 2\lambda_1^3 x^3 y^3 + 6\lambda_1^4 x^3 y^3 z \\
&+ 6\lambda_1^5 x^3 y^3 z^2 + 2\lambda_1^6 x^3 y^3 z^3 + \left(\frac{1}{2}\lambda_1^4 + \frac{1}{2}\lambda_2^2\right) x^3 y^4 + \left(\frac{3}{2}\lambda_1^5 + \frac{1}{2}\lambda_1\lambda_2^2\right) x^3 y^4 z \\
&+ \left(\frac{3}{2}\lambda_1^6 + \frac{1}{2}\lambda_2^3\right) x^3 y^4 z^2 + \left(\frac{1}{2}\lambda_1^7 + \frac{1}{2}\lambda_1\lambda_2^3\right) x^3 y^4 z^3 + \cdots.
\end{aligned} \qquad (41)$$

The coefficients of $x^m y^n z^p$ in (41) are polynomials in λ1 and λ2 each having one or two terms, despite the fact that (35) suggests three terms. This is true for every m, n, p since $b_{m,n,p}$ and $C_{m,n,p}$ cannot be both non zero, due to the fact that $\frac{p-1}{2}$ and $\frac{p}{2}$ cannot be both integral in (39) and (40).

Example 3.2. Counting linear sentences with given parameters.

Corollary 3.2 is particularly useful when one wants to compute an individual term in the inventory of linear sentences. For example, consider the 3-letter alphabet A={a,b,c} and take $\mathbb{L}=\mathbb{N}^+$. This means that we impose no restrictions on the lengths of the words that are assigned to each arrow or loop in linear sentences. Suppose that we want to know the number of such linear sentences having m=7 vertices, n=10arrows, p=5 loops which are made of 7 times the letter a, 6 times the letter b and 12 times the letter c. In this case, the coefficient of $x^7 y^{10} z^5$ is given by

$$2\binom{6}{4}\binom{7}{5}\lambda_1^{15}+\binom{3}{2}\binom{3}{2}\lambda_1\lambda_2^7=630\lambda_1^{15}+9\lambda_1\lambda_2^7 \quad , \qquad (42)$$

where

$$\lambda_1=\frac{t(\mathrm{a}+\mathrm{b}+\mathrm{c})}{1-t(\mathrm{a}+\mathrm{b}+\mathrm{c})},\ \lambda_2=\frac{t^2(\mathrm{a}^2+\mathrm{b}^2+\mathrm{c}^2)}{1-t^2(\mathrm{a}^2+\mathrm{b}^2+\mathrm{c}^2)} \qquad (43)$$

The required number of linear sentences equals the coefficient of t^{25} a^7 b^6 c^{12} in (42). Using Maple, this number is equal to

11026064192581225368 (44)

If we take $\mathbb{L}=\{1,2\}$, then the words that are assigned to each arrow or loop in linear sentences are of length 1 or 2. In this case, (43) is replaced by

$$\lambda_1=t(\mathrm{a}+\mathrm{b}+\mathrm{c})+t^2(\mathrm{a}+\mathrm{b}+\mathrm{c})^2,\ \lambda_2=t^2(\mathrm{a}^2+\mathrm{b}^2+\mathrm{c}^2)+t^4(\mathrm{a}^2+\mathrm{b}^2+\mathrm{c}^2)^2 \qquad (45)$$

and (44) goes down to 16882686796464720.

Example 3.3. Manipulating the inventory of linear sentences.

As we have seen in Section 1, manipulations of inventories (specialization of variables, expansions, etc) can be made to analyze various parameters in graphical sentences. Proposition 3.1 is generally more suitable for such manipulations than Corollary 3.2. For example, let ℓ be the number of letters in alphabet A, $\mathbb{L}=\mathbb{N}^+$, and assign the value 1 to x, y and to each letter $a \in A$ in $\mathbf{w}\mathfrak{L}$ given by (31). Then $\lambda_1 = \ell t/(1-\ell t)$, $\lambda_2 = \ell t^2/(1-\ell t^2)$, and, by (7)

$$\frac{1}{2}\left(\frac{(1-\ell t)^2}{1-4\ell t+2\ell^2 t^2}+\frac{1-\ell t^2}{1-4\ell t^2+2\ell^2 t^4}\right)=\sum_q e_q t^q \qquad ,(46)$$

where e_q is the number of linear sentences without loops that are made of q letters. Note that (46) is a rational function of t, so that the sequence $(e_q)_{q\geq 0}$ satisfies a linear recurrence with constant coefficients and the asymptotic expansion of e_q, as $q \to \infty$, can be established using standard classical methods. The first few terms in expansion (46) are given by

$$\begin{aligned}\sum e_q t^q &= 1+\ell t+\frac{1}{2}\left(7\ell^2+3\ell\right)t^2+12\ell^3 t^3+\left(41\ell^4+5\ell^2\right)t^4+140\ell^5 t^5\\ &+\left(478\ell^6+17\ell^3\right)t^6+1632\ell^7 t^7+\left(5572\ell^8+58\ell^4\right)t^8+19024\ell^9 t^9\\ &+\left(64952\ell^{10}+198\ell^5\right)t^{10}+221760\ell^{11}t^{11}+\left(757136\ell^{12}+676\ell^6\right)t^{12}+\cdots\end{aligned} \qquad (47)$$

Example 3.4. Fibonacci numbers versus ordinary sentences with loops.

Consider now the class $\mathcal{O}^{\circlearrowleft}$ of ordinary sentences with possible loops. Let $y=z$ and assign the value 1 to x and to each letter $a \in A$ in $\mathbf{w}\mathcal{O}^{\circlearrowleft}$ given by (30). Then we have

$$\mathbf{w}\mathcal{O}^{\circlearrowleft}\left(1,z,z,t,(1)_{a\in A}\right)=\frac{1+\lambda_1 z}{1-\lambda_1 z-\lambda_1^2 z^2}=\sum_{p\geq 0}F_{p+2}\lambda_1^p z^p=\sum_{p,q\geq 0}f_{p,q}z^p t^q \qquad ,(48)$$

where F_k are the Fibonacci numbers[11] and $f_{p,q}$ is the number of ordinary sentences with possible loops made of p words and q letters. Now, fix $q\geq 1$ and consider the finite class $\mathcal{O}_q^{\circlearrowleft}$ of all $S \in \mathcal{O}^{\circlearrowleft}$ made of q letters. Since, $\lambda_1 = \ell t/(1-\ell t)$, then collecting the coefficient of t^q in (48) we have the polynomial inventory

$$\mathbf{w}\mathcal{O}_q^{\circlearrowleft}\left(1,z,z,1,(1)_{a\in A}\right)=\sum_{1\leq p\leq q}f_{p,q}z^p,\quad f_{p,q}=\binom{q-1}{p-1}F_{p+2}\,\ell^q \qquad (49)$$

Finally, making use of (10) and invoking Binet's formula

$$F_k = \frac{1}{\sqrt{5}}\left(\phi^k - \psi^k\right) \tag{50}$$

where $\phi = \frac{1+\sqrt{5}}{2}$ *is the golden number,* $\psi = \frac{1-\sqrt{5}}{2} = -1/\phi$*, we find that the expected number of words in a random ordinary sentence with possible loops made of* $q \geq 1$ *letters is*

$$1 + (q-1)\frac{F_{2q}}{F_{2q+1}} \sim \frac{q}{\phi}, \quad q \to \infty \tag{51}$$

The reader can check that if we do not allow loops, then (51) is replaced by $\frac{1}{2}(q+1)$.

A multitude of other similar examples can be obtained using Proposition 3.1 and Corollary 3.2.

3.2. General Tree-Like Sentences

We say that a digraph $g=(g_0,g_1)$ is tree-like if its underlying simple graph is a simple tree or a simple rooted tree. A graphical sentence is tree-like if it comes from a tree-like digraph. Figure 5shows a tree-like digraph, its underlying simple tree and a tree-like sentence over alphabet$A\{0,1\}$.

The tree-like structures of Figure 5 are free in the sense that they are not restricted to be embedded in the plane and no other constraints are assumed on the vertices, arrows and loops. More generally, by allowing such constraints, one can consider, for example, the above linear sentences (see Figure 3 and Figure 4), one way free binary rooted tree sentences (see Figure 6left), one way free full binary rooted tree sentences (see Figure 6 right), plane tree sentences (see Figure 5 right) where, this time, the underlying tree is considered as being embedded in the plane), etc. We shall deal with these cases in a uniform manner by adding extra structure on the underlying trees or rooted trees. More precisely, the underlying trees or rooted trees will be enriched according to the following definition.

Definition 3.1. [5] Let $R=R(X)$ be any given one-sort species.

1) A R-enriched rooted tree is a rooted tree in which the set of immediate descendants (away from the root) of every vertex is equipped with a R-structure (see Figure 7 left, in which each dotted arc represents a R-structure).

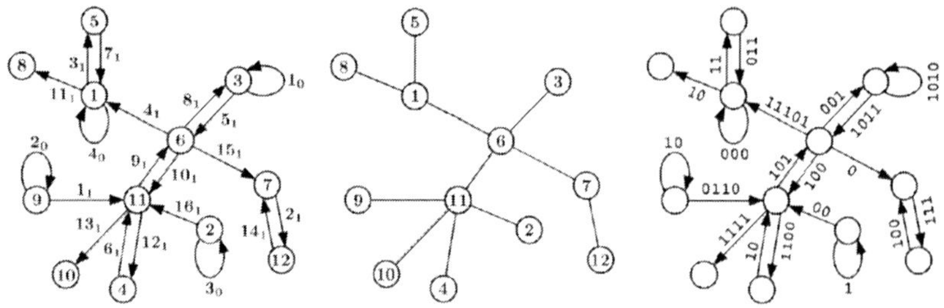

Figure 5. A tree-like digraph, its underlying simple tree, a tree-like sentence over $A = \{0, 1\}$.

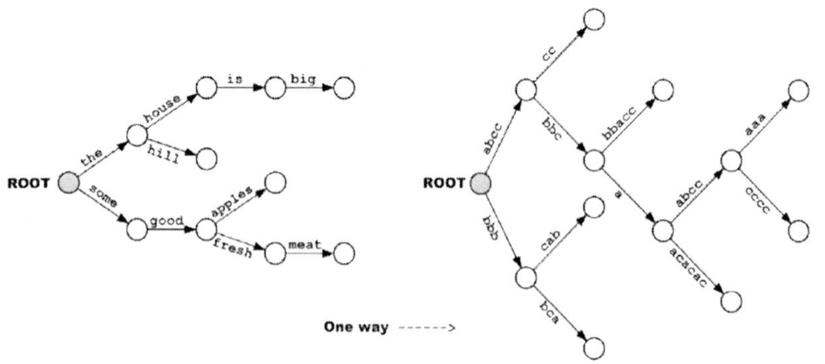

Figure 6. A one way free binary rooted tree sentence and a full one.

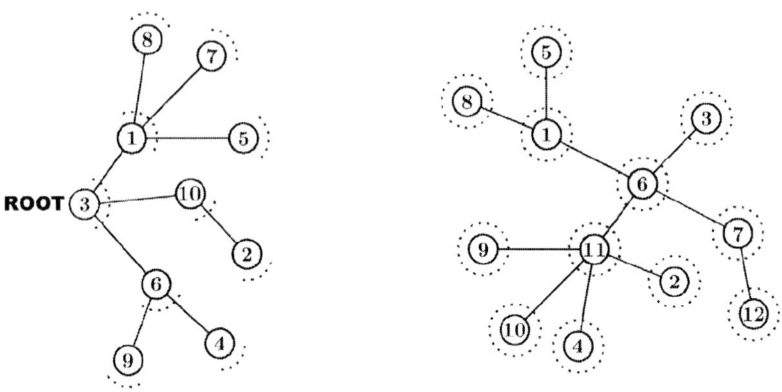

Figure 7. A R-enriched rooted tree and a R-enriched tree.

2) A R-enriched tree is a tree in which the set of immediate neighbors of each vertex is equipped with a R-structure (see Figure 7 right, in which each dotted circle represents a R-structure).

Lemma 3.3 [1] *The species* $A_R = A_R(X)$ *of R-enriched rooted trees is characterized recursively by the combinatorial equation*

$$A_R = XR\left(A_R\right), \tag{52}$$

and the species a_R=$a_R(X)$of R-enriched trees satisfies the combinatorial equality[12]

$$XR\left(A_{R'}\right)+E_2\left(A_{R'}\right)=a_R+A_{R'}^2 \tag{53}$$

where $A_{R'}$ = $A_{R'}(X)$ is the species of R'-enriched rooted trees (R' being the combinatorial derivative of the species R). □

It is easy to see that the species of ordinary rooted trees (resp. ordinary trees) corresponds to the species A_R (resp. a_R) with the choice $R=E$, the species of all finite sets. The species of binary rooted trees (resp. full binary rooted trees) corresponds to the species A_R with the choice $R=1+X+E_2$ (resp.$R=1+E_2$), where 1 denotes, as usual, the species of the empty set and E_2, the species of 2-element sets. The species of all plane trees corresponds to the species a_R with the choice $R=1+C$, where C is the species of cyclic permutations (see Example 3.8 below), etc.

For the computation of the inventories of various classes of enriched tree-like graphical sentences, we will make use of the following 3-sort extension of Lemma 3.3 which includes a new extension, (56) below, of the dissymmetry formula (53).

Lemma 3.4. *The species $\vec{A}_R^{\circlearrowleft}=\vec{A}_R^{\circlearrowleft}(X,Y,Z)$ of one-way R-enriched rooted trees and $A_R^{\circlearrowleft}=A_R^{\circlearrowleft}(X,Y,Z)$ of R-enriched rooted trees on the sorts X of vertices, Y of arrows and Z of loops are characterized recursively by the combinatorial equations*

$$\text{a) } \vec{A}_R^{\circlearrowleft}=(1+Z)XR\left(Y\vec{A}_R^{\circlearrowleft}\right),\ \text{b) } A_R^{\circlearrowleft}=(1+Z)XR\left(\Omega A_R^{\circlearrowleft}\right), \tag{54}$$

where $\Omega=\Omega(Y)=2Y+Y^2$. They can also be expressed explicitly in terms of the 1-sort species $A_R(X)$ as follows

$$\text{a) } \vec{A}_R^{\circlearrowleft}=\frac{1}{Y}A_R\left((1+Z)XY\right),\ \text{b) } A_R^{\circlearrowleft}=\frac{1}{\Omega}A_R\left((1+Z)X\Omega\right) \tag{55}$$

The species $\mathfrak{a}_R^{\circlearrowleft}=\mathfrak{a}_R^{\circlearrowleft}(X,Y,Z)$ of R-enriched trees on sorts X of vertices, Y of arrows and Z of loops satisfies the combinatorial equality (extended dissymmetry formula)

$$\left(1+Z\right)XR\left(\Omega\mathcal{A}_{R'}^{\circ}\right)+E_2\left(Y\mathcal{A}_{R'}^{\circ}\right)=\mathfrak{a}_R^{\circ}+\left(Y+Y^2\right)\cdot\left(\mathcal{A}_{R'}^{\circ}\right)^2 ,\tag{56}$$

where $\mathcal{A}_{R'}^{\circ}=\mathcal{A}_{R'}^{\circ}\left(X,Y,Z\right)$ *is the species of R'-enriched rooted trees on sorts X, Y, Z.*

Proof. See Section 5. □

In our analysis of tree-like graphical sentences, we will use of the following useful compact "plethystic notation" which is classical in the theory of species and cycle index series.

Notation 3.5. *Let* $\mathbf{S}=\mathbf{S}\left(v_1,v_2,v_3,\cdots\right)$ *be a (formal) power series in the variables* $v_1,v_2,v_3,\cdots$*. For any integer k≥1,* S_k *denotes the series S in which each variable is raised to the power k:*

$$\mathbf{S}_k=\mathbf{S}_k\left(v_1,v_2,v_3,\cdots\right)=\mathbf{S}\left(v_1^k,v_2^k,v_3^k,\cdots\right) \tag{57}$$

In particular, S_1*=S. Furthermore, given power series,*

$$\mathbf{f}=\mathbf{f}\left(s_1,s_2,s_3,\cdots\right),\ \mathbf{g}=\mathbf{g}\left(s_1,s_2,s_3,\cdots;t_1,t_2,t_3,\cdots\right),\ etc ,\tag{58}$$

$$\mathbf{S}=\mathbf{S}\left(v_1,v_2,v_3,\cdots\right),\ \mathbf{T}=\mathbf{T}\left(v_1,v_2,v_3,\cdots\right),\ etc \ ,\tag{59}$$

Then $\mathbf{f}\left[\left[\mathbf{S}\right]\right],\ \mathbf{g}\left[\left[\mathbf{S};\mathbf{T}\right]\right]$ *etc, denote the series*

$$\mathbf{f}\left[\left[\mathbf{S}\right]\right]=f\left(\mathbf{S}_1,\mathbf{S}_2,\mathbf{S}_3,\cdots\right),\ \mathbf{g}\left[\left[\mathbf{S};\mathbf{T}\right]\right]=g\left(\mathbf{S}_1,\mathbf{S}_2,\mathbf{S}_3,\cdots;\mathbf{T}_1,\mathbf{T}_2,\mathbf{T}_3,\cdots\right),\ etc ,\tag{60}$$

in the variables $v_1,v_2,v_3,\cdots$. □

For example, taking the variables $x,y,z,t,\left(a\right)_{a\in A}$ for the *vi*'s, then formula (23) of Proposition 2.1 takes the compact form

$$\mathbf{w}\mathfrak{S}\left(x,y,z,t,\left(a\right)_{a\in A}\right)=\mathbf{Z}_{\mathcal{G}}\left[\left[x;\lambda y;\lambda z\right]\right] \tag{61}$$

where $\lambda=\lambda_1=\Lambda\left(t\Sigma a\right),\ \Lambda\left(u\right)=\sum_{i\in\mathbb{L}}u^i$, and $\Sigma a=\sum_{a\in A}a$ is the formal sum of the letters in A.

We now describe how to compute the inventory of classes of tree-like sentences.

Proposition 3.6 *Given an arbitrary species R=R(X)*, let

$$\vec{\mathfrak{T}}_R^{\circlearrowleft} \text{ be the set of one way } R\text{-enriched rooted trees sentences with possible loops,} \quad (62)$$

$$\mathfrak{T}_R^{\circlearrowleft} \text{ be the set of } R\text{-enriched rooted trees sentences with possible loops} \quad ,(63)$$

$$\mathrm{t}_R^{\circlearrowleft} \text{ be the set of } R\text{-enriched trees sentences with possible loops} \quad ,(64)$$

over an alphabet A and a set $\mathbb{L} \subseteq \mathbb{N}^+$ of allowed word-lengths. Then, using Notation 3.5, the inventories $\mathbf{w}\vec{\mathfrak{T}}_R^{\circlearrowleft}$ and $\mathbf{w}\mathfrak{T}_R^{\circlearrowleft}$ can be computed recursively as follows

$$\mathbf{w}\vec{\mathfrak{T}}_R^{\circlearrowleft} = (1+\lambda z)xZ_R\left[\left[\lambda y\left(\mathbf{w}\vec{\mathfrak{T}}_R^{\circlearrowleft}\right)\right]\right], \quad (65)$$

$$\mathbf{w}\mathfrak{T}_R^{\circlearrowleft} = (1+\lambda z)xZ_R\left[\left[(2+\lambda y)\lambda y\left(\mathbf{w}\mathfrak{T}_R^{\circlearrowleft}\right)\right]\right], \quad (66)$$

They can also be expressed explicitly in terms of the cycle index series $Z_{A_R}(x_1, x_2, x_3, \cdots)$ c of the 1-sort species $A_R(R)$ by

$$\mathbf{w}\vec{\mathfrak{T}}_R^{\circlearrowleft} = \frac{1}{\lambda y} Z_{A_R}\left[\left[(1+\lambda z)\lambda xy\right]\right], \quad (67)$$

$$\mathbf{w}\mathfrak{T}_R^{\circlearrowleft} = \frac{1}{(2+\lambda y)\lambda y} Z_{A_R}\left[\left[(1+\lambda z)(2+\lambda y)\lambda xy\right]\right] \quad (68)$$

Where $\lambda = \Lambda(t\Sigma a),\ \Lambda(u) = \sum_{i\in\mathbb{L}} u^i,\ Z_R = Z_R(x_1, x_2, x_3, \cdots)$.
Moreover, let R' be the combinatorial derivative of the species R. Then, the inventory $\mathbf{wt}_R^{\circlearrowleft}$ has the form

$$\mathbf{wt}_R^{\circlearrowleft} = (1+\lambda z)xZ_R\left[\left[(2+\lambda y)\lambda y\left(\mathbf{w}\mathfrak{T}_{R'}^{\circlearrowleft}\right)\right]\right] - \frac{1}{2}(2+\lambda y)\lambda y\left(\mathbf{w}\mathfrak{T}_{R'}^{\circlearrowleft}\right)^2 + \frac{1}{2}\lambda_2 y^2\left(\mathbf{w}\mathfrak{T}_{R'}^{\circlearrowleft}\right)_2, \quad (69)$$

In the case of the corresponding sets $\vec{\mathfrak{T}}_R$, $\mathfrak{T}_R$, t_R, in which no loops are allowed, we have

$$\mathbf{w}\vec{\mathfrak{T}}_R = \mathbf{w}\vec{\mathfrak{T}}_R^{\circlearrowleft}\Big|_{z=0},\quad \mathbf{w}\mathfrak{T}_R = \mathbf{w}\mathfrak{T}_R^{\circlearrowleft}\Big|_{z=0},\quad \mathbf{w}\mathfrak{t}_R = \mathbf{w}\mathfrak{t}_R^{\circlearrowleft}\Big|_{z=0}. \tag{70}$$

Proof. (sketch) Apply Proposition 2.1, taking into account Lemma 3.4.

Note. When written explicitly, (65) takes the form

$$\mathbf{w}\vec{\mathfrak{T}}_R^{\circlearrowleft} = (1+\lambda_1 z)xZ_R\left(\lambda_1 y\left(\mathbf{w}\vec{\mathfrak{T}}_R^{\circlearrowleft}\right), \lambda_2 y^2\left(\mathbf{w}\vec{\mathfrak{T}}_R^{\circlearrowleft}\right)_2, \lambda_3 y^3\left(\mathbf{w}\vec{\mathfrak{T}}_R^{\circlearrowleft}\right)_3, \cdots\right) \tag{71}$$

where,

$\mathbf{w}\vec{\mathfrak{T}}_R^{\circlearrowleft} = \mathbf{w}\vec{\mathfrak{T}}_R^{\circlearrowleft}\left(x, y, z, t, (a)_{a\in A}\right)$, $\left(\mathbf{w}\vec{\mathfrak{T}}_R^{\circlearrowleft}\right)_2 = \mathbf{w}\vec{\mathfrak{T}}_R^{\circlearrowleft}\left(x^2, y^2, z^2, t^2, (a^2)_{a\in A}\right)$, $\cdots$, $\lambda_1 = \lambda = \Lambda\left(t\sum_{a\in A} a\right)$, $\lambda_2 = \Lambda\left(t^2\sum_{a\in A} a^2\right), \cdots$ Formulas (65) and (66) give rise to iterative schemes for the computation of the inven- tories $\mathbf{w}\vec{\mathfrak{T}}_R^{\circlearrowleft}$, $\mathbf{w}\mathfrak{T}_R^{\circlearrowleft}$. See, for example, [7] -[10] for descriptions of efficient ways to do such computations, including adaptations of quadratically convergent Newtonian methods. Formula (69) reduces the computation of $\mathbf{w}\mathfrak{t}_R^{\circlearrowleft}$ to that of $\mathbf{w}\mathfrak{T}_{R'}^{\circlearrowleft}$. Recall that $Z_{R'}(x_1, x_2, x_3, \cdots) = \dfrac{\partial}{\partial x_1} Z_R(x_1, x_2, x_3, \cdots)$.

Sample of explicit examples of computations.

Example 3.5. One way free binary rooted tree sentences.

Consider the set $\vec{\mathfrak{B}}$ of one way free binary rooted tree sentences without loops (Figure 6 left, shows such a tree-like sentence over the 26-letter alphabet $A = \{\mathrm{a}, \mathrm{b}, \cdots, \mathrm{z}\}$) and the set $\vec{\mathfrak{B}}^{\circlearrowleft}$ of such sentences where loops are allowed. These tree-like sentences correspond to R-enriched rooted trees with $R = 1 + X + E_2$. Since $Z_R = 1 + x_1 + \left(x_1^2 + x_2\right)/2$, formula (65) of Proposition 3.6 immediately gives the following recursive scheme for the computation of the inventory $\mathbf{w}\vec{\mathfrak{B}}^{\circlearrowleft}$

$$\mathbf{w}\vec{\mathfrak{B}}^{\circlearrowleft} = (1+\lambda z)x\left(1 + \lambda y \mathbf{w}\vec{\mathfrak{B}}^{\circlearrowleft} + \frac{1}{2}\left(\lambda^2 y^2\left(\mathbf{w}\vec{\mathfrak{B}}^{\circlearrowleft}\right)^2 + \lambda_2 y^2\left(\mathbf{w}\vec{\mathfrak{B}}^{\circlearrowleft}\right)_2\right)\right), \tag{72}$$

and, since $\mathbf{w}\vec{\mathfrak{B}} = \mathbf{w}\vec{\mathfrak{B}}^{\circlearrowleft}\Big|_{z=0}$, (70) leads to

$$\mathbf{w}\vec{\mathfrak{B}} = x\left(1 + \lambda y \mathbf{w}\vec{\mathfrak{B}} + \frac{1}{2}\left(\lambda^2 y^2\left(\mathbf{w}\vec{\mathfrak{B}}\right)^2 + \lambda_2 y^2\left(\mathbf{w}\vec{\mathfrak{B}}\right)_2\right)\right) \tag{73}$$

Of course, as many terms as we want in (72) and (73) can be computed using a computer algebra system. For a more specific application, let ℓ be the number of letters in alphabet A and let $\vec{f}^{\circlearrowleft}(t)=\mathbf{w}\vec{\mathfrak{B}}^{\circlearrowleft}\left(1,1,1,t,(1)_{a\in A}\right)$ and $\vec{f}(t)=\mathbf{w}\vec{\mathfrak{B}}\left(1,1,1,t,(1)_{a\in A}\right)$ as in (7). Then $\lambda=\ell t/(1-\ell t)$, $\lambda_2=\ell t^2/\left(1-\ell t^2\right)$ and (72), (73) give, after some symbolic manipulation,

$$\vec{f}^{\circlearrowleft}(t)=\frac{1-\ell t}{1-3\ell t+\ell^2t^2}\left(1+\frac{1}{2}\left(\frac{\ell^2t^2}{(1-\ell t)^2}\vec{f}^{\circlearrowleft}(t)^2+\frac{\ell t^2}{1-\ell t^2}\vec{f}^{\circlearrowleft}\left(t^2\right)\right)\right) \tag{74}$$

$$=1+2\ell t+\frac{1}{2}\ell(11\ell+1)t^2+\ell^2(17\ell+1)t^3+\frac{3}{2}\ell^2\left(37\ell^2+2\ell+1\right)t^4$$
$$+\frac{3}{2}\ell^3\left(125\ell^2+7\ell+2\right)t^5+\frac{1}{8}\ell^2\left(5197\ell^4+310\ell^3+73\ell^2+34\ell+2\right)t^6+\cdots \tag{75}$$

$$\vec{f}(t)=\frac{1-\ell t}{1-2\ell t}\left(1+\frac{1}{2}\left(\frac{\ell^2t^2}{(1-\ell t)^2}\vec{f}(t)^2+\frac{\ell t^2}{1-\ell t^2}\vec{f}\left(t^2\right)\right)\right) \tag{76}$$

$$=1+\ell t+\frac{1}{2}\ell(5\ell+1)t^2+\frac{1}{2}\ell^2(13\ell+1)t^3+\frac{1}{2}\ell^2\left(35\ell^2+3\ell+2\right)t^4$$
$$+\frac{1}{2}\ell^3\left(97\ell^2+9\ell+2\right)t^5+\frac{1}{8}\ell^2\left(1101\ell^4+110\ell^3+25\ell^2+18\ell+2\right)t^6+\cdots. \tag{77}$$

The coefficient of t^q in (75) (resp. (77)) is the number of one way free binary rooted tree sentences with loops (resp. without loops) on a ℓ-letter alphabet A that are made of q letters. As an illustration, for the usual 26-letter alphabet $A=\{a,b,\cdots,z\}$, series (75) and (77) read as follows up to $q=10$:

$$\vec{f}^{\circlearrowleft}(t)=1+52t+3731t^2+299468t^3+25415910t^4+2232608976t^5$$
$$+201144059077t^6+18486682363084t^7+1726830972452898t^8$$
$$+163481547244045680t^9+15652457718105041990t^{10}+\cdots, \tag{78}$$

$$\tilde{f}(t) = 1 + 26t + 1703t^2 + 114582t^3 + 8024120t^4 + 578320704t^5 + 42679370357t^6 + 3210985966578t^7 + 245445450450824t^8 + 19011557820684416t^9 + 1489108237113437148t^{10} + \cdots. \tag{79}$$

Example 3.6. Ordinary tree and rooted tree sentences.

Let $R = R(X) = E(X)$, be the species of finite sets. Then, by Definition 3.1, the species $A_E = A_E(X)$ of E-enriched rooted trees coincides with the species $A = A(X)$ of ordinary (free) rooted trees and the species $a_E(X)$ of E-enriched trees coincides with the species $a=a(X)$ of ordinary (free) trees. Lemma 3.3 produces the familiar combinatorial equations,

$$A = XE(A), \quad A + E_2(A) = a + A^2, \tag{80}$$

the second equation being the classical dissymmetry formula of Leroux. Taking cycle index series in (80) and using $R' = E' = E$ and $Z_E = \exp\left(\sum_{k\geq 1} x_k / k\right)$, we obtain the classical formulas

$$Z_A = x_1 \exp\left(Z_A + \frac{1}{2}(Z_A)_2 + \frac{1}{3}(Z_A)_3 + \cdots \right) \tag{81}$$

$$= x_1 + x_1^2 + \frac{1}{2}\left(3x_1^2 + x_2\right)x_1 + \frac{1}{3}\left(8x_1^3 + 3x_1x_2 + x_3\right)x_1 + \frac{1}{24}\left(125x_1^4 + 54x_1^2x_2 + 16x_1x_3 + 15x_2^2 + 6x_4\right)x_1 + \cdots \tag{82}$$

$$Z_a = Z_A - \frac{1}{2}Z_A^2 + \frac{1}{2}(Z_A)_2 \tag{83}$$

$$= 1 + \frac{1}{2}\left(x_1^2 + x_2\right) + \frac{1}{2}\left(x_1^2 + x_2\right)x_1 + \frac{1}{6}\left(4x_1^4 + 3x_1^2x_2 + 2x_1x_3 + 3x_2^2\right) + \frac{1}{24}\left(25x_1^4 + 18x_1^2x_2 + 8x_1x_3 + 15x_2^2 + 6x_4\right)x_1 + \cdots, \tag{84}$$

from which $Z_A(x_1, x_2, x_3, \cdots)$ and $Z_a(x_1, x_2, x_3, \cdots)$ can be computed to arbitrary degree[13].

Now, let

$\vec{\mathfrak{T}}^{\circ}$ be the set of one way free rooted tree sentences with possible loops ,(85)

$\mathfrak{T}^{\circ}$ be the set of free rooted tree sentences with possible loops ,(86)

$\mathfrak{t}^{\circ}$ be the set of free tree sentences with possible loops ,(87)

Then, by (67)-(69),

$$\mathbf{w}\vec{\mathfrak{T}}^{\circ} = \frac{1}{\lambda y} Z_A\left[\left[(1+\lambda z)\lambda xy\right]\right], \tag{88}$$

$$\mathbf{w}\mathfrak{T}^{\circ} = \frac{1}{(2+\lambda y)\lambda y} Z_A\left[\left[(1+\lambda z)(2+\lambda y)\lambda xy\right]\right], \tag{89}$$

$$\mathbf{w}\mathfrak{t}^{\circ} = \mathbf{w}\mathfrak{T}^{\circ} - \frac{1}{2}(2+\lambda y)\lambda y\left(\mathbf{w}\mathfrak{T}^{\circ}\right)^2 + \frac{1}{2}\lambda_2 y^2\left(\mathbf{w}\mathfrak{T}^{\circ}\right)_2 \tag{90}$$

For more specific applications, let ℓ be the number of letters in the alphabet and consider the specializations $F^{\circ}(t) = w\mathfrak{T}^{\circ}\left(1,1,1,t,(1)_{a\in A}\right)$ and $f^{\circ}(t) = \mathbf{w}\mathfrak{t}^{\circ}\left(1,1,1,t,(1)_{a\in A}\right)$. Then $\lambda_k = \ell t^k/\left(1-\ell t^k\right),\ \ k=1,2,\cdots$ and by (66) and (69) with $R=E$, we have

$$F^{\circ}(t) = \frac{1}{1-\ell t}\exp\left(\sum_{k\geq 1}\frac{1}{k}\frac{\left(2-\ell t^k\right)\ell t^k}{\left(1-\ell t^k\right)^2}F^{\circ}\left(t^k\right)\right) \tag{91}$$

$$= 1+3\ell t+\ell(14\ell+1)t^2+\frac{1}{3}\ell\left(223\ell^2+15\ell+2\right)t^3$$
$$+\frac{1}{6}\ell\left(2539\ell^3+198\ell^2+50\ell+3\right)t^4$$
$$+\frac{1}{30}\ell\left(75873\ell^4+6790\ell^3+1470\ell^2+95\ell+12\right)t^5+\cdots \tag{92}$$

$$f^{\circ}(t) = F^{\circ}(t) - \frac{1}{2}\frac{(2-\ell t)\ell t}{(1-\ell t)^2}F^{\circ}(t)^2 + \frac{1}{2}\frac{\ell t^2}{1-\ell t^2}F^{\circ}(t^2) \tag{93}$$

$$= 1 + 2\ell t + \frac{1}{2}\ell(13\ell+3)t^2 + \frac{1}{6}\ell(79\ell^2+9\ell+2)t^3$$
$$+\frac{1}{2}\ell(241\ell^3+28\ell^2+18\ell+1)t^4$$
$$+\frac{1}{30}\ell(17993\ell^4+2230\ell^3+760\ell^2+65\ell+12)t^5+\cdots \quad .(94)$$

The coefficient of t^q in (92) (resp. (94)) is the number of free rooted tree sentences (resp. free tree sentences) with loops on a ℓ-letter alphabet A that are made of q letters. For example, in the case of a 4-letter alphabet, say$A$={A,C,G,T} series (92) and (94) read as follows up to$q$=9:

$$F^{\circ}(t) = 1+12t+228t^2+4840t^3+110578t^4+2650928t^5+65889652t^6$$
$$+1683125528t^7+43926578519t^8+1166284622820t^9+\cdots, \tag{95}$$

$$f^{\circ}(t) = 1+8t+110t^2+1736t^3+31890t^4+634848t^5+13484444t^6$$
$$+300435640t^7+6950286407t^8+165682482716t^9+\cdots. \tag{96}$$

Consider now the 2-letter alphabet A={a,b} and make the substitutions x=1, y=1, z=0, t=1, then $\lambda = \dfrac{a+b}{1-a-b}$, no loops are allowed, and (89)-(90) take the forms

$$\mathbf{w}\mathfrak{T} = \frac{(1-a-b)^2}{(2-a-b)(a+b)} Z_A\left[\left[\frac{(2-a-b)(a+b)}{(1-a-b)^2}\right]\right] = \sum_{i+j>0} g_{i,j}a^ib^j \quad ,(97)$$

$$\mathbf{w}t = \mathbf{w}\mathfrak{T} - \frac{(2-a-b)(a+b)}{2(1-a-b)^2}(\mathbf{w}\mathfrak{T})^2 + \frac{(a^2+b^2)}{2(1-a^2-b^2)}(\mathbf{w}\mathfrak{T})_2 = \sum_{i+j>0} h_{i,j}a^ib^j \quad ,(98)$$

where $g_{i,j}$ (resp.$h_{i,j}$) is the number of rooted tree sentences (respected tree sentences) without loops that contain exactly *i* times the letter a and *j* times the letter *b*.

If we choose $\mathbb{L}=\{1,2\}$ (only words of length 1 or 2 are allowed), then $\lambda=(a+b)+(a^2+b^2)$, and (89)-(90) begin with the terms

$$\mathbf{w}\mathfrak{T}=1+2a+2b+10a^2+18ab+10b^2+46a^3+128a^2b+128ab^2+46b^3$$
$$+237a^4+874a^3b+1291a^2b^2+874ab^3+237b^4+1258a^5+5870a^4b$$
$$+11402a^3b^2+11402a^2b^3+5870ab^4+1258b^5+\cdots, \tag{99}$$

$$\mathbf{w}\mathfrak{t}=1+a+b+5a^2+7ab+5b^2+15a^3+39a^2b+39ab^2+15b^3$$
$$+66a^4+218a^3b+327a^2b^2+218ab^3+66b^4+276a^5$$
$$+1234a^4b+2366a^3b^2+2366a^2b^3+1234ab^4+276b^5+\cdots, \tag{100}$$

etc. Again, all the above series, and many variants, can be expanded to arbitrary orders.

Example 3.7. Back to linear sentences.

Since linear sentences are special kinds of tree-like sentences, it is interesting to look at the dissymmetry formula (56) in the context of path-shaped graphs. Take the 1-sort species $R=R(X)=1+X+E_2(X)$. Then a R-structure is either void, a singleton, or an unordered pair of singletons. This means that a R-enriched tree is a simple path (see Definition 3.1, Figure 7right and Figure 3 middle). Hence, the 2-sort species *P*(*X,Y*) of all path-shaped digraphs without loops (see Figure 9) coincides with the 2-sort species $\mathfrak{a}_R(X,Y)=\mathfrak{a}_R^{\circ}(X,Y,0)$ of *R*-enriched trees. Moreover, since $R'=1+X$, a *R'*-enriched rooted tree is a simple path pointed at an extremity (see Definition 3.1 and Figure 7, left). Hence, the 2-sort species *K*(*X,Y*) of all path-shaped digraphs without loops pointed at an extremity (see Figure 9) coincides with the 2-sort species $\mathcal{A}_{R'}(X,Y)=\mathcal{A}_{R'}^{\circ}(X,Y,0)$ of *R'*-enriched rooted trees. In this setting, the dissymmetry formula (56), with *Z*=0, becomes

$$X\cdot\left(1+\Omega K+E_2(\Omega K)\right)+E_2(YK)=P+\left(Y+Y^2\right)K^2 \tag{101}$$

where $\Omega = \Omega(Y) = 2Y + Y^2$ Now, using $K = X \cdot (1 + \Omega K)$, we can solve (101) for P as follows,

$$P = K + XE_2(\Omega K) + E_2(YK) - Y(1+Y)K^2 \quad (102)$$

This formula coincides with formula (124) which is used in the proof of Proposition 3.1.

Example 3.8. Plane tree sentences.

A plane tree is a (unrooted) tree that is embedded in a plane. Such tree-structures have fewer automorphisms than free trees. Take any vertex p of a plane tree τ and draw a vector starting at p which is perpendicular to the plane in which τ is embedded. This gives an orientation to that plane and the vertices that are adjacent to p are cyclically turning around p according to that orientation (see Figure 7). In other words, the set of immediate neighbors of p is equipped with a (1+C)-structure, where C is the species of non-empty oriented cycles (the empty set species, 1, corresponds to the special case where the tree is reduced to one point, $\tau=\{p\}$, for which the the set of immediate neighbors of p is empty). Since p is arbitrary, this shows that the species a_{plane} of plane trees coincides with the species a_{1+c} of (1+C)-enriched trees.

Take R=1+C, then $R' = 1' + C' = 1/(1-X) = L$ is the species of linear orders. The species A_L of L- enriched rooted trees coincides with the species A_0 of linearly ordered rooted trees (the set of immediate descendants, away from the root, of every vertex is linearly ordered). Using the classical formulas,

$$Z_R = Z_{1+C} = 1 + \sum_{k\geq 1} \frac{\phi(k)}{k} \log\left(\frac{1}{1-x_k}\right)$$ (where $\phi(k)$ denotes Euler function) and $Z_{R'} = Z_L = \dfrac{1}{1-x_1}$ in Lemma 3.3, we have

$$Z_{A_o} = \frac{1-\sqrt{1-4x_1}}{2} = \sum_{m\geq 1} \frac{(2m-2)!}{m!(m-1)!} x_1^m, \quad (103)$$

$$Z_{a_{\text{plane}}} = x_1\left(1 - \sum_{k\geq 1} \frac{\phi(k)}{k} \log\left(\frac{1+\sqrt{1-4x_k}}{2}\right)\right) + \frac{1}{2}x_1 + \frac{1}{4}\left(\sqrt{1-4x_1} - \sqrt{1-4x_2}\right) \quad (104)$$

Using the expansion[14] $-\log\left(\frac{1}{2}\left(1+\sqrt{1-4x}\right)\right)=\sum_{i\geq 1}\frac{(2i-1)!}{i!^2}x^i$, we get, from (68) and (69), the following explicit expressions for the inventory $\mathbf{w}\mathfrak{T}_o^{\circlearrowleft}$ (resp., $\mathbf{t}_{\text{plane}}^{\circlearrowleft}$) of linearly ordered rooted tree sentences with loops (resp., plane tree sentences with loops)

$$\mathbf{w}\mathfrak{T}_o^{\circlearrowleft}=\frac{1-\sqrt{1-4\theta}}{2(2+\lambda_1 y)\lambda_1 y}=\frac{1}{(2+\lambda_1 y)\lambda_1 y}\cdot\sum_{m\geq 1}\frac{(2m-2)!}{m!(m-1)!}\theta^m \tag{105}$$

$$wt_{\text{plane}}^{\circlearrowleft}=(1+\lambda_1 z)x\left(1-\sum_{j\geq 1}\frac{\phi(j)}{j}\log\left(\frac{1+\sqrt{1-4\theta_j}}{2}\right)\right)-\frac{1-\sqrt{1-4\theta}-2\delta}{4(2+\lambda_1 y)\lambda_1 y}+\frac{1-\sqrt{1-4\theta_2}}{4(2+\lambda_2 y^2)} \tag{106}$$

$$=(1+\lambda_1 z)x\left(1+\sum_{i,j\geq 1}\frac{(2i-1)!\phi(j)}{i!^2 j}\theta_j^i\right)-\frac{\sum_{i\geq 2}\frac{(2i-2)!}{i!(i-1)!}\theta^i}{2(2+\lambda_1 y)\lambda_1 y}+\frac{\sum_{i\geq 1}\frac{(2i-2)!}{i!(i-1)!}\theta_2^i}{2(2+\lambda_2 y^2)} \tag{107}$$

where

$$\theta=\theta_1=(1+\lambda_1 z)(2+\lambda_1 y)\lambda_1 xy,\ \theta_j=\left(1+\lambda_j z^j\right)\left(2+\lambda_j y^j\right)\lambda_j x^j y^j \tag{108}$$

Note that since linearly ordered rooted tree structures are asymmetric structures, no λ_k except λ_1 appear in (105). As before, let ℓ be the number of letters in alphabet A and let $F^{\circlearrowleft}(t)=\mathbf{w}\mathfrak{T}_o^{\circlearrowleft}\left(1,1,1,t,(1)_{a\in A}\right)$ and $f^{\circlearrowleft}(t)=\mathbf{wt}_{\text{plane}}^{\circlearrowleft}\left(1,1,1,t,(1)_{a\in A}\right)$. Then

$$F^{\circlearrowleft}(t)=1+3\ell t+16\ell^2t^2+105\ell^3t^3+765\ell^4t^4+5951\ell^5t^5+\cdots$$

,(109)

$$f^{\circlearrowleft}(t)=1+2\ell t+\frac{1}{2}\ell(13\ell+3)t^2+\frac{1}{3}\ell\left(83\ell^2+3\ell+4\right)t^3+\frac{1}{6}\ell\left(859\ell^3+6\ell^2+53\ell+6\right)t^4$$
$$+\frac{1}{30}\ell\left(25667\ell^4+30\ell^3+205\ell^2+30\ell+48\right)t^5+\cdots. \tag{110}$$

This time, the coefficient of t_q in (109) (resp. (110)) is the number of linearly ordered rooted tree sentences (resp. plane tree sentences) with loops on a ℓ-letter alphabet A that are made of q letters.

As a final illustration, fix $m \geq 1$ and consider the inventory $\mathbf{w}\mathfrak{T}_{o,m}$ of *linearly ordered rooted tree sentences without loops having exactly m vertices.* Letting z=0 in (105) and (108), we have

$$\mathbf{w}\mathfrak{T}_{o,m} = \frac{(2m-2)!}{m!(m-1)!}\left(2\lambda y + \lambda^2 y^2\right)^{m-1} x^m \tag{111}$$

Now, let ℓ be the number of letters in the alphabet and assume that the length of the word on each arrow is at most k. Then, making the substitutions, x=y=1, a=1, $\forall a \in A$, we get

$$\mathbf{w}\mathfrak{T}_{o,m}\left(1,1,1,t,(1)_{a\in A}\right) = \frac{(2m-2)!}{m!(m-1)!}\left(2\lambda + \lambda^2\right)^{m-1} = \sum_{m-1\leq q\leq 2m-2} v_q t^q \tag{112}$$

which is a polynomial in t, since $\lambda = \dfrac{\ell t\left(1-\ell^k t^k\right)}{1-\ell t}$ has k terms. Differentiation gives

$$\left[\frac{\frac{d}{dt}\mathbf{w}\mathfrak{T}_{o,m}\left(1,1,1,t,(1)_{a\in A}\right)}{\mathbf{w}\mathfrak{T}_{o,m}\left(1,1,1,t,(1)_{a\in A}\right)}\right]_{t=1} = \left[\frac{2(m-1)(1+\lambda)\frac{d}{dt}\lambda}{(2+\lambda)\lambda}\right]_{t=1} = \frac{\sum_q q v_q}{\sum_q v_q} = e_{m,\ell,k}. \tag{113}$$

where $e_{m,\ell,k}$ is the expected total number of letters in random m-vertex linearly ordered rooted tree sentence without loops on a ℓ-letter alphabet in which the word on each arrow has at most k letters. Further computations give

$$e_{m,\ell,k} = 2(m-1)\frac{\left(\ell^{k+1}-1\right)\left(k\ell^{k+1}-(k+1)\ell^k+1\right)}{(\ell-1)\left(\ell^k-1\right)\left(\ell^{k+1}+\ell-2\right)} \sim 2k(m-1)\left(1-\frac{1}{\ell}\right), \quad \ell\to\infty \tag{114}$$

Again, all the above inventories can be manipulated in a great number of ways.

4. CONCLUDING REMARKS

It would be interesting to extend the above analysis to other classes of graphical sentences arising from other families of 3-sort species of connected digraphs. As said before, this is generally a very difficult task. However, the analysis can be done, for example, for the class of cyclic graphical sentences (for which the underlying simple graphs are unoriented cycles) by making use of (3-sort) cycle index series related to subgroups of the dihedral groups. The analysis can also be done for the whole class of all graphical sentences since the cycle index series of the 3-sort species, $\mathcal{D}^{\circ} = \mathcal{D}^{\circ}(X,Y,Z)$, of all digraphs (with labelled vertices of sort X, arrows of sort Y and loops of sort Z) turns out to be tractable[15]. In fact,

$$\mathcal{D}^{\circ}(X,Y,Z) = \mathcal{D}\big((1+Z)X,Y\big) \tag{115}$$

where $\mathcal{D} = \mathcal{D}(X,Y)$ is the 2-sort species of all digraphs with vertices of sort X, arrows of sort Y and no loop, which, in the spirit of [1] [11], can be expressed in terms of simpler species by making use of a 2-sort version of the more advanced operation, $\square$, called functorial composition of species.

Another direction of investigation would be to replace digraphs by dimultigraphs (directed multigraphs) and study associated inventories of classes of multigraphical sentences. For example, in the case of the 3-sort (resp. 2-sort) species $\mathcal{M}^{\circ}(X,Y,Z)$ (resp. $M(X,Y)$) of all dimultigraphs, equation (115) must be replaced by $\mathcal{M}^{\circ}(X,Y,Z) = \mathcal{M}\big(E(Z)X,Y\big)$ and operation $\square$ can be used.

5. PROOFS OF THE MAIN RESULTS

Proof of Proposition 2.1. First, consider the weighted species $\mathcal{G}^{*}(X,Y,Z) = \mathcal{G}(xX,yY,zZ)$ of all $\mathcal{G}$ -struc- tures in which each vertex is given a weight x, each arrow a weight y and each loop a weight z. Then

$$\mathbf{Z}_{\mathcal{G}^{*}}\left(x_1,x_2,\cdots;y_1,y_2,\cdots;z_1,z_2,\cdots\right) = \mathbf{Z}_{\mathcal{G}}\left(xx_1,x^2x_2,\cdots;yy_1,y^2y_2,\cdots;zz_1,z^2z_2,\cdots\right) \tag{116}$$

Next, let $\Lambda(Y)=\sum_{k\in\mathbb{L}} Y^k$ be the species of all k-lists of arrows, where $k\in\mathbb{L}$. Figure 8(a) shows an unla-

belled $\Lambda(Y)$ -structure with k=5. Now, define the 2-sort species, $\widehat{\mathcal{G}^\star}(X,Y)$ by substituting $\Lambda(Y)$ for Y and for Z in $\mathcal{G}^\star(X,Y,Z)$:

$$\widehat{\mathcal{G}^\star}(X,Y)=\mathcal{G}^\star(X,\Lambda(Y),\Lambda(Y))=\mathcal{G}(xX,y\Lambda(Y),z\Lambda(Y)) \tag{117}$$

Figure 8(b) shows an unlabelled $\widehat{\mathcal{G}^\star}(X,Y)$-structure of weight $x^4y^6z^2$ on a set of vertices and arrows.

Taking the cycle index series of (117) we get

$$\mathbf{Z}_{\widehat{\mathcal{G}^\star}}(x_1,x_2,\cdots;y_1,y_2,\cdots)=\mathbf{Z}_{\mathcal{G}}(xx_1,x^2x_2,\cdots;y\Lambda(y_1),y^2\Lambda(y_2),\cdots;z\Lambda(y_1),z^2\Lambda(y_2),\cdots) \tag{118}$$

Next, assigning a weight ta, $a\in A$, to every arrow of every $\widehat{\mathcal{G}^\star}$ c-structure, gives the species

$$\mathcal{S}(X,Y)=\widehat{\mathcal{G}^\star}\left(X,\left(\sum_{a\in A}ta\right)Y\right), \tag{119}$$

whose cycle index, $\mathbf{Z}_{\mathcal{S}}(x_1,x_2,\cdots;y_1,y_2,\cdots)$, is obtained by the substitutions $y_k:=\left(\sum_{a\in A}t^k a^k\right)y_k=t^k\left(\Sigma a^k\right)y_k$, $k=1,2,\cdots$, in (118). Note that an unlabelled S-structure can be canonically identified with a graphical sentence. So that (23) follows by the substitutions $x_k:=1, y_k:=1, k=1,2,\cdots$, in $\mathbf{Z}_{\mathcal{S}}$ $\mathbf{Z}_S$ that unlabels the arrows and vertices in S -structures.

Proof of Proposition 3.1. The inventories (30) of the two special kinds of linear graphical sentences, $\mathcal{O}$ and $\mathcal{O}^{\circlearrowleft}$, are very easy to compute since directed paths are sequences of very simple structures with trivial automorphism groups. More precisely, let $D(X,Y)$ be the species of dipaths without loops and $D^{\circlearrowleft}(X,Y,Z)$ be the species of dipaths with possible loops. Since XY-structures are of the form $\circ\!\!\rightarrow$ and ZXY-structures are of the form $\overset{\circ\!\!\longrightarrow}{\circlearrowleft}$, we have

$$D(X,Y)=\sum_{k\geq 0}(XY)^k X=\frac{X}{1-XY}, \tag{120}$$

$$D^{\circlearrowleft}(X,Y,Z)=\sum_{k\geq 0}(XY+ZXY)^k(1+Z)X=\frac{(1+Z)X}{1-(1+Z)XY} \tag{121}$$

and, since $\mathbf{Z}_X=x_1,\mathbf{Z}_Y=y_1,\mathbf{Z}_Z=z_1$, Proposition 2.1 immediately gives (30).

On the other hand, the inventories of the sets $\mathcal{L}$ and $\mathcal{L}^{\circlearrowleft}$ of linear graphical sentences are more difficult to compute since a path-shaped digraph can have a nontrivial automorphism (as we saw above). We first analyze the species *P*(*X,Y*) of all path-shaped digraphs without loops (see Figure 9 top). Introduce the auxiliary species $\Omega=\Omega(Y)=Y+Y+Y^2=Y(2+Y)$. An ΩX -structure is of the form

$$\rightarrow\!\circ \quad \text{or} \quad \leftarrow\!\circ \quad \text{or} \quad \rightleftarrows\!\circ \tag{122}$$

Let *K*=*K*(*X,Y*) be the species of all *P*-structures pointed at an extremity (see Figure 9bottom).

This pointing induces a global orientation to these pointed structures (see dotted arrow) and implies that the species *K* is a species of sequences:

$$K=X\sum_{k\geq 0}(\Omega X)^k=\frac{X}{1-\Omega X}=\frac{X}{1-Y(2+Y)X} \tag{123}$$

As a consequence of the general dissymmetry formula (56) the species *P* can be expressed in terms of *K* and Ω as follows (see details in Example 3.7)

$$P=K+XE_2(\Omega K)+E_2(YK)-Y(1+Y)K^2 \tag{124}$$

where E_2 denotes the species of 2-element sets. Formula (31) then follows from Proposition 2.1 by taking the cycle index series of (124) and using the fact that $\mathbf{Z}_{E_2(X)}=\left(x_1^2+x_2\right)/2$. Finally, let $P^{\circlearrowleft}=P^{\circlearrowleft}(X,Y,Z)$ be the species of all path-shaped digraphs possible with loops, then the following combinatorial equation holds

$$P^{\circlearrowleft}(X,Y,Z) = P((1+Z)X,Y), \tag{125}$$

since every $P^{\circlearrowleft}$ -structure is obtained from a *P*-structure by adding a loop to each vertex (that is,*X*:=*ZX*) or doing nothing to the vertex (that is, *X*:=*X*). So that (32) follows by substituting (1+*Z*)*X* for *X* in (124). The computations are elementary but long and are left to the reader. □

Proof of Lemma 3.4. Consider an $\vec{\mathcal{A}}_R^{\circlearrowleft}$ -structure τ and look at its underlying R-enriched rooted tree t (see Figure 7 left). To reconstruct τ from *t*, one must replace the root of t by adding a possible loop (that is by replacing the root by an (1+*Z*)*X* -structure) and by replacing each edge adjacent to the root of t by an (outward)

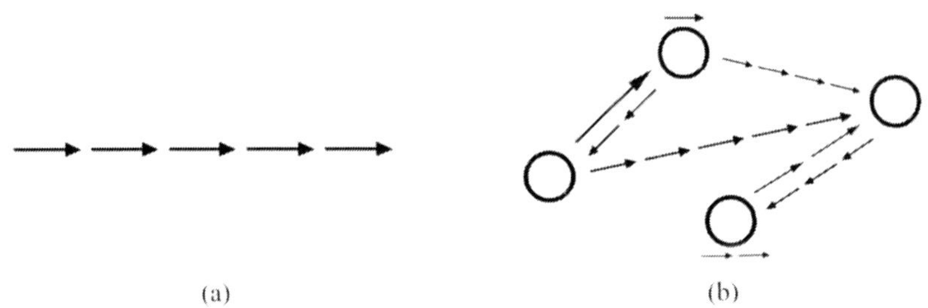

Figure 8. (a) Unlabelled $\Lambda(Y)$-structure, (b) Unlabelled $\widehat{\mathcal{G}}^{*}(X,Y)$- structure (weight: $x^4y^6z^2$ $x^4y^6z^2$).

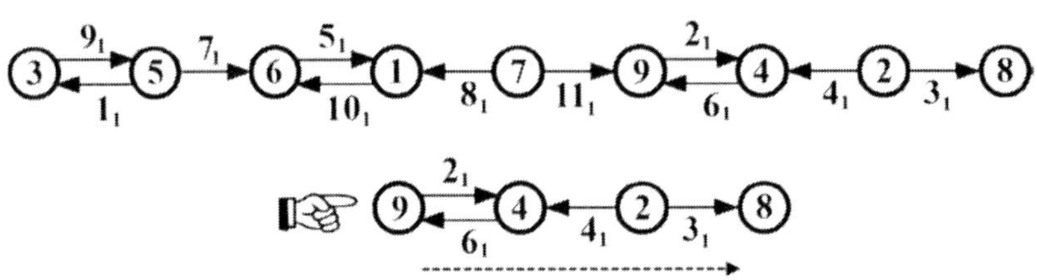

Figure 9. A $P(X,Y)$ *P*(*X*,*Y*)-structure and a *K*(*X*,*Y*)-structure.

arrow (that is, by replacing each such edge by an Y-structure). This establishes (54a). The proof of the combinatorial Equation (54b) is similar, where, this time, each edge adjacent to the root of t is replaced by an outward arrow, an inward arrow or a double arrow (that is, by replacing each such edge by a $(\Omega = 2Y + Y^2)$-structure). To obtain the explicit formula (55a), multiply first both sides of (54a) by *Y*. This gives

$$Y\vec{\mathcal{A}}_R^{\circlearrowright} = (1+Z)XYR\left(Y\vec{\mathcal{A}}_R^{\circlearrowright}\right), \tag{126}$$

But, by (52), the species $A_R((1+Z)XY)$ also satisfies (126). Hence, by the unicity of solution in the implicit species Theorem of Joyal [2], we must have $Y\vec{\mathcal{A}}_R^{\circlearrowright} = A_R((1+Z)XY)$, and (55a) follows by factoring out Y. A similar argumentation can be used to prove (55b) from (54b).

The dissymmetry formula (56) is much more difficult to establish since more automorphisms are involved in enriched trees. To prove this combinatorial equality, we express in two ways the auxiliary species $\mathcal{F}_R = \mathcal{F}_R(X,Y,Z)$ of $\mathfrak{a}_R^{\circlearrowright}$-structures which are pointed either at a single vertex or at two adjacent vertices:

$$\mathcal{F}_R = \left(\mathfrak{a}_R^{\circlearrowright}\right)^{\bullet} + \left(\mathfrak{a}_R^{\circlearrowright}\right)^{\mathrm{adj}\{\bullet\bullet\}} \tag{127}$$

- *The first expression for $\mathcal{F}_R$ reads as follows*

$$\mathcal{F}_R = (1+Z)XR\left(\Omega\mathcal{A}_{R'}^{\circlearrowright}\right) + \left(E_2\left(Y\mathcal{A}_{R'}^{\circlearrowright}\right) + Y\left(\mathcal{A}_{R'}^{\circlearrowright}\right)^2\right) \tag{128}$$

To prove it, consider a $\mathcal{F}_R$-structure ϕ

and look at its underlying pointed or bipointed R-enriched tree, f. We have two cases to consider:

1) If ϕ is a $\left(\mathfrak{a}_R^{\circlearrowright}\right)^{\bullet}$-structure, then by Figure 10 we see that ϕ is canonically equivalent to a

$(1+Z)XR\left(\Omega\mathcal{A}_{R'}^{\circlearrowright}\right)$-structure since to recover ϕ from f, the vertices of f must be replaced by $(1+Z)X$-struc- tures and the edge adjacent to the pointed vertex (and subsequently, all other edges) must be replaced by an Ω-structure.

2) If ϕ is a $\left(\mathfrak{a}_R^{\circlearrowright}\right)^{\mathrm{adj}\{\bullet\bullet\}}$ -structure, then Figure 11 shows that ϕ is canonically equivalent to a

$\left(E_2\left(Y\mathcal{A}_{R'}^{\circlearrowleft}\right)+Y\left(\mathcal{A}_{R'}^{\circlearrowleft}\right)^2\right)$-structure since to recover ϕ from f, the edge of f between the two adjacent pointed vertices must be replaced either by a double arrow (i.e., ϕ is equivalent to a $E_2\left(Y\mathcal{A}_{R'}^{\circlearrowleft}\right)$-structure) or by a single arrow (i.e., ϕ is equivalent to a $Y\left(\mathcal{A}_{R'}^{\circlearrowleft}\right)^2$-structure). This establishes (128).

- *The second expression for $\mathcal{F}_R$ reads as follows*

$$\mathcal{F}_R = \mathfrak{a}_R^{\circlearrowleft} + \left(2Y + Y^2\right)\cdot\left(\mathcal{A}_{R'}^{\circlearrowleft}\right)^2 \tag{129}$$

To prove it, we first split the species $\mathcal{F}_R$ into two subspecies according to whether the pointing(s) coincides exactly with the center or not:

$$\mathcal{F}_R = \mathcal{F}_R^{\text{pointing(s)=center}} + \mathcal{F}_R^{\text{pointing(s)}\neq\text{center}} \tag{130}$$

Since the center of a tree is a canonical object, pointing a tree exactly at its center is naturally equivalent to doing nothing to the tree and we have

$$\mathcal{F}_R = \mathfrak{a}_R^{\circlearrowleft} + \mathcal{F}_R^{\text{pointing(s)}\neq\text{center}} \tag{131}$$

$$= \mathfrak{a}_R^{\circlearrowleft} + \left(\left(\mathfrak{a}_R^{\circlearrowleft}\right)^{\bullet\neq\text{center}} + \left(\mathfrak{a}_R^{\circlearrowleft}\right)^{\text{adj}\{\bullet\bullet\}\neq\text{center}}\right) \tag{132}$$

Now, consider a $\mathcal{F}_R$-structure ϕ that is not pointed at its center. We have two cases to consider:

1) If ϕ is an $\left(\mathfrak{a}_R^{\circlearrowleft}\right)^{\bullet\neq\text{center}}$-structure, then Figure 12 shows that ϕ is canonically equivalent to a $\left(2Y + Y^2\right)\cdot\left(\mathcal{A}_{R'}^{\circlearrowleft}\right)^2$-structure ϕ_1. Indeed, the pointing induces an orientation on the edge from the pointed vertex of f in the direction of the center (see dotted arrow) giving rise to an ordered pair of rooted trees. Moreover, to

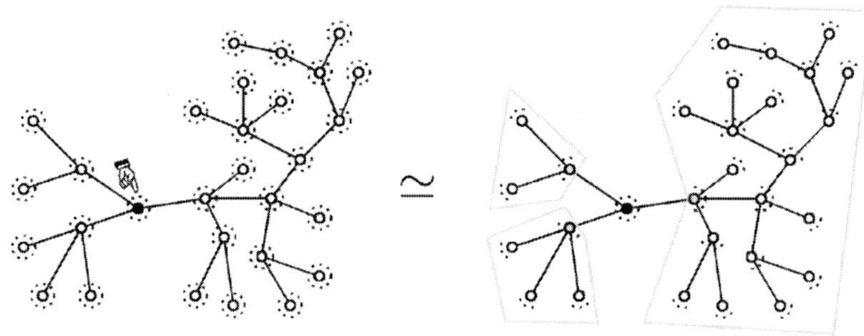

Figure 10. Underlying structures for $\left(\mathfrak{a}_R^{\circlearrowleft}\right)^{\bullet} = (1+Z)\,XR\left(\Omega\mathcal{A}_{R'}^{\circlearrowleft}\right)$.

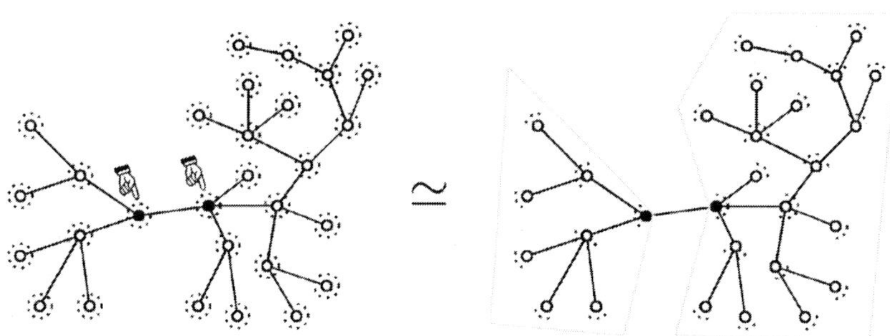

Figure 11. Underlying structures for $\left(\mathfrak{a}_R^{\circlearrowleft}\right)^{\mathrm{adj}\{\bullet\bullet\}} = E_2\left(Y\mathcal{A}_{R'}^{\circlearrowleft}\right) + Y\left(\mathcal{A}_{R'}^{\circlearrowleft}\right)^2$.

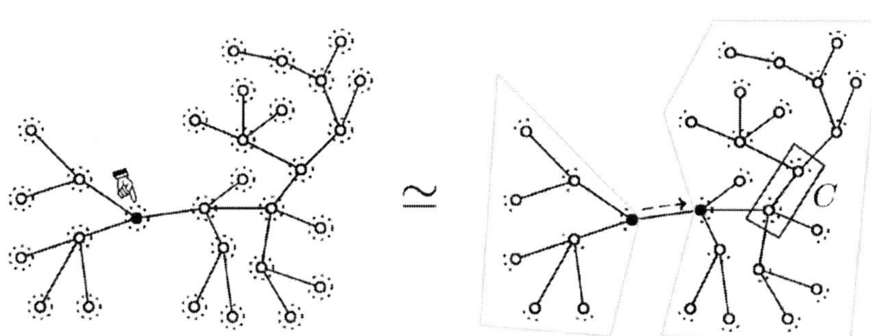

Figure 12. Underlying structures for $\left(\mathfrak{a}_R^{\circlearrowleft}\right)^{\bullet\neq\mathrm{center}} \subset \left(2Y+Y^2\right)\left(\mathcal{A}_{R'}^{\circlearrowleft}\right)^2$.

recover ϕ from f, that edge must be replaced by an arrow going in same direction, or in the opposite direction of the dotted arrow, or by a double arrow. That is, the edge must be replaced by a Ω-structure. Note that ϕ_1 is not an arbitrary $\left(2Y+Y^2\right)\cdot\left(\mathcal{A}_{R'}^{\circlearrowleft}\right)^2$-structure, since the global center is on the side pointed by the dotted arrow.

2) If ϕ is an $\left(\mathfrak{a}_R^{\circlearrowleft}\right)^{\mathrm{adj}\{\bullet\bullet\}\neq\mathrm{center}}$-structure, then Figure 13 shows that ϕ is canonically equivalent to a $\left(2Y+Y^2\right)\cdot\left(\mathcal{A}_{R'}^{\circlearrowleft}\right)^2$-structure $\phi_2 \neq \phi_1$. Indeed, the bi-pointing induces an orientation on the edge between the pointed vertices of f in the direction opposite to the center (see dotted arrow) giving rise to an ordered pair of rooted trees. To recover ϕ from f, that edge must be replaced, as above, by an Ω-structure. Note that ϕ_2 is not

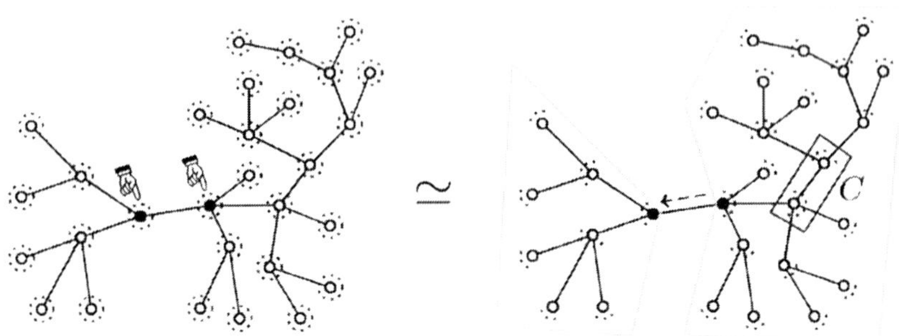

Figure 13. Underlying structures for $\left(\mathfrak{a}_R^{\circlearrowleft}\right)^{\mathrm{adj}\{\bullet\bullet\}\neq\mathrm{center}} \subset \left(2Y+Y^2\right)\left(\mathcal{A}_{R'}^{\circlearrowleft}\right)^2$.

an arbitrary $\left(2Y+Y^2\right)\cdot\left(\mathcal{A}_{R'}^{\circlearrowleft}\right)^2$-structure, since the global center is now on the side of the source of the dotted arrow.

This establishes (129) since any $\left(2Y+Y^2\right)\cdot\left(\mathcal{A}_{R'}^{\circlearrowleft}\right)^2$-structure is either of the form ϕ_1 or of the form ϕ_2. The general 3-sort dissymmetry formula (56) follows by cancelleing the common term $Y\cdot\left(\mathcal{A}_{R'}^{\circlearrowleft}\right)^2$ in the right-hand- sides of the expressions (128) and (129) for $\mathcal{F}_R$. □

FOOTNOTES

[1]In the sense that both its vertices, arrows and loops are unlabelled.

[2]Entry (i, j) being defined as the word on the arrow or loop from vertex i to vertex j, etc.

[3]Variables x, y, z and t are taken distinct from the letters in alphabet A.

[4]This means that $\mathcal{G}$ is a class of connected digraphs on arbitrary finite sets, V of vertices, V_1 of arrows and V_0 of loops, which is closed under arbitrary isomorphisms.

[5]More precisely, X is the species of *singleton vertices*, Y is the species of *singleton arrows* and Z is the species of *singleton loops*.

[6]Or total weight, in the case of weighted digraphs.

[7]Not to be confused with $\sigma(i)$ which is the image of i under σ.

[8]In particular, the empty sequence, 0, has no component and satisfies $0 \vdash 0$.

[9]Where only the vertices are labelled.

[10]However, for the 3-sort species Dig(X, Y, Z) of *all* digraphs, Z_{Dig} can be computed explicitly (see Section 4).

[11] $F_0 = 0, F_1 = 1, F_k = F_{k-1} + F_{k-2} \quad k \geq 2 \quad z/(1 - z - z^2) = \sum_{k\geq 0} F_k z^k$.

[12]Equality (53) is called the dissymmetry formula for trees. It is due P. Leroux [6] for $R = E$ and to the first author [7] for general R.

[13]Explicit expressions for the individual coefficients of Z_A and of Z_a have also been obtained, see [9].

[14]Which can be proved by taking the derivative of both sides.

REFERENCES

1. Bergeron, F., Labelle, G. and Leroux, P. (1998) Combinatorial Species and Tree-Like Structures (Encyclopedia of Mathematics and Its Applications). Cambridge University Press, Cambridge.
2. Labelle, G. (1992) Counting Asymmetric Enriched Trees. Journal of Symbolic Computation, 14, 211-242. http://dx.doi.org/10.1016/0747-7171(92)90037-5
3. Pólya, G. and Read, R.C. (1987) Combinatorial Enumeration of Groups, Graphs and Chemical Compounds. Springer-Verlag, Berlin, Heidelberg, and New-York.
4. Décoste, H., Labelle, G. and Leroux, P. (1982) Une approche combinatoire pour l'itération de Newton-Raphson. Advances in Applied Mathematics, 3, 407-416. http://dx.doi.org/10.1016/S0196-8858(82)80013-4

5. Joyal. A. (1981) Une théorie combinatoire des séries formelles. Advances in Mathematics, 42, 1-82. http://dx.doi.org/10.1016/0001-8708(81)90052-9
6. Labbé, J.-P. and Labelle, G. (2013) Counting Types of Runs in Classes of Arborescent Words. Open Journal of Discrete Mathematics, 3, 7-15. http://dx.doi. org/ 10.4236/ojdm.2013.31002
7. Leroux, P. and Miloudi, B. (1992) Généralisations de la formule d'Otter. Annales des Sciences Mathématiques du Québec, 16, 53-80.
8. Labelle, G. (1986) Some New Computational Methods in the Theory of Species. In: Labelle, G. and Leroux, P., Eds., Combinatoire Enumérative, Lecture Notes in Mathematics 1234, Springer Berlin Heidelberg, Berlin, 192-209. http://dx.doi.org/ 10.1007/bfb0072517
9. Labelle, G. (1981) Une nouvelle démonstration combinatoire des formules d'inversion de Lagrange. Advances in Mathematics, 42, 217-247. http://dx. doi.org/10.1016/0001-8708(81)90041-4
10. Pivoteau, C., Salvy, B. and Soria, M. (2012) Algorithms for Combinatorial Structures: Well-Founded Systems and Newton Iterations. Journal of Combinatorial Theory, Series A, 119, 1711-1773. http://dx.doi.org/10.1016/j.jcta.2012.05.007
11. Décoste, H., Labelle, G. and Leroux, P. (1992) The Functorial Composition of Species, a Forgotten Operation. Discrete Mathematics, 99, 31-48. http://dx.doi. org/ .1016/0012-365X(92)90363-K

CHAPTER 13

On the Signed Domination Number of the Cartesian Product of Two Directed Cycles

Ramy Shaheen

Department of Mathematics, Faculty of Science, Tishreen University, Lattakia, Syria

ABSTRACT

Let D be a finite simple directed graph with vertex set V(D) and arc set A(D). A function f: $V(D) \to \{-1,1\}$ is called a signed dominating function (SDF) if $f\left(N_D^-[v]\right) \geq 1$ for each vertex $v \in V$. The weight $\omega(f)$ of f is defined by $\sum_{v \in V} f(v)$. The signed domination number of a digraph D is D **is** $\gamma_s(D) = \min\{\omega(f) \mid f \text{ is an SDF of } D\}$. **Let** $C_m \times C_n$. Let $C_m \times C_n$ denotes the cartesian product of directed cycles of length m and n. In this paper, we determine the exact values of $g_s(C_m \times C_n)$ for m = 8, 9, 10 and arbitrary n. Also, we give the exact value of $g_s(C_m \times C_n)$ when m, n ≡ 0 (mod 3) and bounds for otherwise.

Keywords: Directed Graph, Directed Cycle, Cartesian Product, Signed Dominating Function, Signed Domination Number

1. INTRODUCTION

Throughout this paper, a digraph $D(V, A)$ always means a finite directed graph without loops and multiple arcs, where $V = V(D)$ is the vertex set and $A = A(D)$is the arc set. If uv is an arc of D, then say that v is an out-neighbor of u and u is an in-neighbor of v. For a vertex $v \in V(D)$, let $N_D^+(v)$ and $N_D^-(v)$ denote the set of out-neighbors and in-neighbors of v, respectively. We write $d_D^+(v)=|N_D^+(v)|$ and $d_D^-(v)=|N_D^-(v)|$ for the out-degree and in-degree of v in D, respectively (shortly $d^+(v)$, $d^-(v)$,. A digraph D is r-regular if $d_D^+(v)=d_D^-(v)=r$ for any vertex $v \in D$. Let $N_D^+[v]=N_D^+(v)\cup\{v\}$ and $N_D^-[v]=N_D^-(v)\cup\{v\}$. The maximum out-degree and in-degree of D are denoted by $\Delta^+(D)$ and $\Delta^-(D)$, respectively (shortly D^+, D^-). The minimum out-degree and in-degree of D are denoted by $\delta^+(D)$ and $\delta^-(D)$, respectively (shortly δ^+, δ^-). *A* signed dominating function of D is defined in [1] as function f: $C \rightarrow \{-1, 1\}$ such that $f(N_D^-[v])\geq 1$ for every vertex $v \in V$. The signed domination number of a directed graph D is $\gamma_s(D)=\min\{\omega(f)|f \text{ is an SDF of } D\}$. Also, a signed k-dominating function (SKDF) of D is a function

f: $V \rightarrow\{-1,1\}$ such that $f(N_D^-[v])\geq k$ for every vertex $v \in V$. The k-signed domination number of a di-

graph D is $\gamma_{ks}(D)=\min\{\omega(f)|f \text{ is an SKDF of } D\}$. Consult [2] for the notation and terminology which are not defined here.

The Cartesian product $D_1 \times D_2$ of two digraphs D_1 and D_2 is the digraph with ver tex set $V(D_1 \times D_2) = V(D_1) \times V(D_2)$ and $((u_1,v_1) \in A(D_1 \times D_2)$ if and only if either $u_1 = v_1$ and $(u_2,v_2) \in A(D_2)$ or $u_2 = v_2$ and $(u_1,v_1) \in A(D_1)$.

In the past few years, several types of domination problems in graphs had been studied [3] -[7], most of those belonging to the vertex domination. In 1995, Dunbar et al. [3], had introduced the concept of signed domination number of an undirected graph. Haas and Wexler in [1], established a sharp lower bound on the signed domination number of a general graph with a given minimum and maximum degree and also of some simple grid graph. Zelinka [8] initiated the study of the signed domination numbers of digraphs. He studied the signed domination number of digraphs for which the in-degrees did not exceed 1, as well as for acyclic tournaments and the circulant tournaments. Karami et al. [9] established lower and upper bounds for the signed domination number of digraphs. Atapour et al. [10]

presented some sharp lower bounds on the signed k-domination number of digraphs. Shaheen and Salim in [11], were studied the signed domination number of two directed cycles $C_m \times C_n$ when $m = 3, 4, 5, 6, 7$ and arbitrary n. In this paper, we study the Cartesian product of two directed cycles C_m and C_n for $mn \geq 8n$. We mainly determine the exact values of $\gamma_s (C_8 \times C_n)$, $\gamma_s(C_9 \times C_n)$, $\gamma_s (C_{10} \times C_n)$ and for some values of m and n. Some previous results:

Theorem 1.1 (Zelinka [8]). Let D be a directed cycle or path with n vertices. Then $\gamma_s (D) = n$.

Lemma 1.2 (Zelinka [8]). Let D be a directed graph with n vertices. Then $\gamma_s (D) \equiv n(\text{mod } 2)$.

Corollary 1.3 (Karami et al. [9]). Let D be a directed of order n in which $d^+(v) = d^-(v) = k$ for each $v \in V$, where k is a nonnegative integer. Then $\gamma_s(D) \geq \frac{n}{1+k}$.

In [11], the following results are proved.

Theorem 1.4 [11]:

$\gamma_s(C_3 \times C_n) = n : n \equiv 0(\text{mod } 3)$, otherwise $\gamma_s(C_3 \times C_n) = n+2$. $\gamma_s(C_4 \times C_n) = 2n$. $\gamma_s(C_5 \times C_n) = 2n : n \equiv 0(\text{mod } 10)$, $\gamma_s(C_5 \times C_n) = 2n+1 : n \equiv 3,5,7(\text{mod } 10)$, $\gamma_s(C_5 \times C_n) = 2n+2 : n \equiv 2,4,6,8(\text{mod } 10)$, $\gamma_s(C_5 \times C_n) = 2n+3 : n \equiv 1,9(\text{mod } 10)$. $\gamma_s(C_6 \times C_n) = 2n : n \equiv 0(\text{mod } 3)$, otherwise $\gamma_s(C_6 \times C_n) = 2n+4$. $\gamma_s(C_7 \times C_n) = 3n$

2. MAIN RESULTS

In this section we calculate the signed domination number of the Cartesian product of two directed cycles C_m and C_n for $m = 8, 9, 10$ and $m \equiv 0(\text{mod } 3)$ and arbitrary n.

The vertices of a directed cycle C_n are always denoted by theintegers $\{1,2,\ldots,n\}$ considered modulo n. The ith row of $V(C_m \times C_n)$ is $R_i = \{(i,j) : j = 1,2,\cdots,n\}$ and the jth column $K_j = \{(i,j) : i = 1,2,\cdots,m\}$. For any vertex $(i,j) \in V(C_m \times C_n)$, always we have the indices i and j are reduced modulo m and n, respectively.

Let us introduce a definition. Suppose that f is a signed dominating function for $C_m \times C_n$, and assume that $1 \leq j, h \leq n$. We say that the hth column of $f(C_m \times C_n)$ is a t-shift of the jth column if $f(i,j) = f(i + t,h)$ for each vertex $(i, j) \in K$, where the indices i, t, $i + t$ are reduced modulo m and j, h are reduced modulo n.

Remark 2.1: Let f is a $\gamma_s(C_m \times C_n)$-function. Then $f[(r,s)] \geq 1$ for each $1 \leq r \leq m$ and each $1 \leq s \leq n$.

Since $C_m \times C_n$ is 2-regular, it follows from $f((i,j)) = -1$ that $f((i \pm 1,j)) = 1$ $f((i,j \pm 1)) = 1$ because $f[(i, j)] \geq 1$, $f((i + 1,j - 1)) = 1$ because $f\left[(i+1, j)\right] \geq 1$ and $f\left[(i, j+1)\right] \geq 1$ $f\left((i-1, j+1)\right) = 1$ because $f\left[(i, j+1)\right] \geq 1$. On the other hand, if $f\left((i \pm 1, j)\right) = f\left((i, j \pm 1)\right) = 1$, $f\left((i+1, j-1)\right) = 1$, and $f\left((i-1, j+1)\right) = 1$, then we must have $f\left((i,j)\right) = -1$ since f is a minimum signed dominating function.

Remark 2.2. Since the case $f\left((i,j)\right) = f\left((i+1,j)\right) = -1$ is not possible, we get $s_j \geq 0$. Furthermore, s_j is odd if m is odd and even when m is even.

Let f be a signed dominating function for $C_m \times C_n$, then we denote $f(K_j) = \sum_{i=1}^{m} f((i,j))$ of the weight of a column K_j and put $s_j = f(K_j)$. The sequence $(s_1, s_2, \cdots, s_n)$ is called a signed dominating sequence corresponding to f. We define

$$X_i = \left|\left\{j : s_j = i\right\}\right|, i = 0, 1, \cdots, m$$

Then we have

$$X_0 + X_1 + \cdots + X_m = n.$$

$$\omega(f) = X_1 + 2X_2 + \cdots + mX_m$$

For the remainder of this section, let f be a signed domination function of $C_m \times C_n$ with signed dominating sequence $(s_1, s_2, \cdots, s_n)$. We need the following Lemma:

Lemma 2.3. If $s_j = k$ then $s_{j-1}, s_{j+1} \geq m - 2k$. Furthermore, $s_{j-1} + s_j \geq m - k$ and $s_j + s_{j+1} \geq m - k$.

Proof. Let $s_j = k$, then there are $(m - k)/2$ of vertices in K_j which get value -1. By Remark 2.1, $K_j + 1$ include at least $2(m - k)/2$ of vertices which get the value 1 and at most $m - (m - k) = k$ of vertices which has value -1. Hence,

s_j+1 ≥ $m - 2k$. Furthermore, $s_j + s_j + 1 \geq m - k$. By the same argument, we get $s_{j-1} \geq m - 2k$ and $s_{j-1} + s_j \geq m - k$. □

Theorem 2.4.

$$\gamma_s(C_8 \times C_n) = \begin{cases} 3n & : n \equiv 0 \pmod{16}, \\ 3n+1 : n \equiv 3,13 \pmod{16}, \\ 3n+2 : n \equiv 6,10 \pmod{16}, \\ 3n+3 : n \equiv 5,7,9,11 \pmod{16}, \end{cases}$$

$$3n+2 \leq \gamma_s(C_8 \times C_n) \leq 3n+4 : n \equiv 2,4,8,12,14 \pmod{16},$$

$$3n+3 \leq \gamma_s(C_8 \times C_n) \leq 3n+5 : n \equiv 1,15 \pmod{16},$$

Proof. We define a signed dominating function f as follows:

$f((i,2j-1)) = f((i+2,2j-1)) = f((i+5,2j-1)) = -1$ for and, $i \equiv (7j-6) \pmod 8$ for $1 \leq j \leq \lfloor n/2 \rfloor$ and $i \equiv (7j-3) \pmod 8$, and $f((i,j)) = 1$ otherwise. Also we define $f_n((i,n)) = 1$ for $i = 1, \cdots, 8$.

By the definition of f, we have $s_j = 2$ for j is odd and $s_j = 4$ for j is even. Notice, f is a SDF for $C_8 \times C_n$ when $n \equiv 0 \pmod{16}$. Therefore, there is a problem with the vertices of K_1 when$n \equiv 1, \ldots, 15 \pmod{16}$.

Now, let us define the following functions:

$$f_1((i,j)) = \begin{cases} f((i,j)) & \text{if } j \neq n, \\ +1 & \text{if } i = 1,2,3,4,5,6,7,8,\ j = n. \end{cases}, \quad f_2((i,j)) = \begin{cases} f((i,j)) & \text{if } j \neq n. \\ -1 & \text{if } i = 5,8,\ j = n, \\ +1 & \text{if } i = 1,2,3,4,6,7,\ j = n, \end{cases}$$

$$f_3((i,j)) = \begin{cases} f((i,j)) & \text{if } j \neq n, \\ -1 & \text{if } i = 5,\ j = n, \\ +1 & \text{if } i = 1,2,3,4,6,7,8,\ j = n. \end{cases}, \quad f_4((i,j)) = \begin{cases} f((i,j)) & \text{if } j \neq n. \\ -1 & \text{if } i = 8,\ j = n, \\ +1 & \text{if } i = 1,2,3,4,5,6,7,\ j = n, \end{cases}$$

We note:

f_1 is a SDF of $C_8 \times C_n$ when $n \equiv 1,2,4,8,12,14,15 \pmod{16}$.

f_2 is a SDF of $C_8 \times C_n$ when $n \equiv 3,13 \pmod{16}$.

f_3 is a SDF of $C_8 \times C_n$ when $n \equiv 6,9,11 \pmod{16}$.

f_4 is a SDF of $C_8 \times C_n$ when n $\equiv 5,7,10 \pmod{16}$.

Hence,

$$\gamma_s(C_8 \times C_n) \le 3n : n \equiv 0 \pmod{16}$$
$$\gamma_s(C_8 \times C_n) \le 3n+1 : n \equiv 3,13 \pmod{16}$$
$$\gamma_s(C_8 \times C_n) \le 3n+2 : n \equiv 6,10 \pmod{16}$$
$$\gamma_s(C_8 \times C_n) \le 3n+3 : n \equiv 5,7,9,11 \pmod{16}$$
$$\gamma_s(C_8 \times C_n) \le 3n+4 : n \equiv 2,4,8,12,14 \pmod{16}$$
$$\gamma_s(C_8 \times C_n) \le 3n+5 : n \equiv 1,15 \pmod{16}$$

(1)

For example, f_1 is a SDF of $C_8 \times C_{12}$, where $\gamma_s(C_8 \times C_{12}) \le 40 = 3(12) + 4$, see Figure 1.

{Here, we must note that, for simplicity of drawing the Cartesian products of two directed cycles $C_m \times C_n$, we do not draw the arcs from vertices in last column to vertices in first column and the arcs from vertices in last row to vertices in first row. Also for each figure of $C_m \times C_n$, we replace it by a corresponding matrix by signs – and + which descriptions −1 and +1 on figure of $f(C_m \times C_n)$, respectively}.

By Remark 2.2, for any minimum signed dominating function f of $C_8 \times C_n$ with signed dominating sequence $(s_1,\ldots,s_n)$, we have $s_j = 0, 2, 4, 6$ or 8 for $j = 1,\ldots,n$. By Lemma 2.3, if $s_j = 0$ then $s_j - 1, s_j + 1 \ge 8$, and if $s_j = 2$ then $s_j - 1, s_j + 1 \ge 4$. This implies that

$$\omega(f) = \sum_{j=1}^{n} s_j \ge 3n \quad \text{for } n \equiv 0 \pmod{2}$$

(2)

$$\omega(f) = \sum_{j=1}^{n} s_j \ge 3n+1 \quad \text{for } n \equiv 1 \pmod{2}$$

(3)

Hence, by (1), (2) and (3) we get

$$\gamma_s(C_8 \times C_n) = 3n \quad \text{for } n \equiv 0 \pmod{16}.$$

$$\gamma_s(C_8 \times C_n) = 3n+1 \quad \text{for } n \equiv 3,13 \pmod{16}$$

Assume that $n \not\equiv 0,3,13 \pmod{16}$.

Let f' ba a signed dominating function with signed dominating sequence $\left(s'_1, s'_2, \cdots, s'_n\right)$.

If m, $n \leq 7$, then by Theorem 1.4 is the required (because $C_m \times C_n \cong C_n \times C_m$). Let $m, n \geq 8$. We prove the following claim:

Claim 2.1. For $k \geq 2$, we have $\sum_{d=j+1}^{j+k} s'_d \geq 3k$ if k is even and $\sum_{d=j+1}^{j+k} s'_d \geq 3k-1$ when k is odd.

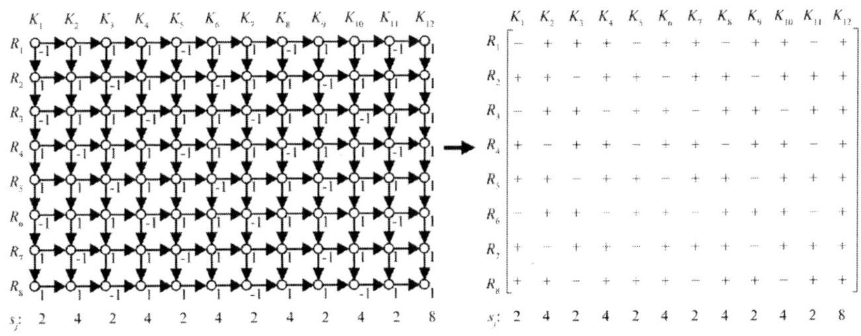

Figure 1. (a) A signed dominating function of $C_8 \times C_{12}$; (b) A corresponding matrix of a signed dominating function of $C_8 \times C_{12}$.

Proof of Claim 2.1. We have the subsequence $\left(s'_{j+1}, \cdots, s'_k\right)$ is including at least two terms. Then, immediately from Remark 2.2 and Lemma 2.3, gets the required. The proof of Claim 2.1 is complete. □

Now, if $s'_j = 0$ for some j, then $s'_{j-1} = s'_{j+1} = 8$. Without loss of generality, we can assume that $s'_2 = 0$. Then Claim 2.1, imply that

$$\omega(f') = \sum_{j=1}^{n} s'_j = \sum_{j=1}^{3} s'_j + \sum_{j=4}^{n} s'_j \geq 16 + 3(n-3) - 1 = 3n + 7 \quad (4)$$

Assume that $s'_j \geq 2$ for all $j = 1, \cdots, n$. We have three cases:

Case 1. If $s'_j = 8$ for some j. Let $s'_1 = 8$. Then from Claim 2.1, we get

$$\omega(f') = \sum_{j=1}^{n} s'_j = s'_1 + \sum_{j=2}^{n} s'_j \geq 8 + 3(n-1) - 1 = 3n+4, \text{ when } n \equiv 0 \pmod 2 \quad . (5)$$

$$\omega(f') = \sum_{j=1}^{n} s'_j = s'_1 + \sum_{j=2}^{n} s'_j \geq 8 + 3(n-1) = 3n+5, \text{ when } n \equiv 1 \pmod 2 \quad . (6)$$

Case 2. Let $2 \leq s'_j \leq 6$. If $(s'_1, \cdots, s'_n)$ include at least two terms which are equals 6, then

$$\omega(f') = \sum_{j=1}^{n} s'_j \geq 3n+4 \quad (7)$$

For $n \equiv 1 \pmod 2$, then 8n is even. By Lemma 1.2, $\gamma_s(C_8 \times C_n) = \omega(f')$ must be even number. Hence, from (7) is

$$\omega(f') = \sum_{j=1}^{n} s'_j \geq 3n+5 \quad (8)$$

Assume that $2 \leq s'_j \leq 4$ for all $j = 1, \cdots, n$ except once which equals 6. Thus,

$$\omega(f') = \sum_{j=1}^{n} s'_j \geq 3n+2 \quad \text{for } n \equiv 0 \pmod 2 \quad . (9)$$

$$\omega(f') = \sum_{j=1}^{n} s'_j \geq 3n+3 \quad \text{for } n \equiv 1 \pmod 2 \quad . (10)$$

For the case 3, we need the following claim:

Claim 2.2. Let f' be a minimum signed dominating function of $C_8 \times C_n$ with signed dominating sequence $(s'_1, s'_2, \cdots, s'_n)$. Then for $(s'_j, s'_{j+1}, s'_{j+1}, s'_{j+2}) = (2,4,2,4)$, and up to isomorphism, there is only one possible configuration for f'', it is shown in Figure 2. The prove is immediately by drawing. □

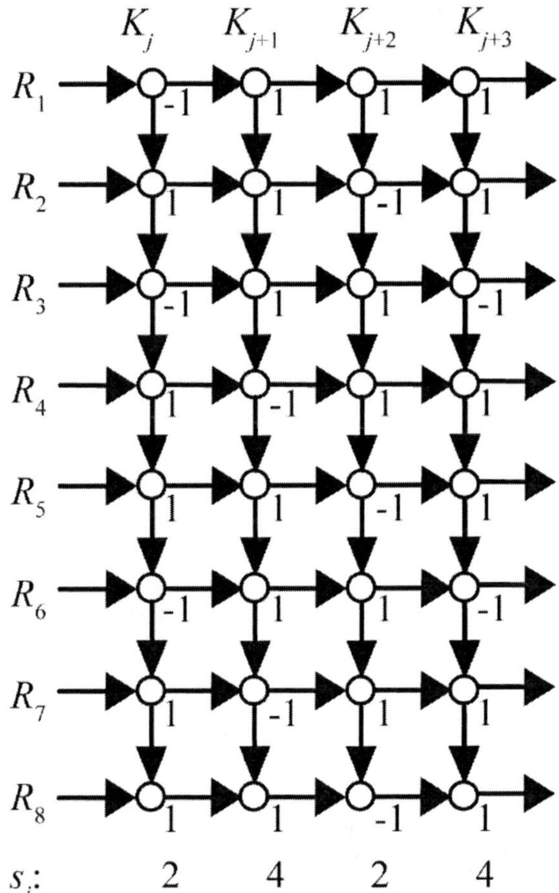

Figure 2. The form $(s'_j, s'_{j+1}, s'_{j+1}, s'_{j+2}) = (2,4,2,4)$.

Case 3. Let $2 \le s'_j \le 4$ for all $j = 1, \cdots, n$. We define

$X_i = \left| j : s'_j = i \right|, i = 2, 4$.

Then we have

$$X_2 + X_4 = n.$$

$$\omega(f') = 2X_2 + 4X_4$$

Since the case $(s'_j, s'_{j+1}) = (2,2)$ is not possible, we have $X_4 \geq X_2$.

If $X_4 \geq \lceil n/2 \rceil + 2$. Then $\omega(f') \geq 2(n - \lceil n/2 \rceil - 2) + 4(\lceil n/2 \rceil + 2) = 2n + 2\lceil n/2 \rceil + 4$. Thus

$$\omega(f') = \sum_{j=1}^{n} s'_j \geq 3n + 4 \quad \text{for } n \equiv 0 \pmod 2 \qquad (11)$$

$$\omega(f') = \sum_{j=1}^{n} s'_j \geq 3n + 5 \quad \text{for } n \equiv 1 \pmod 2 \qquad (12)$$

If $X_4 = \lceil n/2 \rceil + 1$. Then $\omega(f') \geq 2(n - \lceil n/2 \rceil - 1) + 4(\lceil n/2 \rceil + 1) = 2n + 2\lceil n/2 \rceil + 2$. Hence

$$\omega(f') = \sum_{j=1}^{n} s'_j \geq 3n + 2 \quad \text{for } n \equiv 0 \pmod 2 \qquad (13)$$

$$\omega(f') = \sum_{j=1}^{n} s'_j \geq 3n + 3 \quad \text{for } n \equiv 1 \pmod 2 \qquad (14)$$

Let $X_4 = \lceil n/2 \rceil$ and $X_2 = \lfloor n/2 \rfloor$.

Then we have one possible is as the form $(s'_1, s'_2, \cdots, s'_n) = (2,4,2,4,\cdots,2,4,\cdots)$. This implies that $\omega(f')$ = $3n$ for $n \equiv 0$(mod 2) and $\omega(f')$ = $3n$ +1 for $n \equiv$ 1(mod 20). By Claim 2.2, we have f' is as the function f, which defined in forefront of Theorem 2.4. However, f is not be a signed dominating function for $C_8 \times C_n$ when $n \not\equiv 0,3,13 \pmod{16}$. Thus

$$\gamma_s(C_8 \times C_n) > 3n \quad \text{for } n \equiv 0 \pmod 2.$$

$$\gamma_s(C_8 \times C_n) > 3n + 1 \quad \text{for } n \equiv 1 \pmod 2.$$

By Lemma 1.2, and above arguments, we conclude that

$$\gamma_s(C_8 \times C_n) \geq 3n + 2 \quad \text{for } n \equiv 0 \pmod 2 \qquad (15)$$

$$\gamma_s(C_8 \times C_n) \geq 3n + 3 \quad \text{for } n \equiv 1 \pmod 2 \qquad (16)$$

Hence, from (1), (15) and (16), deduce that

$$\gamma_s(C_8 \times C_n) \geq 3n+2 \quad \text{for } n \equiv 6,10 \pmod{16}.$$

$$\gamma_s(C_8 \times C_n) \geq 3n+3 \quad \text{for } n \equiv 5,7,9,11 \pmod{16}.$$

$$3n+2 \leq \gamma_s(C_8 \times C_n) \leq 3n+4 \quad \text{for } n \equiv 2,4,8,12,14 \pmod{16}.$$

$$3n+3 \leq \gamma_s(C_8 \times C_n) \leq 3n+5 \quad \text{for } n \equiv 1,15 \pmod{16}.$$

Finally, we result that:

$$\gamma_s(C_8 \times C_n) = 3n \quad \text{for } n \equiv 0 \pmod{16}.$$

$$\gamma_s(C_8 \times C_n) = 3n+1 \quad \text{for } n \equiv 3,13 \pmod{16}.$$

$$\gamma_s(C_8 \times C_n) = 3n+2 \quad \text{for } n \equiv 6,10 \pmod{16}.$$

$$\gamma_s(C_8 \times C_n) = 3n+3 \quad \text{for } n \equiv 5,7,9,11 \pmod{16}.$$

$$3n+2 \leq \gamma_s(C_8 \times C_n) \leq 3n+4 \quad \text{for } n \equiv 2,4,8,12,14 \pmod{16}.$$

$$3n+3 \leq \gamma_s(C_8 \times C_n) \leq 3n+5 \quad \text{for } n \equiv 1,15 \pmod{16}.$$

□

Theorem 2.5.

$$\gamma_s(C_9 \times C_n) = \begin{cases} 3n & : n \equiv 0 \pmod{3}, \\ 3n+6 & : n \equiv 1,2 \pmod{3}. \end{cases}$$

Proof. We define a signed dominating function f as follows: $f((i,j)) = f((i+3,j)) = f((i+6,j)) = -1$ for $1 \leq j \leq n$ and $i = j \pmod 9$, and $f((i,j)) = 1$ otherwise. Also, let us define the following function:

$$f_1((i,j)) = \begin{cases} f((i,j)) & \text{if } j \neq n, \\ +1 & \text{if } i = 1,2,3,4,5,6,7,8,9,\ j = n. \end{cases}$$

By define f, we have $s_j = 3$ for $1 \le j \le n$. Notice, f is a SDF for $C_9 \times C_n$ for $n \equiv 0 (\text{mod } 3)$. And f_1 is a SDF of $C_9 \times C_n$ for . For an illustration γ $(C_9 \times C_6)$, see Figure 3. Hence,

$$\gamma_s\left(C_9 \times C_n\right) \le 3n \quad \text{for } n \equiv 0(\bmod 3). \tag{17}$$

$$\gamma_s\left(C_9 \times C_n\right) \le 3n + 6 \quad \text{for } n \equiv 1,2(\bmod 3). \tag{18}$$

From Corollary 1.3 is γ_s $(C_9 \times C_n) \ge 3n$. Then by (17), $\gamma_s(C_9 \times C_n)$ for $n \equiv 0(\text{mod } 3)$.

For $n \equiv 1,2(\text{mod } 3)$.

If $4 \le n \le 8$, then by Theorems 1.4 and 2.4, gets the required. Assume that $n \ge 9$.

By Remark 2.2, we have s_j = 1, 3, 5, 7 or 9. By Lemma 2.3, if $s_j = 1$ then $s_{j-1}, s_{j+1} \ge 7$, $s_j = 3$, s_j = 3 then $s_{j-1}, s_{j+1} \ge 3$ and s_j = 5 then $s_{j-1}, s_{j+1} \ge 3$ (because if $s_{j-1}, s_{j+1} < 3$, then we need $s_j \ge 7$). By Lemma 2.3, the cases $\left(s_{j+1}, s_{j+2}\right) = (1,3),(3,1)$ are not possible. Hence, $\sum_{d=j+1}^{j+k} s_d \ge 3k$, for $k \ge 2$, for k ≥ 2. This implies that,

$$\sum_{d=1}^{n-1} s_d \ge 3(n-1) \tag{19}$$

We define

$$X_i = \left|\left\{j : s_j = i\right\}\right|, i = 1,3,5,7,9$$

Then we have

$$\begin{array}{c|cccccc|} & K_1 & K_2 & K_3 & K_4 & K_5 & K_6 \\ R_1 & - & + & + & - & + & + \\ R_2 & + & - & + & + & - & + \\ R_3 & + & + & - & + & + & - \\ R_4 & - & + & + & - & + & + \\ R_5 & + & - & + & + & - & + \\ R_6 & + & + & - & + & + & - \\ R_7 & - & + & + & - & + & + \\ R_8 & + & - & + & + & - & + \\ R_9 & + & + & - & + & + & - \\ s_j: & 3 & 3 & 3 & 3 & 3 & 3 \end{array}$$

Figure 3. A corresponding matrix of a signed dominating function of $C_9 \times C_6$.

$$X_1 + X_3 + X_5 + X_7 + X_9 = n.$$

$$\omega(f) = X_1 + 3X_3 + 5X_5 + 7X_7 + 9X_9.$$

If we have one case from the cases $X_9 \geq 1$, $X_7 \geq 2$, $X_5 + X_7 \geq 2$ or $X_5 \geq 3$. Then by (19) is $\omega(f) \geq 3n + 16$.

Assume the contrary, i.e., ($X_9 = 0, X_7 < 2, X_5 + X_7 < 2$ and $X_5 < 3$).

Hence, $\omega(f)$ $X_1 + 3X_3 + 5X_5 + 7X_7$. We consider the cases $X_7 < 2$ and $X_5 < 3$, which are including the remained cases, i.e., $X_7 = 1$ and $X_5 = 2$. First, we give the following Claim:

Claim 2.3. There is only one possible for $(s_j, s_{j+1}) = (3,3)$ is

$f((i,j)) = f((i+3,j)) = f((i+6,j)) = f((i+1,j+1)) = f((i+4,j+1)) = f((i+7,j+1)) = -1$ and $f((i,j)) = f((i,j+1)) = 1$, otherwise for $1 \le i \le 9$.

The proof comes immediately by the drawing. □

Case 1. $X_7 = 1$ and $X_5 = X_9 = 0$. Without loss of generality, we can assume s_n = 7. Then we have the form (3,3,...3,7). By Claim 2.3, for $j < n - 1$, each column K_{j+1} is 1-shift of K_j, K_{j+2} is 2-shift of K_j and K_{j+1} is 3-shift = (0-shift) of K_j. Without loss of generality, we can assume $f((1,1)) = f((4,1)) = f((7,1)) = -1$ and $f((i, 1))$ otherwise. We consider two subcases:

Subcase 1.1. For $n \equiv$ (mod 3). Then K_{n-1} is $(n - 2)$-shift = (2-shift) of K_1. This implies that $f((3,n-1)) = f((6,n-1)) = f((9,n-1)) = -1$. Hence, we need $f((i,n)) = 1$ for all $i = 1,...,9$. This is a contradiction with $\omega(f(K_n)) = 7$ Thus, $\omega(f) \ge 3X_3 + 9X_9 = 3(n-1) + 9 = 3n + 6$.

Subcase 1.2. For n $\equiv$ 2(mod 3). Then K_{n-1} is $(n - 2)$-shift = (0-shift) of K_1. This implies that $f((1,n-1)) = f((4,n-1)) = f((7,n-1)) = -1$. So, we need $f((i,n))$ = 1 for all $i = 1....,9$. Again, we get a contradiction with $\omega(k(K_n)) = 7$. Thus, $\omega(f) \ge 3X_3 + 9X_9 = 3(n - 1) + 9 = 3n + 6$.

Case 2. $X_5 = 2$ and $X_7 = X_9 = 0$. Here we have $s_k = s_{k+d} = 5$ and $s_j = 3$ otherwise. By the same argument similar to the Case 1, we have K_j is $(j - 1)$-shift of K_1. Thus, if $j \equiv 1$(mod 3), then $f((1,j)) = f((4,j)) = f((7,j))$ and for $j \equiv$ 2(mod 3) is $f((2,j)) = f((5,j)) = f((8,j)) = -1$. Also, for position the vertices of K_1, we always have $f((1,n)) = f((2,n)) = f((4,n)) = f((5,n)) = f((7,n)) = f((8,n)) = 1$. We consider four Subcases:

Subcase 2.1. $d = 1$, without loss of generality, we can assume $s_{n-1} = s_n = 5$.

For $n \equiv 3$(mod 3), $f((2n - 1)) = f((5,n - 1)) = f((7, n - 1)) = f((8, n - 1)) = 1$. Then

$f((1,n-1)) = f((2,n-1)) = f((4,n-1)) = f((5,n-1)) = f((7,n-1)) = f((8,n-1)) = 1$.

The three remaining vertices from each K_{n-1} and K_n, most including two values −1, and this is impossible. The same arguments is for n $\equiv$ 2(mod 3).

Subcase 2.2. d = 2, let $s_{n-2} = s_n = 5$. Then we have the form $(s_1, s_2, \cdots, s_n) = (3,3,\cdots,3,5,3,5)$.

If $n \equiv 1(\text{mod } 3)$, then $n - 3 \equiv 1(\text{mod } 3)$. This implies that K_{n-1} is 0-shift of K_1. Therefore, $f((1,n-3)) = f((4,n-3)) = f((7,n-3)) = -1$. Hence, the three columns K_{n-2}, K_{n-1}, K_n must be including seven values of −1, two in K_{n-2}, three in K_{n-1} and two in K_n and this impossible. The same argument is for $n \equiv 2(\text{mod } 3)$.

Subcase 2.3. $d = 3$, let $s_n - 3 = s_n = 5$. We have the form $(s_1,s_2,\ldots,s_n) = (3,3,\ldots,3,5,3,3,5)$. Then for $n \equiv 1(\text{mod } 3)$, K_{n-4} is 2-shift of K_1. Therefore $f((3,n-4)) = f((6,n-4)) = f((9,n-4)) = -1$. Also, $s_{n-2} = s_{n-1} = 3$. Therefore, two vertices of $\{(1,n-3), (4,n-3), (7,n-3)\}$ must has value −1. By symmetry, let $f((1,n-3)) = f((4,n-3)) = -1$. Then by Claim 2.3, there is one case for (s_{n-2}, s_{n-1}). Hence, $f((2,n-2)) = f((5,n-2)) = f((8,n-2)) = f((3,n-1)) = f((6,n-1)) = f((9,n-1)) = -1$. Therefore, we need two vertices from K_n with value −1. This is a contradiction, (because the vertices of the first column must be a signed dominates by the vertices of the last column). The same argument is for $n \equiv 2(\text{mod } 3)$.

Subcase 2.4. $d \geq 4$, let $s_{n-d} = s_n = 5$ (by symmetry is $s_1,s_2,\ldots,s_n$).

We have the form $(s_1, s_2, \cdots, s_n) = (3,3,\cdots,3,5,3,\cdots,3,5)$. By Claim 2.3, if $(s_j, s_{j+1}, \cdots) = (3,3,\cdots)$ then for each two vertices $f((i,j)) = f((q,j)) = -1$ we must have $|i - q| = 3$ and so for $K_{j+1},\ldots,K_{n-d-1}$. Since $s_j = 3\ (j \leq n-d-1)$ and $s_{n-d} = 5$, then K_{n-d} including two vertices with value −1 by 1-shift of two vertices in K_{n-d-1}. Also, K_{n-d+1} including two vertices with value −1 by 1-shift of vertices in K_{n-d} and the third ver- tex must be distance 3 from any one has value −1 (Since $s_{n-d+1} = \ldots = 3$, Claim 2.3). Thus, the order of the values −1 or 1 of the vertices $K_{n-d+1},\ldots, K_{n-1}$ does not change. Hence the vertices of K_{n-1} has the same order of K_{n-1} when we have the signed dominating sequence (3,3,...,3,3) and this impossible is signed dominating sequence of $C_9 \times C_n$ for $n \equiv 1,2(\text{mod } 3)$. In Subcases 2.1, 2.2, 2.3 and 2.4 there are many details, we will be omitted it.

Finally, we deduce that does not exist a signed dominating function f of $C_9 \times C_n$ for $n \equiv 1,2(\text{mod } 3)$ with $\omega(f) \leq 3n+4$. Hence,

$$\gamma_s(C_9 \times C_n) \geq 3n + 6 : n \equiv 1,2(\text{mod}\, 3) \tag{20}$$

From (18) and (20) is $\gamma_s(C_9 \times C_n) = 3n + 6 : n \equiv 1,2(\text{mod}\, 3)$. □

Theorem 2.6. $\gamma_s(C_9 \times C_n) = 3n + 6 : n \equiv 1,2(\text{mod } 3)$.

Proof. We define a signed dominating function f as follows:

$f((i,j)) = f((i+3,j)) = f((i+6,j)) = -1$ for $1 \le j \le n$ and $i \equiv j \pmod{10}$, and $f((i,j)) = 1$ otherwise. Also, we define

$$f_{n-7}((3,n-7)) = f_{n-7}((7,n-7)) = f_{n-7}((10,n-7)) = -1,$$

$$f_{n-6}((1,n-6)) = f_{n-6}((5,n-6)) = f_{n-6}((8,n-6)) = -1,$$

$$f_{n-5}((3,n-5)) = f_{n-5}((6,n-5)) = f_{n-5}((9,n-5)) = -1,$$

$$f_{n-4}((1,n-4)) = f_{n-4}((4,n-4)) = f_{n-4}((7,n-4)) = -1,$$

$$f_{n-3}((2,n-3)) = f_{n-3}((5,n-3)) = f_{n-3}((9,n-3)) = -1,$$

$$f_{n-2}((3,n-2)) = f_{n-2}((7,n-2)) = f_{n-2}((10,n-2)) = -1,$$

$$f_{n-1}((1,n-1)) = f_{n-1}((5,n-1)) = f_{n-1}((8,n-1)) = -1,$$

$$f_n((3,n)) = f_n((6,n)) = f_n((9,n)) = -1,$$

and

$f_j((i,j)) = 1$ otherwise for $j = n-5, n-4, n-3, n-2, n-1, n$.

By define f and $f_{n-7}, f_{n-6}, f_{n-5}, f_{n-4}, f_{n-3}, f_{n-2}, f_{n-1}, f_n$ we have $s_j = 4$ for all $1 \le j \le n$. Notice that: f is a SDF for $C_{10} \times C_n$ when n $\equiv$ 0,3,(mod 10).

$$\{f \setminus \{f(K_{n-5}) \cup f(K_{n-4}) \cup f(K_{n-3}) \cup f(K_{n-2}) \cup f(K_{n-1}) \cup f(K_n)\}\}$$
$$\cup \{f_{n-5} \cup f_{n-4} \cup f_{n-3} \cup f_{n-2} \cup f_{n-1} \cup f_n\}$$

is a SDF for $C_{10} \times C_n$ when $n \equiv 1 \pmod{10}$.

$$\{f \setminus \{f(K_{n-2}) \cup f(K_{n-1}) \cup f(K_n)\}\} \cup \{f_{n-2} \cup f_{n-1} \cup f_n\}$$

is a SDF for $C_{10} \times C_n$ when $n \equiv 2 \pmod{10}$.

$$\{f \setminus \{f(K_{n-6}) \cup f(K_{n-5}) \cup f(K_{n-4}) \cup f(K_{n-3}) \cup f(K_{n-2}) \cup f(K_{n-1}) \cup f(K_n)\}\}$$
$$\cup \{f_{n-6} \cup f_{n-5} \cup f_{n-4} \cup f_{n-3} \cup f_{n-2} \cup f_{n-1} \cup f_n\}$$

is a SDF for $C_{10} \times C_n$ when $n \equiv 4 (\text{mod } 10)$.

$$\{f \setminus \{f(K_{n-3}) \cup f(K_{n-2}) \cup f(K_{n-1}) \cup f(K_n)\}\} \cup \{f_{n-3} \cup f_{n-2} \cup f_{n-1} \cup f_n\}$$

is a SDF for $C_{10} \times C_n$ when $n \equiv 5 (\text{mod } 10)$.

$$\{f \setminus \{f(K_n)\}\} \cup \{f_n\}$$

is a SDF for $C_{10} \times C_n$ when $n \equiv 6 (\text{mod } 10)$.

$$\{f \setminus \{f(K_{n-7}) \cup f(K_{n-6}) \cup f(K_{n-5}) \cup f(K_{n-4}) \cup f(K_{n-3}) \cup f(K_{n-2}) \cup f(K_{n-1}) \cup f(K_n)\}\}$$
$$\cup \{f_{n-7} \cup f_{n-6} \cup f_{n-5} \cup f_{n-4} \cup f_{n-3} \cup f_{n-2} \cup f_{n-1} \cup f_n\}$$

is a SDF for $C_{10} \times C_n$ when n $\equiv 7 (\text{mod } 10)$.

$$\{f \setminus \{f(K_{n-4}) \cup f(K_{n-3}) \cup f(K_{n-2}) \cup f(K_{n-1}) \cup f(K_n)\}\} \cup \{f_{n-4} \cup f_{n-3} \cup f_{n-2} \cup f_{n-1} \cup f_n\}$$

is a SDF for $C_{10} \times C_n$ when $n \equiv 8 (\text{mod } 10)$.

$$\{f \setminus \{f(K_{n-1}) \cup f(K_n)\}\} \cup \{f_{n-1} \cup f_n\}$$

is a SDF for $C_{10} \times C_n$ when $n \equiv 9 (\text{mod } 10)$.

For an illustration $\gamma_s(C_{10} \times C_{11})$ see Figure 4, (here for $n \equiv 1 (\text{mod } 10)$, we are changing the functions of the columns: $K_{n-5}, K_{n-4}, K_{n-3}, K_{n-2}, K_{n-1}, K_n$). In all the cases we have $\gamma_s(C_{10} \times C_n) \leq 4n$.

By Remark 2.2, we have $s_j = 0, 2, 4, 6, 8$ or 10. Also by Lemma 2.3, if $s_j = 0$, then $s_{j-1}, s_{j+1} \geq 10$ and when $s_j = 2$, is and $s_j = 4$ is $s_{j-1}, s_{j+1} \geq 4$ (because if $s_{j-1} = 2$ or $s_{j+1} = 2$, then $s_j \geq 6$). This implies that

$$\gamma_s(C_{10} \times C_n) = \sum_{j=1}^{n} s_j \geq 4n$$

So, we get $\gamma_s(C_{10} \times C_n) = 4n$. □

Corollary 2.7. For m $\equiv 0 (\text{mod } 3)$, we have

$$\gamma_s\left(C_m \times C_n\right)=\frac{mn}{3} \quad \text{if } n \equiv 0 \pmod 3.$$

$$\frac{mn}{3} \le \gamma_s\left(C_m \times C_n\right)=\frac{mn}{3}+\frac{2m}{3} \quad \text{if } n \equiv 1,2 \pmod 3.$$

Proof. By Corollary 1.3 we have

$$\gamma_s\left(C_m \times C_n\right) \ge \frac{mn}{3} \tag{21}$$

	K_1	K_2	K_3	K_4	K_5	K_6	K_7	K_8	K_9	K_{10}	K_{11}
R_1	−	+	+	+	−	+	−	+	+	−	+
R_2	+	−	+	+	+	+	+	−	+	+	+
R_3	+	+	−	+	+	−	+	+	−	+	−
R_4	−	+	+	−	+	+	−	+	+	+	+
R_5	+	−	+	+	−	+	+	−	+	−	+
R_6	+	+	−	+	+	−	+	+	+	+	−
R_7	−	+	+	−	+	+	−	+	−	+	+
R_8	+	−	+	+	−	+	+	+	+	−	+
R_9	+	+	−	+	+	−	+	−	+	+	−
R_{10}	+	+	+	−	+	+	+	+	−	+	+
s_j:	4	4	4	4	4	4	4	4	4	4	4
						f_{n-5}	f_{n-4}	f_{n-3}	f_{n-2}	f_{n-1}	f_n

Figure 4. A corresponding matrix of a signed dominating function of $C_{10} \times C_{11}$.

Let us a signed dominating function f as follows: $f((3i-2,3j-2))=-1$ for $1 \le i \le m/3$, $1 \le j \le n/3$, $f((3i-1,3j-1))=-1$ for $1 \le i \le m/3$, $1 \le j \le n/3$ and $f((3i,3j))=-1$ for $1 \le i \le m/3$, $1 \le j \le n/3$.

By define f, we have s_j = m/3 for $1 \le j \le n$. Notice, f is a SDF for $C_m \times C_n$ for $m,n \equiv 0(\text{mod } 3)$. Hence, $n \equiv 1,2(\text{mod } 3)$. Then from (21), is $\gamma_s(C_m \times C_n) = mn/3$ for $m,n \equiv 0(\text{mod } 3)$.

For $n \equiv 1, 2(\text{mod } 3)$.

Let $f_n((i,n))$ = 1 for $1 \le i \le m$. Notice, $\{f \setminus \{f(K_n)\} \cup \{f_n\}\}$ is a SDF for $C_m \times C_n$ for $n \equiv 1,2(\text{mod } 3)$.

Thus, $\gamma_s(C_m \times C_n) \le m(n-1)/3 + m = mn/3 + 2m/3$. Hence, by (21) is $mn/3 \le \gamma_s(C_m \times C_n) \le mn/3 + 2m/3$ if $n \equiv 1,2(\text{mod } 3)$. □

3. CONCLUSIONS

This paper determined that exact value of the signed domination number of $C_m \times C_n$ for m = 8, 9, 10 and arbitrary n. By using same technique methods, our hope eventually lead to determination $\gamma_s(C_m \times C_n)$ for general m and n.

Based on the above (Lemma 2.3 and Theorems 1.4, 2.4, 2.5 and 2.6), also by the technique which used in this paper, we again rewritten the following conjecture (This conjecture was mention in [11]):

Conjecture 3.1.

$$\gamma_s(C_m \times C_n) = \left\lceil \frac{m}{3} \right\rceil n \quad \text{when } n \equiv 0(\text{mod } 2m) \text{ or } n \equiv 1(\text{mod } 3)$$

REFERENCES

1. Haas, R. and Wexler, T.B. (1999) Bounds on the Signed Domination Number of a Graph. Discrete Mathematics, 195, 295-298. http://dx.doi.org/10.1016/S0012-365X(98)00189-7
2. West, D.B. (2000) Introduction to Graph Theory. Prentice Hall, Inc., Upper Saddle River.
3. Dunbar, J.E., Hedetniemi, S.T., Henning, M.A. and Slater, P.J. (1995) Signed Domination in Graphs, Graph Theory, Combinatorics and Application. John Wiley & Sons, Inc., Hoboken, 311-322.

4. Cockayne, E.J. and Mynhart, C.M. (1996) On a Generalization of Signed Domination Functions of Graphs. Ars Combinatoria, 43, 235-245.
5. Hattingh, J.H. and Ungerer, E. (1998) The Signed and Minus k-Subdomination Numbers of Comets. Discrete Mathematics, 183, 141-152. http://dx.doi.org /10. 1016/S0012-365X(97)00051-4
6. Xu, B. (2001) On Signed Edge Domination Numbers of Graphs. Discrete Mathematics, 239, 179-189. http://dx.doi.org/10.1016/S0012-365X(01)00044-9
7. Broere, I., Hattingh, J.H., Henning, M.A. and McRae, A.A. (1995) Majority Domination in Graphs. Discrete Mathematics, 138, 125-135. http://dx.doi.org/10.1016/0012-365X(94)00194-N
8. Zelinka, B. (2005) Signed Domination Numbers of Directed Graphs. Czechoslovak Mathematical Journal, 55, 479-482. http://dx.doi.org/10.1007/s10587-005-0038-5
9. Karami, H., Sheikholeslami, S.M. and Khodkar, A. (2009) Lower Bounds on the Signed Domination Numbers of Directed Graphs. Discrete Mathematics, 309, 2567-2570. http://dx.doi.org/10.1016/j.disc.2008.04.001
10. Atapour, M., Sheikholeslami, S., Hajypory, R. and Volkmann, L. (2010) The Signed k-Domination Number of Directed Graphs. Central European Journal of Mathematics, 8, 1048-1057. http://dx.doi.org/10.2478/s11533-010-0077-5
11. Shaheen, R. and Salim, H. (2015) The Signed Domination Number of Cartesian Products of Directed Cycles. Submitted to Utilitas Mathematica.

CHAPTER 14

Signed Tilings by Ribbon L n-Ominoes, n Even, via Gröbner Bases

Kenneth Gill, Viorel Nitica

Department of Mathematics, West Chester University, West Chester, USA

ABSTRACT

Let T_n be the set of ribbon L-shaped n-ominoes for some $n \geq 4$ even, and let T_n^+ be T_n with an

extra 2× 2 square. We investigate signed tilings of rectangles by T_n and T_n^+. We show that a rectangle has a signed tiling by T_n if and only if both sides of the rectangle are even and one of them is divisible by n, or if one of the sides is odd and the other side is divisible by $n\left(\frac{n}{2}-2\right)$. We also show that a rectangle has a signed tiling by $T_n^+, n \geq 6$ even, if and only if both sides of the rectangle are even, or if one of the sides is odd and the other side is divisible by $n\left(\frac{n}{2}-2\right)$.Our proofs are based on the exhibition of explicit Gröbner bases for the ideals generated by polynomials associated to the tiling sets. In particular, we show that some of the regular tiling results in *Nitica, V.* (2015) *Every tiling of the first quadrant by ribbon L n-ominoes follows the rectangular pattern. Open Journal of Discrete Mathematics*, 5, 11-25, cannot be obtained from coloring invariants.

Keywords: Polyomino, Replicating Tile, *L*-Shaped Polyomino, Skewed *L*-Shaped Polyomino, Signed Tilings, Gröbner Basis, Tiling Rectangles, Coloring Invariants

1. INTRODUCTION

In this article, we study tiling problems for regions in a square lattice by certain symmetries of an *L*-shaped polyomino. Polyominoes were introduced by Golomb in [1] and the standard reference about this subject is the book *Polyominoes* [2]. The *L*-shaped polyomino we study is placed in a square lattice and is made out of n, $n \geq 4$, unit squares, or *cells* (see **Figure 1(a)**). In a $a \times b$ rectangle, a is the height and b is the base. We consider translations (only!) of the tiles shown in **Figure 1(b)**. They are ribbon *L*-shaped n-ominoes.

A *ribbon polyomino* [3] is a simply connected polyomino with no two unit squares lying along a line parallel to the first bisector $y = x$. We denote the set of tiles T_n.

Related papers are [4] and [5], investigating tilings by T_n, n even. In [4], we look at tilings by T_n in the particular case $n = 4$. The starting point was a problem from recreational mathematics. We recall that a replicating tile is one that can make larger copies of itself. The order of replication is the number of initial tiles that fit in the larger copy. Replicating tiles were introduced by Golomb in [6]. In [7], we study replication of higher orders for several tiles introduced in [6]. In particular, we suggested that the skewed *L*-tetromino showed in **Figure 2(a)** was not replicating of order k^2 for any odd k. The question is equivalent to that of tiling a k-inflated copy of the straight *L*-tetromino using only the ribbon orientations of an *L*-tetromino. The question is solved in [4], where it is shown that *L*-tetromino is not replicating of any odd order. This is a consequence of a stronger result: a tiling of the first quadrant by T_4 always follows the rectangular pattern, that is, the tiling reduces to a tiling by 4×2 and 2×4 rectangles, each tiled in turn by two tiles from T_4.

The results in [4] are generalized in [5] to T_n, n even. The main result shows that any tiling of the first quadrant by T_n reduces to a tiling by $2\times n$ and $n \times 2$ rectangles. An application is the characterization of all rectangles that can be tiled by T_n, n even: a rectangle can be tiled by T_n, n even, if and only if both sides are even and at least one side is divisible by n. The rectangular pattern persists if one adds an extra 2×2 tile to T_n, n even. The new tiling set is denoted $\boldsymbol{T}_n^+$. A rectangle can be tiled by $\boldsymbol{T}_n^+$ if and only if it has both sides even. The main result also implies that a skewed *L*-shaped n-omino, n even, (see **Figure 2(b)**) is not a replicating tile of order k^2 for any odd k. This development shows that the limitation of the orientations of the tiles can be of interest, in particular when investigating tiling problems in a skewed lattice.

Signed tilings (see [8]) are also of interest. These are finite placements of tiles on a plane, with weights +1 or −1 assigned to each of the tiles. We say that they tile a region R if the sum of the weights of the tiles is 1 for every cell inside R and 0 for every cell elsewhere. The existence of a regular tiling clearly implies the existence of a signed tiling. Many times solving a tiling problem can be reduced to a coloring argument. It was shown in [8] that the most general argument of this type is equivalent to the existence of a signed tiling. Consequently, different conditions for regular versus signed tilings can be used to show that certain tiling arguments are

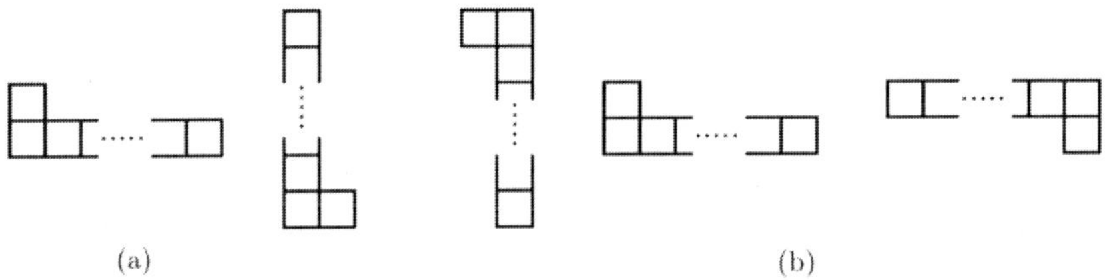

Figure 1. (a) An Ln-omino with n-cells. (b) The set of tiles T_n.

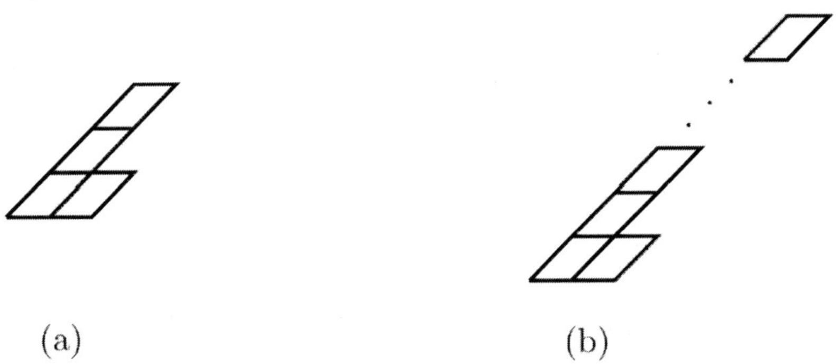

Figure 2. (a) Skewed L-tetromino. (b) Skewed L n-omino.

stronger then coloring arguments. By looking at signed tilings of rectangles by T_n and T_n^+, n even, we show that some of the results in [5] cannot be obtained via coloring arguments.

A useful tool in the study of signed tilings is a Gröbner basis associated to the polynomial ideal generated by the tiling set. See Bodini and Nouvel [9]. One can associate to any cell in the square lattice a monomial in the variable x,y. If the coordinates of the lower left corner of the cell are (a, b), one associates $x^a y^b$. This correspondence associates to any bounded tile a Laurent polynomial with all coefficients 1. The polynomial associated to a tile P is denoted f_P. The polynomial associated to a tile translated by an integer vector (c, d) is the initial polynomial multiplied by the monomial x^c

y^d. If the region we tile is bounded and the tile set consists of bounded tiles, then the problem can be translated in the first quadrant via a translation by an integer vector, and one can work only with regular polynomials in x, y. See Theorem 10 below.

Signed tilings by ribbon L n-ominoes, n odd are studied in [10], where we show that a rectangle can be signed tiled by ribbon L n-ominoes, n odd, if and only if it has a side divisible by n.

The main results of the paper are the following:

Theorem 1. *A rectangle can be signed tiled by T_n, $n \geq 6$, even, if and only if both sides of the rectangle are even and one of them is divisible by n, or one of the sides is odd and the other is divisible by $n(n/2-2)$.*

Theorem 1 is proved in Section 5, after finding a Gröbner basis for T_n, n even, in Section 3. A summary of Gröbner basis theory is shown in Section 2. Theorem 1 shows that some tiling results for T_n, $n \geq 6$ even, in [10] cannot be found via coloring arguments. We recall that it is shown in [4] that a rectangle is signed tiled by T_4 if and only if the sides are even and one side is divisible by 4.

Theorem 2. *A rectangle can be signed tiled by T_4^+ if and only if both sides are even. A rectangle can be signed tiled by $T_n^+, n \geq 6$ even, if and only if it has both sides even or one side is odd and the other side is divisible by $n(n/2-2)$.*

Theorem 2 is proved in Section 6, after finding a Gröbner basis for T_4^+ in Section 4. Theorem 2 shows that some tiling results for $T_n^+, n \geq 6$ even, in [5] cannot be found via coloring arguments.

Due to the Gröbner basis that we exhibit for T_n, $n \geq 6$ even, we also have:

Proposition 3. *A k-inflated copy of the ribbon L n-omino, $n \geq 6$ even, has a signed tiling by T_n if and only if k is even or k is odd and divisible by $n/2-2$.*

The proof of Proposition 3 is shown in Section 7.

Barnes [11] [12] developed a method for solving signed tiling problems with complex number weights. Applied to our tiling sets, the method gives:

Theorem 4. *If complex number weights are used, a rectangle can be signed tiled by T_n, $n \geq 6$ even, if and only if it has a side divisible by n. If only integer weights are used, a rectangle that has a side divisible by n and all cells labeled by the same multiple of $n/2 - n$ can be signed tiled by T_n, $n \geq 6$ even.*

Theorem 4 is proved in Section 8. A Gröbner basis for the tiling set helps even if Barnes method is used.

Theorem 5. *If complex number weights are used, a rectangle can be signed tiled by* $T_n^+, n \geq 6$ *even, if and only if it has an even side, and a rectangle can be signed tiled by* T_4^+ *if and only if both sides are even.*

Theorem 5 is proved in Section 9. It is not clear to us if last statement in Theorem 4 implies Theorem 1 and if Theorem 5 implies Theorem 2. Guided by the work here, we conclude that Gröbner basis method for solving signed tiling problems with integer weights is sometimes more versatile and leads to stronger results then Barnes method.

The methods we use in this paper are well known when applied to a particular tiling problem. Here we apply them uniformly to solve an infinite collection of problems. Our hope was to see some regularity in the Gröbner bases associated to other infinite families of tiling sets, such as the family $T_{m,n,p}$ investigated in [5]. We recall that if m, p are odd and n is even, tilings of the first quadrant by this family follow the rectangular pattern. Nevertheless, our hopes were not validated. The subfamily $T_{3,n,3}$, n even, has a wide variety of Gröbner bases, making difficult to state a general result. Thus, for this particular family, we understand regular tilings of rectangles due to [5], but cannot decide if the results follow from coloring invariants.

2. SUMMARY OF GRÖBNER BASIS THEORY

Let $R[X] = R[X_1, \cdots, X_k]$ be the ring of polynomials with coefficients in a principal ideal domain (PID) R. A *term* in the variables $x_1, \cdots, x_k$ is a power product $x_1^{a_1} x_2^{a_2} \cdots x_k^{a_k}$ with , $a_i \in N$, $1 \leq i \leq n$; in particular $1 = x_1^0 x_2^0 \cdots x_k^0$ is a term. A term with an associated coefficient from R is called *monomial*. We endow the set of terms with the *total degree-lexicographical order*, in which we first compare the degrees of the monomials and then break the ties by means of lexicographic order for the order $x_1 > x_2 > \cdots > x_k$ on the variables. If the variables are only x, y and $x > y$, this gives the total order:

$$1 < y < x < y^2 < xy < x^2 < y^3 < xy^2 < x^2y < x^3 < y^4 < \cdots \quad (1)$$

For $P \in R[X]$ we denote by $HT(P)$ the leading term and by $HM(P)$ the highest monomial in P with respect to the above order. We denote by $HC(P)$ the coefficient of the leading monomial in P. We denote by $T(P)$ the set of terms appearing in P and by $M(P)$ the set of monomials in P. For a given ideal I in $R[X]$ an associated Gröbner basis is introduced as in Chapters 5, 10 in [13].

If G is a finite set in $R[\mathbf{X}]$, we denote by $I(G)$ the ideal generated by G in $R[\mathbf{X}]$.

Definition 1. *Let f, g, $p\in R[\mathbf{X}]$. We say that f D-reduces to g modulo p and write $f\rightarrow_p g$ if there exists $m\in M(f)$ with $HM(p)/m$, say $m = mHM(p)$, and $g = f - mp$. For a finite set Gin $R[\mathbf{X}]$, we denote by $G \rightarrow$ the reflexive-transitive closure of $\rightarrow_p$, $p\in G$. We say that g is a normal form for f with respect to G if $f\rightarrow_G g$ and no further D-reduction is possible. We say that f is D-reducible modulo G if $f\rightarrow_G 0$.*

If $f\rightarrow_G 0$, then $f\in I(G)$. The converse is also true if G is a Gröbner basis.

Definition 2. *A D-Gröbner basis is a finite set G of $R[\mathbf{X}]$ with the property that all D-normal forms modulo G of elements of $I(G)$ equal zero. If $I\in R[\mathbf{X}]$ is an ideal, then a D-Gröbner basis of I is a D-Gröbner basis that generates the ideal I.*

Proposition 6. *Let G be a finite set of $R[\mathbf{X}]$. Then the following statements are equivalent:*

1) *G is a Gröbner basis.*
2) *Every $I\in f\neq 0, f\in I(G)$, is D-reducible modulo G.*

We observe, nevertheless, that if R is only a (PID), the normal form associated to a polynomial f by a finite set G of $R[\mathbf{X}]$ is not unique. That is, the reminder of the division of f by G is not unique.

We introduce now the notions of S-polynomial and G-polynomial that allows to check if a given finite set G of $R[\mathbf{X}]$ is a Gröbner basis for the ideal it generates. As usual, *lcm* is the notation for the least common multiple and *gcd* is the notation for the greatest common divisor.

Definition 3. *Let $0\neq g_i\in R[X]$, $i = 1, 2$, $HC(g_i) = a_i$, $HT(g_i) = t_i$. Let $a = a_i b_i = lcm(a_1, a_2)$ with $b_i\in R$, and $t = s_i t_i = lcm(t_1, t_2)$ with $s_i\in T$. The S-polynomial of g_1, g_2 is defined as:*

$$S(g_1, g_2) = b_1 s_1 g_1 - b_2 s_2 g_2 \tag{2}$$

If $c_1, c_2\in R$ such that $\gcd(a_1, a_2) = c_1a_1 + c_2a_2$. *Then the G-polynomial of g_1, g_2 is defined as:*

$$G(g_1, g_2) = c_1 s_1 g_1 + c_2 s_2 g_2 \tag{3}$$

Theorem 7. *Let G be a finite set of $R[\mathbf{X}]$. Assume that for all $g_1, g_2\in G$, $S(g_1, g_2)\rightarrow_G 0$ and $G(g_1, g_2)$ is top-D-reducible modulo G. Then G is a Gröbner basis.*

Assume now that R is an Euclidean domain with unique reminders (see page 463 [13]). This is the case for the ring of integers Z if we specify reminders upon division by $0\neq m$ to be in the interval $[0,m)$.

Definition 4. *Let f, g, p∈ R[**X**]. We say that f E-reduces to g modulo p and write* $f \to_{E,p} g$ *if there exists m = at ∈M (f) with HM (p) | t , say t = s HT (p), and g = f − qsp where 0 ≠ q∈ R is the quotient of a upon division with unique reminder by HC (p) .*

Proposition 8. *E-reduction extends D-reduction, i.e., every D-reduction step in an E-reduction step.*

Theorem 9. *Let R be an Euclidean domain with unique reminders, and assume G subset of R[**X**] is a D-Gröbner basis. Then the following hold:*

1) $f \to_{E,p} 0$ *for* $f \in I(G)$.
2) *E-reduction modulo G has unique normal forms.*

The following result connects signed tilings and Gröbner bases. See [9] and [14] for a proof.

Theorem 10. *A polyomino P admits a signed tiling by translates of prototiles* $P_1, \cdots, P_k$ *if and only if for some (test) monomial* $x^a y^b$ *the polynomial* $x^a y^b f_p$ *is in the ideal generated by* $f_{P_1}, \cdots, f_{P_k}$.

3. GRÖBNER BASIS FOR *TN*, *N* EVEN

We show first Gröbner bases for the ideals generated by 4 6 *T* ,*T* , as these are different from the general case.

Proposition 11. *The polynomials* $C_1(2)=x^2+x+y+1$, $C_2(2)=y^2+x+y+1$ *form a Gröbner basis for the ideal generated by* T_4.

Proof. The polynomials corresponding to the tiles in T_4 are

$C_1(2)$, $C_2(2)$, xy^2+xy+y^2+x, $x^2y+xy+x^2+y$. The last two can be generated by $C_1(2)$, $C_2(2)$:

$$xy^2+xy+y^2+x=-C_1(2)+(x+1)C_2(2), x^2y+xy+x^2+y=-C_2(2)+(y+1)C_1(2) \quad (4)$$

It remains to show that the *S*-polynomial associated to $C_1(2)$, $C_2(2)$ can be reduced. One has:

$$S(C_1(2),C_2(2))=y^2(x^2+x+y+1)-x^2(y^2+y+1+x)=-(x+y+1)C_1(2)+(x+y+1)C_2(2) \quad (5)$$

Proposition 12. *A Gröbner basis for the ideal of polynomials generated by* T_6 *is given by:*

$$C_1(3)=x^3+x^2+x+y^2+y+1, C_2(3)=y^3+y^2+y+x^2+x+1, C_3(3)=xy-1 \quad (6)$$

Proof. The polynomials associated to T_6 are:

$$H_1(k) = y^4 + y^3 + y^2 + y + 1 + x,\ H_2(k) = y^4 + xy^4 + xy^3 + xy^2 + xy + x,$$
$$H_3(k) = y + y^4 + x^3 + x^2 + x + 1,\ H_4(k) = x^4y + x^3y + x^2y + xy + y + x^4 \tag{7}$$

Similar to what is done in [10] the presence of $C_3(3)$ in the Gröbner basis allows to reduce the algebraic proofs to combinatorial considerations. We leave most of the details of this proof to the reader. The proof that polynomials in Formula (7) are in the ideal generated by $C_1(3)$, $C_2(3)$, $C_3(3)$ is similar to that of Proposition 5 in [10]. The proof that $C_1(3)$, $C_2(3)$, $C_3(3)$ are in the ideal generated by the polynomials in Formula (7) is similar to that of Proposition 6 in [10]. A geometric proof that $C_3(3)$ belongs to the ideal generated by $H_1(3)$, $H_2(3)$, $H_3(3)$, $H_4(3)$ is shown in **Figure 3**.

For the rest of this section $n = 2k$, $k \geq 4$. The polynomials associated to the tiles in T_n are:

$$H_1(k) = \frac{y^{2k-1} - 1}{y-1} + x,\ H_2(k) = y^{2k-2} + \frac{x\left(y^{2k-1} - 1\right)}{y-1},\ H_3(k) = \frac{x^{2k-1} - 1}{x-1} + y,\ H_4(k) = x^{2k-2} + \frac{y\left(x^{2k-1} - 1\right)}{x-1} \tag{8}$$

We show that a Gröbner basis for the ideal generated by the polynomials in Formula (8) is given by:

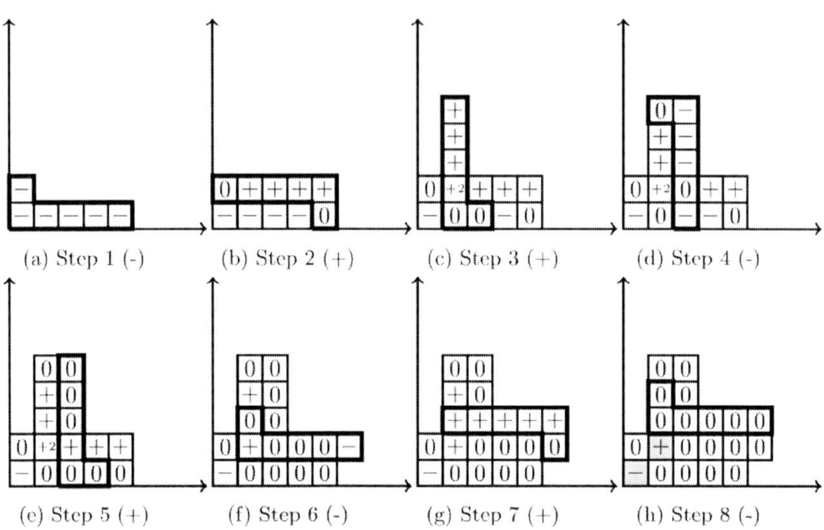

Figure 3. The polynomial $C_3(3)$ is generated by $\{H_1(3), H_2(3), H_3(3), H_4(3)\}$.

$$C_1(k) = \frac{y^{k+1}-1}{y-1} + x\frac{x^{k-1}-1}{x-1} + \left\lfloor \frac{k-1}{2} \right\rfloor xy - \left\lfloor \frac{k-1}{2} \right\rfloor,$$
$$C_2(k) = \frac{x^{k+1}-1}{x-1} + y\frac{y^{k-1}-1}{y-1} + \left\lfloor \frac{k-1}{2} \right\rfloor xy - \left\lfloor \frac{k-1}{2} \right\rfloor$$
$$C_3(k) = x^2y + xy - x - 1, C_4(k) = xy^2 + xy - y - 1, C_5(k) = (k-2)xy - (k-2),$$

where $\lfloor x \rfloor$ is the integer part of x. It is convenient to visualize the elements of the basis as tiles with cells labeled by integers, see **Figure 4**.

Proposition 13. *The polynomials* $H_i(k)$, $1 \le i \le 4$, *belong to the ideal generated by* $C_i(k)$, $1 \le i \le 5$.

Proof. Due to the symmetry, we only show that $H_1(k)$, $H_2(k)$ are in the ideal. The polynomials $C_3(k)$, $C_4(k)$ allow to translate a horizontal domino with both cells labeled by the same sign, respectively a vertical domino, along a vector parallel to the line $y = x$. They also allow to translate horizontally or vertically a block of two cells adjacent at a vertex and labeled by different signs into a similar block. If the length of the translation is even, the signs stay the same. If the length of the translation is odd, all signs are changed. See **Figure 5**.

We show how to build $H_1(k)$. There are two cases to be considered, k odd and k even.

The steps of a geometric construction for k odd are shown in **Figure 6**. To reach Step 1, we add several times multiples of $C_4(k)$, as in **Figure 5(b)**. To reach Step 2, we add several times multiples of $C_3(k)$, as in **Figure 5(a)**. To reach Step 3, first we subtract $C_5(k)$, then add several times multiples of $C_3(k)$, $C_4(k)$ as in **Figure 5(c)** and **Figure 5(d)**. To obtain now $H_1(k)$ in the initial position, we multiply the tile in Step 3 by x^{k-2}, which

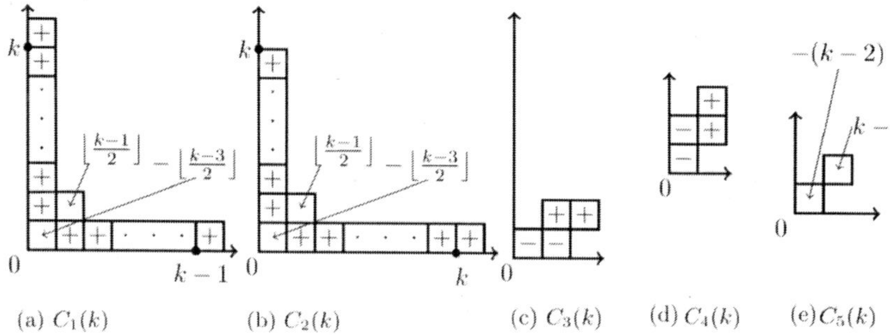

Figure 4. The Gröbner basis $\{C_1(k), C_2(k), C_3(k), C_4(k), C_5(k)\}$.

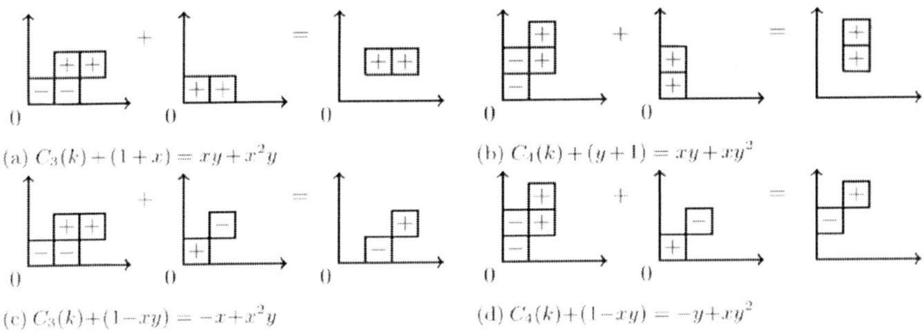

Figure 5. Tiles arithmetic.

will translate the tile $k-2$ cells up, and then add multiples on $C_3(k)$, $C_4(k)$, as in **Figure 5(c)** and **Figure 5(d)**.

The steps of a geometric constructions for k even are shown in **Figure 7**. To reach Step 1, we add several times multiples of $C_4(k)$, as in **Figure 5(b)**. To reach Step 2, we add several times multiples of $C_3(k)$, as in **Figure 5(a)**. To reach Step 3, first we subtract $C_5(k)$, then add several times multiples of $C_3(k)$, $C_4(k)$ as in **Figure 5(c)** and **Figure 5(d)**. To obtain now $H_1(k)$ in the initial position, we multiply the tile in Step 3 by x^{k-2}, which will translate the tile $k-2$ cells up, and then add multiples on $C_3(k)$, $C_4(k)$, as in **Figure 5(c)** and **Figure 5(d)**.

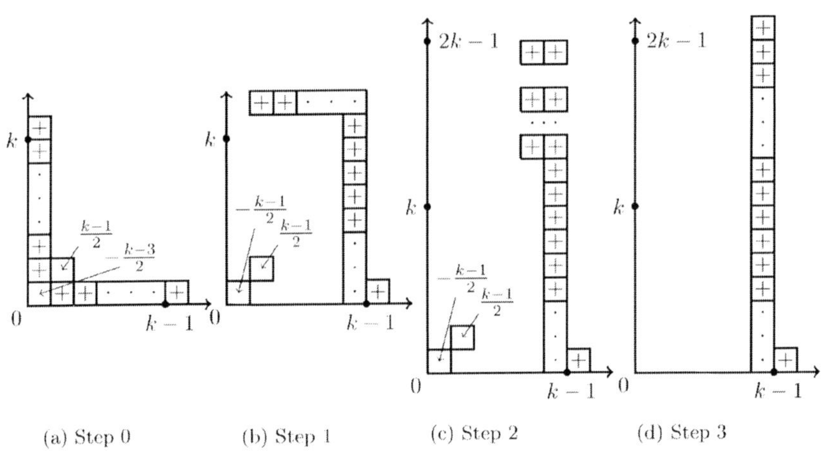

Figure 6. Building $H_1(k)$, k odd, out of $\{C_1(k), C_2(k), C_3(k), C_4(k), C_5(k)\}$.

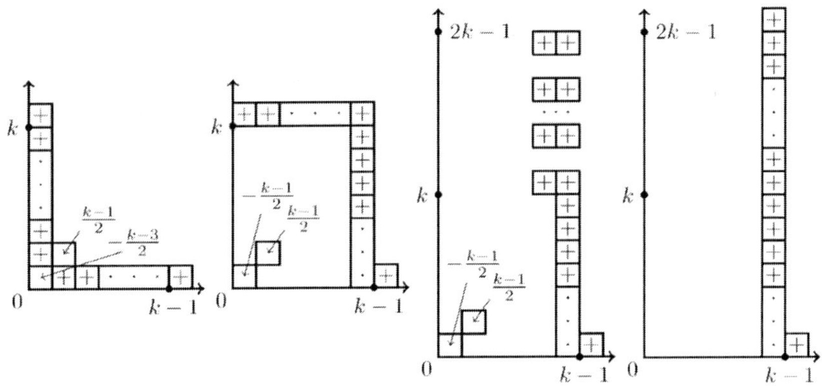

(a) Step 0 (b) Step 1 (c) Step 2 (d) Step 3

Figure 7. Building $H_1(k)$, k even, out of $\{C_1(k), C_2(k), C_3(k), C_4(k), C_5(k)\}$.

We show how to build $H_2(k)$. There are two cases to be considered, k odd and k even.

The steps of a geometric constructions for k odd are shown in **Figure 8**. To reach Step 1, we add several times multiples of $C_4(k)$, as in **Figure 5(b)**. To reach Step 2, we add several times multiples of $C_3(k)$, as in **Figure 5(a)**. To reach Step 3, first we subtract $C_5(k)$, then add several times multiples of $C_3(k)$,$C_4(k)$ as in **Figure 5(c)** and **Figure 5(d)**. To obtain now $H_2(k)$ in the initial position, we multiply the tile in Step 3 by x^{k-2}, which will translate the tile $k - 2$ cells up, and then add multiples on $C_3(k)$, $C_4(k)$, as in **Figure 5(c)** and **Figure 5(d)**. The steps of a geometric constructions for k even are shown in **Figure 9**. To reach Step 1, we add

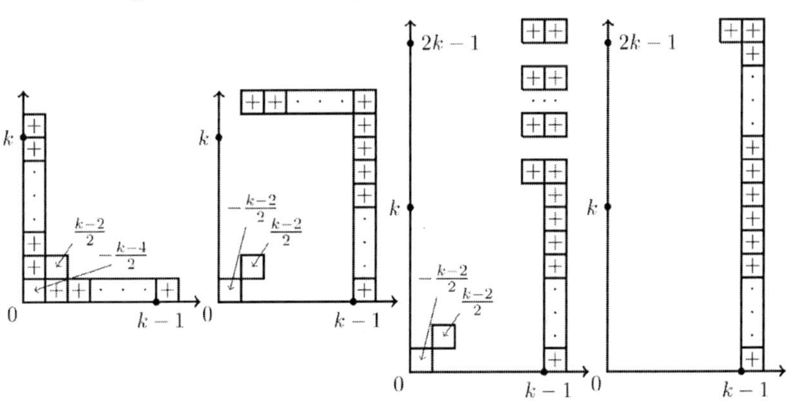

(a) Step 0 (b) Step 1 (c) Step 2 (d) Step 3

Figure 8. Building $H_2(k)$, k odd, out of $\{C_1(k), C_2(k), C_3(k), C_4(k), C_5(k)\}$.

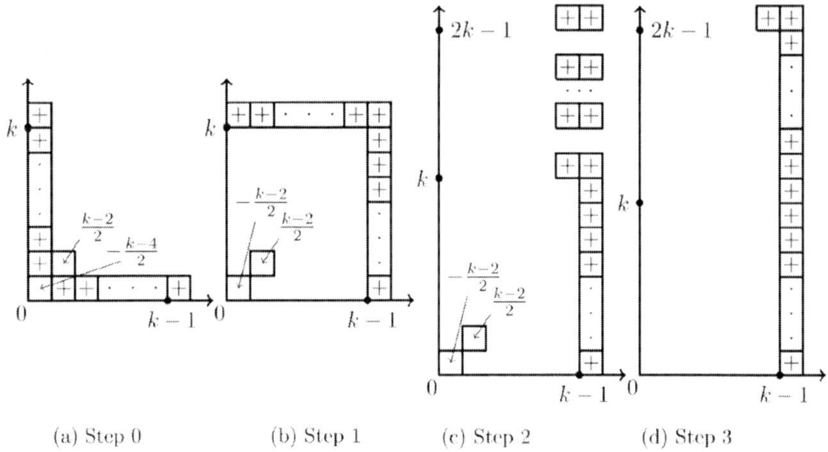

Figure 9. Building $H_2(k)$, k even, out of $\{C_1(k), C_2(k), C_3(k), C_4(k), C_5(k)\}$.

several times multiples of $C_4(k)$, as in **Figure 5(b)**. To reach Step 2, we add several times multiples of $C_3(k)$, as in **Figure 5(a)**. To reach Step 3, first we subtract $C_5(k)$, then add several times multiples of $C_3(k)$, $C_4(k)$ as in **Figure 5(c)** and **Figure 5(d)**. To obtain now $H_2(k)$ in the initial position, we multiply the tile in Step 3 by x^{k-2}, which will translate the tile $k-2$ cells up, and then add multiples on $C_3(k)$, $C_4(k)$, as in **Figure 5(c)** and **Figure 5(d)**.

Proposition 14. *The polynomials $C_i(k)$, $1\le i\le 5$ belong to the ideal generated by $H_i(k)$, $1\le i\le 4$.*

Proof. Due to the symmetry, it is enough to show that $C_1(k)$, $C_3(k)$ and $C_5(k)$ belong to the ideal. We show how to generate $C_3(k)$, $C_5(k)$ (and consequently $C_4(k)$). To generate $C_1(k)$ we can reverse the process in Proposition 13. For $C_3(k)$, one has $C_3(k)=(xy + x - 1)\, H_3(k) - xH_4(k)$. To generate $C_5(k)$ we first show how to obtain a configuration in which all nontrivial cells, 4 of them, are located on the main diagonal. See **Figure 10**. Then we use the tiles arithmetic shown in **Figure 11** to pull the cells in positions $(k-1, k-1)$ and $(2k-2, 2k-2)$ in positions $(1, 1)$ and $(2, 2)$. This gives $C_5(k)$.

Proposition 15. *The sets* $\{C_i(k), 1\le i\le 5\}$ and $\{H_i(k), 1\le i\le 4\}$ *generate the same ideal.*

Proof. This follows from Propositions 13, 14.

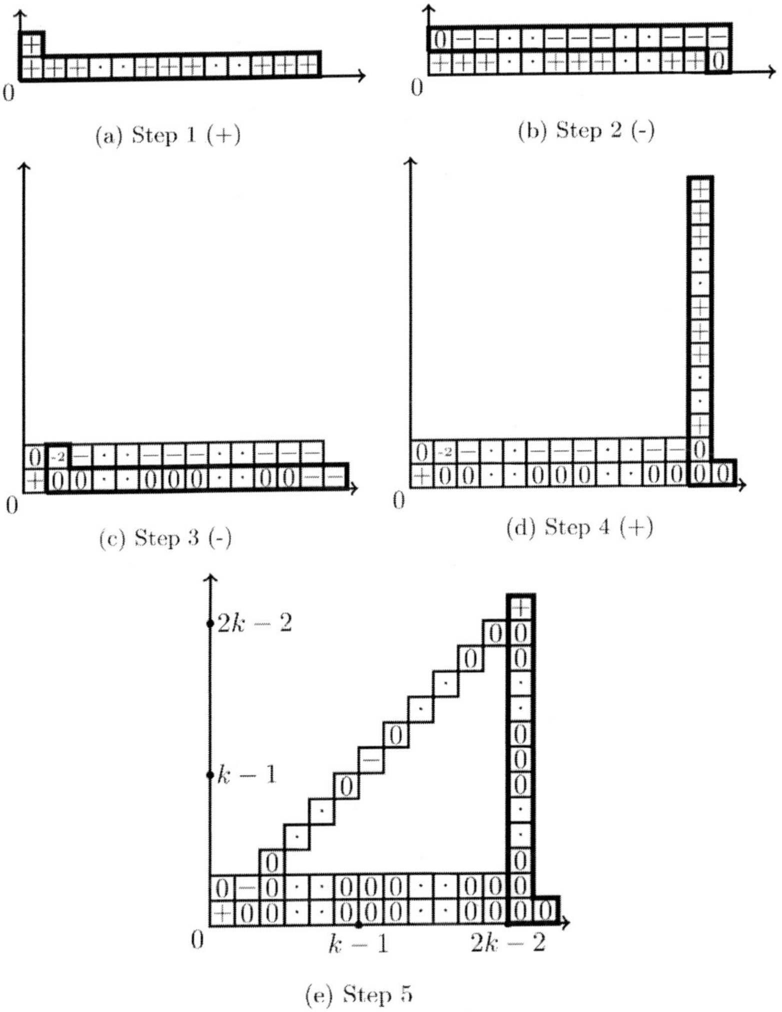

Figure 10. Building $C_5(k)$ out of $\{H_1(k), H_2(k), H_3(k), H_4(k)\}$.

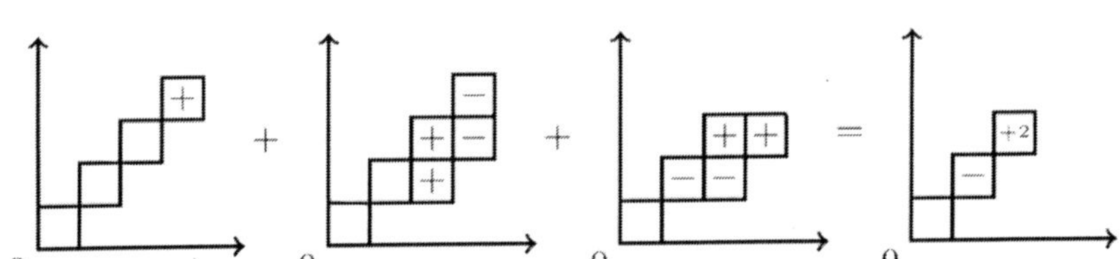

Figure 11. Tiles arithmetic: $x^3+y^3-x^2yC_4(k)+xyC_4(k)=2x^2y^2-xy$.

Proposition 16. *One has the following formulas*:

$$S\left(C_1(k),C_2(k)\right)=-y^{k-1}C_1(k)+x^{k-1}C_2(k)-y^{k-1}\left(1+x^2+\cdots+x^{k-3}\right)C_3(k)$$
$$+x^{k-1}\left(1+y^2+\cdots+y^{k-3}\right)C_4(k)+y^{k-1}C_5(k)$$
$$-y^{k-1}\left\lfloor\frac{k-1}{2}\right\rfloor C_4(k)-x^{k-1}C_5(k)+x^{k-1}\left\lfloor\frac{k-1}{2}\right\rfloor C_3(k)$$

$$S\left(C_1(k),C_3(k)\right)=xC_2(k)-y^{k-2}C_4(k)+y^{k-2}C_3(k)+\left(xy^{\{k-4\}}+xy^{\{k-6\}}+\cdots+xy+\varepsilon(k)\right)C_4(k)$$
$$-xC_5(k)+\left\lfloor\frac{k-1}{2}\right\rfloor xC_3(k),$$

$$S\left(C_1(k),C_4(k)\right)=C_2(k)+\left(x^{k-4}+x^{k-6}+\cdots+x^2+\varepsilon(k)\right)C_4(k)-C_5(k)+\left\lfloor\frac{k-1}{2}\right\rfloor C_3(k),$$

$$S\left(C_1(k),C_5(k)\right)=(k-2)C_2(k)+(k-2)C_3(k)\left(1-\varepsilon(k)+y+y^3+\cdots+y^{\{k-3\}}\right)$$
$$+\left(2\left\lfloor\frac{k-3}{2}\right\rfloor+\varepsilon(k)\right)C_5(k)+(k-2)\left\lfloor\frac{k-1}{2}\right\rfloor C_3(k),$$

$$S\left(C_1(k),C_5(k)\right)=(k-2)C_2(k)+(k-2)C_3(k)\left(1-\varepsilon(k)+y+y^3+\cdots+y^{\{k-3\}}\right)$$
$$+\left(2\left\lfloor\frac{k-3}{2}\right\rfloor+\varepsilon(k)\right)C_5(k)+(k-2)\left\lfloor\frac{k-1}{2}\right\rfloor C_3(k),$$

$$S\left(C_2(k),C_3(k)\right)=C_1(k)+\left(y^{k-4}+y^{k-6}+\cdots+y^2+\varepsilon(k)\right)C_3(k)-C_5(k)+\left\lfloor\frac{k-1}{2}\right\rfloor C_4(k),$$

$$S\left(C_2(k),C_4(k)\right)=yC_1(k)-x^{k-2}C_3(k)+x^{k-2}C_4(k)+\left(x^{k-4}y+x^{k-6}y+\cdots+x^2y+\varepsilon(k)y\right)C_3(k)$$
$$-yC_5(k)+\left\lfloor\frac{k-1}{2}\right\rfloor yC_4(k).$$

$$S\left(C_2(k),C_5(k)\right)=(k-2)C_1(k)+(k-2)C_4(k)\left(1-\varepsilon(k)+x+x^3+\cdots+x^{k-3}\right)$$
$$+\left(2\left\lfloor\frac{k-3}{2}\right\rfloor+\varepsilon(k)\right)C_5(k)+(k-2)\left\lfloor\frac{k-1}{2}\right\rfloor C_4(k).$$

$$S\left(C_3(k),C_4(k)\right)=-C_3(k)+C_4(k)$$

$$S\left(C_3(k),C_5(k)\right)=C_5(k),S\left(C_4(k),C_5(k)\right)=C_5(k),$$ *where* $\varepsilon(k)=\frac{1-(-1)^k}{2}$, *which are given by D-reductions. Therefore,* $\{C_i(k),\ 1\le i\le 5\}$ *form a Gröbner basis.*

Proof. We observe that we can always choose one of the coefficients C_1, C_2 in Definition 3 to be zero. So in order to check that we have a Gröbner basis, we do not need to use G-polynomials. Due to the symmetry, some formulas above follow immediately from others: (14) follows from (12), (15) follows from (11), (16) follows from (13), and second formula in (18) follows from the first. For the rest, note that the leading monomial in

$C_1(k\ )$ is y^k, the leading monomial in $C_2(k)$ is x^k, the leading monomial in $C_2(k)$ is x^2y, the leading monomial in $C_4(k)$ is xy^2, and the leading monomial in $C_5(k)$ is $(k-2)\,xy$.

The D-reduction of $S(C_1(k), C_2(k))$ is shown in **Figure 12**. $S(C_1(k), C_2(k))$ consists of two disjoint symmetric tiles. The reduction of them is similar and it is shown in parallel in **Figure 12**. We start with

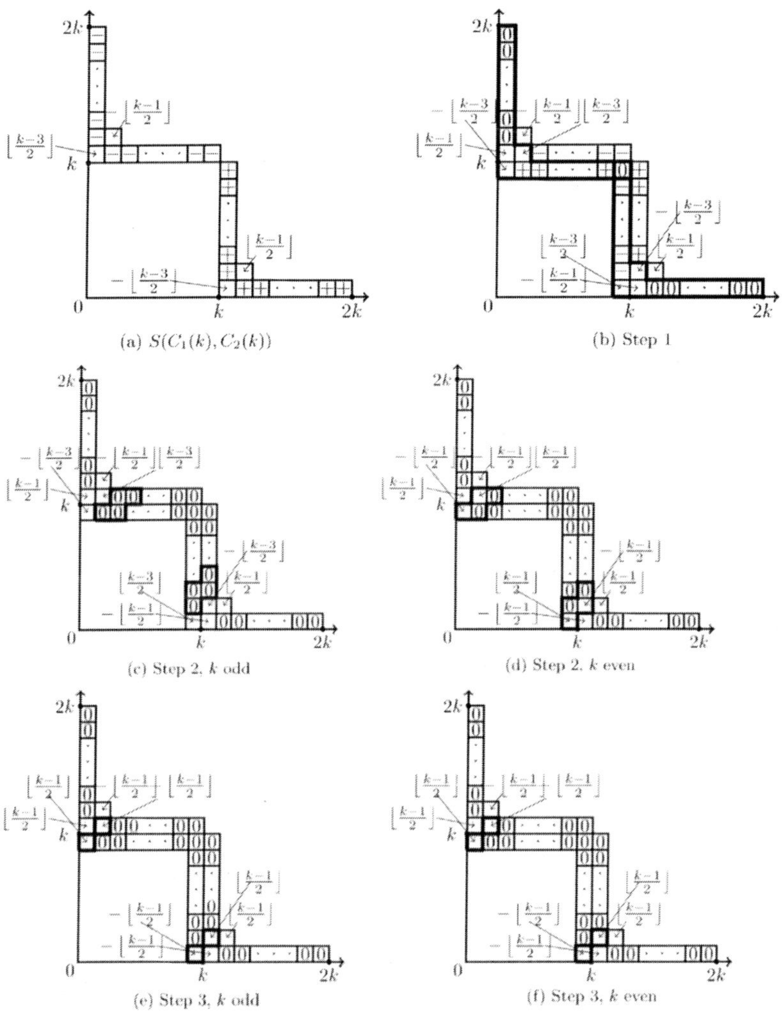

Figure 12. The D-reduction of $S(C_1(k), C_2(k))$.

$$S\left(C_1(k), C_2(k)\right) = x^k C_1(k) - y^k C_2(k) \tag{19}$$

The D-reduction of $S(C_1(k), C_2(k))$ is shown in **Figure 13**. We start with

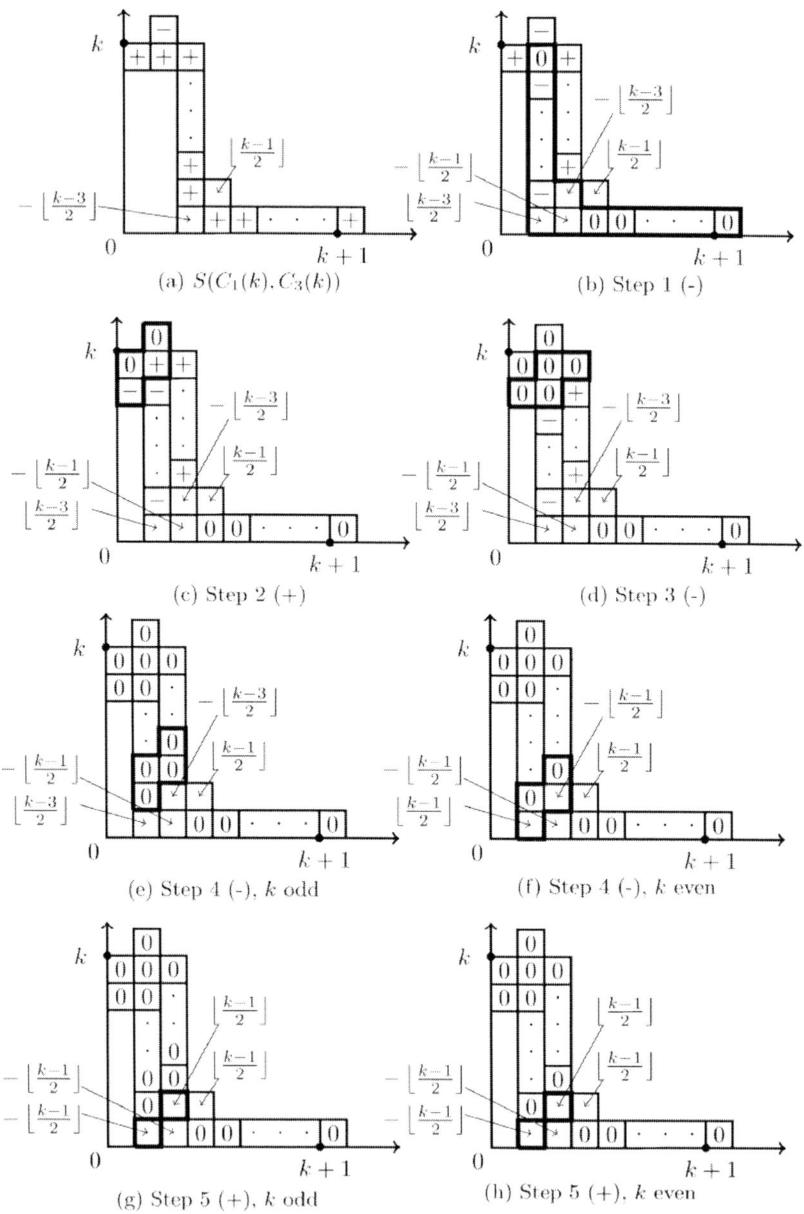

Figure 13. The D-reduction of $S(C_1(k), C_3(k))$.

$$S\left(C_1(k),C_3(k)\right)=x^2C_1(k)-y^{k-1}C_3(k) \quad (20)$$

From Step 3 to Step 4 we subtract $\left(xy^{k-4}+xy^{k-6}+\cdots+xy\right)C_4(k)$ or $\left(xy^{k-4}+xy^{k-6}+\cdots+xy^2+x\right)C_4(k)$, depending on k odd or even. From Step 4 to Step 5 we use the following formulas: $(k-2)-\left\lfloor\frac{k-3}{2}\right\rfloor=\left\lfloor\frac{k-1}{2}\right\rfloor$, if is odd, $(k-2)-\left\lfloor\frac{k-1}{2}\right\rfloor=\left\lfloor\frac{k-1}{2}\right\rfloor$ (21)

The D-reduction of $S(C_1(k), C_4(k))$ is shown in **Figure 14**. We start with

$$S\left(C_1(k),C_4(k)\right)=xC_1(k)-y^{k-2}C_4(k) \quad (22)$$

From Step 1 to Step 2 we subtract $\left(x^{k-4}+x^{k-6}+\cdots+x\right)C_4(k)$ or $\left(x^{k-4}+x^{k-6}+\cdots+x^2+1\right)C_4(k)$, depending on k odd or even. From Step 2 to Step 3 we use Formulas (21).

The D-reduction of $S(C_1(k), C_5(k))$ is shown in **Figure 15**. We start with

$$S\left(C_1(k),C_5(k)\right)=(k-2)xC_1(k)-y^{k-1}C_5(k) \quad (23)$$

To reach Step 1, we subtract (k-2) $C_2(k)$. To reach Step 2, we subtract $\left(1+y^2+\cdots+y^{k-3}\right)C_4(k)$ if k is odd and $\left(y+\cdots+y^{k-3}\right)C_4(k)$ if k is even. To reach Step 3, we add $2\left\lfloor\frac{k-1}{2}\right\rfloor C_5(k)$ if k is odd and $\left(2\left\lfloor\frac{k-3}{2}\right\rfloor+1\right)C_5(k)$ if k is even.

The D-reduction of $S(C_3(k), C_4(k))$ is:

$$S\left(C_3(k),C_4(k)\right)=yC_3(k)-xC_4(k)=-C_3(k)+C_4(k) \quad (24)$$

The D-reduction of $S(C_3(k), C_5(k))$ is:

$$S\left(C_3(k),C_5(k)\right)=(k-2)C_3(k)-xC_5(k)=C_5(k) \quad (25)$$

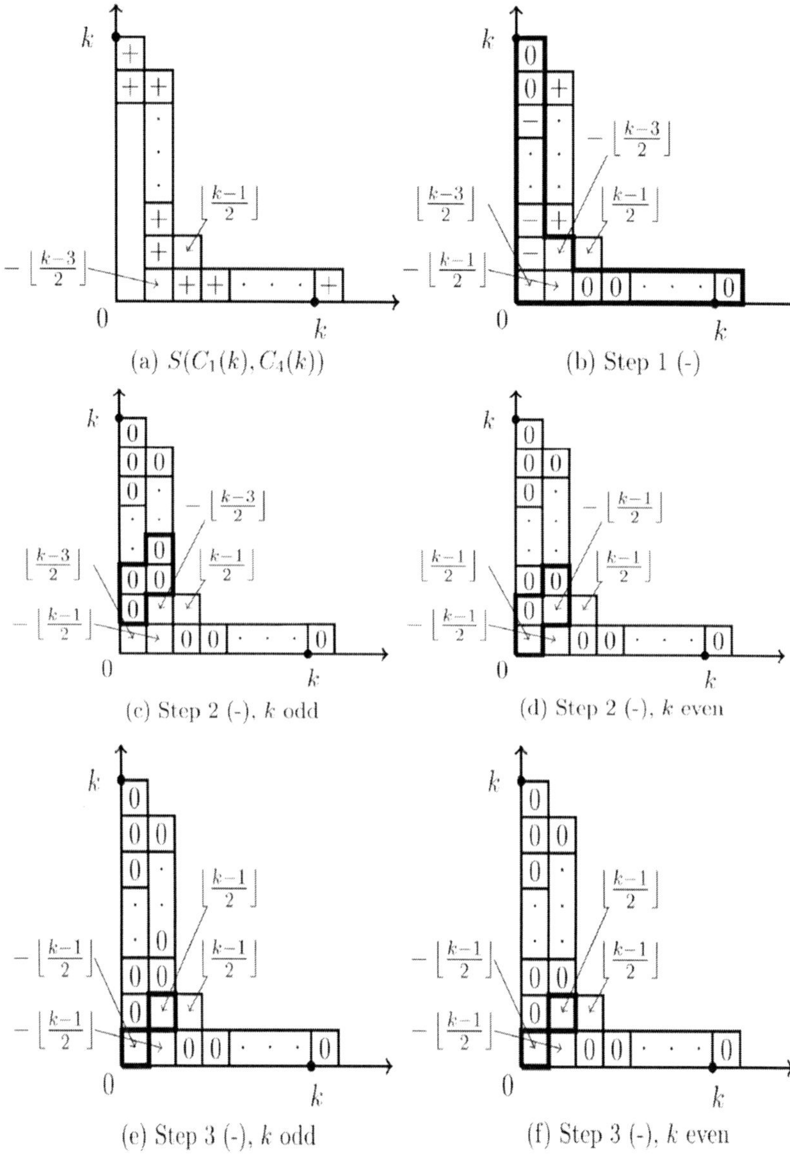

Figure 14. The D-reduction of $S(C_1(k), C_4(k))$.

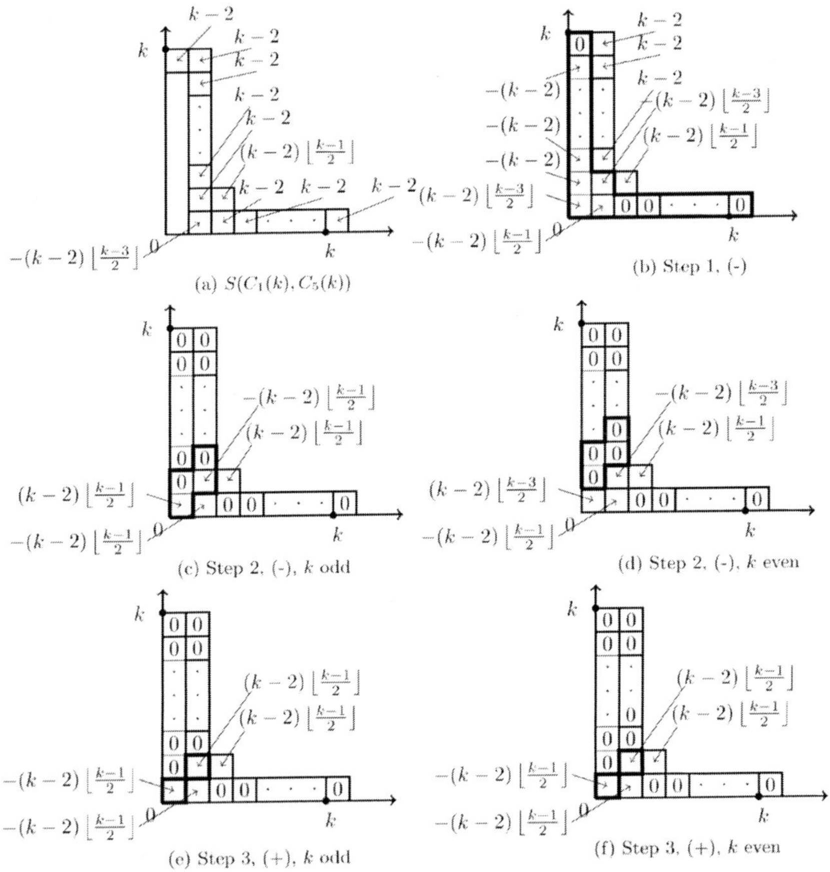

Figure 15. The D-reduction of $S(C_1(k), C_5(k))$.

4. GRÖBNER BASIS FOR T_n^+, *N* EVEN

We consider first the case $n = 4$.

Proposition 17. *A Gröbner basis for the ideal generated by* T_4^+ *is:* $D_1=y^2+2y+1$, $D_2=x-y$.

Proof. The pictures for tiles corresponding to the basis are shown in **Figure 16**. One has:

$$y^2+2y+1=(y^2+y+1+x)+(x^2+y+xy+x^2y)-x(xy+x+y+1),$$
$$x-y=-(x^2+y+xy+x^2y)+x(xy+x+y+1),$$

(26)

thus the Gröbner basis can be generated by the polynomials in T_4^+. Conversely, one has:

$$y^2+y+1+x=(y^2+2y+1)+(x-y),$$
$$y^2+xy^2+xy+x=y(y^2+y+1)+(y^2+y+1)(x-y),$$
$$y+1+x+x^2=(y^2+2y+1)+(x+y+1)(x-y),$$
$$x^2+y+xy+x^2y=y(y^2+2y+1)+(xy+y^2+x+2y)(x-y),$$
$$xy+x+y+1=(y^2+2y+1)+(y+1)(x-y),$$

(27)

thus T_4^+ is generated by the Gröbner basis.

The *S*-polynomial S(D_1, D_2) is reduced as follows:

$$S(D_1,D_2)=x(y^2+2y+1)-y^2(x-y)=2xy+x+y^3=y(y^2+2y+1)+(2y+1)(x-y)$$

(28)

Let now $n = 2k$, $k \geq 3$. Recall that T_n^+ is the set T_n plus a 2× 2 tile with polynomial $H_5(k)=xy+x+y+1$. We show that the Gröbner basis for the ideal generated by T_n^+ is given by:

$$D_1(k)=y^2+2y+1,$$
$$D_2(k)=[(k-2)(k-1)-1]y+x+(k-2)(k-1),$$
$$D_3(n)=k(k-2)(y+1).$$

(29)

As T_n is a subset of T_n^+, T_n^+ generate the Gröbner basis $C_i(k)$ for T_n. Next formula shows how to generate $D_1(k)$:

$$H_5(k)y-C_4(k)=(y+xy+y^2+xy^2)-(-1-y+xy+xy^2)=1+2y+y^2=D_1(k)$$

(30)

Lemma 18. *The polynomial* $1+x+(k-1)(y+y^2)$ *is generated by* $H_1(k)$ and $D_1(k)$, and $H_1(k)$ is *generated by* $1+x+(k-1)(y+y^2)$ and $D_1(k)$.

Proof. First produce $1+y-y^2-y^3=D_1(k)(1-y)$. Adding copies of this tile to $H_1(k)$ gives the sum:

$$H_1(k)+\left[y^{n-5}+2y^{n-7}+3y^{n-9}+\cdots+(k-2)y\right]\left(1+y-y^2-y^3\right) \tag{31}$$

Expanding Formula (31) gives a telescopic sum that reduces to (y^2+y) $(k-1)+1+x$.

Proposition 19. *The polynomials $D_1(k)$, $D_2(k)$, $D_3(k)$ belong to the ideal generated by* T_n^+.

Proof. We showed in Formula (30) how to generate $D_1(k)$. By Lemma 18, we can start from $H_1(k)$ to produce (y^2+y) $(k-1)+1+x$. Then, subtract as follows:

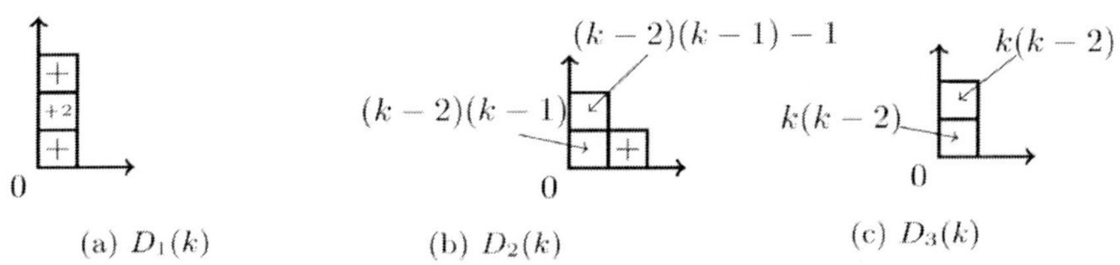

Figure 16. The Gröbner basis $\{D_1(k), D_2(k), D_3(k)\}$.

$$1+x+(k-1)\left(y+y^2\right)-D_1(k)(k-1)=-(k-2)+x-y(k-1) \tag{32}$$

By the symmetry of T_n^+ about $x=y$, we can also generate $-(k-2)-x(k-1)+y$. Combining the two:

$$(k-1)\left[-(k-2)+x-y(k-1)\right]+\left(-(k-2)-x(k-1)+y\right)=-D_3(k) \tag{33}$$

Finally, $D_2(k)$ is produced from $D_3(k)$ and $-(k-2)+x-y(k-1)$, which we have from Formula (32):

$$\left[-(k-2)+x-y(k-1)\right]+(y+1)k(k-2)=D_2(k) \tag{34}$$

Lemma 20. *The polynomials* $\bar{D}_1(k)=1+2x+x^2$, $\bar{D}_2(k)=(k-1)(k-2)+x\left[(k-1)(k-2)-1\right]+y$, $\bar{D}_3(k)=(1+x)k(k-2)$ *belong to the ideal generated by $D_1(k)$, $D_2(k)$, $D_3(k)$.*

Proof. We show below independently in Proposition 21 that $H_5(k)$ is also in this ideal. Then one can easily check that:

$$\overline{D}_1(k)=(1+x)D_2(k)-[(k-1)(k-2)-1]H_5(k)$$
$$\overline{D}_2(k)=[(k-1)(k-2)-1]D_2(k)-(k-1)(k-3)D_3(k)$$
$$\overline{D}_3(k)=k(k-2)D_2(k)-[(k-2)(k-1)-1]D_3(k).$$

Proposition 21. *The members of* T_n^+ *belong to the ideal generated by* $D_1(k)$, $D_2(k)$, $D_3(k)$.

Proof. One has after calculations:

$$H_5(k)=(1+y)D_2(k)-[(k-2)(k-1)-1]D_1(k) \tag{36}$$

To obtain $H_1(k)$, begin with

$$D_2(k)-D_3(k)+(k-1)D_1(k)=1+x+(k-1)(y+y^2) \tag{37}$$

By Lemma 18, this tile may be transformed into $H_1(k)$ using only $D_1(k)$, $D_2(k)$, $D_3(k)$. We also get $H_3(k)$ by symmetry in the following way. Swap the variables x, y in Lemma 18. Then we have that $H_3(k)$ and $1+y+(k-1)(x+x^2)$ can each be produced from the other using either T_n^+, which is symmetric about $x = y$, or the tiles $\overline{D}_1(k),\overline{D}_2(k),\overline{D}_3(k)$. Then swapping x, y in Formula (37) allows to obtain $1+y+(k-1)(x+x^2)$ from the basis $\overline{D}_i(k)$, which in turn can be obtained from the Gröbner basis itself by Lemma 20. Therefore the Gröbner basis also generates $H_3(k)$.

The polynomial $C_4(k)=-1-y+xy+xy^2$ can be used to change $H_1(k)$ into $H_2(k)$:

$$H_1(k)+(-1-y+xy+xy^2)(1+y^2+y^4+\cdots+y^{n-4})=H_2(k) \tag{38}$$

By symmetry, the same process will change $H_3(k)$ and $H_4(k)$ using $C_3(k)$. It remains, then, to show that the Gröbner basis for T_n^+ can generate $C_3(k)$ and $C_4(k)$. Start with

$$D_1(k)+y(D_3(k)-D_2(k))=(y^2+y)k+1-xy \tag{39}$$

Then, multiply by $y+1$ and add $-kyD_1(k)$:

$$(y+1)[(y^2+y)k+1-xy]-ky(y^2+2y+1)=1+y-xy-xy^2=-C_4(k) \tag{40}$$

Once again, symmetry gives us a procedure for $C_3(k) = 1 + x - xy - x^2y$, and the proof is complete.

Proposition 22. *The sets* $\{D_i(k), 1 \le i \le 3\}$ *and* $\{H_i(k), 1 \le i \le 5\}$ *generate the same ideal.*

Proof. This follows from Propositions 19, 20.

Proposition 23. *We have the following formulas*:

$$S(D_1(k), D_2(k)) = k(k-2)D_2(k) + (1-x)D_3(k),$$
$$S(D_1(k), D_3(k)) = D_3(k),$$
$$S(D_2(k), D_3(k)) = k(k-2)\left[1-(k-1)(k-2)\right]D_1(k) + k(k-2)D_2(k) - D_3(k), \tag{41}$$

which are given by D-reductions. Therefore, $\{D_i(k), 1 \le i \le 3\}$ *forms a Gröbner basis.*

Proof. We start with

$$3S(D_1(k), D_2(k)) = k(k-2)\left[(k-2)(k-1)-1\right]D_1(k) - k(k-2)yD_2(k)$$
$$S(D_1(k), D_3(k)) = k(k-2)D_1(k) - (y+1)D_3(k)$$
$$S(D_2(k), D_3(k)) = -k(k-2)yD_2(k) + xD_3(k). \tag{42}$$

The reader may easily check that the given reductions are valid for these *S*-polynomials.

5. PROOF OF THEOREM 1

The case $n = 6$ follows as in paper [4]. We assume for the rest of this section $n \ge 8$. Consider a rectangle $q \times p$, $q \ge p \ge 1$. Using the presence of $C_3(k)$ and $C_4(k)$ in the Gröbner basis, the rectangle can be reduced to one of the configurations in **Figure 17(a)** and **Figure 17(b)**. Configuration (b) appears when q, p are both even. The number of cells labeled by p is $q - p + 1$ in a) and $q - p$ in (b).

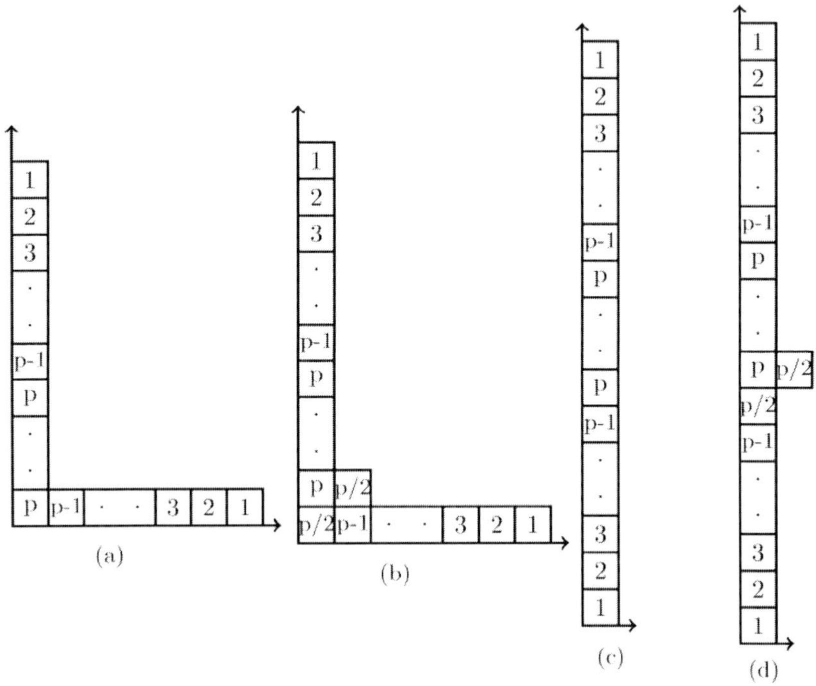

Figure 17. *D*-reductions of a rectangle.

In what follows the signed tile $B = xy - 1$ will play an important role. We recall that it can be moved horizontally/

vertically as shown in **Figure 5**. The tile B does not belong to the ideal generated by T_n. Other signed

tile of interest in the sequel is $D = y^{n+1} + y^n + y^{n-1} + \cdots + y + 1 - xy$, which is the concatenation of a vertical bar of length n and B. The tile $D = y^{n+1} + y^n + y^{n-1} + \cdots + y + 1 - xy$, which is the concatenation of a vertical bar of length n and B. The tile $D = yH_1(k) - C_4(k)$\$ belongs to the ideal generated by T_n.

Multiplying the polynomial associated to the rectangle by y^p, we can assume that the configurations in **Figure 17** are at height $p - 1$ above the x-axis. Using the tiles $C_3(k)$, $C_4(k)$ and an amount of tiles B ($p/2$ if p is even and zero if p is odd), they can be reduced further to the configurations shown in **Figure 17(c)** and **Figure 17(d)**. We observe that (b) is the sum of (a) with $p/2$ copies of B.

Reducing further the configurations in **Figure 17(c)** and **Figure 17(d)**, with copies of D, the existence of a signed tiling for the $q \times p$ rectangle becomes equivalent to deciding when the following two conditions are both true:

1) The polynomial $Q(x) = 1 + y + y^2 + \cdots + y^{n-1}$ divides:

$$P_{pq}(y) = 1 + 2y + 3y^2 + \cdots + py^{p-1} + py^p + \cdots + py^{q-1} + (p-1)y^q + (p-2)y^{q+1} + \cdots + 2y^{p+q-3} + y^{p+q-2}$$

(43)

2) The extra tiles B that appear while doing tile arithmetic for 1), including those from **Figure 17**, can be

cancelled out by $C_5(k)$.

If $p + q - 1 < n$, then $degQ > degP_{pq}$, so divisibility does not hold. If $p + q - 1 \geq n$, we look at P_{pq} as a sum of p polynomials with all coefficients equal to 1:

$$\begin{aligned} P_{pq}(y) &= 1 + y + y^2 + y^3 + \cdots + y^{p-1} + y^p + \cdots + y^{q-1} + y^q + y^{q+1} + \cdots + y^{p+q-4} + y^{p+q-3} + y^{p+q-2} \\ &+ y + y^2 + y^3 + \cdots + y^{p-1} + y^p + \cdots + y^{q-1} + y^q + y^{q+1} + y^{p+q-4} + \cdots + y^{p+q-3} \\ &+ y^2 + y^3 + \cdots + y^{p-1} + y^p + \cdots + y^{q-1} + y^q + y^{q+1} + \cdots + y^{p+q-4} + \cdots + y^p + \cdots + y^{q-1}. \end{aligned} \tag{44}$$

We discuss first 1) and show that it is true when p or q is divisible by n. Then, assuming this condition satisfied,

we discuss 2).

1) Let $p + q - 1 = nm + r, 0 \leq r < n$, and $p = ns + t, 0 \leq t < n$. The remainder $R_{pq}(y)$ of the division of $R_{pq}(y)$ by $Q(y)$ is the sum of the remainders of the division of the p polynomials above by $Q(y)$. If r is odd, one has the sequence of remainders, each remainder written in a separate pair of parentheses:

$$\begin{aligned} R_{pq}(y) &= \left(1 + y + y^2 + \cdots + y^{r-1}\right) + \left(y + y^2 + \cdots + y^{r-2}\right) + \left(y^2 + \cdots + y^{r-3}\right) + \cdots \\ &+ \left(y^{\frac{r-1}{2}}\right) - \left(y^{\frac{r-1}{2}}\right) - \cdots - \left(y + y^2 + \cdots + y^{r-2}\right) - \left(1 + y + y^2 + \cdots + y^{r-1}\right) \\ &+ \left(y^{r+1} + y^{r+2} + \cdots + y^{n-3} + y^{n-2}\right) + \left(y^{r+2} + \cdots + y^{n-3}\right) + \cdots + \left(y^{\frac{r+n-1}{2}}\right) \\ &- \left(y^{\frac{r+n-1}{2}}\right) - \cdots - \left(y^{r+2} + \cdots + y^{n-3}\right) - \left(y^{r+1} + y^{r+3} + \cdots + y^{n-3} + y^{n-2}\right) + \cdots \end{aligned} \tag{45}$$

If $p \geq n$, the sequence of remainders above is periodic with period n, given by the part of the sequence shown above, and the sum of any subsequence of n consecutive remainders is 0. So if p is divisible by n, $P_{pq}(y)$ is divisible by $Q(y)$. If p is not divisible by n, then doing first the cancellation as above and then using the symmetry present in the sequence of remainders, the sum of the sequence of remainders equals 0 only if $r+1=t$, that is, only if q is divisible by n.

If r is even, one has the sequence of remainders, each remainder written in a separate pair of parentheses:

$$\begin{aligned} R_{pq}(y) = {} & \left(1+y+y^2+\cdots+y^{r-1}\right)+\left(y+y^2+\cdots+y^{r-2}\right)+\left(y^2+\cdots+y^{r-3}\right)+\cdots \\ & +\left(y^{\frac{r-2}{2}}+y^{\frac{r}{2}}\right)+(0)-\left(y^{\frac{r-2}{2}}+y^{\frac{r}{2}}\right)-\cdots-\left(y+y^2+\cdots+y^{r-2}\right) \\ & -\left(1+y+y^2+\cdots+y^{r-1}\right)+\left(y^{r+1}+y^{r+3}+\cdots+y^{n-3}+y^{n-2}\right) \\ & +\left(y^{r+2}+\cdots+y^{n-3}\right)+\cdots+\left(y^{\frac{r+n-1}{2}}+y^{\frac{r+n+1}{2}}\right)+(0)-\left(y^{\frac{r+n-1}{2}}+y^{\frac{r+n+1}{2}}\right)-\cdots \\ & -\left(y^{r+2}+\cdots+y^{n-3}\right)-\left(y^{r+1}+y^{r+3}+\cdots+y^{n-3}+y^{n-2}\right)-\cdots \end{aligned} \tag{46}$$

If $p \geq n$, the sequence of remainders above is periodic with period n, given by the part of the sequence shown above, and the sum of any subsequence of n consecutive remainders is 0. So if p is divisible by n, $P_{pq}(y)$ is divisible by $Q(y)$. If p is not divisible by n, then doing first the cancellation as above and then using the symmetry present in the sequence of remainders, the sum of the sequence of remainders equals 0 only if $r+1=t$, that is, only if q is divisible by n.

2) We assume now that n divides p or q and count the extra tiles B that appears. They are counted by the coefficients of the quotient, call it $S(y)$, of the division of $P_{qp}(y)$ by $Q(y)$. We need to compute the sum 1 S of the coefficients in $S(y)$ of the even powers of y and the sum 2 S of the coefficients in $S(y)$ of the odd powers of y. The difference S_1-S_2 gives the number of extra tiles B that we need to consider. We use the equation relating the derivatives:

$$P'_{pq}(y) = Q'(y)S(y) + Q(y)S'(y) \tag{47}$$

Note that $Q(-1) = 0$, $Q'(-1) = n/2$, $S(-1) = S_1 - S_2$. Plugging in $x = -1$ gives:

$$S_1 - S_2 = S(-1) = \frac{2P'_{pq}(-1)}{n} \tag{48}$$

Differentiating the equation of $P_{pq}(y)$ one has:

$$\begin{aligned} P'_{pq}(y) = {} & 2\times 1+3\times 2y+4\times 3y^2+\cdots+(p-1)(p-2)y^{p-3}+p(p-1)y^{p-2}+\cdots+p(q-1)y^{q-2} \\ & +(p-1)qy^{q-1}+(p-2)(q+1)y^{q}+\cdots+2(p+q-3)y^{p+q-4}+(p+q-2)y^{p+q-3} \end{aligned} \tag{49}$$

While computing $P_{pq}(-1)$ we recall that n is even and distinguish the following cases: *Case A.* p even, q odd, *Case B.* p odd, q even, *Case C.* p even, q even.

We need the following formulas:

$$2\times 1-3\times 2+4\times 3-\cdots-(p-1)\times(p-2)=-\frac{p(p-2)}{2}$$

$$p\times(p-1)-p\times(p)+p\times(p+1)-\cdots+p\times(q-2)-p\times(q-1)=-\frac{p(q-p+1)}{2}$$

$$(p-1)\times q-(p-2)\times(q+1)+\cdots+3\times(p+q-4)-2\times(p+q-3)+(p+q-2)=\frac{pq}{2}$$

Case A. One has:

$$P'_{pq}(-1)=2\times 1-3\times 2+4\times 3-\cdots-(p-1)\times(p-2)+p\times(p-1)-\cdots-p\times(q-1)$$
$$+(p-1)\times q-(p-2)\times(q+1)+\cdots-2\times(p+q-3)+(p+q-2)=\frac{p}{2}.$$

The number of extra B tiles is $-\frac{p}{2}+\frac{p}{n}=\frac{p(1-k)}{n}$. To have a complete reduction, the number of B tiles has to be a multiple of $k-2$. As $k-1$ and $k-2$ are relatively prime, p has to be a multiple of $n(k\text{-}2)$.

Case B. One has:

$$P'_{pq}(-1)=2\times 1-3\times 2+4\times 3-\cdots+(p-1)\times(p-2)-p\times(p-1)+\cdots+p\times(q-1)$$
$$-(p-1)\times q+(p-2)\times(q+1)+\cdots-2\times(p+q-3)+(p+q-2)=\frac{q}{2}. \tag{52}$$

The number of extra B tiles is $\frac{q}{n}$. We have the condition that q is a multiple of $n(k\text{-}2)$.

Case C. One has:

$$P'_{pq}(-1)=2\times 1-3\times 2+4\times 3-\cdots-(p-1)\times(p-2)+p\times(p-1)-\cdots+p\times(q-1)$$
$$-(p-1)\times q+(p-2)\times(q+1)+\cdots+2\times(p+q-3)-(p+q-2)=0. \tag{53}$$

The number of extra B tiles is $-\frac{p}{2}+\frac{p}{2}=0$. In this case a signed tiling is always possible.

6. PROOF OF THEOREM 2

Let $k \geq 3$. Consider a $q \times p$, $q \geq p \geq 1$, rectangle. Using the presence of $D_1(k)$ and $\bar{D}_1(k)$ in the ideal, the rectangle can be reduced to one of the configurations in **Figure 18**, where the integers *s,t, st* represent the weights of the corresponding cells. The configuration in (a), and its copies appearing in (b), (c), (d), are multiples of $H_5(k)$ and can be reduced to zero. The remaining region in (b) can be reduced to $\left[1-(k-2)(k-1)\right](1+y)$. As 1− (*k* − 2) (*k* −1) is never a multiple of *k* (*k* − 2) for $k \geq 4$, this configuration can be reduced further by ( ) 2 *D k* to zero only if *s* is a multiple of *k* (*k* − 2) . Same reasoning works for (c). The remaining region in (d) can be reduced further to $\left[(k-2)(k-1)s-t\right]y+\left[(k-2)(k-1)s+s-t+1\right]$ which is never a multiple of $D_2(k)$, thus cannot be reduced to zero.

Assume now $k = 2$. The proof is similar, but one observes that only configuration (a) in **Figure 18** can be

reduced to zero using the Gröbner basis for T_4^+ .

7. PROOF OF PROPOSITION 3

If *k* is even, finding a signed tiling for a k-inflated copy of the *Ln*-omino can be reduced, via reductions by $C_3(k)$, $C_4(k)$tiles, to finding a signed tiling for a $nk \times k$ rectangle. From Theorem 1 follows that such a tiling always exists. If *k* is odd, a reduction to a $kn \times k$ rectangle can be done only modulo a *B* tile, which does not belong to the ideal generated by T_n.

8. THE METHOD OF BARNES FOR T_N, N EVEN

In this section we give a proof of Theorem 4 following a method of Barnes. We assume familiarity with [11] [12]. We apply the method to T_n, $n \geq 6$ even. Consider the polynomials (8) associated to the tiles in *n T* and denote by *I* the ideal generated by them. We show that the complex algebraic variety *V* defined by (8) consists only of the points $\left(\varepsilon, 1+\varepsilon+\varepsilon^2+\cdots+\varepsilon^{n-1}\right)$, where ε is an *n*-th root of identity different from 1.

Separating *x* from $H_1(k)=0$, replacing in $H_2(k)=0$ and factoring the resulting polynomial gives:

$$\left(y^{2k-1}+y^{2k-2}+\cdots+y^2+y+1\right)\left(y^{2k-3}+y^{2k-4}+\cdots+y^2+y+1\right)=0 \tag{54}$$

Denote the polynomial on the left hand side by $f_2(y)$, and denote the corresponding polynomial in the variable x (obtained from $H_3(k)$ and $H_4(k)$ by $f_1(x)$. Their roots are roots of unity of order $2k-1$ and $2k-3$. Using the system of equations that defines V, the roots of order $2k-3$ can be eliminated. Moreover, the only solutions of the system are as above.

We show now that I is a radical ideal. We use an algorithm of Seidenberg which can be applied to find the radical ideal of a zero dimensional algebraic variety over an algebraically closed field. See Lemma 92 in [15]. Compare also with Theorem 7.1 in [11]. As V is zero dimensional, one can find square free polynomials $\overline{f}_1(x)$ and $\overline{f}_2(x)$ that belong to the radical ideal. We take these to be $\overline{f}_1(x)=F(x), \overline{f}_2(x)=F(y)$, where:

$$F(x)=(k-2)\left(x^{2k-1}+x^{2k-2}+\cdots+x^2+x+1\right) \tag{55}$$

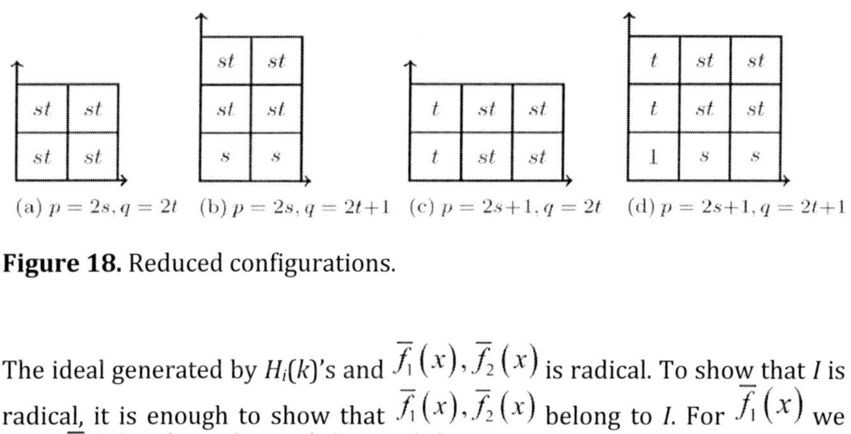

Figure 18. Reduced configurations.

The ideal generated by $H_i(k)$'s and $\overline{f}_1(x), \overline{f}_2(x)$ is radical. To show that I is radical, it is enough to show that $\overline{f}_1(x), \overline{f}_2(x)$ belong to I. For $\overline{f}_1(x)$ we have $\overline{f}_1(x)=(k-2)xH_3(k)+C_5(k)\$$ and $\overline{f}_2(x)$ follows by symmetry.

We apply Lemma 3.8 in [11]: a region R is signed tiled by n T if and only if the polynomial $f_R(x, y)$ evaluates to zero on V. If R is a $p \times q$ rectangle, then $f_R(x, y)=[(x^{q-1})/(x-1)][(y^p-1)/(y-1)]$, which evaluates to zero on V if and only if one of p, q is divisible by n. This gives the first statement in Theorem 4. For the second statement in Theorem 4 we use the method described in the proof of Theorem 4.2 in [11]. A set of generators over the rationals for the rectangles that have a side divisible by n is given by the set $H=\{H_1(k), H_2(k), H_3(k), H_4(k)\}$ and the polynomial with rational coefficients $(k-2)^{-1}C_5(k)$.

As $C_5(k)$ is already generated by H, this implies that (k-2) multiples of the elements in H can signed tile with integer coefficients any (k-2) multiple of a rectangle with a side divisible by n.

9. THE METHOD OF BARNES FOR T_n^+ ,N EVEN

Let $k \geq 3$. Adding the extra polynomial $1+x+y+xy$ to the set of generators, reduces the variety V to the

point $(-1, -1)$. Proceeding as before, the ideal I is radical and the square free polynomials $\overline{f_1}(x), \overline{f_2}(x)$ can

be chosen to be $\overline{f_1}(x) = k(k-2)(x+1), \overline{f_2}(x) = k(k-2)(y+1)$. The second one belongs to the Gröbner basis for T_n^+ and the first one can be generated as well as our set of generators $H_i(k)$ is symmetric in the variables x, y . The statement in Theorem 5 follows now from Lemma 3.8 in [11].

Assume now $k = 2$. In this case the ideal I it is not radical. This follows using the theory developed in [11] about colorings. It is shown in [11] that 4 T has 4 colorings, three standard and one nonstandard due to the differential operator $x\,y\,\partial + \partial$. It is easy to check that one of the standard colorings (see **Figure 1** in [11]) and the nonstandard coloring (see **Figure 4** in [11]) are the only colorings for T_4^+ . One can check that a rectangle fits these colorings only if and only if it has both sides even, so it follows from Theorem 5.3 in [11] that a rectangle is signed tiled by T_4^+ if and only if it has both sides even.

10. CONCLUSIONS

Understanding tilings of rectangles by particular, even simple, polyominoes is a difficult combinatorial problem with a long history. Among the pioneering contributions, we mention those of Klarner [16]. In particular, Klarner on page 113 of [16] emphasizes the difficulty of the problem of classifying rectangles tileable by L-shaped n-ominoes of order two, that is, those for which two copies can be assembled in a rectangle: *It seems impossibly difficult to characterize the rectangles which can be packed with an n-omino of order* 2. *A theorem of this kind restricted to the L-shaped n-ominoes of order* 2 *would probably still be too difficult to formulate.*

Similar problems can be studied for parallelograms. As already mentioned in the introduction, the problem of tiling a general parallelogram positioned on a skewed lattice by all symmetries of a single *skewed* tile that has all sides parallel to the sides of the parallelogram is equivalent to the problem of tiling a rectangle by a polyomino (the straightened tile) allowing only a reduced set of orientations for that polyomino. This new problem seems to be more amenable to a solution and considerable progress has been done in the case of *L*-shaped *n*-ominoes of order two in several recent papers of one of the authors and collaborators [4] [5] [10] [17]. The results are consequences of more general tiling results for quadrants, showing that many of these tilings can be reduced to tilings by rectangles. Several general conjectures about the solution of this problem are formulated in [17].

The main contribution of the present paper is a strengthening of the results in [5]. We show that the regular tiling results obtained in [5] for the tiling sets T_n, T_n^+, n even, cannot be obtained from coloring invariants. The approach we used is via computations of explicit Gröbner basis for the ideals of polynomials generated by the tiling sets T_n, T_n^+, n even. In particular, we are able to classify signed tilings with integer weights of rectangles by T_n, T_n^+, n even. Our tools are graphic combinatorics and algebra. We also revisit some previous results of Barnes [11] [12], relevant for signed tilings with complex/rational weights, and explain what they imply for T_n, T_n^+, n even.

ACKNOWLEDGEMENTS

V. Nitica was partially supported by Simons Foundation Grant 208729. While working on this project, K. Gill

was undergraduate student at West Chester University.

REFERENCES

1. Golomb, S.W. (1954) Checker Boards and Polyominoes. *American Mathematical Monthly*, **61**, 675-682. http://dx.doi.org/10.2307/2307321
2. Golomb, S.W. (1994) Polyominoes, Puzzles, Patterns, Problems, and Packings. Princeton University Press, Princeton.
3. Pak, I. (2000) Ribbon tile Invariants. *Transactions of the American Mathematical Society*, **352**, 5525-5561. http://dx.doi.org/10.1090/S0002-9947-00-02666-0
4. Chao, M., Levenstein, D., Nitica, V. and Sharp, R. (2013) A Coloring Invariant for Ribbon *L*-Tetrominoes. *Discrete Mathematics,* **313**, 611-621. http://dx.doi.org/10.1016/j.disc.2012.12.007

5. Nitica, V. (2015) Every Tiling of the First Quadrant by Ribbon *L* *n*-Ominoes Follows the Rectangular Pattern. *Open Journal of Discrete Mathematics*, **5**, 11-25. http://dx.doi.org/10.4236/ojdm.2015.52002
6. Golomb, S.W. (1964) Replicating Figures in the Plane. *Mathematical Gazette*, **48**, 403-412. http://dx.doi.org/10.2307/3611700
7. Nitica, V. (2003) Rep-Tiles Revisited, in the Volume MASS Selecta: Teaching and Learning Advanced Undergraduate Mathematics. American Mathematical Society.
8. Conway, J.H. and Lagarias, J.C. (1990) Tilings with Polyominoes and Combinatorial Group Theory. *Journal of ombinatorial Theory, Series A*, **53**, 183-208. http://dx.doi.org/10.1016/0097-3165(90)90057-4
9. Bodini, O. and Nouvel, B. (2004) *Z*-Tilings of Polyominoes and Standard Basis, in Combinatorial Image Analysis. Springer, Berlin, 137-150. http://dx.doi.org/10.1007/978-3-540-30503-3_11
10. Nitica, V. (2015) Signed Tilings by Ribbon *L* *n*-Ominoes, *n* Odd, via Gröbner Bases. arXiv:1601.00558v2.
11. Barnes, F.W. (1982) Algebraic Theory of Brick Packing I. *Discrete Math*, **42**, 7-26. http://dx.doi.org/10.1016/0012-365X(82)90049-8
12. Barnes, F.W. (1982) Algebraic Theory of Brick Packing II. *Discrete Math*, **42**, 129-144. http://dx.doi.org/10.1016/0012-365X(82)90211-4
13. Becker, T. and Weispfenning, V. (In Cooperation with Krendel, H.) (1993) Gröbner Bases. Springer-Verlag, Berlin.
14. Dizdarevic, M.M., Timotijevic, M. and Zivaljevic, R.T. (2016) Signed Polyominotilings by *n*-in-Line Polyominoesand Gröbner Bases. *Publications de l'Institut Mathematique, Nouvelle Série*, **99**, 31-42. http://dx.doi.org/10.2298/PIM1613031M
15. Seidenberg, A. (1974) Constructions in Algebra. *Transactions of the American Mathematical Society*, **197**, 273-313. http://dx.doi.org/10.1090/S0002-9947-1974-0349648-2
16. Klarner, D.A. (1969) Packing a Rectangle with Congruent *N*-Ominoes. *Journal of Combinatorial Theory*, **7**, 107-115. http://dx.doi.org/10.1016/S0021-9800(69)80044-X
17. Calderon, A., Fairchild, S., Nitica, V. and Simon, S. (2015) Tilings of Quadrants by *L*-Ominoes and Notched Rectangles. *Topics in Recreational Mathematics*, **5-7**, 39-75.

CHAPTER 15

Extensions of CM Elliptic Curves and Orbit Counting on The Projective Line

Julian Rosen[1] and Ariel Shnidman[2*]

[1]*Department of Mathematics, University of Michigan, 530 Church St., Ann Arbor, MI 48109, USA, 2Department of Mathematics, Boston College, Chestnut Hill, MA 02647, USA.*

[2]*Department of Mathematics, Boston College,*

ABSTRACT

There are several formulas for the number of orbits of the projective line under the action of subgroups of GL2GL2. We give an interpretation of two such formulas in terms of the geometry of elliptic curves, and prove a more general formula for a large class of congruence subgroups of Bianchi groups. Our formula involves the number of walks on a certain graph called an isogeny volcano. Underlying our results is a complete description of the group of extensions of a pair of CM elliptic curves, as well as the group of extensions of a pair of lattices in a quadratic field.

Keywords: Kleinian groups Elliptic curves Complex multiplication

1 INTRODUCTION

We begin with some motivation from two well-known and elegant formulas. The first formula is for the number of orbits of $P^1(Q)$ under the action of the congruence subgroup $\Gamma_0(N) \subset GL_2(Z)$,[1] acting by fractional-linear transformation [3, §II.1]:

$$\#\Gamma_0(N)\backslash\mathbb{P}^1(\mathbb{Q}) = \sum_{d|N} \phi_u\left(\gcd(d, N/d)\right) \tag{1.1}$$

Here ϕ_u is the *reduced totient function*

$$\phi_u(n) = \#\left(\frac{(\mathbb{Z}/n\mathbb{Z})^\times}{\{\pm 1\}}\right) = \begin{cases} \frac{\phi(n)}{2} & \text{if } n \geq 3, \\ 1 & \text{if } n = 1, 2. \end{cases}$$

If we replace ϕuϕu by ϕϕ, then the right hand side of (1.1) also counts the number of cusps on the modular curve X0(N)X0(N).

The second formula is for the number of orbits of P1(K)P1(K) under the action of GL2(OK)GL2(OK), where *K* is a number field and OKOK is its ring of integers:

$$\#GL_2(OK)\backslash P^1(K)=h. \tag{1.2}$$

Here, $h=h(K)$ is the size of the class group Pic(OK). When *K* is an imaginary quadratic field, (1.2) counts the number of cusps on the corresponding Bianchi orbifold and is due to Bianchi himself. A formula counting the number of orbits of $P^1(K)$ under the action of congruence subgroups analogous to $\Gamma 0(N)$ is given in [2].

We can connect these two formulas using the theory of elliptic curves. For this, we let *E* and E' be elliptic curves over C, and we consider the number $N(E,E')$ of elliptic curves on the abelian surface $A=E\times E'$ up to the action of Aut(A). This number is finite by [5], and in fact $N(E,E')=2$ unless there exists an isogeny $\lambda:E\rightarrow E'$. This raises the question: how do we compute $N(E,E')$ if *E* and E' are isogenous?

If End(E)=Z, then Hom(E,E')=Zλ, for a certain minimal isogeny λ, whose degree we will denote by *N*. Then Aut(A) $\simeq$ Γ0(N) and we have $N(E,E')=\#\Gamma_0(N)\backslash P^1(Q)$ [9, Prop. 3.7]. So the number $N(E,E')$ is given by (1.1). On the other hand, if *E* has complex multiplication (CM) by an imaginary quadratic field *K*, we may think of Aut(A) as a subgroup of $GL_2(K)$. As before, we have $N(E,E')=\#Aut(A)\backslash P^1(K)$; see Lemma 3.1. In the special case where $E=E'$ and End(E)=O_K, we have Aut(A) $\simeq$ $GL_2(OK)$ and Bianchi's formula (1.2) gives $N(E,E)=h$.

Our main result is a formula for $N(E,E')$ for any two elliptic curves E,E' with CM by *K*. Equivalently, we compute $\#Aut(M)\backslash P^1(K)$ for any lattice $M\subset K^2$. To state the result, we follow Kani [4] and define the *e-conductor* of an elliptic curve *E* with CM by *K* to be the index [O*K*:End(E)]. Thus, if *E* has e-conductor *c*, then End(E) is isomorphic to O_c, the unique subring of index *c* inside O_K. Concretely, if $E\simeq C/a$, for some lattice aain *K*, then *c* is the index of the ring of multipliers $\{\alpha\in K:\alpha a\subset a\}$ in O*K*.

Theorem 1.1

Let K be an imaginary quadratic field whose only roots of unity are ±1.[2] Suppose E and E' are elliptic curves with complex multiplication by K and with e-conductors c and c', respectively. Define f=lcm(c,c') and f'=gcd(c,c'). Then

$$N(E,E') = \sum_{d|f} h_d \cdot \phi_u(f/d) \sum_{k|d} 2^{\omega(d/k)} r_K(f', k, f/d)$$

Here,

- $$h_d = \#\text{Pic}(\mathcal{O}_d) = \#\text{Pic}(\mathcal{O}_K) \cdot d \prod_{p|d} (1 - \chi_K(p)/p),$$
- χ_K is the quadratic Dirichlet character associated to K,
- $\omega(n)$ is the number of distinct prime factors of n, and
- $r_K(a,b,N)$ is the number of cyclic subgroups of order N of a fixed elliptic curve of e-conductor a such that the quotient has e-conductor b.

The mysterious quantity in our formula is the function $r_K(a,b,N)$. This is a 3-variable multiplicative function which is made completely explicit in Corollary 4.3. For each prime p, the values of r_K on powers of p depend only on how p splits in K, so the number $N(E,E')$ depends only on the two integers c and c' and the values of χ_K on primes dividing cc'. The explicit formulas for $r_K(a,b,N)$ are derived by interpreting these numbers as counting certain walks on graphs called *p-isogeny volcanoes*. The structure of the p-isogeny volcano then allows one to compute $r_K(a,b,N)$ simply by looking at the graph (see Theorem 4.2). We also provide a Sage script for computing the numbers $N(E,E')$ [10].

As a corollary of Theorem 1.1, we obtain orbit counting formulas for a large class of subgroups of $GL_2(K)$ which are commensurable with the Bianchi group $GL_2(O_K)$. Explicitly, if $E=\mathbb{C}/a$ and $E'=\mathbb{C}/a'$, for lattices a and a' in K, then Aut($E\times E'$) is isomorphic to the group:

$$\Gamma(\mathfrak{a},\mathfrak{a}') := \left\{ \gamma \in \begin{pmatrix} \mathcal{O}_c & c\mathfrak{a}(f'\mathfrak{a}')^{-1} \\ c'\mathfrak{a}'(f'\mathfrak{a})^{-1} & \mathcal{O}_{c'} \end{pmatrix} : \det\gamma = \pm 1 \right\}.$$

We then have the following orbit counting formula:

Corollary 1.2

$$\#\Gamma(\mathfrak{a},\mathfrak{a}')\backslash\mathbb{P}^1(K) = \sum_{d|f} h_d \cdot \phi_u(f/d) \sum_{k|d} 2^{\omega(d/k)} r_K(f', k, f/d).$$

In favorable cases, the formula in Corollary 1.2 simplifies. For example, if $a=a'=O_f$ for some $f \geq 1$, then $\Gamma(a,a') = GL_2(O_f)$ and the right hand side becomes a simple Dirichlet convolution:

Corollary 1.3 For any $f \geq 1$,

$$\#GL_2(\mathcal{O}_f)\backslash\mathbb{P}^1(K) = \sum_{d|f} h_d \cdot \phi_u(f/d).$$

Proof See Sect. 4.

These results have application to other counting problems in geometry. For example, $\#\Gamma(a,a')\backslash P^1(K)$ is the number of cusps on the hyperbolic 3-manifold $\Gamma(a,a')\backslash H^3$. It is also the number of equivalence classes of contractions of the abelian surface $A=E\times E'$, in the sense of the minimal model program. Our formula can be used to study the asymptotics of these quantities as A varies, and should be helpful in individual computations as well.

The proof of Theorem 1.1 involves a careful study of ExtExt-groups in the category of products of elliptic curves with complex multiplication by K, or equivalently, in the category of lattices in imaginary quadratic fields. These results, found in Sect. 2, are interesting in their own right and should find other applications.

2 EXTENSIONS OF CM ELLIPTIC CURVES

Fix an imaginary quadratic number field $K\subset C$. A *K-lattice of rank n* is a free abelian subgroup $L\subset K^n$ of rank $2n$. The quotient C^n/L is a complex torus, which is known to be algebraic.

2.1 Singular abelian surfaces

For context, we recall a basic fact about the abelian surfaces we are considering.

Proposition 2.1 [6] Let A be a complex be an abelian surface over CC. Then the following are equivalent:

1) *(1)* $A \simeq C^2/\Lambda$, with Λ a K-lattice of rank 2.

2) *(2)* A is isomorphic to a product of two elliptic curves, both having CM by K.

3) *(3)* A is isogenous to a product of two elliptic curves, both having CM by K.

If these equivalent conditions hold, we say that A is *singular* and that A has *CM by K*.

2.2 Rank 1 *K*-lattices

Recall that an *order* in K is a subring $R \subset O_K$ such that Frac(R)=K. Every order has the form R=Z+fO_K for a unique positive integer f, called the *conductor* of R. For each lattice $L \subset K^n$, the *ring of multipliers* $R(L)$ is the order $\{\alpha \in K{:}\alpha L \subset L\}$. The conductor of L is defined to be the conductor of $R(L)$.

If a is a rank 1 K-lattice, then a is projective as an $R(a)$-module. Two rank 1 K-lattices a and a' are *homothetic* if $a=\gamma a'$ for some $\gamma \in K^{\times}$. The set of homothety classes of lattices of conductor f forms a group under multiplication of lattices, which is denoted Pic(O_f). The set of homothety classes of lattices in K is therefore in bijection with $\coprod_{f\geq 1}$Pic(O_f). If E=C/a, then we define the e-conductor of E to be the conductor of a.

Proposition 2.2 Let A be a singular abelian surface with CM by K. Then there is a positive integer f and a lattice $a \subset K$ of conductor f' dividing f such that

$$A \cong \mathrm{C}/O_f \oplus \mathrm{C}/a.$$

Moreover, the integers f, f' and the class $[a] \in$ Pic($O_{f'}$) are uniquely determined by these conditions. In particular,

$$A \mapsto (f/f', [a])$$

is a bijection between the set of isomorphism classes of singular abelian surfaces with CM by K and pairs $(g,[a])$, where $g \geq 1$ and $[a]$ is a homothety class of lattices in K.

Proof Since A is singular with CM by K, we may write $A=E\times E'$ with E=C/a and E'=C/a' for rank 1 K-lattices a and a'. Let c and c' be the e-conductors of these elliptic curves. Then $E\times E' \simeq \mathrm{C}/O_f \times \mathrm{C}/aa'$, where f=lcm(c_1,c_2). Moreover, if two lattices $a,b \subset K$ have multiplication by O_f, then $\mathrm{C}^2/(O_f \oplus a) \simeq \mathrm{C}^2/(O_f \oplus b)$ if and only if a and b are homothetic (see [6] or [4]).

2.3 Extensions of rank 1 lattices

Let $L_1,L_2 \subset K$ be K-lattices of conductors f_1 and f_2, and let E_i=C/L_i (for i=1,2) be the corresponding elliptic curves. Suppose also that L_1=Z+Z$_{\tau 1}$ and L_2=Z+Z$_{\tau 2}$ for elements τ_1 and τ_2 of K; up to homothety, we may always choose such a basis.

We wish to classify the *extensions* of E_1 by E_2, i.e. the short exact sequences

$$0 \to E_2 \to S \to E_1 \to 0,$$

where S is a complex torus. Another extension

$$0 \to E_2 \to S' \to E_1 \to 0,$$

is said to be equivalent to the first if there is an isomorphism $S \to S'$ which induces the identity on E_2 and E_1. There is a natural group operation on the set of equivalences classes of extensions, and this group is denoted $\mathrm{Ext}^1_{\mathrm{an}}(E_1, E_2)$. The subgroup of extensions with S an abelian surface, i.e. an algebraic complex torus, is denoted $\mathrm{Ext}^1_{\mathrm{alg}}(E_1, E_2)$. We refer to [1, §I] for more details regarding extensions of complex tori.

The following result shows that the extensions of E_1 by E_2 are controlled by a third elliptic curve $\tilde{E} := \mathbb{C}/L_1L_2$.

Proposition 2.3 [1, I.5.7 and I.6.2] The association

$$z \longmapsto \frac{\mathbb{C}^2}{\begin{pmatrix} \tau_2 & 1 & z & 0 \\ 0 & 0 & \tau_1 & 1 \end{pmatrix}}$$

induces an isomorphism between the group $\tilde{E} = \mathbb{C}/L_1L_2$ and the group $\mathrm{Ext}^1_{\mathrm{an}}(E_1, E_2)$. Under this bijection, the torsion points in $\tilde{E}$ corresponds to the subgroup $\mathrm{Ext}^1_{\mathrm{alg}}(E_1, E_2)$ of algebraic extensions.

Remark There is a similar result in [8, Thm. 6.1], attributed to Lichtenbaum, but the result is not stated correctly there.

Proposition 2.3 shows that $\mathrm{Ext}^1_{\mathrm{alg}}(E_1, E_2) \simeq (\mathbb{Q}/\mathbb{Z})^2$ as a group. But it is not clear which (or how many) extension classes correspond to some fixed abelian surface S. The following theorem gives this extra information.

Theorem 2.4 Let $P \in \tilde{E}$ be a torsion point of order n, and let

$$0 \to E_2 \to S \to E_1 \to 0$$

be the corresponding extension given by Proposition 2.3. Then

$$S \simeq \mathbb{C}/O_{nf} \times /\tilde{E}\langle P \rangle,$$

where $f = \mathrm{lcm}(f_1, f_2)$.

In the proof, we consider several different notions of extension.

- For any integer c divisible by f, the lattices L_1 and L_2 can be considered as O_c-modules, and we have the group $\mathrm{Ext}^1_{\mathcal{O}_c}(L_1, L_2)$ of extensions of O_c-modules;

- The group $\mathrm{Ext}^1(L_1,L_2)$ of extensions of K-lattices;
- The group $\mathrm{Ext}^1_{\mathrm{alg}}(\mathbb{C}/L_1,\mathbb{C}/L_2)$ of extensions of abelian varieties;
- The group $\mathrm{Ext}^1_{\mathrm{an}}(\mathbb{C}/L_1,\mathbb{C}/L_2)$ of extensions of complex tori.

An extension of modules determines an extension of K-lattices, which determines an extension of abelian varieties, which determines an extension of complex tori, so there is a sequence of group homomorphisms

$$\begin{aligned}\mathrm{Ext}^1_{\mathcal{O}_c}(L_1,L_2) &\xrightarrow{\gamma_1} \mathrm{Ext}^1(L_1,L_2) \xrightarrow{\gamma_2} \mathrm{Ext}^1_{\mathrm{alg}}(\mathbb{C}/L_1,\mathbb{C}/L_2)\\ &\xrightarrow{\gamma_3} \mathrm{Ext}^1_{\mathrm{an}}(\mathbb{C}/L_1,\mathbb{C}/L_2).\end{aligned} \tag{2.1}$$

It is not hard to see that γ_1, γ_2, and γ_3 are injective. Moreover, γ_2 is an isomorphism, as there is an equivalence of categories between the category of K-lattices and the category of abelian varieties isogenous to a product of elliptic curves with CM by K. The latter equivalence of categories can be deduced from the main results of [4] or [7]. Finally, Proposition 2.3 implies the image of γ_3 is the torsion subgroup of $\mathrm{Ext}^1_{\mathrm{an}}(\mathbb{C}/L_1,\mathbb{C}/L_2)$.

Proposition 2.5 For any c divisible by f=lcm(f_1,f_2), there is an isomorphism of O_c-modules

$$\mathrm{Ext}^1_{\mathcal{O}_c}(L_1,L_2)\cong\frac{\mathrm{Hom}_{\mathcal{O}_c}(L_1,L_2)}{\frac{c}{f}\mathrm{Hom}_{\mathcal{O}_c}(L_1,L_2)}.$$

Proof Choose an algebraic integer ω with $O_K=\mathbb{Z}+\omega\mathbb{Z}$; then $O_c=\mathbb{Z}+c\omega\mathbb{Z}$. Tensoring with O_{f_1} induces a surjection $\mathrm{Pic}(O_c)\to\mathrm{Pic}(O_{f_1})$, so we can find a sublattice $\widetilde{L}_1\subset L_1$ with ring of multipliers O_c such that $\mathcal{O}_{f_1}\cdot\widetilde{L}_1=L_1$. Consider the following resolution of L_1 by projective O_c-modules:

$$0\longleftarrow L_1\xleftarrow{\varphi_0}\widetilde{L}_1^2\xleftarrow{\varphi_1}\widetilde{L}_1^2\xleftarrow{\varphi_2}\dots,$$

with

$$\varphi_0=[1,-k\omega]\quad\text{and}\quad\varphi_1=\varphi_2=\cdots=\begin{bmatrix}c\omega & -f_1c\omega^2\\ \frac{c}{f_1} & -c\omega\end{bmatrix}.$$

We apply the functor $\mathrm{Hom}(-,L_2)$ and examine the first coordinate to obtain

$$\operatorname{Ext}^1_{\mathcal{O}_c}(L_1, L_2) = \frac{\left\{\alpha \in \operatorname{Hom}_{\mathcal{O}_c}(\widetilde{L}_1, L_2) : \mathcal{O}_{f_1}\alpha \subset \operatorname{Hom}_f(\widetilde{L}_1, L_2)\right\}}{\left\{\frac{c}{f_1}\mathcal{O}_{f_1} \cdot \operatorname{Hom}_{\mathcal{O}_c}(\widetilde{L}_1, L_2)\right\}}$$
$$= \frac{\operatorname{Hom}_{\mathcal{O}_c}(L_1, L_2)}{\frac{c}{f}\operatorname{Hom}_{\mathcal{O}_c}(L_1, L_2)},$$

where the second line follows from [4, Lem. 15].

Remark Proposition 2.5 holds if K is a *real* quadratic field as well.

Corollary 2.6 The map γ_1 of (2.1) takes $\operatorname{Ext}^1_{\mathcal{O}_c}(L_1, L_2)$ isomorphically onto the $(c\,/\,f)$-torsion in $\operatorname{Ext}^1(L_1,L_2)$.

Proof Note that as an abelian group, $\operatorname{Hom}_{\mathcal{O}c}(L_1,L_2)$ is free of rank 2. Hence, by Proposition 2.5, $\operatorname{Ext}^1_{\mathcal{O}_c}(L_1, L_2)$ is an $(c\,/\,f)$-torsion group of size $(c/f)^2$ and γ_1 maps into the $(c\,/\,f)$-torsion in $\operatorname{Ext}^1(L_1,L_2)$. Proposition 2.3 implies that the $(c\,/\,f)$-torsion in $\operatorname{Ext}^1_{\mathrm{alg}}(\mathbb{C}/L_1, \mathbb{C}/L_2)$ has cardinality $(c/f)^2$, hence the $(c\,/\,f)$-torsion in $\operatorname{Ext}^1(L_1,L_2)$ also has cardinality $(c/f)^2$ because γ_2 is an isomorphism. As γ_1 induces an injective map between two sets of the same cardinality, it must be an isomorphism onto the $(c\,/\,f)$-torsion of $\operatorname{Ext}^1(L_1,L_2)$.

Lemma 2.7 Suppose

$0 \to L_2 \to L \to L_1 \to 0$

is an extension of K-lattices, where L_i has conductor f_i (for i=1,2). If the corresponding element of $\operatorname{Ext}^1(L_1,L_2)$ has order n, then L has conductor nf.

Proof The conductor of L is the minimal c such that the class in $\operatorname{Ext}^1(L_1,L_2)$ representing L is in the image of $\operatorname{Ext}^1_{\mathcal{O}_c}(L_1, L_2)$. Corollary 2.6 implies that this value is nf.

Proof of Theorem 2.4 By Proposition 2.2, the abelian surface S is isomorphic to $\mathbb{C}/O_N \times \mathbb{C}/a$ for some integer N and some lattice $a \subset K$ with $O_N \subset R(a)$. Since the conductor of $O_N \oplus a$ is N, we must have $N=nf$ by the previous lemma. Thus, it remains to show that $\mathbb{C}/a \simeq \tilde{E}/\langle P\rangle$. We do this by computing the exterior power of the lattice corresponding to S in two different ways.

On the one hand, we can recover a as the quotient of the group

$$\bigwedge^2_{\mathcal{O}_N}(\mathcal{O}_N \oplus \mathfrak{a}) \cong \bigwedge^2_{\mathcal{O}_N}\mathfrak{a} \oplus \bigwedge^2_{\mathcal{O}_N}\mathcal{O}_N \oplus (\mathcal{O}_N \otimes_{\mathcal{O}_N} \mathfrak{a}) \cong \left(\bigwedge^2_{\mathcal{O}_N}\mathfrak{a}\right) \oplus \mathfrak{a}$$

modulo its torsion. On the other hand, we have $S \simeq C^2/L$, where L is the Z-span of the period matrix in Proposition 2.3, with $z \in C$ any lift of the order n torsion point $\in \tilde{E} = C/L_1L_2$. The torsion-free part of $\bigwedge^2_{\mathcal{O}_N} L$ is spanned by the 2×2-minors of the period matrix. Thus, a is the lattice in K generated by these minors, i.e. the lattice generated by L_1L_2 and the element z. This is precisely the lattice corresponding to the elliptic curve $\tilde{E}/\langle P\rangle$.

3 PROOF OF THEOREM 1.1 AND COROLLARY 1.2

Let $A=E\times E'$ be a product of two elliptic curves with CM by the same imaginary quadratic field K. Then we may think of Aut(A) as a subgroup of $GL_2(K)$. Explicitly, choose an isogeny $\lambda:E\rightarrow E'$, and identify $\mathrm{Hom}(E,A)\otimes_Z Q$ with K^2 via the basis (1, 0) and $(0,\lambda)$. Then Aut(A) acts faithfully on this 2-dimensional K-vector space by post-composition. On the other hand, $GL_2(K)$ acts on $P^1(K)$ by fractional linear transformation.

Lemma 3.1 The orbits of $P^1(K)$ under the action of $\mathrm{Aut}(A)\subset GL_2(K)$ are in bijection with the Aut(A)-orbits of elliptic curves contained in A.

Proof Every elliptic curve on $A=E\times E'$ is isogenous to E, so is the image of some non-zero map $E\rightarrow A$. Two maps $a,b:E\rightarrow A$ have the same image if and only if there are non-zero $x,y \in$ End(E) with $ax=by$. In other words, a and b have the same image if and only if they determine the same class in the K-projectivization $P_K(\mathrm{Hom}(E,A)\otimes_Z Q)=P^1(K)$. So we have constructed an injective map from the set of elliptic curves on A to $P^1(K)$, which is also surjective since every non-zero element of $\mathrm{Hom}(E,A)\otimes_Z Q$ is a K-multiple of an element of Hom(E,A). This bijection is evidently compatible with the pushforward action of Aut(A) on elliptic curves and the action of Aut(A) on $P^1(K)$ defined above, so the lemma follows.

Corollary 1.2 follows from Lemma 3.1 and Theorem 1.1, so we focus for the rest of this section on the proof of Theorem 1.1. In other words, we will count the number $N(E,E')$ of Aut(A) -equivalence classes of elliptic curves on the abelian surface $A=E\times E'$. By Proposition 2.2, we may assume $E=C/O_f$ and $E'=C/a$, where a has conductor f' dividing f. If $E_0\subset A$ is an elliptic curve contained in A, then both E_0 and the quotient A/E_0 are elliptic curves with CM by K, and one has a short exact sequence of abelian varieties:

$$0\rightarrow E_0\rightarrow A\rightarrow A/E_0\rightarrow 0. \tag{3.1}$$

If $E_1\subset A$ is another elliptic curve with corresponding sequence

$$0\rightarrow E_1\rightarrow A\rightarrow A/E_1\rightarrow 0, \tag{3.2}$$

then $E_0 \subset A$ and $E_1 \subset A$ are Aut(A)-equivalent if and only if (3.1) and (3.2) are isomorphic as short exact sequences.

Lemma 3.2 If $E_1 \subset A$ is an elliptic curve, then the e-conductor of E_1 divides f.

Proof Choose a map $\iota: E \to A$ such that $E_1 = \iota(E)$. Write $\iota = (\iota_1, \iota_2)$ for maps $\iota_1: E \to E$ and $\iota_2: E \to \mathbb{C}/\mathfrak{a}$. Since $\mathcal{O}_f$ acts on both E and $\mathbb{C}/\mathfrak{a}$, we have $\ker \iota_1 = E[I_1]$ and $\ker \iota_2 = E[I_2]$ for certain ideals I_1 and I_2 of $\mathcal{O}_f$, by a result of Kani [4, Thm. 20(b)]. Here, $E[I]$ is the subgroup of points that lie in the kernel of $\alpha: E \to E$, for every $\alpha \in I$. Thus,

$$\ker \iota = \ker \iota_1 \cap \ker \iota_2 = E[I_1] \cap E[I_2] = E[I_1 + I_2].$$

It follows that $E_1 \simeq E/\ker \iota \simeq E/E\,[I_1 + I_2]$, and so E_1 has complex multiplication by $\mathcal{O}_f$.

Proposition 3.3 $N(E,E')$ is equal to the number of isomorphism classes of short exact sequences

$$0 \to E_1 \to A \to E_2 \to 0,$$

with E_1 and E_2 elliptic curves of e-conductor dividing f.

Proof The previous lemma shows that any elliptic curve $E_1 \subset A$ has e-conductor dividing f. Dualizing, we see that $E_2 = A/E_1$ is also an elliptic curve in $\hat{A} \simeq A$, so has e-conductor dividing f as well.

We continue the proof of Theorem 1.1. Suppose that E_1 and E_2 are elliptic curves with e-conductor dividing f. Observe that if $\gamma_1, \gamma_2 \in \mathrm{Ext}^1(E_2, E_1)$ correspond to extensions

$$0 \to E_1 \to A_1 \to E_2 \to 0, \qquad (3.3)$$

$$0 \to E_1 \to A_2 \to E_2 \to 0, \qquad (3.4)$$

then (3.3) and (3.4) are isomorphic as short exact sequences if and only if γ_1 and γ_2 are in the same orbit of $\mathrm{Aut}(E_1) \times \mathrm{Aut}(E_2)$ on $\mathrm{Ext}^1(E_2, E_1)$. By assumption, K contains no non-trivial roots of unity, so $\mathrm{Aut}(E_1) = \mathrm{Aut}(E_2) = \{\pm 1\}$. Combining this observation with Proposition 3.3, we obtain

$$N(E,E') = \#\mathrm{Aut}(A)\ \mathbb{P}^1(K) = \sum_{f_1, f_2 | f} \sum_{\substack{[L_1] \in \mathrm{Pic}(\mathcal{O}_{f_1}) \\ [L_2] \in \mathrm{Pic}(\mathcal{O}_{f_2})}} \#\mathrm{Ext}^1_{\mathcal{O}_f}(L_2, L_1)_A / \{\pm 1\},$$

where $\mathrm{Ext}^1_{\mathcal{O}_f}(L_2, L_1)_A$ is the set of extensions classes $0 \to L_1 \to L \to L_2 \to 0$ in $\mathrm{Ext}^1_{\mathcal{O}_f}(L_2, L_1)$ such that $A \simeq \mathbb{C}^2/L$.

Now fix integers f_1 and f_2 dividing f, and set $k = \gcd(f_1, f_2)$, $d = \mathrm{lcm}(f_1, f_2)$. Recall that an isogeny of elliptic curves is *cyclic* if its kernel is a cyclic group. We call a cyclic isogeny *based* if the kernel is equipped with a distinguished generator. By Theorem 2.4, classes in $\mathrm{Ext}^1_{\mathcal{O}_f}(L_2, L_1)_A$ correspond to based cyclic isogenies $\mathbb{C}/L_1L_2 \to \mathbb{C}/a$ of degree f/d. Dualizing, we find that these are equinumerous to based cyclic (f/d)-isogenies $\mathbb{C}/a \to \mathbb{C}/L_1L_2$.

Recall that $r_K(a,b,N)$ is the number of cyclic subgroups of order N of a fixed elliptic curve of e-conductor a such that the quotient has e-conductor b. Thus, there are $r_K(f',k,f/d)$ cyclic subgroups $G \subset \mathbb{C}/a$ of order f/d such that the quotient has e-conductor k. For each such subgroup, there are $h_{f1}h_{f2}/h_k = h_d$ pairs $[L_1] \in \mathrm{Pic}(O_{f1})$, $[L_2] \in \mathrm{Pic}(O_{f2})$ with $(\mathbb{C}/a)/G \cong \mathbb{C}/L_1L_2$. Given such G, L_1, and L_2, there are $\phi_u(f/d)$ based cyclic (f/d)-isogenies $\mathbb{C}/a \to \mathbb{C}/L_1L_2$ with kernel G, up to the action of $\{\pm 1\}$. So we conclude that

$$\#\mathrm{Aut}(A)\backslash\mathbb{P}^1(K) = \sum_{\substack{f_1,f_2 \mid f \\ k:=\gcd(f_1,f_2) \\ d:=\mathrm{lcm}(f_1,f_2)}} h_d \cdot \phi_u(f/d) \cdot r_K(f', k, f/d)$$

$$= \sum_{k \mid d \mid f} 2^{\omega(d/k)} \cdot h_d \cdot \phi_u(f/d) \cdot r_K(f', k, f/d),$$

completing the proof of Theorem 1.1.

4 COMPUTATION OF $R_K(A,B,C)$

To make Theorem 1.1 explicit, we need to compute the number $r_K(a,b,c)$ of cyclic subgroups C of order c of a fixed CM elliptic curve E of e-conductor a such that E/C has e-conductor b. We will see that this does not depend on the choice of the elliptic curve E of e-conductor a. First we reduce to the case where a, b, and c are powers of a prime p:

Lemma 4.1

The function $r_K(a,b,c)$ is multiplicative:: if we factor a, b, and c into prime powers $a = \prod_p p^{a_p}$, $b = \prod_p p^{b_p}$, *and* $c = \prod_p p^{c_p}$, then

$$r_K(a,b,c) = \prod_p r_K\left(p^{a_p}, p^{b_p}, p^{c_p}\right).$$

Proof If E is an elliptic curve, then giving a cyclic subgroup of order c is the same as giving a cyclic subgroup of order p^{c_p} for every prime p. Furthermore, if $E \to E'$ is an isogeny of CM elliptic curves of p-power degree, the ratio of the e-conductors of E and E' is a power of p. To see this, it suffices to show that if $a \subset b$ is an inclusion of rank 1 K-lattices of index papa, then the conductors of a and b are off by a power of p. And indeed, if ℓ is any prime, then the ℓ-part of the conductor of a lattice can be computed locally at ℓ. In particular, if $\ell \neq p$, then $a \hookrightarrow b$ induces an isomorphism $a \otimes Z_\ell \simeq b \otimes Z_\ell$. It follows then that

$$r_K(a,b,c) = \prod_p r_K\left(a, p^{b_p} \prod_{\ell \neq p} \ell^{a_\ell}, p^{c_p}\right).$$

To prove the lemma it now suffices to show that

$$r_K\left(a, p^{b_p} \prod_{\ell \neq p} \ell^{a_\ell}, p^{c_p}\right) = r_K(p^{a_p}, p^{b_p}, p^{c_p}),$$

or, equivalently, that $r_K(p^a m, p^b m, p^c) = r_K(p^a, p^b, p^c)$, for any integer m, prime to p.

First note that if E has e-conductor $p^a m$, then there exists an elliptic curve E' of e-conductor p^a and an isogeny $\phi: E \to E'$ of degree prime to p. Indeed, it is enough to show that E admits an isogeny to $\mathbb{C}/\mathcal{O}_{p^a m}$ of degree prime to p. For this, write $E = \mathbb{C}/a$ and choose an ideal λ of $\mathcal{O}_{p^a m}$ in the class of a−1a−1 in Pic($\mathcal{O}_{p^a m}$), and whose norm is prime to p. Then the map

$$E \to E/E[\lambda] \simeq \mathbb{C}/\mathfrak{a}\lambda^{-1} \simeq \mathbb{C}/\mathcal{O}_{p^a m}$$

is such an isogeny. The equality $r_K(p^a m, p^b m, p^c) = r_K(p^a, p^b, p^c)$ now follows since cyclic subgroups of E size p^c are in bijection with cyclic subgroups of E' of size p^c, via $C \mapsto \phi(C)$. Moreover, if E / C has e-conductor $p^b m$, then $E'/\phi(C)$ has e-conductor p^b, by the argument in the first paragraph of this proof.

Next, we fix a prime p and show how to compute $r_K(p^a, p^b, p^c)$. We will use p-isogeny volcanoes [11], a tool typically used in the study of elliptic curves over finite fields. We define a graph G_p, whose vertex set is the set of isomorphism classes of elliptic curves E/C with CM by K (we omit the dependence on K from the notation). The edge set of G_p is the set of isomorphism classes of p-isogenies between two such elliptic curves. For each isogeny $\phi: E \to E'$ there is a dual isogeny $\hat{\phi}: E' \to E$, so we identify these two edges and consider G_p as an undirected graph.

For our purposes we need only consider connected components of G_p containing a vertex representing a curve of e-conductor 1. If H_p is such a component, then the vertices of H are partitioned into *levels* $H_{p,0}$, $H_{p,1}, \ldots$,

$H_{p,k}$,..., one for each $k \geq 0$, where $H_{p,k}$ consists of the curves in H_p with *e*-conductor p^k. We call the subgraph on $H_{p,0}$ the *surface*. The key fact is that all such connected components are isomorphic, and each is a *volcano*. In particular, if H_p is the connected component of a curve of *e*-conductor 1, then H_p has the following properties [11, §2]:

- H_p is $(p+1)$-regular.
- The surface $H_{p,0}$ is a $(1+\chi_K(p))$ -regular graph on t vertices, where t is the order of the class $[p]$ in $\mathrm{Pic}(O_K)$, for any prime ideal p above p.
- For $k \geq 1$, each vertex in $H_{p,k}$ has a unique edge leading to $H_{p,k-1}$, and this accounts for every edge not in the surface.

We note that for this description of H_p to be accurate, we must require that K contain no roots of unity other than $\{\pm 1\}$.

Figures 1, 2 and 3 depict examples of H_p in the case $p=3$, for each of the three splitting types. In p is inert in Kthen every H_p is of the form in the Fig. 1. If p is ramified, each $H_{p,0}$ either has two surface vertices with a unique edge connecting them (as in the Fig. 2), or a single vertex with a self-loop. If p is split, each component either has an n-cycle at the surface with $n \geq 2$ (the case drawn in Fig. 3 is $n=3$), or there is a unique vertex with two self loops.

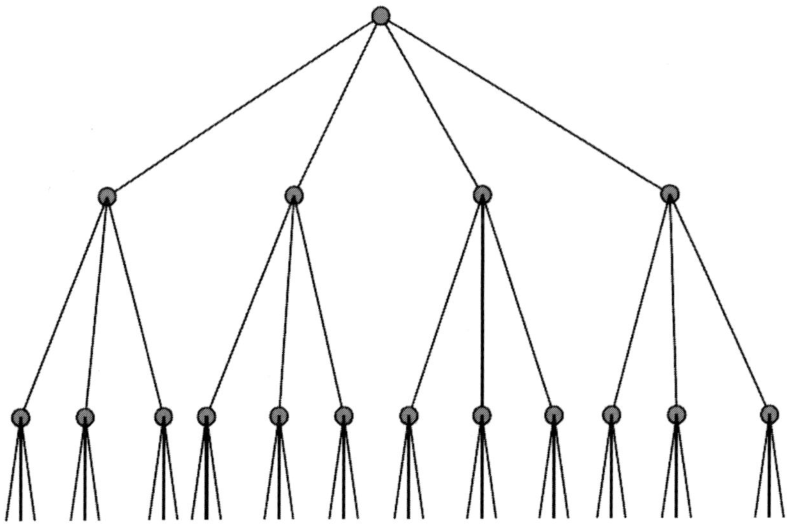

Figure. 1 Connected component of H_3, inert case

Recall that if $G=(V,E)$ is a graph, and $v,v'\in V$ are vertices, then a *walk from v to v′ of length c* is a sequence

$$v=v_0,e_1,v_1,e_2,\ldots,v_{c-1},e_c,v_c=v',$$

where each $e_i\in E$ is an edge connecting v_i-1 and v_i.

Definition A walk is called *non-backtracking* if $e_i\neq e_i+1$ for all i.

Theorem 4.2 Let a, b, c be non-negative integers. Then $r_K(p^a,p^b,p^c)$ is the number of non-backtracking walks of length c in H_p starting at a *fixed* vertex of level a and ending at a vertex of level b.

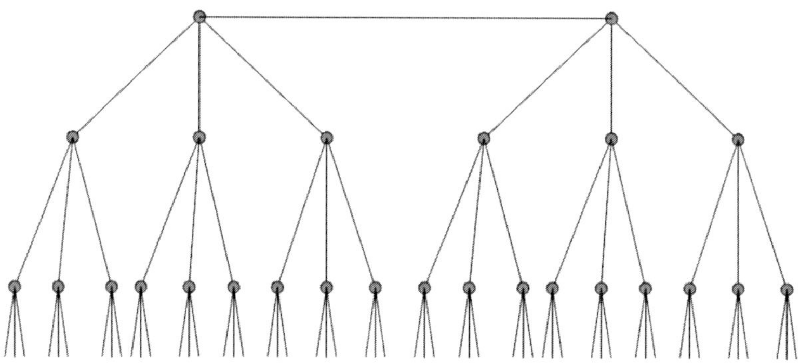

Figure. 2 Connected component of H_3, ramified case

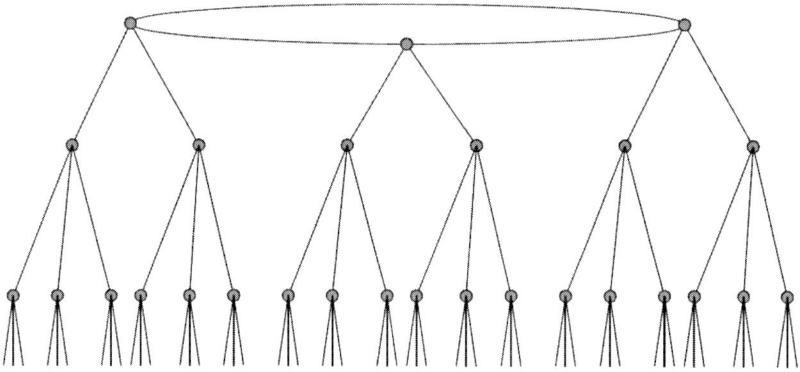

Figure. 3 Connected component of H_3, split case

Proof Suppose E is an elliptic curve with CM by K of e-conductor p^a. Then isomorphism classes of p-isogenies out of E are in bijection with subgroups of E of order p, since we work in characteristic 0, and since $\mathrm{Disc}(K)<-4$. It follows that subgroups of E of order p are in bijection with length 1 walks on H_p starting at E. An isogeny of degree p^c is a composition of p-isogenies, and thus corresponds to a walk along H_p of length c. Backtracking amounts to composing with the dual of the previous p-isogeny, which would make the the composite isogeny divisible by p and hence not cyclic. The theorem follows.

Theorem 4.2 allows us to compute $r_K(p^a,p^b,p^c)$ explicitly:

Corollary 4.3 Let $\chi_K(p)$ equal −1, 0 , or 1, depending on whether p is inert, ramified, or split in K. Then

$$r_K(p^a,p^b,p^c) = \begin{cases} 1 & \textit{if } a=b=c=0 \\ 1+\chi_K(p) & \textit{if } a=b=0 \textit{ and } c>0 \\ 0 & \textit{if } a<b \textit{ and } c<b-a \\ p^c & \textit{if } 0<a\le b \textit{ and } c=b-a \\ (p-\chi_K(p))\,p^{c-1} & \textit{if } 0=a<b \textit{ and } c=b \\ (p-1)p^{(b-a+c)/2-1} & \textit{if } a\le b \textit{ and } b-a<c<b+a \\ (p-\chi_K(p)-1)p^{b-1} & \textit{if } a\le b \textit{ and } c=b+a>0 \\ (1+\chi_K(p))\,(p-\chi_K(p))\,p^{b-1} & \textit{if } a\le b \textit{ and } c=b+a+1>1 \\ (\chi_K(p)+|\chi_K(p)|)\,(p-1)p^{b-1} & \textit{if } a\le b \textit{ and } c>b+a+1>1 \\ r_K(p^b,p^b,p^{c-a+b}) & \textit{if } a>b. \end{cases}$$

Here, we use the convention that p^s=0 if s is not an integer. Note that the formula in the last case reduces to one of the earlier cases.

Proof The key point is that non-backtracking walks on H_p are very constrained: once a non-backtracking walk begins to descend down the volcano, it cannot reascend. Moreover, the only 'horizontal' edges on H_p are at the surface. It follows that a non-backtracking walk on H_p consists of (at most) three stages in the following order: ascend up the volcano, move horizontally at the surface, and then descend down the volcano.

We give the proof of the corollary in the case $a\le b$ and $c>b+a+1$, and leave the proofs of the other cases to the reader, as they are similar. We must choose a vertex in level a and count the number of non-backtracking walks of length c to a vertex of level b. Note that $b\ge1$ in this case. Since $c>a+b$, any non-backtracking walk from level a to level b of length c on H_p must begin by ascending via the unique path to the surface. Then the walk must take $c-b-a\ge2$ horizontal steps along the surface before descending down to level b. This is only possible, without backtracking, if p splits in K, and in that case there are exactly two ways to traverse across the n-cycle at the surface. There are then $(p-1)p^{b-1}$ ways to descend to level b. Thus, in the split case there are a total of $2(p-1)p^{b-1}$ non-backtracking walks, whereas in the non-split cases there are $0=\chi_K(p)+|\chi_K(p)|$ such walks, as claimed.

As an example of how to compute these numbers "by eye", we now give the proof of Corollary 1.3. This is the special case where $A = \mathbb{C}^2/\mathcal{O}_f^2$, and hence $\mathrm{Aut}(A)=\mathrm{GL}_2(O_f)$.

Proof of Corollary 1.3 We need to show that if $k|d|f$, then

$$r_K(f, k, f/d) = \begin{cases} 1 & \text{if } k = d \\ 0 & \text{otherwise.} \end{cases}$$

By Lemma 4.1, we may assume $f=p^a$, $k=p^b$, and $d=p^c$, with $a\geq c\geq b$. Then if $b<c$, a vertex in level a cannot walk to level b in $a-c$ steps. But if $b=c$, there is a unique (ascending) walk from a vertex of level a to a vertex of level b of length $a-b$, as claimed.

In the general case, the formula for $N(E,E')$ obtained by combining Theorem 1.1 and Corollary 4.3 does not seem to simplify much further.

ACKNOWLEDGEMENTS

AS thanks Andrew Snowden for a helpful conversation and Andrew Sutherland for help with the figures. AS was partially supported by NSF Grant DMS-0943832.

FOOTNOTES

1 Our definition of Γ0(N)Γ0(N) is not the traditional one, as we allow elements of determinant −1.

2 There are analogous formulas for the two imaginary quadratic fields with more roots of unity, but we omit these cases for simplicity.

REFERENCES

1. Birkenhake, C., Lange, H.: Complex Tori. Springer, Berlin (1999)
2. Cremona, J. E., Aranés, M. T.: Congruence subgroups, cusps and manin symbols over number fields. In: Boeckle, G., Wiese, G. (eds.) Computations with Modular Forms. Contributions in Mathematical and Computational Sciences, vol. 6, pp. 109–127. Springer International Publishing (2014)

3. Gross, B.H., Zagier, D.: Heegner points and derivatives of LL-series. Invent. Math. **84**, 225–320 (1986)
4. Kani, E.: Products of CM elliptic curves. Collect. Math. **62**(3), 297–339 (2011)
5. Lenstra Jr., H., Oort, F., Zarhin, Y.: Abelian subvarieties. J. Algebra **180**(2), 513–516 (1996)
6. Mitani, T., Shioda, T.: Singular abelian surfaces and binary quadratic forms. In: Classification of Algebraic Varieties and Compact Complex Manifolds. Lecture Notes Mathematics, vol. 412, pp. 259–287 (1974)
7. Oort, F., Zarhin, Y.: Complex tori. Indag. Math. **7**, 473–487 (1996)
8. Papanikolas, M., Ramachandran, N.: A Weil–Barsotti formula for Drinfeld modules. J. Number Theory **98**(2), 407–431 (2003)
9. Rosen, J., Shnidman, A.: Néron-Severi groups of product abelian surfaces. arXiv: 1402.2233 (2014)
10. Sage script for computing orbit counts. arXiv:1608.01390v1
11. Sutherland, A.: Isogeny volcanoes. Open Book Ser. **1**(1), 507–530 (2013)

Electrical Methods in Geophysical Exploration of Deep Sedimentary Basins

S.H. Yungul

Geophysical Consultant
Formerly Senior Research Associate
Chevron Oil Field Research Co.

CHAPMAN & HALL
London · Weinheim · New York · Tokyo · Melbourne · Madras

Published by Chapman & Hall, 2–6 Boundary Row, London SE1 8HN, UK

Chapman & Hall, 2–6 Boundary Row, London SE1 8HN, UK

Chapman & Hall GmbH, Pappelallee 3, 69469 Weinheim, Germany

Chapman & Hall USA, 115 Fifth Avenue, New York, NY 10003, USA

Chapman & Hall Japan, ITP-Japan, Kyowa Building, 3F, 2-2-1 Hirakawacho, Chiyoda-ku, Tokyo 102, Japan

Chapman & Hall Australia, 102 Dodds Street, South Melbourne, Victoria 3205, Australia

Chapman & Hall India, R. Seshadri, 32 Second Main Road, CIT East, Madras 600 035, India

---

First edition 1996

Typeset in Palatino 10/12 pt. by Thomson Press (India) Ltd, Madras
Printed in Great Britain by St Edmundsbury Press, Bury St Edmunds, Suffolk

ISBN 0 412 73710 8

A catalogue record for this book is available from the British Library

Library of Congress Catalog Card Number: 95-83574

∞ Printed on permanent acid-free text paper, manufactured in accordance with ANSI/NISO Z39.48-1992 and ANSI/NISO Z39.48-1984 (Permanence of Paper).

Contents

Preface

The objective of this book is to provide a working knowledge of the deep-penetrating electrical methods of exploration, with specific reference to sedimentary basins. The emphasis is on concepts, methods, exploration strategy and interpretation. Only a limited amount of material on mathematical equation derivation, instrumentation, field procedures and data processing is included. Well logging and those techniques which are designed specifically for prospecting for metallic sulfides are not included.

The material covered here is applicable primarily, but not exclusively, to:

- general exploration of deep or shallow sedimentary basins;
- exploration of the Earth's deep crust; and
- prospecting for oil and gas, geothermal reservoirs and groundwater resources.

Some of these methods are used in earthquake prediction research, and some lend themselves to deep and shallow environmental studies and geotechnical work.

This book may serve practising geologists and geophysicists, especially those in the petroleum industry, and particularly in survey planning and data interpretation. It may also be used as a textbook. There is continuity throughout the book which can be understood by non-geophysicists.

The list of references was kept to an optimum length for such a small book; however, thousands of investigators have contributed to this very diversified field in the past.

Field examples and case histories are given as part of the methods to which they are connected (not in separate sections). Following an introductory chapter and one on electrical properties of sedimentary rocks, the main body of the text starts with Chapter 3, Direct current resistivity methods, not only because these heralded the sounding techniques in applied geophysics before the advent of the practical seismic methods, but also because they constitute an accommodating, all-purpose tool which has evolved through developments and field applications since the 1920s; they are viable and are being further developed.

At the present time the subject of Chapter 4, Natural-source electromagnetic methods, is probably the most important electrical method used in the exploration of deep sedimentary basins. Consequently, this chapter is the most comprehensive one in the book.

Frequency- and time-domain controlled-source electromagnetic methods (including the controlled-source magnetotelluric method) are discussed in Chapter 5. So far as deep sedimentary basins are concerned, these methods have been the subject of recent developments, but they need improvement to be well established in petroleum exploration. They may have a better future in petroleum work than the other electrical methods.

Finally, Chapter 6, Possibilities of direct detection of petroleum, is essential simply because most of the geophysical work done in sedimentary basins pertains to oil and gas. The endeavors in this field span from the 1930s to the present (1994). There is much mystery and controversy associated with this subject.

My thanks are due to Drs James Combs, Lewis H. Cohen, Shawn Biehler, and Tien-Chang Lee of University of California, at Riverside, for having invited me to teach a graduate-level course on electrical methods at their university that eventually led to this book, to Dr Stanley B. Jones of Chevron Oil Field Research Company for making it possible to organize and teach this course, and to my son Martin, who learned wordprocessing to type this book in his spare time.

S.H. Yungul
Mill Valley, California
September 1995

Glossary of symbols and abbreviations

AB	Grounded-wire dipole or bipole
a	Radius of a circle (loop)
$\boldsymbol{B}$	Magnetic induction vector, newton/A × m, tesla, T
CSEM	Controlled-source electromagnetic
CSMT	Controlled-source magnetotelluric
d	depth to top of a geologic layer
DC	direct current
$\boldsymbol{E}$	Electric field vector, V/m; mV/km in practice
EM	Electromagnetic
f	Frequency, cycles/s, hertz, Hz
FDEM, FEM	Frequency-domain electromagnetic
h	Thickness of a geologic layer
$\boldsymbol{H}$	Magnetic field vector, A/m
H_z	Vertical component of $\boldsymbol{H}$
Hz	Hertz, unit of f, cycle/s
I	Electric current, ampere, A
IP	Induced polarization
$\boldsymbol{J}$	Current density vector, A/m^2
J	Jacobian of telluric transfer function
k	Wave number
LOTEM	Long-offset time-domain EM
MT	Magnetotelluric
MTS	Magnetotelluric sounding
n	Number of turns in a loop
nT	Nanotesla, 10^{-9} T
R	Source–receiver separation
s	Second
S	Area of a loop, m^2
S	Conductance of a geologic layer, h/ρ
SP	Spontaneous potential
t	Time, s
T	Period, $1/f$, s/cycle

T	tesla, unit of $\boldsymbol{B}$
TDEM, TEM	Time-domain electromagnetic
uniform	Homogeneous and isotropic
V	Electric potential, EMF, volt, V
x, y, z	Cartesian coordinates, z positive down
Z	Mutual impedance, FEM
Z_0	Free-space mutual impedance, FEM
Z_0	MT impedance at $z=0$
γ	Wave number (propagation constant)
γ	Practical unit of $\boldsymbol{B}$ in cgs-emu, 1 nT
δ	Skin depth, m, $(2\rho/\mu\omega)^{1/2} \cong 503(\rho T)^{1/2}$
ε	Electric inductive capacity, farad/m
ε_0	ε of free space, 8.854×10^{-12}
λ	Wavelength, $2\pi\delta$
λ	Phantom variable of integration
μ	Magnetic inductive capacity, henry/m
μ_0	μ of free space, $4\pi \times 10^{-7}$ henry/m
ρ	Resistivity, ohm × m, Ω m
ρ_a	Apparent resistivity
ρ_l	Longitudinal resistivity (parallel to bedding)
ρ_t	Transversal resistivity (perpendicular to bedding)
ρ_e	Effective ρ_l combined for several layers
σ	Conductivity, $1/\rho$, mho/m, siemens/m
ϕ	Phase, or phase difference, radian, rad
ϕ	Azimuth angle in cylindrical coordinates
ω	Angular frequency, rad/s, $2\pi f$
Ω	ohm, resistance
Ω m	ohm × m, resistivity

1

The place and role of electrical methods

The following is a partial list of common fields of exploration and the most favored methods in these fields.

1. Solid Earth physics (general geophysics)	Earthquake seismology
2. Mining	Electrical
3. Geothermal	Electrical
4. Ground water	Electrical
5. Well logging (borehole geophysics)	Electrical
6. Oil and gas exploration	Reflection seismograph

1, solid Earth physics, is mostly in the hands of non-profit organizations, and the electrical methods have a fair share in it. In items 2–5 the electrical methods are, and have been for a long time, the main exploration tools.

It is item 6, oil and gas exploration, that creates arguments as to what extent the electrical methods are cost effective. Obviously it is proper that seismic methods are the main tool used, but it is sometimes argued that the electrical methods should be utilized more than they are at present.

In what immediately follows, the place and role of the electrical methods in items 1–6 will be discussed in broad terms, without going into individual techniques.

1.1 SOLID EARTH PHYSICS

Geophysical exploration in the study of the internal constitution of the Earth is carried out mainly with earthquake seismology. Electrical methods are also used in this field, especially magnetotellurics, to explore the Earth's crust and upper mantle. For instance, heat anomalies in the upper mantle are associated with resistivity anomalies which show up in magnetotelluric measurements. For this reason, and also due to lithological differences, continental rift zones represent quite different resistivity structures than crustal shields or crust under deep basins (e.g. Madden and Swift, 1969; Madden, 1971; Mitchell and Landisman, 1971). Some of these rifts become sedimentary basins. Direct current resistivity sounding has been used to explore down to the upper mantle to determine resistivity structure of the crust in detail (e.g. Van Zijl and Joubert, 1975).

This field, which is also called general geophysics, is not confined to large-scale studies of the crust and upper mantle, but includes academic studies in sedimentary structural, lithological and depositional problems, and earthquake prediction. The electrical methods discussed in this book lend themselves to these problems as well as to the crustal and upper mantle studies.

1.2 MINING

In mining geophysics electrical methods are the most important. Airborne magnetic surveys are also extensively used in mining exploration, but these are primarily for regional surveys. Detection of metallic sulfides is done primarily by means of electrical methods, disseminated sulfides mostly by induced polarization (IP), which is very briefly discussed in Chapter 6, and massive sulfides mostly by means of the so-called electromagnetic induction (EMI) methods, ground and airborne. These EMI methods are designed primarily to search for shallow buried conductors, such as metallic massive sulfides. They are used mostly in the airborne mode. There are a great variety of methods differing in source–receiver types and configurations, both of which may be coils or electric dipoles. Depth of exploration is of the order of 100 m. Commonly they operate in the audio-frequency range, about 300–3000 Hz. Lower frequencies penetrate deeper, but since the currents induced in the conducting ore are proportional to frequency, they produce lower-amplitude anomalies. Also, the deeper the ore the lower the anomaly at the surface. Thus, the effects of lower frequency and deeper anomaly source are compounded, and lower frequencies become unrewarding. Because of extreme diversification, these EMI methods constitute a field of specialization by themselves. Most books on applied geophysics contain large sections on these methods (e.g. Keller and Frischknecht, 1966; Telford *et al.*, 1976; Nabighian, 1988, 1991). These methods do not have much application in the exploration of sedimentary basins and they are not discussed in this book.

1.3 GEOTHERMAL

Most geothermal reservoirs in the category of convective hydrothermal systems are associated with low-resistivity anomalies, whether they are in sedimentary basins or elsewhere. This is because higher temperature lowers the resistivity of the water contained in rock pores (up to the boiling point), and, water at higher temperature usually carries higher concentrations of dissolved salts, so that the combined effect on resistivity is usually quite large. Such reservoirs may be detected directly, that is, by prospecting for the hot waters themselves, and not for the geology (Lumb, 1981). There are also the effects of alteration minerals which may either lower or raise the host rock resistivity. The consensus is that the most useful geophysical prospecting tools in this field are heat flow measurements and surface electrical surveys of which a few geothermal applications are discussed in this book. An extensive evaluation of geothermal application of the electrical methods is given by Goldstein, Norris and Wilt (1978).

Convective hydrothermal systems in volcanic environments do not display many common features; each one is very different to the others. Those in sedimentary basins are more susceptible to generalizations, as discussed below.

In the Imperial Valley basin of southern California, in an area where the regional resistivity is about 5 Ω m, in a typical geothermal system the uppermost 1 km section may be about 1 Ω m, much lower than the regional value, but below this the resistivity may be about 20 Ω m, much higher than the regional value, all the way down to the basement complex which may be 5 km deep. The upper low-resistivity section is due to higher temperature and higher salinity. Some of it may be lithified and behave like a high-resistivity cap rock. The high-resistivity section below is due to hydrothermally metamorphosed (indurated) rocks, especially shales. The horizontal extent of this anomalous region may be 4–6 km. There may be a hot water plume in fractured rocks, in the high-resistivity, indurated zone, that may yield commercial production.

An example of the type discussed above is the Salton Sea geothermal field in the Imperial Valley. A cross-section is shown in Figure 1.1 (Kasameyer, 1976). This is the resistivity structure obtained from direct current resistivity soundings. The shallow high-resistivity zone (shaded) is the cap rock, anhydrite-rich evaporites in a carbonate–clay matrix; it has undergone self-sealing. The deeper high-resistivity zone is the hydrothermally indurated, low-porosity rocks. Outside these high-resistivity zones lie the low-resistivity zones. The total conductance of the section above the deeper high-resistivity zone indicates that the effective resistivity over the geothermal field is less than one-third of the regional value. These resistivity regions cut across the sedimentary bedding.

A producing field at Cerro Prieto, Mexico, in the southward extension of the Imperial Valley, has been the scene for a special study using surface electrical methods to monitor changes in reservoir conditions due to production, such as formation of a vapor zone or invasion of cold water.

Figure 1.2 shows a two-dimensional resistivity model obtained from resistivity profiling (Wilt and Goldstein, 1981). Such work is discussed in section 3.3. The dashed curve represents the top of hydrothermal metamorphism (the indurated zone) in the production zone, based on borehole data. This is near the top of the high-resistivity, 4 Ω m, zone in

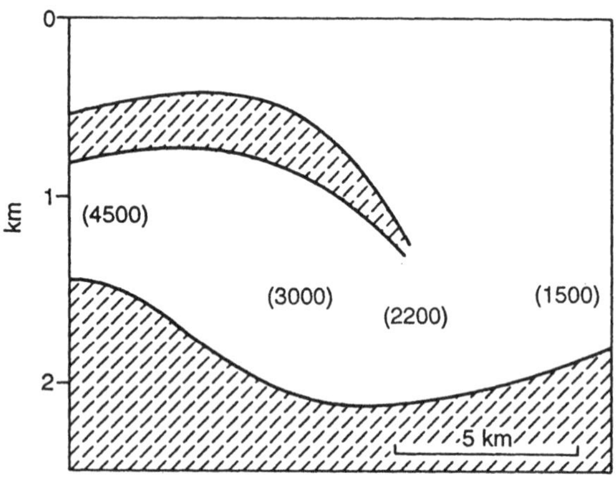

Figure 1.1 Cross-section of resistivity distribution, Salton Sea geothermal field, Imperial Valley, California, obtained from resistivity soundings. Shaded regions are the high-resistivity cap rock (shallow) and the indurated zone (deep). Total conductance, in mho, above the deep indurated zone is shown in parentheses (after Kasameyer, 1976).

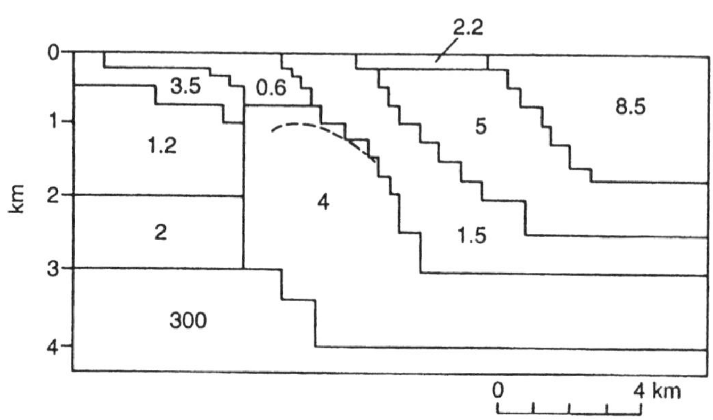

Figure 1.2 Cross-section of resistivity model, Cerro Prieto geothermal field, Mexico, from dipole–dipole resistivity profiling; resistivities are in Ωm; dashed curve represents top of the indurated rocks in the production zone (after Wilt and Goldstein, 1981).

Figure 1.2; the regional average resistivities are less than 2 Ω m. The shallow low-resistivity zone (0.6 Ω m) is also present.

Most of the electrical methods used in geothermal work are discussed fairly extensively in this book.

1.4 GROUND WATER

In the early 20th century, use of the 'divining rod' for finding water was superseded by electrical methods, which have since enjoyed the top spot among geophysical methods, and they still have a large potential for growth. When all the oil prospectors are gone, the water prospectors will be there.

Resistivity of a geologic formation depends primarily on pore water content and on salinity (ion concentration) in that water. High-porosity and water-saturated sands have lower resistivity than low-porosity or unsaturated sands of the same salinity. If a sand layer dipping under a clay layer is followed by means of resistivity sounding, one may locate water, an aquifer or an artesian source. However, the resistivity of a sand may be reduced also when a clean one becomes shaley.

An alluvial water table is the simplest kind of target. Other targets include a perched water table, impermeable basement, and fresh- and saltwater boundaries. In general, electrical exploration of a sedimentary basin yields information on structure, lithology and formation water conditions that leads to finding water.

Several case histories throughout the world showing the results of direct current resistivity methods are given by Breusse (1963). The controlled-source electromagnetic (CSEM) sounding methods may sometimes be better suited than the direct current methods, because CSEM methods have a theoretical advantage in locating low-resistivity layers (Fitterman and Stewart, 1986).

There are some field examples on application to water in this book (Figures 3.15a and 3.15d), which give wide coverage to most of the electrical methods used in this field.

A special type of application of electrical methods to water problems is to determine salinity, to distinguish between salt and fresh water and to map interfaces between the two. This is a use for which electrical methods have no competitor. Figure 3.15a shows an example. Structures on such interfaces may have implications other than freshwater prospecting, such as fault mapping and deep basinal information. Volker and Dijkstra (1955) describe a case history in which the salinity distribution in a deep aquifer under the sea floor (Zuiderzee) was determined by means of resistivity sounding. This is also a field example of the use of electrical methods at sea.

1.5 WELL LOGGING (BOREHOLE GEOPHYSICS)

Electrical well logging is the main branch in this field; it is second only to seismic exploration in geophysical expenditure.

Electrical well logging grew out of surface electrical methods. Basically, it is resistivity profiling (section 3.3) over subcropping, steeply dipping geologic formations. The original pole–dipole array (section 3.1) became the **lateral device** and the pole–pole array became the **normal device** in well logging. Petroleum geophysicists were not very happy with electrical methods after the advent of the seismic reflection method but, when they turned the horizontal electrode arrays into vertical, they liked the results. The reason for this success is probably because resistivities of most sedimentary layers range from 1 to 1000 Ω m, a contrast of 1000, whereas the seismic velocity contrast is only about 3 or 4. This field is now very large, with many specialists and much literature; it is not discussed in this book.

A new field of electrical borehole geophysics is being developed that is directed to exploration of targets far away from boreholes. Such measurements include:

- sounding from one borehole laterally (horizontally) to map flanks of piercement salt domes;
- borehole-to-borehole measurements, to correlate formations, for instance, or to monitor the progress of assisted recovery in petroleum production;
- borehole-to-surface measurements for higher resolution in target delineation.

A special issue of *Geoexploration* (Vol. 28, Nos. 3/4, October 1991) is devoted to this subject. Asch and Morrison (1989) provide an extensive list of references on borehole-to-surface measurements. This field also is not discussed in this book.

1.6 OIL AND GAS (SURFACE EXPLORATION)

In this field there is argument as to what extent the electrical methods are cost effective. The major portion of all geophysical expenditure in the world is in oil and gas exploration, primarily using the seismic reflection method. Needless to say, it is proper that the seismic method is the primary tool in finding oil because it provides far better resolution in deep sedimentary layers than any other method. With the electrical methods, resolution decreases rapidly with depth; there is the problem of equivalence (non-uniqueness), and the lateral resolution is poor. However, all these disadvantages may become silver linings in regional work, in basin evaluation, where one is interested in gross properties. One deep resistivity or magnetotelluric sounding in a suspected deep basin may be more valuable than one deep borehole, at a nominal relative cost. Figures 3.14 and 4.8 show such examples.

In areas where seismic methods are not providing an appreciable amount of information, electrical methods are the only alternative to obtain the kind of structural and lithological information that normally comes from seismic reflection, but with far less detail than seismic results (Figure 3.15d). For instance, sedimentary sections under thick layers of volcanics or high-velocity carbonates may not be effectively explored by seismic methods. In such areas magnetotelluric or controlled-source electromagnetic surveys may be used.

Another field of use is delineating lithological zones favorable to oil accumulation, such as stratigraphic trap zones in sedimentation transition fronts between zones consisting mostly of sandstone and those mostly of shale, zones of highly anisotropic resistivity and buried delta fronts (Keller, 1968). Large oil companies have experimented little with this kind of work.

There are also geological targets that are exclusively electrical, such as detection of oil and gas through detection of geochemically anomalous regions. Chapter 6 discusses such 'direct detection' methods, including case histories.

There are many ways to express the usefulness of electrical methods in prospecting for oil and gas: post-gravity–pre-seismic reconnaissance, post-seismic lithologic information for seismic interpretation, etc. In short, electrical methods can provide a different view to the subsurface, the resistivity view, in addition to the density and velocity views.

Probably the most cost-effective application is in the evaluation of meagerly known basins prior to expensive operations. In the early stages of geophysical work, possibly following an aeromagnetic survey, deep electrical soundings, such as magnetotelluric, made at judiciously selected locations can yield enough information about the major lithologic units down to the basement complex to guide future work.

2

Electrical properties of sedimentary rocks

2.1 GENERAL

The subject of electrical properties of rocks is extensive but not all of it is well understood. There are complete books (e.g. Parkhomenko, 1967), but most of the broad-scope literature is in the form of long chapters in books (e.g. Keller, 1971, 1982, 1988; Olhoeft, 1980).

This chapter is short and of limited scope in that it is concerned primarily with the electrical resistivities of common sedimentary rocks *in situ*, that is, water-bearing clastic (detrital) rocks and limestones.

The electrical property of primary importance in the study of deep sedimentary basins by electrical methods is resistivity, ρ, and its reciprocal, conductivity, σ, which depend primarily on the water component of a buried sedimentary rock: the amount, the salinity and the configuration of the water are the major factors. All subsurface rocks are normally partially saturated above the water table and fully saturated below it. In other words, the resistivity behavior of a water-bearing rock is essentially that of an aqueous electrolyte. Consequently, each rock type, such as shale, sandstone and limestone, displays a wide range of resistivities within itself, by factors of hundreds or even thousands. On the other hand, statistical averages of these common rock types indicate that shales have usually lower resistivities than sandstones, which in turn have lower resistivities than limestones. However, it is understood that shales may have higher resistivities than sandstones, and even higher than limestones in rare cases. This does not create an unsurmountable drawback, because the primary role of electrical methods is to determine the variation of resistivity in the subsurface; the geologic meaning of these variations comes from interpretation which combines all geologic data available in that locality.

The major portion of all sedimentary basins consists of clastic (detrital) rocks (e.g. shales, sandy shales, shaly sands and sandstones) and of limestones.

Electrical surveys frequently involve an 'electrical basement', which is a high-resistivity medium assumed to be a half-space at the base of the sedimentary rocks of interest. It may be composed of igneous rocks, metamorphic schists, evaporites or massive carbonates. In exploring deep basins one tries to obtain an accurate determination of the depth to the basement but not its resistivity. Usually, there is no need to extend the depth of exploration to obtain more accurate estimates of the basement resistivity. In other words, the resistivities in the basement are of no major concern in the study of sedimentary basins; however, they are important in crustal studies. Sometimes high-resistivity rocks, such as basalt flows or evaporites, may be present above or within the sedimentary section of interest.

2.2 THE PHENOMENA OF ELECTRIC CURRENT CONDUCTION

Electric charge in motion constitutes an electric current. The charge flow can occur in the following ways:

Electronic (metallic) conduction takes place by the movement of free electrons in the outermost orbits of atoms, under the influence of an electric field $\boldsymbol{E}$. This kind of conduction occurs in metallic conductors (metals) and, to a lesser degree, in electronic semiconductors (e.g. dry minerals). Native metals and metallic sulfide minerals have high conductivities, but these do not constitute a significant volume in sedimentary basins. The common rock-forming minerals have very high resistivities when dry.

Electrolytic conduction in aqueous solutions is the most important conduction phenomenon. The charge is transported by ions which move under the influence of an electric field $\boldsymbol{E}$ at a constant speed proportional to the magnitude of $\boldsymbol{E}$; anions ($-$) move in the direction opposite to that of $\boldsymbol{E}$ and cations ($+$) move in the direction of $\boldsymbol{E}$; there is transportation of matter. The current density $\boldsymbol{J}$ is proportional to the magnitude of $\boldsymbol{E}$, and Ohm's law is valid for this case also, but the linearity of electrolytic conductors is frequently masked by other effects; these will be discussed in Chapters 3 and 6. This is the type of conduction that is of primary importance in sedimentary rocks, via the waters contained in their interstices, fractures and vugular openings. There is always some salinity in these waters, largely because most sedimentary rocks were deposited in marine environments. Pure water is a good electrical insulator, but a very small amount of salt goes a long way in increasing the conductivity. Clay particles have an additional conduction ability due to their bound water, which will be discussed later.

Solid electrolytic conduction takes place by the motion of ions through a crystal lattice. Most rock-forming minerals act as solid electrolytes, also called **dielectrics.** The level of this type of conduction is several orders of magnitude lower than that in aqueous electrolytes.

Displacement (or dielectric) conduction takes place only in time-varying conditions. In most semiconductors the number of free electrons is too small to constitute appreciable amounts of charge flow. However, alternating current conduction can take place in such media by the displacement of the nuclear and electron patterns. No such current flows in the direct current regime. These currents are significant only in high-frequency work in high-resistivity environments (section 4.3.2).

Maxwell's equations, together with constitutive relations are valid for all types of current conduction cited above.

2.3 ELECTRICAL RESISTIVITY OF ROCKS

For a physical definition of the resistivity, consider a wire made of a homogeneous, isotropic and linear material, whose resistance is R ohms, length L meters, and cross-section S m^2. The resistance R is proportional to L/S, and the experimentally determined proportionality factor ρ is the resistivity,

$$R = \rho L/S.$$

If the wire is in the form of a cube of dimension L, then we have $\rho = RL$, which says that the unit of ρ is ohms times meter, denoted by Ω m. If $L = 1$, then $\rho = R$. This states that ρ is numerically equal to the resistance, in ohms, between two opposite faces of a 1-m cube.

The conductivity, denoted by σ, is the reciprocal of ρ; its unit is siemens/m or mho/m, where mho is ohm spelled backward.

The differential form of Ohm's law for a homogeneous, isotropic and linear medium is

$$\mathbf{E} = \rho \mathbf{J}, \tag{2.1}$$

where $\mathbf{E}$ is electric field intensity (V/m), $\mathbf{J}$ is current density (A/m^2), and ρ is in Ω m; this equation is known as one of the constitutive relations that go with Maxwell's equations, and it establishes the definition of ρ. The term linear implies that the proportionality factor ρ is independent of the magnitudes of $\mathbf{J}$ and $\mathbf{E}$, although it may depend on the direction, in which case the medium is said to be linear but anisotropic. In an anisotropic medium the vectors $\mathbf{J}$ and $\mathbf{E}$ are no longer in the same direction and equation (2.1) takes the form of a symmetric tensor with six different resistivities, ρ_{xx}, ρ_{xy}, ρ_{xz}, ρ_{yy}, ρ_{yz} and ρ_{zz}, which can be reduced to three principal resistivities, ρ_{xx}, ρ_{yy}, ρ_{zz}, by rotating the coordinate axes until the cross terms are zero,

$$E_x = \rho_{xx} J_x, \text{ etc.} \tag{2.2}$$

2.3.1 Resistivity of water-bearing rocks

The role of saline waters in rocks in lowering their resistivities has already been discussed. There are several other parameters governing resistivity. The significant ones are **porosity** (the fraction per unit volume of rock that is void space that may be filled with fluids – water, air, etc.), **saturation** (the fraction of unit volume of porosity that is filled with a fluid – water or oil), **salinity** (ion concentration in the water), **texture** (pore structure), **shaliness** (shale content, degree of ion-exchange capacity) and **temperature**. The effect of these parameters is particularly important to the petroleum industry and has been extensively studied. The results of decades of investigations are summarized in Keller (1971).

With regard to texture (pore structure), porosity is originally intergranular in clastic rocks; it is in the form of joints, fractures and vesicles in igneous rocks, and vugs and fractures in limestones. The pore volume may consist of larger voids and finer connecting pores. Larger voids have to be interconnected and filled with water to conduct electricity. Most of the electrical resistance is due to the connecting pores. In other words, porosity alone does not determine the resistivity, especially in limestones and basalts.

It has been observed that porous rocks fully saturated with water and not containing much clay minerals obey the well-known empirical formula called **Archie's law**,

$$\rho_r = a\rho_w / P^m, \tag{2.3}$$

where ρ_r is the bulk resistivity, ρ_w is the pore water resistivity, P is the porosity expressed as a fraction per unit volume of rock, and a and m are empirical parameters that are experimentally determined. Another parameter derived from this equation is the formation factor, denoted by F,

$$F = \rho_r / \rho_w = a / P^m. \tag{2.4}$$

Figure 2.1 is a plot of F and P for various rock types, each segment representing a rock type and yielding different values for a and m. Equation (2.4) states that F is independent of salinity. If a porous rock is saturated with an electrolyte, the resistivity contrast between the rock and the electrolyte is equal to F. This is also useful in electrical scale model studies.

When pores are partially saturated with electrolyte and partially by insulating fluids, e.g. oil, gas or air, Archie's equation is modified to

$$\rho_r = \rho_w / P^m S^n = F\rho_w, \tag{2.5}$$

where S is the fraction of unit pore space filled with electrolyte, and n is another experimentally determined parameter which is usually close to 2. Since m is also close to 2 (Figure 2.1), a first-order approximation is

$$\rho_r = \rho_w / P^2 S^2. \tag{2.6}$$

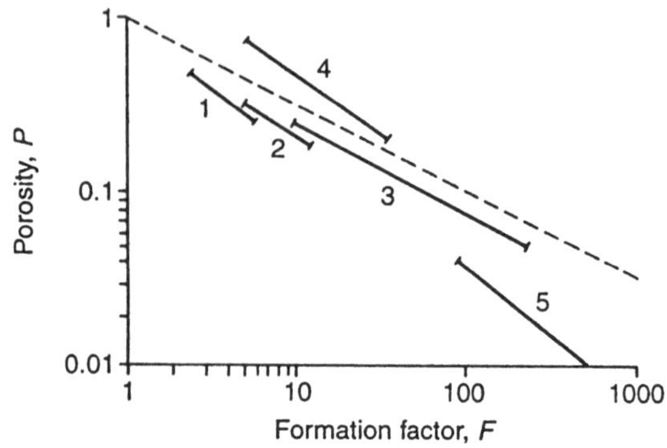

Figure 2.1 Archie's law for various rock types. Formation factor, F, is the ratio of rock (bulk) resistivity, ρ_r, to water resistivity, ρ_w. The average resistivities of porosity groups have been plotted. 1, Tertiary sedimentary rocks, $P=0.25$–0.45, $a=0.88$, $m=1.37$; 2, Mesozoic sedimentary rocks, $P=0.18$–0.35, $a=0.62$, $m=1.72$; 3, Paleozoic sedimentary rocks, $P=0.05$–0.25, $a=0.62$, $m=1.95$; 4, porous volcanic rocks, $P=0.20$–0.80, $a=3.5$, $m=1.44$; 5, dense igneous and metamorphic rocks, $P<0.04$, $a=1.4$, $m=1.58$. The dashed line is for $a=1$ and $m=2$ (after Keller, 1971).

Equation (2.5) is valid provided that the electrolyte saturation S is greater than a critical value, which is about 0.25 for sandstones.

The resistivity of a NaCl solution, in Ω m, at 20°C, is approximately

$$\rho_w \approxeq 5/c, \tag{2.7}$$

where c is NaCl concentration in grams per liter. Combining equations (2.6) and (2.7), we obtain an approximate but easy to remember formula,

$$\rho_r = 5/c\, P^2 S^2. \tag{2.8}$$

(a) Shaliness

In accordance with Archie's law, equation (2.4) states that the formation factor F should be independent of the water resistivity. However, it is observed that F is lower than that predicted by equation (2.4) when a rock is saturated with a water of high salinity. This may be explained by considering that clay particles contribute exchange ions to the electrolyte, thereby lowering the resistivity of the formation (Waxman and Smits, 1968). Clay minerals absorb certain ions in an exchangeable state. Actually all fine-grained minerals, including quartz, have ion exchange capabilities, and usually the finer the particles, the larger the ion exchange and the lower the resistivity. Thus, shaliness can be envisioned as the amount of fine-grained particles, and it results in lower resistivities than equation (2.5) predicts.

(b) Temperature effect

In the Earth the normal gradients of pressure and temperature are such that the electrolytes remain below the boiling point, except under abnormal conditions such as those found in geothermal reservoirs. Under normal conditions a rise in temperature lowers the resistivity of the electrolyte by lowering the viscosity of the water, and thereby increasing the mobility of the ions. Resistivities of NaCl solutions within a wide range of concentrations drop by a factor of ~ 4 when the temperature is raised from 20°C to 100°C.

(c) Resistivities of rocks, minerals and waters

As discussed above, resistivities of water-bearing sedimentary rocks vary widely, even for one rock type of a certain age, because the resistivity depends primarily on the electrolyte. On the other hand, electrical surveys deal with the gross properties of substantial thicknesses and volumes of sedimentary formations. Consequently, the foregoing discussion (Archie's law, etc.) constitutes only a general and qualitative guidance in planning a survey and in interpreting the field data. For these reasons, some well log data obtained in or near the survey area are needed for quantitative planning and realistic geologic interpretation. The analysis of well logs for this purpose is discussed in detail by Keller (1968).

The following is an extremely simplified table which is the result of my experience and of a visual averaging of the published data. The commonly observed ranges are in Ω m. The figures in parentheses indicate the most commonly observed ranges.

Sedimentary rocks	
Carbonates	$50–10^4$ (100–2000)
Sandstones	$10–10^3$ (50–1000)
Shales	1–100 (5–20)
Evaporites	$100–10^5$ (2000)
Igneous rocks	
Granites	$500–10^6$ ($10^3–10^4$)
Basalts	$100–10^4$ (200–1000)
Rock-forming minerals	As high as 10^{15}
Metallic ore minerals	As low as 10^{-6}
Waters	
Pure water	$10^4–10^5$
Rain water	$30–10^3$
Tap water	10–150
Ocean water	0.27

2.4 DIELECTRIC CONSTANT OF ROCKS

The definition of a pure dielectric material is that it does not permit the passage of conduction (galvanic or ohmic) currents; it is an insulator, but it permits the passage of a different kind of current, called displacement current, only in the time-varying regime in which charge separation (polarization) arises and an electric field passes.

The mechanism of displacement (or dielectric) conduction in semiconductors was briefly discussed above, in section 2.2, the phenomena of electric current conduction.

One of Maxwell's equations governing the macroscopic electromagnetic field is

$$\text{curl } \boldsymbol{H} = \boldsymbol{J} + \partial \boldsymbol{D}/\partial t, \tag{2.9}$$

where $\boldsymbol{H}$ is the magnetic field intensity (curl is a function of spatial derivatives), $\boldsymbol{J}$ is the conduction current density, $\boldsymbol{D}$ is the electric displacement (coulomb/m^2) and its time-derivative, $\partial \boldsymbol{D}/\partial t$, is the displacement current density. $\boldsymbol{D}$ is due to an electric field intensity $\boldsymbol{E}$ via a constitutive relation which is, for a linear and isotropic medium,

$$\boldsymbol{D} = \varepsilon \boldsymbol{E} \tag{2.10}$$

which constitutes the definition of ε, **electric inductive capacity**, also called **permittivity**, measured in farad/m. In practice, for a material medium ε is normalized to that of free space, ε_0,

$$\boldsymbol{K} = \varepsilon/\varepsilon_0; \tag{2.11}$$

K is called the **dielectric constant**; $\varepsilon_0 = 8.854 \times 10^{-12}$ farad/m for a free space.

With reference to equation (2.9), water-bearing sedimentary rocks have low resistivities and the geophysical methods used in the exploration of sedimentary basins employ low frequencies. Consequently, the displacement current is usually negligible and $\boldsymbol{J}$ therefore overrides, except in special cases such as in measurements on or in dry massive carbonates and rock salt; these rocks have very high resistivities and must be explored by means of high-frequency sources. Otherwise, the dielectric constant drops out of the picture (section 4.3.2).

On the other hand geophysicists, especially those dealing with mining exploration, have been interested in the possibility of using the dielectric constant as a distinct rock or mineral property, analogous to resistivity. However, this yielded discouraging results. At radio frequencies most dry rocks and rock-forming minerals have about the same dielectric constants, mostly between 5 and 7, and this cannot be used to distinguish one rock type from another. At low frequencies the dielectric constant is extremely variable, especially in water-bearing rocks, even for one small unit of rock, and again it does not serve as a meaningful characteristic property. It is possible that at low frequencies the dielectric polarization is obscured by other types of polarization (Olhoeft, 1985; Chapter 6). This is probably why the dielectric constants of water-bearing rocks, obtained in laboratories and in the field, turn out to be very large, even though the dielectric constant of pure water is about 60–80. 'It is likely that none of the published data on dielectric constants of water-bearing rocks are meaningful' (Keller, 1971).

3

Direct current resistivity methods

3.1 AN OVERVIEW OF THE METHODS

In general, the procedure with the resistivity methods is to inject a direct current (DC) into the Earth by means of a pair of electrodes and to measure the potential difference between another pair of electrodes at a multiplicity of locations. These measurements are inverted into a distribution of electrical resistivity in the subsurface. The resistivity boundaries are interpreted in terms of lithological boundaries. The foundation of this is simply Ohm's law.

It can be said that such measurements constituted the first quantitative geophysical exploration technique whereby downward information could be obtained and contour maps of sedimentary structures could be constructed, before the advent of seismic reflection methods. Probably the most significant contributor to the developments in this subject was Professor Conrad Schlumberger, whose field experiments date back to 1912. His book, *Étude sur la Prospection Électrique du Sous-Sol*, written in 1920, was largely responsible for the early economic applications of resistivity methods. A survey in 1923 led to a gas discovery in Romania. Salt domes were discovered and delineated in France in 1926, while in the United States such work began in 1925. Field applications were fairly well developed between the two World Wars, but effective interpretation methods had to wait until the 1960s, primarily for digital computers.

According to the terminology the expression 'resistivity methods' implies that the condition of steady-state direct current flow (simply called DC regime) is met in the practical sense; it is not necessary to specify that the source is DC. However, the field procedures are such that, even though the source current may not be DC, the final product practically meets the condition of DC regime (see The near field in section 5.2.2). Under these conditions, in a linear, isotropic but inhomogeneous medium the current flow obeys the simple form of Ohm's law:

$$\boldsymbol{E} = \rho \boldsymbol{J}, \tag{3.1}$$

where $\boldsymbol{E}$ is the electric field intensity (potential gradient), V/m, and $\boldsymbol{J}$ is the current density, A/m^2; it states that $\boldsymbol{E}$ is proportional to $\boldsymbol{J}$ via an experimentally determined proportionality factor ρ, called resistivity; it also constitutes the definition of ρ, ohm m (Chapter 2). All these quantities are functions of space coordinates only. The media which obey this relationship are called linear, or ohmic, meaning that ρ is independent of the magnitude of $\boldsymbol{J}$.

On the other hand, the principle of conservation of electric charge (since charge can neither be created nor destroyed) implies, in the DC regime, that the **divergence** of $\boldsymbol{J}$ is zero,

$$\nabla \boldsymbol{J} = 0; \tag{3.2}$$

this is none other than a mathematical expression of 'what goes out is equal to what comes in'. Equations (3.1) and (3.2) lead to Laplace's equation:

$$\nabla^2 V = 0, \tag{3.3}$$

where V is the electric potential, a scalar quantity, measured in volts, the gradient of which is $\boldsymbol{E}$,

$$\boldsymbol{E} = -\nabla V; \tag{3.4}$$

$\boldsymbol{E}$ is said to be derived from V.

Direct current cannot be introduced into the earth inductively; it must be introduced by means of physical contact of electrodes, that is, galvanically; one must have two earthed electrodes, A and B, connected to a current source of I amperes by means of an insulated wire. This is called the source dipole, source bipole or transmitter. In theory, the definition of a dipole of this type is that the product (AB) $\times I$, called dipole moment, remains finite as (AB) approaches zero. This is called an ideal dipole. In some resistivity work it is assumed that (AB) = 0 to facilitate the calculation. Some workers proposed that the term dipole be confined to (AB) = 0 assumption and the term bipole be used for finite (AB). However, there is no uniformity in this usage at present. Here we shall use the term dipole for finite (AB) as well as for the assumption (AB) = 0, unless otherwise specified. If we assume that electrode A constitutes a source of $+I$ amperes, then we must assume that B is a source of $-I$ amperes. This way the potential theory works fine; electric potential V, measured in volts, can be calculated at any point as the algebraic sum of the two potentials due to $+I$ at A and $-I$ at B. This is called the principle of superposition.

The potential can be measured only in terms of the potential difference between a pair of electrodes which are usually designated by points M and N. This is called the probe dipole. The system of four electrodes, ABMN, is called a **quadrupole**.

Let us now consider a hypothetical case in which there is only one current source of $+I$ amperes at point A, and the potential V is measured in volts at M, which is at a distance of r meters from A, as shown in Figure 3.1. In practice this means that points B and N are so far away as not to perturb the measurements appreciably. This electrode array is called **pole–pole**. It is also known as a **normal device** in electric well logging, which came into existence merely as a vertical application of the methods developed for surface exploration. If the subsurface is 'uniform' (homogeneous and isotropic), and of resistivity $\rho\ \Omega$ m, we have

$$V = \frac{\rho I}{2\pi} \frac{1}{r}, \quad \rho = 2\pi r \frac{V}{I}. \tag{3.5}$$

Equation (3.5) constitutes a solution to Laplace's equation for this simple model.

If the subsurface is not uniform, ρ in equation (3.5) has no meaning other than the definition of a fictitious quantity called apparent resistivity, represented by ρ_a, which is used as an intermediate parameter. Now ρ_a depends on the locations of A and M with respect to the resistivity boundaries; it does not represent the average resistivity of the environment; it can be lower than the lowest and higher than the highest resistivity, except when the subsurface is made of horizontal layers. Referring to Figure 3.1, various resistivity methods can be defined in terms of the independent variables x, y, r and θ as follows.

1. Homogeneous and isotropic subsurface: ρ_a is independent of x, y, r and θ and is equal to ρ.
2. Horizontally layered subsurface (each layer may have θ-independent anisotropy): ρ_a is a function of r only; $\rho_a(r)$ is measured and interpreted in terms of downward resistivity distribution, $\rho(z)$; it yields well log type data when mapping structures in areas with gentle tectonics. This is called resistivity sounding.
3. Arbitrary, three-dimensional resistivity boundaries: r and θ are kept constant, and x and y are varied. Apparent resistivity values are contoured. Different selections of r and θ yield different maps. By

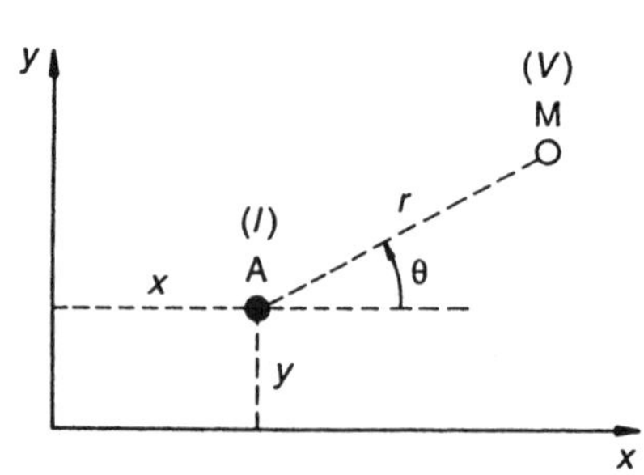

Figure 3.1 Variables of resistivity measurements (after Lasfargues, 1957).

satisfying a number of such maps simultaneously, one may obtain knowledge about $\rho(x, y, z)$. This is called resistivity mapping.

4. Arbitrary two-dimensional resistivity boundaries: A and M are always kept on a straight line perpendicular to the geologic strike, that is r, A, B are always on the same straight line. Different values of r yield different ρ_a profiles. By satisfying a number of such profiles simultaneously, one may obtain a geologic cross-section. This technique is called resistivity profiling.

In order to see a target, one has to cast light on it, so to speak; one has to establish a workable amount of current density at the depth where the target is located. Consider a source dipole AB placed on the surface of a uniform (homogeneous and isotropic) half-space. The center of AB line is point O where the current density (A/m^2) is

$$J_0 = \frac{4I}{\pi(\mathrm{AB})^2}. \tag{3.6}$$

The current density on the vertical line downward from O, namely on the z-axis, is

$$J(z) = \frac{J_0}{\{1+[4z^2/(\mathrm{AB})^2]\}^{3/2}}, \tag{3.7}$$

which shows that one has to increase the length of AB to increase the current density at depth relative to that at the surface. Now we have to 'focus' our 'eyes' to see a target at a depth of interest. This is done by properly selecting the distance of the probe M (or MN) from the source A (or AB). The proof follows.

Consider the model shown in Figure 3.2, which involves the pole–pole device as in Figure 3.1. The potential at M is (Keller and Frischknecht, 1966)

$$V = \frac{\rho_1 I}{2\pi r}\left[1+2\sum_{n=1}^{\infty}\frac{k^n}{\left[1+\left(\frac{2nh}{r}\right)^2\right]^{1/2}}\right], \tag{3.8}$$

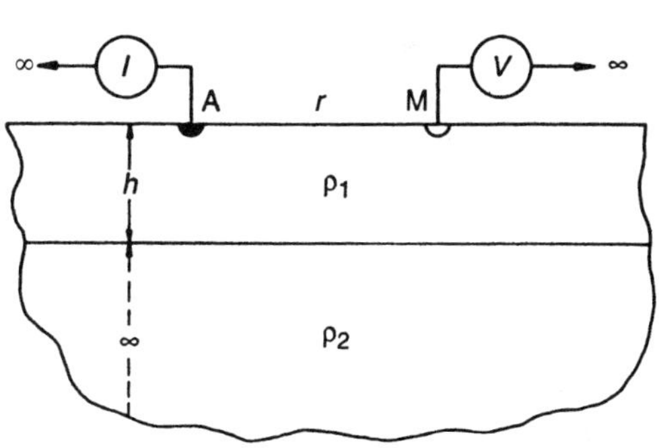

Figure 3.2 A two-layer subsurface and a pole–pole device.

where k, called the reflection coefficient, is

$$k = \frac{\rho_2 - \rho_1}{\rho_2 + \rho_1}.$$

A comparison of equations (3.8) and (3.5) shows that the summation term is the perturbation due to the substratum. As $r \to 0$ the perturbation term vanishes; the subsurface behaves as a half-space of resistivity ρ_1, that is, the substratum has no influence on V, no matter what the current density is at depth h. As $r \to \infty$ the limit of equation (3.8) approaches

$$V = \frac{\rho_2 I}{2\pi r},$$

which means that when r is much larger than h the measurement is not appreciably influenced by the overburden; the subsurface appears to be practically a half-space of resistivity ρ_2. Therefore, one focuses on deeper targets by increasing r. As r is increased from very small to very large values, the measured V varies gradually from a value for a ρ_1 half-space to that for a ρ_2 half-space. For any finite value of r all the subsurface media make their presence felt on V. This constitutes the principle of the method of resistivity sounding, which lends itself to the determination of depths, thicknesses and resistivities of geologic formations.

Resistivity measurements can be made by means of an arbitrarily arranged quadrupole shown in Figure 3.3. Applying equation (3.5) to the source-and-probe pairs the superposition of potentials yields

$$\rho_a = \frac{2\pi}{\left(\frac{1}{AM} - \frac{1}{BM}\right) - \left(\frac{1}{AN} - \frac{1}{BN}\right)} \frac{\Delta V}{I}, \tag{3.9}$$

where ΔV is the potential difference between M and N. This would be the true resistivity of the subsurface if it was a uniform half-space. However, in the presence of inhomogeneities, it would be very difficult

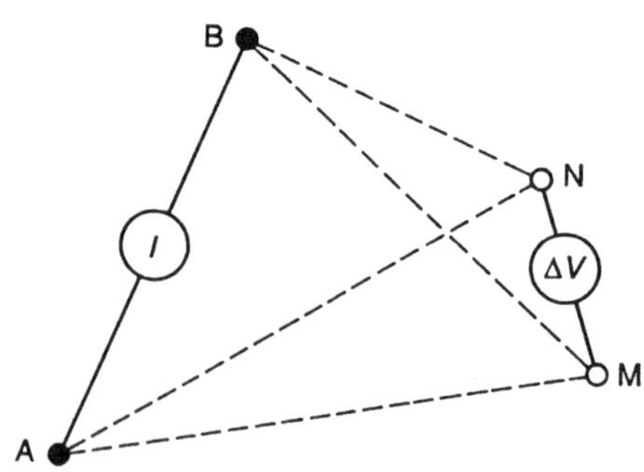

Figure 3.3 Arbitrarily arranged quadruple, plan view.

to interpret values of ρ_a obtained by means of such an arbitrary quadrupole. One would rather use certain regular forms of electrode arrays. There is a large number of types of electrode array in use. The selection depends mainly on the types of geologic targets, on surface conditions and on accessibility. We shall describe only a few types of electrode array that are of particular interest in deep work in sedimentary basins, e.g. exploration related to petroleum. These are (1) Schlumberger quadrupole, (2) azimuthal dipole, (3) equatorial dipole, (4) axial dipole (also known as dipole–dipole, or polar dipole), (5) Wenner quadrupole, and (6) rotating bipole–dipole. Types 1, 2 and 3 are suitable for resistivity sounding, 4 and 5 are suitable for resistivity profiling, and type 6 for resistivity mapping.

One more important item of general interest is the principle of reciprocity. With reference to the arbitrarily arranged quadrupole, and for an arbitrarily inhomogeneous subsurface, the source dipole and the probe dipole are interchangeable. If a current I_0 at AB creates ΔV_0 at MN, a current I_0 at MN would create ΔV_0 at AB, and ρ_a would also remain the same.

3.2 RESISTIVITY SOUNDING

3.2.1 General

The objective of resistivity sounding is to obtain depths, thicknesses and resistivities of geologic formations in a multilayered subsurface. This allows one to map structures and lithologic changes. The procedure is analogous to that of seismic refraction. With reference to Figure 3.4, consider the pole–dipole array and the horizontally layered (1-D) subsurface;

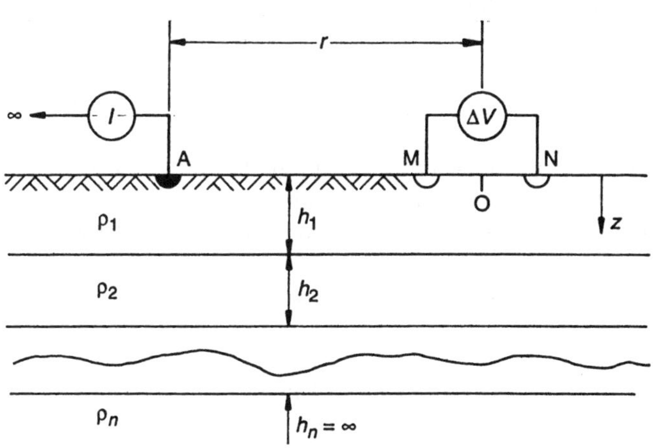

Figure 3.4 Pole–dipole and 1-D subsurface, cross-section.

points A, M, N are on a straight line, and MN is much smaller than AO. The current I creates an electric field $\boldsymbol{E}$ at point O that is in the radial direction r; it is the gradient of V and is measured in volt/m,

$$E_{\mathrm{r}} = -\frac{\partial V}{\partial r} \cong \frac{\Delta V}{\mathrm{MN}}. \tag{3.10}$$

If a well log, $\rho(z)$, representing a 1-D subsurface is given, then $E_{\mathrm{r}}(r)$ can be calculated. Conversely, if $E_{\mathrm{r}}(r)$ is measured, one can calculate $\rho(z)$ for a simple 1-D model, and this is what resistivity sounding attempts to do. The pole–dipole of Figure 3.4 is called the **lateral device** in electric well logging. Although E_{r} is a point function, it is neither practical nor desirable to measure it as such; its value averaged over a desirable length MN is more meaningful. The desirable length is such that MN/r is small but not so small as to create instrumental and geologic (near-surface) noise problems. One proceeds by computing the apparent resistivities $\rho_{\mathrm{a}}(r)$ from the $E_{\mathrm{r}}(r)$ measurements. For the pole–dipole arrangement and for $(\mathrm{MN}/r) \to 0$, we have

$$\rho_{\mathrm{a}} = 2\pi r^2 \frac{E_{\mathrm{r}}}{I}, \tag{3.11}$$

which follows from equations (3.5) and (3.10). The process of inversion, or interpretation, takes the $\rho_{\mathrm{a}}(r)$ data and converts it to information about $\rho(z)$.

Let us now clarify again the significance of separation r in terms of the depth of exploration by means of an anology to the electromagnetic wavelength λ in the magnetotelluric (MT) method. (The longer the wavelength the deeper the EM wave penetrates.) In MT the resistivity ρ of a uniform half-space is determined by

$$\rho = \frac{1}{2\pi\mu_0} T \left| \frac{E_x}{H_y} \right|, \tag{3.12}$$

where E_x and H_y are the electric and magnetic fields respectively and T is the period. H_y is a measure of the total current in the 'unit rectangle' in the subsurface (equation (4.33) in section 4.3.3(e)). Let us represent this current by a symbol I^*. We can also represent the period T in terms of wavelength λ, and equation (3.12) becomes

$$\rho = \frac{1}{\pi 8^{1/2}} \lambda \left| \frac{E_x}{I^*} \right|. \tag{3.13}$$

It is clear that r^2 in equation (3.11) plays the role of λ in equation (3.13); both are the 'yard sticks' of depth.

3.2.2 Electrode arrays

(a) Wenner array

The most commonly used electrode array during the early days of the resistivity method was the Wenner array. This is the collinear equidistant quadrupole, with (AB) = 3(MN), the center of MN being at that of AB. The apparent resistivity is

$$\rho_a = 2\pi a(\Delta V/I),$$

where a = (MN). The Wenner array pioneered the method because of its simplicity in field operations, but it has since been largely replaced by the Schlumberger array for sounding, except for shallow work. However, the Wenner array is still used for profiling involving moderate depths, because the fieldwork proceeds along equally spaced markers in the field.

(b) The Schlumberger array

The pole–dipole array is not practical, except for very shallow depths, because one cannot effectively place electrode B at infinity. Instead, B is placed symmetrically with respect to O, as shown in Figure 3.5. This doubles the magnitude of the electric field E_r at O, and the apparent resistivity becomes

$$\rho_a = \pi(\mathrm{AO})^2 \frac{E_r}{I}, \tag{3.14}$$

which is called the Schlumberger apparent resistivity. For a 1-D subsurface the Schlumberger and the pole–dipole apparent resistivities are identical for AO = r. In practice, MN remains constant and AO is varied by expanding the AB dipole outward from O along a straight line. The interpretation theory and equation (3.14) are based on the assumption that (MN/AB) → 0. However, the field operations require a finite dimension of MN to measure the apparent resistivity, and the exact equation for apparent resistivity is

$$\rho_a = \pi \frac{(\mathrm{AO})^2 - (\mathrm{MO})^2}{\mathrm{MN}} \frac{\Delta V}{I}, \tag{3.15}$$

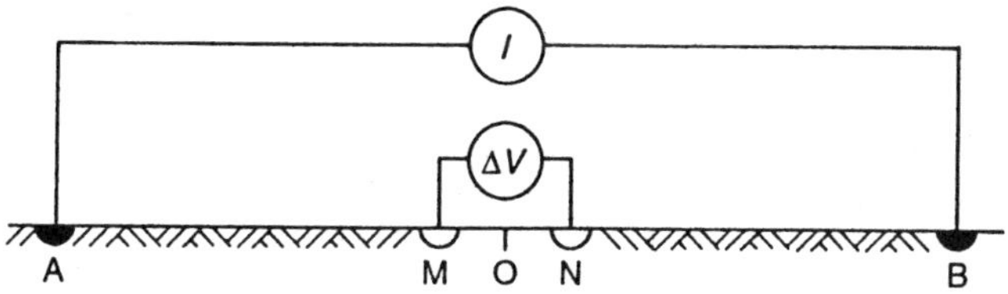

Figure 3.5 The Schlumberger electrode array.

which follows from equation (3.9). Although equation (3.16) is exact for a uniform subsurface, it deviates from the Schlumberger apparent resistivity when inhomogeneities are present, because it is calculated by using ΔV across a finite distance MN instead of the electric field at point O. The deviations may be negative or positive, and they can be taken into account during data processing.

The field data are obtained in the form of overlapping segments, each segment being for a fixed MN dipole. Figure 3.6 shows an actual sounding data set. The two segments, each corresponding to a fixed MN, are displaced due to two effects: (1) deviation from the condition $MN = 0$, and (2) inhomogeneities near electrodes M and N. The first effect is smooth and predictable (Mundry, 1980). No corrections would be required if $(MN/AB) \leqslant 0.1$. As a rule, this ratio is kept smaller than 0.2. The correction of the second effect can be made by shifting a segment vertically.

The data of Figure 3.6 were obtained in a sedimentary basin, at a location where the basement complex is known to be at a depth of about 2500 ft. The right-hand rising branch represents the high-resistivity basement. In this case the basement depth can be determined from the portion of the data corresponding to an AB expansion of about 5000 ft; the depth of exploration is approximately AB/2 for this model. Other types of models may require larger expansions for the same relative depth of exploration which may be less than AB/4.

The Schlumberger array is a time-tested, well-developed, versatile and very useful tool; it has been extensively published in special monographs with theoretical curves (Compagnie Générale de Géophysique, 1963;

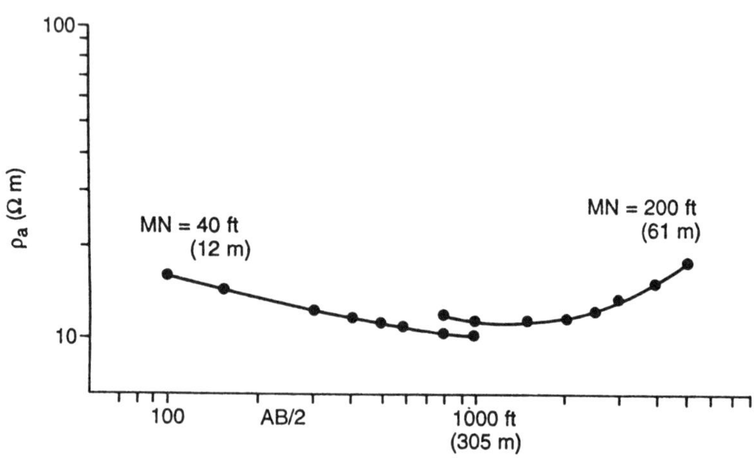

Figure 3.6 Schlumberger apparent resistivity field data, Jasmine area, Bakersfield, California.

Orellana and Mooney, 1966; Rijkswaterstaat, 1969), in books on direct current methods (Lasfargues, 1957; Kunetz, 1966; Bhattacharya and Patra, 1968; Orellana, 1972; Koefoed, 1979), and in books on electrical methods in general (Tarkhov, 1963; Keller and Frischknecht, 1966). However, deep exploration with this array is cumbersome and is frequently impractical, because the necessity of expanding the AB line along a straight line about three times the depth of exploration creates problems in accessibility, safety and labor. It also suffers from reduced horizontal resolution. Consequently, its general usage has been confined to depths of about 1 km and to areas with suitable surface conditions. In order to get around the requirement of long AB lines, the so-called **dipole** arrays were developed in the 1950s, especially in the USSR. Among these, the azimuthal dipole and its special form called the equatorial dipole proved to be practical for deep sounding in sedimentary basins.

(c) The azimuthal dipole array

Figure 3.7 shows how an azimuthal dipole array is used to perform a sounding. The source dipole AB remains stationary, while the probe dipole MN moves along a road, which may be curved, to expand the dipole separation *R*. The depth of exploration no longer depends on the length of AB, which can now be as small as noise problems allow. An azimuthal distance AB is usually <0.1 of that of a Schlumberger array for the same depth of exploration, but now a larger power generator is required. The theory and practice of azimuthal sounding are given by Berdichevskii (1958).

Assuming that the subsurface is horizontally layered (1-D), resistivity sounding measurements obtained with one array can be transformed into those that would have been obtained with another array. A review of the transformation methods is given by Koefoed (1979). It can be

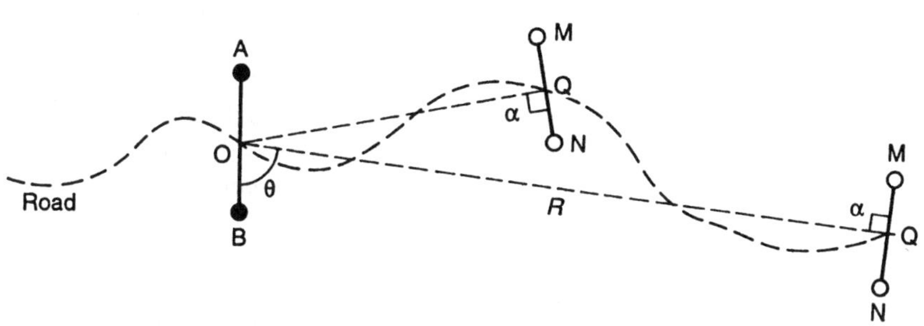

Figure 3.7 Azimuthal dipole array. Requirements are $\alpha=90°$, $60° \leqslant \theta \leqslant 120°$, $AB \leqslant 0.6\,R$, $MN \leqslant 0.2\,R$.

achieved by digital linear filtering (Başokur, 1983), or by a simple regressional analysis (Banerjee and Sengupta, 1987). However, a transformation of field data may actually compound the problem of interpretation unless the subsurface lends itself to the 1-D assumption. Fortunately the azimuthal array is such that the field measurements are practically Schlumberger resistivities, and no laborious transformation process is needed provided that

1. $\alpha = 90°$;
2. $60° \leqslant \theta \leqslant 120°$;
3. $(\mathrm{MN}/R) \leqslant 0.2$;
4. $(\mathrm{AB}/R) \leqslant 0.6$;

To calculate the apparent resistivity one follows the procedure of equation (3.5) and makes the approximation that $(\Delta V/\mathrm{MN})$ is the azimuthal component of electric field at point Q. The justification follows from condition 3 above. The azimuthal apparent resistivity is

$$\rho_{\mathrm{a}} = \frac{\Delta V}{I} \frac{R^3}{(\mathrm{AB})(\mathrm{MN})} \frac{4\pi}{\sin\theta} \times \frac{1}{\left(1+\frac{(\mathrm{AB})^2}{4R^2}+\frac{\mathrm{AB}}{R}\cos\theta\right)^{-3/2}+\left(1+\frac{(\mathrm{AB})^2}{4R^2}-\frac{\mathrm{AB}}{R}\cos\theta\right)^{-3/2}}. \tag{3.16}$$

If the survey plan is such that $(\mathrm{AB}/2R)$ is smaller than 0.1, equation (3.16) may be simplified as

$$\rho_{\mathrm{a}} = \frac{\Delta V}{I} \frac{R^3}{(\mathrm{AB})(\mathrm{MN})} \frac{2\pi}{\sin\theta}. \tag{3.17}$$

Equation (3.17) implies that the array consists of an ideal source dipole, $\mathrm{AB} \to 0$, whose dipole moment $(\mathrm{AB})I$ remains finite, and an ideal probe dipole, $\mathrm{MN} \to 0$. It is easy to show that, for a 1-D subsurface, the azimuthal ρ_{a} of equation (3.17) for an arbitrary R is equal to the Schlumberger ρ_{a} of equation (3.14) for $(\mathrm{AO}) = R$ (Bhattacharya and Patra, 1968, p. 36; Koefoed, 1979, p. 46). However, for real azimuthal ρ_{a} data, as expressed by equation (3.16), a transformation to Schlumberger ρ_{a} is preferred. This transformation consists of correcting the dipole separation R to obtain an effective dipole separation $\overline{R}$, so that $\overline{R}$ is equivalent to AB/2 of Schlumberger ρ_{a}; the correction factor is designated by p, which is close to unity and depends on the array parameters θ and $(\mathrm{AB})/2R$,

$$\overline{R} = pR; \tag{3.18}$$

p is read off graphs that have been constructed by taking the average values from theoretical Schlumberger curves. Figure 3.8 is the graph for p (Berdichevskii, 1958). The error in p as read from Figure 3.8 is less than

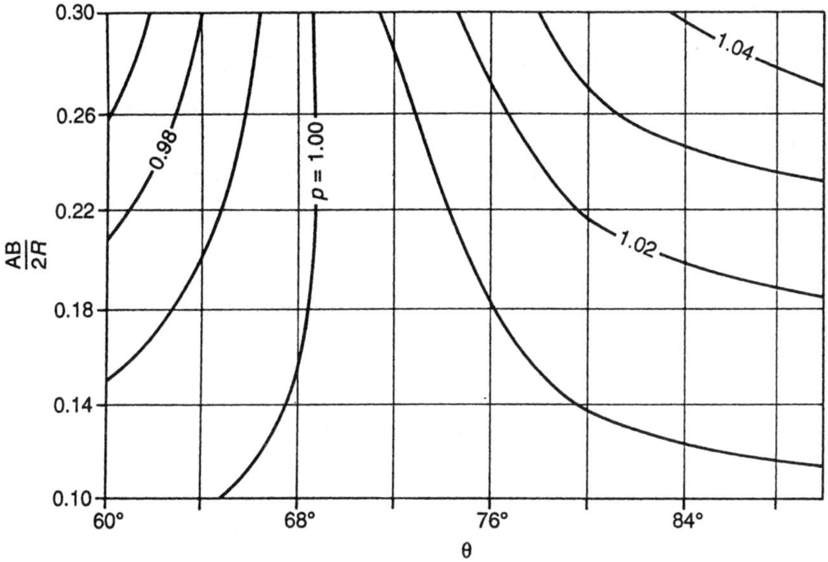

Figure 3.8 Azimuthal dipole sounding, graph of correction factor p for transformation of R to effective dipole separation $\overline{R}$ (from Berdichevskii, 1958).

2%. Note that for $(\mathrm{AB}/2R)=0.1$ the correction factor p is nearly unity; this is the ideal dipole condition. In summary, azimuthal $\rho_a(\overline{R})$ data are the same as Schlumberger $\rho_a(\mathrm{AB}/2)$ data when the subsurface is 1-D. It is not practical to use an azimuthal array for the very shallow part of a sounding; one normally starts with a Schlumberger array and switches to an azimuthal array when a desirable AB length and an AB/R ratio has been reached. Thus, the initial portion of a field curve is derived from Schlumberger array data, $\rho_a(\mathrm{AB}/2)$, as shown in Figure 3.6.

(d) The equatorial dipole array

The azimuthal dipole provides flexibility in coping with accessibility, topography and other surface obstacles, but it necessitates laborious land surveying of R, θ and α. It also puts a restriction on the length of AB, $(\mathrm{AB}/R)\leqslant 0.6$. When surface conditions allow one to expand R along a straight line, it is advantageous to keep the azimuth angle θ equal to 90° all the time. This special type is called an equatorial dipole array; the theory and practice are given by Berdichevskii and Petrovskii (1956). In Figure 3.9 the quadrupole ABMN constitutes an equatorial dipole; θ is always 90° and MN is perpendicular to R. For a horizontally layered (1-D) subsurface, it is easy to show, with the aid of the circle in Figure 3.9, that AQ $(=\overline{R})$ is exactly equivalent to $(\mathrm{A'B'}/2)$ of the Schlumberger array. In other words,

$$\text{Dipole } \rho_a(\overline{R}) = \text{Schlumberger } \rho_a(\text{A}'\text{B}'/2),$$

and no array transformation is needed. There is no restriction on the ratio AB/R. The restriction on the dimension of MN is about the same as in the case of the Schlumberger array, that is $(\text{MN}/\overline{R}) \leqslant 0.4$, because, for this regularly shaped array, it is simple to derive ρ_a exactly from equation (3.9) which is for an arbitrary quadrupole. In other words, the potential difference across a finite MN is used instead of the electric field approximation at point Q. The exact apparent resistivity is

$$\rho_a = \pi \frac{\left[R^2+\left(\frac{\text{AB}}{2}+\frac{\text{MN}}{2}\right)^2\right]^{1/2}\left[R^2+\left(\frac{\text{AB}}{2}-\frac{\text{MN}}{2}\right)^2\right]^{1/2}}{\left[R^2+\left(\frac{\text{AB}}{2}+\frac{\text{MN}}{2}\right)^2\right]^{1/2}-\left[R^2+\left(\frac{\text{AB}}{2}-\frac{\text{MN}}{2}\right)^2\right]^{1/2}} \frac{\Delta V}{I}. \tag{3.19}$$

However, with proper facilities, one can and should keep MN as small as near-surface irregularities permit, because each measurement requires a new location of MN, and a shorter value of MN is more economical. While a short MN receives a smaller signal, it also picks up less telluric and cultural noise; and a short MN yields measurements which are closer to theoretical Schlumberger apparent resistivities. If $(\text{MN})/\overline{R} \leqslant 0.2$, one is allowed to use the azimuthal dipole equation (3.16) with $\theta = 90°$; this yields

$$\rho_a = \frac{2\pi\overline{R}^3}{(\text{AB})(\text{MN})} \frac{\Delta V}{I}, \tag{3.20}$$

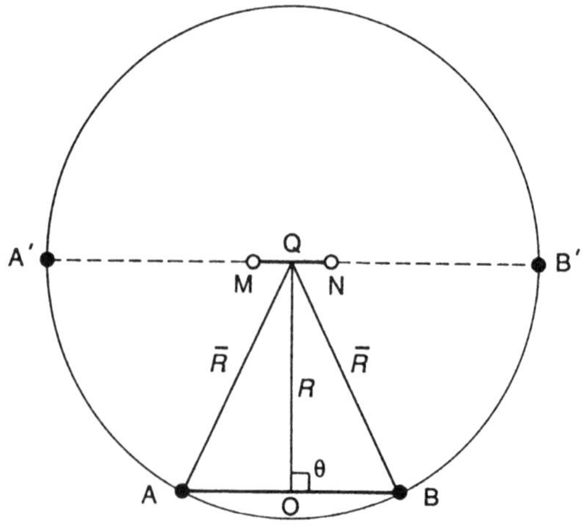

Figure 3.9 Equatorial dipole array.

where

$$\overline{R} = AQ = \left[R^2 + \frac{(AB)^2}{4}\right]^{1/2}. \qquad (3.21)$$

It should be noted that in equation (3.20) there is no restriction on the size of AB. As in the case of the azimuthal dipole array, the usual procedure is to start the fieldwork with a Schlumberger expansion and then to switch to the equatorial dipole mode, but, in this case, this is done merely for fieldwork economy and not for any theoretical requirement.

(e) The axial (polar) dipole array

In this array electrodes AB and MN are in a straight line, shown as 'Ax.' in Figure 3.10, which is a compilation of all the dipolar arrays that have been used for deep sounding. In certain cases accessibility may necessitate the use of an axial array, or availability of a straight road may render motorized cable laying advantageous. Another logistical advantage is that it sometimes enables one to obtain two additional soundings from each setup of a source dipole, as shown in Figure 3.11. Here we see a pair of bilateral equatorial and a pair of bilateral axial soundings, all four measured from one AB.

An axial dipole does not yield the Schlumberger apparent resistivity; it necessitates a different set of theoretical curves and computer programs for interpretation. Its depth of exploration is such that OQ has to be expanded to about 1.5 times that which would be needed with the equatorial or azimuthal arrays. This is not only an economic and logistical

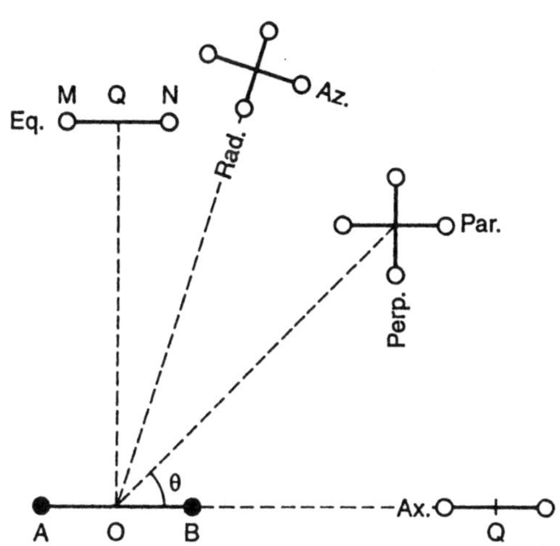

Figure 3.10 Dipole sounding arrays. Eq.: equatorial; Az.: azimuthal; Rad.: radial; Par.: parallel; perp.: perpendicular; Ax.: axial.

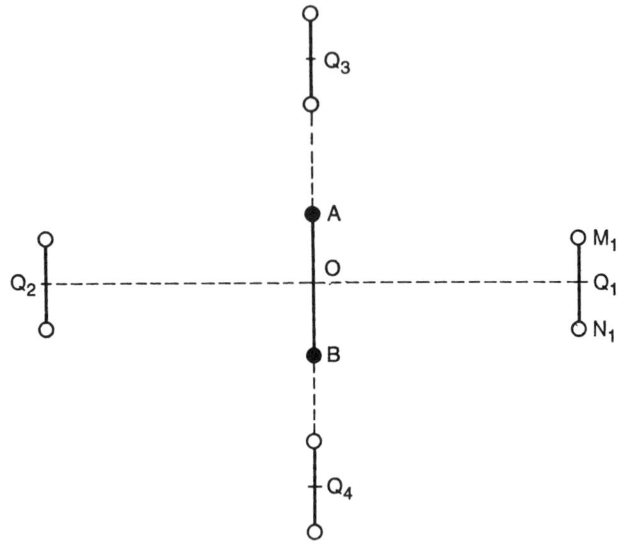

Figure 3.11 Pairs of bilateral equatorial and bilateral axial soundings obtained from one source dipole.

drawback, but it also decreases horizontal resolution. Futhermore, its rigid (straight line) geometry is restrictive. In short, it is seldom used for sounding. However, it is the best array for deep profiling for exploration of two-dimensional structures, as we shall discuss in section 3.3.

The exact apparent resistivity is given by equation (3.9) which is for an arbitrary quadrupole. If it is assumed that both AB and MN are ideal dipoles, as for equation (3.17), the apparent resistivity is

$$\rho_a = \frac{\Delta V}{I} \frac{\pi R^3}{(AB)(MN)}, \tag{3.22}$$

where $R = (OQ)$.

(f) Other types of dipole arrays

Figure 3.10 shows all the major types of dipole array, among which we have discussed the azimuthal (Az.), equatorial (Eq.) and axial (Ax.) types. The remainder, namely the radial (Rad.), parallel (Par.), and perpendicular (Perp.) arrays are seldom used in sounding, and will not be discussed here. An English translation of a collection of papers written in Russian (Al'pin *et al.*, 1966) has extensive information on all the types of dipole array.

In petroleum work a combination of the Schlumberger and azimuthal arrays is the most useful resistivity sounding technique.

3.2.3 Field operations

As mentioned before, fieldwork normally starts with a Schlumberger array and turns into a dipole array when an appropriate AB length has been reached. If the geologic strike is approximately known, the Schlumberger AB line should be parallel to the strike and, in the case of a fault, it should be offset from the fault by a distance of at least one-third of the maximum AB.

When one has a choice in the orientation of the *R* axes of a dipole array, they should be perpendicular to the geologic strike. Fortunately, this conforms to the requirements of the Schlumberger array associated with each dipole sounding. However, if the *R* axes cross outcrops of alternately high and low resistivities, they should be parallel to the strike.

Each setup of the source dipole is normally used for at least two soundings, one on each side, as shown in Figure 3.11. Such a doublet is called a bilateral dipole sounding. Its purpose is twofold. It increases the reliability of the results, so far as the vertical information at one location is concerned. The divergence between the two field curves is a measure of the applicability of the assumption of a 1-D subsurface, and an average of the two is a better representation of a 1-D subsurface. The divergence supplies the dip information more reliably than could have been obtained by correlating the results of separate soundings.

One source dipole can supply the signal to many MN measuring crews working simultaneously on either the same or opposite sides of AB.

An important phase of the fieldwork is the preparation of the A and B groundings associated with large dipole separations. This is because the source voltage is usually restricted to about 1000 V due to safety reasons, and, regardless of how high the power may be, one has to provide low electrode resistance to obtain the required current. Figure 3.12 shows the outputs of two 1000-V generators. If the resistance is, for instance, 200 Ω, both the 5-kW and 20-kW generators can deliver only 5 A. Fieldwork in sedimentary basins of average type usually necessitates about 20 A at the source dipole. If the voltage is 1000 V, the total circuit resistance must be brought down to 50 Ω for a 20-kW generator.

The MN electrodes are of the nonpolarizable (reversible) type; their electrical resistance is usually of no consequence.

The source dipole current is usually a commutated direct current (a square wave). The commutation is necessary to cancel out the MN electrode potentials and the telluric noise, and to prevent polarization at the AB electrodes. The commutation interval (half-period of the square wave) has to be longer than the recovery time (the time required for a practical establishment of the steady-state direct current regime at MN); the lower the resistivity of the subsurface and the larger the array dimensions, the longer the recovery time. One usually determines the recovery time experimentally in the field; it can also be estimated very

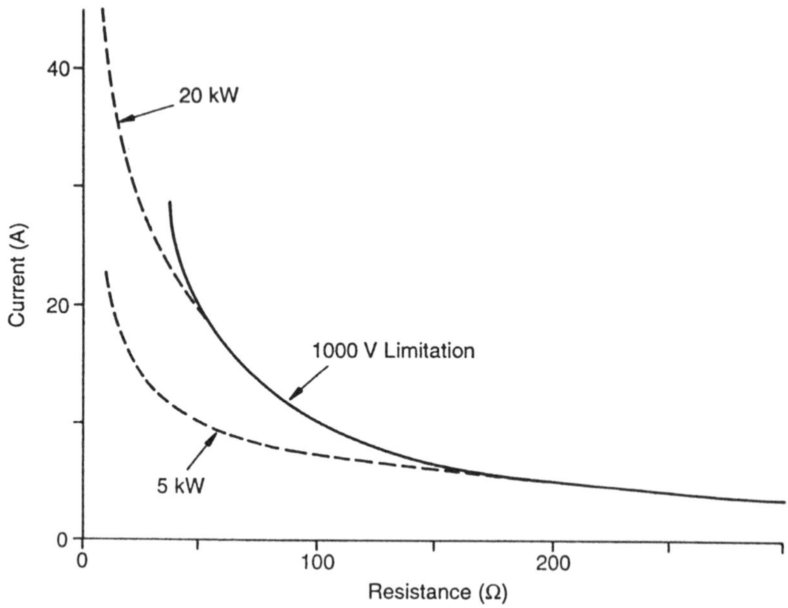

Figure 3.12 Amperages obtainable from two 1000-V generators. To obtain 20A with a 20-kw, 1000-V generator, the resistance has to be 50 Ω.

approximately, for the Schlumberger, azimuthal and equatorial arrays, for a horizontal sedimentary section underlain by an insulating basement complex, by means of the following formula (Tarkhov, 1963):

$$t \cong 0.002\, S\overline{R}, \tag{3.23}$$

where t is recovery time in seconds, S is the longitudinal conductance of the sedimentary section in mhos and $\overline{R}$ is the effective dipole separation in kilometers; $\overline{R} = \text{AB}/2$ for the Schlumberger array.

Figure 3.13 is an actual strip chart record of ΔV measurement at an azimuthal MN site. A visual separation of the signal from the telluric noise becomes difficult if the commutation interval is longer than about 8 s, because 20 s is usually an outstanding period for the telluric noise. Recording and data processing may be accomplished by paper strip chart recording and visual analysis in simple cases (Figure 3.13), and by digital recording and noise cancellation. The visual analysis technique by reconstructing the telluric signal can be very challenging; it is an art.

3.2.4 One-dimensional (1-D) interpretation

The field data, which are in the form of smoothed $\rho_a(\overline{R})$ curves, are to be interpreted in terms of thicknesses h_i and resistivities ρ_i of a 1-D subsur-

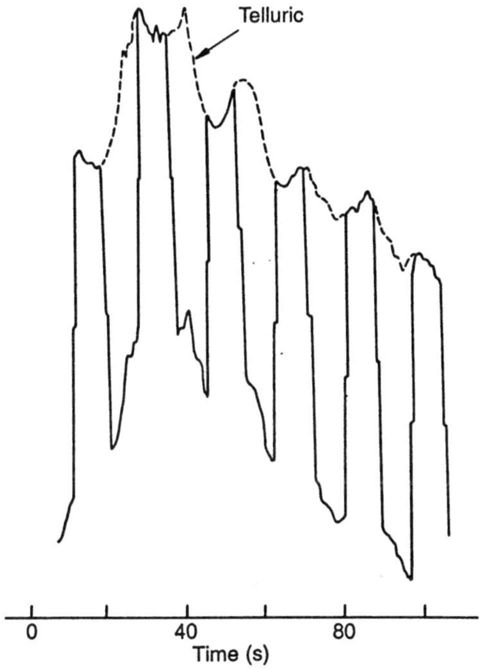

Figure 3.13 A strip chart record of ΔV measurement at an azimuthal dipole MN site.

face. Here we shall assume that the validity of the 1-D assumption has been established by means of the procedure described in section 3.2.3.

There are three major methods of interpretation: (1) approximate, by manual curve matching, (2) iterative, by forward modeling, and (3) direct (inversion).

(a) Interpretation by manual curve matching

The manual curve-matching technique is useful in survey planning, in work progress control, and in providing the initial models for the final computerized interpretation. This time-tested technique plays an important role in coping with the problem of non-uniqueness (section 3.2.4(f)). This is because catalogs of theoretical curves readily provide visual inspection of many possible models simultaneously.

A catalog of standardized theoretical three-layer apparent resistivity curves is needed for this technique. The top three layers are interpreted by means of a three-layer theoretical curve. The upper two layers are combined into one layer of a fictitious thickness (replacement thickness, h_r) and a fictitious resistivity (replacement resistivity, ρ_r) by means of an auxiliary-point graph. Another three-layer theoretical curve is used to interpret the fourth layer, and so on. This is a stripping operation. A considerably less reliable approximation can be obtained by means of two-layer theoretical curves, but at present several excellent catalogs of three-layer curves are available in the literature, and there is no need to discuss how this is done by means of two-layer curves only. It is not practical to possess a catalog of four-layer curves, because the number of curves becomes too large and unmanageable, since both ρ and h are independent parameters.

The standardization of the theoretical curves is performed by normalizing ρ_a, $\overline{R}$, ρ_i, h_i to the parameters of the top layer, namely ρ_1 and h_1. Thus, all of the variables occur as dimensionless ratios. A plot of ρ_a/ρ_1 versus $\overline{R}/h_1$ constitutes a standardized (dimensionless) theoretical curve; the units of $\overline{R}$ and h_i are immaterial. One does not have the electromagnetic similitude problem with direct current (section 4.3.3(d)). The reflection coefficient k is also dimensionless. Furthermore, these curves are plotted on bilogarithmic papers. Thus, a theoretical curve can be matched, by translation, to a field curve ρ_a ($\overline{R}$), regardless of the units of the latter. For instance, if the field curve is a plot of ρ_a in Ω m and $\overline{R}$ in feet, the curve matching will yield ρ_i in Ω m and h_i in feet.

Figure 3.14 shows how an actual azimuthal dipole sounding curve is interpreted by means of the auxiliary-point method and three-layer theoretical Schlumberger curves. The solid-line curve represents the smoothed field measurements. The portion from $\overline{R}=80$ ft to $\overline{R}=2000$ ft was measured employing a Schlumberger array, and the rest by an azimuthal dipole array.

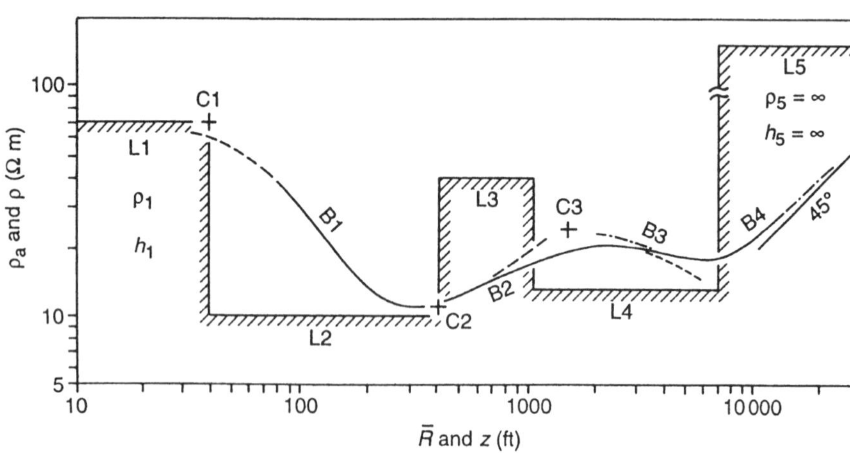

Figure 3.14 The auxiliary-point method of curve matching. The solid-line curve represents smoothed field measurements with Schlumberger and azimuthal dipole arrays.

At the time this work was done the only three-layer catalog available was the 49-set collection published by Compagnie Générale de Géophysique (1963). Subsequently, extensive catalogs were published by Orellana and Mooney (1966), and Rijkswaterstaat (1969).

The auxiliary-point curves used here were published by Ebert (1943). They consist of four sheets, each being for a different echelon (sequence) of layer resistivities in the three layers from the top downward: high–low–high (H-type); descending (VH-type); ascending (A-type); and, low–high–low (VA-type). Reviews of auxiliary-point methods and curve sets are given by Orellana and Mooney (1966), and Koefoed (1979).

With reference to Figure 3.14, the field curve displays four 'branches' (B1–B4). Therefore, the simplest model will consist of five layers (L1–L5), the deepest one (L5) being a half-space. The layer parameters, as determined progressively through interpretation, have been plotted on the same scale as the field curve. Note that if one shifts the field curve to the left, its lows and highs correspond qualitatively to the lows and highs of the layer resistivities. This explains the meaning of the 'branches' of the field curve. Each field curve can be interpreted in terms of as many layers as there are data points on it, but now we are discussing an approximate method of interpretation in terms of a very simple model. The interpretation procedure is described step by step below.

1. Fit a three-layer curve to B1 and B2. This curve, shown as a dashed line, is for $\rho_1=1$, $\rho_2=1/7$, $\rho_3=1$, and $h_2/h_1=9$, where h_i is the thickness of a layer. This yields the first 'cross point' C1. The cross point is the position on the field curve of the intersection of the $\rho_a=1$ and $\overline{R}=1$ lines of the theoretical curve; it reads, on the field curve coordinates, the parameters of the first layer, $h_1=40$ ft and $\rho_1=70\ \Omega$ m. The dimensionless parameters of the theoretical curve yield $\rho_2=10\ \Omega$ m, $h_2=9h_1=380$ ft, and $h_1+h_2=420$ ft.
2. Now that the parameters of the upper two layers are known, one can combine them into one fictitious layer with a 'replacement resistivity' ρ_r (1, 2) and 'replacement thickness' h_r (1, 2). For the parameters of the Ebert graphs, one has $\rho_2/\rho_1=1/7$, and $h_2/h_1=9$. Take the H-type Ebert graphs; the intersection of the curves with ratios 1/7 and 9 is at C2. Transfer C2 to the field curve coordinates; it now reads on the field curve coordinates h_r (1, 2) = 420 ft and ρ_r (1, 2) = 11 Ω m.
3. Take the three-layer curves set symbolically described by the ratio $\rho_1 : \rho_2 : \rho_3=1:4:1$. Place the cross point of this set on C2 and find a curve that matches B2 and B3. This curve is shown as a dotted line in Figure 3.14; it yields $h_3=1.5\,h_r\ (1,2)=630$ ft, $\rho_3=44\ \Omega$ m, and $h_1+h_2+h_3=1050$ ft.
4. Combine the upper three layers into one layer by means of VA-type Ebert graphs; $\rho_3/\rho_r\,(1,2)=4$, $h_3/h_r\,(1,2)=1.5$. Obtain C3 as before; this

reads h_r (1, 2, 3) = 1500 ft and ρ_r (1, 2, 3) = 24 Ω m on the field curve coordinates.

5. Take a three-layer curve set for $\rho_1 : \rho_2 : \rho_3 = 1 : 0.54 : \infty$. Place the cross point of this set on C3 and find a curve that fits B3 and B4. This curve is shown as the dot–dash line in Figure 3.14; it yields $h_4 = 4 \times 1500 = 6000$ ft, $\rho_4 = 0.54 \times 24 = 13$ Ω m, and $\rho_5 = \infty$.

Obviously the foregoing technique is highly subjective, but sometimes it is more reliable than a strictly objective (direct) interpretation, because the curve-matching technique allows one to honor any kind of additional data as the interpretation progresses from the top downward into less and less known environments.

Total conductance

With reference to Figure 3.14, B4 is approximately asymptotic to a straight line having a slope of 45° with the abscissa. The significance of this is explained below.

If it is assumed that a stack of horizontal layers is underlain by a substratum of infinite resistivity, called an electrical 'basement', it can be shown that for large separations, $\overline{R}$, the asymptotic value of ρ_a for the Schlumberger and azimuthal arrays is (e.g. Bhattacharya and Patra, 1968, p.49)

$$\rho_a = \overline{R}/S, \tag{3.24}$$

where S is the total longitudinal conductance of the layers overlying the basement,

$$S = \sum (h_i/\rho_i).$$

Taking the logarithm of equation (3.24), we get

$$\log \rho_a = \log \overline{R} - \log S.$$

This is the equation of a straight line having a 45° slope with the abscissa. It follows that a 45° asymptote constructed to fit B4 intersects the $\rho_a = 1$ abscissa line at an $\overline{R}$ value which is numerically equal to S.

A map of S values is of geologic significance by itself. It can also serve to check whether an interpretation of any kind has been done satisfactorily. For instance, with reference to Figure 3.14, the value of S computed from the layers determined by means of the curve matching interpretation is 515 ft/Ω m. On the other hand, the asymptote yields 520 ft/Ω m. In this case the discrepancy is negligible. Were the discrepancy large, the interpretation results would have to be modified, because the asymptotic value of S is usually more reliable than the sum of the layer conductances, except for the effects of anisotropy (section 3.2.4(g)).

In Figure 3.14 the top of L5 (the basement) is thought to be the top of massive limestones which are considered to be close to the Precambrian rocks.

(b) Interpretation by forward modeling

Having accomplished an approximate interpretation by means of curve matching, as shown in Figure 3.14, one can compute a theoretical curve for the approximate model. The difference between the field curve and the theoretical curve is decreased by altering the layer parameters and by adding more layers. This procedure is repeated until a satisfactory match is obtained. For this, a digital computer program that can rapidly generate theoretical curves for models having relatively large numbers of layers (say 20) is required. The modification and matching can be done visually by human judgement, or automatically by the computer, for example by a nonlinear regression subroutine using least-squares.

The automatic iterative procedure converges slowly, or frequently fails to converge, unless the initial model is a good approximation. Another cause of non-convergence is that field data may not be compatible with the assumption of a 1-D subsurface, that is, the subsurface may be 2-D or 3-D, or experimental errors may be large. The major drawback of automatic iteration is non-uniqueness caused by the existence of equivalent geologic models. In other words, widely differing 1-D models will yield practically the same theoretical curve. We shall return to this problem later. It is first necessary to input additional data into the process. The results may be greatly improved if some of the resistivities are known from well log data. Some computer programs provide options to assign fixed values either to resistivities or to thicknesses, or to assign upper and lower limits to these.

Automatic iterative interpretation methods are extensively reviewed by Koefoed (1979), who also gives several computer programs.

Pous, Marcuello and Queralt (1987) proposed an iterative interpretation algorithm based on a probabilistic treatment of the *a priori* information that allows constraints between the layer parameters. This treatment comes closer to obtaining a unique solution among several equivalent ones.

(c) Computation of theoretical apparent resistivity

We wish to calculate the distribution of electrical potential V on the surface of a horizontally layered earth due to a point source of current also placed on the surface. The subsurface consists of an arbitrary number of horizontal, homogeneous and isotropic layers of resistivity ρ_i and thickness h_i, the deepest medium (the nth layer) being a half-space. Due to the cylindrical symmetry, V is a function of the scalar distance r from the source; it is not dependent on the azimuth angle. Therefore, the problem is to solve Laplace's equation in cylindrical coordinates, and with cylindrical symmetry,

$$\frac{\partial^2 V}{\partial r^2} + \frac{1}{r}\frac{\partial V}{\partial r} + \frac{\partial^2 V}{\partial z^2} = 0. \tag{3.25}$$

The z-axis is positive downward and the source is at the origin. By using the method of separation of variables, one obtains the following solution:

$$[V(r)]_{z=0} = \frac{\rho_1 I}{2\pi}\left\{\frac{1}{r} + 2\int_0^\infty K(\lambda)\,J_0(\lambda r)\mathrm{d}\lambda\right\}, \tag{3.26}$$

where ρ_1 is the resistivity of the uppermost layer, I is the source strength (amperes), λ is the phantom variable (dummy) of integration, J_0 is a Bessel function and $K(\lambda)$ is called the **kernel function** which depends on the resistivities and thicknesses of the layers; $K(\lambda)$ can be determined numerically by means of the boundary conditions.

By definition, and according to equation (3.14), the Schlumberger array apparent resistivity is

$$\rho_a = -\frac{\pi(\mathrm{AO})^2}{I}\frac{\partial V}{\partial r}, \tag{3.14}$$

where V is given by equation (3.26), $\mathrm{AO}=\mathrm{AB}/2=\overline{R}$, and $\partial V/\partial r$ is the electric field E_r at the center point O of the array and in the direction AB. Equations (3.26) and (3.14) lead to the expression for the Schlumberger apparent resistivity

$$\rho_a = \rho_1\left\{1 + 2\overline{R}^2\int_0^\infty K(\lambda)\,J_1(\lambda\overline{R})\,\lambda\mathrm{d}\lambda\right\}. \tag{3.27}$$

As stated before, equation (3.26) is also valid for the equatorial and azimuthal dipole arrays; $\overline{R}=\mathrm{AB}/2$ the Schlumberger, $\overline{R}=\mathrm{AQ}$ for the equatorial and $R=p(\mathrm{OQ})$ for the azimuthal array. To obtain theoretical $\rho_a(\overline{R})$ curves, equation (3.27) can be evaluated by several methods: (1) numerical integration, (2) the image-point method, (3) decomposition into partial fractions and (4) linear-filter methods. At present the linear-filter method (Gosh, 1971) is universally preferred. This and other methods were reviewed by Koefoed (1979), who also provided computer programs. The linear-filter method requires extensive numerical computations. There are some computer programs that simplify the computation through approximations. Nivas and Israil (1986) proposed a simple method using an exponential approximation of the kernel function.

In section 3.2.2 and in Figure 3.11 I have briefly mentioned the use of the axial dipole array, which requires a different set of theoretical apparent resistivity curves. Catalogs of such curves are not as readily available as the Schlumberger catalogs. However, computer programs have been published (Laine and Lytle, 1976; Nivas and Israil, 1987). Also, because the axial dipole is the most popular array in induced polarization measurements, some mining geophysics contractors sell theoretical curves computed with this array for 1-D and 2-D models.

(d) Direct interpretation (inversion)

In the previously described iterative method of interpretation, one obtains a match between a field curve and a computer-generated curve, based on an assumed model, whether the matching is done visually or automatically by the computer. The computer generates the apparent resistivity curves for assumed models by means of equation (3.27), that is, the assumed layer parameters determine the kernel function, and ρ_a is computed from this kernel function. On the other hand, the principle of direct interpretation is that the field curve of apparent resistivities is directly converted to the kernel function which yields the layer parameters. The expression for the kernel function in terms of the field curve of apparent resistivities is (Koefoed, 1968)

$$K(\lambda) = \int_0^{\infty} \frac{\rho_a(\overline{R}) - \rho_1}{2\rho_1 \overline{R}} J_1(\lambda \overline{R})\, d\overline{R}, \tag{3.28}$$

where $\rho_a(\overline{R})$ is the Schlumberger apparent resistivities measured in the field. Equation (3.28) can in principle be solved by numerical integration, which requires the separate numerical evaluation of an infinite integral for each point of the kernel curve. The first direct interpretation method was published in 1940. However, practical methods were developed much later, the main problem being the magnification of observational errors that occur during the process of transforming to the kernel domain.

A conversational FORTRAN program is given by Koefoed (1979) who also applied this method to interprete resistivity sounding data obtained in The Netherlands for groundwater studies; one of his examples follows.

In Figure 3.15a the dots are the field data obtained with a Schlumberger array. The solid line is the theoretical curve derived from the interpreted layer parameters. The layer parameters are also plotted in Figure 3.15a as a dashed line, in the form of a well log, that is, true resistivity ρ against depth z. The interpretation resolved five layers. These are mainly Pliocene and Pleistocene sandy formations. The lowest (fifth) layer, which was assumed to be a half-space, represents a formation saturated with salt water; its resistivity is 5.4 Ω m. This example effectively illustrates the usefulness of resistivity sounding in mapping an interface between freshwater- and saltwater-saturated formations.

Continuous inversion

A unique inversion method by Zohdy (1989) produces as many fictitious layers as there are digitized points on the field curve, ρ_a (AB/2). It is a theoretically simple but ingenious method which works well. The iterations are guided somewhat empirically; they are based on the observed morphological relationships between resistivity versus depth data on the one hand and electrode spacing versus apparent resistivity data on the other. For instance, both sets of data are approximately parallel to each other, but the characteristic points are shifted horizontally and vertically with respect to each other, as shown in Figure 3.15b. The iterations modify the model in the direction of such anticipated shifts. The initial model is that in which depth is equal to AB/2, using the Schlumberger array, and ρ is equal to ρ_a; there is no need for an initial guess.

The initial model is successively modified until the computed curve matches the field curve, utilizing a least-squares optimization. Figure 3.15c shows an application. This is the step-function model; each digitized point on the field curve yields a step in the model, too many layers

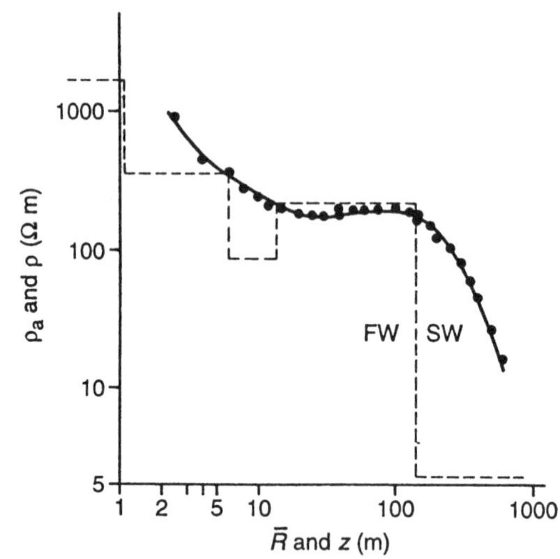

Figure 3.15a

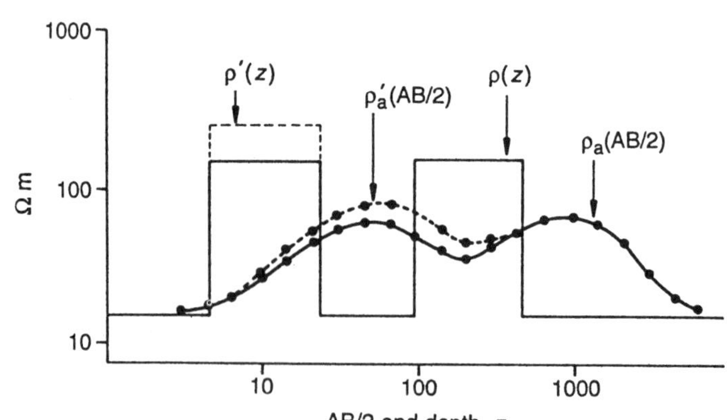

Figure 3.15b

to correlate. Therefore, for each step-function model either a continuous variation or a simpler but equivalent step-function model with fewer layers is constructed by means of a computer routine.

Figure 3.15d shows a cross-section obtained by Schlumberger soundings near Jean, Nevada (Zohdy, 1989). It is a 1-D automatic interpretation in the form of continuous variation of resistivity as described above.

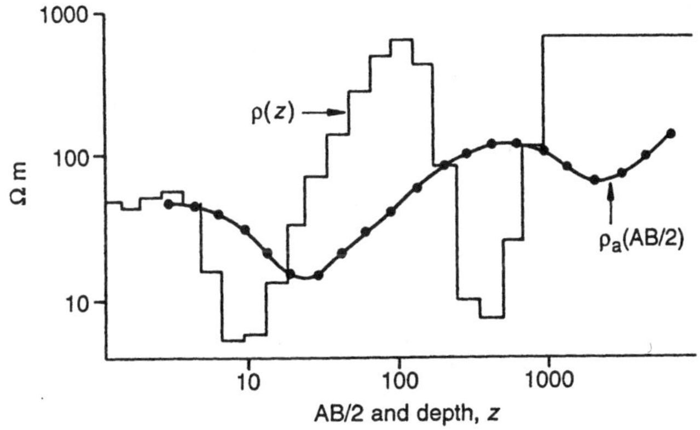

Figure 3.15c

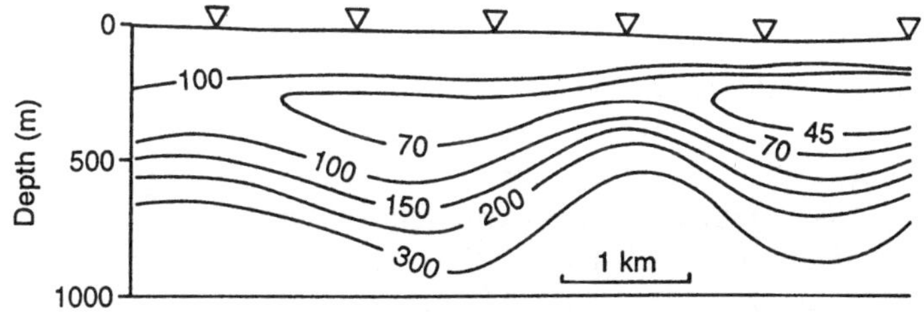

Figure 3.15d

Figure 3.15a–d (a) Direct interpretation of apparent resistivity field data. Dots are the field data; dashed line represents the results of interpretation; solid line is the theoretical curve derived from the interpreted layers; FW and SW are fresh- and saltwater formations respectively (from Koefoed, 1979). (b) Morphological and spatial relations between the subsurface model (ρ) and the Schlumberger sounding curve (ρ_a) computed from the model. If the model resistivity is raised to the dotted line (ρ'), the sounding resistivity also rises by a comparable amount. Depth, z, is assumed to be equal to (AB/2) and ρ to ρ_a. Assumed depths are shifted to the left and amplitudes up, until they are brought in phase (after Zohdy, 1989; the abscissa unit is not given but is immaterial). (c) Automatic interpretation of an actual Schlumberger sounding in the form of step-function model (after Zohdy, 1989; the abscissa unit is not given). (d) Resistivity cross-section from Schlumberger soundings at stations indicated by triangles, near Jean, Nevada, for groundwater exploration. Contours are in Ω m. Inversion is in the form of continuous variation of resistivity (after Zohdy, 1989).

Zohdy (1989) states that these results 'provided new information that was later supported by drilling'. More information about the Jean, Nevada, work can be found in a United States Geological Survey report by Zohdy (1988).

(e) Dipping layers

The routine 1-D interpretation requires mild tectonics and low dips, less than about 5 °. For steeper dips, up to about 20°, the interpretation is amenable to the 1-D case if the direction of the geologic strike is known beforehand, if the sounding is of the bilateral equatorial type with updip and downdip expansions of *R* from the AB line, which is set parallel to the strike, and if the dip is uniform within the array expansion. The average of the updip and downdip $\rho_a(\overline{R})$ curves is nearly the same as that which would have been obtained with a Schlumberger array all the way along the same AB direction, and it would yield the approximate depth underneath point O. With the Schlumberger array, one tries to minimize the dip effect by expanding AB along the strike, whereas with the bilateral equatorial array one tries to maximize the dip effect by placing AB in the strike direction and then estimating the dip from the divergence between the updip and downdip curves. We shall see later how this is done.

The deviation of the downdip dipole curve from that due to a 1-D model is less than that of the updip curve. Consequently, if a sounding consists of a one-sided expansion (not a bilateral one), it is better to expand it in the downdip side. For a one-sided dipole sounding, point O is not the reference point if there is any dip effect. If the existence of a layer becomes clear at a separation *R*, then the reference point for the top of this layer is at a distance approximately $R/2$ from point O.

The dip of a monoclinal electrical basement can be estimated by means of the set of theoretical curves shown in Figure 3.16. The AB line of a bilateral equatorial array is placed parallel to the strike, and *R* is expanded updip (R_+) and downdip (R_-), as shown in Figure 3.16. The two field curves, $\rho_+(R)$ and $\rho_-(R)$, are plotted on the same coordinates. The average of these curves is interpreted as a 1-D case, and the result is referred to point O. As explained in section 3.2.4(a), the replacement thickness and the replacement resistivity of the layers overlying the basement are determined from the 1-D interpretation results. This yields the final cross point which is placed on the intersection of the axes of the graph in Figure 3.16 ($R/d=1$ and $\rho_a/\rho_1=1$); the ρ_{a-} and ρ_{a+} curves read the dip angle on the theoretical curves.

The following field example shows the effect of dip in an area where the subsurface is known in detail from borehole and seismic data; it is also a comparison between the Schlumberger, equatorial dipole and azimuthal dipole soundings at the same location. Figure 3.17 shows the

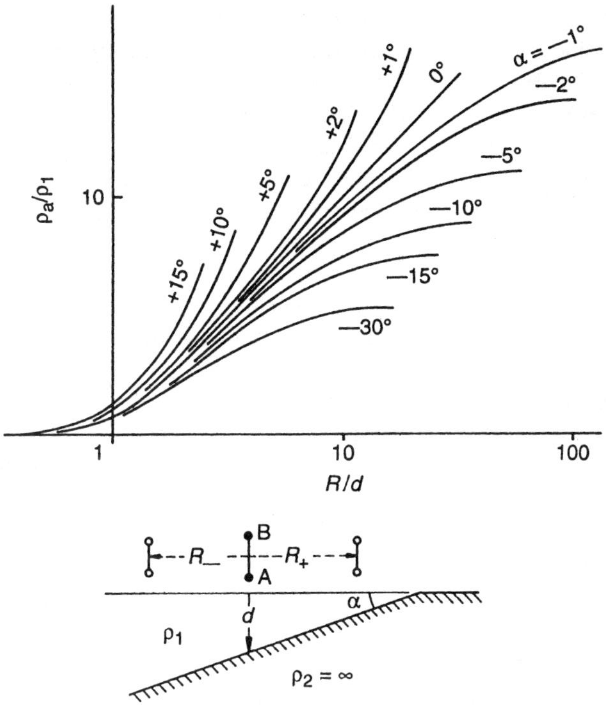

Figure 3.16 Theoretical curves for an inclined basement for a bilateral equatorial dipole array (from Berdichevskii and Zagarmistr, 1958).

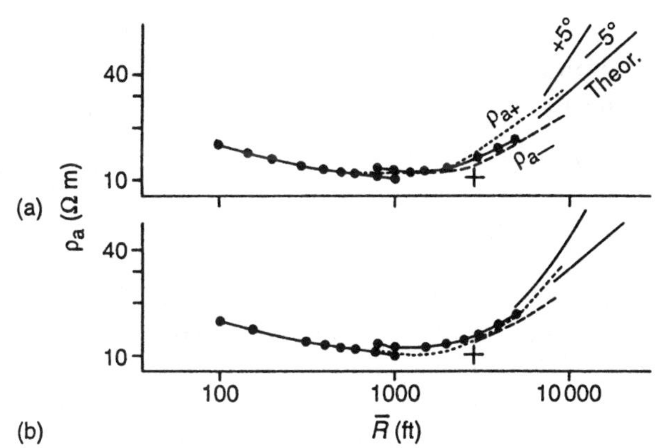

Figure 3.17 Comparison of resistivity sounding data obtained by different arrays at the same location where the dips are known. The Schlumberger data are shown as dots on solid lines. Updip (ρ_{a+}) and downdip (ρ_{a-}) curves for (a) bilateral azimuthal dipole data and (b) bilateral equatorial dipole data. Theoretical curve segments for 5° basement dip (solid lines).

data set. The Schlumberger data (dots) are the same as in Figure 3.6. They were obtained by expanding AB parallel to the strike.

Figure 3.17a shows the bilateral azimuthal dipole data superposed on the Schlumberger data. A 3000-ft (914-m) long portion of the Schlumberger AB line was used as source dipole. The azimuth angle θ varied between 65° and 105°. The updip and downdip expansions yielded the ρ_{a+} and ρ_{a-} curves respectively; both have been smoothed in Figure 3.17. The right-hand rising branch represents the basement complex at a depth of about 2500 ft (760 m). Also shown are the last cross point representing the replacement resistivity and the replacement thickness of the sedimentary section above the basement, and the theoretical curves for a basement inclined at 5°, transferred from Figure 3.16. The dip of the basement is known to be about 5°.

Figure 3.17b shows the bilateral equatorial data under the same conditions, but with $\theta = 90°$. In this case the dip information obtained from the bilateral dipole data is only semiquantitative; the deviation from the theoretical value is probably primarily due to faulting, which is also well known from other data.

(f) Layer resolution

In order to resolve a certain layer, the ratio of its depth z to its thickness h must be smaller than a certain value which may range from about 4 to about 20, depending on the particular case. Very generally, in order to recognize the presence of a layer, its thickness must be greater than 0.1 times its depth. Furthermore, when the presence of a relatively thin layer is recognized, it does not mean that its resistivity and thickness can be computed separately and uniquely. If $h_n \ll z_n$, the layer can be determined only in terms of

1. $S_n = h_n/\rho_n$ for a conductive layer, that is, if $\rho_n < \rho_{n-1}, \rho_{n+1}$, and in terms of
2. $R_n = h_n\,\rho_n$ for a resistive layer, that is, if $\rho_n > \rho_{n-1}, \rho_{n+1}$.

For proof of this, see, for instance, Bhattacharya and Patra (1968, pp.32–3 and 61–4). Thus the interpretation yields the values of S or R, but not ρ_n and h_n, and the solution is not unique. This aspect is called the 'principle of equivalence', which is due to the fact that the problem is an 'ill-posed' one, largely, but not totally, because of experimental noise and geological irregularities. For a unique interpretation of a thin layer, its resistivity must be known from well logs. It is obvious that one cannot resolve a large number of discrete layers, usually less than about six, excluding the very shallow and insignificant layers. The deeper the exploration, the larger the electrode array and the larger the volume of the subsurface influencing the measurements. Consequently, the vertical and horizontal resolutions decrease rapidly as the depth increases.

The lack of a high degree of resolution at depth becomes an advantage in some cases; it provides information on average properties of major rock formations. One resistivity sounding by itself may have more regional significance in basin evaluation than a borehole.

(g) *Effects of anisotropy*

Sedimentary formations are normally anisotropic. Resistivity may be assumed to be independent of direction parallel to bedding. Thus, each layer possesses two resistivities: longitudinal resistivity, ρ_l parallel to bedding, and transverse resistivity, ρ_t, perpendicular to bedding; ρ_t is always larger than ρ_l. Resistivity of a discrete layer computed from a resistivity sounding is neither ρ_l nor ρ_t, but ρ_m which is called the mean resistivity,

$$\rho_m = (\rho_t \, \rho_l)^{1/2}.$$

If the true thickness of a layer is h, its thickness computed from resistivity sounding is λh, where λ is called the coefficient of anisotropy,

$$\lambda = (\rho_t / \rho_l)^{1/2}.$$

The proof is given by Grant and West (1965, pp. 413–5). Anisotropy includes microanisotropy, which is an intimate property of rocks, and macroanisotropy due to stacking of thin layers making up one layer which is thick enough to be resolved by resistivity sounding. The magnitude of λ is usually in the neighborhood of 1.1. This means that computed depths are systematically too large by about 10%.

In the case of a deep, thin layer, that is, for $h_n \ll z_n$, combining the expressions for S_n and R_n, defined in section 3.2.4(f), with the expressions for ρ_m and λ, one finds that resistivity sounding yields either

$$S = h/\rho_l$$

for a thin, low-resistivity layer, or

$$R = h\rho_t$$

for a thin, high-resistivity layer.

In summary, if a layer is thick enough to have its thickness and depth resolved separately, its depth and thickness are exaggerated due to anisotropy, and the interpreted layer resistivity is the mean resistivity.

Electrical and induction well logs yield only ρ_l. If such logs and resistivity sounding measurements are combined, λ, ρ_t and ρ_l can be computed separately. The depth discrepancy between a well log and a sounding is an effective method of determining the bulk anisotropy coefficient, which is a significant stratigraphical parameter in its own right in oil exploration; it is thought to be related to the degree of mixing of sand, shale and limestone in the subsurface. This in turn is useful in locating the ancient shore lines and isolating areas favorable for oil accumulation (Keller, 1968).

It was stated (Total conductance, in section 3.2.4(a)) that the 45° asymptote to the right-hand rising branch of a resistivity sounding field curve yields the total conductance, S, of the sedimentary section above the electrical basement,

$$S=\sum(h_i/\rho_i),$$

which can be expressed as

$$S=\left(\sum h_i\right)/\rho_l,$$

where ρ_l is the replacement resistivity (or effective resistivity) of the whole section above the basement, and it is the longitudinal resistivity, since it corresponds to very large electrode separations and horizontal current flow. Thus, interpretation of sounding data in terms of discrete layers yields ρ_m, and the asymptotic S yields the overall ρ_l.

On the other hand, the magnetotelluric and loop-source electromagnetic sounding methods yield ρ_l for a discrete layer. It is possible that a combination of one of these methods with resistivity sounding can enable one to obtain the anisotropy coefficient and resolve ρ_t and ρ_l for discrete layers.

(h) Depth of exploration

It was experimentally shown in section 3.2.2(b) that, in determining the depth to the basement, the depth of exploration is about AB/2 for the Schlumberger array. This is AQ for the equatorial and pR for the azimuthal dipole arrays. In other words the depth of exploration in this case is about equal to $\overline{R}$ for these three arrays. It is less for the axial array; it is about 0.7 R, where $R=\text{OQ}$.

For a discrete layer, the concept of depth of exploration becomes very vague. Many authors have proposed a multitude of definitions and have reached many different conclusions. These are reviewed by Roy and Elliot (1981). In reality there can be no generalized correspondence between depth and electrode separation. Each model has to be considered separately by means of computer modeling or curve catalogs. Here I shall use the definition given by Roy and Elliot (1981) and give a rule of thumb.

The limiting depth of exploration is defined as the maximum depth of the layer whose contribution to the measurement becomes significant for a given type and dimension of array. The limiting depth is:

- for the Schlumberger array: from AB/8 to AB/2, average AB/4;

- for the azimuthal and equatorial dipoles: average $\overline{R}/2$, where $\overline{R}$ is p(OQ) or AQ respectively,
- for the axial dipole: average $R/3$, where R is OQ.

3.3 RESISTIVITY PROFILING AND MAPPING

3.3.1 Introduction

The classical definitions of resistivity profiling and mapping were given in section 3.1.

The purpose of profiling is to explore 2-D structures. The electrode array is kept constant both in dimension and in orientation, and is moved in a straight line perpendicular to the geologic strike. Before the advent of the dipole methods, the Wenner array was used extensively, especially in the 'longitudinal' mode in which the line of electrodes moves along the line of profile, along equally spaced markers in the field (section 3.2.2). For special problems, such as determining a steeply dipping layer or dike, or for precise determination of a fault, it may be desirable to move the line of electrodes broadside; this is called 'transverse profiling' which is considerably more expensive than the longitudinal method and may be cost effective only for shallow work. In the longitudinal mode, Wenner profiling is still used for moderate depths, primarily because of its simplicity. For deep work, and especially with mechanized equipment, the preferred array is the axial dipole (section 3.2.2); this is the only kind of profiling that will be discussed here.

Resistivity mapping is carried out, in general, to explore 3-D structures. This can be envisaged as a series of parallel profiles in which the apparent resistivities for each fixed array dimension and orientation are displayed in map form. A number of such maps, each for a different array dimension, and possibly some for different orientations, may be qualitatively, and sometimes quantitatively, interpreted. This method has been used extensively since the 1930s, especially to obtain detailed subcrop geology. Three such case histories are found in a book by Poldini (1947). Mapping can also be envisaged as one way of displaying the data obtained from a number of resistivity soundings made on a square grid. These are valid methods but are frequently not cost effective for deep work, unless one is doing a multiplicity of profiles or a grid of soundings anyway.

A special type is called quadripole (or rotating dipole) resistivity mapping. Two orthogonal dipoles constitute the source, which stays fixed in order to survey the surroundings, which may be a fairly large area. The receiver moves around arbitrarily and is used to measure two orthogonal components of the electric field. Apparent resistivities are assigned to the receiver locations. This method has been used in prospecting for geothermal reservoirs in a sedimentary basin by Harthill (1978).

3.3.2 Resistivity profiling with an axial dipole array

An axial dipole array (also called polar dipole) consists of two collinear dipoles (Figure 3.10). In profiling, it is logistically advantageous to make the lengths of the source and receiver dipoles equal, and the distance between the dipole centers a multiple of dipole lengths, that is $MN = AB = a$, and $OQ = na$, as shown in Figure 3.18. In this special form it is called the **dipole–dipole array**, with which the apparent resistivity is

$$\rho_a = \pi n\, a(n-1)\,(n+1)\,\Delta V/I; \tag{3.29}$$

this is derived directly from equation (3.9). The reader is cautioned that in some other literature the distance between the two inner electrodes (BM) is taken as na.

The most favored style of data display is shown in Figure 3.18. The apparent resistivities are plotted at the intersections of diagonals at 45° to the line between the centers of the dipoles, and then contoured. The result is a pseudo-section. It should be remembered that this is merely a style of data display, that it does not convey any quantitative notion about the depth or the shape of the subsurface model.

With a mechanized crew, the fieldwork proceeds in the following fashion. Assume that the transmitter (Tx) is fixed and the receiver (Rx) starts at a large value of n at the right-hand side of Figure 3.18, and moves toward Tx, making measurement intervals a distance a apart, and passes Tx by. Then, at an appropriate distance on the left side of Tx, the Rx dipole becomes a Tx dipole.

Figure 3.19 is a field example from the Cerro Prieto geothermal area, Mexico (Wilt, Goldstein and Raso, 1980). Subsequently, a profile of magnetotelluric soundings were done here and its results were compared to those of dipole–dipole profiling (Gamble *et al.*, 1981); the authors state

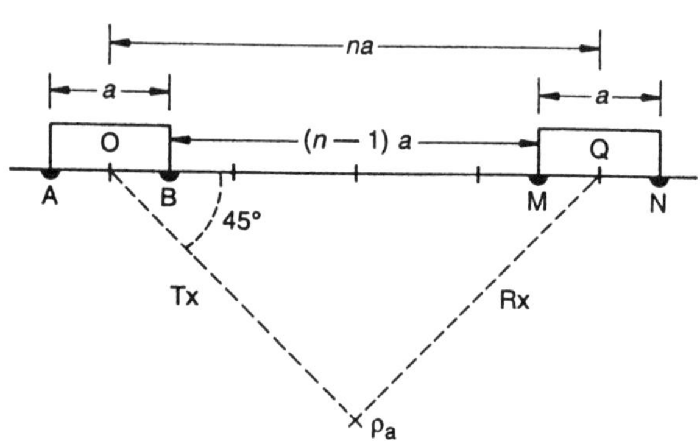

Figure 3.18 A cross-section showing the axial dipole (dipole–dipole) array for resistivity profiling. Tx is transmitter (source dipole) and Rx is receiver (probe dipole).

that the only substantial disagreements between the two models occur at depths greater than below 2 km, that is, below the model shown in Figure 3.19B.

(a) Interpretation

Interpretation in terms of arbitrary 2-D models of any complexity is at present performed by numerical forward modeling: (1) finite difference, (2) network anology, (3) finite element and (4) integral equation.

Trial-and-error data fitting with complex models requires many time-consuming iterations; it is very difficult to choose the proper alterations to the model. There is much need for an automated inversion. However, the consensus is that it is doubtful whether strictly automated inversion with arbitrary complex models can be achieved in the foreseeable future, because of non-uniqueness, geological noise and the large amount of computer time required.

Forward modeling

A finite-difference matrix inversion technique was developed by Dey and Morrison (1979). They state that this has significantly less attendant computational costs than with finite-element or network-solution

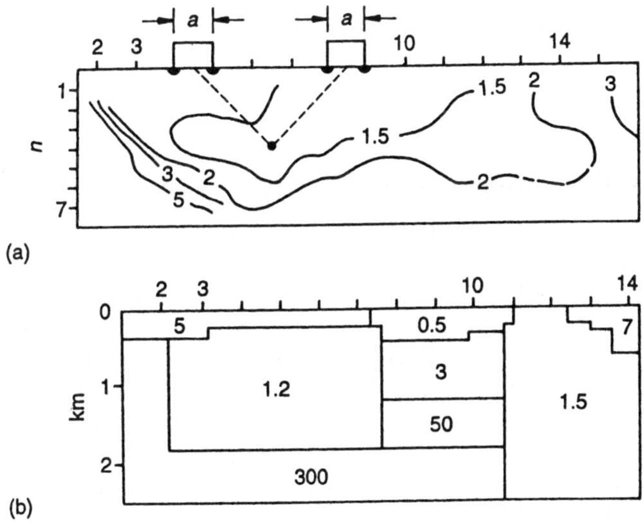

Figure 3.19 Dipole–dipole resistivity profiling, cross-section, Cerro Prieto geothermal field, Mexico (data are in Ω m). (a) Field data, apparent resistivity pseudo-section, $a = 1$ km, (b) 2-D interpretation by forward modeling (after Wilt, Goldstein and Raso, 1980).

techniques. Dey (1976) published the algorithm and the user's guide for a computer program. The network analogy (transmission-surface) was published by Madden (1971) and Pelton, Rijo and Swift (1978). A generalized treatment of 2-D EM numerical modeling, including DC, is found in Hohmann (1988).

Inversion

Tripp, Hohmann and Swift (1984) describe a nonlinear inversion technique that estimates the resistivities of cells of predetermined geometry through a method of least squares. The forward modeling component is based on the transmission-surface analogy.

Smith and Vozoff (1984) use a technique that is similar to the one above, except that they use a straightforward finite-difference calculation. A generalized treatment of 2-D EM inversion is given by Hohmann and Raiche (1988).

Joint inversion

Methods of 2-D joint inversion of EM and resistivity data have been published since the 1970s. Sasaki (1989) describes a 2-D joint inversion of magnetotelluric and dipole–dipole resistivity data, with the emphasis on a computer algorithm. The resulting model is composed of a large number of rectangular blocks, each having a constant resistivity. The algorithm determines iteratively only the resistivities. The forward solutions for both the dipole–dipole and the magnetotelluric problems are based on the finite-element method. Examples of field and synthetic data are used to show the improvements obtained with joint inversion over the single inversion method.

3.3.3 Resistivity mapping with a quadripole array

This method grew out of the **dipole mapping** method which employs one stationary source dipole as the transmitter, and a wandering receiver to measure the total electric field. The apparent resistivity, which is assigned to the receiver location, varies considerably with the orientation of the transmitter dipole. The **quadripole** array was developed to alleviate the source orientation effects that were bothersome in dipole mapping.

The quadripole array was developed in stages in the 1970s. Bibby's (1977) theory unifies the previous developments by others. Harthill (1978) presented the results of a large survey based on Bibby's paper. The following procedure is that described by Harthill (1978).

Figure 3.20 shows the quadripole layout. Two approximately orthogonal dipoles, no. 1 and no. 2, constitute the transmitter, Tx. The receiver, Rx, consists of two orthogonal dipoles to measure the components of the electric field *E*. Each transmitter dipole is excited alternately. Up to 700 A

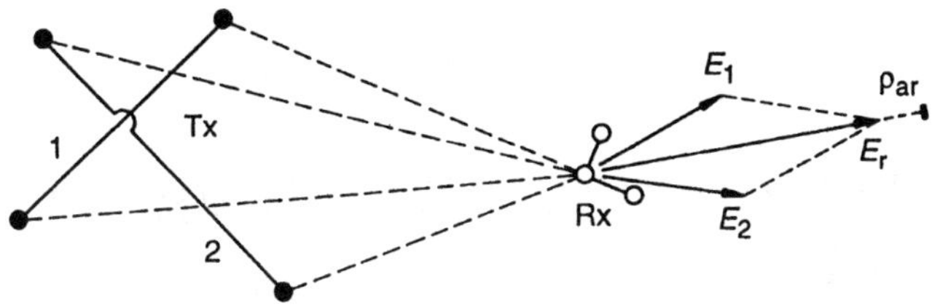

Figure 3.20 The quadripole resistivity mapping array. Tx and Rx are the transmitter and receiver respectively. $\boldsymbol{E}_1$ is the total electric field due to no. 1 transmitter dipole; likewise for $\boldsymbol{E}_2$ (schematic).

of square wave current is used. The Tx–Rx separation is from 1 to 10 km. When dipole no. 1 is excited the total electric field at Rx is $\boldsymbol{E}_1$, and when no. 2 is excited the total field is $\boldsymbol{E}_2$. The vector sum of $\boldsymbol{E}_1$ and $\boldsymbol{E}_2$ is the resultant field $\boldsymbol{E}_r$, which is the total of the totals. This yields a kind of tensor apparent resistivity, ρ_{ar}. By changing the ratio of $|\boldsymbol{E}_1|$ to $|\boldsymbol{E}_2|$ the resultant $\boldsymbol{E}_r$ can be caused to rotate through 360°. At every Rx location, for any value of $\boldsymbol{E}_r$ at any azimuth a value of ρ_{ar} is calculated. As $\boldsymbol{E}_r$ rotates, ρ_{ar} values fall on an ellipse. This is why this method is also called **rotating dipole tensor resistivity mapping**. The data display can be in a variety of forms. The arithmetic mean of the maximum and minimum axes of the ellipse of resistivity is mapped as the **tensor invariant resistivity**.

Having made measurements in the surroundings of the transmitter, the latter moves to a new location. Some of the new receiver locations overlap with those measured with the previous transmitter. Usually, there are appreciable discrepancies between the overlapping data, and subjective judgement is used to smooth them out. Also, a form of spatial filtering may be used to make contouring possible where the data overlap.

The primary objective of the survey presented by Harthill (1978) was to locate low-resistivity zones that might be associated with geothermal reservoirs in the Imperial Valley sedimentary basin in California. It appears that useful results were obtained.

Quadripole mapping is based on the premise that measurements of the electric field at each receiver station represent the subsurface resistivity conditions in the immediate vicinity of that station only. In reality, this assumption violates the theorem of reciprocity, that the transmitter and receiver electrodes are interchangeable; each measured apparent resistivity cannot be assigned only to that receiver station; it represents the earth resistivities in the vicinity of the whole array, down to depths governed by the distance between the transmitter and the receiver.

In conclusion, this method may be cost-effective only in the hands of those explorationists who are highly experienced in this particular technique, and only under certain geologic conditions.

4

Natural-source electromagnetic methods

4.1 INTRODUCTION

As the name implies, these methods utilize natural electromagnetic fields whose origins are mostly outside the atmosphere. The study of the variation of the Earth's magnetic field with time began about four centuries ago. The frequency spectrum of these variations is very broad; it ranges from hundreds of years per cycle to more than 30 megacycles per second. Time variations of any magnetic field are associated with an electric field which induces electric currents in conducting media, such as sedimentary rocks. Although the existence of such currents could have been inferred in the past, their existence was proven by Barlow in 1847 by means of measurements made using telegraph lines; they are now called telluric currents.

At any location at the surface of the Earth, the magnetic field, electric field and electric current are related to each other and to the physical properties of the subsurface rocks in the vicinity of that location via Maxwell's equations. For instance, Figure 4.1 shows the variation with time of the natural magnetic (H_y) and electric (E_x) fields measured simultaneously at one location. The magnetic field is measured by means of a magnetometer, and the electric field measurement is derived from the voltage difference between two grounded electrodes. Either one of the traces could be computed synthetically from the other by means of Maxwell's equations if the properties of the subsurface rocks were known. Conversely, the two traces together supply geological information. This is the magnetotelluric method of exploration, meaning that time variations of the magnetic field and the telluric electric field are measured simultaneously at one location where geologic information is sought.

Now let us consider another form of natural-field measurements called the telluric method. Figure 4.2 shows two electric field traces measured simultaneously at two stations 43 km apart, for 30 min. These are

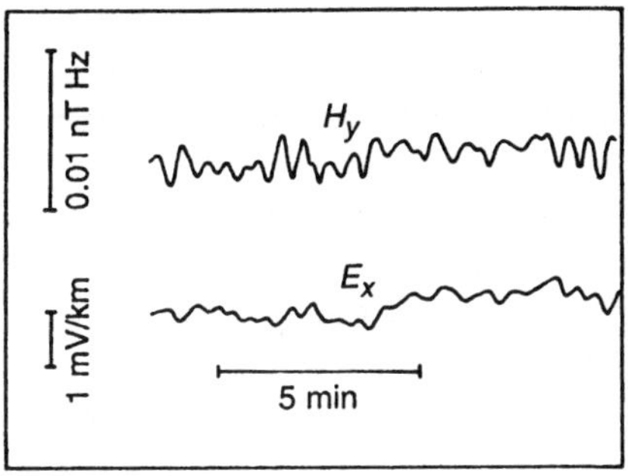

Figure 4.1 Time variations of the horizontal components of the natural magnetic (H_y) and electric (E_x) fields, simultaneously measured at one station (from Yungul, 1982).

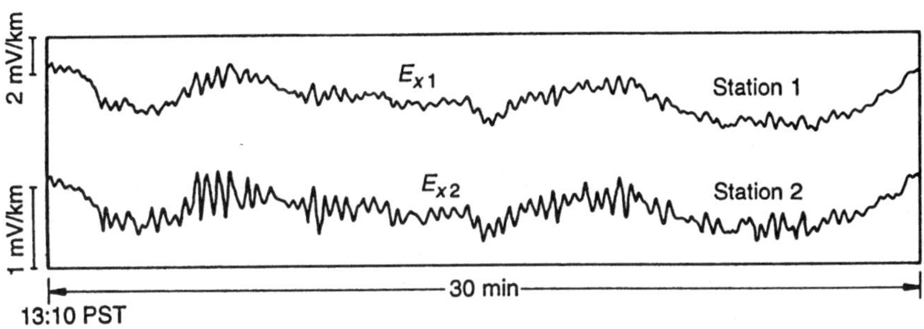

Figure 4.2 Two electric field traces (tellurograms) simultaneously measured at two stations 43 km apart for 30 min, in the direction of station separation, San Joaquin Valley, California (from Yungul, 1982).

components of the electric field in the direction of station separation. The two traces are very similar morphologically but different quantitatively; different frequency components have different amplitude ratios. These differences between the two traces are due almost totally to the differences in the geologic conditions at the two stations. If the geology in the vicinity of one station is known, the two traces yield geologic information at the other station. For instance, a simple visual inspection shows that the amplitudes at station 2 are smaller than those at station 1. This suggests that the thickness of the low-resistivity sedimentary rocks at station 2 is larger than that at station 1; the lower the amplitudes, the thicker the low-resistivity section. Futhermore, the amplitude ratio varies with frequency. Such data, which can be obtained by either visual or spectral analysis, yields quantitative knowledge on resistivities as well as thicknesses of sedimentary sections. This is the telluric method; the tellluric electric field is measured simultaneously at two or more stations, without measuring the magnetic field, whereas in the magnetotelluric method the electric and magnetic fields are measured simultaneously at one station.

Actually, the difference between the two methods is not as clear-cut as expressed above; there is overlap and gradation between various techniques of the two methods; and the two methods are basically equivalent. The choice between the two methods, or rather two classes of techniques, depends on the type and/or value of geologic targets.

4.2 NATURAL ELECTROMAGNETIC FIELDS

4.2.1 Definitions and units

Telluric current is the worldwide electric current of natural origin (*Tellus* means the Earth in Latin). The unit of electric current is the ampere (A); it is a scalar quantity but it must be related to a particular surface across which the current flows. In dealing with natural currents a more significant quantity is the current density vector, $\boldsymbol{J}$, measured in amperes per square meter (A/m^2).

Telluric field. Telluric electric field intensity, $\boldsymbol{E}$, measured in volts per meter (V/m), is associated with telluric current density via Ohm's law. This unit, V/m, is too large for telluric measurements; it is customary to use millivolts per kilometer (mV/km).

A **tellurogram** is a strip chart record of time variations of one component of telluric field, e.g. $E_x(t)$, as shown in Figure 4.2.

Magnetic Field intensity, $\boldsymbol{H}$, is measured in amperes per meter (A/m). Another quantity is magnetic induction $\boldsymbol{B}$. For a linear and isotropic medium free of permanent magnetization,

$$\boldsymbol{B} = \mu \boldsymbol{H},$$

where μ is the magnetic inductive capacity. It has become customary in geophysical prospecting that, when one says magnetic field intensity, designated by $\boldsymbol{H}$, one actually means magnetic induction, designated by $\boldsymbol{B}$ and measured in (newton/ampere $\times$ meter), which is called tesla (T), because $\boldsymbol{B}$ is what most magnetometers measure. There is much confusion among geophysicists as to what magnetometers measure, $\boldsymbol{B}$ or $\boldsymbol{H}$, and in what units it should be given, A/m or T. The problem has not yet been resolved on an international basis. This, however, does not necessarily create a quantitative error, because the numbers that are used are in the units of $\boldsymbol{B}$. Also, for commonly occurring rocks we are allowed to assume that μ is equal to that of free space, μ_0, and $\mu_0 = 1$ in cgs emu, so that in emu, $\boldsymbol{B}$ is numerically equal to $\boldsymbol{H}$. The unit of $\boldsymbol{B}$ in cgs emu is the gauss (Gs) which is equal to 10^{-4} T.

Both gauss or tesla are too large for natural field measurements. It is customary to use either γ or nT, as shown below:

$$\text{nanotesla, nT} = 10^{-9}\,\text{T} = 10^{-5}\,\text{gauss} = \gamma\text{, gamma.}$$

A **magnetogram** is a strip chart record of time variations of one component of $\boldsymbol{H}$ or $\boldsymbol{B}$, e.g. $H_y(t)$, as shown in Figure 4.1. Here the scale is given in nanotesla times frequency in hertz, because $\boldsymbol{H}_y$ was measured with a coil–core magnetometer whose output is the time derivative of $\boldsymbol{H}_y$.

The **magnetotelluric field** is the natural electromagnetic field when $\boldsymbol{E}$, $\boldsymbol{H}$ and f are considered together so as to interrelate them.

4.2.2 Types and characteristics of natural fields

The amplitude of $\boldsymbol{E}$ depends strongly on local geology; it can vary by a factor of 20 over a short distance, say a kilometer. On the other hand, $\boldsymbol{H}$ depends much less on local geology (section 4.3.3(e)). Generally speaking, the amplitude of $\boldsymbol{H}$ seldom varies by a factor of more than 1.5 over a distance of a few kilometers. Therefore, in the study of the worldwide characteristics of the natural EM fields, it is customary to consider the time variations of $\boldsymbol{H}$ only. Obviously, in geophysical exploration, $\boldsymbol{E}$ has much more diagnostic capabilities than $\boldsymbol{H}$. The primary role of $\boldsymbol{H}$ is to provide the normalization for $\boldsymbol{E}$; it is like the source strength in amperes in the resistivity method. In the terminology of magnetic observatories 'variations' mean time variations of $\boldsymbol{H}$.

There are various types, sources and frequencies of the variations. Figure 4.3 is an approximate and schematic representation of origins and

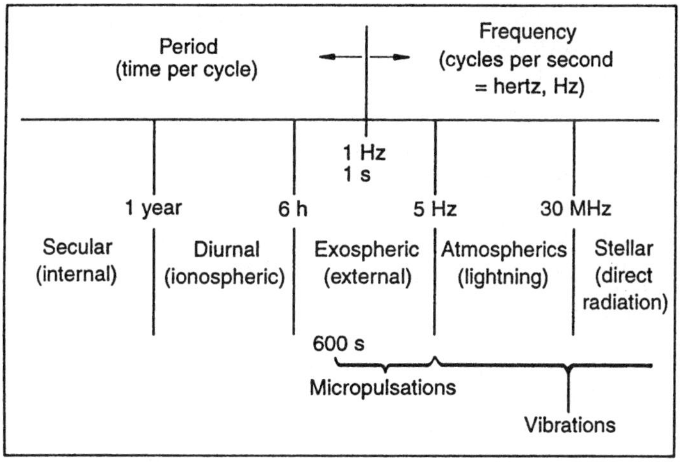

Figure 4.3 Approximate and schematic frequencies and origins of the natural electromagnetic fields (from Yungul, 1982).

frequencies. In the exploration of sedimentary basins, e.g. in petroleum work, the frequency spectrum of primary importance is from about $T=600$ s to about $f=100$ Hz. This range comprises all of the phenomena called **geomagnetic micropulsations** and some of the atmospherics; these will be discussed later.

(a) Secular variations

These are entirely of internal origin, thought to be caused by motion in the electrically conducting fluid core of the Earth. The periods are measured in tens of years, which are of no practical use in petroleum work.

(b) Diurnal (daily) variations

Thermal effects of the Sun and attraction effects of the Moon, which cause the ionosphere, which has a high electrical conductivity, to oscillate. Thus, a conductor moves in the Earth's main magnetic field and thereby forms an 'atmospheric dynamo' whose output has periods from about 6 to about 25 h. The lunar day is about 50 min longer than the solar day, so that the solar and lunar day changes are shifting in phase. This causes the monthly cycles. The amplitudes are large, of the order of tens of nanotesla. These events have substantial application in the exploration of the lower crust and upper mantle, but the frequencies they contain are too low for petroleum work.

(c) Variations of exospheric origin

Variations of exospheric origin are the most significant for exploration, and are called **micropulsations**; the most intense phenomena called magnetic storms fall into this category.

Micropulsations occur within the approximate period range from 0.2 to 600 s (5–0.0017 Hz). They occur almost continuously as a background noise. The amplitudes depend on latitude, solar activity, frequency, local and universal times, the season and local geology. The worldwide and long-term statistical average amplitudes of the fields are of the order of a few millivolts per kilometer and a few tenths of one nanotesla (Figure 4.4a). A large number of papers and several books has been written on micropulsations, especially in connection with research in space physics. These variations of exospheric origin are generated through the interaction of ionized gas clouds coming from the Sun, called the **solar wind**, with the main magnetic field and the atmosphere of the Earth. These interactions finally set up large loops (sheets) of electric current in the ionosphere, called 'overhead currents'; the diameter of such a loop is a few thousand kilometers. From there on electromagnetic propagation

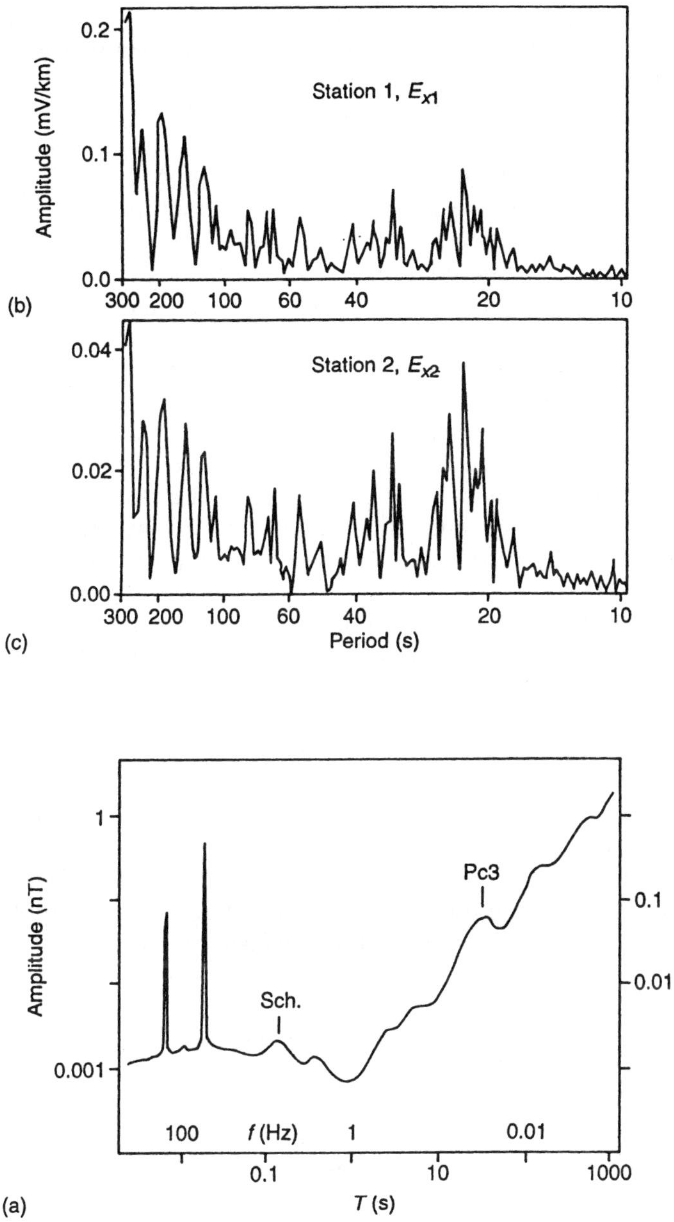

Figure 4.4 (a)Example of a magnetic field amplitude spectrum, for a horizontal component. Spikes to the left are power line harmonics, Sch = Schumann resonance (after Malergue *et al.*, 1986). (b), (c) Amplitude spectra of the tellurograms shown in Figure 4.2, at discrete frequencies. Record section is 1800 s; sampling interval = 1.875 s (from Yungul, 1982).

(or induction) takes place in the lower atmosphere, which is a very poor conductor. Thus, it appears to us as though the ionosphere was the source of the micropulsations.

Micropulsations are generally classified into two groups: continuous pulsations (Pc) and irregular pulsations (Pi). Continuous pulsations occur mostly during daytime. They form wave trains lasting tens of minutes. Their amplitude is at a maximum at about 10 local mean time (LMT). Annual maxima are at the equinoxes and a minimum occurs in December. Pcs are subdivided into several kinds and frequency ranges. Pc3s, whose period range is 10–45 s, are the most useful events in exploring sedimentary basins.

Figure 4.2, which was discussed in section 4.1, shows two typical tellurograms. They represent normal activity on a magnetically quiet day at mid-latitudes. They display well developed Pc3 wave trains. Figures 4.4 b,c show the amplitude spectra of the tellurograms shown in Figure 4.2. The high amplitudes in the period range of about 20–40 s are those of Pc3 wave trains. For periods longer than about 100 s the amplitudes rise again, as is frequently observed. The ratio of the amplitude at station 1 to that at station 2 is about 5 at a period of 300 s, and about 2.5 at 25 s. This is because the sedimentary basin is much deeper at station 2 than at station 1.

Irregular pulsations occur mostly during night time. The wave trains are of limited duration. Their period range is usually 40–120 s and the amplitude is of the order of 1 nT. These are sometimes very useful, especially in areas where industrial EM noise becomes prohibitive during daylight hours.

Magnetic storms are also of exospheric origin. These are very intense and prolonged disturbances that occur about once a month on average. They are caused by large-scale bursts of ionized clouds accelerated away from the Sun. Their amplitudes may reach hundreds of nanotesla and hundreds of millivolts per kilometer. They last from a few hours to a few days and include an abundance of micropulsations. Observatories regularly issue storm forecasts. These storms are useful in exploration, while they last, especially in areas with industrial noise.

(d) Vibrations

This is a general name for variations whose frequencies are higher than 5 Hz. From about 5 Hz to about 10 kHz the major cause is lightning, which occurs almost continuously in Indonesia, in central Africa and in the Amazon region; these are the so-called atmospherics which are mostly in the audiofrequency range. Resonance in the Earth–ionosphere cavity, called Schumann resonance, has a fundamental frequency of about 8 Hz; it is usually present and is useful in exploration. Above 30 MHz the dominant source is the direct radiation of EM waves from the Sun. In petroleum work, in sedimentary basins, there is seldom need for frequencies higher than about 200 Hz.

4.2.3 Source conditions

For the resistivities encountered in sedimentary basins and for the frequencies employed in petroleum work, the signals (micropulsations and some of the atmospherics) can be considered to be plane waves (Madden and Nelson, 1964; Srivastava, 1965). Furthermore, the Earth's surface may be assumed to be an infinite plane. Also, the resistivity contrast between air and the Earth is so large that micropulsation waves impinging on the Earth at all angles of incidence are transmitted into the Earth very nearly vertically. Thus, one can assume that the signals are vertically incident plane waves. This implies that in a horizontally layered subsurface the telluric currents flow horizontally, that ***E*** and ***H*** are horizontally uniform (but amplitude and phase vary with depth), and that variations with horizontal distance are entirely due to lateral changes in geology.

The fact that micropulsations and atmospherics lend themselves to the assumption of vertically incident plane waves is a very substantial advantage in exploration over the controlled-source methods, because it facilitates interpretation by not involving source geometries with respect to geological geometries.

Electric currents associated with micropulsations scan the subsurface from all directions, usually displaying a statistical elliptic polarization if a narrow band of frequencies is taken, as shown in Figure 4.5. This is another advantage over the controlled-source signals. However, on rare occasions the polarization may turn out to be somewhat linear and may create difficulties.

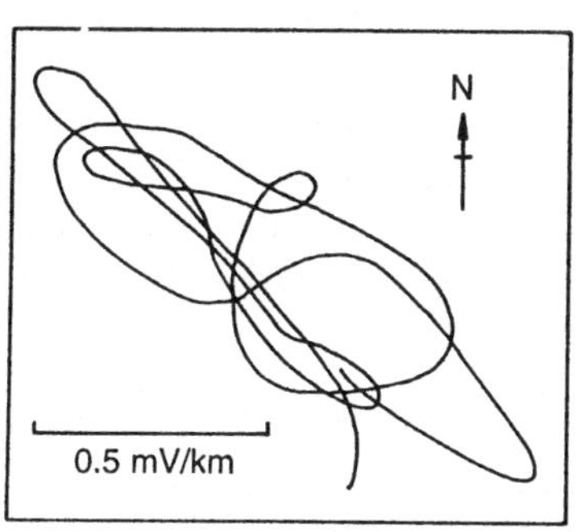

Figure 4.5 A vector diagram representing a few minutes of recording of the horizontal electric field vector of micropulsations. The bandpass filter peaked at a period of 20 s (from Yungul, 1982).

A major difficulty with natural-field methods is the abundance of noise due to those industrial currents which do not meet the condition of plane waves. To cope with the noise one may resort to judicious selection of site and time, or to **referential** measurements that will be discussed in section 4.3.7. However, if the noise source is sufficiently far away, at a distance R, such that $R(f/\rho)^{1/2}$ is sufficiently large, where ρ is the resistivity of the subsurface and f is frequency, then one is justified in using industrial noise as though it were a micropulsation signal. We shall return to this subject in section 5.2.

The character of the natural fields and the frequencies of interest are such that, if records of $\boldsymbol{E}(t)$ and $\boldsymbol{H}(t)$ are obtained over long periods, say of the order of one hour, the fields can be decomposed into their Fourier components, and each component can be treated separately as a harmonic (sinusoidal) wave.

If, at a measuring site, the subsurface is effectively two-dimensional on a local basis (piecewise two-dimensional) along the principal axes of anisotropy x and y, a wave can be decomposed into x and y polarizations. Each one can be treated separately and combined at the end. Therefore, the theory is reduced to a linearly polarized, vertically incident, harmonic plane wave.

4.3 MAGNETOTELLURIC (MT) METHODS

The magnetotelluric (MT) method was developed and used in the investigation of the Earth's deep crust by Kato, Kikuchi and Rikitake in Japan and by Tikhonov in the USSR, all in 1950. It appears that Tikhonov published the first definitive paper in 1950 on this subject. The subject became popular in theory among petroleum geophysicists of the western hemisphere following the publication of Cagniard's (1953) paper in *Geophysics*. However, the method remained an academic subject until about 1967, due to lack of proper instruments and computer programs. It gained a fair amount of practical acceptance in the petroleum industry following Vozoff's (1972) comprehensive paper; the subject of this paper has been updated and expanded in Vozoff (1991). There are several books on MT, published in different countries, e.g. a book by Kaufman and Keller (1981), and a volume of 54 reprinted MT papers, edited by Vozoff (1986).

4.3.1 An introductory overview of MT methods

At each observation point (station) at the surface, two horizontal components of the electric field $\boldsymbol{E}$ and the three orthogonal components of the magnetic field $\boldsymbol{H}$ are measured simultaneously in the form of time series. Analysis of these time series in the frequency domain yields information concerning the electrical resistivity distribution in the subsurface.

The main objective of MT, insofar as its petroleum applications are concerned, is to obtain highly simplified 'resistivity logs' (without drilling) in areas of mild dips. Under such conditions, measurements made at each observation location yield downward information at that location, by assuming that the subsurface is locally horizontally layered (1-D). This particular procedure is called **magnetotelluric sounding** (MTS). This type of information can also be obtained by controlled-source methods (frequency-domain and time-domain electromagnetic, and direct current), but each method is a different tool and has different advantages.

In other cases, where the dips are large, the subsurface may be assumed to be 2-D. This necessitates a number of closely spaced MT stations along a profile perpendicular to the geologic strike. Geologic interpretation of this kind of data is achieved in a way that is analogous to gravity modeling: the MT data at all of the stations must satisfy the 2-D model simultaneously. In this case, the result is not in the form of 'resistivity logs' but a cross-section, usually a simple one.

If the subsurface must be treated as a 3-D model, such as an isolated reef, quantitative interpretation (simultaneous inversion of all the data) becomes very cumbersome and expensive, except for very simple geologic models which are of marginal significance in petroleum work. In such cases, at the present time, the telluric methods discussed in section 4.4 may prove to be more cost effective than 3-D MT.

The MT methods, in their application to the exploration of sedimentary basins, usually make use of electromagnetic signals within the frequency range from about 100 Hz down to about 0.002 Hz. The signals in the frequency range from 5 Hz down to about 0.0017 Hz are called **micropulsations**, which are generated by ionized gas clouds coming from the Sun. The major cause of the signals within the frequency range from about 5 Hz to about 10 kHz is lightning occurring almost continuously in Central Africa, South-east Asia, and in the Amazon region. These are called atmospherics.

So far as exploration of sedimentary basins is concerned, the selected micropulsations and the atmospherics of distant origin can be considered to be plane waves. Also, the resistivity contrast between air and the Earth (sedimentary rocks in particular) is so large that waves at all angles of incidence are transmitted into the Earth very nearly vertically. These considerations lead to the assumption that the signal is a vertically incident plane wave. This is what makes MT so attractive: no source dimensions enter into the interpretation, in contrast to controlled-source methods in which the source proximity is a factor in interpretation. Another major advantage of MT is that signals at practically all frequencies of interest are available.

The amplitude of the magnetic field is of the order of 0.1 **gamma** (nT) while the amplitude of the electric field depends largely on subsurface geology and is of the order of 1 mV/km.

With reference to Figure 4.6, a vertically incident electromagnetic plane wave sets up a horizontal electric current in a uniform (homogeneous and isotropic) subsurface of resistivity ρ. This current is manifested by a horizontal electric field $\boldsymbol{E}$ and a horizontal magnetic field $\boldsymbol{H}$. These two fields are at right angles to each other; E_x component is associated with component H_y. A grounded dipole serves to measure E_x in the form of a time series. A magnetometer measures H_y time series simultaneously with E_x. The magnetometer may be a coil with a metal core. The record sections have to be sufficiently long that they can be assumed to be stationary time series, meaning that statistical properties are independent of time; thus they can be represented by Fourier sine–cosine series. One now has $E_x(t)$ and $H_y(t)$; these yield, through Fourier analysis, the amplitudes $|E_x|$ and $|H_y|$ at a number of frequencies. The resistivity of the subsurface is determined by

$$\rho = \frac{0.2}{f}\left|\frac{E_x}{H_y}\right|^2, \tag{4.1}$$

where E_x and H_y are amplitudes in mV/km and gammas (γ) respectively at frequency f, and ρ is in ohm × meters (Ω m). If the subsurface is uniform, ρ remains the same, independent of frequency. The attenuation of the downgoing wave is frequency-dependent; the higher the frequency the faster it attenuates; high frequencies are associated with shallow currents and therefore yield shallow geologic information. If the amplitude of E_x is A at the surface, it becomes 0.37 A at a depth δ, in meters, which is called **skin depth**:

$$\delta \cong 503\,(\rho/f)^{1/2}. \tag{4.2}$$

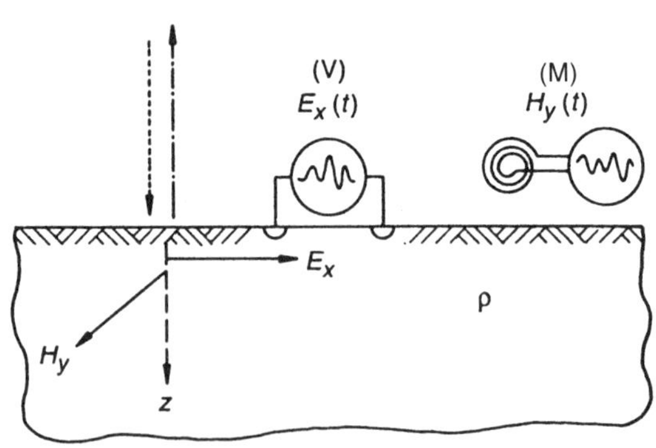

Figure 4.6 MT field setup. Dotted curve = normally incident plane wave; dot–dash curve = reflected wave; dashed curve = transmitted wave; V = voltmeter; M = magnetometer (from K. Vozoff, personal communication 1970).

The skin depth for a subsurface of $\rho = 10\ \Omega$ m and for $f = 10$ Hz is 503 m. In this case most of the electric current is confined to the upper 0.5 km.

Let us now suppose that the subsurface is horizontally layered (1-D) as shown in Figure 4.7. Let us measure ρ by means of equation (4.1) by employing a high frequency f_1 (short period T_1), such that for ρ_1 and f_1 the skin depth δ_1 is a small fraction of the thickness of the first layer. This means that the subsurface appears to this signal as a homogeneous half-space of resistivity ρ_1, except for minor oscillations which are part of the frequency-domain representation; they are not indicative of reflection of electromagnetic waves.

Next let us consider equation (4.1) for a frequency f_2, such that the wave penetrates deeper, and the effects of deeper layers are felt. Now equation (4.1) has no meaning, and the concept of skin depth loses its physical meaning. However, let us still use equation (4.1) to arrive at a fictitious parameter called **apparent resistivity**, ρ_a, with the stipulation that we will get rid of it eventually and obtain the true layer parameters (ρ_1, z_1, etc.).

Finally, let us consider ρ_a for a very low frequency f_3. In this case the wave will penetrate to great depths and sample a very large volume in

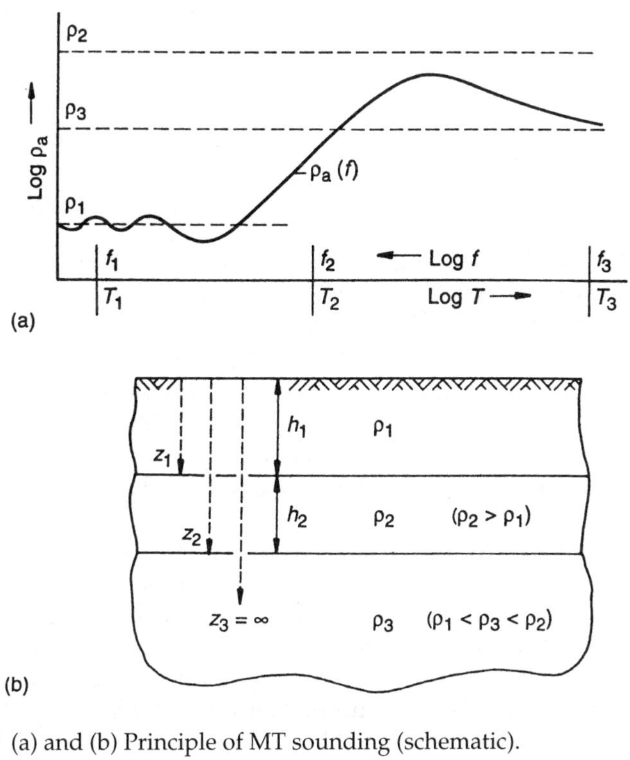

Figure 4.7 (a) and (b) Principle of MT sounding (schematic).

which the upper layers will be insignificant. For practical purposes the subsurface will appear as a homogeneous half-space of resistivity ρ_3, and equation (4.1) will yield a value which will be asymptotic to ρ_3.

Field data plotted in the form of Figure 4.7, apparent resistivity versus frequency, $\rho_a(f)$, can be interpreted automatically (by inversion), or by trial-and-error calculations (forward modeling), or by curve matching. This yields values of ρ and z in the form of a 'resistivity log', so to speak.

Phase differences between the electric and magnetic components at different frequencies, $\phi(f)$ are also measured. A horizontally layered model interpreted from a $\rho_a(f)$ curve must also satisfy the $\phi(f)$ curve.

The subsurface is never exactly horizontally layered. It may be approximately two-dimensional on a local basis (piecewise two-dimensional). Further, the source field direction varies all the time. These conditions necessitate tensor measurements. At each frequency there are four values each of ρ_a and ϕ. These tensor measurements must be analytically rotated to the principal axes of anisotropy (roughly dip and strike directions). To do this, the vertical component of the magnetic field is also needed. The procedure in the field is to measure E_x, E_y, H_x, H_y and H_z simultaneously. The rotation yields two $\rho_a(f)$ curves: $\rho_{\|}$, for which $\boldsymbol{E}$ is in the strike direction, $\rho_{\perp}$, for which $\boldsymbol{E}$ is in the dip direction, and two phase curves associated with them, $\phi_{\|}$ and $\phi_{\perp}$. When 1-D interpretation is justified, in most cases $\rho_{\|}$ yields better results than $\rho_{\perp}$.

The problem of equivalence (non-uniqueness) in MT interpretation is a serious drawback. This difficulty is common to all electrical methods in differing degrees. Consequently, automatic interpretation (inversion) often yields unrealistic results, unless initial estimates are not too far from the correct model, and constraints are imposed on some parameters.

A rapid method of obtaining an approximate 1-D interpretation (an estimate) is by matching the MT data to standardized theoretical curves. One can interpret a subsurface consisting of several layers by means of three-layer standard curves, by 'stripping' the upper two layers and combining them into one, and repeating this downward. Small catalogs of three-layer curves were published by Yungul (1961), Srivastava (1967) and others. Whittall (1986) discusses a method to combat the non-uniqueness by imposing physical constraints derived from external sources.

A method of inversion that is a necessity in 1-D MT interpretation, especially in petroleum work, is an extremely simple and somewhat empirical one. This is the continuous inversion method of Bostick (1977). In some sedimentary basins where resistivities do not display thick discrete layers but vary gradually, the discrete layer inversion may not yield satisfactory results while the continuous inversion may; it yields approximate results that can be correlated like well logs.

A field example of typical basin evaluation by MT is a traverse of widely spaced stations in South Texas (Vozoff, 1972). Figure 4.8 shows the cross-section obtained by 1-D trial-and-error interpretation of the MT data at six stations. The numbers on formations are the ranges of resistivities in ohm × m. This cross-section conforms fairly well to the one obtained by smoothing existing well logs (Vozoff, 1972, Figure 31, not shown here). Figure 4.9 shows the $\rho_{\perp}$ and $\rho_{\parallel}$ data at station G2, and Figure 4.10 is the 1-D interpretation of $\rho_{\perp}$, superposed on a smoothed induction log from a drill hole in the vicinity of G2. The solid curve is the induction log. The dashed curve is a portion of the four-layer model which was obtained by trial-and-error modeling of $\rho_{\perp}(f)$ data. The dot–dash curve is the result of automatic inversion of the same data; six-layer models were used and both layer thicknesses and their resistivities were permitted to vary freely.

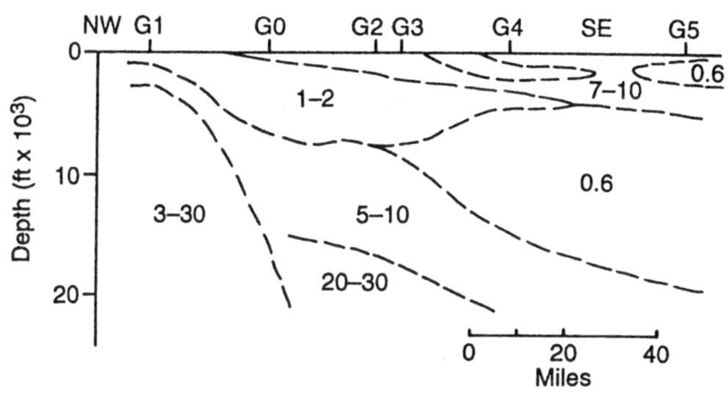

Figure 4.8 One-dimensional interpretation of MT survey, South Texas. Data are in Ω m. (from Vozoff, 1972).

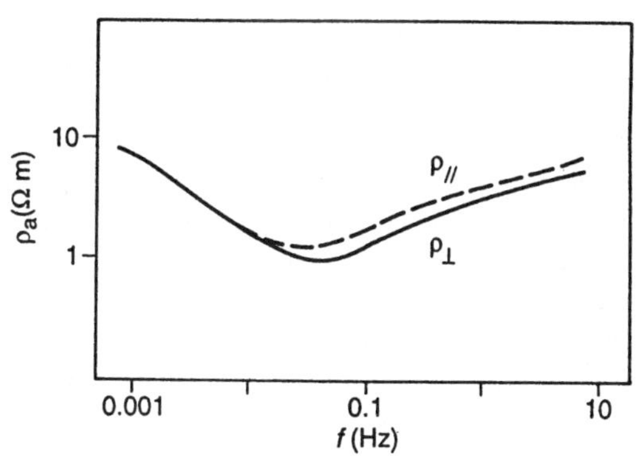

Figure 4.9 Rotated MT apparent resistivities at station G2, South Texas (from Vozoff, 1972).

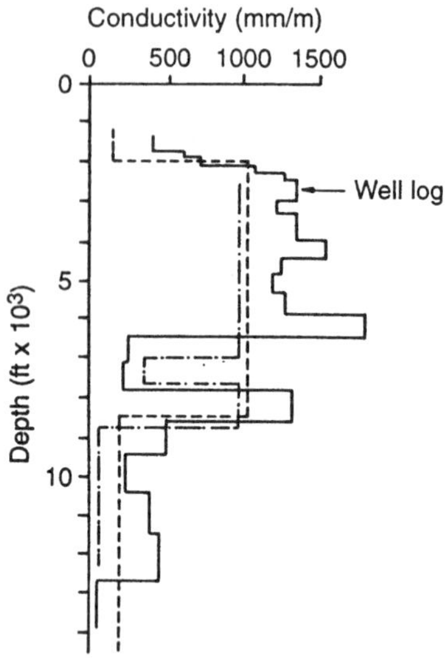

Figure 4.10 Comparison of well log with MT one-dimensional interpretation at station G2, South Texas. Solid curve = induction log; dashed curve = trial-and-error model; dot–dash curve = automatic inversion model (from Vozoff, 1972).

4.3.2 Theoretical principle of MT methods

We shall first assume that a harmonic (sinusoidal), one-dimensional (plane) electromagnetic (EM) wave is propagating in an unbounded, linear, isotropic and homogeneous medium in which there is no electric polarization, no free charges and no permanent magnetization. The constitutive relations in such a medium are

$$\boldsymbol{J} = \sigma \boldsymbol{E}, \tag{4.3}$$

$$\boldsymbol{H} = \boldsymbol{B}/\mu, \tag{4.4}$$

$$\boldsymbol{D} = \varepsilon \boldsymbol{E}, \tag{4.5}$$

where $\boldsymbol{J}$ is current density (A/m^2), $\boldsymbol{E}$ is electric field intensity (V/m), σ is conductivity ($1/\rho$, mho/m), $\boldsymbol{H}$ is magnetic field intensity (A/m), $\boldsymbol{B}$ is magnetic induction (tesla, T), μ is magnetic inductive capacity (henry/m), $\boldsymbol{D}$ is electric displacement (coulomb/m^2), and ε is electric inductive capacity (farad/m). Equation 4.3 is the differential form of Ohm's law. The term 'linear' implies that the proportionality factors between $\boldsymbol{J}$ and $\boldsymbol{E}$, etc., are independent of the magnitudes of $\boldsymbol{J}$ and $\boldsymbol{E}$, etc., although they may depend on the direction, in which case the medium is said to be linear but anisotropic. Earth materials are fairly linear. Electromagnetic theory in nonlinear media is extremely complicated, and very complicated in linear but anisotropic media. In geophysical work one usually assumes that media are linear and isotropic, and deals with inhomegeneity alone.

Under these circumstances, Maxwell's equations yield the following field equations for a plane wave traveling in the z-direction:

$$\frac{\partial^2 \boldsymbol{E}}{\partial z^2} = \gamma^2 \mathbf{E}, \tag{4.6}$$

$$\frac{\partial^2 \boldsymbol{H}}{\partial z^2} = \gamma^2 \boldsymbol{H}, \tag{4.7}$$

where $\boldsymbol{E}$ and $\boldsymbol{H}$ are dependent on only one space coordinate, z, and on time t, and where

$$\gamma^2 = i\omega\mu\sigma - \omega^2 \varepsilon\mu; \tag{4.8}$$

γ is called the **propagation constant**, or **wave number**; ω is angular frequency in radians per second. Components of a field can be treated separately and combined at the end. Therefore, we shall deal with one component, say E_x. It follows from Maxwell's equation that E_x is associated with H_y only. The general solutions to equations (4.6) and (4.7) yield these field equations:

$$E_x(z, t) = \mathrm{A}e^{i\omega t + \gamma z} + \mathrm{B}e^{i\omega t - \gamma z}, \tag{4.9}$$

$$H_y(z, t) = -\frac{\gamma}{i\omega\mu}[\mathrm{A}e^{i\omega t + \gamma z} - \mathrm{B}e^{i\omega t - \gamma z}], \tag{4.10}$$

where A and B are arbitrary constants. It should be remembered that these equations are for an unbounded homogeneous medium. If the subsurface is divided into a number of homogeneous subregions, each having different properties, different equations must be used for each subregion, and these equations are related to each other at the boundaries via the boundary conditions. A simultaneous solution of the field equations and the boundary conditions constitutes the unique solution. We shall see how this is done in connection with a horizontally layered subsurface.

For the exploration of sedimentary basins, the generalized field equations (4.9) and (4.10) can be simplified. In such work σ is of the order of 0.1 mho/m, ω is of the order of 0.1 radian/s, and ε is of the order of 10^{-10} farad/m. Consequently, σ is several orders of magnitude larger than $\omega\varepsilon$; the term with ε may be neglected, and γ in equation (4.8) becomes

$$\mp \gamma = (i\,\omega\mu\sigma)^{1/2}, \tag{4.11}$$

as was done by other authors, e.g. Keller and Frischknecht (1966).

By dropping the term with ε we are neglecting the displacement currents without which the propagation can no longer be called a real wave, but rather diffusion, like heat flow. The field equations without the ε term will represent a **quasi-static** electromagnetic field.

For earth materials μ is approximately the same as that for free space, $\mu_0 = 4\pi \times 10^{-7}$ henry/m. Thus, putting $T = 2\pi/\omega$ and designating the real part of γ by α,

$$\alpha = Re\ \gamma = \frac{2\pi}{10^4}\left(\frac{10}{\rho T}\right)^{1/2}. \tag{4.12}$$

Let us assume that the x–y plane represents the surface of a flat Earth which is a homogeneous half-space, and that the z-axis is positive downward. Consider a hormonic plane wave normally incident on the surface. Part of it is reflected upward and part is transmitted downward along z. The downgoing part is described by equations (4.9) and (4.10) in which ρ pertains to the subsurface. With reference to equations (4.9) and (4.10), E_x cannot be infinite as $z \to \infty$. Consequently, $A = 0$. Taking the real parts, and using

$$e^{i\omega t} \equiv \cos \omega t + i \sin \omega t,$$

equations (4.9) and (4.10) become

$$E_x(z, t) = Be^{-\alpha z} \cos(\omega t - \alpha z), \tag{4.13}$$

$$H_y(z, t) = \left(\frac{1}{\omega\rho\mu_0}\right)^{1/2} Be^{-\alpha z} \cos\left(\omega t - \alpha z - \frac{\pi}{4}\right). \tag{4.14}$$

Note that there is a 45° phase lag in H_y; this comes from $i^{1/2} \equiv e^{i\pi/4}$. At the surface ($z = 0$) the ratio of the amplitudes is

$$|Z_0| = \left|\frac{E_x}{H_y}\right|_{z=0} = (\omega\rho\mu_0)^{1/2} = \frac{2\pi}{10^3}\left(\frac{\rho}{5T}\right)^{1/2}, \tag{4.15}$$

where Z is called **impedance**; its unit is ohms. Obviously the resistivity ρ of the subsurface can be measured via equation (4.15),

$$\rho = \frac{5 \times 10^6}{4\pi^2}\, T \left|\frac{E_x}{H_y}\right|^2_{z=0}, \tag{4.16}$$

where the units are in MKS. In geophysical work it is customary to employ the practical units of gamma (γ), or nanotesla (nT), for $\boldsymbol{H}$, millivolt/km for $\boldsymbol{E}$ and ohm × m (Ω m) for ρ. These units for $\boldsymbol{H}$ are actually for magnetic induction, $\boldsymbol{B}$, but this confusion does not cause any problem in practice (section 4.2.1). With these practical units,

$$\rho = 0.2T\left|\frac{E_x}{H_y}\right|^2_{z=0}, \tag{4.17}$$

where ρ is an invariant, regardless of period T.

We shall now investigate what happens to different frequency components of a downgoing wave, in terms of attenuation and phase lag with depth. With reference to equation (4.13) one observes the following relations.

- Amplitude of E_x at $z=0$ is B.
- Amplitude of E_x at $z=1/\alpha$ is $B/e \cong 0.37$ B.
- Amplitude of E_x at $z=n/\alpha$ is B/e^n.
- Phase of E_x at $z=1/\alpha$ lags 1 rad with respect to E_x at $z=0$.
- Phase of E_x at $z=n/\alpha$ lags n rad with respect to E_x at $z=0$.
- The depth $z=1/\alpha$ is called a **radian wavelength**, or a **skin depth**, designated by δ(m),

$$\delta=\frac{1}{\alpha}=\left(\frac{2\rho}{\omega\mu_0}\right)^{1/2}\cong 503(\rho T)^{1/2}. \tag{4.18}$$

- At a depth $z=2\pi/\alpha$ the phase lags by 2π rad. Consequently, the full wavelength (m) is

$$\lambda=2\pi\delta=2\pi\left(\frac{2\rho}{\omega\mu_0}\right)^{1/2}=10^4\left(\frac{\rho T}{10}\right)^{1/2}. \tag{4.19}$$

- When z is increased by δ, the amplitude is attenuated by a factor of $1/e$ and the phase lags by one rad. This is why α is called the **attenuation constant**.

An interesting observation concerns the velocity of a downgoing wave in a sedimentary basin of say $\rho=10\ \Omega$ m and for $T=10$ s.

$$\text{Velocity}=f\lambda=10^4(\rho/10T)^{1/2}\cong 3163\ \text{m/s},$$

which is of the order of seismic wave velocities.

It is clear that there is a relation between the depth of penetration (attenuation) on one hand and T and ρ on the other. One can be selective about exploring a certain section of the subsurface (from the surface down) by choosing the proper Fourier components. This suggests the possibility of downward exploration, 'sounding', at one location. The strategy in accomplishing this has already been discussed in section 4.3.1 by means of Figure 4.7.

The theory of primary importance is the MT response of a 1-D subsurface containing an arbitrary number of layers that would be used in a computer program suitable for forward modeling or inversion in an economical way.

Let us first briefly describe the outline of Cagniard's (1953) procedure. The subsurface consists of $n-1$ horizontal homogeneous and isotropic layers underlain by a half-space of resistivity ρ_n. Figure 4.7 is such a model where $n=3$. Let us rewrite the general field equations (4.9) and (4.10),

$$E_x=\mathrm{A}e^{i\omega t+\gamma z}+\mathrm{B}e^{i\omega t-\gamma z}, \tag{4.9}$$

$$H_y=-\frac{\gamma}{i\omega\mu}[\mathrm{A}e^{i\omega t+\gamma z}-\mathrm{B}e^{i\omega t-\gamma z}], \tag{4.10}$$

where $\gamma^2=(i\omega\mu\sigma-\omega^2\varepsilon\mu)$, and A and B are arbitrary constants. These equations are for an unbounded medium. Each layer has a different γ

and therefore different constants A and B. With n layers there are $2n$ constants A and B. E_x cannot approach infinity as $z \to \infty$. Therefore $A_n = 0$ in the half-space. Any one constant can be arbitrary. Thus, there are $2(n-1)$ constants and that many boundary conditions to determine these constants, since there are $(n-1)$ interfaces at which E_x and H_y have to be continuous. A simultaneous solution of the boundary conditions yields the constants A and B in each medium. At the end, all we need is A_1 and B_1, so that we can express Z_0,

$$\dot{Z}_0(\omega) = \dot{E}_x(\omega)/\dot{H}_y(\omega), \tag{4.20}$$

where dots represent complex amplitudes.

$$\dot{Z}_0 = |Z_0| e^{i\phi} = \frac{|E_x| e^{i\phi_x}}{|H_y| e^{i\phi_y}} = \frac{|E_x|}{|H_y|} e^{i(\phi_x - \phi_y)}. \tag{4.21}$$

Thus, the theoretical $\rho_a(T)$ curve is obtained [equations (4.15), (4.16), (4.17)]. The phase of $\dot{Z}_0$ is the phase difference between $\dot{E}_x$ and $\dot{H}_y$.

4.3.3 One-dimensional (1-D) interpretation

We shall now assume that the subsurface consists of horizontal, homogeneous and isotropic layers. Given the experimental data in the form of $\rho_a(T)$ and $\phi(T)$, we shall try to determine the parameters of the subsurface h_i and ρ_i, as shown in Figure 4.7. This can be done only if the experimental data indicate that 1-D assumption is justified (section 4.3.4). Theoretically each solution is unique, but geologic and natural electromagnetic field irregularities and experimental errors render the problem 'ill-posed'. That is, each experimental curve can satisfy various models. A monograph by Whittall and Oldenburg (1992) is devoted to 1-D MT inversion. It includes an in-depth discussion of non-uniqueness.

(a) Interpretation by forward modeling

One starts with an initial model, an estimate of the probable solution which is usually in the form of a small number of discrete (parameterized) layers, say five. Such an estimate model may be obtained by the method of curve matching (section 4.3.3(d)). The theoretical ρ_a and ϕ curves for the model are computed and compared with the experimental curves. The model is personally modified and the theoretical curves are recomputed until they fit the experimental curves. A variety of techniques for the calculation of theoretical ρ_a and ϕ curves have been published (e.g. Madden and Swift, 1969).

(b) Automatic interpretation

This is the so-called 'inversion'. In one form of such interpretation, as before, one starts with an initial estimate model of discrete layers, but now the trial-and-error computations (the iterations) are accomplished through an automated algorithm (Rankin, Mozeson and Nebatani, 1974; Jupp and Vozoff, 1975).

If the algorithm is such that the iterations with large numbers of layers are inexpensive, such as with transmission line analogies, the user may fix the thickness of a large number of thin layers and determine the resistivities. This yields a stepwise continuous resistivity structure. In a development in the opposite direction of this approach, Smith and Booker (1988) use a program which yields a minimum number of layers for a satisfactory fit.

In an inversion method developed by Oldenburg (1979) the assumption of discrete layers is dropped and the 1-D model is allowed to be an arbitrary continuous function of depth. This is achieved by applying smoothing functions to the layer parameter distribution at each iterative stage.

An alternative to Oldenburg's (1979) method was developed by Hobbs (1982). This also yields a continuous and smooth resistivity structure, but it does not require an initial estimate; it starts with a uniform half-space.

(c) Continuous inversion

There are several somewhat empirical techniques whereby experimental curves $\rho_a(T)$ and $\phi(T)$ can be inverted point by point, that is, each point on ρ_a(T) yields a depth–resistivity point (Jones, 1983). Obviously the results cannot be accurate, regardless of data quality and subsurface parameters, because real solution necessitates that all of the data points be satisfied simultaneously. One such method developed by Bostick (1977) is very simple and extensively used in Europe and North America. We shall call it the 'Bostick inversion'. It is commonly used to obtain an initial estimate before the start of iterative interpretation. It may also be considered as a style of data display.

The Bostick inversion is based on the asymptotic behavior of the $\rho_a(T)$ curves at low frequencies. Figure 4.11 shows two theoretical $\rho_a(T)$ curves. The one with an ascending 45° asymptote is for a horizontal overburden in which resistivity varies with depth, $\rho(z)$; it is underlain by a basement, a half-space, of infinite resistivity. The one with a descending 45° asymptote is for the same overburden but the basement resistivity is zero. The depth to the basement is D; the total conductance of the overburden is defined by

$$S = \int_0^D \mathrm{d}z/\rho(z). \qquad (4.22)$$

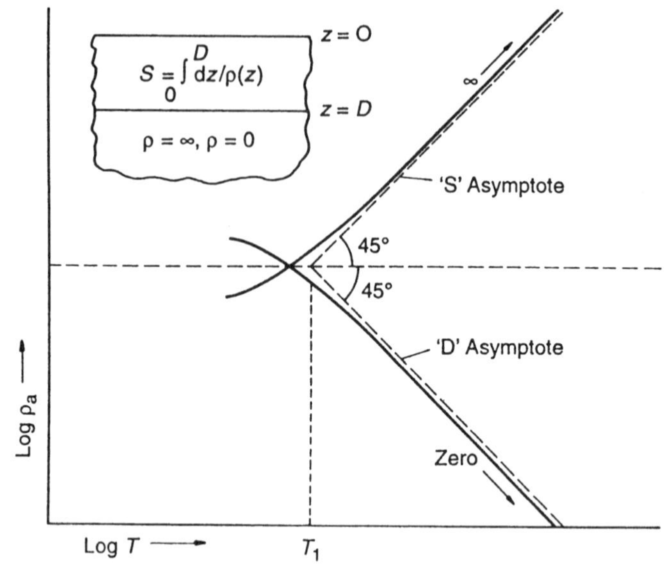

Figure 4.11 Principle of the Bostick inversion. Solid line curves are long-period parts of two-layer MT theoretical curves, one with a basement of infinite resistivity ($\rho_B = \infty$), and the other with zero resistivity ($\rho_B = 0$). Dashed lines are asymptotes to the ρ_a curves.

The asymptotes have these well-known properties:

The equation of the ascending asymptote is

$$\rho_a = T/2\pi\mu_0 S^2; \tag{4.23}$$

it depends only on S, regardless of the form of $\rho(z)$; and that of the descending one is

$$\rho_a = 2\pi\mu_0 D^2/T; \tag{4.24}$$

it depends only on D. If we imagine an experimental apparent resistivity value for a period shorter than T_1 at which the asymptotes intersect, this value will not know which way it is going to go, up to ∞ or down to zero; it therefore depends only on resistivities shallower than D. The method is based on expressing each apparent resistivity value in terms of asymptotic intersection at that resistivity value.

A simultaneous treatment of equations (4.22), (4.23), (4.24) yields

$$D = (\rho_a T/2\pi\mu_0)^{1/2}, \tag{4.25}$$

and

$$\rho = \rho_a\left(1 + \frac{\mathrm{d}\log\rho_a}{\mathrm{d}\log T}\right)\bigg/\left(1 - \frac{\mathrm{d}\log\rho_a}{\mathrm{d}\log T}\right), \tag{4.26}$$

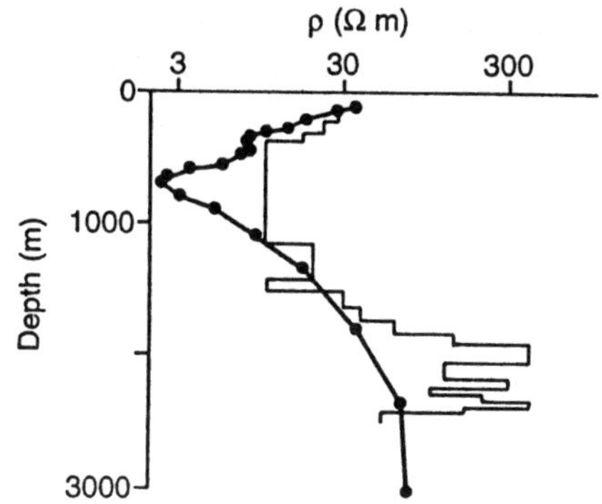

Figure 4.12 A comparison of the Bostick inversion of MT data and resistivities from borehole data, Kiskunhalas, Hungary. Curve with dots = inversion of E-polarization ($\rho_{\parallel}$) (from Ádám, Nagy and Varga, 1989).

where ρ_a is a function of T, and the derivative terms are the slope of the $\rho_a(T)$ curve plotted on a log–log scale. Thus, given a point on the curve, at T, the values of ρ_a and the slope of the curve are known, and a corresponding resistivity ρ at depth D is computed. The slope values obtained from the $\rho_a(T)$ curve frequently cause scatter in the inversion; it may be better to obtain the value of the slope from the phase data, $\phi\,(T)$, if they are available. Using the $\phi\,(T)$ data, and a Hilbert transform operation, equation (4.26) is transformed to

$$\rho \cong \rho_a\left(\frac{\pi}{2\phi} - 1\right), \tag{4.27}$$

where ϕ is in radians. A field example of the Bostick inversion, given by Ádám, Nagy and Varga (1989), is shown in Figure 4.12. The borehole data were obtained from resistivity logs. Another example is given by Goldberg and Rotstein (1982).

(d) Interpretation by manual curve matching

A catalog of standardized three-layer theoretical curves can be used to interpret a multilayer case by 'stripping' from the surface down. The procedure with, and the benefits of, curve matching in MT are similar to those for resistivity sounding (section 3.2). Unfortunately, an adequate three-layer catalog is not readily available. Those published by Yungul (1961) and Srivastava (1967) are not adequate for general usage. A family of two-layer curves aided by auxiliary curves will not do at all. A catalog of four-layer curves will not do either, because the volume would be intolerably large.

A convenient way of standardizing theoretical curves, to render them dimensionless, so that they can be superposed on experimental curves, is to make use of the law of similitude in electrodynamics. Neglecting the displacement currents as before, the wavelength, in meters, is

$$\lambda_i = 10^{3.5}\,(\rho_i T_i)^{1/2}. \tag{4.19}$$

Let h_i represent any one geometric dimension of the real model in the Earth. Then, the quantity

$$N_i = \rho_i T_i / h_i^2 \tag{4.28}$$

is dimensionless, a pure number. Therefore, every system which has the same values of N must produce the same response, regardless of the actual values of, ρ, T and h, because the solution of Maxwell's equations is the same for all models having the same values of N.

Let us consider an Earth model with geometric dimensions h_i, resistivities ρ_i, and signal period T_i. The electrodynamic scale model of it has the dimensions

$$\overline{h}_i = h_0 h_i, \quad \overline{\rho}_i = \rho_0 \rho_i, \quad \overline{T}_i = T_0 T_i, \tag{4.29}$$

where h_0, ρ_0, and T_0 are scaling factors, constants which are interdependent. By equating an N in the Earth model to that in the scale model we find

$$h_0^2 = \rho_0 T_0. \tag{4.30}$$

Two of the scale factors can be selected arbitrarily, but the third one must obey equation (4.30). The scale factor of ρ_a is the same as that of ρ; the phase ϕ is the same in both models. If we know the solution for the scale model, we can obtain the solution for the Earth model via equations (4.29). It follows that if we plot ρ_a (T) for the Earth model and $\overline{\rho}_a$ $(\overline{T})$ for its scale model on the same kind of logarithmic paper, the two curves can be superimposed, because multiplications turn into translations in logarithmic plotting. Standardization of scale-model curves is done by fixing the resistivity and thickness of the uppermost layer, $\overline{\rho}_1$ and $\overline{h}_1$, at 1 Ω m and 1 km respectively. Catalogs of three-layer curves are prepared in such a way that each sheet contains a family of curves for a certain combination of $\overline{\rho}_2$ and $\overline{\rho}_3$; each curve pertains to one value of $\overline{z}_2$, the depth to the third layer which is a half-space (Figure 4.7).

Two-layer curves are special forms of three-layer curves in which $z_2 \to 1$ or ∞, but these are collected in one sheet to facilitate the interpretation of two-layer cases or asymptotic branches.

With reference to Figure 4.13, let us suppose that the solid line curve represents a three-layer Earth model and that we wish to interpret it by means of a three-layer catalog. The specifications of the matched stan-

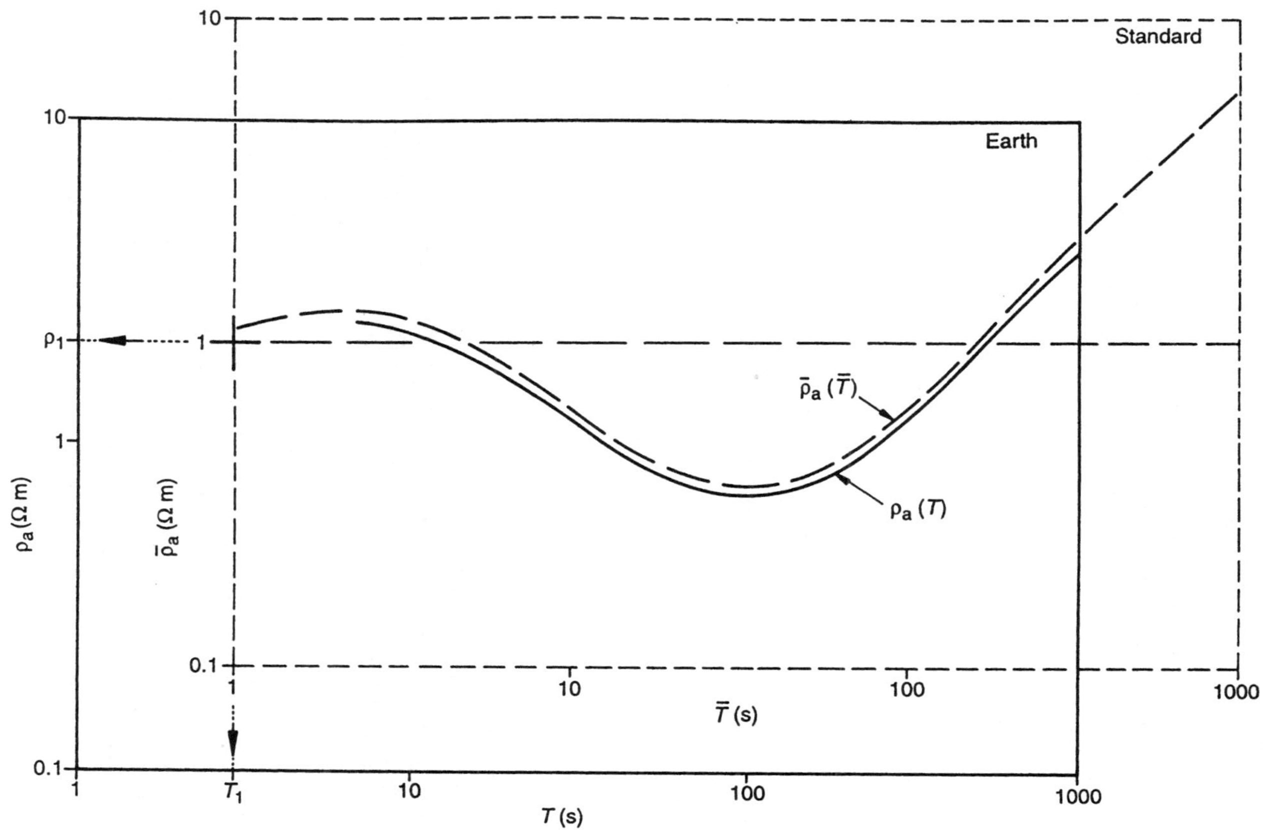

Figure 4.13 The matching of a three-layer standard curve to a three-layer Earth model curve.

dard curve yield the ratios

$$\rho_2/\rho_1, \quad \rho_3/\rho_1, \quad z_2/z_1,$$

where z_1 and z_2 are the respective depths to the second and third layers. The horizontal line at $\bar{\rho}_a = 1$ reads ρ_1 on the ρ_a axis. The vertical line at $\bar{T} = 1$ reads T_1 on the T-axis. According to equation (4.30) we have

$$z_1 = (\rho_1 T_1)^{1/2}. \tag{4.31}$$

Thus, all of the Earth model parameters are determined by means of the above ratios.

Let us now interpret a curve for a four-layer Earth model. Having determined the parameters of the two uppermost layers as described above, we can combine these two layers into one replacement layer of thickness z_2 having a replacement resistivity of ρ^* as described below:

Telluric currents flow nearly horizontally, especially where the use of a 1-D model is justified. To determine z_3, the depth to the fourth layer, one must use longer periods than in determining z_2 and, for these longer periods, the skin effect in the upper two layers can be neglected. One is then justified in assuming that the upper two layers behave as resistors in parallel, in accordance with Kirchhoff's law,

$$\frac{z_2}{\rho^*} = \frac{z_1}{\rho_1} + \frac{z_2 - z_1}{\rho_2}. \tag{4.32}$$

Equation (4.32), known as Hummel's relation, has been used since the 1930s in connection with resistivity sounding.

Figure 4.14 is given to show the degree of validity of combining two layers into one. Theoretical MT sounding (MTS) curves were computed once for model 1 (five layers) and once for model 2 (two layers) in which the top four layers were combined into one $\rho^* = 10.4$. The two models are quite similar for long periods even for this model which has large resistivity contrasts.

Figure 4.15 shows a four-layer MTS curve (solid line). This is a synthetic curve computed from the model shown at the upper left-hand corner. A standard curve marked (1:10:1) for $z_2 = 6$, shown as a dotted line, is fitted to the part of the MTS curve that corresponds to the upper three layers. Note that the two curves diverge considerably around $T = 10$ s. This is because the right-hand ascending (asymptotic) branch of any MTS curve is preceded by an undershoot. Likewise, a right-hand descending branch would be preceded by an overshoot.

The cross point of the dotted curve ($\rho_a = 1$, $T = 1$), indicated as X_1, reads $\rho_1 = 8$ and $T_1 = 0.031$ on the coordinates of the MTS curve. Therefore, $\rho_1 = 8\ \Omega$ m; $\rho_2 = 10\ \rho_1 = 80$; $\rho_3 = \rho_1 = 8$. The above value for ρ_3 is a tentative

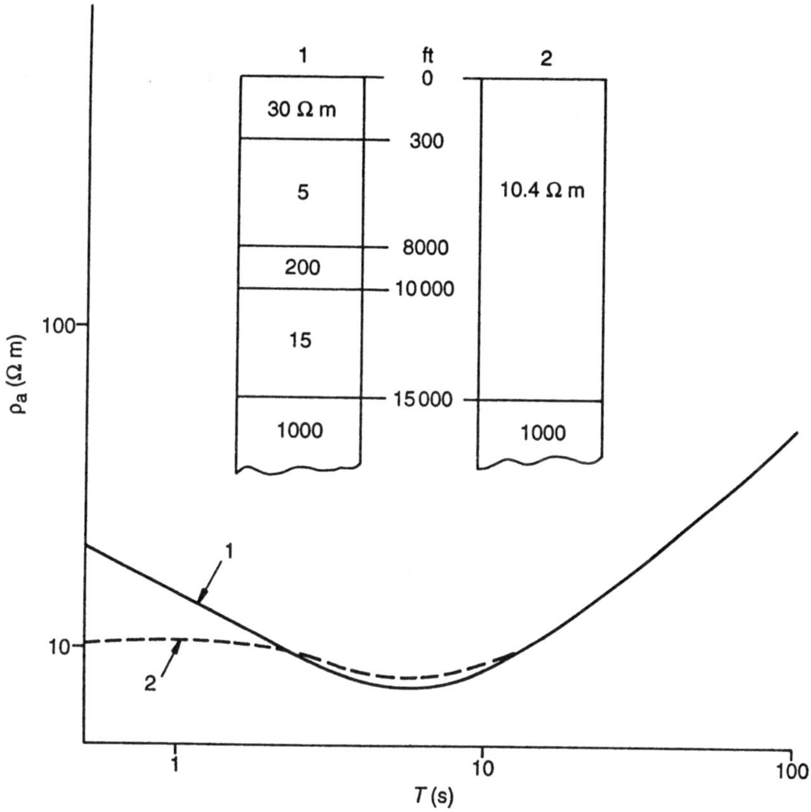

Figure 4.14 Equivalence of two-layer and five-layer models for long periods.

one and not needed at this stage. Equation (4.31) leads to

$$z_1 = (\rho_1 T_1)^{1/2} = (8 \times 0.031)^{1/2} \cong 0.5 \text{ km},$$

$$z_2 = 6\, z_1 = 3 \text{ km}.$$

Combining the upper two layers into one replacement layer, we find

$$\rho^* = 32\ \Omega \text{ m}.$$

The cross point ($\rho_a = 1$, $T = 1$) of the second set of standard curves that we will fit to the remaining portion of the MTS curve, indicated as X_2, will be placed at the following coordinates of the MTS curve:

$$\rho_a = \rho^* = 32\ \Omega \text{ m},$$

$$T = T^* = \frac{z_2^2}{\rho^*} = \frac{9}{32} = 0.281 \text{ s},$$

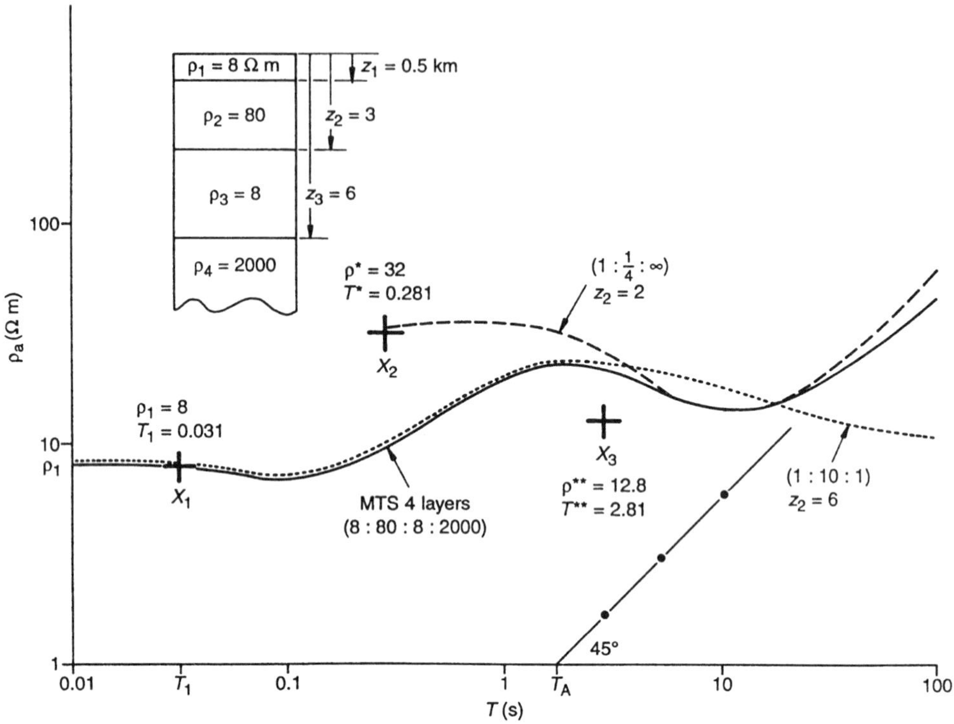

Figure 4.15 Interpretation of a four-layer Earth model (MTS) curve by means of three-layer standard curves.

the unit of z being 1 km. A fit is obtained by $(1:1/4:\infty)$, the dashed-line curve, for $z_2 = 2$. This yields

$$\rho_3 = \frac{1}{4}\rho^* = \frac{32}{4} = 8\ \Omega\ \text{m},$$

$$z_3 = 2z_2 = 6\ \text{km}.$$

The third-layer resistivity of the dashed-line curve is ∞; it diverges upward from the MTS curve. An estimate of ρ_4 is obtained by combining the replacement layer obtained above with the third layer. This gives a second replacement layer, whose thickness is 6 km and resistivity, ρ^{**}, is 12.8 Ω m. The cross point, X_3, of a two-layer standard curve that we shall fit to the right-hand rising branch of the MTS curve will be placed at the following coordinates of the MTS curve:

$$\rho_a = \rho^{**} = 12.8\ \Omega\ \text{m},$$

$$T = T^{**} = z_3^2/\rho^{**} = 36/12.8 = 2.81\ \text{s}.$$

The two-layer standard curve with $\rho_2 \cong 160\ \Omega\ \mathrm{m}$ fits the right-hand rising branch of the curve. Therefore,

$$\rho_4 \cong 160 \times 12.8 \cong 2000\ \Omega\ \mathrm{m}.$$

Total conductance

Let us now check the overall results obtained above to make sure that a gross error has not been made somewhere. With reference to Figure 4.15, the asymptote to the right-hand ascending branch of the dashed curve, for the hypothetical case of $\rho_4 = \infty$, is shown as the line with slope of 45°. The line equation of the asymptote, as before, is

$$\rho_a = T/2\pi\mu_0\, S^2, \tag{4.23}$$

where S is the total conductance of the section above the $\rho_4 = 2000\ \Omega\ \mathrm{m}$ half-space. Our interpretation results yield

$$S = \sum h_i/\rho_i = 468.75\ \text{mhos}.$$

A simplified form of equation (4.23) is

$$S = 356\ T_A^{1/2},\ \text{in mhos},$$

where T_A is the intercept of the asymptote with the $\rho_a = 1$ axis; it is about 1.7 s in this case. The two values of S, one from the layered interpretation, 468.75 mhos, and the other from T_A, 464 mhos, are very close.

If the last medium (the half-space approximation) was of very low resistivity, equation (4.24) could be used to check the results in terms of total depth.

(e) Layer resolution

Theoretically, each solution to the 1-D inversion problem may be unique (Whittall and Oldenburg, 1992), but geologic irregularities (deviations from 1-D), a non-planar source field, and experimental errors render the problem 'ill-posed'. That is, each experimental (field) curve can satisfy various models. Anticipated model response can be readily determined, but the geologic irregularities and the experimental noises cannot effectively be taken into account. Thus, the resolution is a moot question. However, there are some theoretical shortcomings, as explained below.

Making the same assumptions as in section 4.3.2, i.e. neglecting the displacement currents, Maxwell's equations yield the amplitudes of a pair of components H_y and E_x at the surface, at one frequency ω,

$$H_y(0) = \int_{\infty}^{0} \sigma(z) E_x(z)\,\mathrm{d}z, \tag{4.33}$$

$$E_x(0) = -i\omega\mu_0 \int_0^{\infty} H_y(z)\,\mathrm{d}z. \tag{4.34}$$

Equation (4.33) is none other than a special form of Ampere's circuit law. It says that $H_y(0)$ measures the total current in amperes flowing in the x-direction through a vertical unit rectangle, in the y–z plane, extending from the surface down to infinity, regardless of the form of $\sigma(z)$. In a way, $H_y(0)$ measures the source strength to which $E_x(0)$ is normalized, analogous to direct current in resistivity work. Thus, this is a cumulative effect.

Equation (4.34) also indicates that $E_x(0)$ depends on cumulative $H_y(z)$. Therefore, the impedance $z_0 = E_x(0)/H_y(0)$, and hence $\rho_a(T)$, is not sensitive to the details of the $\sigma(z)$ distribution. This is not really an electromagnetic reflection method. Futhermore, since current density very rapidly decreases with depth (the skin effect), because a sedimentary basin is a medium subject to excessive current loss to heat, due to its high conductivity, the cumulative effects of deeper layers also decrease rapidly and resolution deteriorates. Deep layers have to be thick to be resolved.

For thin layers, that is, if the layer thickness h is much less than the skin depth in that layer, the resolution is as follows (Madden, 1971).

1. If the layer has low resistivity, only its unit conductance, h/ρ, can be determined, just as in resistivity sounding.
2. If the layer has high resistivity, only its thickness, h, can be determined. On the other hand resistivity sounding yields only the unit resistance, $h\rho$. This fact created interest in combining MT and resistivity measurements to enable one to determine ρ and h separately for a high-resistivity layer.

(f) Depth of exploration

Depth of exploration is practically unlimited; however, the use of long periods renders surveys more expensive. It is necessary to know beforehand what period range is needed. There is no substitute for modeling for this purpose, for ρ_a and for ϕ. However, there is a crude rule of thumb based on visual inspection of standard curves: the depth of exploration is about $\delta/2$, one-half the skin depth, for the ρ_a curve. For instance, using the $\delta/2$ rule, considering a sedimentary section having a thickness of 5000 m and an effective resistivity of 20 Ω m, the basement will be felt at a period of about 20 s on the ρ_a curve but it will show up at shorter periods on the ϕ curve.

4.3.4 Data analysis

Many of the basic developments in this subject took place in the 1960s. Vozoff (1972) cites many references to these basic developments, and his papers (1972, 1991) largely influenced the following discussion.

The primary objective is to determine the smoothed amplitude and phase spectra of the real-time series, $E_x(t)$ etc., so that $\rho_a(T)$ and $\phi(T)$

curves can be constructed. With the advent of the fast Fourier analysis techniques it became possible to obtain the smooth spectra directly by Fourier analysis, rather than using autocorrelation and cross-correlation techniques described in the older literature.

Let us suppose that we are interested in the micropulsations within the frequency range of 10^{-3} to 10 Hz. Superimposed on these there will be very low-frequency noise consisting of diurnal variations, electrode and instrument drifts, and high-frequency noise due to instruments, atmospherics and culture. Most of the high-frequency noise can be removed through various stages of electronic and/or digital filtering. The low-frequency noise can be removed by means of detrending, that is, by subtracting a least-squares fit polynomial curve. Frequently, a cubic function will do. However, sometimes detrending can cause more harm than good. Consequently, some people do not use detrending and are satisfied with an effective windowing after having removed a linear trend (ramp) and a DC shift (pedestal). The windowing must be done anyway to prevent truncation errors, whether detrending is done or not. For this, Hanning, Hamming, half-cosine, trapezoid and triangular windows may be used. Several overlapping windows are analyzed and averaged to obtain smooth amplitude and phase spectra.

One may obtain a total of about 2 h of record sampled at about five to ten samples per second to resolve the period range from 1 s to 1000 s. A sampling rate of exactly 10 samples per second would result in the 60 Hz noise causing aliasing by folding back to DC and is not recommended. This is the 'low-frequency' record. The window may be 30 min wide and shifted by 15-min steps. Another record is needed to resolve the period range from 0.1 s to 1 s, preferably overlapping to 10 s. A record of 5-min duration sampled at about 100 samples per second may suffice. This is the 'high-frequency' record.

Recording at two or more frequency bands, as described above, may be a necessity governed by the dynamic range of the instruments on hand; it is also done to reduce the number of data points needed to resolve a frequency range of four decades.

Having sampled at considerably higher speeds than twice the Nyquist frequencies, a digital low-pass filtering is carried out and the data are decimated. Of course this procedure is not essential in every case. Under favorable signal-to-noise ratio circumstances one can get along by recording a minimum of two samples per period and not decimating.

Let us consider first the case of a 1-D subsurface. We can envisage the time series $E_x(t)$ and $H_y(t)$ for a single-frequency signal and express the impedance at that frequency.

$$\frac{\dot{E}_x(t)}{\dot{H}_y(t)} = \dot{Z}_{xy}, \tag{4.35}$$

where dots signify complex quantities, because magnitudes of E_x and H_y fields are interrelated at different times due to the phase difference between them. Equation (4.35) does not have as much physical meaning as mathematical meaning. $\dot{E}_x(t)$ and $\dot{H}_y(t)$ can be replaced by the complex Fourier transforms of real-time series.

$$\frac{|E_x(\omega)|\,\mathrm{e}^{i\phi_x(\omega)}}{|H_y(\omega)|\,\mathrm{e}^{i\phi_y(\omega)}} = |Z_{xy}(\omega)|\,\mathrm{e}^{i\,[\phi_x(\omega)-\phi_y(\omega)]}, \tag{4.36}$$

where

$$|E_x(\omega)|\,\mathrm{e}^{i\phi_x(\omega)} = \int_{-\infty}^{+\infty} E_x(t)\,\mathrm{e}^{-i\omega t}\,\mathrm{d}t. \tag{4.37}$$

We take the modulus to form the $\rho_a(T)$ curve and the argument to form the $\phi\,(T)$ curve. For instance, using E_x and H_y components, we have, in practical units,

$$\rho_{xy} = 0.2\,T\,|Z_{xy}|^2.$$

If the subsurface is exactly 1-D, the direction of the x-axis is immaterial and the ρ_{xy} curve constitutes all of the data necessary for interpretation. The assumption of a 1-D model is justified if the following conditions are met simultaneously:

1. $\rho_{xy}\,(T)$ is practically the same for different directions of the measuring axes, that is, $\rho_{xy} \cong \rho_{yx}$.
2. Likewise for $\phi\,(T)$.
3. H_z/H_x is negligible for all directions of x.
4. $\rho_a(T)$ and $\phi\,(T)$ curves are simultaneously satisfied by the same 1-D model. Note that 1, 2 and 3 are not sufficient for the data obtained at the epicenter of a circularly symmetrical structure.

Let us now consider a 3-D subsurface. As in equation (4.35), for one frequency, the complex field components ($\dot{E}_x(t)$, etc.) are linearly interrelated. However, for linear polarization of the source in the x-direction, the induced current is now curved and has both x and y components. A pair of total horizontal field vectors **E** and **H** are no longer orthogonal. An E_x field is associated with H_y and H_x components, and therefore with Z_{xy} and Z_{xx}; the former is called **principal impedance** and the latter **cross impedance**. We thus have, for a single-frequency plane wave (Cantwell, 1960),

$$\dot{E}_x(t) = \dot{Z}_{xy}\,\dot{H}_y(t) + \dot{Z}_{xx}\,\dot{H}_x(t),$$

$$\dot{E}_y(t) = \dot{Z}_{yx}\,\dot{H}_x(t) + \dot{Z}_{yy}\,\dot{H}_y(t), \tag{4.38}$$

where Z represents the components of a surface impedance tensor Z_{ij}; they depend upon the resistivity distribution and the direction of the measuring axes. As before, the field components can be replaced by the complex Fourier transforms of the real-time series, and equation (4.38) can be transformed into

$$E_x(\omega) = Z_{xy}(\omega)H_y(\omega)e^{i\phi_{xy}(\omega)} + Z_{xx}(\omega)H_x(\omega)e^{i\phi_{xx}(\omega)},$$

$$E_y(\omega) = Z_{yx}(\omega)H_x(\omega)e^{i\phi_{yx}(\omega)} + Z_{yy}(\omega)H_y(\omega)e^{i\phi_{yy}(\omega)}, \tag{4.39}$$

where all values of E, H and Z are real numbers. According to equation (4.39) we now have four $\rho_a(T)$ curves and four $\phi(T)$ curves. In addition we now have a vertical component $H_z(T)$, which can be defined in terms of ratios

$$H_z/H_x \text{versus } T \quad \text{and} \quad H_z/H_y \text{ versus } T,$$

and their phases. Thus, one now has 12 curves at one measurement site, that is, 12 quantities for each value of T. In order to interpret the subsurface, we must possess such data at a number of stations and find a model which would satisfy all of these data simultaneously. This, of course, is an extremely complicated problem.

We now fall back to a 2-D subsurface and see what we can do. The subsurface may be approximately two dimensional on a local basis (piecewise two-dimensional). This is to say that if the system of measuring axes is rotated, at a particular direction the cross impedances will be either zero or small enough to be neglected. The new axes aligned in this particular direction are called the **principal axes of anisotropy**. If the measurements are made along these axes, equation (4.39) reverts to

$$E_x(\omega) = Z_{xy}(\omega)H_y(\omega)e^{i\phi_{xy}(\omega)},$$

$$E_y(\omega) = Z_{yx}(\omega)H_x(\omega)e^{i\phi_{yx}(\omega)}. \tag{4.40}$$

For each value of ω, two equations with two unknowns (Z) would yield the values of Z. However, if the geologic strike is unknown and the measuring axes x and y are arbitrarily oriented all four elements of the impedance tensor Z_{ij} are nonzero, as in equation (4.39). If the coordinate axes are rotated physically or by computation until one of them is parallel to the geologic strike, the cross impedances would be zero and equation (4.40) would be satisfied in the new coordinate system, which would coincide with the principal axes of anisotropy. The principal impedances would be nonzero and unequal. In reality, the cross impedances are never zero, because neither a really two-dimensional subsurface nor plane waves exist. All that one could do is to minimize the cross impedances. It should be noted that the rotation may be different for each frequency.

Instead of rotating the axes of the real-time series by computation, it is more advantageous to compute the four Z_{ij} components first and then rotate the axes as follows. The four impedances are computed by solving equations (4.39), which relates the Fourier transforms in the original measuring coordinate system. It appears that there are two equations and four unknowns. The solution is achieved by taking advantage of the fact that the Z_{ij} components change very slowly with ω, and consequently can be calculated as averages over frequency bands with each band including many values of E and H. One such process is described by Madden and Nelson (1964).

Having computed the Z_{ij} values in the original x, y system, they can be rotated to any other x', y' system by an angle θ. Frequently y' coincides with the strike when Z'_{xy} becomes the maximum and Z'_{yx} the minimum, but sometimes x' coincides with the strike (Z_{xy} is obtained from E_x and H_y). Generally the x' and y' axes coincide with the principal axes of anisotropy, although which one is the strike is undetermined. One way of finding the x' and y' axes is to compute the principal impedances for many values of θ and interpolate to find the maximum and minimum for each value of ω. However, an analytical technique can be used to reduce computation. Swift (1967) developed a technique whereby the rotation angle $\theta°$ required to maximize the sum of the squares of the principal impedances is determined. The same value of $\theta°$ yields a minimum value for the sum of the squares of the cross impedances. All four impedances are computed for each value of ω but only the principal apparent resistivities, ρ'_{xy} and ρ'_{yx}, are plotted, because the cross apparent resistivities would be interpretable only in terms of a three-dimensional model. It will be assumed that the subsurface is piecewise two-dimensional under each station and that the cross apparent resistivities are negligible, although their magnitudes are used to find the degree to which the subsurface departs from a two-dimensional model, as explained below.

According to Sims and Bostick (1969), both $|Z_{xx}+Z_{yy}|$ and $|Z_{xy}-Z_{yx}|$ are independent of θ. The magnitude of the ratio of these quantities, also independent of θ, is called the **skewness**, S,

$$S=\frac{|Z_{xx}+Z_{yy}|}{|Z_{xy}-Z_{yx}|}. \tag{4.41}$$

If S is small, it is concluded that the subsurface is practically 1-D or 2-D at that site and at that frequency, otherwise it is 3-D. Obviously, if the subsurface is really 1-D then S is zero. A practical but not infallible rule of thumb is as follows:

- 1-D if $S \leqslant 0.05$;
- 3-D if $S \geqslant 0.2$;
- 2-D if $0.2 > S > 0.05$.

As stated above, the rotation of the coordinates renders one of the principal impedances maximum and the other minimum; the rotation angle $\theta°$ being different at each frequency. These two impedances yield

ρ_{max} and ρ_{min} curves. In most cases when the x'-axis is perpendicular to the strike, Z'_{xy}, which is derived from E_x, is maximum, and thus the y'-axis is the strike. However, this is not always so. The axes x' and y' are the principal axes, in the direction of dip and strike, but which one is the strike may be determined as follows. Under ideal source conditions, H_z is totally due to currents flowing parallel to the strike. Therefore, the horizontal direction in which H is most coherent with H_z is the dip axis, the direction perpendicular to the strike. The procedure for determining this direction was developed by T.R. Madden (1968, unpublished) and described by Vozoff (1972). It is assumed that H_z is linearly related to H_x and H_y, and at each frequency their Fourier transforms can be written as

$$H_z = T_x H_x + T_y H_y, \qquad (4.41a)$$

where T_x and T_y are unknown complex coefficients that can be calculated. If the structure is 2-D with the strike in the y' direction, equation (4.41a) becomes

$$H_z = T_{x'} H_{x'}, \qquad (4.41b)$$

where x' is in the direction perpendicular to the strike (dip axis). $T_{x'}$ is called the **tipper**, meaning that a part of the horizontal component of $\boldsymbol{H}$ is tipped into the vertical. The modulus of $T_{x'}$ is usually of the order of 0.1.

Having plotted the ρ_{min} and ρ_{max} curves in which each frequency has been rotated separately, some people prefer to re-rotate all frequencies of each curve to one average angle $\bar{\theta}$. It might be advisable to see that $\bar{\theta}_{max}$ and $\bar{\theta}_{min}$ are 90° apart. The tipper information allows one to determine whether ρ_{min} or ρ_{max} coincides with the strike. The one that coincides with the strike is designated as $\rho_{\parallel}$, meaning that it corresponds to the component of $\boldsymbol{E}$ parallel to the strike. This is also called ρ_a for E-polarization; also called TE mode, meaning that E is transversal to the paper on which the 2-D model cross-section is drawn. The other is designated as $\rho_{\perp}$, likewise, H-polarization, and TM mode. Sometimes the $\rho_{\parallel}$ and $\rho_{\perp}$ curves cross each other. In other words, part of ρ_{min} and part of ρ_{max} must be joined together to form the $\rho_{\parallel}$ curve; likewise for $\rho_{\perp}$.

Skewness rules for dimensionality, and strike determination rules that were discussed above, do not function satisfactorily in the presence of 3-D near-surface inhomogeneities which distort the impedance tensor, and prevent the separation of the two modes, $\rho_{\parallel}$ and $\rho_{\perp}$ (Park and Livelybrooks, 1989).

If the objective of a survey is primarily 1-D interpretation, one normally does not obtain a station density to resolve the 3-D and 2-D near-surface inhomogeneities, which may cause spectacular, somewhat frequency-independent **static shifts** between the $\rho_{\parallel}(T)$ and $\rho_{\perp}(T)$ curves.

Figure 4.16 shows the effects of a 2-D near-surface inhomogeneity on ρ_a. The model is shown in the upper left-hand corner, and also below

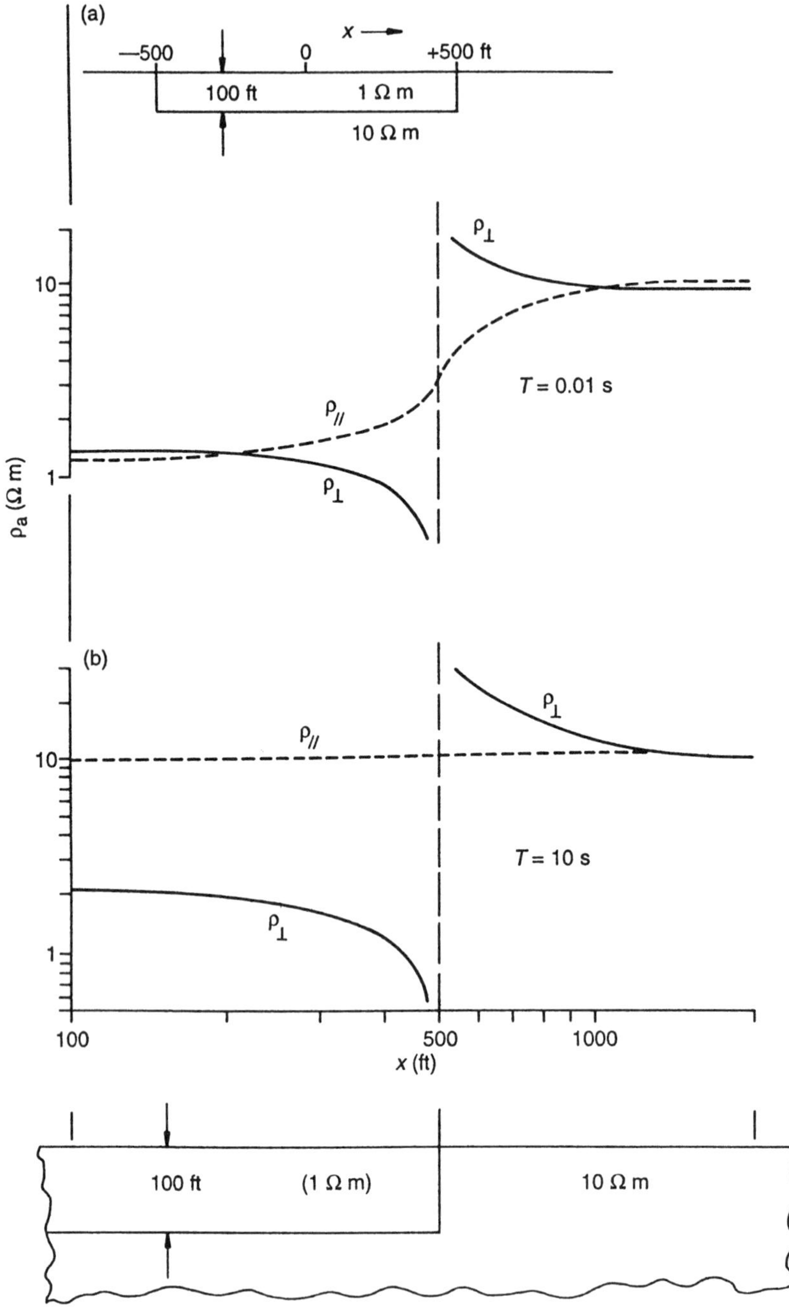

Figure 4.16 Effects of a near-surface inhomogeneity on magnetotelluric sounding measurements, with the electric field taken as a point function. The inhomogeneity is a two-dimensional horizontal slab of infinite extent in the direction perpendicular to the profile.

with logarithmic distance, x, from the center of the slab which extends to infinity in the $\mp y$ directions. There are large overshoots in $\rho_\perp$ and large differences (shifts) between $\rho_\parallel$ and $\rho_\perp$ at all frequencies. These will create large shifts between $\rho_\parallel$ (T) and $\rho_\perp(T)$ curves: about a factor of ten in this case. At low frequencies (curves for $T = 10$) $\rho_\parallel$ is not affected appreciably by the slab, off or on it; it is non-anomalous for all practical purposes. This is why in most cases $\rho_\parallel$ is used for 1-D interpretation; this is the TE mode; but it is ρ_{min} off the high-resistivity slab and ρ_{max} on the slab. This rule is useful in interpretation near fault scarps. However, in presence of 3-D features $\rho_\perp$ may yield more reliable 1-D interpretation than $\rho_\parallel$.

Figure 4.17 shows the results of a model computation for an MT sounding over a 2-D near-surface inhomogeneity. The static shift between $\rho_\parallel(T)$ and $\rho_\perp(T)$ curves amounts to a factor of about 8. Another synthetic example of static shifts is shown in Figure 4.20. Figure 4.9 shows an actual case of a mild shift.

Most of the shift arises from the sensitivity of the electric field to the near-surface features. This can be subdued by using long measuring lines, say 500 m. Such a procedure is laborious and does lower the spatial

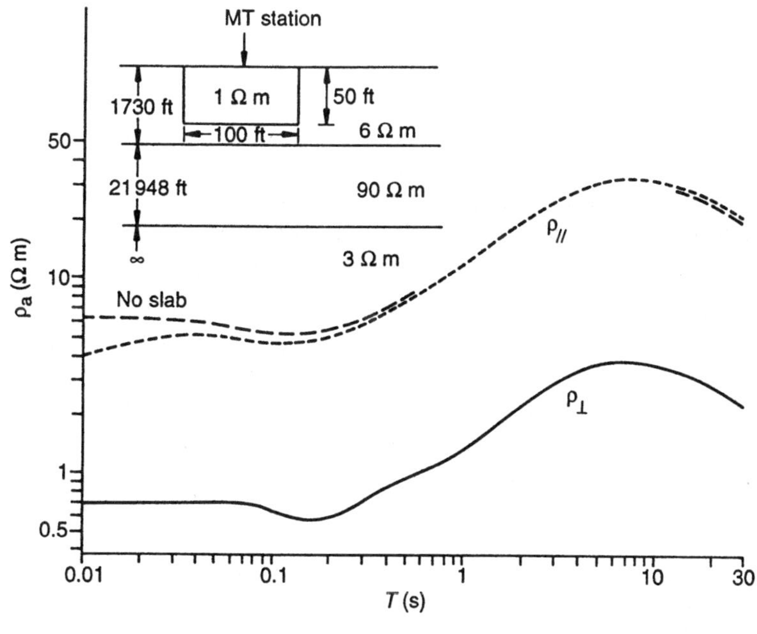

Figure 4.17 Static shift in magnetotelluric sounding curves $\rho_\parallel(T)$ and $\rho_\perp(T)$ due to a 2-D near-surface inhomogeneity, a slab which extends to $\mp\infty$ perpendicular to the schematic cross-section shown in the upper left corner. (Point-function electric field.)

resolution. To improve the static shift in the data, some geophysicists use the arithmetic averages of Z_{max} and Z_{min}. Others use the logarithmic averages.

Pellerin and Hohmann (1990) suggest that a central-loop transient electromagnetic sounding (TEM) be used to shift the shallow part of an MT sounding curve back to its proper position, if one does not mind the substantial additional expenditure.

Bostick (1986) developed a continuous profiling technique (section 4.3.7(f)) to attenuate the 2-D and 3-D near-surface effects, thereby reducing the static shift.

4.3.5 Two-dimensional interpretation

The MT relationship at each station over a two-dimensional model is defined by six different curves as functions of frequency: $\rho_{\parallel}$, $\rho_{\perp}$, $\phi_{\parallel}$, $\phi_{\perp}$, H_z/H_α and phase of H_z/H_α, where α is the direction perpendicular to the strike. The first step is to construct an approximate 2-D model based on independent information and on 1-D interpretation at each MT station. To do this, each of the $\rho_{\parallel}$ and $\rho_{\perp}$ curves at each station is independently interpreted by assuming 1-D models directly below the station. A compromise is made between the two results, usually putting more emphasis on the $\rho_{\parallel}$ curve results, except in certain cases, such as over the apex of an anticline, and sometimes in areas with 3-D influence. Having interpreted all the $\rho_{\parallel}$ and $\rho_{\perp}$ curves at all stations in this fashion, a two-dimensional model is constructed. This is where the two-dimensional modeling starts. It was stated above that the complete MT relationship at each station requires six independent curves. Some of these curves are used qualitatively sometimes, but the $\rho_{\parallel}$ and $\rho_{\perp}$ curves constitute the essential data at each station; these may be displayed in the following ways.

1. Apparent resistivity cross-sections (**pseudo-sections**), one for $\rho_{\parallel}$ and one for $\rho_{\perp}$. The *x*-axis represents the distance (station locations) and the *z*-axis represents the period, increasing downward, on a logarithmic scale. Apparent resistivities are written against the period and distance coordinates and then contours of equal apparent resistivity are drawn. This is the practice favored by most workers (Figure 4.18).
2. Apparent resistivity profiles are drawn across the structure, one profile for each period, for $\rho_{\parallel}$ and $\rho_{\perp}$.
3. Small-scale MT sounding curves are drawn on top of each station.
4. The profiles of $E_{\parallel}$,$E_{\perp}$,$H_{\parallel}$ and $H_{\perp}$ are drawn across the structure, one profile for each period.

The interpretation is usually performed by trial-and-error forward modeling. Three types of modeling methods have been popular: (1) the network solution using transmission-line or -surface analogy, (2) finite

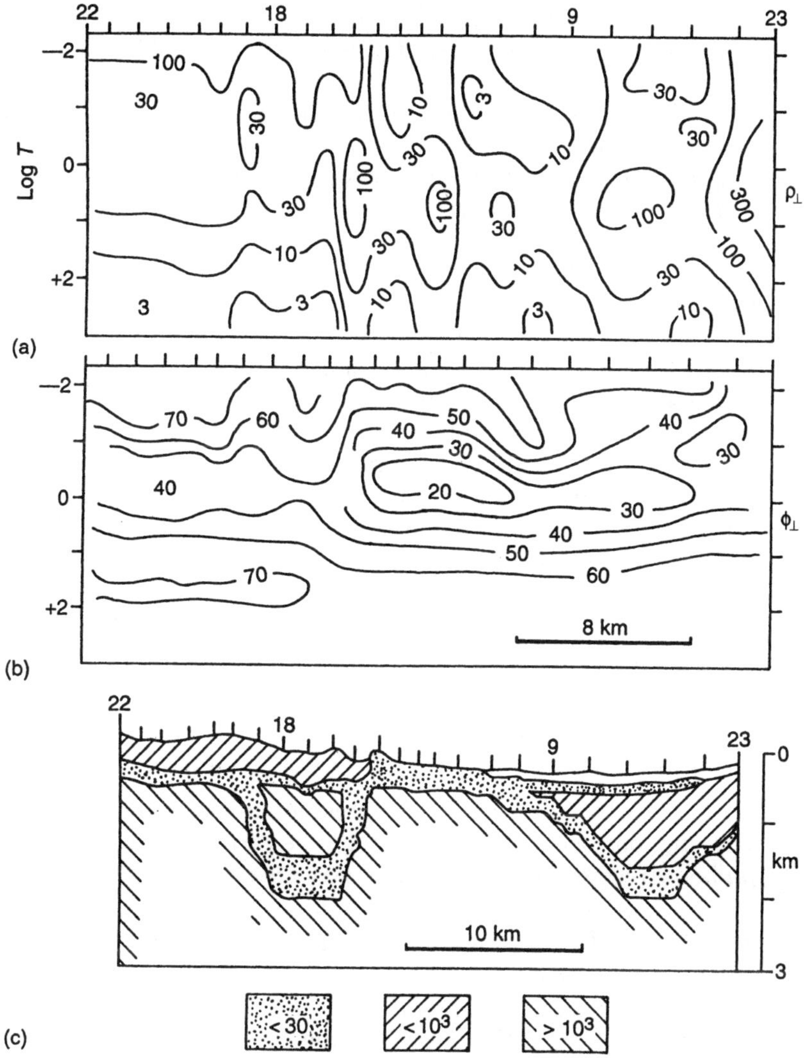

Figure 4.18 MT survey of Long Valley caldera, California. Pseudo-sections of $\rho_\perp$ and $\phi_\perp$, in (a) Ω m and (b) in degrees. (c) Two-dimensional interpretation by finite-element forward modeling (after Wannamaker *et al.*, 1991).

difference and (3) finite element. The network solution method is discussed by Swift (1971). The listing of a program using the finite-difference method was published by Jones and Pascoe (1971). A finite-element program was developed at the University of Utah, Salt Lake City, that uses triangular subdomains. The 1981 version is called *MT2D: An*

Interactive, Two-Dimensional MT and Line-Source Modeling Program. These numerical modeling methods are discussed extensively by Hohmann (1988).

The modeling results are highly sensitive to the accuracy of initial model estimates and to the degree of parameterization. It can be said that the geologic information obtained from such modeling is likely to show what it is not rather than what it is, unless the initial model is simple enough to be in accord with the resolving power of the MT method and is a realistic one. DeGroot-Hedlin and Constable (1990) developed an approach whereby an extremely simple structure that would fit the data is sought. This is done by relaxing the misfit criterion by a small amount.

Figure 4.18 shows a field example (Wannamaker *et al.*, 1991) in which finite-element forward modeling was used. The data consist of a profile of 24 MT soundings across the Long Valley caldera in California. A 2-D model was constructed and interpreted in terms of geothermal potential. The use of $\rho_\perp$ and $\phi_\perp$ has been emphasized in the modeling, because experience in this area indicated that these are less influenced by the 3-D effects than $\rho_\parallel$ and $\phi_\parallel$. Figure 4.18c is the result of forward modeling. Only the part of the model shallower than a depth of 3 km is shown here. Some of the shallow information came from time-domain EM soundings and drill hole data.

4.3.6 Concluding remarks about 1-D, 2-D and 3-D interpretations

The state of the art with regard to interpretation in terms of dimensionality is such that one should be careful not to be carried away beyond 1-D if the probable returns do not seem to be cost effective.

1. If the MT data at each station justify the assumption of a 1-D model, the results can be very rewarding.
2. If the subsurface has to be treated via 2-D modeling the MT method is relatively less likely to be rewarding; the cost and the interpretation difficulties increase considerably. Model alterations for each trial are difficult to choose, and many time-consuming trials are necessary, unless the subsurface and the near-surface features are very simple. An effective direct interpretation (inversion) method in terms of arbitrary cross-sections will not be available in the foreseeable future.
3. If the subsurface has to be treated as 3-D on a local basis, this can be done in terms of extremely simple and regular geometric forms., e.g. prisms. This may be useful in mining geophysics, but is of limited use in the study of sedimentary basins, except for educational purposes. In this case one should perhaps favor the telluric methods instead; they are more cost effective in such cases (section 4.4).

4.3.7 Varieties of magnetotelluric and quasi-magnetotelluric methods

Beginning with Cagniard's (1953) paper, a number of variations and modes of operation of MT have been developed that suit, in different degrees, various types of geologic problems, facilities and monetary budgets. Some of these are complete MT, referential MT, remote MT, telluric sounding, telluric MT, and electromagnetic array profiling (EMAP).

The method envisaged by Cagniard (1953) was based on the assumptions of perfect 1-D subsurface and negligible H_z. This would involve the measurements of horizontal components, two of $\boldsymbol{E}$ and two of $\boldsymbol{H}$, along a coordinate system whose direction was to be selected in the field. No rotation would be performed. Such apparent resistivities are now called **Cagniard resistivities,** or scalar resistivities, to distinguish them from rotated tensor resistivities, and such procedure is now called **simple MT**. This method is still being used by organizations having limited facilities, and for appropriate geologic problems, by judiciously selecting locations and directions of measuring axes. If this is done properly, Cagniard resistivities may be satisfactory as final data; they might even be more valuable than rotated tensor resistivities derived from data obtained along arbitrarily oriented axes. Such surveys can be carried out by means of paper chart recorders, and even without digital processing for some very limited objectives, such as determining the depth to the electrical basement and the total conductance of the sedimentary section. Subsequently, the following methods were developed.

(a) Complete (Conventional) MT

This method requires the procedure described in section 4.3.4. The data have to be in digital form so that measuring axes can be analytically rotated to obtain $\rho_{\parallel}$ and $\rho_{\perp}$ and their phases. Also useful in interpretation is the 'tipper' information that comes from the three components of $\boldsymbol{H}$. Thus, complete MT implies five-component, digital measurements over a broad band of frequencies. Vozoff (1972) describes such operations. This type of survey is expensive on a cost-per-station basis, but represents a bargain for regional work, such as basin evaluation in petroleum exploration. The cost is reduced if one is not interested in very low frequencies, and if the signal-to-noise ratio is favorable.

(b) Referential MT

Two complete MT systems are employed and measurements are made simultaneously at two stations between about 3 and 15 km apart. Measurements at one station are telemetered to the other station where recording is done for both stations, or recorded and clocked individually at each station. In processing the data, cross-products between horizontal magnetic fields at two stations are used to compute impedance tensors at both stations. A field evaluation of this procedure was published by Gamble, Goubau and Clarke (1979). This method reduces various types of noise that do not meet the specifications of MT-grade distant-origin signals.

Referential MT can be accomplished in the following ways.

1. Two complete MT measurements, with magnetic reference, are made and the two systems move to another pair of stations. One can use the electric field as reference (telluric reference), but the consensus is that magnetic reference is better, primarily because the electric field is much more sensitive than the magnetic field to the local geology.
2. One complete MT system, plus one two-component magnetic reference unit at a base station to which a number of MT stations are referred. This mode requires only one MT crew.

Needless to say, referential MT data are normally better than single-station complete MT data for the same duration of field recording, and they do not always cost more. The comparative cost advantage can go either way, depending on the number and density of stations, the signal-to-noise ratio and logistics. However, in areas of high cultural noise one may not have any choice but to use referential MT, which rejects all non-planar waves.

This method probably constitutes the most important development since complete MT became available, because locations free of cultural noise are becoming rare. Clarke *et al.* (1983) describe the equipment and procedures used.

(c) Remote (wandering, roving) MT

It has been stated above that the diagnostic power resides mainly in the electric field; the dependence of the horizontal components of the magnetic field on local geology is relatively small. Therefore, the horizontal magnetic field may be assumed to be uniform spatially, and the magnetometer may be displaced from the telluric measuring lines by some distance. How far this distance can be depends on survey objectives and on subsurface resistivities; it may be desirable to keep it within 100 m in some instances and tens of kilometers in others. Cagniard (1953), recognizing this option, suggested that offshore MT data may be obtained by measuring the magnetic field on land and the telluric field on sea bottom. However, the horizontal magnetic field can change by a factor of about 1.5 over a distance of about 7 km if the subsurface has major lateral variations. This effect is frequency dependent and can lead to serious errors. On the other hand, interpretation of remote MT data can be made without assuming spatial uniformity of the horizontal magnetic field, by calculating actual remote MT responses (Stodt, Hohmann and Ting, 1981), but this would largely reduce the incentive in choosing the remote MT method in the first place.

It has been common practice, ever since the MT method became operational, to take this option in areas with accessibility difficulties, because telluric field measuring units can be portable, or simply to reduce cost. With the recent developments in telemetering and referential real-time data processing, this method became more popular.

A complete MT base station and about three remote telluric stations are set up. Data are measured simultaneously at four sites by telemetering the teluric stations to the MT base. Telluric reference is used. Then the whole setup is moved to another location. Thus, about one-fourth of the stations are of the complete MT type and provide tipper information. On a statistical basis, one crew with only one MT system can obtain about six stations per day. The cost per station is usually one-half to one-third of that of single-station complete MT.

This method may be used for surveys with high station density, for detection of targets having small lateral extent, such as isolated reefs, or along profiles with closely spaced stations in order to reduce the effects of near-surface inhomogeneities, and to increase the subsurface resolution in 2-D interpretation.

A large number of remote telluric stations can be referred to one MT base by recording in batches of two or three telluric stations, but, for similar results, one should consider using a different technique, namely telluric MT, that will be discussed in section 4.3.7(e) below.

Some organizations possess MT systems with a large number of channels. For instance, a 16-channel system could be used to obtain two complete MT stations with magnetic reference plus three remote telluric stations, or one four-component MT base plus six remote telluric stations. Some contractors have been offering two such 16-channel units working in synchronization, so that one can obtain up to 15 stations of remote MT at one recording episode. This would lend itself to long-duration recording, day and night, until a satisfactory signal-to-noise ratio is accomplished, and to recording along high station density profiles for spatial filtering of the near-surface inhomogeneities.

(d) Telluric sounding

This method yields MT-type data in the form of scalar resistivities without making any magnetic measurements whatsoever (Yungul, 1966). It was developed in the late 1950s when prospecting-grade micropulsation magnetometers did not exist. In principle, it is like remote MT if magnetic measurements at a base station are replaced by synthetic magnetic data obtained from telluric measurements plus subsurface resistivity data gathered there. However, the practice is simpler; it consists of transferring a base station resistivity model to remote telluric stations via simultaneous telluric data at the base. Resistivity data at the base may be obtained by means of direct current soundings expanded in two azimuths, or from well logs.

The method is particularly useful if the telluric measurements are confined to the components which approximate $E_\perp$ (perpendicular to the average strike of major geologic features), because for a strictly 2-D subsurface $H_\parallel$ is exactly the same everywhere at the surface if the air resistivity is assumed to be infinite. All that is necessary is to obtain amplitude ratios of $E_\perp$ at moving stations to those at the base station. Because the cost per station is very low, the moving stations can be closely spaced, even continuous, so as to achieve spatial filtering.

In the early stages of development, I used only paper chart recorders and portable radio communication, and obtained good results in deep sedimentary basins. It can be used by organizations which do not possess MT facilities.

The shortcomings of the method are obvious, but it does have some theoretical advantages as well. In MT, long telluric measuring lines are employed, say 400 m, primarily to suppress the effects of 2-D and 3-D near-surface inhomogeneities, whereas magnetic measurements are made as point functions. Thus, telluric and magnetic measurements are not really the components of a point-function electromagnetic field. This situation is further aggravated when impedance axes are analytically rotated, especially at locations where near-surface lateral inhomogeneities exist. With telluric sounding, still longer telluric lines can be used at the base and direct current soundings can be obtained along them, thereby rendering the averaging compatible. In addition, there is no magnetic noise.

(e) Telluric MT

The principle of this method is the same as that of remote MT, but the simultaneous MT measurements at the base station are replaced by an MT impedance tensor obtained from one MT measurement there (Hermance and Thayer, 1975). It requires MT-type data processing, but it costs less than remote MT.

With reference to equation (4.38), the components Z_{ij} of the surface impedance tensor at the base station, $[Z_b]$, can be determined from one set of MT data (Vozoff, 1972). Likewise, components of $\boldsymbol{E}$ at base and wandering stations define a 'telluric transfer tensor', $[T]$, whose components are T_{ij}. If it is assumed that $\boldsymbol{H}$ is the same everywhere, the impedance tensor at the wandering station, $[Z_w]$, is equal to the product of $[T]$

and $[Z_b]$; this is the MT data obtained at the wandering station. An additional assumption is that there exists an exact telluric transfer tensor at all frequencies of interest.

It may seem that, so far as field procedures are concerned, the only difference between telluric MT and remote MT is that with telluric MT the magnetometer is returned home having made one measurement at the base station, whereas with remote MT it stays there throughout the survey. However, for large surveys there is a difference in the planning. Telluric MT is planned such that MT measurements are made first at a network of evenly distributed base stations, and the remainder of the fieldwork is all telluric, by reoccupying the previously established telluric electrode locations at base stations.

(f) Electromagnetic array profiling (EMAP)

The primary purposes of this method are to reduce the ill-effects of static shift due to near-surface inhomogeneities, and those of spatial aliasing due to sparse sampling (Bostick, 1986; Dobrin and Savit, 1988; Vozoff, 1991; Torres-Verdin and Bostick, 1992a, 1992b; Warren and Srnka, 1992).

The basic field layout is shown in Figure 4.19f. The dipoles measuring the electric field, each about 300 m long, are positioned end to end along a continuous survey path which is assumed to be perpendicular to the strike of a 2-D subsurface. These E-fields are assumed to be $E_\perp$ components (perpendicular to the strike) which, when combined with $H_\parallel$ (paralleled to the strike) at the magnetic base station, yield TM-mode apparent resistivities. Measurements are made at many contiguous *E*-field-measuring dipoles simultaneously. Two horizontal components (x,y) or all three components (x, y, z) of the magnetic field are measured at a fixed base station. Torres-Verdin and Bostick (1992a, 1992b) state that *E*-field measurements are not referred to the local *H*-field at the base, but to the 'primary plane-wave magnetic field within the survey area', which is estimated by areal averaging of *H*-field measurements taken along and about the survey line. However, a field study at White Pine County, Nevada (Torres-Verdin and Bostick, 1992b), was made by employing a single magnetic base station.

Torres-Verdin and Bostick (1992a) state that an analysis of the characteristics of EMAP data, when acquired along oblique traverses over a 2-D subsurface, as well as a complex 3-D subsurface, will be published in the future.

Simultaneous measurement of the *E*-fields in many contiguous dipoles reduces spatial aliasing and lends itself to a spatial adaptive low-pass wavenumber filtering. The benefits of this kind of filtering are shown by means of three synthetic 2-D models in Torres-Verdin and Bostick (1992b).

In EMAP fieldwork, magnetic remote reference measurements are sometimes made in addition to the measurements at the magnetic base

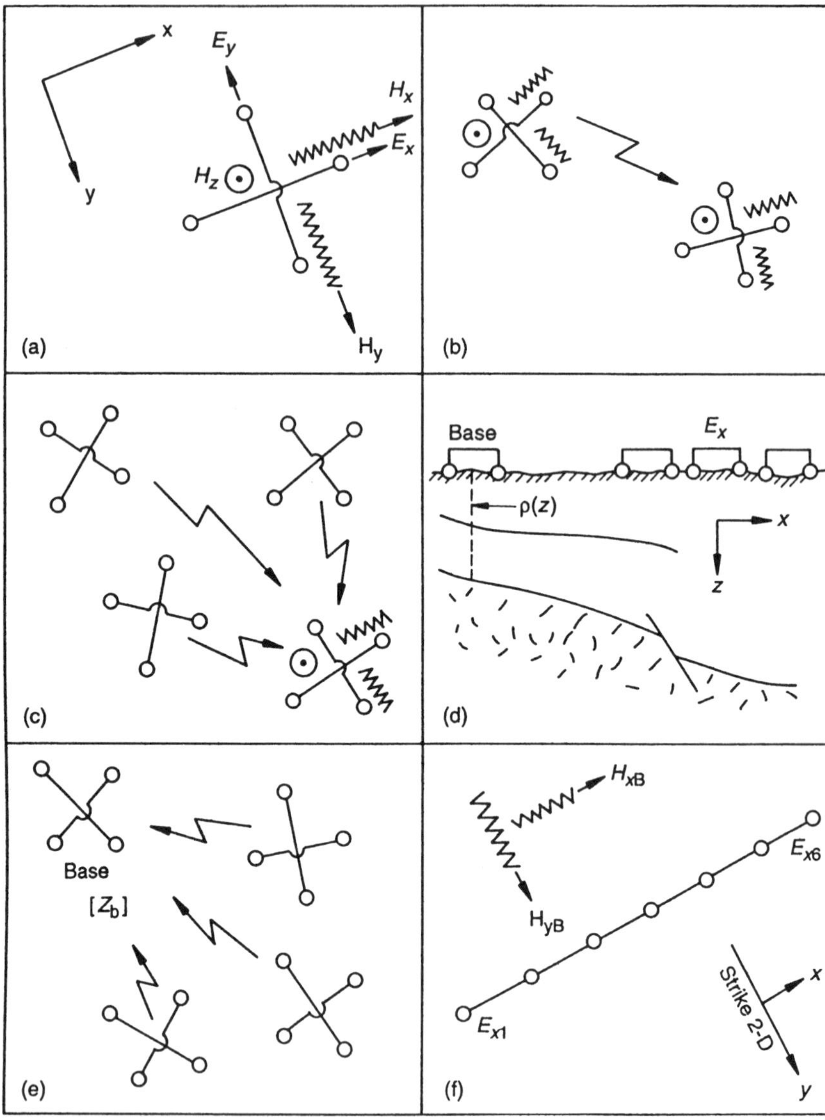

Figure 4.19 Varieties of MT and quasi-MT methods: (a) complete MT; (b) referential MT; (c) remote MT; (d) telluric sounding; (e) telluric MT; (f) electromagnetic array profiling. Small circles are telluric electrodes, coils are horizontal magnetic sensors and large circles are vertical magnetic sensors.

station. These serve to cancel out the cultural noise as usual, but also serve to sum up the adjacent E dipoles in a phase-coherent manner.

Warren and Srnka (1992) describe an EMAP test survey at three basalt-covered locations on the Columbia Plateau in Washington State. Three 16-km long separate survey lines were located such that each passed close to one of three previously drilled boreholes. The lines were separated from each other by about 20 km. Magnetic fields were measured at one base and at one remote reference site. The $E_{\perp}/H_{\parallel}$ impedances were spatially filtered and inverted to resistivities by assuming a 1-D subsurface under each dipole. Resistivity cross-sections are included in the article. These are at Saddle Mountain, Whiskey Mountain and Page. The first two seem to agree with the borehole data down to depths of about 4 km, but not that at Page. The authors state that the line orientation at Page is not suitable for the assumption of a 2-D subsurface.

4.3.8 Concluding remarks about the various methods

The foregoing discussion treated a broad spectrum of methods in terms of the value, quality and cost of similar types of data. Each method is a viable tool if it is employed judiciously.

At the bright end of the spectrum we have referential MT; this yields high-quality data at high cost. The other end is somewhat open; one can keep reducing the cost and quality for a long way. A survey may cost ten times as much as another, for the same number and density of stations. To reduce the cost still further, one could use the telluric methods, discussed in section 4.4, which routinely utilize the Pc3 signals (10–45 s period range), which are strong and abundant and are easy to measure and to interpret, if the problem lends itself to such data of the quasi-potential field type. Given a specific geologic problem and a limited budget, one has to make a difficult choice between closely spaced low-cost and widely spaced high-cost stations. The decision is not dictated solely by the geometric dimensions of geologic targets and wavelengths of signals, but also by logistics, and by surface and near-surface conditions.

4.3.9 Measuring the electric and magnetic fields

A horizontal component of the electric field is measured in essentially the same way as is done in resistivity work,

$$E_r(t) \cong \frac{\Delta V(t)}{l},$$

where $\Delta V(t)$ represents the time variations of the voltage difference between two grounded nonpolarizing electrodes at the ends of an insulated wire, the direction and distance between the electrodes being r and l respectively. This is the average E_r along l, measured in mV/km. The voltage from the electrodes is DC-coupled to differential amplifiers. We assume that this average is a point-function value of E_r at the center of l; however, it is not a rigorous representation of the electric field intensity even as an average. The length l has to be long enough to override noises from both the electronic system and the electrodes. It has also to be long enough to cope with the surface and near-surface irregularities, to act as a spatial filter. A significant noise problem can arise from wind causing the wire to move (the dynamo effect).

Lead, silver and copper cells are used as nonpolarizing electrodes. Their management is a lengthy subject which has been extensively described in the literature ever since spontaneous potential (SP) methods were developed (e.g. Clarke *et al.*, 1983). This is a more serious problem in MT work than it is in resistivity, telluric or SP methods, because here one deals with signals in a broad band of frequencies, and some of them, such as those in the 0.1–5 Hz band, are usually of very low amplitude.

The most widely used magnetic sensors at present are (1) induction coils with an alloy core of high magnetic permeability and (2) Josephson junction-effect superconducting magnetometers called SQUID; it seems that the latter is more widely used for deep work.

An induction coil works on the basis of Faraday's law; the electromotive force in the coil is proportional to the time derivative of $\boldsymbol{H}$. Consequently, the output voltage amplitude increases in proportion to frequency. On the other hand, natural field amplitudes generally decrease with increasing frequency (Figure 4.4a). The result is that the coil output is prewhitened; this extends the dynamic range of the system. Induction coils have lower noise than SQUIDs at higher frequencies, above 100 Hz, and may be preferred for MT work that requires frequencies up to 1 kHz. The resolution is about 10^{-5} nT at 100 Hz and about 0.01 nT at 0.01 Hz. However, they are heavy and long, and three of them must be buried separately to prevent interaction and wind vibration noise; the H_z sensor may necessitate an auger hole about 1 m deep. The reader will find more information and many references on induction coils in Vozoff (1991) who is one of the pioneers in their development.

The SQUID magnetometer contains three sensors, mounted orthogonally in one cryostat containing liquid helium. The resolution is 10^{-4} to 10^{-5} nT; it does not prewhiten the natural field, but its dynamic range is large. Except with its shortcomings at the higher end of MT frequencies, and the problems associated with procurement, transportation and management of the liquid helium, the SQUID magnetometer is more versatile, less laborious and has higher resolution at frequencies lower than 10 Hz, compared to the induction coil. More information is found in Clarke *et al.* (1983) and Vozoff (1991).

4.4 TELLURIC METHODS*

4.4.1 Introduction

The telluric methods have already been defined in section 4.1. Theoretically speaking, telluric and MT methods are equivalent, that is, measuring spatial variations of $\boldsymbol{E}$ (telluric) is equivalent to measuring $\boldsymbol{E}$ and $\boldsymbol{H}$ at discrete points (MT). However, this is where the equivalence ends. The telluric methods, at their most effective, are lateral exploration tools, analogous to gravitational methods, and are similar to resistivity mapping with an immensely long rotating source dipole. The spatial variations in the measured quantities are almost entirely due to lateral variations in resistivity distribution in the subsurface; the more pronounced the lateral resistivity variations (the structures) in sedimentary basins, the more rewarding the results. MT is, at its most effective, a vertical exploration tool similar to resistivity sounding, which yields well log type information; the milder the structures, the more rewarding the results. Thus, in general the two methods may complement each other if used in a cost-effective fashion. However, some geophysicists not familiar with telluric methods believe that they have been replaced by MT.

The first practical telluric techniques were developed by Schlumberger (1936, 1941). The **relative ellipse area** method by Kunetz (1952a) constituted a milestone, in that scalar telluric anomalies could be assigned to measurement sites. This method is described by Boissonnas and Leonardon (1948); their article is probably the earliest significant English literature on this subject. A comprehensive paper by Migaux (1951) describes the principles and geologic significance of the telluric method. A review paper by Kunetz (1958) contains many interesting case histories concerning petroleum exploration. Another milestone was a book by Berdichevskii (1960), which contributed substantial quantitative information on the telluric method. This was done on the basis of then newly developed MT theory.

Telluric methods preceded MT by about two decades and led to it, but a quantitative understanding of the former was a by-product of the latter.

4.4.2 Telluric and magnetotelluric methods compared

Figure 4.20(a) is a cross-section of a 2-D model representing a salt ridge in a deep, unusually low-resistivity sedimentary basin, similar to the Texas Gulf Coast. The two near-surface inhomogeneities could represent freshwater sands.

* This section is an abridged version of the paper: Yungul, S.H. (1977) The telluric methods in the study of sedimentary structures – a survey. *Geoexploration*, **15**, 207–38. With permission of Elsevier Science Publishers B.V.

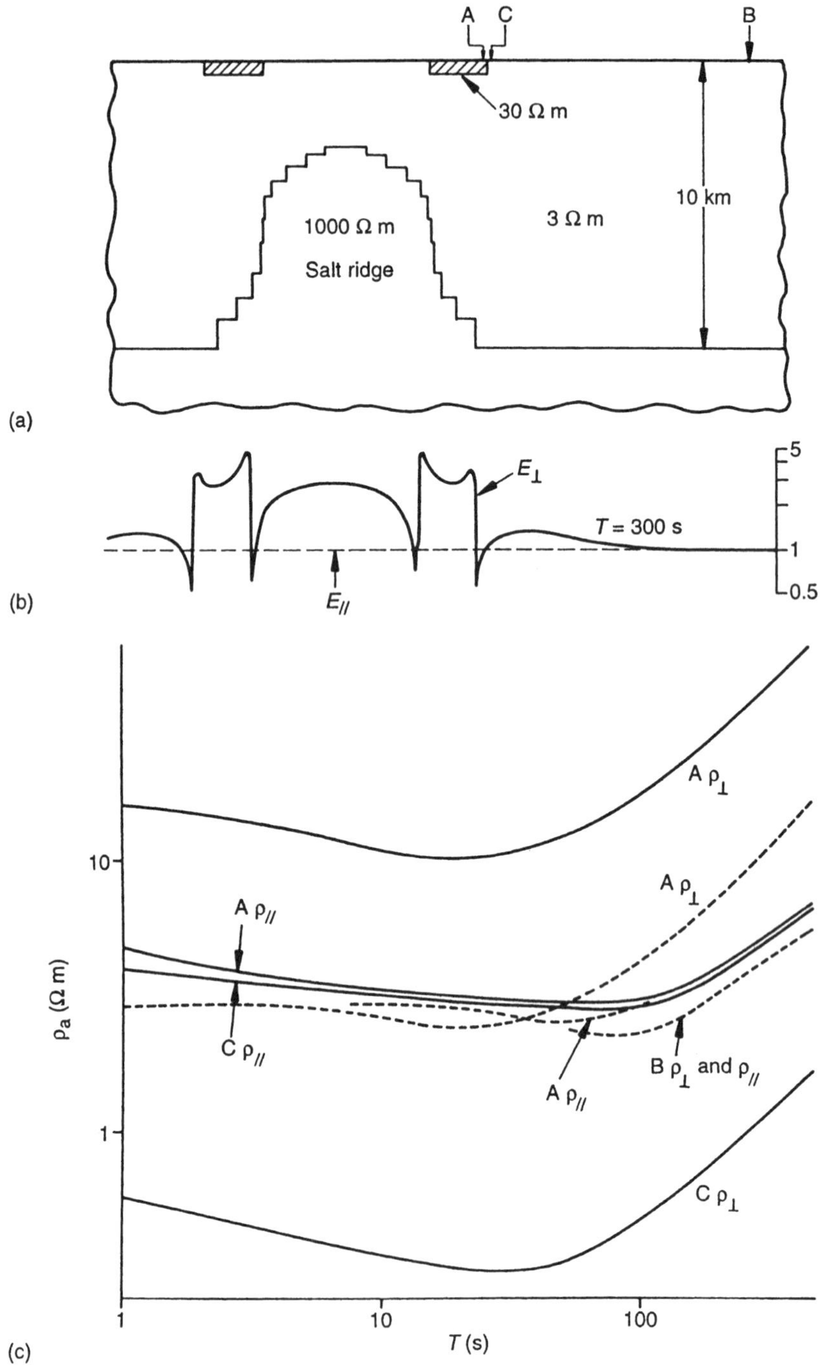

Figure 4.20 Numerically computed magnetotelluric and telluric responses of a 2-D model. Dotted magnetotelluric sounding curves would have been obtained if the near-surface inhomogeneities were not present. Solid-line curves are for the complete model (from Yungül, 1977; with permission of Elsevier Science Publ.).

Figure 4.20c shows the computed MT sounding curves at stations A, B and C. The electric fields are point functions. The dotted curves represent the condition without the near-surface inhomogeneities. As before, $\rho_{\parallel}$ is derived from $E_{\parallel}$, parallel to the strike. The solid line curves have been computed for the complete model. The dotted curves at A, without the near-surface inhomogeneities, cannot be interpreted by themselves; the survey and the interpretation have to be carried out as explained in section 4.3.5.

The case with the near-surface inhomogeneities included (the solid line curves) is much worse; the $\rho_{\perp}$ curve is shifted by a factor of 10 if the station is displaced 200 m. The $\rho_{\parallel}$ curve is not shifted much, but this case occurs only if the near-surface inhomogeneities and the deep structure are 2-D and have the same strike, which is unusual; otherwise the situation would be hopeless. A space-domain filtering of the data is not practical, because the per-station cost of such MT work is high.

Figure 4.20b shows a profile of normalized $E_{\perp}$ and $E_{\parallel}$ components across the complete model for a period of 300 s. Such data measured through band-pass filters, that constitute essentially telluric data, are extracted from a survey made without knowing anything about the subsurface and near-surface conditions. The effects of the inhomogeneities can now be filtered out if there are sufficient data points, and the salt ridge anomaly can be isolated and interpreted. One can have a high station density because the cost is much lower than MT. Two or three MT soundings at sites judiciously selected through the telluric data may be necessary for a quantitative interpretation of the telluric data. This example is very unusual in that a 300-s period was used; this is because the geologic section of interest is thick and of extremely low resistivity. Typical cases involve signals in the period range 15–40 s.

Another consideration is the search for relatively small 3-D features such as salt domes, reefs and geothermal reservoirs. Such surveys require many closely spaced stations; Figure 4.31 shows 287 stations. An MT survey with the same number of stations would be very expensive. Also, most of the wealth of MT data thus obtained could not have been used effectively, because general 3-D interpretation is not practical at present.

4.4.3 Principle of telluric methods

With the telluric methods we measure the spatial variations of $\boldsymbol{E}$. It is stated above that these measurements are also theoretically equivalent to measurements of $\boldsymbol{H}$, hence telluric and MT are the same thing in theory, as the following shows.

One of Maxwell's equations is

$$\operatorname{curl} \boldsymbol{E} = \mu \frac{\partial \boldsymbol{H}}{\partial t}; \tag{4.42}$$

it states that a function of the space derivatives of ***E*** at one point measures the time derivative of ***H*** there. Equation (4.42) is the differential form of Faraday's law. However, curl ***E*** includes vertical derivative of ***E*** which cannot practically be measured, but this shortcoming can be compensated for by some independent data, e.g., well logs or resistivity soundings at a base station for telluric measurements.

The structure analyzed in Figure 4.21, shown in the upper left-hand corner, is a horizontally layered sedimentary section overlying a high-resistivity half-space (the basement). Below it are shown the computed MT apparent resistivity (ρ_a) and phase (ϕ) curves. The ρ_a curve rises on the right to become asymptotic to the resistivity of the basement.

The section of the rising branch between T_1 and T_2 is close to a straight line with a slope of 45° in the logarithmic plot; let us assume that it is exactly so. This branch has to satisfy the MT relationship (in the practical

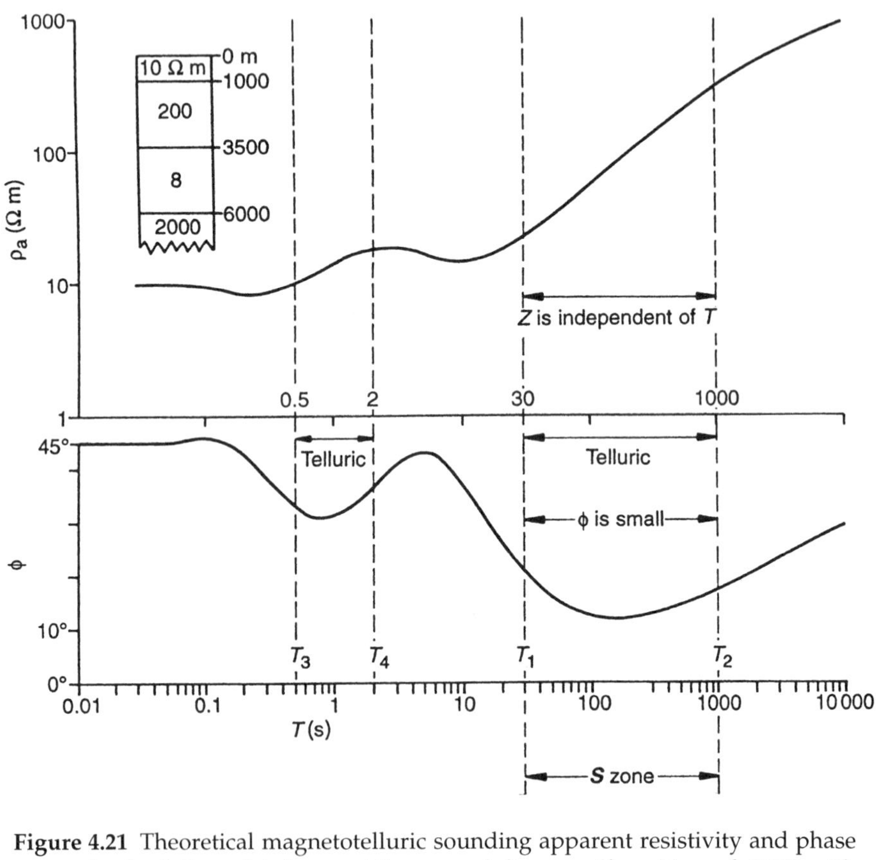

Figure 4.21 Theoretical magnetotelluric sounding apparent resistivity and phase curves for the 1-D model shown at the upper left corner (from Yungul, 1977; with permission of Elsevier Science Publ.).

units mV/km, nT and Ω m):

$$\rho_a = 0.2T\,|\,E_x/H_y\,|^2. \tag{4.1}$$

A 45° straight-line relationship requires that in

$$\log \rho_a = \log T + \log\,[0.2\,|\,E_x/H_y\,|^2] \tag{4.43}$$

the second term be independent of T. That is, within the T_1–T_2 range the impedance $Z = E_x/H_y$ is independent of T. The T_1–T_2 interval is called the 'S zone' (Berdichevskii, 1960); S is the horizontal conductance, in mhos, of the sedimentary section:

$$S = \sum_{i=1}^{n} (h_i/\rho_i) = D/\rho_e, \tag{4.44}$$

where n is the number of sedimentary layers, h_i are layer thicknesses and D is the depth to basement, both in meters. This formula also defines ρ_e, called the **effective resistivity** of the sedimentary section. The 45° branch does not change substantially so long as S remains constant, no matter how many layers and what sequential combinations are considered (equation (4.23) and Figure 4.14).

Berdichevskii (1960) developed an asymptotic equation for Z for periods longer than about T_1 shown in Figure 4.21. His equation (in MKS units) is:

$$\dot{Z} = \frac{\dot{E}_x}{\dot{H}_y} = 1\bigg/\left(S + \frac{10^3}{4\pi}\left(\frac{T}{5\rho_b}\right)^{1/2} e^{i\pi/4}\right), \tag{4.45}$$

where ρ_b is basement resistivity, and dots represent complex quantities. Because Z is independent of T in the S zone, equation (4.45) must then take the form (in MKS)

$$Z = 1/S, \tag{4.46}$$

or (in practical units: mV/km, nT and Ω m)

$$Z \cong 796/S.$$

On the other hand the phase curve in Figure 4.21 shows that the phase difference between E_x and H_y is small within the S zone and can be assumed to be zero. It follows that the impedance is approximately a real number (i.e. purely resistive); for instantaneous values the relationship between $\boldsymbol{E}$ and $\boldsymbol{H}$ is that of a DC electromagnetic field. The telluric field can be assumed to be an instantaneous DC field within the window T_1–T_2.

Figure 4.21 shows that another approximately DC window exists within the T_3–T_4 range, for which the top of the 200 Ω m layer would behave like a basement. The S zone periods are such that the skin depth for ρ_e is between about 1.5 and 10 times the depth to the basement.

So far we have based our argument on models of a horizontally layered subsurface. This would be valid if the measuring sites were far from lateral changes, or if the lateral changes were not abrupt. However, an important role of the telluric method is the study of abrupt changes, such as faults and subcrops. In reality, from the viewpoint of exact analysis, in such cases the phase relationships and the dependence of Z on T, become very complicated. However, theory and practice show that so long as one operates in the S zone, in terms of the average characteristics of the sedimentary section, the instantaneous DC assumption is still met satisfactorily, usually good enough for quantitative interpretation, as will be demonstrated later. When such is not the case, qualitative interpretation can be made. However, if the DC condition is met, the interpretation of measurements is drastically simplified.

As before, in the general case, for an arbitrary single-frequency plane wave incident to an arbitrary subsurface, the components of $\boldsymbol{E}$ and $\boldsymbol{H}$ are related by (Cantwell, 1960):

$$\dot{E}_x(t) = \dot{Z}_{xy}\,\dot{H}_y(t) + \dot{Z}_{xx}\,\dot{H}_x(t),$$

$$\dot{E}_y(t) = \dot{Z}_{yx}\,\dot{H}_x(t) + \dot{Z}_{yy}\,\dot{H}_y(t), \tag{4.38}$$

where dots signify complex quantities, and Z represents the components of a surface impedance tensor. If we assume that within the S zone the components of Z are independent of frequency and the phases are negligibly small, equation (4.38) reverts to:

$$E_x(t) = C_1 H_y(t) + C_2 H_x(t),$$

$$E_y(t) = C_3 H_x(t) + C_4 H_y(t), \tag{4.47}$$

where all quantities are real numbers, C represents constants that depend only on the subsurface geology and the direction of the measuring axes, and $E(t)$ represents real-time series that include all frequencies within the S zone.

It follows that the components of $\boldsymbol{E}$ at two separate locations, measured along the x- and y-axes at a base station B and along the u- and v-axes at a field station F, are also linearly related:

$$E_u(t) = \mathrm{a}\,E_x(t) + \mathrm{b}\,E_y(t),$$

$$E_v(t) = \mathrm{c}\,E_x(t) + \mathrm{d}\,E_y(t), \tag{4.48}$$

where constants a, b, c and d depend only on the directions of the coordinate systems and subsurface geology. However, the Jacobian of the transformation from the x–y system at B to the u–v system at F,

$$J = |\mathrm{ad} - \mathrm{bc}|, \tag{4.49}$$

is an invariant, for the pair of stations B and F, independent of measuring axes directions, depending only on the geology at B and F. This is the same as the **telluric transfer tensor** discussed in section 4.3.7(e), except that now the phases are zero. *J*, a real and scalar number, is contoured and interpreted in a way that is analogous to the interpretation of gravitational data.

The following methods pertain primarily to the procedures of extracting the *J* values.

4.4.4 The relative ellipse and triangle methods

These methods were developed by Kunetz (1952a). In equation (4.48) the instantaneous *E*-values can be replaced by synchronous increments between time intervals, as shown in Figure 4.22:

$$U = \mathrm{a}X + \mathrm{b}Y \quad \text{and} \quad V = \mathrm{c}X + \mathrm{d}Y, \tag{4.50}$$

where capital letters signify components of a pair of conjugate variation vectors, $\Delta \boldsymbol{E}_\mathrm{F}$ at F and $\Delta \boldsymbol{E}_\mathrm{B}$ at B. These correspond to an arbitrary time interval, $\Delta t = t_2 - t_1$, as schematically shown in Figure 4.22. Such strip chart records are called tellurograms.

The x–y system at B and the u–v system at F are oriented arbitrarily. If a number of pairs of such conjugate variation vectors are determined, and those at F are normalized to the magnitudes of their conjugates at B,

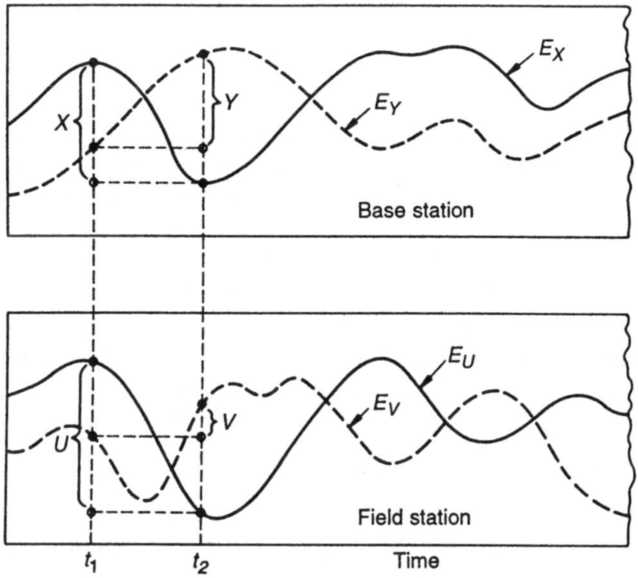

Figure 4.22 Hypothetical tellurograms representing simultaneous recordings along orthogonal measuring axes x, y at base station and u, v at field station (from Yungul, 1968).

the normalized $\Delta \boldsymbol{E}_{\mathrm{F}}$ traces an ellipse whose area is equal to

$$\pi J = \pi \,|\, \mathrm{ad} - \mathrm{bc} \,|\,, \tag{4.51}$$

where J is the Jacobian of the transformation from the x–y system to the u–v system, as in equation (4.49). Normalization is equivalent to assuming that $\Delta \boldsymbol{E}_{\mathrm{B}}$ traces a circle; and a circle at B transforms into an ellipse at F through equation (4.50).

Experience shows that fairly regular ellipses are obtained, an indirect proof of the validity of equation (4.50). The ellipse, and therefore J, is an invariant for a pair of measuring stations, depending only on the geology at F and B. The ellipse at F is called the **relative ellipse**, and J is the **relative ellipse area**.

Theoretically, two pairs of conjugate vectors, determined by eight components, are sufficient to solve for J from equation (4.50). The ratio of the areas of two conjugate triangles formed by these four vectors is

$$J = \frac{|\,U_1 V_2 - U_2 V_1\,|}{|\,X_1 Y_2 - X_2 Y_1\,|}. \tag{4.52}$$

This is called the **triangle method**, which yields J without constructing an ellipse.

Usually, many triangles are constructed at each site, and a statistical value is determined among the selected ones. There is considerable scattering; one of the reasons is that equation (4.50) is not rigorously satisfied. Several triangles must be used to obtain one J-value. The procedure for the triangle method, described in detail by Berdichevskii (1960), is tedious and requires considerable skill. It does not readily lend itself to digital computer processing, perhaps primarily because the phases are not really zero. Favorable time intervals are visually selected first and then the resulting variation vectors are subjected to another visual selection.

One could use telluric MT, discussed in section 4.3.7(e) and measure a, b, c and d with their phases and at a number of frequencies. This would require digital recording and Fourier analysis, and is using an MT technique. Here we are interested in a narrow-band (at frequencies of the *S*-zone window), rapid, simple and low-priced method.

4.4.5 The intrinsic ellipse method

Approximate values of J can be obtained by an automated version of the intrinsic ellipse method of Kunetz (1952b, 1957). As with the triangle method, tellurograms are obtained at F and B along the u–v and x–y axes; a schematic one is shown in Figure 4.23a. A synthetic tellurogram is computed for the median direction of the measuring axes, such as along the w-axis (Figure 4.23b). The median direction tellurogram can also be

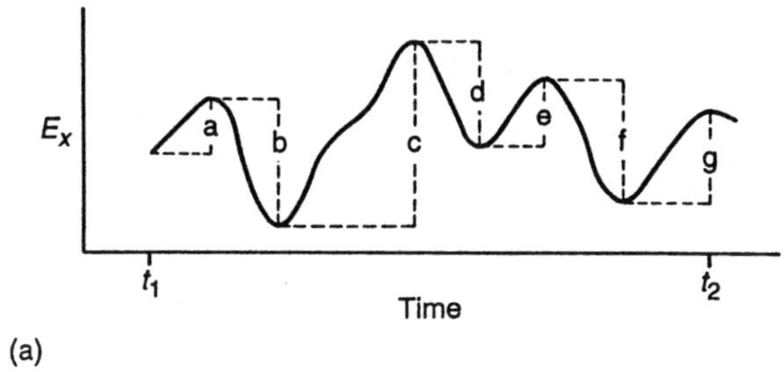

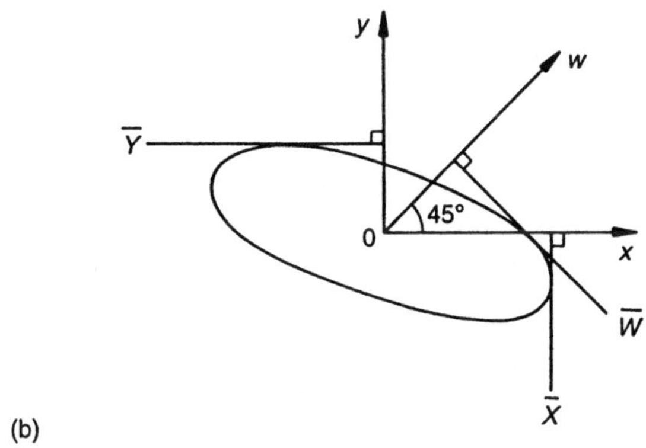

Figure 4.23 The intrinsic ellipse method: (a) a hypothetical tellurogram; (b) determination of the intrinsic ellipse by means of its tangents (after Kunetz, 1952b).

obtained directly in the field, by means of a third measuring line. The total variations on the tellurograms are computed for a selected time interval, say 15 min. On Figure 4.23a the total variation along the x-axis is the numerical summation of the peak-to-trough excursions.

$$\overline{X} = a + b + c + d + e + f + g.$$

A straight line perpendicular to the x-axis, drawn at distance $\overline{X}$ from the origin, constitutes a tangent to the intrinsic ellipse (Figure 4.23b). The ellipse is determined by its origin and three tangents. Its area at F is (Kunetz, 1952b):

$$A_F = \frac{\pi}{2}\{[(\overline{X}+\overline{Y})^2 - 2\overline{W}^2][2\overline{W}^2 - (\overline{X}-\overline{Y})^2]\}^{1/2}. \tag{4.53}$$

The ratio of A_F to A_B is approximately equal to J. A comparison of the intrinsic ellipse and triangle (or relative ellipse) methods by Kunetz (1957) shows a mean deviation of 7% between the values of J obtained.

The intrinsic ellipse method can readily be processed by digital computer. The total variations can also be obtained directly in the field, by means of counters (Kunetz, 1957). I have found that for computerized automatic processing of 20-min tellurograms, the accuracy of J deteriorates as ellipticity increases. Sometimes equation (4.53) yields imaginary values. This could be helped by using longer record sections and also by measuring the third component along the w-axis (Figure 4.23) in the field. The procedures of this method are described in detail by Thieme (1963).

The azimuth and ellipticity of the intrinsic ellipse have some geological significance, but this and the subject of absolute ellipse methods will not be discussed here.

4.4.6 The vectogram method

I find that the most economical method of determining J, compared with the ellipse, triangle, and intrinsic ellipse methods, is a polar diagram recording of $\mathbf{E}$ on X–Y recorders (Yungul, 1968).

Figure 4.24 shows a pair of vectograms, obtained approximately simultaneously at F and B, in about 1 min of recording time. The band-

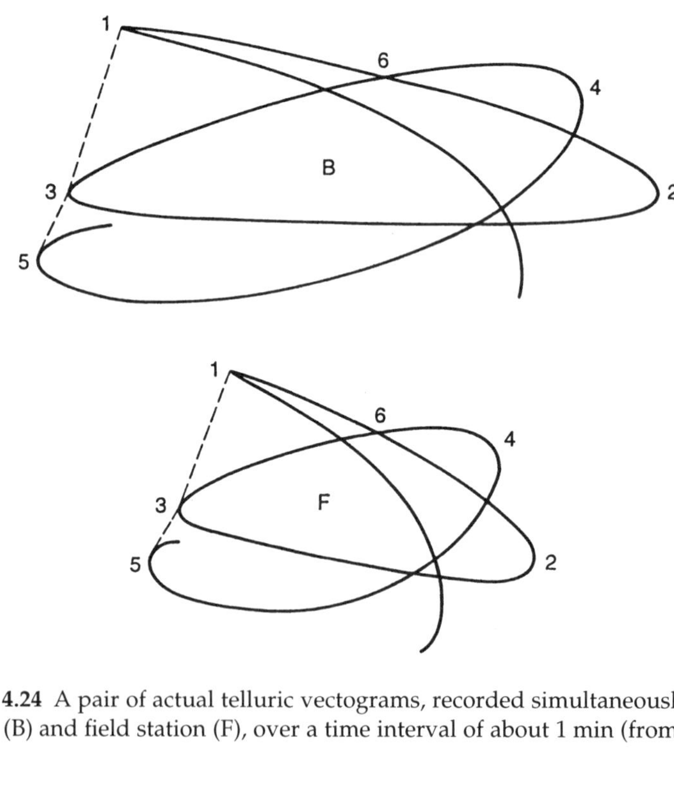

Figure 4.24 A pair of actual telluric vectograms, recorded simultaneously at base station (B) and field station (F), over a time interval of about 1 min (from Yungul, 1968).

pass filter was peaked at T = 20 s. The ratio of any single or combination of closed areas at F, including the tangents, to the conjugate at B is equal to *J*. This is in accordance with equation (4.50). About a dozen values of *J* can be measured from this pair, but three, obtained from the contours (1, 2, 3), (3, 4, 5) and (2, 3, 6), will suffice.

Normally, a few such vectogram pairs are obtained at each field station. Many fewer values of *J* are needed than in the triangle method, because each closed area is equivalent to the average of several triangles. The measurement of closed areas effectively integrates the undesirable signals and noise of relatively higher frequencies.

I have found that the standard deviation for the vectogram method is about one order of magnitude lower than for the triangle method, at the same 20-min recording interval (Yungul, 1968).

Approximately simultaneous vectograms can be obtained at F and B through oral communication by means of portable radio transmitters. One person at B can manage three or four F-crews. The closed areas on the vectograms can be measured with a planimeter, without the need for accurate time-tying.

At locations where ***E*** is somewhat linearly polarized the accuracy of *J* becomes low and is sometimes unusable with the triangle method. This is not so with the vectogram method; one can wait for perhaps several minutes until one or two nonlinear loops occur.

4.4.7 The split-spread method

Figure 4.25 shows a field setup having a collinear and approximately equally spaced three-electrode array. The two potential differences, each one being between the center electrode and one outer electrode, are

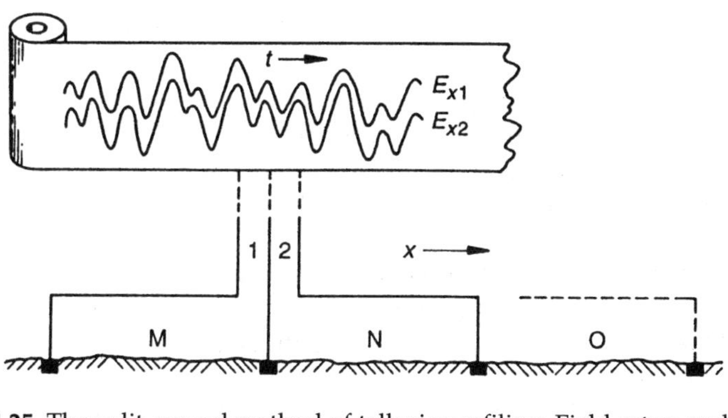

Figure 4.25 The split-spread method of telluric profiling. Field setup and a hypothetical tellurogram (from Yungul, 1977; with permission of Elsevier Science Publ.).

recorded simultaneously on the same strip of paper for about 20 min. The result is a pair of approximately meshing tellurograms at each setup.

Proceeding with continuous profiling (overlapping spreads), hence the terminology **split spread**, one can determine values of J at the midpoints of consecutive electrode pairs; e.g. at N with respect to M, and at O with respect to N, all normalized to M. Even though it appears that the determination of J requires two-dimensional (x,y) data, whereas this procedure provides only one-dimensional (x) data, the J-values are obtainable if $\Delta \boldsymbol{E}$ scans the subsurface from various directions, and if it is assumed that the subsurface is piecewise but arbitrarily two-dimensional only in the immediate vicinity of each spread.

On each pair of tellurograms the peak-to-trough amplitudes of roughly meshing events are measured with an expanding-scale ruler, and the ratios, R, of these amplitudes are computed:

$$R = \Delta E_{x1} / \Delta E_{x2}.$$

About 30 R-values are obtained and plotted, either as dot histograms or sequentially (Figure 4.26). In general, R has a large amount of dispersion. Then J-values are picked by visual inspection of these histograms, as explained below.

Suppose that the horizontal acute angle between the geologic dip direction and the spread is θ, which is arbitrary and unknown . In theory R varies, in general, from $-\infty$ to $+\infty$ for each setup. However, if values of ΔE_x are taken from roughly meshing events, the following theoretical and approximate relations are obtained.

- If θ is small, say 10°, dispersion in R is small; J equals the average value of R.
- If θ is large, say 20–70°, dispersion in R is large; it varies from 1 to J; therefore J equals the maximum deviation of R from 1. This is the crucial aspect of the method.
- If $\theta = 90°$ (spread along the strike), there is no dispersion in R; R and J equal 1.

Figure 4.26 shows four actual cases and how values of J are picked from R-histograms. To reduce office work, the histograms are plotted on a logarithmic R-scale by applying the calibration factor to only selected J-values.

The measurement of J is repeatable within about 5% if cumulative errors are adjusted by tilting long profiles, because random errors accumulate as $n^{1/2}$, where n is the number of setups (stations).

The split-spread method has high resolution for local and small anomalies associated with such features as minor faults, as will be shown in section 4.4.8 (Figure 4.29); however, it suffers from cumulative errors over long-distances, hence it is not recommended for regional surveys.

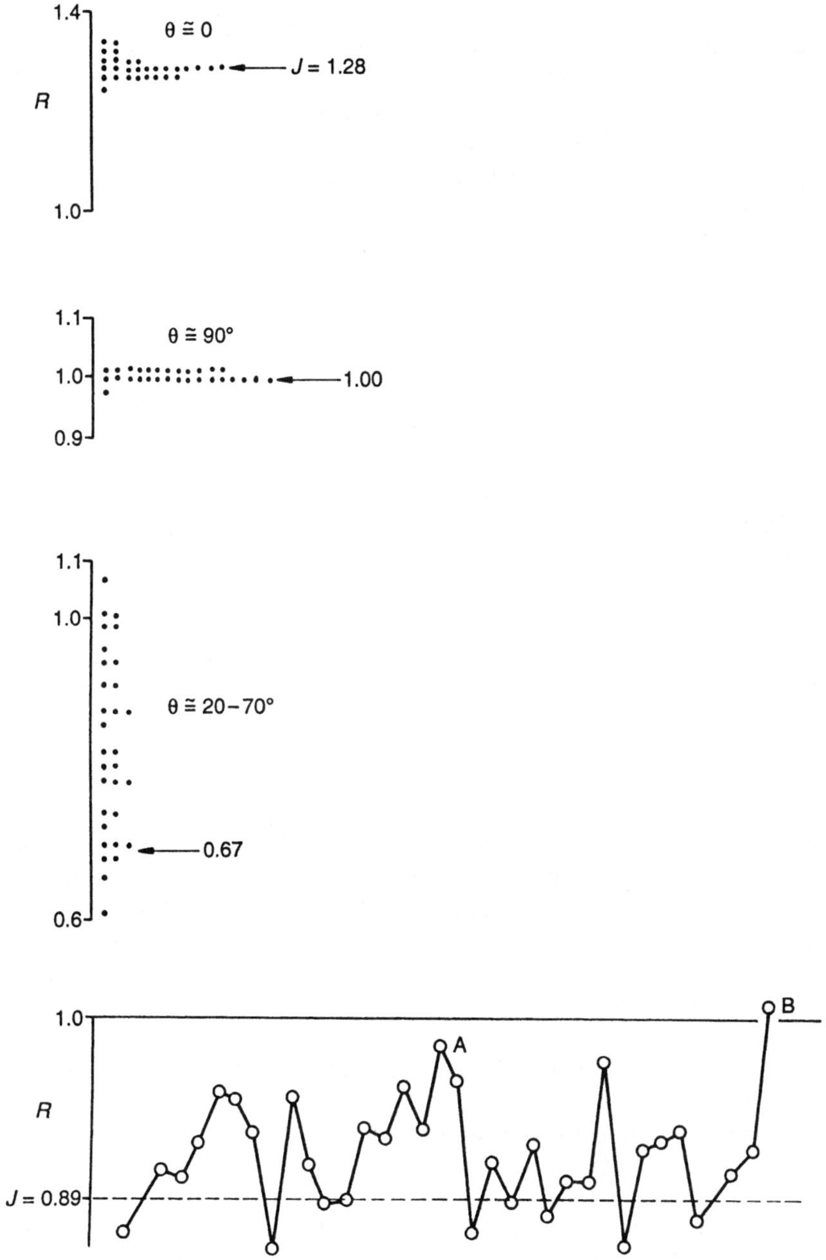

Figure 4.26 Actual examples of processed data with the split-spread telluric method. R is the amplitude ratio: θ is the horizontal acute angle between geologic dip bearing and the line of electrodes (from Yungul, 1977; with permission of Elsevier Science Publ.).

Data processing at the office is extremely simple, but R-values can be obtained directly in the field, by means of special instruments.

The leapfrogging collinear three-electrode procedure was used by Neuenschwander and Metcalf (1942) and Dahlberg (1945) in the frequency range from 0.5 Hz to several kilo hertz. Their analyses were confined to extracting relative noise levels of average amplitude ratios, which would yield repeatable and geologically analytically significant data only when θ is either 0° or 90°.

(a) Dip and strike determination

It is sometimes desirable to obtain dip and strike information at an isolated field station. Here dip and strike pertain to the gross electrical anisotropy (macro- and micro-anisotropy combined). This information can be obtained by making two consecutive split-spread measurements, obtaining J_1 first, and then J_2 from a spread formed by rotating the first spread 90° around the middle electrode. The data analysis is described below.

The boundary conditions show that any one telluric current sheet in the subsurface, say at a unit depth under the center electrode, can be replaced by the top of a fictitious insulating basement without perturbing $\boldsymbol{E}$ at the surface. If the top of this basement immediately under the cross-spread is assumed to be an arbitrarily inclined plane under a homogeneous and isotropic geologic section, then the values of J are approximately inversely proportional to the depth to the basement, as will be explained in section 4.4.8. Thus, it is a simple problem of geometry to relate independently normalized J_1 and J_2 on one hand and the azimuth of the dip of this basement on the other. The basement is determined by a triangle having one corner vertically below the center electrode and two corners below the reference points of J_1 and J_2, at depths 1, D_1 and D_2 respectively, where

$$D_1 = 2/(J_1+1) \quad \text{and} \quad D_2 = 2/(J_2+1). \tag{4.54}$$

Alternatively, if a third channel is available, one can record three spreads laid out in the form of a letter T simultaneously, and obtain dip and strike information readily.

For an arbitrary orientation of the T-spread, the split-spread data will come out as shown at the bottom of Figure 4.26. Here R-values A and B are close to unity, so for A and B the variation vector $\Delta\boldsymbol{E}$ was nearly parallel to the strike. Now the direction of this $\Delta\boldsymbol{E}$ can be determined by means of the third spread, and this will be the strike direction. The dip azimuth is determined by J, as explained above.

4.4.8 Interpretation

The most common problem is the interpretation of contour maps of J, called telluric anomaly maps.

(a) Qualitative interpretation

To facilitate understanding of the geologic significance of J, we shall confine discussion to two-dimensional structures. For a two-dimensional structure whose strike is in the y-direction, J at F is

$$J_F = E_{xF}/E_{xB}. \tag{4.55}$$

For a DC field E_y is the same everywhere, and an $\boldsymbol{E}$ in the x-direction does not generate any y-component. This implies that in equation (4.48) we have $b = c = 0$, $d = 1$, and $J = a$, hence equation (4.55).

Approximately two-dimensional anomalies can be obtained by constructing J-profiles from a telluric anomaly map along directions substantially perpendicular to contours.

We can mentally construct a two-dimensional analog model that replaces: (1) the electric current by a nonturbulent fluid flow, (2) the electrical conductivity ($1/\rho$) distribution by a fluid permeability distribution and (3) $\boldsymbol{E}$ by fluid velocity v. This can be done because both phenomena are governed by the same equations and boundary conditions. This analogy is helpful in visualizing qualitatively what kind of J-distribution to expect at the surface over certain geologic models. For most people water velocity is easier to visualize than $\boldsymbol{E}$.

Figures 4.27 and 4.28 schematically show the J-anomalies associated with typical structures in sedimentary basins. Several of these characteristic effects may be included in one model. Let us now examine a few case histories corresponding to some of these schematic models.

Figure 4.29 shows two telluric profiles over a fault. The subsurface information comes from detailed seismic and drilling data. The sedimentary section (under flat farmland) consists of gently dipping formations that include no major discrete resistivity layers. The effective resistivity is about 4 Ω m. The telluric data were obtained by means of split-spread measurements. Profiles 1 and 2 are 1 mile (1.6 km) apart. A band-pass filter peaked at $T = 20$ s was used. The repeat of profile 1 was obtained at a different time and by different instrumentation to the other two profiles, to check repeatability.

The anomaly associated with the fault would have probably been within experimental error for another telluric method. It corresponds to Figure 4.28b, 'faulted stack'. It is practically impossible using well logs to construct a model of this subsurface that could quantitatively account for this anomaly. This is because all the microscopic and macroscopic

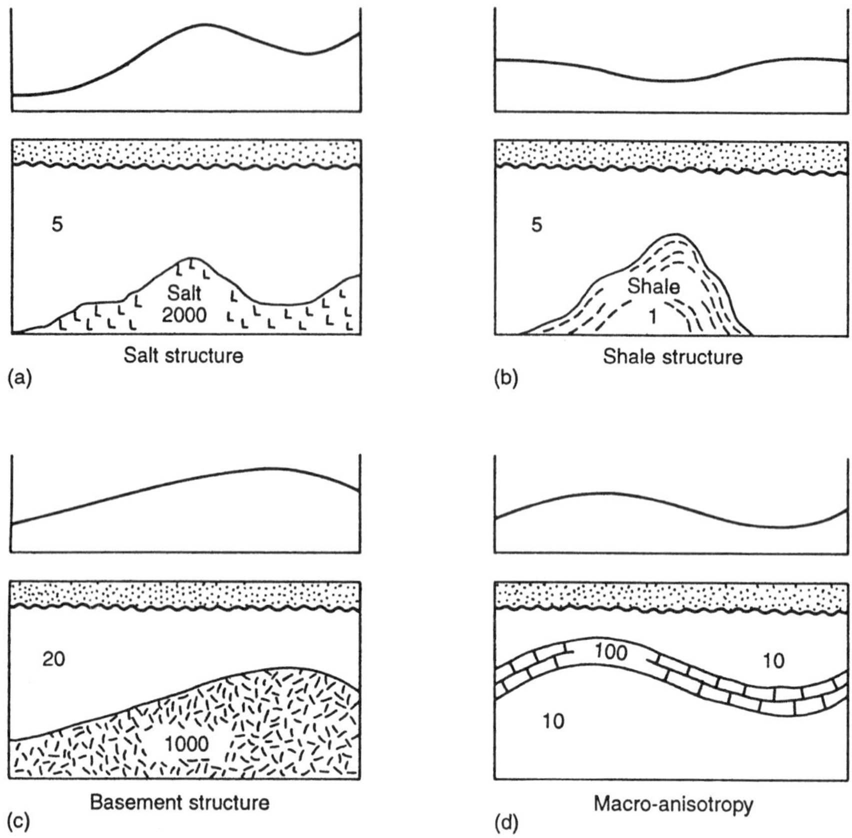

Figure 4.27 Schematic telluric anomalies due to 2-D hypothetical structures. Numbers represent resistivities in Ω m (from Yungul, 1977; with permission of Elsevier Science Publ.).

properties of the sedimentary section enter into the picture. One has to consider this anomaly as a 'signature' and be satisfied with it.

In Figure 4.29 the regional rise in J to the right is primarily due to rising basement. This corresponds to Figure 4.27c, 'basement structure'. If the sedimentary section was homogeneous and isotropic, one would obtain the dotted curve; this will be explained in section 4.4.8(b). The steeper rise in J is due to increasing resistivity to the right, toward the basin margin.

The telluric anomaly shown in Figure 4.30 was obtained over the Haynesville piercement salt dome in Texas (Boissonnas and Leonardon, 1948). The depth to the top of the salt is about 1200 ft. The salt outline near the top is shown by the dashed contour. This anomaly corresponds to Figure 4.27a, 'salt structure' and will be analyzed in section 4.4.8(c).

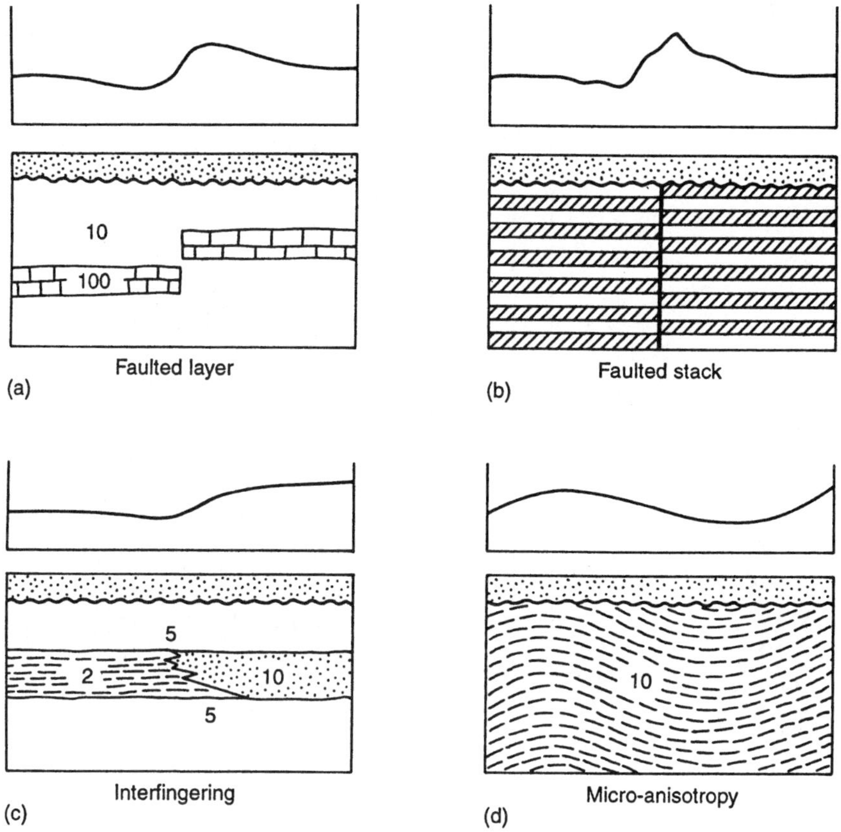

Figure 4.28 Schematic telluric anomalies due to 2-D hypothetical structures. Numbers represent resistivities in Ω m (from Yungul, 1977; with permission of Elsevier Science Publ.).

Figure 4.31 is a telluric anomaly map of an area in the Midland Basin, Texas, comprising three major, isolated, known and deeply buried reefs (Yungul, Hembree and Greenhouse, 1973). These are the Millican (M), Jameson (J) and IAB Reefs. The data were obtained mostly by the vectogram and partly by the split-spread methods. Only the vectogram stations, 287 of them, are indicated on the map by heavy dots.

Figure 4.32 is an east–west cross-section and *J*-profile across the Millican Reef, showing the results of quantitative interpretation (explained in section 4.4.8(c)). The numbers in the resistivity zones are the resistivity ratios. The known reefs are geographically associated with large telluric highs. However, the cause of these anomalies is lateral variation in resistivity within the post-reef sediments, which seem to be of

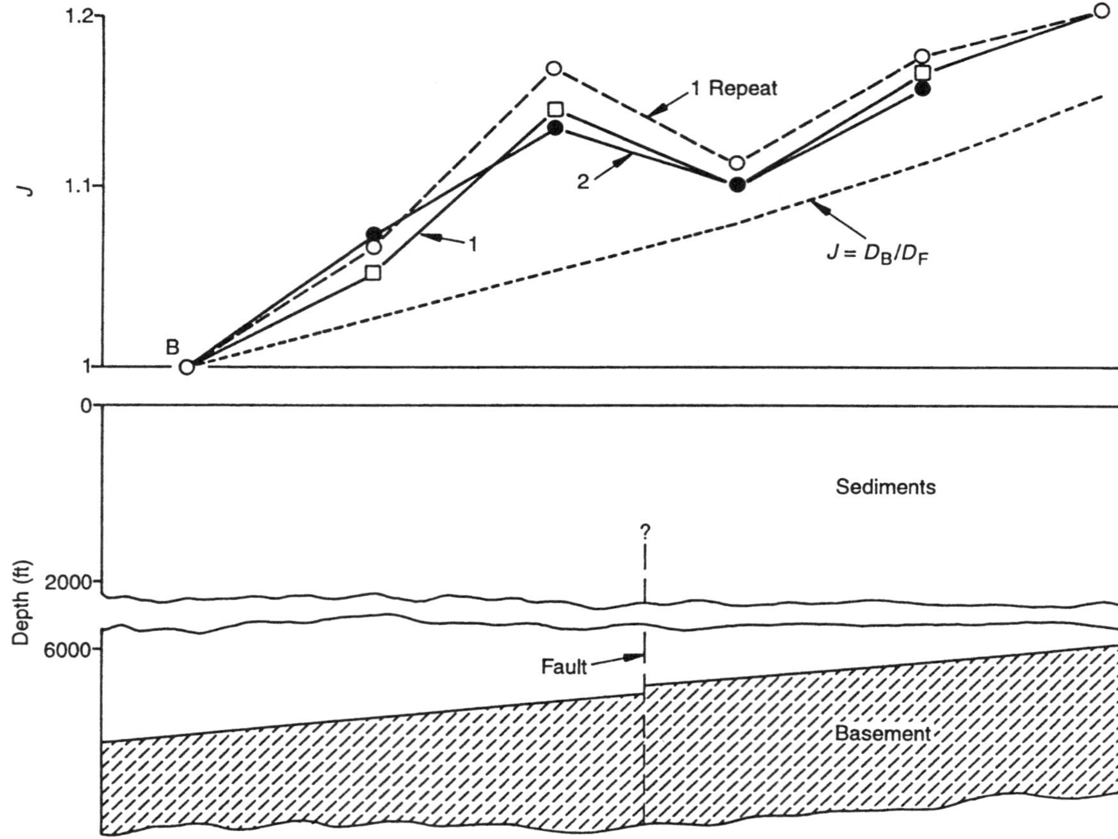

Figure 4.29 A telluric anomaly due to a fault, from split-spread profiling. Subsurface information came from seismic and well-log data. Profiles 1 and 2 are 1 mile (1609 m) apart. The dotted curve would have been obtained if the sediments were homogeneous and isotropic (from Yungul, 1977; with permission of Elsevier Science Publ.).

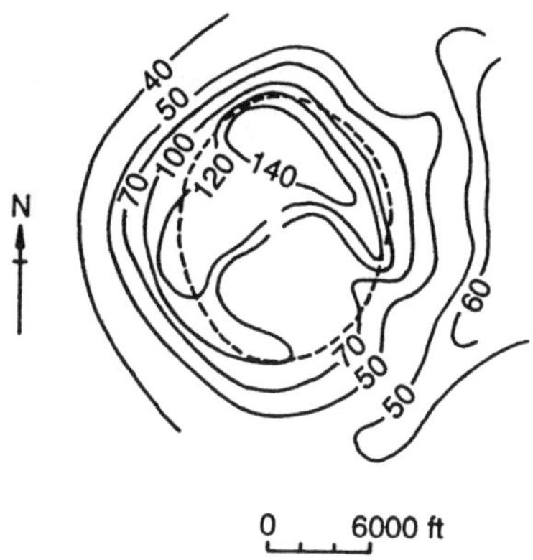

Figure 4.30 Telluric anomaly map (*J*-contours), Haynesville piercement salt dome, Texas (modified from Boissonnas and Leonardon, 1948).

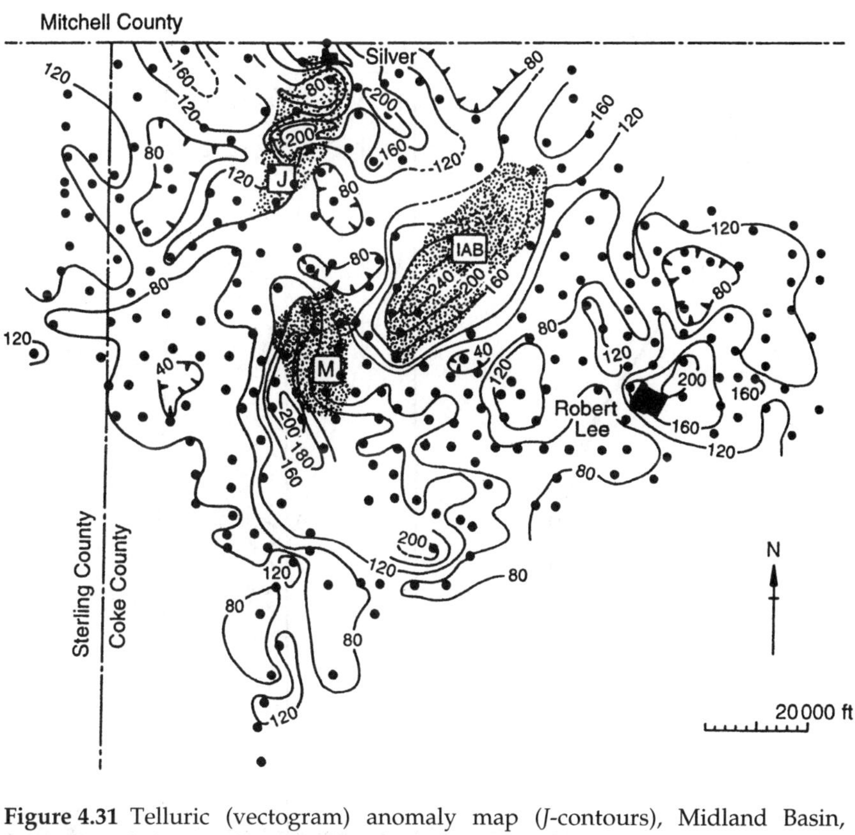

Figure 4.31 Telluric (vectogram) anomaly map (*J*-contours), Midland Basin, Texas. Stippled areas show the known producing reef buildups: Jameson (J), IAB, and Millican (M) (from Yungul, Hembree and Greenhouse, 1973).

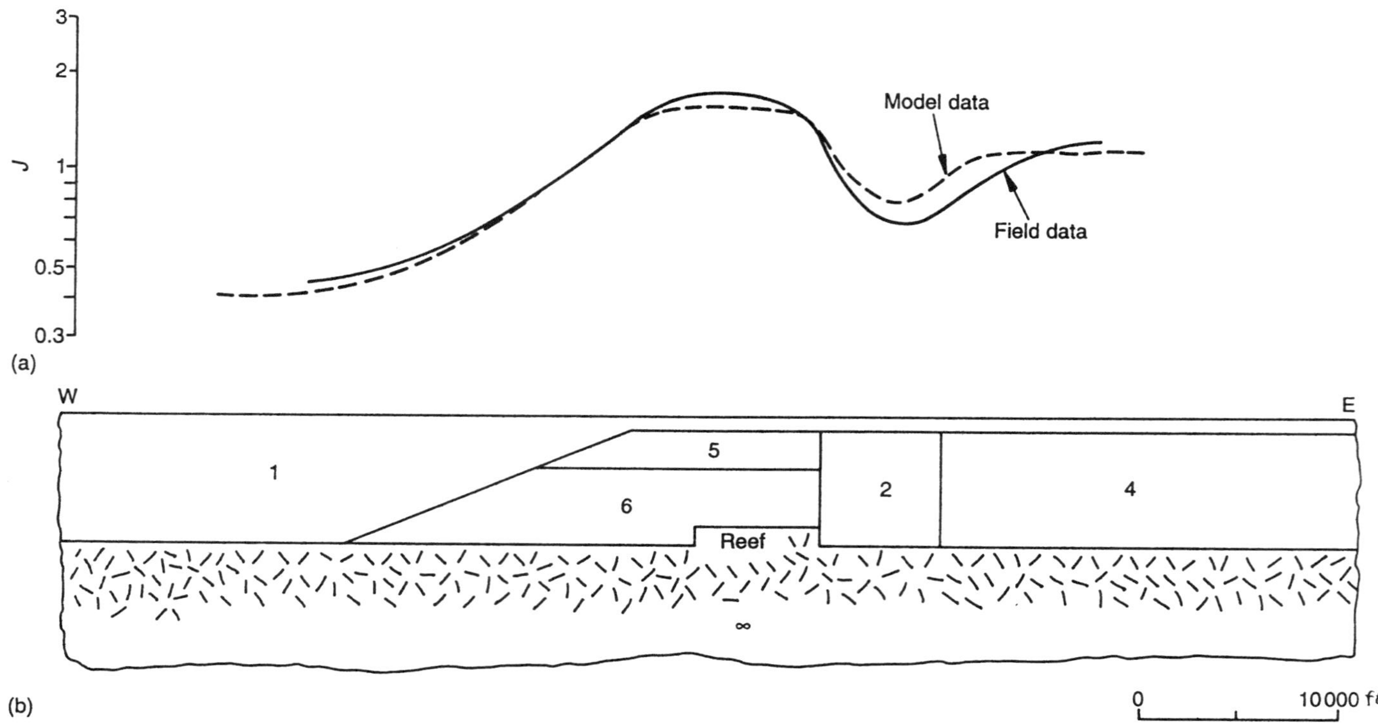

Figure 4.32 Two-dimensional model of Millican Reef, Midland Basin, Texas. East–west cross-section and the theoretical (model) and actual (field) data. Numbers represent resistivity ratios; 1 corresponds to 7 Ω m (from Yungul, Hembree and Greenhouse, 1973).

the ordinary 'layer-cake' types on the basis of extensive drilling data. This anomaly is of the interfingered type of Figure 4.28c. The effect of the reef mass on this very large anomaly is negligible.

Finally, Figure 4.33 shows a classical example of a detail telluric survey in the Aquitaine Basin, southern France (Migaux, 1946). This is a detailed portion of a reconnaissance survey of a large area. Telluric stations are shown as heavy dots. The relative ellipse method was used. An anticline lies unconformably under a horizontal section several hundreds of meters thick.

The subcrop of a limestone layer in the anticline creates 'cuesta'-type minor low and high telluric anomaly belts around the apex of the domal portion of the anticline, as shown by dotted and dashed lines, respectively. The anticline is represented by a telluric high that corresponds to a combination of macro-anisotropy (Figure 4.27d) and micro-anisotropy (Figure 4.28d).

(b) *Approximate interpretation*

Smooth structures with low relief and no layers of very high resistivity within the sediments produce simple relationships between J and either the total horizontal conductance, S, of the sediments above the basement, or depth, D, to the basement.

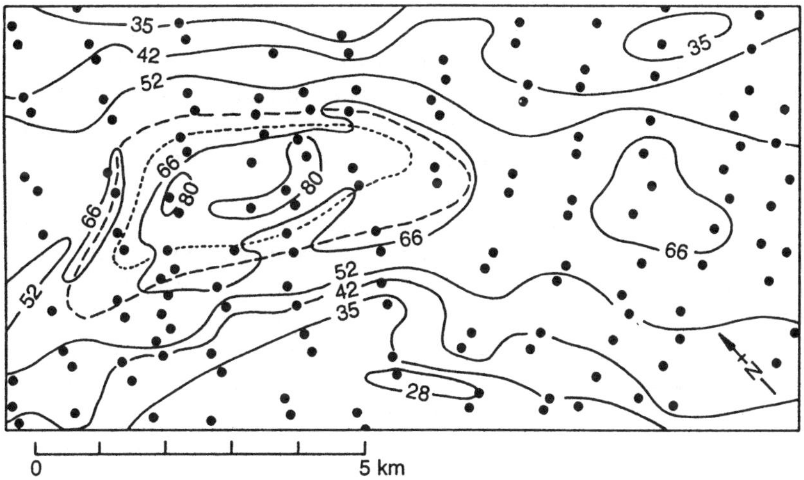

Figure 4.33 Telluric anomaly map (J-contours) of an area in the Aquitaine Basin, France, over the domal part of an anticline. The dotted and dashed curves are the minor low and high axes respectively, associated with a limestone subcrop (modified from Migaux, 1946).

For an arbitrary two-dimensional subsurface whose strike is in the y-direction, we have seen that

$$J_F = E_{xF}/E_{xB}, \tag{4.55}$$

and that H_y is the same everywhere at the surface. It follows that

$$J_F = Z_{xyF}/Z_{xyB} = (\rho_{\perp F}/\rho_{\perp B})^{1/2}, \tag{4.56}$$

where Z_{xy} is the MT impedance obtained from E_x and H_y, and $\rho_\perp$ is the MT transverse apparent resistivity.

If it is assumed that B and F are away from abrupt lateral changes in the geology and overlie nearly horizontal layers, although in different sequences, we have, according to equation (4.46), $Z_B = 1/S_B$, $Z_F = 1/S_F$, in the MKS system. Therefore, equation (4.56) takes the following form

$$J_F = S_B/S_F, \tag{4.57}$$

where, as defined by equation (4.44), $S = D/\rho_e$. Therefore

$$J_F = \frac{D_B}{D_F}\frac{\rho_{eF}}{\rho_{eB}}. \tag{4.58}$$

If the structures are smooth and of low relief, even though three-dimensional, equation (4.58) holds so long as stations are away from abrupt lateral changes. Another limitation is that there should be no high-resistivity screening inclusions in the sedimentary section, such as evaporite and carbonate layers of appreciable thickness. Such a layer accentuates the telluric effect of the section above it at the expense of that below (Figure 4.27d).

Equation (4.57) was derived by Migaux (1951), but the above derivation is similar to that of Berdichevskii (1960, pp. 38–42). Returning to equation (4.57), if S_B is known, then values of J can be converted into an S-map. Evidently this does not resolve D and ρ_e separately. However, if values of ρ_e are obtained at some locations by means of resistivity or MT soundings or well logs, J-values can be converted directly to D-values. This requires empirically modified forms of equation (4.57), from a statistical relationship between J-values and sounding data. Berdichevskii (1960, pp. 197–204) demonstrates the usefulness of such relationships in several case histories involving basin evaluation.

The approximate interpretation technique described above is too restrictive and yields crude quantitative information at best, unless it is combined with an adequate amount of sounding data. It cannot be used at all for interpretation of abrupt lateral discontinuities such as those in Figure 4.32.

(c) Quantitative interpretation

This is achieved by forward modeling, either in the form of DC potential distribution, or 2-D magnetotelluric response for the middle frequency of the pass band (section 4.3.5). MT modeling is not of much help, as yet, in 3-D interpretation. However, since telluric anomalies represent the DC case, simpler solutions can be found for them by either experimental modeling (analog or scale modeling) or calculations.

With the DC assumption, telluric current flow is governed by Ohm's law (section 3.1):

$$\mathbf{J} = \mathbf{E}/\rho, \tag{4.59}$$

where $\mathbf{J}$ is the current density. This leads to Laplace's equation:

$$\nabla^2 V = 0, \tag{4.60}$$

where V is the electric potential of which $\mathbf{E}$ is the gradient. The subsurface model is confined between two half-spaces of infinite resistivity, namely the air and basement. The source is a primary uniform horizontal $\mathbf{E}$. A solution of Laplace's equation that satisfies the boundary conditions of the model yields $\mathbf{E}(x, y)$, from which J-maps can be constructed. However, a straightforward analytical procedure is not practical, and recourse is made to mathematical 'tricks' even for extremely simple models. Two early and significant papers of this nature are due to Baranov (1951) and Kunetz and Chastenet de Géry (1956).

These papers and many others now have only historical value, in view of the developments in MT numerical calculations.

Experimental modeling

A practical technique for three-dimensional interpretation seems to be physical scale modeling in electrolytic tanks in which the surface of the electrolyte represents the surface of the Earth. Utzmann (1954) describes in detail the instrumentation and the techniques of constructing the models and making the measurements for the case of instantaneous DC, although an alternating current source is used.

MT scale modeling facilities can also be used to determine telluric anomalies, but such systems have additional constrictions which are not necessary for telluric measurements. An extensive discussion of electromagnetic physical scale modeling is given by Frischknecht (1988).

In certain cases master curves can be obtained by scale modeling for use in interpreting anomalies due to a certain type of model, but for a wide range of geometric parameters. Such a case is now discussed. A piercement salt dome was modeled by an upright circular paraffin cylinder, placed on the tank bottom, which represented the basement. The electrolyte represented the sediments. The ratio of salt to sediment resistivity is practically infinite in this model. In reality (such as on the Gulf Coast) this is of the order of 1000, but whether it is 1000 or ∞ makes no appreciable difference to the results.

Even this simple model does not lend itself to analytic calculations. Three models were constructed for different ratios of salt height, h, to diameter, *d*, and telluric anomalies for each one were measured for several depths, *z*, to the top. The results were compiled in the form of master curves, as shown in Figure 4.34. The solid curves represent maximum values of *J*, which occur over the cylinder axis, as a function of depth-to-diameter ratio, z/d, for three models characterized by different height-to-diameter ratios, h/d. Similarly, the dashed curves are values of *J* above the salt edge, that is, one radius away from the axis.

Interestingly, the curves are practically the same for models for which the height-to-diameter ratio, h/d, is equal to or larger than unity: a condition that applies to the piercement domes in the Gulf Coast area in general. Therefore, one solid curve and one dashed curve, say for $h/d=2$, will suffice to interpret domes of all geometric parameters so long as $(h/d)\geqslant 1$.

Let us apply this to the interpretation of the anomaly shown in Figure 4.30. If the *J*-contours are smoothed and normalized to the nonanom-

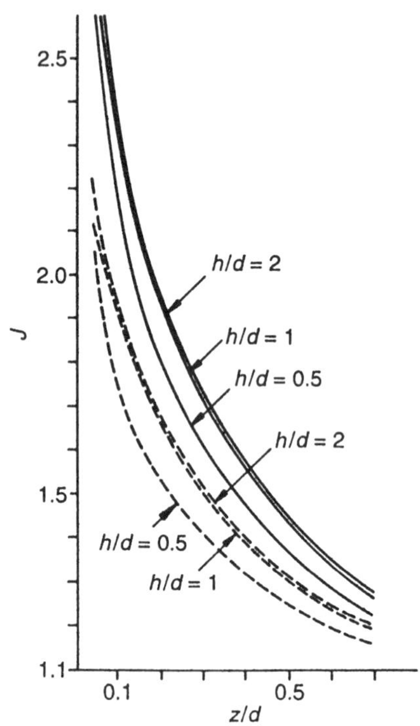

Figure 4.34 Master curves for telluric anomalies due to an insulating upright circular cylinder, compiled from electrolytic tank scale modeling. The cylinder, resting on the tank bottom, represents a piercement salt dome. *d* is a diameter, *h* is height and *z* is depth to the top. Solid curves are maximum anomalies over the axis, and dashed curves are the anomalies at one radius from the axis (from Yungul, 1977; with permission of Elsevier Science Publ.).

alous region and an average radial representative of the anomaly is obtained (not shown), the maximum value of J at the center is 2.4. This yields $(z/d)=0.095$ from the solid curve for $(h/d)=2$, and $J=1.96$ over the edge from the corresponding dashed curve. On the telluric anomaly, $J=1.96$ corresponds to a diameter of 14 500 ft. Therefore, the depth is 1380 ft. According to the map given by Boissonnas and Leonardon (1948), the drill data indicate an average diameter of about 13 800 ft near the top, and an average depth of 1200 ft. The interpreted depth and diameter figures are satisfactorily close to the actual data.

Apparently, the anomaly of Figure 4.30 was obtained by using signals whose periods were shorter than the S zone would require at Haynesville, Texas, if the total sedimentary section is considered. On the other hand, the curves of Figure 4.34 are for the DC case (the S zone). Nevertheless, the interpretation results are good. This is because the anomaly is not sensitive to the entire height of the salt stock, as evinced from the h/d independence of the scale model data. All that matters in this case is that enough of the salt stock lies within the depth of penetration, say less than a skin depth. Evidently, this example shows that quantitatively significant telluric data can be obtained even for periods shorter than those in the S zone.

Numerical modeling

Numerical 2-D telluric modeling consists of computing values of $E_\perp$, normalized to a nonanomalous location, for the frequency with which telluric data had been measured (middle frequency of the filter pass band). This constitutes one of the outputs of MT modeling computer programs. Normalized values of $E_\perp$ are equal to J-values if the frequency is within the S zone, as shown by equation (4.55). Obviously, this requires that the model be in terms of resistivity values and not resistivity ratios.

We have seen that the S zone comprises a broad band of frequencies. According to the law of electromagnetic similitude (section 4.3.3(d)), resistivity and frequency can be changed so long as their ratio is kept constant. Consequently, telluric modeling results will remain practically the same for a wide range of resistivities and frequencies so long as the resistivity ratios are correct.

Numerical modeling methods have already been discussed in section 4.3.5, for the MT methods.

4.4.9 Instruments and equipment

A horizontal component of $\mathbf{E}$ is measured by stretching an insulated wire on the ground, planting two electrodes at its ends and measuring the potential difference between them. Two orthogonal components of $\mathbf{E}$ can be measured simultaneously by using three electrodes and an L-shaped wire.

The length of the wire depends largely on the nature of near-surface irregularities, which must be averaged out by using a sufficiently long wire; otherwise 40-m lengths would be suitable so far as instrument noise is concerned. It is possible to operate with 150-m lengths if the measuring sites are judiciously selected.

Electrodes used include nonpolarizable (reversible) types of copper, cadmium, silver or zinc. Bare copper pipes or lead plates are also used. I prefer weathered, paper-thin lead strips having no fresh surfaces, attached to the wire with an alligator clip. This seems to be the best compromise between precision and economy.

The instrumental system for each channel consists of a variable DC bias, a low-pass filter, a DC amplifier with an input impedance of at least 10 M Ω, a band-pass filter and a strip-chart or X–Y recorder, depending on the technique used.

The amplifiers and recorders are off-the-shelf items; but the filters may be either home-made, to suit the geologic problem on hand, or commercial variable pass-band filters. Overall system noise of less than about 3 μV rms within the frequency band of interest is satisfactory, and not difficult to achieve. A two-channel system can be accommodated in a suitcase, excluding the power source.

5

Controlled-source electromagnetic (CSEM) methods

5.1 GENERALITIES AND COMPARISONS

5.1.1 Introduction to CSEM methods

The controlled-source electromagnetic (CSEM) methods are associated with several acronyms for long names of the techniques, and there is advantage in establishing a short glossary at the beginning.

EM	Generally electromagnetic, specifically all methods based on electromagnetic induction from a controlled source.
CS	Controlled source, e.g. an insulated wire loop or a grounded wire.
LO	Long offset, large separation between source and receiver.
FEM	Frequency-domain EM.
TEM	Time-domain (or transient) EM.
LOTEM	Long-offset time-domain EM.
LOFEM	Long-offset frequency-domain EM.
MT	Magnetotelluric methods (natural-source)
CSMT	Controlled-source MT method.
DC	Direct-current resistivity methods.

More than 60 years ago many geophysicists believed that the future of deep electrical exploration belonged to the method of frequency sounding, what we now call controlled-source frequency-domain electromagnetic sounding, simply denoted by FEM. The magnetotelluric (MT) method, which is no other than a natural-source FEM, had not been invented yet, but its appearance in the 1950s is a credit to the earlier proponents of FEM.

In the 1930s, direct current (DC) resistivity methods employing collinear quadrupoles (e.g. Wenner) were practically the only methods of electrical sounding; they required laborious expansion of a source–receiver electrode array and it was necessary to take measurements at many spacings. It was thought that one fixed source–receiver setup could yield a sounding by making measurements at different frequencies, the lower the frequency the deeper the penetration. This concept alone was attractive enough for the development of FEM. Better yet, one could inject a step function, or a pulse, into the earth, that would contain all the frequencies of interest, and record the decay in time at the receiver. This would yield the sounding data in a short time. This technique is now called time-domain EM (TEM).

Until the 1970s the performance of the deep EM sounding methods was poor; they were very little used, primarily because interpretation was difficult and the instrumentation was complicated. One of the main difficulties is that these methods involve two independent 'yardsticks' in measuring the parameters of the Earth: (1) the geometric dimensions of the source–receiver array and (2) the wavelengths in the source field. Consequently, inversion of the measurements in terms of a 1-D multilayer earth was expensive and 2-D modeling still is expensive, compared with DC which has only the geometric factor, and with MT which has only wavelength dependence. The effects of the geometric dimensions of the array are reduced if source–receiver separation is large, e.g. tens of kilometers. Thus, the source proximity effects are reduced, instrumentation and interpretation are somewhat simplified, and many soundings can be obtained from one source setup, thereby reducing cost per sounding. However, long separations create signal-to-noise ratio problems and necessitate large source equipment, of the order of megawatts, even magnetohydrodynamic generators. Furthermore, when the separation is tens of kilometers, the source and the receiver may lie on entirely different geologic structures, thereby creating a different kind of interpretation problem, namely in the assumption that the measurements represent a 1-D multilayer earth.

In the 1970s major improvements took place in the instrumentation and in the 1-D multilayer interpretation methods. Consequently, the demand for these methods rose significantly in the late 1970s and in the 1980s. This trend gave impetus to the publication of four recent books on EM methods by Patra and Mallick (1980), Kaufman and Keller (1983), Nabighian (1988, 1991) and Strack (1992).

This rapid development influenced some geophysicists to change their minds rather drastically as to the usefulness of EM sounding, TEM in particular, from 'no good' to 'the best' to 'the only one' among the electrical sounding methods.

In 1975 the controlled-source magnetotelluric (CSMT) method made its debut and became popular in the 1980s. This is a hybrid between LOFEM and MT. The source–receiver separation is so large that the measurements at the receiver are treated as though they were due to plane-wave excitation, as in the case of MT.

This rapidly increased interest during the 1980s was primarily in mining, groundwater and geothermal exploration and, in general, in problems involving depths mostly less than about 1 km. Such surveys are being done on a routine basis by contractors, and many case histories have been published. It seems that, for such problems, EM is being used more than DC at present. However, not many results have yet been published on routine application to deep sounding. Thus, it is still too early to evaluate the deep EM sounding methods applicable to sedimentary basins and to compare them with DC and MT.

5.1.2 Planning a survey with EM methods

There is a large variety of EM techniques and equipment, but only those that are popular in deep sounding will be mentioned here. About ten different source–receiver geometries are being used on various problems, deep or shallow (e.g. Spies and Frischknecht, 1991, p. 291), but only two types of source, namely the horizontal loop and grounded wire, will be mentioned.

The planner of an EM survey has to make several choices: FEM or TEM, short-offset or long-offset separation, horizontal loop or grounded wire as the source, strength of the source, sensitivity of the receivers and the quantities to be measured. These decisions are difficult to make and should be based on earth model calculations. The discussions in sections 5.1.3 and 5.1.4 may be helpful in making these choices.

(a) Sources

The choice to be made is between a loop of insulated wire laid on the ground (inductive source) and an insulated wire laid straight on the surface and grounded at both ends (conductive, or galvanic source). The loop may be a large square, hundreds of meters on one side, consisting of a single turn, or of a smaller size made of multi-conductor cable. If the source–receiver separation is more than about five times the loop dimension, the loop may be treated mathematically as a vertical magnetic dipole whose moment (strength) is equal to the product of amperes × number of turns × loop area. This simplifies the mathematics. An obvious operational reason for using a loop instead of a grounded wire is to provide adequate source strength when electrode resistances are high, but there are theoretical and other operational reasons in making this choice, as will be discussed later.

In FEM systems the source waveform is usually sinusoidal or square. With a square waveform, simultaneous measurements can be made of the fundamental frequency and of a few harmonics. The frequency is changed in discrete steps to perform a sounding.

In TEM systems the popular waveform is a train of square pulses, negative and positive, with an off-time between pulses.

Grounded-wire sources also can be treated mathematically as electric dipoles if they are shorter than about 1/5 of the source–receiver separation.

(b) Receivers

The measurements at the receiver station may include, in the most general case, three orthogonal components of the magnetic field ***H***, two horizontal orthogonal components of the electric field ***E***, and phase or time reference to the transmitter. The ***H*** and ***E*** measurements are basically the same as in MT (section 4.3.9). The phase or time reference to the transmitter may be accomplished by fiber optic cable or by radio telemetry. Most FEM systems measure in-phase and quadrature parts of the received signal that yield amplitude and phase information.

Selection regarding the strength of the source has to be made in accordance with the sensitivity of the receiver, with the maximum source–receiver separation and with anticipated subsurface resistivities.

5.1.3 Comparisons among EM techniques

These comparisons of various methods are given below in the form of a quick reference without explanations. They are based on discussions to be found in this book, in other books (e.g. Nabighian, 1988, 1991; Kaufman and Keller, 1983) and in many articles written by the users of electrical methods who are not unanimous about all of these comparisons.

(a) FEM versus TEM

It seems that deep TEM soundings provide better layer resolution than FEM. Furthermore, the problem of the primary field can be avoided with TEM, if the measurements are made during the time when the power is off. This allows the TEM equipment to be simpler than that of FEM, and yet record the secondary signals accurately even though they are far smaller than the primary field. However, narrow-band filtering that is used with FEM to improve the signal-to-noise ratio cannot be used with TEM; therefore, in areas with much industrial EM noise FEM may necessarily be the choice. With TEM, in order to overpower the noise the source equipment has to be much larger than that of FEM. Narrow-band filtering is time consuming in the field and FEM measurements require longer crew time than TEM measurements. For the reasons given below, FEM is preferable for shallow soundings; it may also be preferable in high EM noise environments; it needs lighter but more complicated equipment. TEM is becoming more popular than FEM, but the level of cultural noise is increasing rapidly and the use of FEM may increase.

(b) Long offset versus short offset

With FEM, the secondary field must be separated from the primary field. With short-offset FEM, if the source–receiver separation is smaller than the depth to a target layer, the secondary field associated with this layer will be very small compared with the primary field. Thus, the separation must be at least as large as the maximum depth to be explored; otherwise the primary field must be compensated. If the separation is considerably larger than the depth (LOFEM), the parameters of thin layers at depth cannot be determined accurately.

With short-offset TEM, there is no problem of the primary field, but deep information is only in the tail portion of the decay curve after the receiver voltage is drastically diminished. This requires equipment with a very large dynamic range. Consequently, shallow work that can be accomplished with short offset and modest source equipment can be done more efficiently with FEM. The popular usage of TEM is in the long-offset mode, LOTEM, for deep work.

In general, with long-offset measurements, with separations of tens of kilometers, many soundings can be made from one source; this is considerably less expensive than short-offset soundings which necessitate relocation of the source for each sounding. This, however, has two disadvantages: (1) large separations require powerful source equipment, and (2) when the source and the receiver are located over dissimilar geologic structures the conventional concept of 'sounding', that is, the assumption of 1-D multilayer earth, becomes less viable than in the case of short offset.

In summary, long-offset (LOFEM, LOTEM and CSMT) measurements are simpler to make and easier to process (to interpret) than short-offset measurements, but they require heavier equipment and have less horizontal resolution and more geological uncertainty.

(c) Horizontal-loop versus grounded-wire source

A loop may be necessary if the contact resistance of a grounded wire is too high. For deep soundings, fieldwork with a large loop is usually more expensive than that with a grounded wire. On the other hand the setting up of a medium-size loop may be less laborious than that of a grounded wire and loop sources are preferred for shallow work. Generally, a grounded wire is preferred for long offsets and for deep work, because the primary field falls of less rapidly at large separations from a grounded wire than from a loop.

The theory and interpretation of measurements are simpler with a loop source than a grounded-wire source, mainly because a horizontal-loop dipole (vertical magnetic dipole) is horizontally omnidirectional; for a 1-D earth the field at the receiver consists of H_R, H_z and E_ϕ, the radial, vertical and azimuthal components, respectively, whereas the field of a grounded wire has all six components.

The response of an earth model will be different with a loop source than with a grounded-wire source, especially at short offsets and at low frequencies. This is another factor affecting the choice between the two types of sources for a particular earth model.

5.1.4 Comparisons of EM techniques with DC and MT

Comparisons of EM with other methods should be made in combinations of specific earth models and techniques, such as FEM versus DC in shallow aquifer mapping. Generalizations are risky; they tend to pit one method against the other, whereas, in general, they should complement each other. Nevertheless, the following discussion consists largely of generalizations.

(a) EM versus DC

Most of the following statements are well-established facts, but some are debatable.

- EM can be carried out without ground contact in areas where hard rock or ice or dry sand at the surface renders DC fieldwork uneconomical.
- EM involves less labor in the field because each receiver location yields one sounding. This may reduce the cost in certain cases but not always; the capital cost of complicated or heavy equipment may reverse the relative cost advantage.
- With EM, source–receiver separation can be shorter than that in DC; it can even be smaller than the depth of exploration. This provides better spatial resolution than DC. However, the DC dipole techniques (section 3.2) are not as bad, in this respect, as the Schlumberger array which most investigators compare with EM.
- In general EM is less sensitive than DC to lateral near-surface inhomogeneities, but this depends on the size of inhomogeneities and on the specific techniques used. The reduction in sensitivity to inhomogeneities is due primarily to two reasons:

  1. EM sounding measurements are made at a stationary point and the near-surface inhomogeneities create smooth variations, whereas

they create random variations with DC as the measuring electrodes move;

2. EM sounding can be made with the ***H*** field only, and this is not as sensitive to near-surface inhomogeneities as the ***E*** field that DC must use.

- With EM, high-resistivity layers (e.g. massive limestone, volcanics, or evaporites) do not screen deeper layers. With DC, such layers necessitate a larger than usual expansion of array dimensions, thereby reducing horizontal resolution.
- With TEM, non-uniqueness is somewhat less than that with DC.
- On a 1-D multilayer earth, EM measures the horizontal resistivity whereas DC measures a combination of horizontal and vertical resistivities. With a highly anisotropic earth, EM sounding results could yield more accurate depth estimates than DC.
- Parameters ρ and h of low-resistivity layers can be determined more effectively with EM than with DC. This is important in aquifer mapping. However, resistivity of a high-resistivity layer can be determined better with DC.
- DC sounding usually yields more detail than EM. In general, DC responds more to resistivity variations.
- Interpretation usually is easier and more reliable with DC than with EM.
- Instrumentation and equipment are simpler and less expensive with DC than with EM, for the same depth of exploration.
- The practical depth of exploration in routine work is larger with DC than with EM, except LOTEM.

(b) EM versus MT

The following statements concern the conventional type of MT.

- Interpretation of MT sounding data is much easier than that of EM data.
- EM is less sensitive to distant inhomogeneities than MT.
- EM can be done with the ***H*** field only, without the ***E*** field, and is thereby less affected by static shift, which constitutes a serious problem in MT work.
- Shallow EM sounding may cost less than shallow MT, but deep MT sounding costs less than deep EM sounding.
- Layer resolution depends on the problem and on the EM technique used; it is difficult to say whether MT or EM is better in general. MT resolution is generally better than short-offset EM resolution.
- The depth of exploration is virtually unlimited with MT, whereas it is only a few kilometers for EM with practical expenditures.

5.2 CONTROLLED-SOURCE MAGNETOTELLURIC (CSMT) METHODS

5.2.1 An overview

In magnetotelluric (MT) methods, for the exploration of sedimentary basins, one is allowed to assume that the source field is a vertically incident plane wave at all frequencies of interest (section 4.2.3). This is a very substantial advantage over all the controlled-source methods, because it facilitates the interpretation by not involving any source location and geometry complications. On the other hand, a major disadvantage of MT is that the source field is not always readily measurable at all frequencies of interest, especially in the 0.2–5 Hz band. The signals at these frequencies, (the so-called 'dead band') are very useful in most MT work. However, their amplitudes are frequently too low to be measured readily. This gives rise to long waits in the field and expensive crew time. Furthermore, the source field direction varies continuously with time. This necessitates tensor measurements of five components and computational rotation of the measuring axes. (The omnidirectional nature of the source is actually an advantage, but it becomes a disadvantage in low-budget projects.) Still another disadvantage of MT is that the industrial currents (cultural noise) do not meet the condition of plane waves. However, if the noise source is sufficiently far away then one is justified in using the noise as though it was an MT-grade signal (section 4.2.3). Now, if such industrial noise is included in the measurements, one is, in essence, making controlled-source MT (CSMT) measurements.

Goldstein and Strangway (1975) thought that an artificial source could be devised in such a way that the plane-wave advantage of MT could be approximately preserved, while the direction, frequency and strength of the source field could still be chosen. This would combine the advantages of MT and FEM methods and remove the disadvantages of MT. Such a technique would have the potential of providing rapid and low-cost data gathering and processing in certain type of work, such as high-density grid measurements used in ore prospecting.

Figure 5.1 shows a field setup. The source is usually an insulated wire grounded at both ends by electrodes A and B, as in the case of DC methods in which it is called a source dipole (or bipole). This is a galvanic source. One can also employ an inductive source in the form of a horizontal loop of insulated wire if the resistance at electrodes A and B is high, but this mode of operation is expensive. The source transmitter system feeds a signal of selected frequencies into the earth. At locations that are far enough from the source, that is, in the **far field** where the plane-wave assumption is justified, regular MT-type measurements are made, the simplest being one horizontal component of both the electric field, E_x,

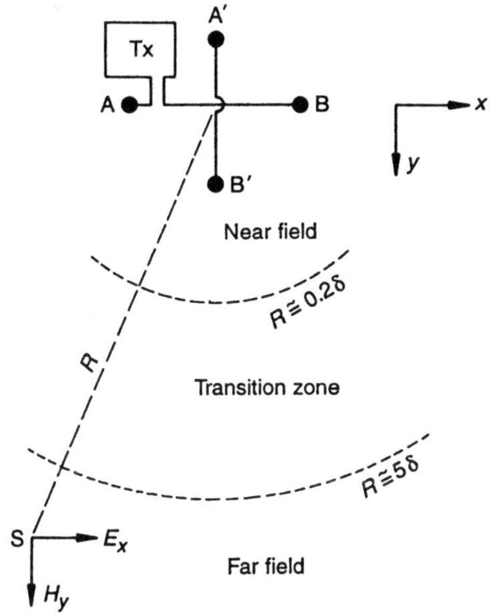

Figure 5.1 Plan view of CSAMT field setup. AB is source dipole grounded at A and B; A′B′ is an additional source dipole used for optional tensor measurements; Tx is transmitter unit with motor generator; S is station where orthogonal components of ***E*** and ***H*** are measured; *R* is separation between Tx and S, preferably larger than five skin depths. Scalar surveys require measurements of E_x and H_y from one source. Tensor surveys require the measurement of 10 components, five from AB and five from A′B′, orthogonal horizontal components of ***E*** and three orthogonal components of ***H***.

and the magnetic field, H_y, for the selected frequencies. These measurements yield MT apparent resistivities which are interpreted as discussed in Chapter 4.

For deep work, the far field where the plane-wave condition is valid may be too far to obtain measurable signals; and the locations where the signals can be measured accurately may be confined to the **near field** and to the **transition zone** where the plane-wave assumption is not valid. Some of the literature constitutes attempts to make corrections to the transition zone measurements to make them acceptable as far-field data (Yamashita, Hallof and Pelton, 1985; Zonge and Hughes, 1991). However, interpretation of transition zone data is difficult; one no longer has the plane-wave advantage; the technique is no longer a CSMT, but a kind of FEM.

The necessary distance of an effective far-field from the source transmitter will be discussed later; a literature survey indicates that it ought to be at least five **skin depths** away, that is, the separation R between the transmitter and the station where $\boldsymbol{E}$ and $\boldsymbol{H}$ measurements are made must be larger than five times the skin depth for the lowest frequency used (Sandberg and Hohmann, 1982; Butterworth, 1988; Vozoff, 1991). The skin depth is

$$\delta \cong 503(\rho/f)^{1/2}, \tag{5.1}$$

where δ is in m, ρ in Ω m, and f is the frequency in Hz (section 4.3.2). For the MT case, the depth of exploration is about $\delta/2$ (section 4.3.3(f)). It follows that the far-field begins at a distance of about 10 times the desired depth of exploration. For instance, considering a sedimentary basin in which depth to the basement is 5 km, the transmitter–receiver separation has to be at least 50 km. The lowest frequency has to correspond to a value of δ of 10 km. According to equation (5.1), this frequency is 0.025 Hz if the effective resistivity is 10 Ω m and 0.05 Hz if it is 20 Ω m. These frequencies (periods of 20 s and 40 s) correspond to Pc3-type natural fields that are abundant and have large amplitudes (section 4.2). Pc3s are the most useful events in exploring sedimentary basins with MT. It is obvious at this point that the CSMT transmitter signal of 0.025–0.05 Hz coming from a distance of 50 km is unlikely to be able to compete with Pc3s. This is a somewhat conservative situation. Usual MT work in deep basins involve frequencies down to about 0.005 Hz; Figure 4.9 shows values as low as 0.001 Hz. If we assume that ρ is 10 Ω m, which is typical, and the lowest frequency is 0.005 Hz, the skin depth becomes 23 km, and the required separation, 5 δ, is 115 km.

Obviously, CSMT is not practical for deep basinal work. What depth then is it practical for? In the foregoing discussion the purpose was to obtain quantitatively significant MT-grade data, say within 10% of Cagniard resistivities. However, if the primary purpose is to locate a target qualitatively, such as prospecting for metallic sulfides, then the separation requirement of five skin depths can be relaxed; one can operate within the transition zone. Mining geophysicists use separations of three skin depths. Let us now return to the MT-grade CSMT data, because this is essential in the interpretation of layered geologic formations.

Let us assume that a practical separation is 10 km, at which δ can be 2 km, and this allows a depth of exploration of about 1 km. If ρ is somewhere between 10 Ω m and 20 Ω m, the lowest frequency ought to be about 1 Hz. Zonge and Hughes (1991) state that they usually collect their data across the 1–8192 Hz range. This technique, which deals only with frequencies higher than about 1 Hz, is called the **controlled-source audio frequency magnetotelluric** (CSAMT) method which is now well established in the exploration industry. It has been applied mostly to mining, geothermal, groundwater and geotechnical problems, and occasionally to shallow petroleum-related problems. Many case histories with an extensive treatment of CSAMT are found in Zonge and Hughes (1991); they utilize the 0.1–10 kHz frequency band, with an ideal separation of 4δ which may be 5–10 km, and a source dipole length of 1–3 km; they state that their system is 'effective in mapping the top 2 to 3 km of the earth's crust'; they accomplish this depth by making the low-frequency measurements in the transition zone and roughly correcting them to the far field, or by interpreting the data qualitatively, or by backing away from the MT assumption and performing an exact impedance modeling as in the case of FEM.

A geothermal case history reported by Sandberg and Hohmann (1982) contains much data and theoretical results about CSMT in general. It concerns delineation of the shallow resistivity pattern above a hydrothermal system at Roosevelt Hot Springs, Utah. Apparent resistivities were measured at four frequencies from 32 Hz to 5208 Hz. The maximum depth of exploration was somewhere between 150 m and 300 m. The authors conclude that CSAMT resistivity mapping at Roosevelt Hot Springs geothermal area delineates the same low-resistivity zones as those shown on a DC dipole–dipole resistivity map, that the CSAMT data were collected more rapidly, and that CSAMT could be more cost effective than dipole–dipole resistivity mapping. They also conclude that the technique has difficulty detecting structure beneath conductive overburden.

As a petroleum exploration application, CSAMT measurements were made over Trap Spring oil field, Nevada, located in the frontier Great Basin of the western USA (Hughes and Carlson, 1987). The field produces oil from fractured volcanics. The CSAMT data delineated the major faulting in the volcanics, which lie at depths of 400–1220 m. A source-to-receiver separation of 8 km and a frequency band of 2 Hz–4 kHz were employed. The overburden resistivities are mostly in the neighbourhood of 20–30 Ω m. The maximum depth of penetration was approximately 1100 m.

The theory of data processing and interpretation with CSMT is essentially the same as that of MT, which is discussed in Chapter 4 and will not be repeated here. In the following the theories and problems related to source proximity will be briefly discussed.

5.2.2 Theoretical principle of CSMT

In so far as one operates well in the far field, that is, when the vertically incident plane-wave assumption is valid, the principle of CSMT is the same as in MT (section 4.3.2). Here our primary concern is with source proximity problems.

The electromagnetic field of a grounded dipole

As before, it will be assumed that the displacement currents are negligible; this leads to a quasi-static electromagnetic field for which the real part of the propagation constant, for a homogeneous earth, is

$$\alpha = (\omega\mu/2\rho)^{1/2} \quad (5.2)$$

$$= 2\pi\,(10f/\rho)^{1/2}/10^4 = 1/\delta, \quad (5.3)$$

and

$$\delta \cong 503(\rho/f)^{1/2} \text{ meters}, \quad (5.1)$$

where δ is the skin depth, which is the depth at which the wave amplitude attenuates to $1/e$ or to 0.37 of what it is at the surface.

Let us assume that the source dipole AB is a horizontal ideal dipole, that is, its length L is infinitesimally small; its dipole moment, current times length, remains finite as its length approaches zero. Figure 5.2 shows a plan view of the coordinate system; the z-axis is into the paper.

Wait (1961) developed the mathematical equations for the components E_R, E_ϕ, H_R, H_ϕ and H_z, at the surface, as a function of R, ϕ, ω, ρ and current I, for a homogeneous earth. These equations are complicated; they contain Bessel functions and cannot be evaluated easily, and they will not be reproduced here; they are also given in Kaufman and Keller (1983), Ward and Hohmann (1988) and Zonge and Hughes (1991). Wait (1961) also derived the limits of his equations for the near field (for $R \ll \delta$) and for the far field (for $R \gg \delta$). These are discussed below.

The near field

The limiting equations for $R \ll \delta$, for a homogeneous earth, are

$$E_R \cong IL\rho \cos\phi/\pi R^3, \quad (5.4)$$

$$E_\phi \cong IL\rho \sin\phi/2\pi R^3, \quad (5.5)$$

$$H_R \cong IL \sin\phi/4\pi R^2, \quad (5.6)$$

$$H_\phi \cong IL \cos\phi/4\pi R^2, \quad (5.7)$$

$$H_z \cong IL \sin\phi/4\pi R^2. \quad (5.8)$$

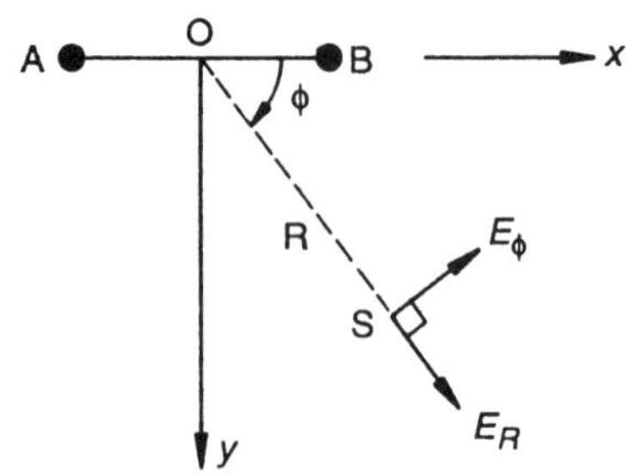

Figure 5.2 Plan view of the coordinate system of CSMT for the electromagnetic field components of a grounded dipole AB. The z-axis is into the paper.

These equations are independent of frequency. The ***H*** components are even independent of ρ; they simply measure the current *I*. There is no phase difference between the ***E*** and ***H*** components. This is the case of an instantaneous DC field, because the wavelength is very large compared to the array dimension *R* which remains the only effective 'yardstick' to measure ρ. Equation (5.5) yields

$$\rho = 2\pi R^3 E_\phi / IL \sin\phi \tag{5.9}$$

which is none other than equation (3.13), for a DC resistivity measurement with an azimuthal dipole array (section 3.2.2). In equation (5.9) E_ϕ is normalized to *I*. This can also be achieved by normalizing it to H_R, as is done with MT, via equations (5.5) and (5.6),

$$\rho = \frac{R}{2}\left|\frac{E_\phi}{H_R}\right|, \tag{5.10}$$

in which case one is not concerned with the current *I* in the source dipole. The DC method, which is the subject of Chapter 3, operates in accordance with equation (5.9) while the **magnetometric resistivity method** operates using equation (5.10) (Edwards and Nabighian, 1991). In other words, the near field is not the proper domain of CSMT; it can be explored more efficiently by other methods.

The far field

The limiting equations for $R \gg \delta$, for a homogeneous earth, are

$$E_R \cong IL \cos\phi / 2\pi R^3 \sigma, \tag{5.11}$$

$$E\phi \cong IL \sin\phi / \pi R^3 \sigma, \tag{5.12}$$

$$H_R \cong IL \sin\phi / \pi R^3 (i\mu\sigma\omega)^{1/2}, \tag{5.13}$$

$$H_\phi \cong -IL \cos\phi / 2\pi R^3 (i\mu\sigma\omega)^{1/2}, \tag{5.14}$$

$$H_z \cong 3IL \sin\phi / 2\pi R^4 \, i\mu\sigma\omega. \tag{5.15}$$

E_R and E_ϕ are independent of frequency, as in the case of the near field, but they are now proportional to ρ. H_R and H_ϕ depend on frequency and on ρ. The phase difference between E_ϕ and H_R is 45°, because $i^{1/2} \equiv e^{i\pi/4}$. Equations (5.12) and (5.13) yield

$$\rho = \frac{1}{\mu\omega}\left|\frac{E_\phi}{H_R}\right|. \tag{5.16}$$

This, in the *x*–*y* system of Figure 5.2, and with the practical units of mV/km, gamma and Ω m, becomes

$$\rho = \frac{1}{5f}\left|\frac{E_x}{H_y}\right|, \tag{5.17}$$

which is the same as equation (4.17) in MT.

The far field is the true domain of CSMT, as the name implies, but how far is far enough? Field geophysicists do not want to go further than 7 δ; this would require impractically large power sources; and they do not think that they can treat the data as MT for less than 4 δ separations. The consensus is that 5 δ is a good compromise which yields MT resistivities within about 10% error if the subsurface is horizontally layered, otherwise the results are unpredictable, because in 2-D and 3-D subsurface conditions ρ depends on the direction of R and on the location and direction of the source dipole.

Figure 5.3 is a comparison of CSMT and MT through model calculations (Butterworth, 1988). The model, Figure 5.3, is a horizontal rectangular prism of 10 Ω m embedded in a half-space of 100 Ω m. Figure 5.4 shows the MT response and Figure 5.5 shows the CSMT response. If we assume that the subsurface is a homogeneous half-space of 100 Ω m, point M in Figure 5.5 represents about 5.5 δ and is about 7% in error relative to MT.

Model studies by Sandberg and Hohmann (1982) also indicate that a separation of 5 δ is good enough for an error of about 10% for a homogeneous half-space.

A major problem is that the 5 δ rule, and as a matter of fact the concept of skin depth, applies only to a homogeneous earth. Otherwise, one would have to perform model calculations for the CSMT and MT responses of an anticipated model to estimate where the far field starts, meaning the region where the error is within $\pm 10\%$. In an inhomogeneous earth the error at a nominal separation of 5δ may be hundreds of percent. Figure 40 of Vozoff (1991, not reproduced here) is a field comparison of MT and CSMT measurements made at the same station with $R = 10$ km at an equatorial position (broadside). Down to 10 Hz, at which $R = 20\,\delta$, the agreement is very good. At 1 Hz, which corresponds to a

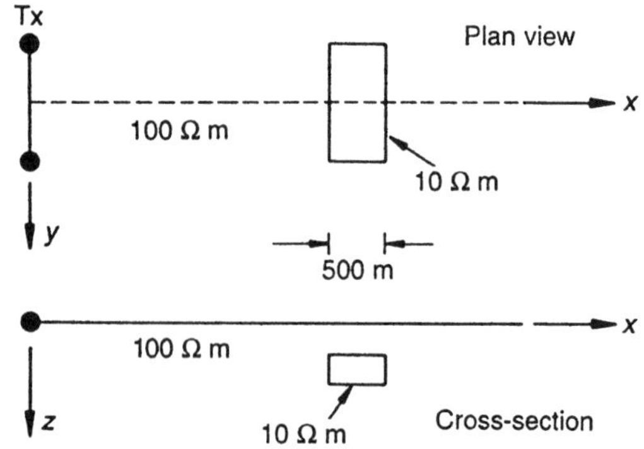

Figure 5.3 Model for comparison of CSMT with MT (from Butterworth, 1988).

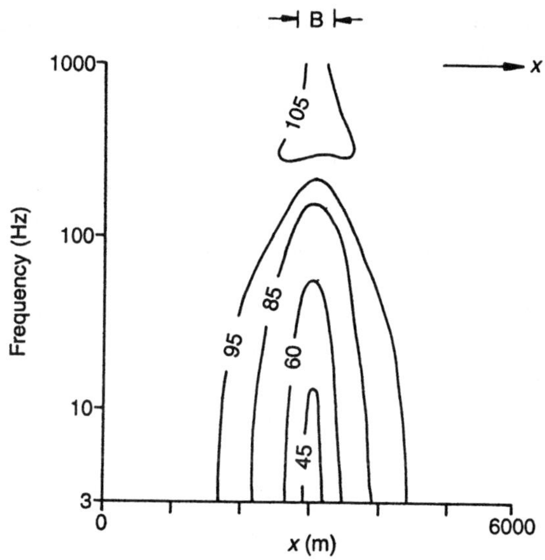

Figure 5.4 MT ρ_{yx} pseudo-section for the model shown in Figure 5.3; ρ_{yx} is the apparent resistivity, in Ω m, derived from E_y and H_x. Space at B is the location of the buried 10 Ω m block (from Butterworth, 1988).

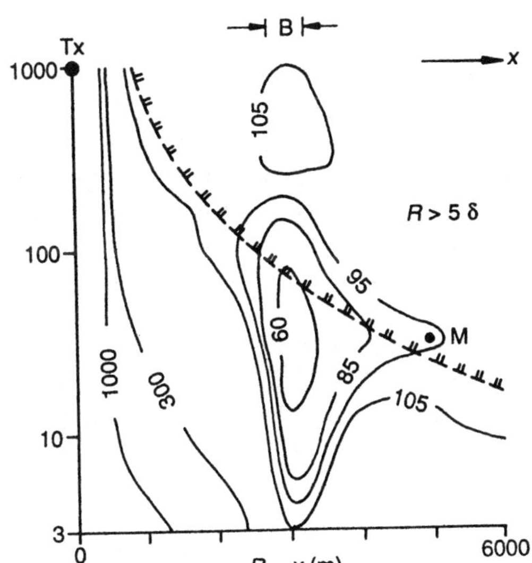

Figure 5.5 CSMT ρ_{yx} pseudo-section for the model shown in Figure 5.3; ρ_{yx} is the apparent resistivity, in Ω m, derived from E_y and H_x. The dashed line is the locus of $R = 5\delta$ points; M represents distortion caused by source proximity at $R = 5.5\delta$. Space at B is the location of the buried 10 Ω m block (from Butterworth, 1988).

nominal skin depth of 1590 m, and $R=6.3\,\delta$, MT ρ is about 22 Ω m and CSMT ρ is about 70 Ω m.

The transition zone

This is the region between the practical boundaries of the near field and the far field, say between $R=0.5\,\delta$ and $R=5\,\delta$. In this region Wait's (1961) equations containing Bessel functions (mentioned above but not reproduced here) express the components of the magnetic and electric fields for a homogeneous earth. Such equations are difficult to evaluate for an inhomogeneous earth, which is the justification for exploration anyway.

When the depth of exploration is extended beyond what is obtainable under the far-field conditions, by employing frequencies that are too low for the separation R, the result is the transition-zone data which are no longer CSMT but transition-zone FEM data. Nevertheless, as mentioned above, these are merged into the far-field data by approximately correcting them for the source proximity effect (Yamashita, Hallof and Pelton, 1985; Yamashita, 1987; Zonge and Hughes, 1991), or they are used directly as transition-zone FEM data.

5.2.3 Field practice

With reference to Figure 5.1, normal practice is to make one set of E_x and H_y measurements at each station for each frequency and from one source dipole AB. This yields the **scalar** apparent resistivities $\rho_s(f)$ which are satisfactory in effectively 1-D geology, or if the geologic strike is approximately known and if the x- or y-axis is chosen parallel to it.

In areas of complex geology a more useful set of data consists of vector apparent resistivities, $\rho_v(f)$. These are obtained from one source dipole but by measuring E_x, E_y, H_x and H_y. These yield the horizontal field vectors $\boldsymbol{E}$ and $\boldsymbol{H}$ from which values of ρ_v are calculated independent of the x and y directions, but these are not tensors, because the source polarization is linear.

If one is making sounding measurements at widely spaced stations, it may be cost effective to measure this complete set at each station: two orthogonal source dipoles AB and A′B′, as shown in Figure 5.1; five components, E_x, E_y, H_x, H_y, H_z, from each source; a total of ten components. These are processed as in MT, and yield **tensor** resistivities, $\rho_t(f)$ (section 4.3.4).

Figure 5.6 shows the results of calculation for a homogeneous earth and for vector resistivities obtained from one source dipole (Zonge and Hughes, 1991). Skin depth δ is 78 m. Broadside measurements reach the far zone condition at about $y=300$ m, at a separation of about $4\,\delta$, while the collinear measurements become so at $x=400$ m, at a separation of more than $5\,\delta$. Furthermore, a broadside beam of 45° on each side is the

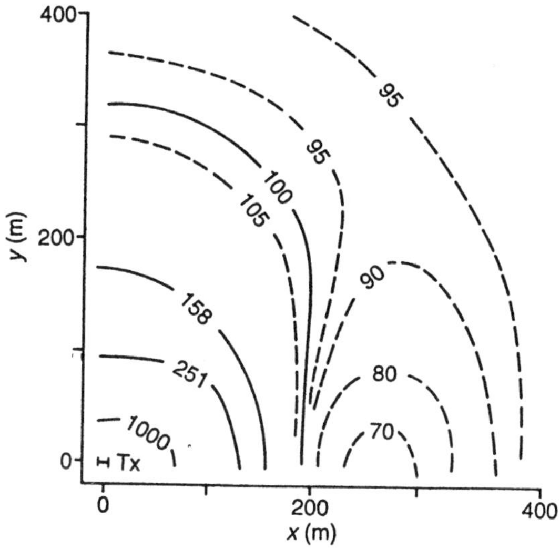

Figure 5.6 Model study results of CSMT vector resistivities, for a homogeneous earth of 100 Ω m; Tx is the grounded dipole source with a center point at the coordinate origin, L = 10 m in the x-direction, I = 1000 A, f = 4096 Hz. Skin depth is 78 m (from Zonge and Hughes, 1991).

preferred zone, because it does not display as much spatial variation ($135° > \phi > 45°$: Figure 5.2).

The source dipole is essentially the same as in DC work. The maximum voltage for safety considerations is about 1000 V, and this necessitates very low electrode resistances to obtain large amounts of current needed to operate in the far field (section 3.2.3). With DC dipole sounding the depth of exploration is from $R/2$ to R, whereas with CSMT, and with the 5-δ rule, the depth of exploration at the lowest frequency is $R/10$; therefore, with CSMT, R is much larger than that with DC dipole work for comparable depths. The dipole moment with CSMT has to be much larger; this requires large power sources, such as 100 kW. With a 100-kW power source at 1000 V, in order to obtain 100 A, one has to lower the total resistance to 10 Ω, and this is a major problem (Figure 3.12); it requires multiple electrodes, much water, even earth moving equipment and metal sheets. Even then, the outputs of the ***E*** and ***H*** sensors are at sub-microvolt levels. These can be computed from equations (5.11)–(5.15), given the magnetic sensor specifications.

The station measurement procedures are also essentially the same as in MT. Core–induction-coil magnetic field sensors are popular in this case, because CSMT usually employs frequencies higher than 1 Hz (section 4.3.9).

5.2.4 Concluding remarks

CSMT data are intended to be interpreted as MT data, but one has to make compromises and accept irregularities and surprises, because the data depend on the location and direction of the source dipole, and on the direction of R, no matter how large R is, 4δ or 7δ, because geologic conditions under the source dipole, and also those between the source and station, substantially influence the measurements at a station. In other words, apparent resistivities at a station, regardless of whether they are scalar, vector or tensor, will change if the source dipole is moved to another location, keeping the dimension of R the same. For instance, in Figure 5.5, the elongation of contours near point M would occur at the left-hand side if Tx is located on the right-hand side of B.

It appears that CSMT is practical only to depths of about 1 km in typical sedimentary basins, so far as far-field data (MT-type) are sought. It may be cost effective for surveys with grid-type or closely spaced stations, but not for long traverses and widely spaced soundings, because of the source logistics.

CSMT may be used in conjunction with deep MT. A combination of many CSMT stations plus a few MT stations that use only low frequency may prove to be cost effective. It can also reduce the static shift errors associated with MT by providing some sort of spatial filtering, although CSMT also suffers from static shift since it measures the electric field.

5.3 FREQUENCY-DOMAIN ELECTROMAGNETIC (FEM) SOUNDING

5.3.1 Principles of FEM sounding

The expression 'sounding' implies, as before, that at each source–receiver setup the earth is assumed to be horizontally layered.

(a) Horizontal loop source

The simplest and the most favored form of FEM sounding is achieved by two horizontal coplanar loops at the surface of the earth which is the x–y plane, as shown in Figure 5.7a. No. 1 is the source and no. 2 is the receiver sensor to measure the vertical component of the magnetic field, H_z. The following assumptions are made.

1. The source–receiver separation, R, is five times larger than the loop dimensions. This allows the assumption that the source is a vertical magnetic dipole of infinitesimal dimensions; its field in free space is uniform across the area of the receiver loop and perpendicular to it, and consists of H_z only; one or both loops may be circular or square shaped.

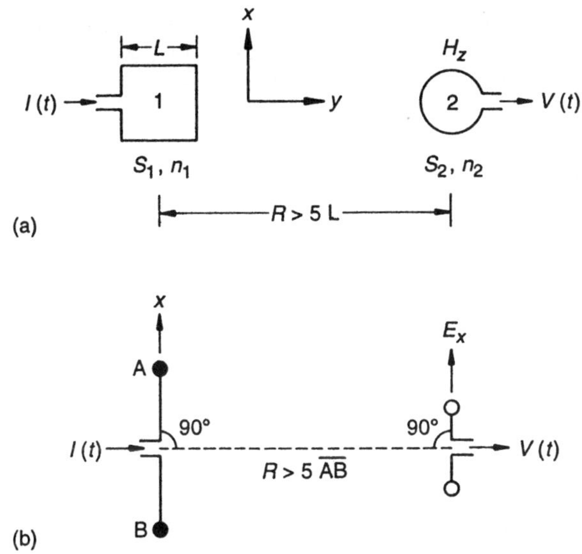

Figure 5.7 Plan views of (a) horizontal coplanar loops and (b) equatorial electric dipoles.

2. The quasi-static approximation is valid; conduction currents dominate over displacement currents which are neglected when the source–receiver separation is much smaller than the wavelengths in free space. This reduces the wave number (propagation constant) to

$$k = (1 - i)\,(\omega\mu\sigma/2)^{1/2}. \tag{5.18}$$

3. The source current I and the fields $\boldsymbol{E}$ and $\boldsymbol{H}$ are harmonic in time, $\boldsymbol{H} = A\mathrm{e}^{-i\omega t}$.

S_1 and S_2 are the areas enclosed by the loops, and n_1 and n_2 are the numbers of turns of wire in the loops.

Let us now assume that the system is in the free space. The magnetic field at the receiver is $\boldsymbol{H}_0 = H_{0z}$, where

$$H_{0z} = -IS_1n_1/4\pi R^3; \tag{5.19}$$

this equation is derived from the Biot–Savart law and is valid for DC or AC. The time-varying H_{0z} induces a voltage V_0 in the receiver loop according to Faraday's law. The ratio of receiver voltage to the source current is the mutual impedance Z_0,

$$Z_0 = V_0/I = i\mu_0\omega S_2n_2S_1n_1/4\pi R^3. \tag{5.20}$$

To facilitate the interpretation, mutual impedance $Z(=V/I)$, measured at the surface of the earth, is normalized by the free-space mutual impedance, Z_0, of the same loops. The mutual impedance ratio, more generally known as the mutual coupling ratio, Z/Z_0, is equal, in this case, to the ratio of the vertical magnetic fields.

$$Z/Z_0 = H_z/H_{0z}, \tag{5.21}$$

where H_{0z}, the free-space value, is called the primary field. For a homogeneous earth the general expression of H_z is (Wait, 1951, 1961, 1962; Ward and Hohmann, 1988)

$$H_z = \frac{-IS_1n_1}{2\pi k^2R^5}[(9+9ikR-4k^2R^2-ik^3R^3)\mathrm{e}^{-ikR}-9]. \tag{5.22}$$

The mutual impedance ratio for a homogeneous earth is

$$Z/Z_0 = (2/k^2R^2)[-9+(9+9ikR-4k^2R^2-ik^3R^3)\mathrm{e}^{-ikR}]. \tag{5.23}$$

Now, in addition to H_z, there is a radial component, H_R, but a horizontal loop receiver does not respond to it.

For a 1-D layered earth the mutual impedance ratio is (Koefoed and Biewinga, 1976)

$$Z/Z_0 = 1 - \int_0^\infty R^3\lambda^2 K^*(\lambda)\, J_0(\lambda R)\, \mathrm{d}\lambda, \tag{5.24}$$

where R is the source–receiver separation, $\mathrm{K}^*(\lambda)$ is a kernel function which depends on thicknesses h_i and resistivities ρ_i of the subsurface layers and on frequency, J_0 is the Bessel function of zero order, and λ is a phantom variable of integration. This equation can be solved numerically. Koefoed, Ghosh and Polman (1972) use a procedure whereby the integral is transformed into a convolution integral and it is computed by the application of a linear filter.

What has been done above with H_z only can also be done with the radial component H_R and with the azimuthal component of the electric field, E_ϕ. Three orthogonal components of $\boldsymbol{H}$ can be measured by means of a magnetometer. Two orthogonal components of $\boldsymbol{E}$ can be measured by means of two orthogonal grounded wire dipoles. Thus, one can obtain three mutual coupling ratios, each normalized to its primary field, H_z/H_{0z}, H_R/H_{0z}, and $E_\phi/E_{0\phi}$. The primary field H_{0z} has already been given in equation (5.19). H_R is also normalized to H_{0z} because $H_{0R}=0$; for a 1-D earth H_R is purely secondary (induced). The primary field of E_ϕ is

$$E_{0\phi} = -i\mu_0\omega IS_1n_1/4\pi R^2. \tag{5.25}$$

These primary fields are those for horizontal coplanar loops situated in free space on the x–y plane. They represent the fields caused only by the current in the source, without any contribution by the induction currents in the conducting earth.

Mathematical expressions for the electromagnetic fields, primary fields, mutual coupling ratios and other relations, for different source–

receiver arrays, and for a uniform earth are compiled in Spies and Frischknecht (1991).

Each receiver location may yield three amplitude and three phase curves. These six curves can be treated independently, leading to a common model.

Operating with horizontal coplanar loops only, the amplitude and phase of H_z/H_{0z} can be measured. However, measurement of phase difference between the transmitter and receiver signals require somewhat complicated instruments, and the simplest operation consists of measuring the amplitude of H_z/H_{0z}. In the case of CSMT, $\boldsymbol{E}$ is normalized to $\boldsymbol{H}$, independent of the phase and amplitude of the source, such as E_ϕ/H_R, but this can be done only in the far field, that is, when $R > 5\,\delta$.

Returning to the general case, with horizontal coplanar loops, Figure 5.8 shows a set of theoretical curves for a two-layer earth, in the case of a conductive substratum, $\rho_2 < \rho_1$, for various resistivity ratios indicated on the curves, and for $R/h_1 = 5$, where h_1 is the thickness of the top layer and R is the source–receiver separation. Amplitudes of Z/Z_0 are plotted on a linear ordinate scale. The frequencies enter as $\rho_1 T/R^2$ values on a logarithmic abscissa scale, where T is the period. As discussed in Chapter 4, this normalized parameter is dimensionless, because it is proportional to $(\lambda_1/R)^2$, where λ_1 is the wavelength in the top layer (equations (4.19) and

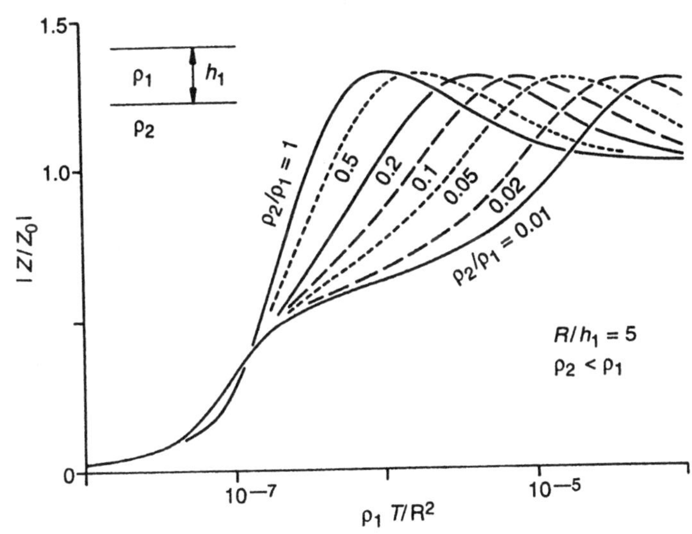

Figure 5.8 Theoretical FEM curves (amplitude of Z/Z_0 versus $\rho_1 T/R^2$) for a two-layer earth with horizontal coplanar loops, for $R/h_1 = 5$, for various values of $\rho_2/\rho_1 < 1$, in the case of a conductive substratum (after Koefoed and Biewinga, 1976).

(4.28)). Another form of normalization is R/δ_1, where δ_1 is the skin depth in the top layer; this ratio is called the induction number.

Figure 5.9 shows the same set of theoretical curves as Figure 5.8, but for a resistive substratum, $\rho_2 > \rho_1$. Separation of the curves from that which is for a uniform half-space ($\rho_2/\rho_1 = 1$) is small. It is much more so for $R/h_1 = 2.2$, as shown in Figure 5.10. The common practice of operating with $R/h_1 \geqslant 1$ is not justified in this case. It thus can be said that the horizontal coplanar loop system does not work well in determining depth to a resistive substratum. As for determining its resistivity, ρ_2, the situation is still worse; there is a saturation effect beyond about $\rho_2/\rho_1 = 5$, and it makes very little difference whether this ratio is 5 or infinity.

In exploring deep sedimentary basins, depth to the basement complex is very important. It would require a separation R of about 20 km to obtain a rather poor resolution to locate a basement at a depth of 5 km, with the system discussed above. A loop source is not practical for a separation R of this magnitude anyway.

(b) Grounded-wire dipole source

Another FEM technique that may solve the basement problem discussed above is the equatorial electric dipole array, the same configuration that is used in DC work, as shown in Figure 5.7b, and also in Figure 3.10. The

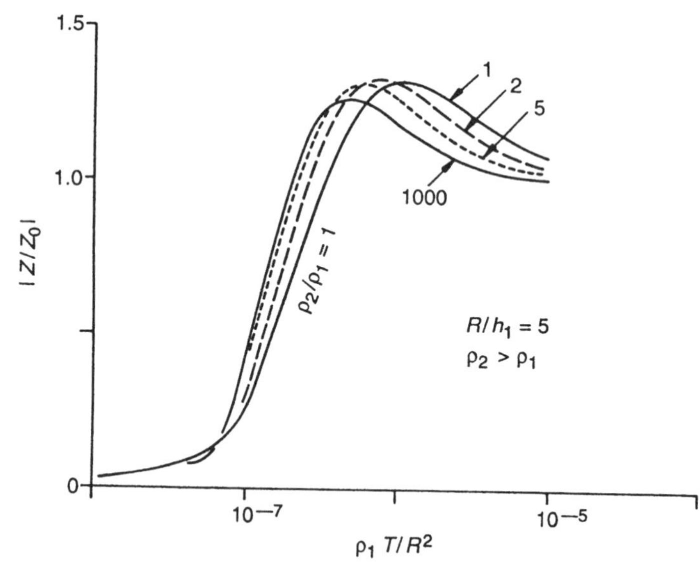

Figure 5.9 Theoretical FEM curves (amplitude of Z/Z_0 versus $\rho_1 T/R^2$) for a two-layer earth with horizontal loops, for $R/h_1 = 5$, for various values of $\rho_2/\rho_1 > 1$, in the case of a resistive substratum (after Keofoed and Biewinga, 1976).

source is a grounded-wire dipole; the receiver is also a grounded-wire dipole in the equatorial position and parallel to the source. This will be called simply the equatorial array.

According to the collection of FEM response curves of Spies and Frischknecht (1991), any combination of grounded-dipole source and measurements of ***H*** field does not provide appreciable improvement over the horizontal loop–loop system in the basement problem. It is only when the receiver is also an electric field dipole that the separations between the curves are enhanced, for a high-resistivity substratum, especially with the equatorial array, but the axial (in-line) electric dipoles are also satisfactory. However these arrays fail to solve the reverse problem, which involves a low-resistivity substratum, such as an aquifer. These points are demonstrated in Figure 5.11 in which the amplitude of the mutual coupling ratio for an equatorial array is plotted against frequency (Hz) for a two-layer earth with $\rho_1 = 4.25\ \Omega$ m, ρ_2 varies, $h_1 = 250$ m and separation $R = 1000$ m.

The general expression for an equatorial array on a uniform earth, for an x-directed source dipole is (Wait, 1961; Ward and Hohmann, 1988)

$$E_x = -E_\phi = \frac{-IL}{2\pi\sigma R^3}[2-(1+ikR)e^{-ikR}], \qquad (5.26)$$

where L is the length of the source and I is the current. The primary field

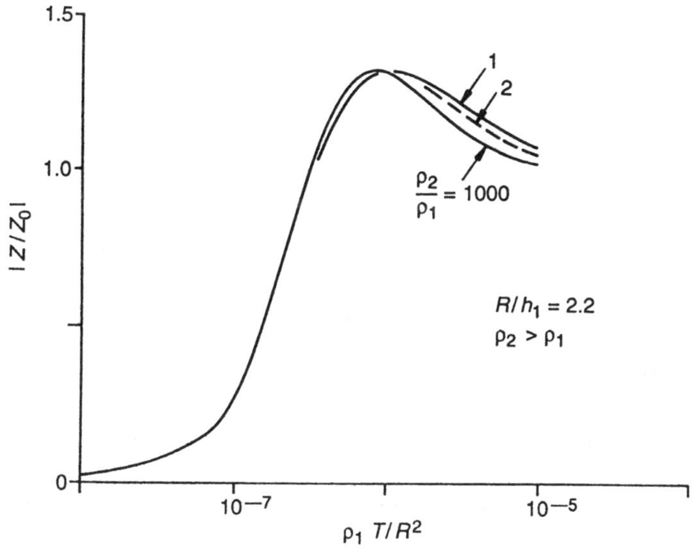

Figure 5.10 Theoretical FEM curves (amplitude of Z/Z_0 versus $\rho_1 T/R^2$) for a two-layer earth with horizontal loops, for $R/h_1 = 2.2$, for various values of $\rho_2/\rho_1 > 1$ (after Koefoed and Biewinga, 1976).

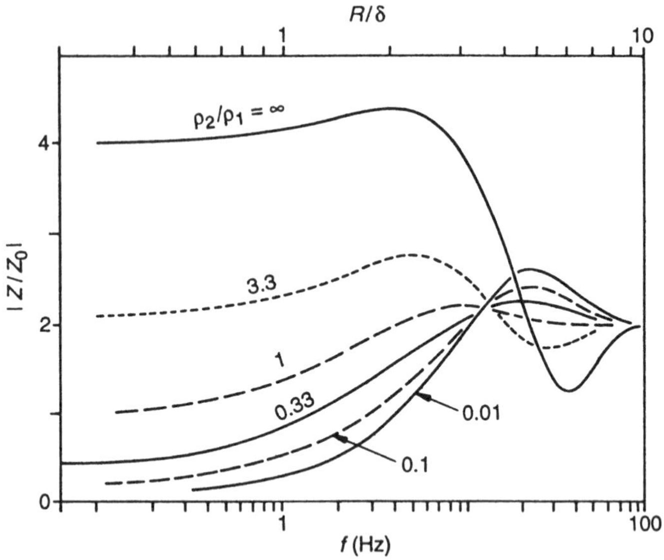

Figure 5.11 Theoretical FEM curves with equatorial electric dipoles, for a two-layer earth, showing amplitude of Z/Z_0 versus frequency (Hz) and induction number R/δ, where δ is the skin depth, for various values of ρ_2/ρ_1; top layer resistivity $\rho_1 = 4.25\ \Omega$ m; ρ_2 varies, top layer thickness $h_1 = 250$ m, and separation $R = 1000$ m (after Spies and Frischknecht, 1991).

for a uniform earth is

$$E_{0x} = -IL/2\pi\sigma R^3. \tag{5.27}$$

The mutual coupling ratio is

$$Z/Z_0 = E_x/E_{0x} = 2 - (1 + ikR)e^{-ikR}. \tag{5.28}$$

In view of the advantage of the equatorial array over the horizontal coplanar loops in sedimentary basin evaluation, as discussed above, Sherwood and Yungul (1968) developed an FEM technique, utilizing the equatorial array, and a coded source signal. The coded signal is a pseudo-random wave train designed in accordance with a frequency spectrum of interest. The simplest form of coded signal is a square wave of variable period. Let $I(t)$ represent the coded-signal current and $V(t)$ the output voltage at the receiver. We can envisage $V(t)$ being the output of a filter which is the earth in this case, plus a background noise consisting of telluric and industrial voltages.

$$V(t) = I(t) * G(t) + N(t), \tag{5.29}$$

where $G(t)$ is the unit impulse response of the earth filter and $N(t)$ is the unpredictable natural and artificial noise. What is needed is to separate the signal

$$S(t) = I(t) * G(t)$$

from the noise $N(t)$. This is done by cross correlation of $I(t)$ with $V(t)$. This yields the cross correlation function $C(t)$. Poorly correlating parts, which represent the unpredictable noise, are discarded. A Fourier analysis of $C(t)$ yields the amplitude and phase spectra. The amplitude spectrum of $C(t)$ is a measure of the product of the amplitude spectra of $I(t)$ and $S(t)$ as a function of frequency. This technique allows one to obtain data for all frequencies in one swoop, instead of making measurements with discrete frequencies one by one in order to carry out the narrow-band filtering. In a way it combines some of the advantages of FEM and TEM; it does it in one swoop like TEM and it provides an effective filtering like FEM. It is advantageous to take H_z measurements as well at the receiver site.

The coded source-signal technique was further developed by Edwards (1976), Duncan *et al.* (1980), Gomez-Trevino and Edwards (1983), Holladay (1987), and Edwards and Holladay (1991). These investigators, except Holladay (1987), measured only the H_z field. Duncan *et al.* (1980) explain the technique effectively and give two case histories, one in the sedimentary basin of southern Ontario, Canada, sounding to a depth of about 500 m, and another for deep crustal investigations at Timmins, Ontario, at a site within the Precambrian shield. Gomez-Trevino and Edwards (1983) describe a case history in the sedimentary basin of southern Ontario, Canada, in which coded source-signal and DC Schlumberger measurements were made and jointly inverted. Up to eight distinct layers down to a depth of about 1 km were fitted to the data. Edwards and Holladay (1991) give an updated version of the theory and instrumentation and two field examples, one in the sedimentary basin of southern Ontario and the other near Orangeville, Ontario, where an axial array was used and E_x (in-line) and H_y were measured.

5.3.2 Interpretation

As discussed in Chapters 3 and 4, 1-D inversion with DC and MT methods is easy; 2-D modeling is simple although laborious. With EM, 1-D inversion is somewhat difficult, and 2-D modeling is difficult and/or very laborious. In what follows only 1-D interpretation will be considered.

Any one style of interpretation falls into one or more of the following categories:

1. to find a model for which the theoretical response curve will match the observed data, which may be achieved by (1) manual curve matching, (2) computerized interactive trial-and-error curve matching and (3) computerized automatic curve fitting;

2. direct inversion;
3. qualitative interpretation using various methods of data display, e.g. pseudo-sections similar to those discussed in Chapters 3 and 4.

(a) 1-D interpretation by manual curve matching

Evaluation of equation (5.24), which is for a 1-D multilayer earth, shows that if the Z/Z_0 curves are plotted on a logarithmic horizontal scale a change in top layer resistivity ρ_1, period T or R results in a horizontal shift of the Z/Z_0 curve, provided that layer resistivities and thicknesses normalized to ρ_1 and R, respectively, remain the same (Koefoed and Biewinga, 1976). This renders the matching of dimensionless theoretical curves to observed curves practical.

For a two-layer earth, each family of standardized theoretical curves is for one measuring array form (such as two horizontal coplanar loops), for one component (such as H_z or E_x) or for other parameters (such as ellipticity), for amplitudes or phases, for one h_1/R ratio, where h_1 is the top-layer thickness, and each curve is for one ρ_2/ρ_1 ratio. Such families of curves are shown in Figures 5.8–5.11. Obviously, an extensive catalog of theoretical curves are required even for a two-layer earth interpretation, whereas MT requires only two families, one for amplitudes and one for phases, and DC requires only one family of curves. The case of a two-layer earth, horizontal coplanar loops and $|H_z|$ will now be considered.

In a family of theoretical curves for a certain h_1/R, the amplitude of the mutual coupling ratio, $|Z/Z_0| = |H_z/H_{0z}|$, is plotted on the vertical axis with a linear scale. The normalized period, $\rho_1 T/R^2$, is plotted on the horizontal axis with a logarithmic scale. Each curve is for a separate ρ_2/ρ_1 value.

The observed data are plotted as $|Z/Z_0|$ versus T. The vertical coordinate axis scale is the same for both theoretical and observed values. An observed data curve can be matched to a theoretical curve by shifting horizontally. R is known. The h_1/R value for the matched curve family yields h_1 directly. Each value of T on the observed curve corresponds to a $\rho_1 T/R^2$ value on the theoretical curve; thus ρ_1 is determined. The ratio ρ_2/ρ_1 is read from the curve. This is extremely simple if one is able to select the proper family of curves from an album. In order to carry out interpretation in terms of 1-D multilayers one needs a catalog of three-layer master curves, as was done with DC and MT. For the reasons discussed above, an effective three-layer catalog would be too voluminous and it would not be practical to search through it, trying to select the proper family of curves. Another reason that renders an extensive master curve catalog largely ineffective is that the theoretical amplitude and phase curves, for a variety of 1-D models, look very similar; they do not visually display the model characteristics, not in the way that the DC and MT curves do. This makes the selection of a proper family of curves difficult, as demonstrated below.

Figure 5.12 shows a field example, an FEM sounding with horizontal coplanar loops. The observed data (crosses) were interpreted as a three-layer earth, and yet they do not diverge much from the dashed curve which represents a uniform earth.

Figure 5.13 shows another field example of observed data (dots) also obtained with horizontal coplanar loops. The data were fitted first to a two-layer model and then to a three-layer model. The differences between the two model curves are hardly discernible. Obviously, such a fine tuning cannot be accomplished by searching through the pages of volumes of an album, and this is only a three-layer model. A DC sounding employing a Schlumberger array was also made at the site of the FEM sounding. It was interpreted (Mundry and Blohm, 1987) in terms of seven layers. Resistivities of the layers from top to bottom, in Ω m, are: 12 607, 20 000, 4295, 395, 73, 119 and 2. Depths to the layer boundaries, in m, are: 1.1, 3.1, 8.4, 39, 65 and 215. As usual, the detail obtained with FEM is considerably less than that obtained by DC. Also, FEM is not as accurate as DC in determining the resistivities of resistive layers in the section, hence the wide discrepancies in layer resistivities between FEM and DC. These observations also suggest that FEM curves would not display much visually observable character.

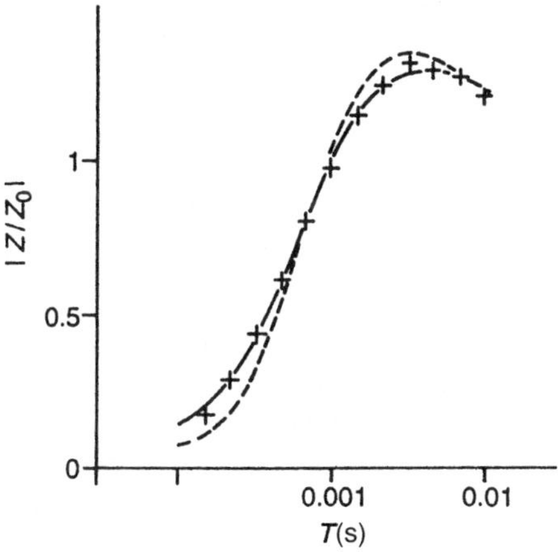

Figure 5.12 FEM sounding with horizontal coplanar loops, $R=200$ m; observed data shown by crosses; theoretical three-layer curve fitted to the observed data (solid line); parameters of the model are $h_1=30$ m, $\rho_1=22.5\ \Omega$ m, $h_2=80$ m, $\rho_2=11.5\ \Omega$ m, $\rho_3=4.5\ \Omega$ m. Dashed line is a theoretical curve for a uniform half-space (modified from Koefoed and Biewinga, 1976).

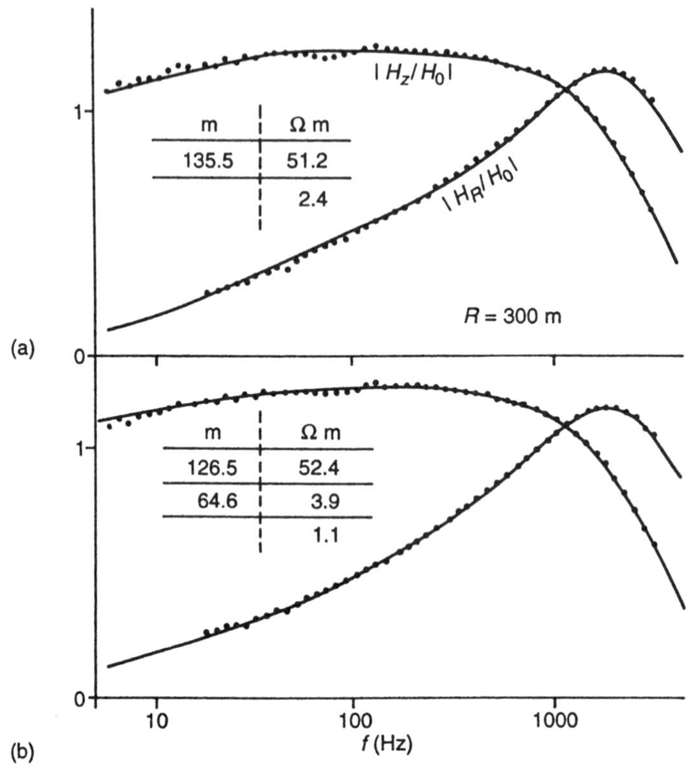

Figure 5.13 FEM sounding with horizontal loop source with H_z and H_R receivers. $R=300$ m. (a) Two-layer interpretation and (b) three-layer interpretation of a set of data (from Mundry and Blohm, 1987).

Due to the reasons discussed above, master curve catalogs for routine interpretation, for the kind of functions that the DC and MT catalogs perform, are not available for FEM sounding interpretation, and perhaps never will be, simply because curve matching by the computer alone, automatic or interactive, is so much more attractive, assuming that the present-day direct inversion methods are not yet entirely satisfactory.

Two-layer catalogs have practical value, and selected compilations of three-layer curves have educational and some practical value. Two-layer tables for Z/Z_0 for horizontal loop–loop and loop–wire systems are given by Frischknecht (1967). Verma (1980a, 1982) published volumes of three- and four-layer master tables and curves for Z/Z_0.

A variety of FEM two- and three-layer master curves and auxiliary curves are given in Kaufman and Keller (1983). Their curves are for two distinct types of EM sounding techniques: parametric sounding, as discussed here so far, whereby one measures values of Z for a number of frequencies at one value of R; and geometric sounding whereby one uses one frequency while expanding the R-distances. In geometric sounding one may use two or more frequencies and obtain a more extensive data set, encompassing both techniques of sounding. Their curves for the parametric mode are plotted as ρ_a/ρ_1 versus λ_1/h_1, where ρ_a is the apparent resistivity and λ_1 is the wavelength in top layer, whose parameters are ρ_1 and h_1. The apparent resistivity is obtained from the limiting value of Z for a uniform earth at the high-frequency end. The geometric sounding curves are plotted as ρ_a/ρ_1 versus R/h_1 for various ρ_2/ρ_1 and ρ_3/ρ_1 ratios and for one frequency.

Spies and Frischknecht (1991) give many two-layer curve families for a variety of source–receiver arrays, including horizontal coplanar loops and equatorial electric dipoles (Figure 5.11 is one of them). In these curves amplitudes and phases of Z/Z_0 are plotted against R/δ and also against frequency in hertz.

(b) Computer-assisted interpretation

The most common interpretation technique at present is curve fitting by computer, by trial and error. This requires a rapid and accurate algorithm for the forward problem (for calculating theoretical response curves). It is effective in finding approximate matches to two- and three-layer models, but it is slow in achieving a good match to four or more layers. It is also highly subjective and requires highly skilled interpreters.

A more advanced method of model fitting is an iterative least-squares nonlinear regression technique optimizing the parameters of the layers. This approach, widely known as Marquardt's technique, was used in the interpretation of the data shown in Figure 5.13 (Mundry and Blohm, 1987). This method computes the parameters of an assumed model; the model may be constrained through independent data or assumptions, thereby alleviating some of the non-uniqueness (Hohmann and Raiche, 1988). The user must select (1) number of layers, (2) first-guesses for layer parameters (h_i and ρ_i), (3) weighting factors and (4) equivalence (section 5.3.3). Much information on the application of computer-assisted interpretation (inversion) is found in Spies and Frischknecht (1991), including the Backus–Gilbert and Monte Carlo methods, and the effects of other than 1-D structures.

(c) Direct inversion

Undoubtedly the most objective method is direct inversion where h_i, ρ_i and the number of layers are calculated directly from data. This problem is very complicated because it is nonlinear. It is not practical at present, because direct inversion results with the existing algorithms are said to be unstable and noisy. This is a subject that needs further research.

5.3.3 Layer resolution

As in DC and MT, the EM response of a thin low-resistivity layer depends only on the conductance of that layer. For a high-resistivity thin layer, EM methods, in general, are sensitive to the thickness but not the resistivity.

Equivalence parameters for a thin layer for the horizontal coplanar loop configuration, based on various investigators' limited observations are $h^{1/2}/\rho$ for a low-resistivity layer, and $h^2\rho$ for a high-resistivity layer. The corresponding parameters for DC are h/ρ and $h\rho$; for MT they are h/ρ and h.

The horizontal coplanar loop configuration seems to be the best to resolve an intermediate layer sandwiched between either higher- or lower-resistivity media, the so-called H-type and K-type three-layer models, respectively (Verma, 1980b).

It is well known that the horizontal coplanar loop configuration works best if resistivity decreases with depth; this is the proper configuration to prospect for aquifers but not to study the basement (Figures 5.8 and 5.9).

The equatorial electric dipoles configuration works well in detection of a thin high-resistivity layer which may be invisible to an H_z measurement. Therefore, combining E_x and H_z measurements in an equatorial array may provide a less ambiguous detection of high-resistivity thin layers such as oil and gas reservoirs. The equatorial electric dipole array also works well in basement studies, but not in detecting low-resistivity layers, such as aquifers (Figure 5.11).

5.3.4 Depth of exploration

This problem is not as straightforward with EM as it is with DC and MT, because with EM the depth of exploration is dependent on the separation R as well as on the frequency. There is no simple and effective rule of thumb, and there is no substitute to modeling for this purpose.

For horizontal coplanar loops the maximum depth of exploration varies roughly from $R/2$ to R, where R is the source–receiver separation, when the proper frequencies are employed. However, this depth is severely restricted by signal-to-noise ratio (S/N), due to three factors which are external to the measuring-system noise.

1. Signal strength falls off rapidly with increasing R (equation (5.22)).
2. The secondary induced EM field decreases as period T increases.
3. As T increases, the geomagnetic micropulsation amplitudes (telluric noise) also increase (Figure 4.4A).

Telluric noise becomes so overpowering for periods larger than about 5 s that a remote-reference noise cancellation system becomes a necessity. Thus, in a typical sedimentary basin, in order to extend the depth of exploration beyond about 2 km, one would have to employ large and powerful source equipment, remote reference (section 4.3.7(b)) and complicated measuring and signal processing systems. Most commercially available systems are routinely capable of exploring down to about 1 km.

5.3.5 Field practice

The following discussion of field practice with an EM system developed by Lawrence Berkeley Laboratory and the University of California in Berkeley will serve the purpose. This is the EM-60 system (Morrison *et al.*, 1978; Morrison and Goldstein, 1991). Development of this system and field results obtained with it have been fairly well published; it is fairly new (post-1978); it was designed as a low-frequency, large-moment horizontal-loop, deep FEM sounding system. It thus fits well into the present subject. Square source-loop sizes of up to 3 km on each side and source–receiver separations of up to 30 km have been used. A maximum current of 400 A peak-to-peak at 250 V can be fed into the loop, with a transmitter moment of up to 10^8 Am^2. The waveform is square. The frequency range is from 10^{-3} to 10^3 Hz, but below 0.05 Hz (above $T=20$ s) data are not reliable due to telluric noise. A three-component SQUID magnetometer is used to measure H_z, H_R and H_ϕ; E_ϕ may also be measured. A remote-reference telluric noise cancellation system is used; it is placed about 10 to 15 km away from the transmitter loop. The remote-reference signal is simply subtracted from the signal at the receiver.

A survey with the EM-60 was carried out at Soda Lakes geothermal area, Nevada (Stark *et al.*, 1980). A total of 13 stations were occupied. Results for one sounding are shown in Figure 5.14. Interpretation was carried out by means of an automatic Marquardt least-squares inversion. The maximum depth of exploration attained was about 1–1.5 km; the authors anticipated deeper penetration and were disappointed. In Figure 5.14 the lowest frequency is 0.1 hz; the system capability is 0.001 Hz.

Wilt *et al.* (1983) describe their experience with the EM-60 system for geothermal exploration in Nevada. System description, data interpretation and field surveys at Panter Canyon, Soda Lakes and McCoy are discussed.

A petroleum-related survey was conducted in the Columbia Basin, Washington (Wilt *et al.*, 1989). With reference to Figure 5.15, the primary purpose was to map the bottom configuration of the basalt volcanics which extend mostly from the surface down to depths of about 1300–1800 m; their average resistivity is about 200 Ω m and that of the sedimentary section underneath is about 10 Ω m. It was also a trial to find out whether subtle configurations at the bottom of the volcanics can

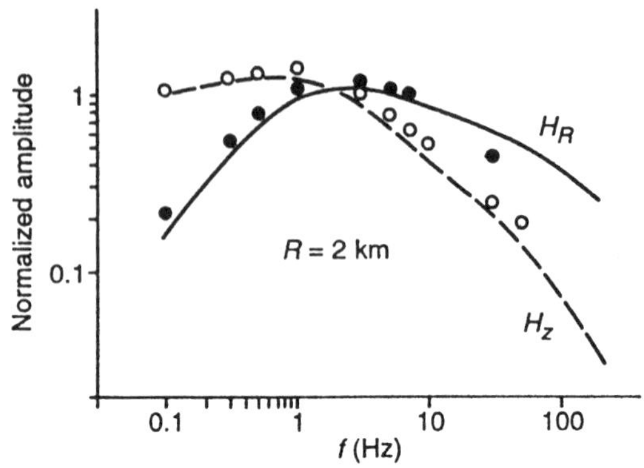

Figure 5.14 FEM sounding with EM-60 at Soda Lakes, Nevada (Line A–Á trending northeast, transmitter no.1, transmitter–receiver separation 2 km, using a horizontal loop source, showing normalized amplitudes. Three-layer interpretation with ρ in Ω m and thickness (h) in m: $\rho_1 = 11$, $\rho_2 = 2.5$, $\rho_3 = 50 \mp 77$, $h_1 = 230$, $h_2 = 850 \mp 50$ (from Stark *et al.*, 1980).

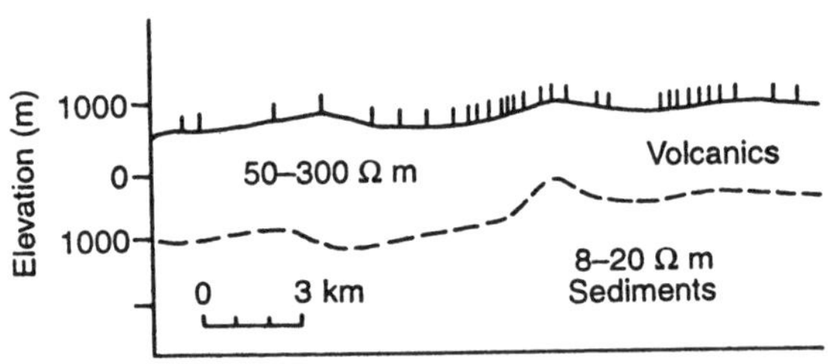

Figure 5.15 Cross-section from 1-D inversion of horizontal loop FEM soundings at locations indicated by vertical bars at the surface, Columbia Basin, Washington. Dashed line represents the bottom of the volcanics (abbreviated from Wilt *et al.*, 1989).

be mapped. Such features could be related to sedimentary structures, folds and faults, which could constitute oil drilling targets. This problem is disadvantageous for seismic methods and advantageous for the EM-60 type of system, as illustrated in Figures 5.8 and 5.9, namely a high-resistivity surface layer and a low-resistivity substratum. Magnetic fields H_z, H_R and H_ϕ were measured with SQUID magnetometers. The separation R ranged from 1 to 5.5 km. Square source loops, 500 m on a side, with a frequency range of 0.1–100 Hz and remote-reference noise cancellation were used. During 15 field days 38 stations were occupied. A comparison of the induction well log at one well and three FEM soundings in its vicinity shows very good correlation down to a depth of about 2 km, about 500 m underneath the volcanics (Figure 5.15). However, to what extent the well log data were used in constraining the Marquardt inversion was not stated. Most of the interpretation consisted of 1-D three-layer inversions. A pieced-together 2-D cross-section was constructed from the 1-D inversions, as shown in Figure 5.15. The validity of doing so was controlled by the $|H_\phi| / |H_R|$ values which were used as 3-D indicators ($H_\phi = 0$ for a 1-D earth). The final 2-D cross-section was also compared to 2-D model data obtained from physical scale modeling.

5.4 TIME-DOMAIN ELECTROMAGNETIC SOUNDING (TEM OR TRANSIENT SOUNDING)

5.4.1 Introduction

The introductory material in section 5.1 also includes the time-domain methods. Only a few points need be emphasized here.

In theory, time-domain and frequency-domain methods (TEM and FEM, respectively) are equivalent; the time-domain data can be transformed to frequency-domain, but this would not be advantageous in practice, considering the shape and length of the records and the noise problems. The main difference between TEM and FEM is that TEM measurements can be made in the absence of the source signal, when the power is off, and the primary field is nil. This usually allows the TEM equipment to be simpler than that of FEM and yet record the secondary signals accurately.

So far as deep EM sounding is concerned, at the present time TEM is more popular than FEM; it provides better layer resolution and is more cost effective than FEM for deep work.

Early investigators in petroleum-oriented exploration in the USA applied a variety of time-domain EM techniques, such as the Eltran (electromagnetic transient) and the Elflex (electromagnetic reflexion) methods, in which oil companies were interested during the years from the 1930s to the 1960s.

The Eltran method employed a dipole–dipole array (collinear grounded-wire source and receiver dipoles). The source signal is an electric surge supplied by a condenser. The receiver signal is recorded simultaneously with the source signal. The method was largely abandoned in the 1950s, because it was not known how to interpret the data quantitatively in terms of a layered earth. Some investigators searched for reflections among thousands of field oscillograms but did not find any valid ones. However, the data were still somewhat helpful in that they showed the lateral variations in gross electrical properties, but this could have been accomplished more effectively by means of DC methods.

Proponents of the Elflex method claimed that their measurements represented EM reflections from oil- or gas-bearing formations and constituted 'direct detection' (Chapter 6). Subsequently, theoretical developments indicated that it would be very unlikely to be able to resolve such reflections coming from depths of about 2000 m (as commonly claimed) in commonly encountered sedimentary basins. This created a disillusionment with the time-domain EM methods within oil companies in the USA.

The petroleum industry became actively interested in TEM, mostly in the 1940s and the 1950s, in a promotional fashion, and largely abandoned it before 1960 as a result of theoretical studies. Both entry into this field and exit from it have been somewhat subjective, because the decisions depended largely on the advice of seismic and gravity operational personnel. On the other hand, the mining exploration industry has been steadily increasing its interest and capability in time-domain measurements.

The modern-day practice of deep TEM has its foundation in the former Soviet Union, back in the 1930s. Keller reviewed these developments in the introduction to Vanyan *et al.* (1967) that he translated. Much of the Russian literature has been translated to English by Professor G.V. Keller, of the Colorado School of Mines, who has been an active investigator in the theory of TEM sounding and in its application to geothermal and petroleum exploration and to crustal studies. The senior author of Kaufman and Keller (1983) is from the former Soviet Union.

The interest of some oil companies in applying deep TEM sounding in sedimentary basins, to depths of several kilometers, is increasing. However, TEM is not as well established in this field as the magnetotelluric (MT) method; not many results have been published yet, more theoretical and instrumentational developments are needed.

5.4.2 Principles of TEM sounding

What has been discussed about EM methods in general and TEM in particular in sections 5.1, 5.2 and 5.3 will not be repeated here.

One of the two most significant advantages of TEM sounding over the FEM and DC methods is that the depth of exploration depends primarily on the duration of the measuring time, t, on instrument specifications (e.g. dynamic range and resolution) and on ambient noise, but not so much on source–receiver separation. This provides a very significant advantage in spatial resolution, logistics, and cost effectiveness. For instance, with central loop sounding, in which the source is a large horizontal loop and the receiver is a small horizontal loop in the center of the large loop, for a two-layer earth the substratum is first detected when the top-layer diffusion depth,

$$\delta_{TD} = (2t\rho_1/\mu_0)^{1/2}, \qquad (5.30)$$

is approximately equal to the thickness of the top layer (Spies and Frischknecht, 1991). The separation (the source–loop radius in this case) does not even enter into equation (5.30), although the larger the source loop the better the signal quality and the deeper the depth of exploration.

The source waveform may be a train of bipolar (negative and positive) square pulses with a zero-power interval between them, or on-and-off square pulses of the same sign, or other waveforms. Measurements at the receiver are made when the power is off; thus the induction coupling decay (the secondary field) is recorded in the absence of a large primary field. This is probably the other most significant advantage over the FEM method.

On the other hand, TEM measurements are difficult to work with. The signal decays very rapidly, the receiver instrumentation requires a large dynamic range, measurements cannot tolerate much ambient noise, the sounding curves do not have much character that would be suggestive of layering in the earth, and this renders interpretation difficult.

A field setup of a typical deep sounding system with a grounded-wire dipole source and a horizontal loop receiver, plus horizontal electric field measuring dipoles, is shown in Figure 5.16. The source dipole current, I, is usually from a few tens to several hundred amperes. A typical square pulse cycle would be one minute on positive, one minute off and one minute on negative. The length of the source dipole, AB, is usually from 1 to 3 km. The separation, R, is from a few to tens of kilometers. The receiver is a multiturn square loop about 100 m on one side. Electric field components, E_x and E_y, are also measured. Vector magnetometers may be used instead of horizontal loop to measure H_z, which is usually converted to $\partial H_z/\partial t$ to make it consistent with $V(t)$ of a horizontal loop receiver.

Figure 5.17 shows a field record, voltage output of a horizontal loop receiver and a grounded-wire dipole source. Tens of such records

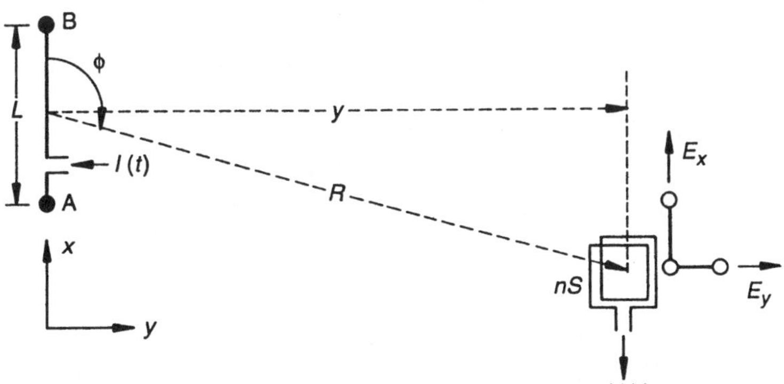

Figure 5.16 A typical deep TEM sounding field setup, grounded-wire dipole source (AB) and horizontal loop receiver. Electric fields E_x and E_y are also measured.

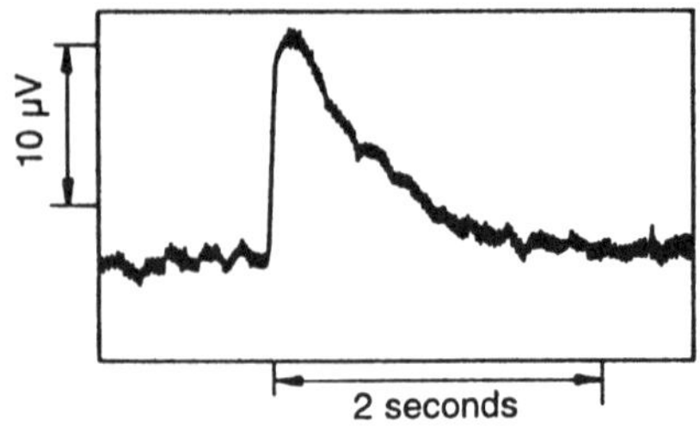

Figure 5.17 A typical field record obtained by means of a system of the type shown in Figure 5.16 (redrawn from Jackson and Keller, 1972; the noise is schematic).

containing much random noise (e.g. natural and cultural EM noise) are synchronously stacked to reduce the noise. Distortion in the recording system is removed by a deconvolution process. Uncorrelated signals are further reduced by special filters. The final $V(t)$ data may be inverted into a layered earth model. However, the common practice at the present is to transform the $V(t)$ curve to an apparent resistivity versus t curve which is discussed below.

(a) Apparent resistivity

It is advantageous to normalize the measurements to a uniform earth, as was described for DC and MT. This is done by converting the measurements (V, H_z, $\partial H_z/\partial t$, etc.) to an apparent resistivity function, $\rho_a(t)$, which is defined as the resistivity of a uniform earth which would produce the measured quantities. There are several advantages in doing so. (1) It renders the measured quantities independent of the geometric and other parameters not related to the subsurface. (2) It reduces the dynamic range of the sounding curves. (3) Before converting to ρ_a, the shapes of sounding curves do not give much clue, when inspected visually, about the subsurface resistivity profile: the measured quantities decrease rapidly with increasing time whether or not the resistivity increases at depth, and the relationship between the experimental curve shape and the vertical resistivity profile is not obvious. Apparent resistivity curves show a similarity to the resistivity profile and facilitate interpretation.

Grounded-wire dipole source and horizontal loop receiver

Let us assume that an x-directed source and a horizontal-loop receiver are placed on the surface of a uniform earth. The voltage output of the receiver is (Kaufman and Keller, 1983; Ward and Hohmann, 1988)

$$V = \frac{nSIL\rho}{2\pi}\,\frac{y}{R^5}\,\left[3\mathrm{erf}(\theta R) - \frac{2}{\pi^{1/2}}\,\theta R(3+2\theta^2 R^2)\mathrm{e}^{-\theta^2 R^2}\right], \tag{5.31}$$

where nS is the turns–area product of the receiver, IL is the dipole moment of the source (A × m), ρ is the resistivity of the uniform earth, y is the coordinate of the receiver center perpendicular to the source dipole which is x-directed, R is the source–receiver separation ($y=R\sin\phi$),

$$\theta^2=\mu_0/4\rho t,$$

and erf is the error function,

$$\operatorname{erf}(\theta R)\equiv 1-\frac{2}{\pi^{1/2}}\int_0^{\theta R} \mathrm{e}^{-v}\,\mathrm{d}v. \tag{5.32}$$

The voltage may be replaced by H_z via Faraday's law,

$$V=-\mu_0\, nS(\partial H_z/\partial t).$$

One would like to obtain a relation between V and ρ via equation (5.31), and this would be the desired $\rho_a(t)$ curve. However, equation (5.31) cannot be solved for ρ and there is no unique relation between V and ρ_a in this case; ρ_a can be computed only under asymptotic conditions, namely the early-time (ET) and the late-time (LT) conditions. Vanyan *et al.* (1967) give an expression for the early-time apparent resistivity (ρ_{ET}) for a short time after a step function is turned off, for the configuration described above,

$$\rho_{ET}=\frac{2R^3y}{3nSIL}\,V. \tag{5.34}$$

This is the asymptotic behavior of equation (5.31) for small values of t. Jacobson (1969) gives an expression for the late-time apparent resistivity (ρ_{LT}), for the same measuring system as described above, for long times after the step function is turned off,

$$\rho_{LT}=\left(\frac{nSILy}{Vt^{5/2}}\,\frac{4}{5}\,\pi\times 10^{-17.5}\right)^{2/3}. \tag{5.35}$$

This is the asymptotic expression for equation (5.31) for very large values of t. Different forms of relations for ρ_{ET} and ρ_{LT} are given in Jackson and Keller (1972), Kaufman and Keller (1983), Keller *et al.* (1984) and Strack (1992). The subject of apparent resistivity in TEM has created much confusion (Spies and Eggers, 1986). With the advent of direct inversion techniques it may not be necessary to convert the field data to apparent resistivity. Meanwhile, differences between various expressions for apparent resistivities and the divergence between ρ_{ET} and ρ_{LT} curves should not bother users as long as these features are consistent with the procedures and master curves used for interpretation.

Figure 5.18(a) shows the early- and late-time apparent resistivities for a uniform earth. Immediately after the source current is turned off, ρ_{ET} is equal to the true resistivity, 10 Ω m, for about 0.6 s [for $\tau/R\leqslant 2$, where $\tau=(8\pi^2\rho t/\mu_0)$]. From then it reduces toward zero. The ρ_{LT} curve

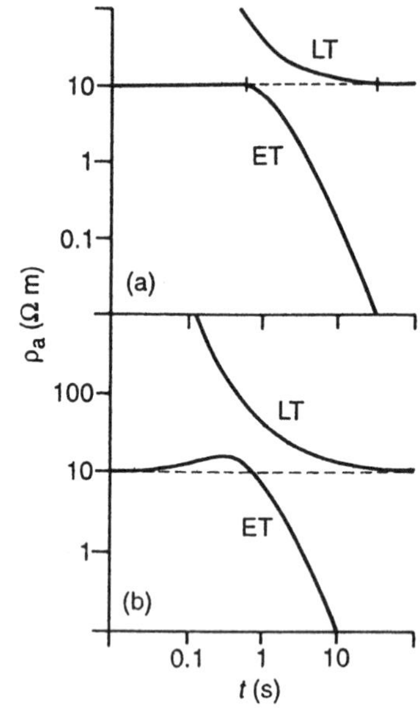

Figure 5.18 Theoretical early-time and late-time apparent resistivity curves, ET and LT respectively, for grounded-wire dipole source and horizontal loop receiver, offset $R = 10$ km; (a) for a uniform earth of 10 Ω m; (b) for a three-layer earth with $\rho_1 = \rho_3 = 10$ Ω m, $\rho_2 = 1000$ Ω m, $h_1 = h_2 = 700$ m (from Strack, 1992).

approaches the true resistivity and practically becomes so for $\tau/R \geqslant 16$ (beyond about 30 s). In a 1-D layered earth, for very small values of t, ρ_{ET} would be equal to the true resistivity of the top layer, ρ_1. The ρ_{LT} curve would approach, for very large values of t, the true resistivity of the basement, ρ_n.

Figure 5.18b shows the ρ_{ET} and ρ_{LT} curves for a three-layer earth. The presence of the resistive layer in the middle is clearly displayed; this would not be so obvious on a $V(t)$ curve.

Central loop sounding

A horizontal loop receiver is placed at the center of a large circular horizontal loop source. The voltage output of the receiver and the corresponding apparent resistivities are given in Spies and Eggers (1986). The voltage is

$$V = \frac{nSI\rho}{a^3}\left[3\mathrm{erf}(\theta a) - \frac{2}{\pi^{1/2}}\theta a(3 + 2\theta^2 a^2)e^{-\theta^2 a^2}\right], \tag{5.36}$$

where a is the source loop radius; the other symbols are the same as in equation (5.31). The apparent resistivities are

$$\rho_{\mathrm{ET}} = a^3 V / 3nSI, \tag{5.37}$$

$$\rho_{\mathrm{LT}} = I^{2/3}\, a^{4/3} (nS)^{2/3}\, \mu_0^{5/3} \Big/ 20^{2/3}\, \pi^{1/3}\, t^{5/3}\, V^{2/3}. \tag{5.38}$$

Note that equation (5.36) for a central loop and equation (5.31) for an x-directed grounded-wire dipole are identical if the receiver loop is placed on the equatorial axis of the x-directed dipole ($y = R$), for the equivalent source moment. This is intuitively explainable by envisaging an integration of the grounded-wire dipole effect along the circumference around the center of the receiver. This is also valid for a 1-D multilayer earth because of circular symmetry. In practice the large source loop is usually a square.

(b) Theory of 1-D multilayer earth

The theory of TEM is basically the same as that of FEM, since equations in one domain can be transformed to the other. Some of the comprehensive monographs on this subject are by Vanyan *et al.* (1967), Kaufman and Keller (1983) and Ward and Hohmann (1988). The coverage in these monographs is necessarily selective, because there are many varieties of source–receiver configuration and many different kinds of measured quantities.

In order to find out which source–receiver configuration to use and which quantities to measure and for how long, for a given geologic problem, one has to analyze the expressions and theoretical curves for the response of multilayer earth models.

A list of formulas for a variety of sources for the turn-off step function, for a variety of receiver types (electric field, magnetic field and its time derivative) and asymptotic apparent resistivities for a uniform earth is given in Spies and Frischknecht (1991); they also give a collection of two-layer theoretical curves with both voltage and magnetic field responses; the earth model is the same as for the FEM theoretical curves discussed in connection with Figure 5.11.

Some theoretical expressions refer to certain operating zones with regard to depth to the basement, D, versus source–receiver separation, R, defined on the assumption of a two-layer earth, as follows:

Near zone: $R \ll D$,
Far zone: $R \gg D$,
Intermediate zone: between the near and far zones.

We also have the following operating zones relative to the maximum depth of exploration d:

Short offset: $\mathrm{R} \leqslant \mathrm{D}$,
Long offset: $\mathrm{R} \geqslant \mathrm{D}$,

There are other operating zones; each has its early-time, late-time and intermediate-time windows.

5.4.3 Interpretation

Much of what has been discussed in FEM interpretation (section 5.3) also applies to TEM 1-D interpretation.

As before, manual curve matching can be done. Extensive catalogs of TEM late-time master curves have been published in the former Soviet Union since the 1960s. These curves are described in Kaufman and Keller (1983), who also give some two- and three-layer curves. In these curves the vertical axis is ρ_a/ρ_1 and the horizontal axis is $2t\rho_1/\mu_0$, which is the square of the diffusion depth in the top layer. Curve matching in TEM is not easy, for two reasons. (1) When visually inspected, curves do not display much character that would suggest the resistivity profile of the subsurface. (2) As mentioned above, both the time and apparent resistivity coordinates of theoretical curves are normalized to ρ_1, the top layer resistivity; therefore, when the matching is accomplished, the position (cross point) must also yield the correct value of ρ_1. This explains why the trial-and-error matching by a computer is also difficult. As discussed in section 5.3.2, parameter optimizing techniques are used. Much information on computer-assisted interpretation (inversion) is found in Hohmann and Raiche (1988) and Spies and Frischknecht (1991).

Kaufman and Keller (1983) give a FORTRAN listing for inversion of data obtained with a grounded-wire dipole source and horizontal loop receiver, for long-offset setups (far zone). It employs a nonlinear regression and Fibonacci search to find a model in terms of an arbitrary number of layers and arbitrary layer parameters.

The inversion of central-sounding (in-loop) data can be accomplished by means of a United States Geological Survey open-file program (Anderson, 1982).

5.4.4 Layer resolution

The two-layer theoretical curves given by Spies and Frischknecht (1991), for FEM and TEM, generally show that the relative differences between curves for various resistivity contrasts are larger for TEM than for FEM. This means that, in general, TEM provides better layer resolution than FEM.

With TEM an interface at a particular depth d can be detected best with a particular source–receiver separation R. For the grounded-wire dipole source and horizontal loop receiver, maximum sensitivity at late time is obtained when $R \leqslant d$. As discussed before, this rule is also applicable to central loop sounding; however, larger sized loops yield better quality data (better signal-to-noise ratio and less influence of near-surface inhomogeneities), and the optimum loop radius is about equal to depth, $a \cong d$.

Layer resolution and equivalence with EM methods, discussed in section 5.3 for FEM, also applies to TEM.

5.4.5 Depth of exploration

It was discussed in section 5.4.2 that the depth of exploration d depends primarily on the duration of the measuring time, t, on instrument specifications (e.g. dynamic range and resolution) and on ambient noise, but not so much on offset distance R or source-loop radius a. It also depends much on the resistivity profile of the subsurface. So many independent parameters render the depth of exploration difficult to quantify. So far as the subsurface parameters are concerned, the easiest way to study this problem is to consider the detection of the basement in a two-layer earth model, and then to consider the latest time that can be measured, given the noise level and the resolution of the measuring system, and how much stacking can be done given the economic limitations. Considering these variables, Spies (1989) finds that, for short-offset and for the voltage of the receiver in central sounding (small horizontal loop receiver at the center of a large horizontal loop source with $a \leqslant d$), the depth of investigation is

$$d \cong 0.5\,(AI\rho_1/N)^{1/5}, \tag{5.39}$$

where A and I are the source-loop area times the number of turns and current, respectively, and N is the threshold voltage noise level detected by the system, which 'typically reduces to about 0.5 nV/m^2 after 15 minutes of stacking'. According to this expression, to double the depth d the source moment IA has to be increased 32-fold! Assuming the two-layer model and the threshold noise level mentioned above, for a 200 m square source loop with a current of 20 A and a sedimentary section of 10 Ω m effective resistivity, the maximum depth of exploration would be 950 m, and the latest sample time would be 58 ms (Spies and Frischknecht, 1991). Note that in equation (5.39) there is no loop radius a which is equivalent to offset distance R with a grounded-wire dipole source. If the receiver is an H_z magnetometer, instead of a horizontal loop that responds to $\partial H_z/\partial t$, the expression for the depth of investigation becomes quite different; it no longer depends on ρ_1, and is proportional to the one-third power of the source loop moment.

There are no rules of thumb between d on one hand and t and R (or a) on the other; every variety of source–receiver configuration has to be treated separately. Theoretically there is no relation between d and R (or a). About the only rule of thumb is that, for central sounding with a horizontal loop receiver, and for a grounded-wire dipole source and horizontal loop receiver, for R or a smaller than d, the basement at a depth d is evident when

$$\delta_{TD} = (2\rho_1 t/\mu_0)^{1/2} \cong d,$$

that is, when the diffusion depth δ_{TD} is in the proximity of the basement depth d. The diffusion depth in the time domain, δ_{TD}, is equivalent to skin depth, δ_{FD}, in the frequency domain; δ_{FD} is proportional to $(\rho T)^{1/2,}$ where T is the period and δ_{TD} is proportional to $(\rho t)^{1/2}$.

The relation between R (or a) on the one hand and d and layer resolution on the other is complicated; it is somewhat enigmatic to casual users. Keller *et al.* (1984) state that for a given threshold noise level 'the greatest depth of investigation is obtained with the greatest separation (R)', and explain the reason why this is so, even though δ_{TD} is independent of R. For this reason, some investigators prefer sources with very large power to accommodate very long offsets; this is not just long offset (LOTEM), which is defined as $R \geqslant d$, but it is operating in the far zone, $R \gg D$, where D is depth to the basement.

5.4.6 Field practice

In what follows, two measuring systems that were developed for deep work, and some field results obtained by means of them, will be briefly described. These are the Colorado School of Mines and the German TDEM systems.

(a) The Colorado School of Mines (CSM) system

This is the so-called 'megasource' CSM TDEM system designed to achieve offset distances (R) of several tens of kilometers that may provide depths of investigation of several kilometers. The system together with some field results are described by Keller *et al.* (1984) and Keller (1991).

The basic elements of the system are essentially the same as in Figure 5.16. Usually, the voltage from the horizontal loop receiver is measured; but sometimes H_z is measured with a vector magnetometer (e.g. SQUID) instead and converted to $\partial H_z/\partial t$ to provide about the same signal as the loop voltage; either one is recorded digitally. Identical precision clocks control the time of the waveform at the source and the time of digitization at the receiver. Tens of successive transmissions are recorded and later a number of them are stacked synchronously.

The grounded-wire dipole source was selected primarily because the transient signal from it decays less rapidly with t and R than does the signal from a horizontal loop source.

The preferred receiver signal is $\partial H_z/\partial t$, from either V or H_z, mainly because the horizontal fields are more susceptible to the ambient EM noise. The electric field components are also measured but these are also more susceptible to the ambient EM noise than the vertical field; they may also be complicated by induced polarization effects. Nevertheless, the electric field may be indispensable in resolving high-resistivity layers.

The source dipole moment, *IL*, is in excess of 10^6 A m. This is provided by a source capacity of 1 MW at 1000 V and 1000 A, and *L* from 1 to 3 km. This source, 2000 A peak-to-peak square wave, allows separations (*R*) from a few kilometers to more than 30 km; thus, a large area containing hundreds of receiver stations can be served by one set up of the source dipole, and several measurement crews can work simultaneously.

One of the case histories given in Keller *et al.* (1984) is the Pacific Northwest survey, near Yakima, Washington, USA, in a search for petroleum-bearing sedimentary sections, under a thick section of volcanic cover. One 1900-m long source dipole served 267 sounding (receiver) locations. The establishment of such a dipole with low resistance constitutes a major portion of the survey work. The dipole location was chosen along an irrigation canal. Ten 5- in boreholes, 20 ft deep, with 4-in aluminum pipes emplaced in the boreholes, constituted each end of the AB dipole. The source signal was a square wave with a period of 80 s and a peak-to-peak amplitude of 610 A. The receiver was a horizontal loop with 102 turns and a perimeter of 166 m. 16 to 48 recordings were made at each receiver station for stacking.

Keller *et al.* (1984) give two profiles of sounding curves and 1-D inversion results, selected from an areal coverage of 267 stations. One of these profiles is close to the equatorial axis of the source dipole ($\phi \cong 90°$, near the *y*-axis in Figure 5.16). Figure 5.19 shows two stations at the near end (offsets of about 8–10 km), and two stations at the far end (offsets of about 30–31 km). My purpose in selecting these end stations is to discuss what happens to the 1-D assumption when *R* varies over such a large range. There is a maximum in the tail portion of the early-time ρ_a curves at the far end of the profile; this maximum disappears somewhat at the near end of the profile; however, the inversion results do not show substantial changes. Keller *et al.* (1984) state that 'The change in shape of the curves does not indicate a change in the geoelectric sequence in this case, but instead, is caused by the change in separation from source to receiver'. The other profile (not shown here) is made of stations that are roughly equidistant (about 30 km) from the source dipole and it is about 30 km long. The shape of the sounding curve remains relatively constant along this profile.

The inversion was done in two stages. In the first stage only the layer thicknesses were free to vary. In the second stage the output of the first stage was taken as the initial model and the first and last layer resistivities, 200 and 100 Ω m, respectively, were fixed. No significant well log data were available to Keller *et al.* (1984) at the time of writing to evaluate the results more effectively.

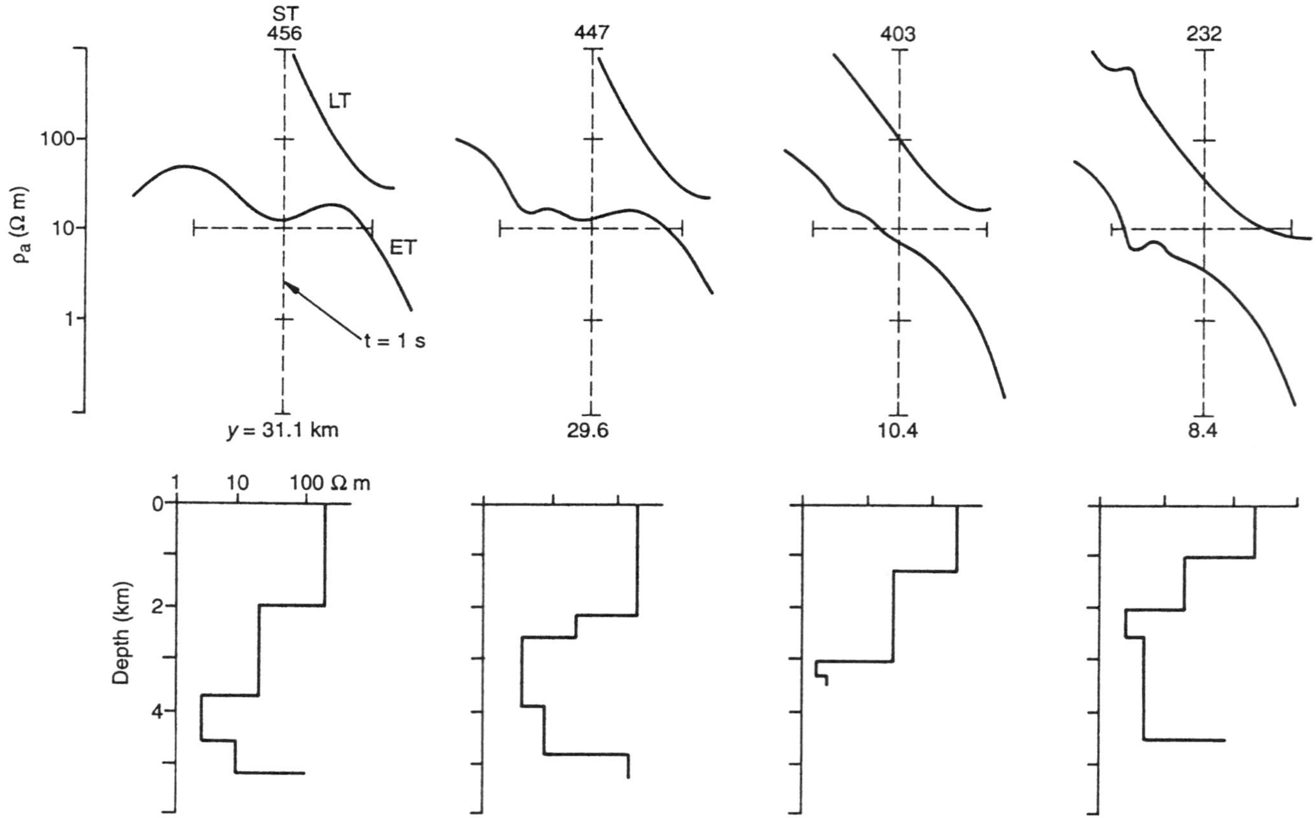

Figure 5.19 LOTEM survey at Yakima, Washington. Offsets for sounding stations (receiver locations) 232 and 403 are about 8–10 km and those for 447 and 456 are about 30–31 km, all are near the equatorial axis of the grounded-wire dipole source (from Keller *et al.* , 1984).

Recent developments in the CSM TDEM systems are discussed in Keller (1991).

(b) The German deep TDEM system

The basic elements of this system are the same as in Figure 5.16. The source dipole length is between 1 and 2 km. Usually, a continuous bipolar square wave current of several tens to several hundreds of amperes is used. The offset distance (*R*) is from 2 to 20 km. It has a multichannel data acquisition system called TEAMEX. Many digital receiver station units are connected together by a line.

This system is described by Strack (1991, 1992); his book (1992) contains much detail about instrumentation, system tests, preparation of the source dipole and safety. Many field examples and interpretation results, including 3-D effects, are given.

As discussed in section 5.3.2 (b), parameter optimizing Marquardt-type inversion techniques are used with *a priori* information from well logs, seismic and other subsurface data. First, all stations on a profile are inverted individually, using the same starting model, usually three or four layers. Then the inversion is repeated by using the inversion results at one station as starting model for the next station. Further stages of inversion are the *soft-bound* and *hard-bound* inversions. In the soft-bound inversion the layer parameter weights are determined from other subsurface data. Parameters with large weights are allowed to change little and those with small weights can change more extensively. In the hard-bound inversion, parameters are strictly constrained to a given range which is derived from other subsurface data.

(c) Siberian platform, Yaraktin survey

This short-offset TEM survey over the Yaraktin oil field in the Siberian platform, Russia, is discussed in section 6.2. It has been reported by Rabinovich, Surkov and Mandelbaum (1977), Kaufman and Keller (1983), and Spies (1983). It comes close to being a 'direct detection' of oil in that it shows the lateral position of oil–water contact in a stratigraphic trap which is at a depth of 2600 m and is 42 m thick.

5.4.7 Concluding remarks

It is well known that MT measurements see too far laterally into distant inhomogeneities that are of no concern to 1-D interpretation at a receiver station. This is the consequence of using a plane-wave source, whereas with controlled-source systems such effects are relatively small because the source field strength rapidly falls off with distance. This statement is certainly correct for short-offset TEM, especially for central sounding, but not for long offsets of several tens of kilometers. Strictly from the point of view of spatial resolution, the short-offset systems are fine but they create an economic problem, because each sounding requires a new source to be set up. It seems that there is need for extensive model studies to learn more about what happens to a 1-D interpretation at a receiver location when the offset distance is several tens of kilometers. Also, improvements can be made by developing small, powerful, even portable, source dipoles, and receivers with higher dynamic range and better signal resolution, so that short-offset measurements can be made economically for both shallow and deep sections.

Another subject of discussion is the degree of parameter optimizing employed in the published case histories concerning petroleum exploration. A profile of individually inverted stations using an initial model (starter) that is normally guessed may yield a highly irregular cross-section. By the time all of the information from other subsurface data is used and the parameters are optimized and constrained, together with the spatial filtering effects of profile inversion, the cross-section may look as smooth as a seismic section. In such circumstances it is difficult to appraise the geologic significance of a log obtained from TEM inversion, plotted right on top of a blocked-off well log. Improvements in direct inversion and more well-documented deep case histories are needed before TEM becomes widely accepted in petroleum exploration.

6

Possibilities of direct detection of petroleum

6.1 INTRODUCTION

This book would not be complete without this chapter, with such a sensational title, simply because by far the most geophysical work carried out in sedimentary basins pertains to oil and gas.

During the decades from about 1933 up to the 1960s many geophysicists nurtured the belief that one day some electrical method would help to detect oil directly; and this would lead to the discovery of vast stratigraphic traps which contain perhaps 75% of the oil ultimately found in the USA. Furthermore, these geophysicists also believed that only electrical methods could accomplish this, because oil-bearing formations frequently display high resistivities and unusual electrochemistries. Since then, seismic methods have come a long way in so-called direct detection, not electrical methods. At present, the consensus is that the electrical methods have turned out to be disappointing in this field. This let-down was partly due to unrealistically high expectations, especially at the beginning, prior to the substantial theoretical developments that took place in the 1960s. However, subsurface electrochemistry is still not well known.

Perhaps one should not even use the familiar expression 'direct detection' at all; this itself constitutes an over-expectation. It may be called 'detection of indications of oil other than structure', or simply 'non-structural prospecting for oil'.

A thorough discussion of the subject, similar to what has been done in the remainder of the book for other methods, including how the promoters on one hand and the scientists on the other have explained the observations, would be too voluminous.

It is almost impossible to produce a satisfactory classification of the methods so far employed, because some of them are known only under trade names which do not effectively represent the basic principles involved. However, several tentative headings may represent the methods used:

- resistivity of oil and gas reservoirs;
- electromagnetic reflection (?);
- induced polarization;
- miscellaneous.

Geophysicists are aware of the fact that present-day techniques in kinds of exploration are always partly empirical; the empiricism varies in degrees from one method to another and also from one case to another. The present subject includes an unusually high dose of empiricism, but we must reserve a degree of respect for the available data for the sake of the complexity of this subject, because some of the unexplainable data may turn out to be useful in the future; some already have done so.

Some of the methods are aiming to detect oil and gas reservoirs directly. Others are aiming to detect shallow indications of oil or gas at depth; complicated geochemical and biological factors are involved. The geochemical phenomena associated with oilfields that create concentrations of minerals in shallow formations above oil reservoirs are outside the scope of this book; the reader will find a list of references in Goldhaber and Reynolds (1991). Such anomalous mineralization includes pyrite, marcasite, pyrrhotite, sedimentary ore deposits, carbonate cementation, silica and salinity anomalies.

6.2 RESISTIVITY OF OIL AND GAS RESERVOIRS

The objective is to locate oil reservoirs directly on the basis of high resistivities associated with them, when oil and/or gas replace formation water (electrolyte) in a porous rock. Oil, gas and dry rock have practically infinite resistivities. For a clean sand with 20% porosity, saturated with a mixture of electrolyte and oil, the bulk resistivity of the formation is relatively low (about 20 Ω m) up to 80% oil saturation (i.e. 80% of pore space is filled with oil and 20% with electrolyte). Above 80% oil saturation the resistivity rises rapidly, to thousands of Ω m for 92% saturation.

On the basis of a simple form of Archie's law, the resistivity of an oil sand may be written as (Keller, 1971)

$$\rho_B = \rho_W / P^2 S^2, \tag{6.1}$$

where ρ_B is the bulk resistivity, ρ_W is the electrolyte resistivity, P is the porosity expressed as a fraction per unit volume of rock, and S is the fraction of pore space filled with electrolyte.

Let us consider a hypothetical formation with $\rho_W = 1\ \Omega$ m and $P = 0.25$. When it is saturated with electrolyte only (no oil), that is, $S = 1$, its bulk resistivity is 16 Ω m. When it is 90% oil saturated ($S=0.10$), its resistivity is 1600 Ω m; a 100-fold rise. Let us assume that such a 90% oil saturated formation is 40 m thick, 1.5 km deep, has a horizontal extent of 4 km^2 and has horizontal uniformity and continuity. Could this oil sand be detected by resistivity sounding?

With a Schlumberger array, AB would have to be expanded up to about 6 km, because depth of exploration is about AB/4. This is too large for the extent of the target. An equatorial dipole array with a maximum separation of about $\bar{R}=3$ km has a depth of exploration of 1.5 km. This might do it.

It is discussed in sections 3.2.4(f) and 4.3.3(e) that in resistivity sounding (DC) such a thin, high-resistivity layer can be determined only in terms of the product $h\rho$ which, in this case, is 40×1600. The response from this layer would be similar to that due to a much thicker layer having a much lower resistivity. This is the so-called shielding (screening) effect that constitutes a serious disadvantage for resistivity sounding, because it necessitates excessive expansion of the array to penetrate through a thin high-resistivity layer. Now it is not necessary to penetrate through the oil sand; all that is needed is to detect the top of the 40×1600 layer. Thus, a disadvantage may be turned into an advantage.

It is shown in section 4.3.3(e) that in magnetotelluric (MT) sounding such a thin high-resistivity layer shows up only in terms of thickness h, simply because such a layer does not carry much current.

It appears that the oil sand would create a large response with DC and a very small one with MT. It might be possible to identify such thin high-resistivity layers embedded in a lower-resistivity environment if the two kinds of measurements are combined. However, a thin layer of carbonate or evaporite rock might give a similar response.

A controlled-source electromagnetic (CSEM) sounding method may be added to the DC–MT combination discussed above, or MT may be replaced by CSEM because, like MT, the CSEM soundings also are sensitive to the thickness, but not the resistivity, of a thin resistive layer (Chapter 5). This could be a time-domain (TEM) or frequency-domain (FEM) short-offset array in which the source–receiver separation (R) is less than the depth of exploration. Alternatively a TEM or FEM central sounding may be used, in which a large horizontal square-loop source is used with a magnetic field receiver at the center. Such systems would have good horizontal resolution and would be less influenced by near-surface inhomogeneities and by distant structures than MT measurements are. Joint inversions of different kinds of data. DC-MT (Jupp and Vozoff, 1977) and DC-TEM (Raiche *et al.*, 1985), may improve the resolution of a thin resistive layer substantially.

Detection of an isolated oil sand, as discussed above, is possible only under very favorable conditions which are rather restrictive. The resistivity contrast of 100 is unusually high; this is for a clean sand. Shaly sands have much lower resistivities for the same oil saturation. Carbonate reservoirs are unpredictable, because the rocks above and below the reservoir may have higher resistivities.

This kind of work seemed to be promising in the 1930s and 1940s. Successes were claimed in certain instances, but the outcome has been generally discouraging (see, for instance, Karcher and McDermott, 1935). However, these trials were done too long ago, when MT and DC dipole sounding methods did not even exist, and TEM and FEM sounding techniques were not developed.

In certain special cases of multiple layers of hydrocarbon saturation whose vertical extent may be comparable to its horizontal extent, the anomalous high-resistivity zone may be detected by many kinds of electrical methods. The Gazli field in Uzbekistan is such a case. The results of an electromagnetic survey were reported by Keller (1969). There was much subsurface control from seismic and drill data to check the results. High gas saturation is found at depths of between 600 and 1200 m. The electromagnetic survey (unspecified) showed a high-resistivity anomaly over the gas field that was double the value measured off the gas field.

Rabinovich, surkov and Mandelbaum, (1977), Kaufman and Keller (1983) and Spies (1983) describe the TEM surveys carried out in the Siberian platform, Russia, that were effective in locating oil and gas reservoirs by identifying the lateral positions of oil–water contacts in stratigraphic traps. In this region the reservoir rock is a deep, thin argillaceous sand which normally is saturated with saline water (0.1 Ω m) and has very low resistivity (1 Ω m) when not saturated with oil or gas. The overlying rocks, carbonates and evaporites, have high resistivities. When the saline water in the reservoir rock is replaced by oil and gas the conductance $S(=\Sigma h_i \sigma_i)$ of the whole section above the basement is strongly modified. The lateral position of the oil–water contact can be identified by means of a short-offset TEM with grounded-wire dipole source and horizontal loop receiver. This configuration is suitable in recording the late-stage transient coupling which is sensitive to changes in S. Figure 6.1 shows a set of theoretical curves for a section similar to that described above. The presence of a single, deep (2600 m), thin (20 m), low resistivity (0.5 Ω m) layer causes a major change in the sounding curve. The Yaraktin oilfield in this area is one case history in which the TEM system described above was effective. The source–receiver separation (R) varied from 1 to 4 km. The oil-bearing formation is a 42-m thick argillaceous sand forming a stratigraphic trap at a depth of 2600 m. Figure 6.1 is a model study for this field. Curve I represents the case in which the reservoir is saturated with oil or gas; the resistivity is so high that the reservoir is lumped together with the basement ($\rho=\infty$). Curves II and III correspond to models with saline water-saturated reservoirs. In other

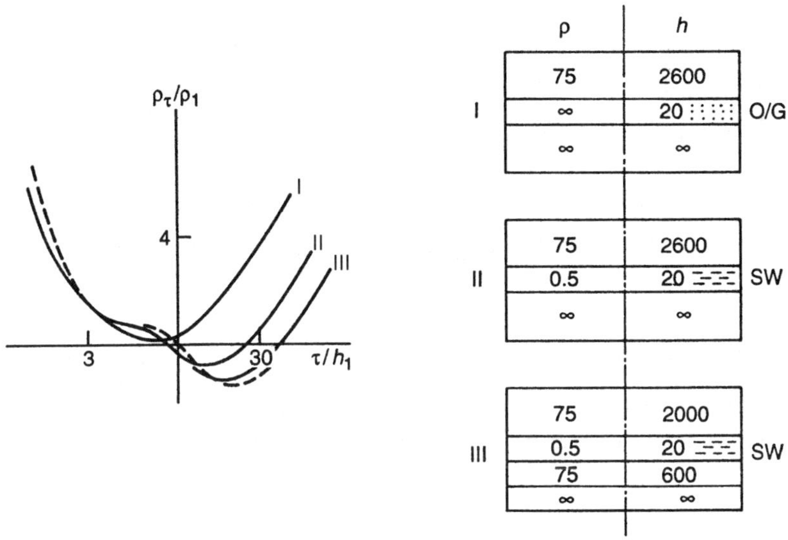

Figure 6.1 Model study using short-offset time-domain electromagnetic (TEM) sounding with a grounded-wire dipole source and horizontal loop receiver. Top layer parameters are ρ_1 and h_1 (thickness); ρ_τ is the late-time apparent resistivity; $\tau = (2\pi\rho_1 t \times 10^7)^{1/2}$. Curves I, II and III are for respective models shown in which thicknesses are in meters and resistivities in Ω m. Curve I represents the case in which the reservoir is saturated with oil or gas (O/G); the resistivity is so high that the reservoir is lumped together with the basement ($\rho = \infty$). Curves II and III correspond to a saline water-saturated reservoir (SW). The dashed line is an actual soundings curve from the Yaraktin oilfield (from Rabinovich, Surkov and Mandelbaum 1977; also from Kaufman and Keller, 1983, and Spies, 1983).

words the TEM data detect the absence of oil or gas. The dashed line is an actual sounding near the oilfield. Profiles of TEM soundings show clear changes in the sounding curves where the oil saturation zone is laterally crossed, as shown in Figure 6.2.

An extensive compilation of data in many regions of the former Soviet Union, involving a study of 950 boreholes, led to a classification of electrical resistivity models of oilfields. These models are represented by six schematic four-layer cross-sections, from 'very large deposits' to 'non-prospective areas' (Spies, 1983). The third layer from the top represents the section which includes the oil- or gas-bearing formation; its average resistivity in the producing zone is from 1.5 to 3 times the values measured off the producing zone for very large deposits (it is a factor of 10 for the Gazli field, as reported by Spies, 1983), and from 0.7 to 1.3 for the productive stratum in non-productive areas. The top layer is usually associated with a resistivity decrease by a factor of 0.8.

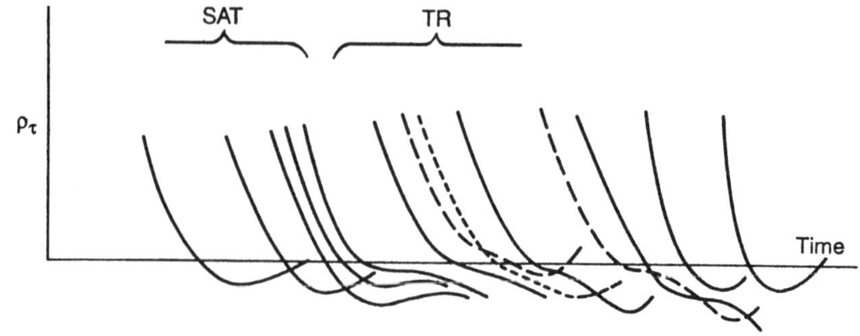

Figure 6.2 A profile of TEM soundings over the Yaraktin oilfield, showing the character change from the oil–water transition zone (TR) to the fully oil-saturated zone (SAT). Notations are the same as in Figure 6.1 (from Rabinovich, Surkov and Mandelbaum 1977; also from Kaufman and Keller, 1983, and Spies, 1983).

6.3 METHODS SAID TO BE ELECTROMAGNETIC REFLECTION

Most geophysicists would dismiss this subject on the basis of a conventional analysis of Maxwell's equations, because in sedimentary basins electromagnetic (EM) waves attenuate very rapidly and they are very dispersive with frequency. Reflections can take place only at high frequencies but these do not have appreciable penetration. Low frequencies do penetrate but their wavelengths are too long to resolve geologic formations.

However, in applying Maxwell's equations we make many assumptions, such as that of linearity (Ohm's law). In addition to the fairly well-known induced polarization (IP) phenomena in relatively shallow sections above oil reservoirs, the reservoirs themselves may display complicated electrochemical phenomena whose responses to an external EM field would not be predictable.

Two of the methods claimed to be based on EM reflections are Radoil and Elflex. The names imply that the former finds oil with radio waves, and the latter is electromagnetic reflection. These methods preoccupied many geophysicists of oil companies, generated much fieldwork and much controversy, especially during the 1950s and 1960s. Here, unfortunately we have to discuss this subject with reference to the trade names, for the reasons mentioned in section 6.1, and for the sake of historical continuity.

The Radoil method was developed during the years following 1936 (see, for example, Pratt, 1953). The transmitted waves have a constant frequency of 1700 kHz and constant amplitude; this is claimed to be in a frequency window of favorable penetration. The fieldwork is carried out in a way that is analogous to seismic refraction surveys. It is claimed that this procedure creates constructive interference of reflected waves and is

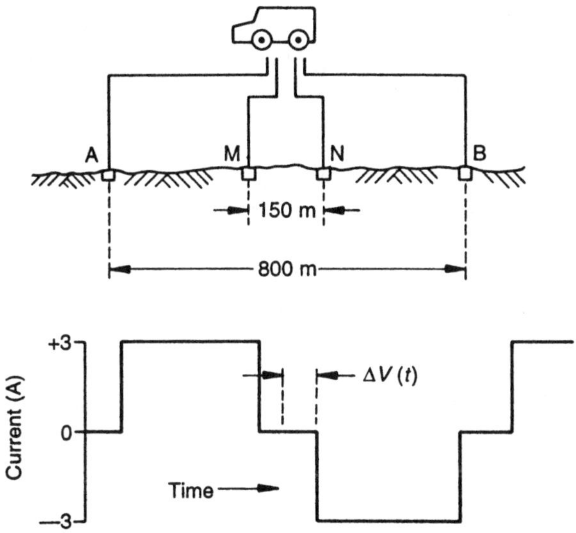

Figure 6.3 Field setup for the Elflex method. The transmitter AB is about 800 m long; the receiver MN is about 150 m long; the current cycle period is from 1 to 10 s; the voltage measurements are made in the interval indicated as $\Delta V(t)$.

the reason for its success. Published data show diagnostic anomalies observed at the edges of oilfields. It is possible that the Radoil measurements are influenced by the near-surface effects of oil at depths.

The other method, Elflex (Evjen, 1948), has taken a much larger share of oil company interest. Different versions of it have been offered in different parts of the world.

The field procedure is shown in Figure 6.3. The electrode array may be Schlumberger (as shown here) or dipole–dipole (axial dipole), or Wenner. At MN (receiver) the time variations of voltage, $\Delta V(t)$, are recorded during the gap while no current is fed into the ground. The integrated voltage at MN divided by the zero-to-peak amperage in AB (transmitter) is the quantity measured at each station. Note, in Figure 6.3, that there is a time lapse between shutting off the transmitter current and measuring the receiver signal. It is claimed that the time lapse allows one to exclude the effects of EM coupling and also to bypass reflections of the high-frequency components from shallow layers which are of no interest in petroleum exploration. The measured signal is attributed to low-frequency reflections from oil sand or saltwater sand at depths such as 2000 m.

Spectacular case histories were published by the proponents (see, for instances, Evjen, 1953), mostly in trade journals.

Even though the consensus on the reported results seemed to be in favor of the Elflex method, geophysicists were reluctant to take this method seriously, because they were unable to reconcile Elflex data with the classical theory of electromagnetic wave propagation in sedimentary basins, in terms of reflections from depths of 2000 m, and also in terms of oil reservoir singularities: why should there not be reflections from a limestone layer?

Another method that is very similar to the Elflex is the Electraflex. Its principles and measuring procedures are explained by Azad (1973). Here also the anomalies are attributed to EM reflections. Azad (1973) gives six field examples in Canada with before-and-after drill results, over pinnacle reefs, structural and stratigraphic traps. Petersen (1974) gives a summary of the results with the Electraflex method.

During the last 50 layers or so, hundreds of crew–months of Elflex-type surveys have been carried out and, according to the case histories mentioned above, some oil was discovered as a result of these surveys. To what extent these surveys were cost effective is not clear.

It appears that the Elflex device shown in Figure 6.3 would measure primarily a combination of induced polarization (IP) and EM coupling responses of a shallow section, down to a few hundred meters. This kind of EM coupling is discussed in section 5.4. The IP effects are discussed below.

The resistivity information obtained from the EM coupling, even though it may pertain to a shallow section, is frequently correlated with petroleum at depth via several geochemical and physical effects, as discussed in section 6.5.1. This is in addition to the familiar IP effects. Also, when measurements were made over known fields to 'calibrate the data', well casings and pipelines might have affected the measurements, mostly via IP effects, and these could have sometimes accounted for some of the anomalies observed over known oilfields.

6.4 INDUCED POLARIZATION (IP) METHODS

The induced polarization (IP) methods of prospecting date back to 1912 when Professor C. Schlumberger of France obtained a German patent on such a method.

The IP response of a subsurface is similar to that of a rechargeable electric battery; after the electric current has been cut off, the subsurface gives back electric current that gradually decays. Schlumberger was concerned with prospecting for metallic sulfides, and still, at present, IP is one of the most cost-effective tools to prospect for disseminated metallic sulfides.

In the 1930s some scientists tried to apply this concept to detect petroleum reservoirs directly, i.e. the oil itself *in situ*, not by means of shallow geochemical by-products of oil. Professor M. Müller of the University of Jena in Germany published a series of articles from 1932 to 1950 (Müller, 1940, 1950) on what he called the electrochemical method. Professor

G. Potapenko of the California Institute of Technology was granted a US patent in 1940 for direct detection of petroleum by means of IP.

In the late 1940s 'IP-type' field measurements appeared under the name of electromagnetic reflection (Elflex) methods, as discussed in section 6.3.

Finally, in the mid-1970s the same types of field procedure were offered by contractors as IP and said to be working on relatively shallow sulfide mineralization caused by oil and gas at depth.

There are several books, e.g. Sumner (1976) and Fink *et al.* (1990), and a great number of articles on IP, since it is one of the major tools in mining geophysics.

When direct current is injected into the earth by means of a transmitter dipole it creates a potential field at the surface that can be measured as a voltage difference by means of a receiver dipole. When the transmitter current is shut off it is found that, in general, the receiver dipole voltage does not vanish immediately but persists for some time and decays gradually, over a few minutes or a few seconds. There are several principal causes of this gradual decay behavior.

For IP it can be due to particles of metallic sulfides. Each particle behaves like an electrode which constitutes a boundary between ionic and electronic conductions. This is called **electrode polarization**. Metallic disseminated sulfides create large anomalies of this kind. IP can also be in the form of **membrane polarization** due to the presence of clay particles, without the presence of any metallic sulfide particles. The two effects are usually present in different proportions.

Electromagnetic (EM) decay, also called electromagnetic coupling, is a cause of decay. This is the ordinary EM phenomenon related to frequency dependent dispersion and attenuation in low-resistivity environments. In sedimentary basins it can be very pronounced and can drown IP signals if one is not careful. The EM coupling can be separated from IP through data processing by assuming certain subsurface models.

There are two basic types of IP measuring procedures: (1) time domain, and (2) frequency domain. With both procedures any of the electrode configurations used in conventional resistivity work can be employed. Most frequently used configurations are either Schlumberger or dipole–dipole (axial dipole). The time- or frequency-domain techniques are equivalent. Figure 6.4 explains the procedure and the quantities measured with time-domain IP. The normalized area under the decay curve at the receiver is the **chargeability**,

$$M = \frac{1}{V_{\mathrm{P}}} \int_{t_1}^{t_2} V_t \, \mathrm{d}t. \tag{6.2}$$

The 'theoretical chargeability' is

$$\mathrm{m} = (V_{\mathrm{S}}/V_{\mathrm{P}}) = (\rho_{\mathrm{DC}} - \rho_{\mathrm{AC}})/\rho_{\mathrm{AC}}, \tag{6.3}$$

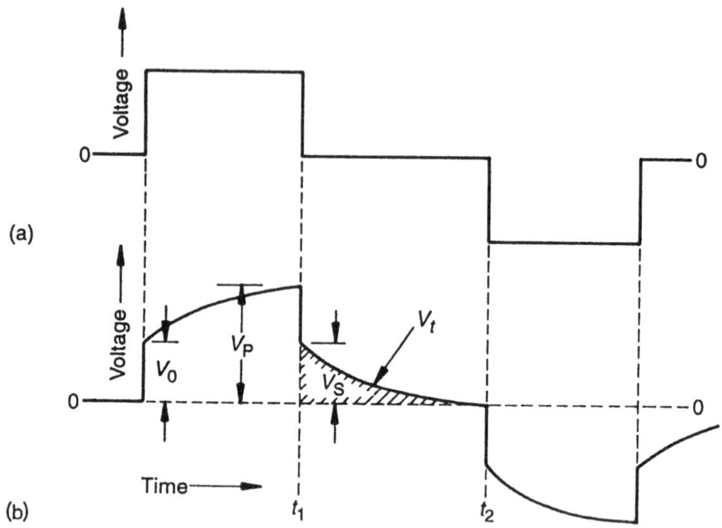

Figure 6.4 Time-domain IP. (a) Transmitter voltage, (+) and (−) are current-on periods, each about 2 min; (b) receiver voltage (schematic).

where ρ_{DC} is the apparent resistivity with DC or very low frequency; ρ_{AC} is the apparent resistivity with high frequency. The above equation states that the time-domain and frequency-domain measurements are equivalent.

Figure 6.5 explains the principle of frequency-domain IP. Since it takes time to build up polarization in the earth, measured apparent resistivity varies with the frequency of transmitter signal; high frequencies show less IP effect than low frequencies. Also, the phase between the transmitted and received signals is another measured IP parameter. The **frequency effect** (FE) is

$$FE = (\rho_2 - \rho_1)/\rho_1 \cong (\rho_{DC} - \rho_{AC})\ \rho_{AC} \cong -1.8\ \phi, \tag{6.4}$$

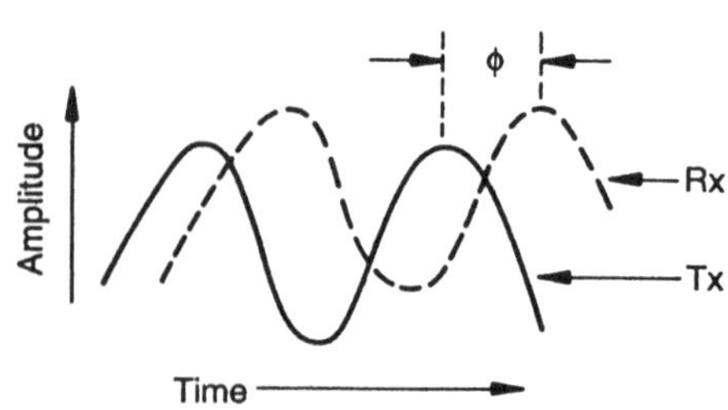

Figure 6.5 Frequency-domain IP, showing transmitter voltage (Tx), receiver voltage (Rx) and phase difference (ϕ) (schematic).

where ρ_2 is the apparent resistivity with low frequency; ρ_1 is the apparent resistivity with high frequency; ρ_{DC} and ρ_{AC} are the same as in equation (6.3); ϕ is the phase shift in radians, as shown in Figure 6.5. Phase from one frequency yields the FE; the same information comes from apparent resistivities at two frequencies. As implied in equation (6.4), phase is approximately independent of frequency (within the frequency range of about 0.05–1.2 Hz for metallic mineralization). On the other hand the phase shift of the EM coupling effect increases approximately linearly with frequency. This allows the separation of the IP phase from the EM coupling phase.

Another development in IP is what is called the complex resistivity (CR). This is a multifrequency IP, also called spectral IP (SIP), that measures apparent resistivities and phase shifts at a large number of discrete frequencies. It yields a curve showing imaginary apparent resistivities versus real apparent resistivities for a number of frequencies. This enables one to remove the EM coupling, and may permit one to identify host rocks and to discriminate minerals. The decoupled EM data, called remanent EM (REM), also constitute valuable information in their own right, since they constitute frequency-domain EM data. There are simple types of IP methods (not really CR methods) which allow one to accomplish a first-order EM coupling removal, such as the one that measures the response for a square wave and for its third harmonic simultaneously. These will also be considered as CR here, so long as they allow EM decoupling to some degree.

Probably the first significant application of CR to oilfield study was made at the Lambert Field in northern Texas (Snyder *et al.*, 1981). This small field was discovered early in 1979 and was surveyed in the summer of 1979 before the completion of development drilling and the placement of permanent production facilities, in order to avoid their IP effects. The production depth is 6500 ft (1980 m). The field appears to be associated with a structure draped over a paleotopographic high on the basement. Broadband data covering the frequency range from 0.02 Hz to 110 Hz were acquired by using four fundamental frequencies of square waves.

Figure 6.6a is a pseudo-section plot of the 'DC' phase (phase shift at 0.02 Hz in this case). The data were gathered by means of a dipole–dipole (axial dipole) electrode array (collinear transmitter and receiver dipoles of fixed lengths but expanding separation *R*). This type of data plotting, at the intersections of 45° diagonals, in the form of a cross-section, puts the reference point for each measurement at a 'depth' of $R/2$ (section 3.3.2). However, this is merely a style of data display; it is not supposed to convey any quantitative notion about the depth of exploration or the shape of the subsurface model. The depth of exploration is, on average, $R/3$. It is clear that there is a well-developed IP phase

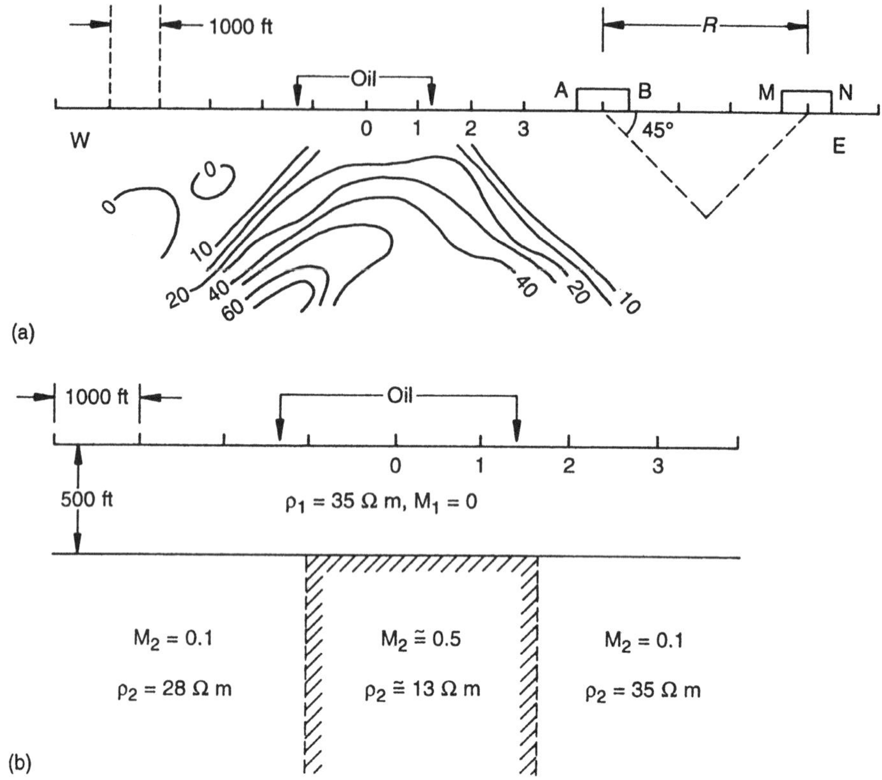

Figure 6.6 (a)Pseudo-section of DC phase (0.02 Hz) of CR data, in milliradians, for the Lambert Field, Texas. Transmitter AB and receiver MN both are 1000 ft (305 m) long. (b) Geologic model from inversion of broadband CR data from the Lambert Field, Texas (after Synder *et al.*, 1981).

anomaly coinciding with the oil–water contact boundaries. Figure 6.6b is a two-dimensional subsurface cross-sectional model which was obtained by inversion of broadband CR data along the profile of Figure 6.6a. The uppermost 500 ft (152 m) is fairly uniform. Below 500 ft, down to an undetermined depth, the chargeability M (Figure 6.4) is laterally anomalous, being high over the reservoir. The resistivity is low over the reservoir. Figure 6.7 shows the normalized complex resistivity plots in the form of an Argand diagram. Phase and amplitude relations yield imaginary apparent resistivity versus real apparent resistivity plots for frequencies increasing counterclockwise on the diagram. Imaginary apparent resistivity comes from the component of receiver signal that is 90° out of phase from the transmitter signal. Figure 6.7a was obtained at a site off the Lambert Field, Texas, away from the oil. At the low end of the frequencies, 0.02 Hz, the imaginary component nearly vanishes. This

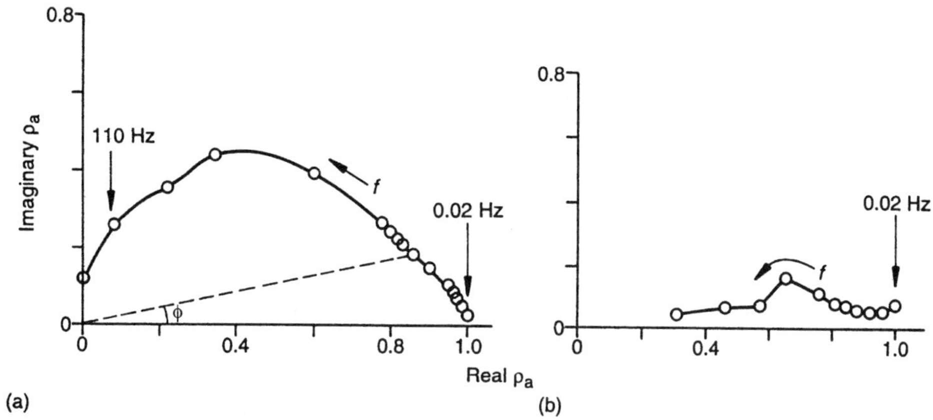

Figure 6.7 Argand diagrams of normalized broadband complex resistivity from the Lambert Field, Texas; (a) at a site off the oilfield; (b) at a site on the oilfield (after Synder *et al.*, 1981).

means that this is a purely EM phenomenon, namely EM coupling without IP effect. Figure 6.7b was obtained at a site over the field. At 0.02 Hz the imaginary component does not vanish. This diagram represents a combination of IP and EM coupling.

Microscopic examination of drill cuttings from the Lambert Field showed micron-sized pyrite, and the authors concluded that the cause of the CR anomaly over the field is IP which is probably generated by the presence of pyrite in the sediments overlying the oil reservoir that was precipitated in the reduced hydrocarbon migration zone which is in the form of a geochemical plume.

Figure 6.8 is a pseudo-section of IP anomaly (DC phase) over the Meridian Field, West Texas (Snyder *et al.*, 1981). This is a stratigraphic and hydrodynamic trap which produced oil and gas at a depth of 5000 ft (1524 m). This reconnaissance CR survey was completed in one day. It shows a well-defined IP anomaly coinciding with the producing area.

Hughes, Zonge and Carlson (1986) summarize their experiences and thoughts concerning the general subject of 'application of electrical techniques in mapping hydrocarbon-related alteration'; they have done valuable pioneering research in mining-related CR, and made several petroleum-related CR and controlled-source MT surveys. They seem to have mixed feelings, and conclude that the occurrence of geochemical alteration due to upward hydrocarbon migration is well established, but much more research is needed in order to see whether or not this alteration produces measurable and consistent electrical responses. They seem to have been influenced in their thoughts by Oehler and Sternberg's (1984) paper describing two IP surveys over two locations;

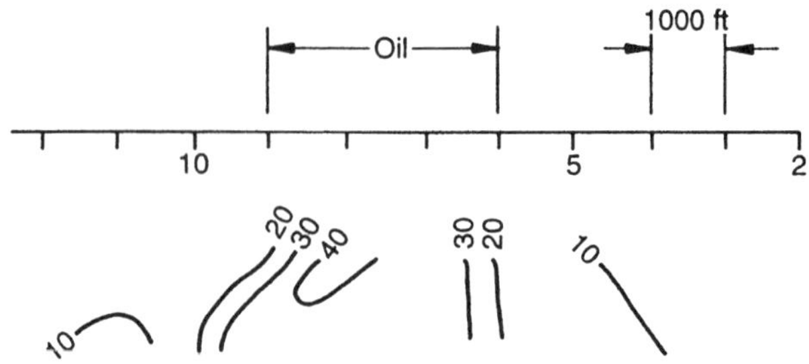

Figure 6.8 Pseudo-section of DC phase of IP anomaly in milliradians, from the Meridian Field, Texas (after Synder *et al.*, 1981).

one, over the Ashland Field, Oklahoma, has a real (petroleum-related) IP anomaly; the other, over a prospect at Salt Draw, Texas, has a 'false' IP anomaly (see also Sternberg and Oehler, 1990).

There is a time-domain IP method with an interesting data plotting technique called the Transiel method. Duprat, Roudot and Spitz (1986) explain the data gathering and processing procedures and present two case histories, one on metallic sulfides the other at a geothermal field. The authors describe the technique as a qualitative IP method that can distinguish deep effects from shallow effects (noise) through digital data processing. It is this discrimination that distinguishes the Transiel method from conventional IP. Otherwise, the field procedure is very similar to that of the Elflex method (see Figure 6.3). Transiel employs a Schlumberger array with a transmitter AB length of 1500 m. The transmitter signal consists of on–off–reverse–off rectangular pulses, of the order of 10 A, each stage being 4 s long (16-s period). Digital recording is done at the receiver (MN) during the entirety of the transmitter-off stages. At each site measurements are repeated many times and decay curves are stacked by microprocessors. Data processing does not allow even an approximate removal of EM coupling. The depth of exploration does not exceed 500 m. Through normalization and averaging of elementary time windows a parameter called **residual amplitude** (RA) is plotted as a function of decay time. This yields RA time sections that look like seismic time sections, for a decay time interval of from zero to 4 s. Anomalies confined to very short times, say from zero to 1 s, are considered as shallow noise, and those between 2 and 4 s are considered as valid RA anomalies.

6.5 MISCELLANEOUS OBSERVATIONS AND IDEAS

6.5.1 Shallow resistivity anomalies

These are among a class of indications called **surface patterns** which includes geochemical and bacteriological analyses and radioactive surveys. It was known in the 1930s that there is an excess of shallow mineralization over deep-seated oil structures. Rosaire (1938) gives an account of the observations. The mineralization forms a stratigraphic 'cap rock'. Its presence is evidenced by shallow resistivity and Eltran (electromagnetic transient) anomalies which coincide with shallow seismic refraction anomalies (velocity highs), at places where there is no structural uplift at depths involved. Drilling rates over structures, in shallow sections, are lower that those over the flanks. Rosaire (1938) gives three shallow Eltran anomalies coinciding with deep-seated oil structures and cites a number of them. The Eltran method, which was used in the 1940s and 1950s, employs a dipole–dipole (axial dipole) array. An electric surge, supplied by a condenser, is fed into the earth at the source dipole. The resulting response voltage is simultaneously recorded at the receiver dipole. This oscillogram is equivalent to the time derivative of the step function response voltage.

6.5.2 Spontaneous potential (SP) anomalies

The issue here is the direct detection of petroleum with the SP method from the surface. It is claimed that oilfield environments create direct electric current flows which can be detected and defined by mapping the equipotential contours at the surface. Since 1940 US patents on such methods have been issued and many surveys have been made. The proponents wrote, in trade journals, about current streams arising from the peripheries of an oil reservoir. They observed various kinds of anomalies and speculated about the mechanism of generation.

Pirson (1980) believes that confining (entrapment) of oil is governed by redox potential, E_H, which is a function of the ratio of reduced to oxidized substance in a rock. He also believes that an oil reservoir is overlain by a chimney of reduced rocks that gives rise to a natural electric current that flows downward to the reservoir, and this creates an SP anomaly.

6.5.3 Possibility of nonlinear behavior of oil and gas reservoirs

In the course of a series of field experiments concerning resistivity sounding, carried out by Lee (1939), the following observations were made.

Resistivity soundings were made upon shallow gas pools in central Michigan. A correlation of the characteristic abrupt changes (kicks) on the curves resulted in regular horizons over a widespread area. Apparent resistivity varied with current density.

It is possible that these correlatable kicks were not caused by near-surface resistivity distribution of IP effects, or by any deep resistivity boundaries. The continuity is against a near-surface origin and the sharpness is against a deep-seated ohmic origin. Lee (1939) believes that these features are due to presence of electrical double layers at interfaces of two chemically dissimilar geologic formations.

A different kind of resistivity measurement made with the Resistolog method (West and Beacham, 1944, 1946) yielded systematic results which were similar to Lee's (1939) observations. The field setup is shown in Figure 6.9a. It is a quasi-pole–dipole (or bipole–dipole) array with two consecutive positions, 1 and 2, of the source electrode. Originally it was intended to be a resistivity sounding method in which the effects of near-surface inhomogeneities would be suppressed.

In Figure 6.9a for each position of MN two measurements of ΔV are made, one with current flow between 1 and 3, and the other with current flow between 2 and 3, which yield two apparent resistivities R_1 and R_2. Keeping 1, 2 and 3 stationary and the length of MN constant, the electrode separation a is expanded. The ratio R_1/R_2 is plotted against a, in the

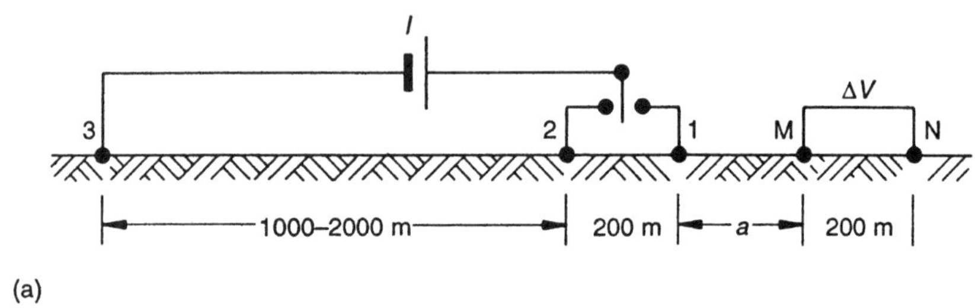

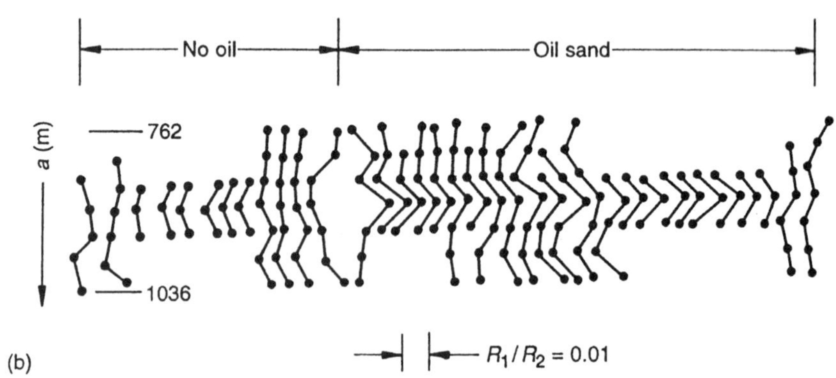

Figure 6.9 The Resistolog method: (a) field setup; (b) the results at Sam Fordyce field, Texas. R_1 and R_2 are apparent resistivities when current electrode is at positions 1 and 2 respectively (after West and Beacham, 1944).

form of cross-sections, as shown in Figure 6.9b. The electrode separation a is empirically converted to depth and the Resistolog data is compared to well log data. (The depth of exploration is of the order of 0.7 a.)

West and Beacham (1944, 1946) discuss the applicability of this method to direct detection of oil sands on the basis of resistivity contrasts. On this basis one should expect broader maxima and minima as the depth to the interface is increased, and this is what the results indicated, up to a certain depth. Beyond that depth the singular points became sharp kicks. Another confusing result was that at Pettus Area, Bee County, Texas, the oil sand at about 1220 m depth was correlated through strong kicks of the Resistolog curves, whereas the electric well logs indicated no appreciable resistivity contrast. In general, correlatable sharp kicks occur at depths deeper than 600 m, up to 1800 m. The most positive results, especially in direct detection of oil, were obtained in this depth range and in areas with little resistivity contrast and with low resistivities. The experiments made by varying the current density did not yield conclusive results.

At a round table discussion of the Society of Exploration Geophysicists (West and Beacham, 1942), it was believed that these probably deep-origin and sharp anomalies are of electrochemical origin rather than being a result of ohmic resistivity variation.

Is it possible that there may be two critical values of current density, J_1 and J_2, at an interface that corresponds to changes from linear to nonlinear and to quasi-linear current conduction? From zero to J_1, for infinitesimally small values, the interface would behave linearly. For values higher than J_1, up to J_2, it would behave nonlinearly. At values higher than J_2 the current would flow freely across the interface and a quasi-linear condition would prevail. If so, the Resistolog success depth range, 600–1800 m, could correspond to the J_1–J_2 interval of nonlinear behavior. However, J due to a point source at the surface would have different directions at different space coordinates; the magnitude of the component normal to an interface would also depend largely on the space coordinates as well as on the strength of the source; and this could lead to inconclusive experimental results.

References

Ádám, A., Nagy, Z. and Varga, G. (1989) Magnetotelluric (MT) research and exploration in Hungary. *Geophysics*, **54**, 795–7.

Al'pin, L.M., Berdichevskii, M.N., Vedrintsev, G.A. and Zagarmistr, A.M. (1966) Dipole methods for measuring earth conductivity. Translated by G.V. Keller. Consultants Bureau.

Anderson, W.L. (1982) Nonlinear least-squares inversion of transient soundings for a central induction loop system (program NLSTCI), *US Geol. Surv. Open File Report* 82–1129.

Asch, T. and Morrison, H.F. (1989) Mapping and monitoring electrical resistivity with surface and subsurface electrode arrays. *Geophysics*, **54**, 235–44.

Azad, J. (1973) Direct oil prospecting with electrical transient reflections. *J. Can. Soc. Expl. Geophys.*, **9**, 1–11.

Banerjee, B. and Sengupta, B.J. (1987) Transformation of dipolar, Wenner and two-electrode curves to Schlumberger apparent resistivity sounding curves. *Geophys. Prosp.*, **35**, 445–53.

Baranov, V. (1951) Quantitative interpretation of measurements in prospecting by telluric currents. Proc. Third World Petroleum Congress, The Hague, Section 1, 646–53 (in French).

Başokur, A.T. (1983) Transformation of resistivity sounding measurements obtained in one electrode configuration to another configuration by means of digital linear filtering. *Geophys. Prosp.*, **31**, 649–63.

Berdichevskii, M.N. (1958) The method of curved electrical probe. *Prikladnaia Geofizika*, **18**, 128–144. [In Russian; English translation in Rast, N. (1962) *Applied Geophysics, USSR*, Pergamon Press, Inc., pp. 223–40.]

Berdichevskii, M.N. (1960) Electrical surveying by means of telluric currents: Gostoptekhizdat. [English translation by Bradley, J.E.S. (1963) National Lending Library; also by Keller, G.V. (1965) *Quarterly of the Colorado School of Mines*, **60**. Referenced pages are those of Keller's translation.]

Berdichevskii, M.N. and Petrovskii, A.D. (1956) Procedures of bilateral equatorial electrical surveying. *Prikladnaia Geofizika*, **14**, 97–114 (in Russian).

Berdichevskii, M.N. and Zagarmistr, A.M. (1958) Interpretation of bilateral electrical soundings obtained with dipole arrays. *Prikladnaia Geofizika*, **19**, 57–108 (in Russian; English translation in Al'pin *et al.*, 1966).

Bhattacharya, P.K. and Patra, H.P. (1968) *Direct Current Geoelectric Sounding*. Elsevier Science Publ.

Bibby, H.M. (1977) The apparent resistivity tensor. *Geophysics*, **42**,1258–61.

Boissonnas, E. and Leonardon, E.G. (1948) Geophysical exploration by telluric currents, with special reference to a survey of the Haynesville salt dome, Wood County, Texas. *Geophysics*, **13**, 387–403.

Bostick, F.X., Jr, (1977) A simple and almost exact method of MT analysis (Abstract), in Workshop on electrical methods in geothermal exploration, Ward, S.H. (ed.) University of Utah, Salt Lake City, pp. 175–83.

Bostick, F.X., Jr. (1986) Electromagnetic array profiling (EMAP). 56th Annual International Meeting Society of Exploration Geophysicists, Expanded Abstracts, pp. 60–1.

Breusse, J.J. (1963) Modern geophysical methods for subsurface water exploration. *Geophysics*, **28**, 633–57.

Butterworth, N.A. (1988) Controlled-source audio-frequency magnetotelluric responses of three-dimensional bodies. MSc Thesis, Univ. of Utah.

Cagniard, L. (1953) Basic theory of the magneto-telluric method of geophysical prospecting. *Geophysics*, **18**, 605–35.

Cantwell, T. (1960) Detection and analysis of low-frequency magnetotelluric signals. PhD Thesis, Massachusetts Institute of Technology.

Clarke, J., Gamble, T.D., Goubau, W.M., Koch, R.H. and Miracky, R.F. (1983) Remote-reference magnetotellurics: equipment and procedures. *Geophys. Prosp.*, **31**, 149–70.

Compagnie Générale de Géophysique (1963) *Master Curves for Electrical Sounding*, 2nd edn: European Association of Exploration Geophysicists.

Dahlberg, R.S. Jr (1945) An investigation of natural earth currents. *Geophysics*, **10**, 494–506.

DeGroot-Hedlin, C. and Constable, S. (1990) Occam's inversion to generate smooth, two-dimensional models from magnetotelluric data. *Geophysics*, **55**, 1613–24.

Dey, A. (1976) Resistivity modeling for arbitrarily shaped two dimensional structures, Part II: user's guide to the FORTRAN algorithm RESIS2D. *Lawrence Berkeley Laboratory Report 5283*, Univ. California, Berkeley.

Dey, A. and Morrison, H.F. (1979) Resistivity modeling for arbitrarily shaped two-dimensional structures. *Geophys. Prosp.*, **27**, 106–36.

Dobrin, M.B. and Savit, C.H. (1988) *Introduction to Geophysical Prospecting*, McGraw-Hill.

Duncan, P.M., Hwang, A., Edwards, R.N., *et al.* (1980) The development and applications of a wide band electromagnetic sounding system using a pseudo-noise source. *Geophysics*, **45**, 1276–96.

Duprat, A., Roudot, M. and Spitz, S. (1986) Testing the Transiel method in mineral and geothermal exploration. *Geophys. Prosp.*, **34**, 445–62.

Ebert, A. (1943) Grundlagen zur Auswertung geoelektrischer Tiefenmessungen. *Beitr. Angew. Geophys.*, **10**, 1–17.

Edwards, R.N. (1976) Electrical methods for the study of regional crustal conductivity anomalies. *Acta Geodaetica, Geophysica et Montanistica, Acad, Sci. Hung.*, **11**, 399–425.

Edwards, R.N. and Holladay, J.S. (1991) Pseudo-random binary sequence (PRBS) techniques, in *Electromagnetic Methods in Applied Geophysics*, (ed. M.N. Nabighian) Vol. 2, Part A, Society of Exploration Geophysicists, pp. 388–97 (Appendix B in Spies and Frischknecht, 1991).

Edwards, R.N. and Nabighian, M.N. (1991) The magnetometric resistivity method, in *Electromagnetic methods in Applied Geophysics*, (ed. M.N. Nabighian) Vol. 2, Part A, Society of Exploration Geophysicists, pp. 47–104.

Evjen, H.M. (1948) Theory and practice of low frequency electromagnetic exploration. *Geophysics*, **13**, 584–94.

Evjen, H.M. (1953) Surface electrical method detects oil directly. *World Oil*, **136**, No. 2, 93–6.

Fink, J.B., McAlister, E.O., Sternberg, B.K. and Wieduwilt, W.G. (eds) (1990) *Induced Polarization*. Society of Exploration Geophysicists, *Investigations in Geophysics*, No. 4.

Fitterman, D.V. and Stewart, M.T. (1986) Transient electromagnetic sounding for groundwater. *Geophysics*, **51**, 995–1005.

Frischknecht, F.C. (1967) Fields about an oscillating magnetic dipole over a two-layer earth, and application to ground and airborne electromagnetic surveys. *Quarterly, Colorado School of Mines*, **62**, 1–326.

Frischknecht, F.C. (1988) Electromagnetic physical scale modeling, in *Electromagnetic Methods in Applied Geophysics* (ed. M.N. Nabighian) Vol. 1, Society of Exploration Geophysicists, pp. 365–441.

Gamble, T.D., Goubau, W.M. and Clarke, J. (1979) Magnetotellurics with a remote magnetic reference. *Geophysics*, **44**, 53–68.

Gamble, T.D., Goubau, W.M., Goldstein, N.E. *et al.* (1981) Magnetotelluric studies at Cerro Prieto. *Geothermics*, **10**, 169–82.

Goldberg, S. and Rotstein, Y. (1982) A simple form of presentation of magnetotelluric data using the Bostick transform. *Geophys. Prosp.*, **30**, 211–6.

Goldhaber, M.B. and Reynolds, R.L. (1991) Relations among hydrocarbon reservoirs, epigenetic sulfidization, and rock magnetization – examples from the south Texas coastal plain. *Geophysics*, **56**, 748–57.

Goldstein, M.A. and Strangway, D.W. (1975) Audio-frequency magnetotellurics with a grounded electric dipole source. *Geophysics*, **40**, 669–83.

Goldstein, N.E., Norris, R.A. and Wilt, M.J. (1978) Assessment of surface geophysical methods in geothermal exploration and recommendations for future research. Lawrence Berkeley Laboratory, Univ. of California, Berkeley, LBL-6815 for US Department of Energy.

Gomez-Trevino, E. and Edwards, R.N. (1983) Electromagnetic soundings in the sedimentary basin of southern Ontario – a case history. *Geophysics*, **48**, 311–30.

Gosh, D.P. (1971) Inverse filter coefficients for the computation of apparent resistivity standard curves for a horizontally stratified earth. *Geophys. Prosp.*, **19**, 769–75.

Grant, F.S. and West, G.F. (1965) *Interpretation Theory in Applied Geophysics*, McGraw-Hill Book Co.

Harthill, N. (1978) A quadripole resistivity survey of the Imperial Valley, California. *Geophysics*, **43**, 1485–500.

Hermance, J.F. and Thayer, R.E. (1975) The telluric–magnetotelluric method. *Geophysics*, **40**, 664–8.

Hobbs, B.A. (1982) Automatic model for finding the one-dimensional magnetotelluric problem. *Geophys. J. Roy. Astr. Soc.*, **68**, 253–64.

Hohmann, G.W. (1988) Numerical modeling for electromagnetic methods of geophysics, in *Electromagnetic methods in applied geophysics* Vol. 1, (ed. M.N. Nabighian) Society of Exploration Geophysicists, pp. 313–63.

Hohmann, G.W. and Raiche, A.P. (1988) Inversion of controlled-source electromagnetic data, in *Electromagnetic Methods in Applied Geophysics*, Vol. 1, Theory (ed M.N. Nabighian) Society of Exploration Geophysicists, pp. 469–503.

Holladay, J.S. (1987) The generalized electrosounding method for sedimentary basin exploration. PhD Thesis, Univ. of Toronto.

Hughes, L.J. and Carlson, N.R. (1987) Structure mapping at Trap Spring oilfield, Nevada, using controlled-source magnetotellurics. *First Break*, **5**, 403–18.

Hughes, L.J., Zonge, K.L. and Carlson, N.R. (1986) The application of electrical techniques in mapping hydrocarbon-related alteration, *Unconventional Methods in Exploration for Petroleum and Natural Gas* (ed. M.J. Davidson), Symposium IV, May 1–2 1985, Southern Methodist Univ. Press, Dallas.

Jackson, D.B. and Keller, G.V. (1972) An electromagnetic sounding survey of the summit of Kilauea volcano, Hawaii. *J. Geophys. Res.*, **77**, 4957–65.

Jacobson, J.J. (1969) Deep electromagnetic sounding technique. DSc Thesis, Colorado School of Mines, Golden, Colorado.

Jones, A.G. (1983) On the equivalence of the 'Niblett' and 'Bostick' transformations in the magnetotelluric method. *J. Geophys.*, **53**, 72–3.

Jones, F.W. and Pascoe, L.J. (1971) A general computer program to determine the perturbation of alternating electric currents in a two-dimensional model of a region of uniform conductivity with an embedded inhomogeneity. *Geophys. J. Roy. Astr. Soc.*, **24**, 3–30.

Jupp, D.L.B. and Vozoff, K. (1975) Stable iterative methods for inversion of geophysical data. *Geophys. J. Roy. Astr. Soc.*, **42**, 957–76.

Jupp, D.L.B. and Vozoff, K. (1977) Resolving anisotropy in layered media by joint inversion. *Geophys. Prosp.*, **25**, 460–70.

Karcher, J.C. and McDermott, E. (1935) Deep electrical prospecting. *Am. Ass. Petrol. Geol.*, **19**, 64–77.

Kasameyer, P.W. (1976) Preliminary interpretation of resistivity and seismic refraction data from the Salton Sea geothermal field. Lawrence-Livermore Laboratory, Univ. of California, RL-52115.

Kaufman, A.A. and Keller, G.V. (1981) *The Magnetotelluric Sounding Method*, Elsevier Science Publ.

Kaufman, A.A. and Keller, G.V. (1983) *Frequency and Transient Soundings*, Elsevier Science Publ.

Keller, G.V. (1968) Electrical prospecting for oil. *Quarterly of the Colorado School of Mines*, **63**, No. 2.

Keller, G.V. (1969) Electromagnetics may be the key to direct oil finding. *World Oil*, Dec., 85–8.

Keller, G.V. (1971) Electrical characteristics of the Earth's crust, in *Electromagnetic Probing in Geophysics* (ed. J.R. Wait), Golem Press, 13–79.

Keller, G.V. (1982) Electrical properties of rocks and minerals, in *Handbook of Physical Properties of Rocks*, Vol. 1 (ed. R.S. Carmichael), CRC Press, pp. 217–94.

Keller, G.V. (1988) Rock and mineral properties, in *Electromagnetic Methods in Applied Geophysics*, vol.1 (ed. M.N. Nabighian), Society of Exploration Geophysicists, pp. 13–51.

Keller, G.V. (1991) Colorado School of Mines TDEM systems, in *Electromagnetic Methods in Applied Geophysics*, Vol. 2, Part A (ed. M.N. Nabighian), Society of Exploration Geophysicists, pp. 408–11 (Appendix F in Spies and Frischknecht 1991).

Keller, G.V. and Frischknecht, F.C. (1966) *Electrical Methods in Geophysical Prospecting*, Pergamon Press.

Keller, G.V. Pritchard, J.I., Jacobson, J.J. and Harthill, N. (1984) Megasource time-domain electromagnetic sounding methods. *Geophysics*, **49**, 993–1009.

Koefoed, O. (1968) *The Application of the Kernel Function in Interpreting Geoelectrical Resistivity Measurements*, Gebrüder Borntraeger.

Koefoed, O. (1979) *Resistivity Sounding Measurements*. Elsevier Science Publ.

Koefoed, O. and Biewinga, D.T. (1976) The application of electromagnetic frequency sounding to groundwater problems. *Geoexploration*, **14**, 229–41.

Koefoed, O., Ghosh, D.P. and Polman, G.J. (1972) Computation of type curve for electromagnetic depth sounding with a horizontal transmitting coil by means of a digital linear filter. *Geophys. Prosp.*, **20**, 406–20.

Kunetz, G. (1952a) Method for the electrical investigation of the ground, US Patent 2 586 667.

Kunetz, G. (1952b) Method for the electrical prospection of the subsoil, US Patent 2 623 697.

Kunetz, G. (1957) Application of statistical properties of earth currents in practical geophysics. *Freiberg. Forschungsh.*, **C32**, Geophysik, 5–19 (in German).

Kunetz, G. (1958) Principles and applications of telluric prospecting. *Erdöl-Z.*, Heft 9, Urban-Verlag, 3–15 (in German).

Kunetz, G. (1966) *Principles of Direct Current Resistivity Prospecting*, Gebrüder Borntraeger.

Kunetz, G. and Chastenet de Géry, J. (1956) Conformal representation and various problems of potential in media of different 'permeability' (electrical conductivity). *Rev. Inst. Fr. Pet.*, **11**(10), 1179–92 (in French).

Laine, E.F. and Lytle, R.J. (1976) A computer program for four probe resistivity measurements in a horizontally layered earth. *IEEE Trans. Geosci. Electronics*, **14**, 232–5.

Lasfargues, P. (1957), *Prospection Électrique par Courants Continus*; Masson & Cie.

Lee, F.W. (1939) The possibility of electrical stratification in the earth as disclosed by surface measurements of currents and potentials. *Trans. Am. Geophys. Union*, Pt. 3, 382–9.

Lumb, J.T. (1981) Prospecting for geothermal resources, in *Geothermal Systems: Principles and Case Histories* (eds L. Raybach and L. J. Muffler), John Wiley & Sons.

Madden, T.R. (1971) The resolving power of geoelectric measurements for delineating resistive zones within the crust, in *The Structure and Physical Properties of the Earth's Crust* (ed. J.G. Heacock), *Geophys Monogr.* **14**, American Geophysical Union, pp. 95–105.

Madden, T. and Nelson, P. (1964) A defense of Cagniard's magnetotelluric method. Massachusetts Institute of Technology, Project NR-371–401.

Madden, T.R. and Swift, C.M., Jr. (1969) Magnetotelluric studies of the electrical conductivity structure of the crust and upper mantle, in *The Earth's Crust and Upper Mantle* (ed. P.J. Hart), *Geophys. Monogr.* **13**, American Geophysical Union, pp. 469–79.

Malergue, G., Aissa, J., Herisson, C. and Rainaud, J. (1986) Recent developments in the magnetotelluric prospecting method. *First Break*, **4**, 23–8.

Migaux, L. (1946) A new method of applied geophysics – prospecting by telluric currents. *Ann. Géophys.*, **2**, 131–46 (in French).

Migaux, L. (1951) Ten years of application of the telluric methods. Third World Petroleum Congress, The Hague, Section 1, pp. 624–45 (in French).

Mitchell, B.J. and Landisman, M. (1971) Electrical and seismic properties of the Earth's crust in the southwestern Great Plains of the USA. *Geophysics*, **36**, 363–81.

Morrison, H.F. and Goldstein, N.E. (1991) Lawrence Berkeley Laboratory EM-60 system, in *Electromagnetic Methods in Applied Geophysics*, Vol. 2, part A (ed. M.N. Nabighian), Society of Exploration Geophysicists, pp. 398–402.

Morrison, H.F., Goldstein, N.E., Hoversten, G.M., *et al.* (1978) Description, field test and data analysis of controlled source EM system (EM-60), Lawrence Berkeley Laboratory, LBL-7088.

Müller, M. (1940) Results of geoelectric measurements of polarization. *Z. Geophys.*, **16**, 274–84 (in German).

Müller, M. (1950) An induction-impulse method for the detection of petroleum. *Geof. Pura et Appl.*, **17**, 54–60 (in German).

Mundry, E. (1980) The effect of a finite distance between potential electrodes on Schlumberger resistivity measurements – a simple correction graph. *Geophysics*, **45**, 1872–5.

Mundry, E. and Blohm, E.-K. (1987) Frequency electromagnetic sounding using a vertical magnetic dipole. *Geophys. Prosp.*, **35**, 110–23.

Nabighian, M.N. (ed.) (1988) Electromagnetic Methods in Applied Geophysics, Vol. 1, *Theory*, 1988 Society of Exploration Geophysicists.

Nabighian, M.N. (ed.) (1991) *Electromagnetic Methods in Applied Geophysics*, Vol. 2, Parts A and B, *Applications*, Society of Exploration Geophysicists.

Neuenschwander, E.F. and Metcalf, D.F. (1942) A study of electrical earth-noise. *Geophysics*, **7**, 69–77.

Nivas, S. and Israil, M. (1986) Computation of apparent resistivities using an exponential approximation of kernel functions. *Geophysics*, **51**, 1594–1692.

Nivas, S., and Israil, M. (1987) A simple method of interpreting dipole resistivity soundings: *Geophysics*, **52**, 1412–17.

Oehler, D.Z. and Sternberg, B.K. (1984) Seepage-induced anomalies, 'false' anomalies, and implications for electrical prospecting. *Bull. Am. Ass. Petrol. Geol.*, **68**, 1121–45.

Oldenburg, D.W. (1979) One-dimensional inversion of natural source magnetotelluric observations. *Geophysics*, **44**, 1218–46.

Olhoeft, G.R. (1980) Electrical properties of rocks, in *Physical Properties of Rocks and Minerals* (eds Y.S. Touloukian, W.R. Judd and R.F. Roy), McGraw-Hill Book Co., pp. 257–330.

Olhoeft, G.R. (1985) Low-frequency electrical properties. *Geophysics*, **50**, 2492–503.

Orellana, E. (1972) *Prospeccion Geoelectrica en Corriente Continua*, Paraninfo.

Orellana, E. and Mooney, H.M. (1966) *Master Tables and Curves for Vertical Electrical Sounding over Layered Structures*, Interciencia.

Park, S.K. and Livelybrooks, D.W. (1989) Quantitative interpretation of rotationally invariant parameters in magnetotellurics. *Geophysics*, **54**, 1483–90.

Parkhomenko, E.I. (1967) *Electrical Properties of Rocks*, Plenum Press (English translation by G.V. Keller).

Patra, H.P. and Mallick, K. (1980) *Time-varying Geoelectric Sounding*. Elsevier Science Publ.

Pellerin, L. and Hohmann, G.W. (1990) Transient electromagnetic inversion – a remedy for magnetotelluric static shifts. *Geophysics*, **55**, 1242–50.

Pelton, W.H., Rijo, L. and Swift, C.M., Jr. (1978) Inversion of two-dimensional resistivity and induced-polarization data. *Geophysics*, **43**, 788–803.

Petersen, E.V. (1974) Direct detection with electric crew will eliminate stratigraphic dusters. *Oilweek*, August **5**, 16–7.

Pirson, S.J. (1980) Oil is confined in the earth by redox potential barriers. *Oil & Gas J.*, July **7**, 153–8.

Poldini, E. (1947) *La Prospection Électrique du Sous-sol*, F. Rouge & C.

Pous, J., Marcuello, A. and Queralt, P. (1987) Resistivity inversion with *a priori* information. *Geophys. Prosp.*, **35**, 590–603.

Pratt, R.B. (1953) New oil finding method tested. *World Oil*, November, 98–105.

Rabinovich, B.I. Surkov, V.S. and Mandelbaum, M.M. (1977) The use of electrical methods in exploration for oil and gas fields in The Siberian platform. *Sovet. Geol.*, No. 2, pp. 314 (in Russian).

Raiche, A.P., Jupp, D.L.B., Rutter, H. and Vozoff, K. (1985) The joint use of coincident loop transient electromagnetic and Schlumberger sounding to resolve layered structures. *Geophysics*, **50**, 1618–27.

Rankin, D., Mozeson, C. and Nabetani, S. (1974) Interactive programming in the inverse method by sequential layering for magnetotelluric analysis. *J. Geophys. Res.*, **79**, 2022–6.

Rijkswaterstaat (1969) *Standard Graphs for Resistivity Prospecting*, European Association of Exploration Geophysicists.

Rosaire, E.E. (1938) Shallow stratigraphic variations over Gulf Coast structures. *Geophysics*, **3**, 96–121.

Roy, K.K. and Elliot, H.M. (1981) Some observations regarding depth of exploration in d.c. electrical methods. *Geoexploration*, **19**, 1–13.

Sandberg, S.K. and Hohmann, G.W. (1982) Controlled-source audiomagnetotellurics in geothermal exploration. *Geophysics*, **47**, 100–16.

Sasaki, Y. (1989) Two-dimensional joint inversion of magnetotelluric and dipole–dipole resistivity data. *Geophysics*, **54**, 254–62.

Schlumberger, M. (1936) Method for prospecting the undersoil. US Patent 2 034 447.

Schlumberger, M. (1941) Method and apparatus for electrical underground prospecting. US Patent 2 240 520.

Sherwood, J.W.C. and Yungul, S.H. (1968) Frequency spectrum analysis of injected coded signal and measured probe signal for geophysical prospecting. US Patent 3 382 428.

Sims, W.E. and Bostick, F.X., Jr (1969) Methods of magnetotelluric analysis. EGRL Technical Report No. 58, Univ. of Texas at Austin.

Smith, J.T. and Booker, J.R. (1988) Magnetotelluric inversion for minimum structure. *Geophysics*, **53**, 1565–76.

Smith, N.C. and Vozoff, K. (1984) Two-dimentional DC resistivity inversion for dipole–dipole data. *Inst. Electr. Electron. Engineers, Trans. Geosci. Remote Sensing*, **22**, 21–8

Snyder, D.D., Kolvoord, R.W., Frangos, W. *et al.* (1981) Exploration for petroleum using complex resistivity measurements, in *Advances in Induced-Polarization and Complex Resistivity*, Univ. of Arizona, pp. 209–53.

Spies, B.R. (1983) Recent developments in the use of surface electrical methods for oil and gas exploration in the Soviet Union. *Geophysics*, **48**, 1102–1112.

Spies, B.R. (1989) Depth of investigation in electromagnetic sounding methods. *Geophysics*, **54**, 872–88.

Spies, B.R. and Eggers, D.E. (1986) The use and misuse of apparent resistivity in electromagnetic methods. *Geophysics*, **51**, 1462–71.

Spies, B.R. and Frischknecht, F.C., (1991) Electromagnetic sounding, in *Electromagnetic methods in Applied Geophysics*, Vol. 2, Part A (ed. M.N. Nabighian) Society of Exploration Geophysicists, pp. 285–425.

Srivastava, S.P. (1965) Method of interpretation of magnetotelluric data when source field is considered. *J. Geophys. Res.*, **70**, 945–54.

Srivastava, S.P. (1967) Magnetotelluric two- and three-layer master curves: *Dom. Obs. Publ.*, **35**, No. 7, Canada Department of Energy, Ottawa.

Stark, M., Wilt, M., Haught, J.R. and Goldstein, N. (1980) Controlled-source electromagnetic survey at Soda Lakes geothermal area, Nevada. Univ. of California, Berkeley, LBL-11221, UC-66b.

Sternberg, B.K. and Oehler, D.Z. (1990) Induced-polarization hydrocarbon surveys – Arkoma Basin case histories, in *Induced Polarization* (eds J.B. Fink, E.O. McAlister, B.K. Sternberg and W.G. Weiduwilt (eds) *Investigations in Geophysics*, No. 4, Society of Exploration Geophysicists.

Stodt, J.A., Hohmann, G.W. and Ting, S.C. (1981) The telluric–magnetotelluric method in two- and three-dimensional environments. *Geophysics*, **46**, 1137–47.

Strack, K.-M., 1991, German deep transient EM systems, *Electromagnetic Methods in Applied Geophysics*, vol. 2, Part A, (ed. M.N. Nabighian) Society of Exploration Geophysicists, Appendix L in Spies and Frischknecht, pp. 422–425.

Strack, K.-M. (1992) *Exploration with Deep Transient Electromagnetics*, Elsevier Science Publ.

Sumner, J.S. (1976) *Principles of Induced Polarization for Geophysical Exploration*, Elsevier Science Publ.

Swift, C.M., Jr. (1967) A magnetotelluric investigation of an electrical conductivity anomaly in the southwestern United States. PhD Thesis, Massachusetts Institute of Technology.

Swift, C.M., Jr. (1971) Theoretical magnetotelluric and TURAM response from two-dimensional inhomogeneities. *Geophysics*, **36**, 38–52.

Tarkhov, A.G. (1963) *Electrorazvedka*, Gostoptekhizdat.

Telford, W.M., Geldart, L.P., Sheriff, R.E. and Keys, D.A. (1976) *Applied Geophysics*, Cambridge University Press.

Thieme, H.G. (1963) Problems and successes of the telluric methods during the investigation of high-resistivity anticline structures in GDR. Deutsch Verl. für Grundstoff-Industrie (in German).

Torres-Verdin, C. and Bostick, F.X., Jr. (1992a) Applications of the Born approximation for the magnetotelluric problem in three-dimensional environments. *Geophysics*, **57**, 587–602.

Torres-Verdin, C. and Bostick, F.X., Jr. (1992b) Principles of spatial surface electric field filtering in magnetotellurics: electromagnetic array profiling (EMAP). *Geophysics*, **57**, 603–22.

Tripp, A.C., Hohmann, G.W. and Swift, C.M., Jr. (1984) Two-dimensional resistivity inversion. *Geophysics*, **49**, 1708–17.

Utzmann, R. (1954) Electrical and telluric prospecting studies on scale models. *Bull. Assoc. Fr. Tech. Pét.*, **107**, 1–61 (in French).

Vanyan, L.L., Bobrovnikov, L.Z., Loshenitzina, V.L. *et al.* (1967) *Electromagnetic Depth Soundings.* Selected and translated by G.V. Keller, Consultants Bureau, New York.

Van Zijl, J.S.V. and Joubert, S.J. (1975) A crustal geoelectrical model for South African Precambrian granitic terraines based on deep Schlumberger soundings. *Geophysics*, **40**, 657–63.

Verma, R.K. (1980a) *Master tables for electromagnetic depth sounding interpretation.* IFI Data Base Library, Plenum Publ. Corp.

Verma, R.K. (1980b) Equivalence in electromagnetic (frequency) sounding. *Geophys. Prosp.*, **28**, 776–91.

Verma, R.K. (1982) *Electromagnetic Sounding Interpretation Data over a Three-layer Earth, 1 and 2*, IFI Data Base Library, Plenum Publ. Corp.

Volker, A. and Dijkstra, J. (1955) Détermination des salinités des eaux dans le sous-sol du Zuiderzee par prospection geophysique. *Geophys. Prosp.*, **3**, 111–25.

Vozoff, K. (1972) The magnetotelluric method in the exploration of sedimentary basins. *Geophysics*, **37**, 98–141.

Vozoff, K. (ed.) (1986) *Magnetotelluric Methods, Geophysics Reprint Series*, No. 5, Society of Exploration Geophysicists.

Vozoff, K. (1991) The magnetotelluric method, in *Electromagnetic Methods in Applied Geophysics*, Vol. 2. (ed. M.N. Nabighian) Society of Exploration Geophysicists, 641–711.

Wait, J.R. (1951) The magnetic dipole over the horizontally stratified earth. *Can. J. Phys.* **29**, 557–92

Wait, J.R. (1961) The electromagnetic fields of a horizontal dipole in the presence of a conducting half space. *Can. J. Phys.* **39**, 1017–28.

Wait, J.R. (1962) *Electromagnetic Waves in Stratified Media.* Pergamon Press.

Wannamaker, P.E., Wright, P.M., Zi-xing, Z. *et al.* (1991) Magnetotelluric transect of Long Valley caldera – resistivity cross-section, structural implications, and the limits of a 2-D analysis. *Geophysics*, **56**, 926–40.

Ward, S.H. and Hohmann, G.W. (1988) Electromagnetic theory for geophysical applications, in *Electromagnetic Methods in Applied Geophysics*, Vol. 1, (ed. M.N. Nabighian), Society of Exploration Geophysicists, pp. 131–311.

Warren, R.K. and Srnka, L.J. (1992) Exploration in the basalt-covered areas of the Columbia River Basin, Washington, using electromagnetic array profiling (EMAP). *Geophysics*, **57**, 986–93.

Waxman, M.H. and Smits, L.J.M. (1968) Electrical conductivities in oil-bearing shaley sand. *Soc. Petr. Eng. J.*, **8**, 107–22.

West, T.S. and Beacham, C.C. (1942) Precise measurement of the electrical resistivity anomaly resulting from oil or gas saturation. *Geophysics*, **7**, 339.

West, T.S. and Beacham, C.C. (1944) Precise measurement of deep electrical anomalies. *Geophysics*, **9**, 494–539.

West, T.S. and Beacham, C.C. (1946) A 'Resistolog' survey of the Loma Alto-Seven Sisters area, Texas. *Geophysics*, **11**, 491–504.

Whittal, K.P. (1986) Inversion of MT data using localized conductivity constraints. *Geophysics*, **51**, 1603–7.

Whittal, K.P. and Oldenburg, D.W. (1992) *Inversion of Magnetotelluric Data for a One-dimensional Conductivity, Geophysical Monograph Series*, No. 5, Society of Exploration Geophysicists.

Wilt, M.J. and Goldstein, N.E. (1981) Resistivity monitoring at Cerro Prieto. *Geothermics*, **10**, 183–91.

Wilt, M.J., Goldstein, N.E. and Raso, M.A. (1980) LBL resistivity studies at Cerro Prieto. *Geothermics*, **9**, 15–26.

Wilt, M. Goldstein, N.E., Stark, M., *et al.* (1983) Experience with the EM-60 electromagnetic system for geothermal exploration in Nevada. *Geophysics*, **48**, 1090–1101.

Wilt, M.J., Morrison, H.F., Lee, K.H. and Goldstein, H.E. (1989) Electromagnetic sounding in the Columbia Basin, Yakima, Wahington. *Geophysics*, **54**, 952–61.

Yamashita, M. (1987) Controlled source audio-frequency magneto-tellurics, Phoenix Geophysics Ltd, Ontario, Canada.

Yamashita, M., Hallof, P.G. and Pelton, W.H. (1985) CSAMT case histories with a multichannel CSAMT system and near-field data correction. 55th Annual International Meeting Society of Exploration Geophysicists Expanded Abstract MT8.

Yungul, S.H. (1961) Magnetotelluric sounding three-layer interpretation curves. *Geophysics*, **26**, 465–73.

Yungul, S.H. (1966) Telluric sounding – a magnetotelluric method without magnetic measurements. *Geophysics*, **31**, 185–91.

Yungul, S.H. (1968) Measurement of telluric relative ellipse area by means of vectograms. *Geophysics*, **38**, 127–31.

Yungul, S.H. (1977) The telluric methods in the study of sedimentary structures – a survey. *Geoexploration*, **15**, 207–38.

Yungul. S.H. (1982) Geoelectricity, in *McGraw-Hill Encyclopedia of Science and Technology*, 5/e, pp. 171–5.

Yungul, S.H., Hembree, M.R. and Greenhouse, J.P. (1973) Telluric anomalies associated with isolated reefs in the Midland Basin, Texas. *Geophysics*, **38**, 545–56.

Zohdy, A.A.R. (1988) Groundwater exploration with Schlumberger soundings near Jean, Nevada. US Geological Survey Open-File Report 88–291, 66.

Zohdy, A.A.R. (1989) A new method for the automatic interpretation of Schlumberger and Wenner sounding curves. *Geophysics*, **54**, 245–53.

Zonge, K.L. and Hughes, L.J. (1991) Controlled source audio-frequency magnetotellurics, in *Electromagnetic Methods in Applied Geophysics*, Vol. 2, Part B (ed. M.N. Nabighian) Society of Exploration Geophysicists, pp. 713–809.

Index

Page numbers appearing in **bold** refer to figures.